PATTY'S INDUSTRIAL HYGIENE AND TOXICOLOGY

Volume I
GENERAL PRINCIPLES

Volume II
TOXICOLOGY

Volume III
THEORY AND RATIONALE
OF INDUSTRIAL HYGIENE
PRACTICE

PATTY'S INDUSTRIAL HYGIENE AND TOXICOLOGY

Volume III
THEORY AND RATIONALE OF
INDUSTRIAL HYGIENE PRACTICE

SECOND EDITION
3B BIOLOGICAL RESPONSES

LEWIS J. CRALLEY, PH.D.
LESTER V. CRALLEY, PH.D.
Editors

Contributors

J. S. Bus	P. J. Gehring	C. H. Powell
J. S. Chapman	J. E. Gibson	K. S. Rao
A. Cohen	A. Glorig	L. E. Scheving
L. J. Cralley	J. C. Guignard	M. J. Smolensky
L. V. Cralley	S. M. Horvath	R. G. Thomas
J. C. Davis	S. M. Michaelson	R. S. Waritz
F. Dukes-Dobos	D. J. Paustenbach	P. G. Watanabe

A WILEY-INTERSCIENCE PUBLICATION

JOHN WILEY & SONS, New York · Chichester · Brisbane · Toronto · Singapore

Copyright © 1985 by John Wiley & Sons, Inc.

All rights reserved. Published simultaneously in Canada.

Reproduction or translation of any part of this work beyond that permitted by Section 107 or 108 of the 1976 United States Copyright Act without the permission of the copyright owner is unlawful. Requests for permission or further information should be addressed to the Permissions Department, John Wiley & Sons, Inc.

Library of Congress Cataloging in Publication Data:

(Revised for volume 3)
Main entry under title:

Theory and rationale of industrial hygiene practice.

   At head of title: Patty's Industrial hygiene and toxicology, volume III.
   "A Wiley-Interscience publication."
   Includes index.
   Contents: 3A. The work environment—3B. Biological responses.
   1. Industrial hygiene—Collected works. 2. Industrial toxicology—Collected works. I. Cralley, Lewis J., 1911– . II. Cralley, Lester Vincent, 1911– III. Patty's Industrial hygiene and toxicology.
RC967.T48   1985   613.6'2   84-25727
ISBN 0-471-82333-3

Printed in the United States of America

10 9 8 7 6 5 4 3 2 1

Contributors

JAMES S. BUS, Ph.D., Scientist, Department of General and Biological Toxicology, Chemical Industries Institute of Toxicology, Research Triangle Park, North Carolina

JOHN S. CHAPMAN, M.D., 3606 Lovers Lane, Dallas, Texas

ALEXANDER COHEN, Ph.D., Chief, Applied Psychology and Ergonomics Branch, Division of Biomedical and Behavioral Science, NIOSH, Cincinnati, Ohio

LESTER V. CRALLEY, Ph.D., 1453 Banyan Drive, Fallbrook, California

LEWIS J. CRALLEY, Ph.D., 7126 Golden Gate Drive, Cincinnati, Ohio

JEFFERSON C. DAVIS, M.D., Hyperbaric Medicine P.A., Methodist Plaza, Sublevel 2, 4499 Medical Drive, San Antonio, Texas

FRANCIS DUKES-DOBOS, M.D., Science Advisor for Environmental Physiology and Ergonomics, Division of Biomedical and Behavioral Science, NIOSH, Cincinnati, Ohio

PERRY J. GEHRING, D.V.M., Ph.D., Director, Toxicology Research Laboratory, Health and Environmental Sciences, DOW Chemical Company, U.S.A., Midland, Michigan

JAMES E. GIBSON, Ph.D., Vice President for Research, Chemical Industries Institute of Toxicology, Research Triangle Park, North Carolina

ARAM GLORIG, M.D., Consultant in Industrial and Forensic Otology, Otology Medical Group, Inc., Assistant Director of Research, House Ear Institute, Los Angeles, California

JOHN C. GUIGNARD, M.B., Ch.B., Research Medical Officer, Naval Medical Research Laboratory, Michaud Station, New Orleans, Louisiana

STEVEN M. HORVATH, Ph.D., Director and Professor, Institute of Environmental Stress, University of California, Santa Barbara, California

SOL M. MICHAELSON, D.V.M., Professor, Department of Radiation Biology and Biophysics, The University of Rochester Medical Center, Rochester, New York

DENNIS J. PAUSTENBACH, Ph.D., Manager, Environmental and Occupational Toxicology, Syntex Corporation, Palo Alto, California

CHARLES H. POWELL, Sc.D., Charan Associates, Inc., Gualala, California

K. S. RAO, D.V.M., Ph.D., Toxicology Research Laboratory, Health and Environmental Sciences, DOW Chemical Company, U.S.A., Midland, Michigan

M. J. SMOLENSKY, Ph.D., Professor of Environmental Physiology, The University of Texas School of Public Health and Graduate School of Biomedical Science, Houston, Texas

ROBERT G. THOMAS, Ph.D., Health Effects Research Division, Office of Environmental Research, Department of Energy, Washington, D.C.

RICHARD S. WARITZ, Ph.D., Senior Toxicologist, Hercules, Incorporated, Wilmington, Delaware

PHILIP G. WATANABE, Ph.D., Toxicology Research, Laboratory, Health and Environmental Sciences, DOW Chemical Company, U.S.A., Midland, Michigan

LAWRENCE E. SCHEVING, Ph.D., Rebsman Professor of Anatomical Sciences, University of Arkansas for Medical Sciences, Department of Anatomy, Little Rock, Arkansas

Preface

The scope and depth in the development of industrial hygiene as a science has continued at an accelerated pace over the past thirty years. Prior to this period, that is, the first half of the twentieth century, the concepts of industrial hygiene began to develop with emphasis on preventive aspects of maintaining a healthful work environment. This encompassed research into obtaining basic knowledge and techniques for the recognition, evaluation, and control of health hazards in the work environment. These programs undertaken by government, industry, academia, insurance carriers, associations, foundations, and labor organizations developed the basic knowledge and techniques needed by the participating multidisciplinary professions to define full-concept programs in preventing occupational disease arising from the workplace. In the United States, these along with the enactment of the Occupational Safety and Health Act of 1970 gave support to the establishment of industrial hygiene as a science and supported its development in the fullest concept in assuring a healthful working environment.

An important approach in the early development of industrial hygiene was the emphasis placed on quantitative excellence wherever this program quality was encountered. This led to the current procedures and techniques used in establishing threshold limit values, now designated in government standards as personal exposure limits, the assessment of exposure levels, and the development of control procedures needed to keep hazardous exposures within safe limits.

An exceedingly high level of professionalism has always existed in the science of industrial hygiene. This is attested by the acceptance by management of control programs as an integral part of industrial processes and of industrial hygienists as important members of the management team. This close interrelationship is essential in the emergence and application of high technology in industry.

It is timely and important that these aspects be strengthened and that the theoretical basis and rationale of industrial hygiene practice be examined thoroughly and continually to restate fundamental facts and to direct attention to areas of weakness before they influence and become incorporated into acceptable practices by virtue of precedence.

Lewis J. Cralley, Ph.D.
Lester V. Cralley, Ph.D.

February 1985
Cincinnati, Ohio
Fallbrook, California

Notation

The subject areas covered in this volume are based on information and interpretation of regulations available in 1983. The practice of industrial hygiene necessitates a continuing updating in these areas.

Advancing industrial technology, including the more recent high technology, is associated with a number of newer types of health stresses with the increasing potential for synergistic interactions. This, along with similar health stresses associated with an ever-expanding life-style, often makes it difficult to distinguish health problems of the workplace from those of off-the-job activities.

The complexity of industrial technology, its application in industrial production, and the associated health stresses at the worksite are such that persons with expertise in the related industrial hygiene and allied specialties are needed for the recognition, evaluation, and control of these stresses. Similarly, persons with the proper expertise are needed in the interpretation of information presented in this volume and in the extrapolation of these data to specific situations.

Contents

1	**Rationale**	1
	Lewis J. Cralley, Ph.D., and Lester V. Cralley, Ph.D.	
2	**Toxicologic Data in Chemical Safety Evaluation**	27
	K. S. Rao, D.V.M., Ph.D., Philip G. Watanabe, Ph.D., and Perry J. Gehring, D.V.M., Ph.D.	
3	**Biological Indicators of Chemical Dosage and Burden**	75
	Richard S. Waritz, Ph.D.	
4	**Body Defense Mechanisms to Toxicant Exposure**	143
	James S. Bus, Ph.D., and James E. Gibson, Ph.D.	
5	**Biological Rhythms, Shiftwork, and Occupational Health**	175
	M. J. Smolensky, Ph.D., Dennis J. Paustenbach, Ph.D., and Lawrence E. Scheving, Ph.D.	
6	**Work Costs and Work Measurements**	313
	Steven M. Horvath, Ph.D.	
7	**Interpreting Exposure Levels to Chemical Agents**	333
	Charles H. Powell, Sc.D.	
8	**Applied Ergonomics**	375
	Alexander Cohen, Ph.D., and Francis Dukes-Dobos, M.D.	
9	**Abnormal Pressure**	431
	Jefferson C. Davis, M.D.	

10	**Biological Agents**	**451**
	John S. Chapman, M.D.	
11	**Hot and Cold Environments**	**481**
	Steven M. Horvath, Ph.D.	
12	**Ionizing Radiation**	**501**
	Robert G. Thomas, Ph.D.	
13	**Noise**	**557**
	Aram Glorig, M.D.	
14	**Nonionizing Electromagnetic Energies**	**579**
	Sol M. Michaelson, D.V.M.	
15	**Vibrations**	**653**
	John C. Guignard, M.B., Ch.B.	
Index		**725**

PATTY'S INDUSTRIAL HYGIENE AND TOXICOLOGY

Volume I
GENERAL PRINCIPLES

Volume II
TOXICOLOGY

Volume III
THEORY AND RATIONALE
OF INDUSTRIAL HYGIENE
PRACTICE

CHAPTER ONE

Rationale

LEWIS J. CRALLEY, Ph.D., and
LESTER V. CRALLEY, Ph.D.

1 BACKGROUND

The emergence of industrial hygiene as a science has followed a predictable pattern. Whenever a gap of knowledge exists and an urgent need arises for such knowledge, dedicated people will gain the knowledge.

The harmful effects from exposures to toxic substances in the workplace, producing disease and death among workers, have been known well for over two thousand years. Knowledge on the toxicity of materials encountered in industry and means for control of exposure to toxic materials as well as techniques for its measurement and evaluation, however, were not available during the earlier period of industrial development. With few exceptions, the earliest attention given to worker health was in applying the knowledge at hand, which concerned primarily the recognition and treatment of illnesses associated with the job.

However, the devotion of prime attention to the preventive aspects of worker health maintenance through controlling job-associated health hazards became quite evident if the best interest of the worker was to be served in preventing occupational diseases. Not until around the turn of the twentieth century, though, did major effort begin to be directed toward the recognition, measurement, evaluation, and control of workplace environmental health stresses in the prevention of occupational diseases.

The aim of this chapter is not to document or present chronologically the major past contributors to worker health and their relevant works or the events and episodes that gave urgency to the development of industrial hygiene as a science. Rather, the purpose of the chapter is to place in perspective the many factors involved in relating environmental stresses to health and the rationale upon which the practice of industrial hygiene is based, including the recognition, measurement, evaluation, and control of workplace stresses, the biological responses to these stresses, the body defense mechanisms involved, and their interrelationships.

The individual chapters in Volume III, A and B, cover these aspects in detail.

Similarly, it is not the intent of this volume, A and B, to present procedures, instrumental or otherwise, for measuring airborne levels of exposure to chemical and other stress agents. This aspect is covered in detail in Patty's Industrial Hygiene, Volume I, *General Principles*. Rather, attention in this volume, A and B, is devoted to other aspects of airborne exposure such as representative and adequate sampling, variations in exposure levels, exposure durations, and the like.

2 INSEPARABILITY OF ENVIRONMENT AND HEALTH

Knowledge is being constantly developed on the ecological balance that exists between the earth's natural environmental forces and the existing biological species and how the effects of changes in either may affect the other. In the earth's early history this balance was maintained by the natural interrelationships of stresses and accommodations between the environment and the existing biological species at that site. This system related as well to the ecological balance within the species, both plant and animal.

Studies of past catastrophic events such as the ice age have shown the effects that changes in this balance can have on the existing biological species. The forces that brought on the demise of the dinosaurs that lived during the Cretaceous and Mesozoic periods are uncertain. Most probably major geological events were involved. Studies have also shown that in the earth's past history other animal species, as well, have originated and disappeared.

The human species, however, has been an exception to the ecological balance that existed in the earth's earlier history between the natural environment and the evolving biological species. The human ability to think, create, and change the natural environment has brought on changes, above and beyond those of the existing natural forces and environment, that have an ever increasing impact upon the previous overall ecology balance.

The capacity of humans to alter the environment to serve their benefits is beyond the bounds of anticipation. In early human history these efforts predictably addressed themselves to a better means of survival, that is, food, shelter, and protection. As these efforts progressed, emphasis was placed on gaining knowledge concerning factors affecting human health and well-being. Thus evolved the medical sciences, including public health. In some instances, these efforts resulted in the intervention of the ecological balance in the control of disease. In other situations the environment may have been altered to make necessary resources available as in the building of dams for flood control and

RATIONALE

for developing hydroelectric power. This type of alteration of the localized natural environment and the associated ecological systems may have an impact in developing additional stresses in readjusting the existing ecological balance.

Of more recent impact on health has been the stresses in living and on health brought on by activities associated with personal gratifications such as life-styles as well as those associated with an ever more complex and advancing technology in all of the sciences.

The quality of indoor environment is receiving increasing attention in relation to good health. This applies to the study and control of factors giving rise to psychological stresses associated with living or working in enclosed spaces, as well as air pollution arising from life-styles and building designs and materials. Examples of the latter include airborne contamination from smoking, insulating materials, and fabrics as well as location of housing sites.

Thus the advantages associated with changes in the environment for human benefit and improving the essential quality of living must be at the same time equated with their cost-effectiveness as well as their potential to deteriorate the environment and present a new scheme of stresses.

That humans, for optimal health, must exist in harmony with the surrounding 24-hour daily environment and its stresses is self-evident.

To better understand the significance of the on-the-job environmental health hazards, an overview of the 24-hour daily stress patterns of workers will be helpful, for this permits a perspective in which the overall component stresses are related to the whole of the worker's health.

Our habitat, the earth and its flora and fauna, is in reality a chemical one, that is, an entity that can be described in terms of an almost infinite number of related elements and compounds, the habitat in which the species was derived and in which a sort of symbiosis exists that supports the survival of the individual species.

The intricacy of this relationship is illustrated in the presently recognized essential trace elements: copper, chromium, fluorine, iodine, molybdenum, manganese, nickel, selenium, silicon, vanadium, and zinc. All are toxic when ingested in excess and all are listed in the standards relating to permissible exposure limits in the working environment. Some forms of several of these trace elements are classified as carcinogens. It is most revealing that certain trace elements essential for survival are under some circumstances capable of destruction. Thus it is a question of how much.

The environment is both friendly and hostile. The friendly milieu provides the components necessary for survival: oxygen, food, and water. On the other hand, the hostile environment constitutes a stress in which survival is constantly challenged.

Although numerous factors are obviously involved in the optimal health of an individual, stresses arising out of the overall environment, that is, the workplace, life-style, and off-the-job activities, are dominant. Thus the stresses over the 24-hour period have an overall impact on an individual's health. Any activity over the same period of time that can be stress relieving will have a

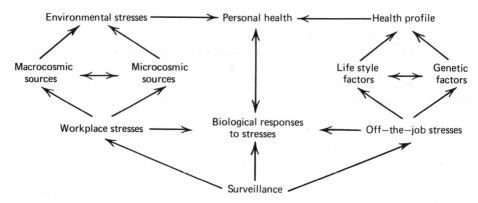

Figure 1.1 Inseparability of environment and health.

beneficial effect in helping the body to adjust to the remaining insults of the day.

The inseparability of the environment and its relation to health is presented graphically in Figure 1.1.

2.1 Environmental Health Stresses

An environmental health stress may be thought of as any agent in the environment capable of significantly diminishing a sense of well-being, causing severe discomfort, interfering with proper body organ functions, or causing disease. These stresses may be chemical, physical, biological, or psychological in nature. They may arise from natural or created sources.

2.1.1 Macrocosmic Sources

Macrocosmic sources of environmental stress agents are those emanating from the solar system or from extensive geographical regions and capable of affecting large geographical areas such as the quality of the air, soil, or water.

Examples of natural sources of these stress agents include ultraviolet, thermal, and other radiations from the sun, volcanic eruptions that release huge quantities of gases and particulates into the upper atmosphere, the changing of the upper air jetstream and other factors that influence the climate, and the movement of the earth's surface plate structure resulting in earthquakes, and tidal waves.

Examples of created macrocosmic stresses include the potential interference of the ozone layer of the upper atmosphere that shields the earth from excessive ultraviolet radiation through the release of large quantities of some organic compounds into the air, the excessive burning of fossil fuels that increase the carbon dioxide level and temperature of the atmosphere, or excessive industrial, community, or life-style pollution over a prolonged period that may affect the quality of the air, water, or soil.

RATIONALE

These stresses may act directly upon the individual as in excessive exposure to ultraviolet and thermal radiation and toxic materials, or indirectly by influencing the earth's climate—sunshine, rain, and temperature—thus affecting vegetation and habitability.

2.1.2 Microcosmic Sources

Microcosmic sources of stress agents are those emanating from localized areas and generally affecting the same region. These are most commonly at the community or regional community level and may also include the home and work environment.

Examples of natural sources of microcosmic stress agents are pollen, which gives rise to sensitization, allergy, and hay fever; water pollution from ground sources having a high mineral or salt content; and the release of methane, radon, sulfur gases and other contaminants from underground and surface areas.

Created sources of health stress agents at the community level are, however, by far the most common source. These include noise from everyday activities as in lawn mowing, motorcycle and truck street traffic, and loud music; air pollution from motor vehicle exhaust, release of industrial emissions into the air, emissions from refuse and garbage land fills and from toxic waste disposal sites, spraying of crops, and life-styles; water pollution through the release of contaminants from home, community and industrial activities into the waste water and seepage into the ground water from garbage and land fills and from toxic waste disposal sites.

These stress agents may have a direct response such as the effects of noise on hearing and toxic exposures on health, or an indirect response such as acid smog and rain affecting vegetation and soil quality downwind from the source.

2.1.3 Life-Style

The life-style of an individual, including habits, nutrition, off-the-job activities, recreation, exercise, and rest, may have beneficial effects as well as innate and insidious stresses that exert a profound influence on that person's health.

Extensive knowledge is being developed constantly on the influence of habits such as excessive smoking, alcohol consumption, and the use of drugs on health. These may have additive, accumulative, and synergistic actions or may exert superimposed responses on other exposures in the workplace. These agents can be an overwhelming cause of off-the-job respiratory, cardiovascular, renal, and other diseases and may create a grave health problem in an individual previous to other exposures encountered in the workplace.

Knowledge is also emerging on the deleterious effects of excessive smoking, alcohol consumption, and drug use by women during pregnancy on the health of their offspring, including malformation and improper functioning of body organs and systems, abnormal weight, and so forth.

The benefits to health of a good, adequately balanced nutrition, that is, vitamins, minerals, and other essential food intake, are gaining increased attention in relation to general fitness, weight control, prevention of disease, supporting the natural body defense mechanisms, recovery from exposure to environmental stresses, and aging. Conversely, malnutrition and obesity are the cause of many illnesses and may have a synergistic effect on exposure to other stresses.

Recreational activities are important aspects of good health practices, releasing tension brought on through both off- and on-the-job stresses. Conversely, many recreational activities may be harmful, such as listening to loud music at home or at discos, which may lead to a loss in hearing, frequent engagements in events and schedules that may interfere with the body's internal rhythmic functions, failing to observe needed precautions while using chemicals in hobby activities, and pursuing activities to the point of exhaustion.

Both exercise and rest are important activities in maintaining good health. Exercise helps maintain proper muscular tone as well as weight control. The exercise should, however, be designed for specific purposes, maintained on a regular basis, and structured to accommodate the body physique and health profile of an individual. Otherwise, more harm than benefit may result. Rest provides time for the body to recuperate from physical and psychological stresses.

The hours between shifts and during weekends provide time for the body to excrete many agents absorbed during the workshift. Work patterns that disturb this recovery period, as in moonlighting, may have an even increased deleterious effect if similar stresses are encountered on the second job. Likewise, smoking, alcohol consumption, and use of drugs may impair the body's proper recuperation from previous stresses.

2.1.4 Off-the-Job Stresses

The worker may encounter a host of innate stresses outside the workplace. These may be of a physical or psychological nature and usually include the eight to ten hour period between the end of the shift and retiring and on weekends. Most any of the off-the-job activities if performed in excess and without regard to necessary precautions are capable of producing stress and injury. Hobby and recreational activities likely account for a good portion of this time. More recently, increasing periods of time are spent sedentarily watching television during which eyestrain, mental and physical fatigue, and general inactivity occur—poor preparation for a refreshing night's sleep.

In hobby activities the participant often encounters many of the environmental exposures experienced on the job, such as in soldering, welding, cleaning, gluing, woodworking, grinding, sanding, and painting, during which the related exposures are well controlled. The home hobbyist, however, will not likely have available the protective devices needed such as local exhaust ventilation, protective clothing, goggles, and respirators, or protective creams. The same hobbyist often does not read or observe the precautionary labels, takes short cuts to save time, and fails to use the basic principles of keeping toxic materials from the skin, thoroughly washing after skin contact, not smoking or eating while working with these materials, and keeping upwind of harmful materials encountered as in insecticide and other spraying in gardening.

The activities previous to and supporting the sleep environment may not augment sleep. Thus the body may be fatigued or otherwise stressed prior to entering the workplace and endanger the worker or those around that person.

The off-the-job gamut of health stresses is wide and formidable. To deal with these stresses satisfactorily requires that a degree of accommodation be reached based on judgment, feasibility, personal option, objectives, and other factors. It is evident that although the components of the total environmental health stresses must be considered on an individual basis, each component cannot stand alone and apart from the others.

2.1.5 Workplace Stresses

The workplace may be a most important source of health stresses if the operations have not been studied thoroughly and the associated health hazards eliminated or the necessary controls incorporated into plant processes.

This was evidenced during the earlier industrial growth when little information was available on the methodology for identifying, measuring, evaluating, and controlling work related stresses. During this period workplace exposures were often severe, leading to a high incidence of disease and related deaths.

At the turn of this century, and especially since the 1950s, management, unions, government, academic, and other groups have taken an increased interest in worker health and in the control of exposures to stresses in the workplace. This, along with the setting of standards of exposure limits, has created an urgency for a broad support of functions and knowledge to prevent on-the-job illnesses.

2.1.6 Biological Response to Environmental Health Stresses

The human body consists of a number of discrete organs and systems derived from the embryonic state, encased in a dermal sheath, and developed to perform specific functions necessary for the functioning of the body as an integral unit. These organs are interdependent so that a malfunction in one may affect the functioning of many others. As an example, a malfunction in the alveoli which hinders the passage of oxygen into the blood transport system may have a direct effect on other organs through their diminished oxygen supply. Similarly, any hormonal imbalance or enzyme aberration may affect the functioning of many other body organs. Once invasion has occurred, toxic agents may selectively target one or more of the organs. Each organ has a varying degree for resisting stress, adjusting to injury, and propensity for repair.

Although the organ structures and functions of the several animal species have many similarities, with some more approximate than others to that of the human body, care has to be taken in extrapolating research data on any one animal species to that of the human body. Similarly, research data obtained within a smaller frame of reference, such as may be done through cell cultures, has to be cautiously interpreted with regard to the whole integrated human body organ system.

2.1.7 Body Protective Mechanisms Against Environmental Stresses

The human body in coexisting with the hostile stresses of the external environment has established a formidable scheme of protection against many of these stresses. This is accomplished in a remarkable manner by the ectodermal and endodermal barriers preventing absorption through inhalation, skin contact, and ingestion, and supported by the backup mesodermal and biotransformation mechanisms once invasion has occurred. These external and internal protective mechanisms against invasion, however, are not absolute and can be overwhelmed by a stress agent to the extent that they are ineffective, with resultant disease and death. Also, these protective mechanisms may become impaired in various degrees from insults associated with life-styles and other daily activities.

In studying the effects of specific stresses on health it is important to be aware of the body protective mechanisms involved. Suitable control of a specific stress should suppliment the body's protective response to that stress.

2.1.8 Coaptation of Health and Environmental Stresses

To survive, the human body must live in balance with the surrounding environment and its concomitant stresses. Since these stresses, singularly or combined, are not constant in value even for short periods of time or over limited geographical areas, the body must have built-in mechanisms for adjusting to differing levels of stresses through adaptation, acclimatization, acclimation, and other regulating systems. There is a limit, however, to which the body can protect itself against these stresses without a breakdown occurring in these protective systems.

Consequently, if humans are to have freedom of geographic movement and of living in an excessively hostile environment, some kind of an additional accommodation must be reached with the environmental insults that are not satisfactorily handled by or overwhelm the related body protective mechanisms. Exceptions include limited living and working in confined spaces where the environment is under absolute control against outside catastrophic stresses as in space travel and underwater activities.

Examples of this coaptive relationship include exposure to ultraviolet radiation from the sun. It is obvious that avoidance of all ultraviolet radiation from this source is impracticable. In addition to the ozone layer of the atmosphere shielding the earth from major levels of ultraviolet radiation emitted by the sun and the body's own protective mechanisms such as pigmentation, further accommodation is reached through the use of skin barriers, eye protection through use of sun glasses, and a managed limitation to exposure.

In the case of allergens in foods, cosmetics, air, fabrics, drugs, and so on, discovery of the offending agents and avoidance of them constitute a prime approach, though medical management is highly important.

In high altitudes where the oxygen content of the air is diminished, the body acclimatizes by increasing the number of red blood cells and hemoglobin that carry oxygen to the tissues. Further accommodation may be provided through the use of supplemental oxygen supply.

In exposures to excessively high and low temperatures, the body has a number of regulating systems to keep the body temperature within normal limits. This in turn permits living in a wide range of environmental temperatures. Further accommodation to extremes in environmental temperatures may be provided through special clothing, protective equipment, and living and working in climate controlled buildings.

Where excessive exposures to environmental stresses exist in the workplace, emphasis is placed on their elimination or lowering the stress levels through recognized control procedures to the point that the body defense mechanisms can adequately prevent injury to health, though some situations may recommend the use of personal protective equipment or other control strategies.

There is a limit to what can be done to alter environmental stresses from natural macrocosmic sources. Thus, these become ubiquitous background stresses upon which other exposures from microcosmic sources are added including community and industrial pollution, off-the-job activities, life-styles, and on-the-job activities.

Rationally, then, it is primarily the created stresses from a few macrocosmic but predominantly microcosmic sources that are amenable to control and must be kept within acceptable limits to permit the human body to reasonably exist in an increasingly complex industrial technological age.

3 INDUSTRIAL TECHNOLOGICAL ADVANCES

In the early history of civilization, the many activities associated with living were at the tribal level where emphasis was placed upon survival, that is, procuring adequate food, clothing, protection, and shelter. Though the tribes were nomadic, they were undoubtedly aware that they must live in accommodation with the environment. This would have been evidenced through the appropriate use of clothing, safe use of fire, and observing climatic patterns.

It was inevitable that the nomadic way of life would give way in most instances to a more settled life-style in which food, shelter, and protection could be more dependable, based upon individual effort and ingenuity. During this period the accommodations to the natural elements were made easier through a more permanent shelter and pattern of living.

The next advancement in industrialization came through the realization that increased production could be attained through specialization of work pursuits wherein a designated work group devoted its entire effort to the making of a single product as clothing, pottery, and tools, or the growing of foods, and in which each group shared their commodity in exchange for the commodity of other groups. This was the beginning of the cottage industries that related primarily to the community level of trade. Even at that level of production many of the health stresses associated with the different pursuits were intensified over that period of nomadic living where every person was a sort of jack-of-all-trades. This was especially true where the operations tended to be restricted to confined and crowded spaces.

As means of communications and transportation improved, trade increased between adjoining communities, and the search continued for better ways of producing commodities at an increased volume with less manpower. Similar production operations tended to expand and be concentrated within the same housing structure. This led to an increase in the health stresses of the whole workforce in instances where the stress agents were cumulative in intensity and response.

This industrialization trend was continued and intensified with the advent of the steam engine, which gave rise to an industrial revolution involving larger factories and newer production techniques along with increased associated health risks. Where in the past, exposures to health risks in any one workplace were dominantly to a few stress agents, the newer industrialization and technology led to a more complex pattern of exposures.

Since the 1950s technological advances and their application to production have expanded at an ever increasing pace. The application in industry of knowledge gained through space technology research has rapidly expanded into the newer electronic high-technology age.

The advent of this high technology and its application to production is having its effect upon both the nature of employment and the concomitant health stresses. While in the past workers needed only special instructions to perform most job operations—and this will continue for some time—the move into higher technology has created a demand for highly trained employees for many job positions, and this trend will increase dramatically. Computers, word processors, video display terminals, radar, and other electronic equipment are becoming commonplace in industry. Robots are rapidly taking over many repetitive operations.

Similarly, the nature and extent of associated health stresses are becoming more complex with the advent of the high technology industry.

The urgent need for knowledge concerning the effects of exposures to health stresses associated with an ever expanding industrial technology, along with the methodology for their recognition, evaluation, and control, gave rise to the science of industrial hygiene. This science must keep attuned to the ever changing applications of technology in industry.

4 EMERGENCE OF INDUSTRIAL HYGIENE AS A SCIENCE

Science may be defined as an organized body of knowledge and facts established through research, observations, and hypothesis. As such, a science may be basic as relating to the physical sciences, or applied in the sense that the principles of other sciences are brought to bear in developing facts and knowledge in a specific area of application.

The lack of knowledge during the early history of industrial development on the effects of health stresses associated with industrial operations and how these could be controlled, with the often concomitant massive exposures to harmful materials in the workplace, led to many serious episodes of illnesses and deaths among workers. An example is the high incidence of silicosis and silicotuberculosis that existed at the turn of the century in the workers of the hard rock mines, granite industry, and tunnel operations where the dust had a high free silica content.

During this early period the major effort on behalf of the workers was to apply the knowledge at hand, which related primarily to the recognition and treatment of occupational illnesses.

It was not until around the turn of this century that specific attention began to be devoted to the preventive aspects of industrial illnesses. Scientists, including engineers, chemists, and physicists, began to apply their knowledge and expertise toward the development of methods and procedures for identifying, measuring, and controlling exposures to harmful airborne dusts and chemicals in the workplace. At that time there were no recognized procedures for accomplishing these objectives.

After trying various potential procedures, the impingement method was judged the most adaptable one at that time for collecting many of the airborne contaminants such as particulates, mists, some fumes, and gases. The light field dust counting technique was developed for enumerating levels of dust in the air; and conventional analytical methods available at that time were adapted for measuring levels of chemicals in these samples.

Even during this early period of developing airborne sample collection and analytical procedures, these scientists realized that the exposure patterns that existed were more complex and complicated than the instrumentation for sample collection and analysis would define. These scientists also knew of many of the deficiencies associated with the proposed procedures being developed. They were aware that the data being collected represented only a segment of the overall exposure pattern. They believed, however, that this segment value could be used as an index which would represent the overall exposure pattern so long as the production techniques and other operational factors remained the same. It must be remembered that at that time information was not available on respiratory deposition and dust size. It was imperative to them, and rightly so, that some method, with whatever deficiencies that may have been incumbent, be developed for indexing airborne levels of contamination in the workplace both for estimating levels of exposure and for use as a benchmark in determining degrees of air quality improvement after controls had been established.

Since the earliest instrumentation for collecting airborne contaminant samples was nonportable, such samples represented general room levels of exposure depending upon where the samples were taken.

The above procedures for airborne sampling and analysis in the workroom, as primitive as they may seem in comparison to those methods available today, served that period of time well. They accounted for the drastic reduction in massive exposures that existed in many work sites and were the methods and procedures upon which future refinements would be made.

These scientists showed that it was feasible to lower the massive workplace airborne contamination that often existed at that time, and by relating the data to the health profile of the worker, observed that lowering of the exposure level also lowered the incidence of the associated disease. Thus began the first field studies that were to have a profound influence on the collection of relevant data upon which to base permissible limits and in developing the rationale upon which the practice of industrial hygiene is predicated. The latter was to be further substantiated through laboratory and clinical research.

The development of the hand-operated midget impinger pump during the 1930s was a decided improvement over the standard impinger pump, since it was portable and permitted taking airborne samples closer to the worker using impingement, fritted glass bubbler, and other collecting techniques. The newer exposure data showed that the worker often had a higher level of exposure than that indicated by the general room airborne level.

Other instrumentation, such as the electrostatic precipitator and evacuated containers, came into use during this period. In the late 1940s the filter paper method for collecting some airborne particulate samples came into use. This procedure was found to be superior in many instances to the impingement method. This method did not fracture particulates or disperse conglomerates, which often accompanied impingement collection, and permitted direct gravimetric measurement of the sample.

Data also began to emerge on the importance of particulate size in relation to retention and deposition in the respiratory tract. The electron microscope made it possible to study the characteristics and response of submicron particle sizes.

Also, a dramatic increase began in toxicologic and epidemiological studies by government, industry, educational institutions, and foundations directed to obtaining data upon which to base exposure standards as well as good industrial hygiene practices.

Another major advancement in developing better methodology for studying occupational diseases occurred at midcentury. Toxicologic and other studies had revealed that lowering the exposure and extending the period of exposure changed the dose–response pattern of many toxic agents. As an example, a high airborne level of exposure to lead produces an acute response over a relatively short period of time. In contrast, lowering the exposure level of the same agent and extending the exposure time shows a different dose–response pattern, a chronic form of lead poisoning. Thus in studying the effects of exposure to health stress agents it is important to obtain relevant dose–response data over an extended period of exposure time. One method of obtaining relevant health profile data on workers is through medical records and special studies. Another and more recent method is through the study of causes of death on death certificates through Social Security, management retirement system, and Union records. These studies have revealed that a lifetime of work exposure to an agent or an extended latency period of twenty or more years from time of initial exposure may be necessary to fully define the wide range of dose–response relationships. This may be especially true for carcinogenic and other long-term types of exposure response.

A more recent advancement relates to chronobiology, the study relating to the body's internal biological rhythms and their effects on organ functions, etc. The workweek schedule can have a direct effect on these rhythmic patterns and health. Also, there is some evidence that the rate of absorption of toxic materials and reaction to stress may relate in some way to an individual's chronobiology.

A surge of improved and sophisticated techniques took place in the 1960s for quantifying worker exposure to health stress agents in the workplace. This applied both to sample procurement and analytical techniques in which much lower levels of exposure to specific agents could be determined.

The establishment of professional associations to support the interests and growth of the profession has played an important role in developing industrial hygiene as a science. In the United States, the American Public Health Association in the 1930s had a section on industrial hygiene that supported the early growth of the profession. The American Conference of Governmental Industrial Hygienists was organized in 1938. The American Industrial Hygiene Association was organized in 1939. The American Board of Industrial Hygiene was created and held its first meeting in 1960. The Board certifies in the comprehensive practice of industrial hygiene as well as in six additional specialties. Industrial hygienists certified by the Board have the status of Diplomates and as such are eligible for membership in the American Academy of Industrial Hygiene.

The American Industrial Hygiene Association has established a laboratory accreditation program with the objective of assisting those laboratories performing industrial hygiene analysis in achieving and maintaining the highest level of professional performance.

The American Industrial Hygiene Foundation was established in 1979 under the auspices of the American Industrial Hygiene Association. The functions of the Foundation are carried out by an independent Board of Trustees. The Foundation provides fellowships to worthy industrial hygiene graduate students, encourages qualified science students to enter the industrial hygiene profession, and entices major universities to establish industrial hygiene graduate programs.

In the United States, a number of occupational health regulations were established in the early 1900s with emphasis, in several, on listing limits of exposure to a relatively few agents. These regulations were effective at the local, state, and federal levels depending on governmental jurisdiction.

The Social Security Act of 1935 and the Walsh-Healy Act of 1936 had an immense impact in giving increased stability, incentive, and expanded concepts in the practice of industrial hygiene. These acts stimulated industry into incorporating industrial hygiene programs as an integral part of management. They also stimulated broad base programs in industry, foundations, educational institutions, insurance carriers, unions, and government into the cause, recognition, and control of occupational diseases. These acts established the philosophy that the worker had a right to earn a living without endangerment to health, and were the forerunners for the passage of the Occupational Safety and Health Act of 1970.

The passage of the latter act, enacted for the purpose of assuring "so far as possible every man and woman in the nation safe and healthful working conditions," had a very broad bearing on the further development and practice in the industrial hygiene profession. The above regulations and Acts greatly supported the establishment of industrial hygiene as a science. It has been necessary to expand the profession in all of its concepts and technical aspects to meet these responsibilities.

Other industrialized countries have had a similar experience in the professional recognition and growth of the science relating to the recognition, measurement, evaluation, and control of work related health stresses.

4.1 Definition of Industrial Hygiene

The American Industrial Hygiene Association defines industrial hygiene as "that science and art devoted to the recognition, evaluation, and control of those environmental factors or stresses, arising in or from the workplace, which may cause sickness, impaired health and well-being, or significant discomfort and inefficiency among workers or among the citizens of a community."

By definition, industrial hygiene is an applied science encompassing the application of knowledge from a multidisciplinary profession including the sciences of chemistry, engineering, biology, mathematics, medicine, physics, toxicology, and other specialties. Industrial hygiene meets the criteria for the definition as a science since it brings together in context and practice an organized body of knowledge necessary for the recognition, evaluation, and control of health stresses in the work environment.

In the early 1900s, the major thrust in the control of workplace health stresses was directed toward those areas in industry having massive exposures to highly toxic materials. The professional talents of the engineer, chemist, physician, physcist, and statistician were those largely used in these programs.

As industrial technology advanced, the complexity of the worker exposure also increased along with the professional talents needed to study the effects, recognition, evaluation, and control of the newer health stress agents. This is especially true with the advent of high technology in the electronic and allied industries. Stress factors such as improper lighting and contrast, glare, posture, need for strict attention for a given task, fatigue, ability to concentrate, tension, and many others arise in the operation of computers, word processors, video display terminals, and radar equipment, which are becoming commonplace in industry. Thus, concerns for the health of the employees above and beyond that of toxicity response arise. The study and control of these newer stress agents point to the need for the occupational health nurse, psychologist, human factors engineer, and others to join the professional team studying the effects and control of the ever widening list of health stress agents in the workplace.

In the early practice of occupational health nursing, emphasis was placed around such activities as the emergency treatment of traumatic injuries stipulated in written orders of a physician and in obtaining records information relating to physical examinations and the like. With the advanced training of the occupational health nurse, this limited role was found to be wasteful of professional talent and resources. The occupational health nurse is often the first interface between the worker and pending health problems and is in a position to gain information on situations and health stresses both on and off-the-job that may, unless addressed, lead to a more serious response. The occupational health nurse has increasingly become a member of the multidisciplinary team needed in the recognition of job associated health stresses.

Similarly, the psychologist, in the study of the effects of strain, tension, and similar factors, and the human factors engineer in designing machines, tools, and equipment to meet the requirements of the worker, are examples of other professionals joining the multidisciplinary team studying the effects and control of the ever increasingly complex health stresses associated with advanced industrial technology.

The complexity of this multidisciplinary profession needed for carrying out the related responsibilities of the Occupational Safety and Health Act of 1970 is further illustrated in the more than 20 technical committees of the American Industrial Hygiene Association and the seven different areas of certification by the American Board of Industrial Hygiene.

4.2 Rationale of Industrial Hygiene Practice

The practice of industrial hygiene is based upon the following tenets:

Environmental health stresses in the workplace can be quantitatively measured and expressed in terms that relate to the degree of stress.

Stresses in the workplace, in main, show a dose–response relationship. The dose can be expressed as a value integrating the concentration and the time duration of the exposure to the stress agent. In general, as the dose increases the severity of the response also increases. As the dose decreases the biological response decreases and may at some time in dose value exhibit a different kind of response, chronic versus acute, depending on the time duration of the stress even though the total stress expressed as a dose–response value may be the same.

The human body has an intricate mechanization of protection, both in preventing the invasion of hostile stresses into the body and in dealing with stress agents once invasion has occurred. For most stress agents there is some point above zero level of exposure which the body can thus tolerate over a working lifetime without injury to health. Levels of exposure to specific stress agents should be kept within prescribed safe limits. Regardless of their type, all exposures in the workplace should be kept within the limits attainable through good industrial hygiene and work practices. Some stress agents, though, may have a biological response at such a low level of exposure that a minimal level of exposure should be observed. Examples of such agents include those that have hypersensitive, hypersusceptible and genotoxic properties.

The elimination of health hazards through process design should be the first objective in maintaining a healthful workplace. Where this is not feasible, recognized engineering methods should be used to keep exposures within acceptable limits. In some instances, supplemental programs such as the use of personal protective equipment and other control strategies have application.

Surveillance of both the work environment and the worker should be maintained to assure a healthful workplace.

4.3 Elements of an Industrial Hygiene Program

The purpose of an industrial hygiene program is to assure a healthful workplace for employees. It should include all the functions needed in the recognition, evaluation, and control of occupational health hazards associated with production, office and other work. This requires a comprehensive program designed around the nature of the operations, documented to preserve a sound retrospective record and executed in a professional manner.

The basic components of a comprehensive program include the following:

1. An ongoing data collection system that provides the essential functions for identifying and assessing the level of health hazards in the workplace.
2. Participation in the periodic review of worker exposure and health records to detect the emergence of health stresses in the workplace.
3. Participation in research, including toxicological and epidemiological studies designed to generate data useful in establishing safe levels of exposure.
4. A data storage system that permits appropriate retrieval to study the long term effects of occupational exposures.
5. Assuring the relevancy of the data being collected.
6. An integrated program capable of responding to the need for the establishment of appropriate controls, both current and those resulting from technological advances and associated process changes.

The industrial hygienist at the corporate or equivalent level should have responsibility of reporting to top management. This involves appropriate input

RATIONALE 17

wherever product, technological, operational, process changes, or other considerations may have an influence on the nature and extent of associated health hazards so that adequate controls can be incorporated at the design stage.

5 HEALTH HAZARD RECOGNITION

An important aspect of a responsive industrial hygiene program is that it is capable of recognizing potential health hazards or, where new materials and operations are encountered, to exercise judicious judgment in maintaining an adequate surveillance program until the associated health hazards have been defined. This should not be a problem in cases involving operations, procedures, or materials where adequate knowledge is available, and it is primarily a matter of application of knowledge and techniques. In operations and procedures involving a new substance or material where relevant information is limited or unavailable, it may be necessary to extrapolate information from other kindred sources and use professional judgment in setting up a control program with a reasonable factor of safety and to incorporate an ongoing surveillance program to further define health hazards that may emerge. In some instances it may be necessary to undertake toxicological research prior to the production stage to better define parameters needed in setting up the control and surveillance program.

One of the basic concepts of industrial hygiene is that the environmental health stress of the workplace can be quantitatively measured and recorded in terms that relate to the degree of stress.

The recognition of potential health hazards is dependent on such relevant basic information as:

1. Detailed knowledge of the industrial process and any resultant emissions that may be harmful.
2. The toxicological, chemical, and physical properties of these emissions.
3. An awareness of the sites in the process that may involve worker exposure.
4. Job work patterns with energy requirements.
5. Other coexisting stresses that may be important.

This information may be expressed in a number of ways depending upon its ultimate use. A most effective form is the material-process flow chart that lists each step in the process along with the appropriate information just noted. This permits the pinpointing of areas of special concern. The effort in whatever form it may take, however, remains only a tool for the use of the industrial hygienist in the actual assessment of the stresses in the workplace. In the quantitation itself, many approaches may be taken depending on the information sought, its intended use, the required sensitivity of measurement, the level of effort and instrumentation available, and the practicality of the procedures.

Aside from the production workplace with the attended toxicological, physical, and other related health stresses, a new area of concern is rapidly gaining special attention where employees may be subjected to a high degree of stress from tension, physical and mental strain, fatigue, excessive concentration, and distraction such as may exist for operators of computers, word processors, video display terminals, and radar equipment. Off-the-job stresses, life-style factors, and the immediate room environment may become increasingly important for such operators. The recognition of associated health stresses and their evaluation require a special battery of psychological and physiological body reaction and response tests to define and measure factors of fatigue, tension, eyestrain, ability to concentrate, and the like.

6 EXPOSURE MEASUREMENTS

As shown in the Table 1.1, both direct and indirect methods may be used to measure worker exposure to stress agents.

6.1 Direct Measurement

To measure directly the quantity of the environmental stress actually received by the body, fluids, tissues, expired air, excreta, and so on must be analyzed to determine the agent per se or a biotransformation product. Such procedures

Table 1.1 Methods for Measuring Worker Exposure to Stress Agents

Direct	Indirect
Body dosage	Environment
Tissues	Ambient air
Fluids	Interface of body and
Blood	stress
Serum	Physiological response
Excreta	Sensory
Urine	Pulse rate and recovery
Feces	pattern
Sweat	Heart rate and recovery
Saliva[a]	pattern
Hair[a]	Body temperature and
Nails[a]	recovery pattern
Mother's milk[a]	Voice masking, etc.
Alveolar air	

[a] Not usually considered to be excreta; see, however, Volume 3 B, Chapter 3, Sections 11 and 12.

may be quite involved, since the evaluation of the data at times depends on previous information gathered through epidemiological studies and animal research. Studies on animals, moreover, may have used indirect methods for measuring exposure to the stress agent, necessitating appropriate extrapolation in the use of such values. Examples are blood levels now in use for the evaluation of lead exposures and urinary fluoride levels for environmental fluoride exposures..

One decided advantage of biological monitoring is that a time-weighted factor is integrated that is difficult to estimate through ambient air sampling when the exposure is highly intermittent or involves peak exposures of varying duration. Conversely, it may fail to reflect adequately peak concentrations per se that may have special meaning. Urine analysis may also provide valuable data on body burden in addition to current exposures when the samples are collected at specific time intervals after exposure, such as at the end of the work shift and before returning to the job, to incorporate a suitable time lapse.

Sampling the alveolar air may be an appropriate procedure for monitoring exposures to organic vapors and gases. An acceleration of research in this area can be anticipated because of the ease with which the sample can be collected.

6.2 Indirect Measurement

The most widely used technique for the evaluation of occupational health hazards is indirect in that the measurement is made at the interface of the body and the stress agent, for example, the breathing zone or skin surface. In this approach the stress level actually measured may differ appreciably from the actual body dose. For example, all the particulates of an inhaled dust are not deposited in the lower respiratory tract. Some are exhaled and others entrapped in the mucous lining of the upper respiratory tract and eventually are expectorated or swallowed. The same is true of gases and vapors of low water solubility. Thus the target site for inhaled chemicals is scattered along the entire respiratory tract, depending on their chemical and physical properties. Another example is skin absorption of a toxic material. Many factors, such as the source and concentration of the contaminant, i.e., airborne or direct contact, and its characteristics, body skin location, and skin physiology, relate to the amount of the contaminant that reacts with or is absorbed through the skin.

The indirect method of health hazard assessment is, nevertheless, a valid one when the techniques used are the same or equivalent to those relied on in the studies that established the standards.

The sampling and analytical procedures must relate appropriately to the chemical and physical properties of the agents assessed, such as particle size, solubility, and limit of sensitivity for analytical procedures. Other factors of importance are weighted average values, peak exposures, and the job requirements with the energy demand which is directly related to respiratory volume and retention characteristics.

7 ENVIRONMENTAL EXPOSURE QUANTIFICATION

Procedures for measuring airborne exposure levels of a stress agent depend to a great degree upon the reasons for making the measurements. Some of these are (1) Obtaining worker exposure levels over a long period of time on which to base permissible exposure limits; (2) Compliance with standards; and (3) Performance of process equipment and controls. It is essential that the data be valid regardless of the purpose for which they were collected and that they be capable of duplication. This is a key factor in establishing exposure limits to be used in standards and in fact finding relating to compliance. Since judgment and action will in some way be passed on the data, validity is paramount if these are to be used as a bona fide basis for action.

7.1 Long-Term Exposure Studies

In an epidemiologic study in which the relationship between the stress agent and the body response is sought, the stress factor must be characterized in great detail. This may require massive volumes of data suitable for statistical analysis and a comprehensive data procuring procedure so that a complete exposure picture may be accurately constructed. The sampling procedures and strategy should be fully described, including number and length of time of samples, and their location, i.e., personal or area samples, and be adequate to cover the full work shift activities of the workers. Any departure from normal activities should be noted. These are important since the data may be used at a later date for a purpose not anticipated at the time of sample collection.

The collection of valid retrospective data may be extremely difficult. If available at all, the data may be scanty; the sample collection and analytical procedure may not have been well documented as to precise methodology and may have been less sensitive and efficient or may have measured different parameters in comparison to current procedures; sampling locations and types may not be well defined; and the job activities of the workers may have changed considerably even though the job designation may be the same. Other factors which need to be considered in securing retrospective data relate to contrasting past and current plant operations, including changes in technology and raw materials, effectiveness of control procedures and their surveillance, and housekeeping and maintenance practices. In many instances an attempt to accommodate these differences has been made through broad assumptions and extrapolations with an unknown degree of validity and without expressing the limitations of such derived data.

The effect of national emergencies may significantly change the nature and extent of worker exposure to associated stress agents. The experience during World War II is an example. The work hours per week were increased by four to eight hours in many industries. Control equipment was allocated to specified industries. Local exhaust ventilation systems at times became ineffective or completely inoperable due to lack of maintenance. Less attention was given to

RATIONALE

plant maintenance, housekeeping, and monitoring procedures. Substitute or lower quality raw material had to be used in many instances.

Although the major impact of World War II upon the levels of exposure to harmful agents occurred from around 1940 to the early 1950s, the effect of many of these exposures may not show up in the older work force and retired workers until the early 1980s.

Thus, expressing exposure levels in the past for more than a few years may be only extrapolated guesses unless factors such as the above can clearly be examined and data validity established.

7.2 Compliance with Standards

In contrast to the collection of data for epidemiological studies, data collected for the purpose of compliance with standards may require relatively few samples if the values are clearly above or below the designated value for that agent. If the values are borderline, the evaluation may call for a more comprehensive sampling strategy and may be a matter for legal interpretation. The nature and type of samples taken should meet the criteria upon which the standards were based. Scientifically, though, the data should be adequate to establish a clear pattern with no one single value being given undue weight and should meet data analysis requirements.

7.3 Spot Sampling

The exposure of a worker may arise from a number of sources, including the ambient levels of the agent in the general room air which in turn may be influenced by ambient levels of the agent in the community air, leaks from improperly maintained operating equipment such as from joints and flanges, the inadequate performance of control equipment, and the care which workers observe in performing job operations. Spot sampling can easily detect the effects of any one of these factors on the overall worker exposure level and point to the direction in which further control action should be taken.

8 DATA EVALUATION

The evaluation of airborne levels of a stress to determine compliance with a standard or to determine specific sources of the stress agent are generally uncomplicated and straightforward.

The evaluation of environmental exposure data that serve as a basis for determining whether a health hazard exists is more complicated and requires a denominator that characterizes a satisfactory workplace. Similarly, the use of environmental exposure data for establishing safe levels of exposure or a permissible exposure level, as in empidemiologic studies, requires their correlation with other parameters such as the health profile of the work force.

As pointed out earlier on, the early field studies of the 1920–1930s showed that when the very high exposures of workers were lowered, there was a corresponding lowering of the related disease incidence in the workers. These and other studies gave support to the dose–response rationale upon which the practice of industrial hygiene is primarily based, i.e., there is a dose–response relationship between the extent of exposure and severity of biological response to most stress agents and in which the response is negligible at some point above zero level.

There is great difficulty, however, in determining lower levels of exposure over a working lifetime to a specific agent and its effect on the health of the worker. Often, this is done through extrapolation of other data or trying to estimate past exposures. In the lower range of the dose–response region, the incidence of disease from exposure to an agent may be so low that it approaches or is within the existing reservoir for that disease in the community outside the industry under study. This results from exposure of the general population to stress agents such as from smoking, alcohol consumption, drug use, hobby activities, community and in-house noise, and the like, which may in many instances be similar to those on the job or may be additive to, accumulative, or synergistic with stresses from on-the-job exposures. Even the best available control studies are often not sensitive enough to give reliable data upon which to further extrapolate data for use in the lower dose–response region. Thus, at some point the effects on health from ever-present, extraneous, off-the-job stresses cannot be distinguished easily from low-level, on-the-job stresses.

It is known, however, that the body has protective mechanisms to guard against the entrance of many environmental agents and to handle those where invasion has occurred. For the vast number of agents encountered in the industrial environment, data on dose–response relationship support the industrial hygiene rationale that the level of the stress agent does not have to be zero over a lifetime of work to prevent injury to the worker's health. Some agents, however—such as those having genotoxic properties in being capable of directly damaging genetic material and those associated with hypersensitivity and hypersusceptibility—may have a biological response at such a low level of exposure that a minimal level of exposure should be observed.

Evaluation of data from exposure stresses relating to tension, fatigue, annoyances, irritation, ability to concentrate and discern, and the like are often subjective and may also involve the personal background, traits, habits, etc., of those being stressed for proper definition and control. This evaluation must be done by a different specialist brought on by a changing technology and society.

Thus, each exposure stress must be considered on an individual basis even though it may not be readily separated from the complex pattern of the 24-hour overall environment.

9 ENVIRONMENTAL CONTROL

The cornerstones of an acceptable industrial hygiene program can be described as follows:

1. Proper identification of on-the-job health hazards.
2. The exposure measurements of such hazards.
3. Data evaluation.
4. Environmental control.

In essence, the success of the entire program depends upon the degree and method of implementation, that is, the control strategy. The technical aspects of the program must encompass sound practices and must be related both to the worker and to the medical preventive program. This constitutes a challenge to the professionalism and to the ultimate contribution of the industrial hygienist. The effort and cost of the program must be commensurate with its effectiveness when examined in context with the other off-the-job health stresses.

The heart of the control program must rest with process and/or engineering controls properly designed to protect the workers' health. The most effective and economic control is that which has been incorporated at the stage of production planning and made an integral part of the process. With new processes this can be accomplished by bringing input at the bench design, pilot, and final stages of process development. It is neither good industrial hygiene practice nor sound economics to design minimal control into a process with the intention of adding supplemental control hardware piecemeal, as indicated by future production, or to comply with regulations. On the other hand, the need may exist for the judicious use of personal protective equipment under unique circumstances—for example, breakdowns, spills, accidental releases, maintenance, housekeeping, and certain repair jobs. The adequate control program must embrace a proper mix of process and/or engineering control hardware, personal protective equipment, and administrative control. No single design can be made to fit all circumstances. Rather, each program must be tailored to fit the individual situation without violating the basic tenets of industrial hygiene practice.

Engineering controls are being supported increasingly with automatic alarm systems to give an alert when the controls are malfunctioning and excessive air contamination is occurring.

The control of stresses associated with high technology in the operation of equipment such as computers, word processors, video display terminals, radar equipment, and the like requires a different engineering approach from that used in the control of toxic stresses. Providing optimal lighting contrast, preventing glare, adjusting the equipment to the operator's stature and capability, and maintaining an overall general room compatibility with the tasks are required where eyestrain, fatigue, tension, and concentration on details are encountered. Additive considerations including special rest periods and designated exercises may be indicated.

10 EDUCATIONAL INVOLVEMENT

In the late 1930s only a few universities in the United States offered courses leading to degrees in industrial hygiene per se at the undergraduate or graduate level. In contrast, in 1983 there were over 55 colleges and universities offering courses leading to undergraduate and graduate degrees in industrial hygiene. This attests to the enormous growth in the profession that has taken place over the past 40 years. The passage of the Occupational Safety and Health Act of 1970 had an impact on this growth.

The American Industrial Hygiene Association has increased in membership from 160 in 1940 to 5954 in 1982. In 1982 the Academy of Industrial Hygiene had 1865 diplomates, 800 industrial hygienists in training, and 126 industrial hygiene technologists.

Professional organizations such as the American Industrial Hygiene Association, the American Conference of Governmental Industrial Hygienists, and the American Academy of Industrial Hygiene offer an excellent opportunity for the interchange of professional knowledge and the continuing education of the industrial hygienist. These professional organizations invite participation through technical publications, lectures, committee activities, seminars, and refresher courses. As an example, the American Industrial Hygiene Conference of 1983 presented 485 technical papers covering a wide range of subjects. The same conference offered 43 professional development courses for the purpose of increasing knowledge and expanding skills in the practice of industrial hygiene. This participation by the experts in the many facets of the profession will enhance the overall performance of the profession and will permit members to keep abreast of newer industrial technology in the recognition, measurement and control of associated work stresses. These associations and the academy support the profession in its fullest concept.

The industrial hygienist is obligated to keep involved in the educational process by making professional information available to other groups having an interest in and a responsibility for the health of workers.

The industrial hygienist should have an active role in educating management concerning environmental stresses in the plant and the programs for their control. An alert management can bring pending situations to the industrial hygienist for study and follow-up and thus prevent the inadvertent occurrence of health problems.

The educational involvement of the worker is extremely important. The worker has a right to know the status of the job environment, the factors that may be deleterious to health if excessive exposures occur, and the control programs that have been instituted. Knowledgeable workers are in a position to enhance their own protection through the proper use of control equipment such as local exhaust ventilation and personal protective devices. A worker is often the first to observe that a control system is not functioning properly and can inform management of this situation. In cases of emergency spills and leaks or equipment breakdown, the worker who is informed of the hazardous nature of the materials involved can better follow prescribed procedures for such situations. The industrial hygienist is in an excellent position to participate in these types of educational programs.

RATIONALE

11 SUMMARY

Gigantic strides have been made during the past four decades in characterizing and controlling environmental health hazards in the workplace. In many industries where full concept industrial hygiene programs are in effect, the off-the-job stresses such as smoking, alcohol consumption, drug use, and hobby activities often have a greater effect on worker health than that of on-the-job activities where the stresses associated with the job have been studied and controlled. These off-the-job stresses offer a tremendous opportunity for the development of relevant preventive programs in which the industrial hygienist can participate.

Industrial technology is rapidly changing the characteristics of many industries in that high technology is increasingly being applied with the emergence of different types of health stresses. The industrial hygienist must keep abreast of these changes along with the procedures for their recognition, evaluation, and control.

It is vital that the techniques used in measuring occupational health stresses cover all the relevant components of each stress and that these are incorporated into the control program strategy. The practice of industrial hygiene rests on proper judgment in evaluating valid data, combined with effective follow-through.

The high quality and relevancy of the industrial hygiene profession have been proven in meeting the challenge of maintaining a healthful working environment in the presence of an ever advancing and increasingly complex industrial technology.

The following chapters cover comprehensively the theoretical basis and rationale for the science of industrial hygiene.

CHAPTER TWO

Toxicologic Data in Chemical Safety Evaluation

K. S. RAO, D.V.M., Ph.D.,
PHILLIP G. WATANABE, Ph.D., and
PERRY J. GEHRING, D.V.M., Ph.D.

1 INTRODUCTION

The ultimate objective of toxicological research on chemicals is to obtain information that will form a sound basis for recommendation of "safe" levels of exposure for humans contacting these substances during manufacture, use, and disposal. The information generally sought in such research is the dose–response function that characterizes the untoward effects produced by selected doses of a chemical. Subsequently the "safe" level of exposure has been selected traditionally by judgment to be $\frac{1}{10}$ to $\frac{1}{5000}$ of the highest dose that produced no discernible effect (1, 2). Safety factors are selected in accordance with the seriousness and persistence of the adverse effects and the shape of the dose–response curve.

"Safe" as used here is a relative term. No matter what adverse effect is of concern, absolute safety to everyone regardless of conditions can never be assured for any chemical, or for that matter any activity in which humans indulge.

In general, data depicting the adverse response of humans to selected doses of a chemical are unavailable. In some instances data from human experimentation are available; however, the adverse effects selected for study are generally transient and mild, such as skin irritation or slight depression of central nervous system (CNS) function. Rarely, epidemiological studies of people exposed during the manufacture and use of a chemical provide dose–response information for more severe effects. Such information, when available, is most important. Because human experimentation is not feasible for elucidation of serious manifestations of toxicity, it is necessary to use experiments in animals to characterize such responses. However, the use of animals to predict the response of humans creates a difficulty. It is not possible to assure absolutely that people will not be more or less sensitive than the animal populations; therefore, the range of doses producing a given effect in humans may be much larger than in the animal species selected for experimentation. Indeed, even experimental results from studies on small numbers of humans are not totally reliable for predicting the response of large population groups for the same reasons.

Nevertheless, to satisfy the ultimate objective, data collected in studies using animals must be extrapolated to predict the response in humans. This chapter provides insight into such extrapolation. The basic premise is that toxicity is manifest as a result of the presence of the toxic agent at a specified concentration at the target area. Species difference in reactions to given exposures to a chemical occur generally because of differences in how the chemical is absorbed into the body, distributed and biotransformed once in the body, and subsequently excreted from the body. A species different in the reaction of the toxicant—the chemical per se or a product formed from it—with the receptor is perceived to be rare.

2 DOSE RESPONSE

Fundamental to the extrapolation process, whether intraspecies or interspecies, is the dose–response function. This concept was enunciated in the sixteenth century by the physician–alchemist Philippus Aureolus Theophrastus Bombast von Hohenheim, better known as Paracelsus. In a Third Defense of the Principle written in 1538, he said: "Was ist das nit gift? Alle ding sind gift und nichts ohn gift. Allenin die dosis macht das ein ding kein gift ist." ("What is it that is not poison? All things are poison and none without poison. Only the dose determines that a thing is poison.")

Figure 2.1 simulates, for a normally distributed population, the percentage of individuals responding to a range of doses for two chemicals. For fictitious chemical I, the range of doses producing a discernible response is considerably smaller than that for chemical II. Curves for both chemicals indicate that at some selected doses on the low side, only a few in a population will respond. At the other extreme are those few individuals in the population requiring large doses to elicit the response. Dose X for both chemicals will cause 50 percent of the population to respond, since the areas under the two curves to the left of this point constitute one-half of the total areas under the respective curves. The dose causing 50 percent of the population to respond is the effective dose, 50 percent or ED_{50}; specifically for lethality, the dose is termed LD_{50}.

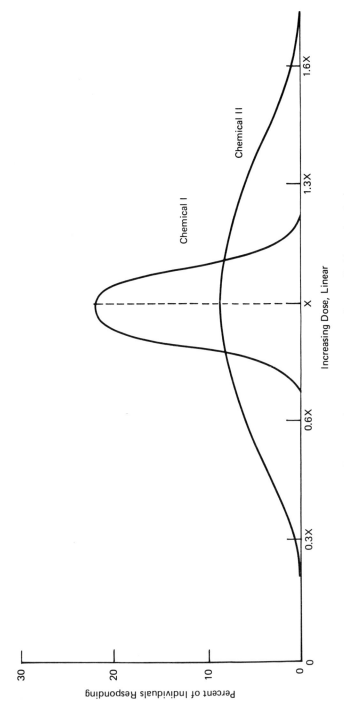

Figure 2.1 Illustrative dose–response curves for two fictitious chemicals.

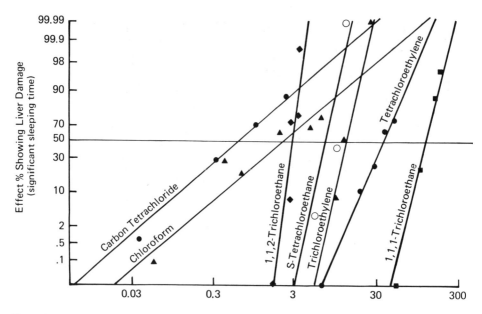

Figure 2.2 Dose–response curves for the effect of each of seven halogenated hydrocarbons on prolongations of pentobarital sleeping time in mice.

The primary reason for showing simulated curves for two chemicals in Figure 2.1 is to emphasize the necessity of using judgment to select a "safe" level of exposure. A sharply defined curve such as that for chemical I allows greater assurity that one-tenth of the dose not producing a response will be "safe." Because of the wide range of doses producing a response to chemical II, a larger safety factor is needed. For this reason the use of a single value such as ED_{50} or LD_{50} for assessment of the relative hazard of a chemical is inadequate.

If the data in Figure 2.1 are plotted as the cumulative percentage of the responding population, a sigmoid curve will result, going from 0 to 100 percent. For greater sharpness of the sigmoid curve, the dose response may be plotted frequently as the percentage responding versus the logarithm of the dose (see Figure 2.7). Since straight lines rather than sigmoid line curves are preferred, the most desirable representation of the data is the use of a logarithmic probability (probit) display (3).

Utilizing a logarithmic probability display of data for the hepatotoxicity incurred from single doses of various chlorinated solvents to mice, Plaa et al. (4) published the results appearing in Figure 2.2. Obviously the ED_{50} values for the various compounds provide some indication of the relative hepatotoxicity. However, with respect to judgment as to what dose will not produce hepatotoxicity to any of the population, a much more conservative safety factor

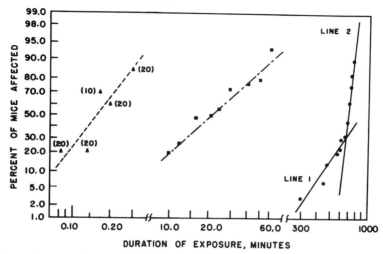

Figure 2.3 Carbon tetrachloride vapor, 8500 ppm. Percentage plotted on probability scale of mice anesthetized (squares) dead (circles), or having a significant serum glutamic acid–pyruvic acid–transaminase (SGPT) elevation (triangles), as a function of the $\log_{10}$ duration of exposure. Each point for anesthesia and lethality was obtained by using a single group of 30 mice; the number in each group used to obtain the points for SGPT activity as given in parentheses.

must be used for carbon tetrachloride and chloroform than for the other materials because of the shallow slope for these compounds.

Another important criterion for selection of a safety factor for a chemical lies in a comparison of the doses producing a subtle unnoticed effect with those producing an effect that can be readily discerned. To illustrate, refer to Figures 2.3 to 2.5, depicting the dose responses for hepatotoxicity, narcosis, and death for carbon tetrachloride, chloroform, and 1,1,1-trichloroethane (5).

For carbon tetrachloride and chloroform, unnoticeable hepatotoxicity can occur with exposure levels less than those that will produce narcosis. For 1,1,1-trichloroethane, however, injury to the liver occurs only at exposure levels in excess of those needed to produce narcosis. Indeed, liver injury occurs only when the exposure levels are sufficient to cause death to some individuals. For 1,1,1-trichloroethane, these results provide considerable confidence that prevention of narcosis, a noticeable effect, will preclude development of liver disease.

For carbon tetrachloride and chloroform, more restrictive judgments for tolerable exposures are needed to prevent the hazard of hepatotoxicity. The shallow curve depicting the hepatotoxicity incurred from various exposures to chloroform indicates greater variability in the population; therefore, a more conservative safety factor is justifiable for chloroform than for carbon tetrachloride. Stated another way, flat dose–response curves indicate that susceptibility to the chemical is highly variable in the population and there is greater probability that a small number of individuals may be adversely affected even

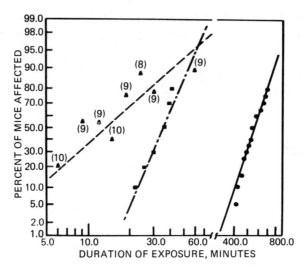

Figure 2.4 Chloroform vapor, 4500 ppm. Percentage plotted on probability scale, of mice anesthetized (squares) dead (circles), or having a significant SGPT elevation (triangles), as a function of the $\log_{10}$ duration of exposure. Each point for anesthesia was obtained by using a single group of 10 mice, and each point for lethality was obtained with a single group of 20 mice; group size used for determining SGPT activity is indicated in parentheses.

though most will be unaffected. Diethylene glycol is another substance that gives rise to a flat dose–response in most species. In humans this flatness has been reflected in human poisoning cases in which some individuals have died from ingestion of small doses, whereas others have survived relatively large doses (6). In contrast, Gonyoulex toxin, a substance occurring occasionally in clams and mussels, has a steep dose–response curve, and one-fourth the LD_{50} can be ingested without measurable risk (7).

3 ROUTE OF EXPOSURE

The toxic effects of a chemical depend on the absorption of a toxic substance into the organism. Under conditions of occupational exposure, a toxic substance can gain entry into the human body through the respiratory tract, the skin, and the gastrointestinal tract. The entry of chemical through these routes is dependent on the chemical and physical properties of the agent. Whether a toxic effect occurs depends on the concentration of the substance at the target site and the susceptibility of the biological system.

3.1 Inhalation

In cases of occupational exposure, inhalation frequently represents an important way of entry of toxic substances into the organism. In almost every manufac-

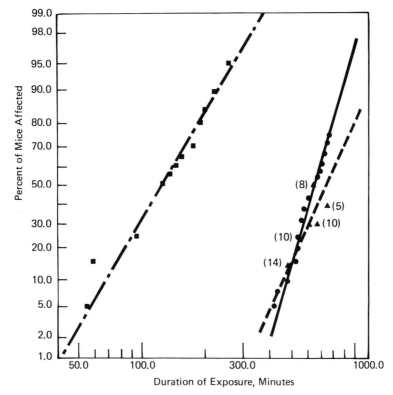

Figure 2.5 1,1,1-Trichloroethane vapor, 13,500 ppm. Percentage plotted on probability scale of mice anesthetized (squares), dead (circles), or having a significant SGPT elevation (triangles), as a function of the $\log_{10}$ duration of exposure. Each experimental point for anesthesia and lethality was calculated by using composite groups of 20 to 135 mice. Individual group sizes used to obtain SGPT activity are indicated in parentheses.

turing process the atmosphere is likely to contain some amounts of vapors, dust particles, mist, and other substances. The penetration of toxic substances by way of the respiratory tract depends on its physical properties.

Gaseous or vaporous substances generally pass into the body mainly by way of the alveoli and are distributed throughout the body by the bloodstream. However, some materials are effectively absorbed in the upper respiratory tract. The degree of retention of such substances in the upper air passages depends on their solubility: the more soluble the substance, the greater the degree of its absorption through the upper air passages, and vice versa. Some are quickly broken down on moist mucosal surfaces.

The retention of inhaled particles (aerosols) occurs throughout the respiratory tract, from the nasal cavities downward depending on the particle size. As the particle size increases, an increasing number of particles tend to be retained in the upper airways. Particles about 1 μm in size settle mainly in the alveoli, whereas those over 5 μm tend to deposit in the upper airways. The site and degree of retention also depends on the density, shape, and electric charge of the particles.

3.2 Cutaneous Absorption

Absorption through the skin is also an important entry for many toxic substances into the body. Too often skin absorption is given inadequate emphasis in assessment of exposure potential. The skin may be considered as a multilayered protective cover for the organism. This barrier, however, is not continuous; it is perforated by hair roots and follicles that penetrate deep into the dermis and subcutaneous tissue. The ability of a substance to penetrate the skin depends mainly on its lipid and water solubilities. Lipid-soluble substances are capable of moving through the fatty layers of the skin, but their further absorption will be hindered if their hydrophobic properties prevent their dissolution in the blood. Skin absorption is also affected by such factors as temperature, contact surface area, and durations of contact.

3.3 Ingestion

Ingestion is not usually considered as a form of occupational exposure (except if it occurs accidentally), and if the basic rules of occupational hygiene are followed, overexposure due to ingestion following contamination of hands, food, beverages, cigarettes, and other materials at the workplace rarely occurs. Accidentally ingested materials are absorbed into the body at one or more sites in the gastrointestinal tract.

Gastric absorption depends on the nature and quantity of gastric contents. Gastric and intestinal secretions may considerably alter the chemical nature of a substance and increase its solubility. Chemicals may also be transformed by intestinal bacteria. For example, aromatic nitro compounds are reduced by bacteria to amines.

After their absorption from the stomach and intestine, many of the gastric contents are carried by the blood to the liver, where they are frequently biotransformed or excreted in the bile. Biotransformation processes in the liver generally degrade a foreign substance to less toxic compounds; however, more toxic entities may be produced by the liver. Some compounds are absorbed into lymphatic pathways and bypass the liver.

4 ACUTE TOXICITY

Acute toxicity of chemicals in animals is evaluated by administration of single doses of various routes of exposure. Although oral administration is uncommon for exposure to most industrial chemicals, it is commonly used for a first assessment of the potential of a chemical to cause toxicity. Too frequently, lethality is considered to be the most important adverse effect, and other signs of adverse effects such as the physical condition of the animals and damage to particular organs are overlooked. Rigorous evaluation of these parameters frequently reveal the organ system affected by the chemical and occasionally the mechanism of toxicity.

For determination of the acute oral toxicity, strict attention should be given to the persistence of the effects. For example, delayed deaths occurring 2 or more days following administration, or depression of body weight gain 1 or 2 weeks after administration, suggests either a persistence of the chemical in the body or slow repair of the damage incurred. In either case more conservative handling of the chemical to preclude adverse effects on health is necessary. It may also be anticipated that repeated exposure to a chemical eliciting persistent manifestations of toxicity subsequent to a single exposure will constitute a greater hazard than exposure to a chemical that does not.

Exposure to industrial chemicals occurs most frequently by way of contamination of the skin or inhalation of the vapor or dust. Contamination of skin may result in local damage and in some instances absorption into the body and systemic toxicity. Adverse local reactions of either skin or eyes are generally extrapolated directly to humans, and appropriate measures are instituted to preclude contamination of the skin and eyes with injurious concentrations of chemicals.

Some chemicals penetrate the skin readily. An indication of rapid penetration is an equivalency or near equivalency of acutely toxic doses, whether given orally or by skin application. Evidence of significant penetration of a chemical through the skin warrants institution of precautions to minimize skin contamination. The rigor of these precautions will be dictated by the probability of dermal exposure and the severity of potential effects as revealed by acute, subchronic, and chronic toxicity, usually revealed in studies relying on other routes for administration.

Exposure to chemicals by way of inhalation is a major concern in an industrial environment. Frequently the acute toxicity of a chemical by way of inhalation is initially assessed by exposing animals to a concentrated atmosphere of the chemical for 6 to 8 hr. If injury or death occurs, the duration of exposure and the exposure concentrations are decreased progressively until the adverse effects disappear. The objective of such experimentation is to characterize the toxicity as a function of duration and concentration of exposure. Frequently, this function can be visualized by plotting the log of the concentration versus the log of the exposure needed to produce a given effect or lack of effects.

To illustrate, consider the data due to Adams et al. (8) depicting the acute toxicity of carbon tetrachloride (Figure 2.6). For rats, line *CD* represents the most severe exposures not resulting in death, and *EF* represents exposures causing no discernible effects. The large difference between exposures represented by these lines carries the same interpretive significance as discussed previously for the data in Figure 2.3.

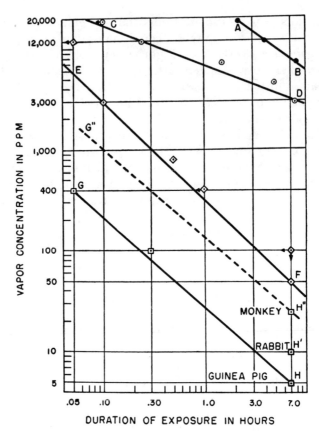

Figure 2.6 Vapor toxicity of carbon tetrachloride. Line *AB* represents the least severe single exposures causing death of all rats tested and line *CD*, the most severe single exposures, permitting survival of all rats tested. Line *EF* represents the most severe single exposures without detectable adverse effect and line *GH*, the most severe repeated exposures without detectable adverse effect in rats. Points, *H*, *H'*, and *H"* represent the most severe repeated exposures without detectable adverse effect in the guinea pig, the rabbit, and the monkey, respectively. Line *G" H"* represents the most severe repeated exposures with little probability of adverse effect in human beings.

In the absence of human experience, tolerable single exposures to carbon tetrachloride would be set at least tenfold less than those represented by line *EF*. Data for additional species—dog, monkey, and rabbit—indicating lesser or equivalent susceptibility, increase confidence that a tenfold margin of safety is adequate.

For many chemicals such as carbon tetrachloride, human experience will augment the judgment. When Adams et al. (8) reported their results, sufficient human experience was available to support the conclusion that single exposures such as those represented by line *EF* did not produce discernible adverse health effects in humans.

5 SUBCHRONIC AND CHRONIC TOXICITY

Only initial judgments are possible from acute toxicity tests. If carefully done, they provide insight into the potential toxicity on repeated exposure to a chemical, but repeated exposure studies are needed to provide a more definitive assessment of subchronic and chronic toxicity.

In studies to characterize the potential subchronic and chronic toxicity, chemicals are administered to animals daily for 5 to 7 days per week for an entire lifetime or a fraction thereof. When administered orally, the chemicals are admixed with the diet when feasible. Inhalation exposure is conducted in chambers with a regimen of 6 to 8 hr per day, 5 days per week. Sometimes other routes of exposure are used (e.g., daily application to the skin or injection). Extensive antemortem and postmortem evaluations are generally made with the use of the most recent clinical, chemical, and pathological methods. Typically, more than 50 to 100 different parameters are assessed.

Adams et al. (8) also reported the subchronic toxicity of carbon tetrachloride. Line *GH* in Figure 2.6 represents daily exposures that did not produce discernible adverse effects in rats over a duration of 6 months. Also shown as single points are no effect exposures for guinea pigs, rabbits, and monkeys exposed 7 hr per day, 5 days per week, for 6 months. In these studies rats and guinea pigs appeared to be most susceptible; rabbits and monkeys were intermediate and least susceptible, respectively. The liver was the target organ. Since human experience indicated that humans were less susceptible than rats, a line, "*G, H*" parallel to *GH* was drawn to represent exposures judged to be safe in humans. As a result, 5 ppm was recommended as the standard for the work environment for 7-hr daily exposures.

In the absence of human experience, exposures represented by line *GH* or possibly even lower exposures may have been selected. To preclude subchronic toxicity, a workroom standard of 5 ppm would have been justified in that this would have protected the most sensitive species, rats and guinea pigs. The volume of air breathed by these species per unit body weight is greater than those for humans. Thus the dose received by humans is less, providing a built-in safety factor.

A key question in subchronic and chronic toxicology studies is, how long should the animals be exposed? Ideally, chronic toxicology studies should involve treatment of the animals for a significant portion of their lifetime, considered to be 2 years for rats and mice. When the chemical is administered by routes other than in diet, administration for a lifetime requires an extensive effort. For example, daily generation of atmospheres containing the vapor of a chemical and chemical analyses to assure the desired concentration are time-consuming and costly, as is handling of the animals. In addition, physical facilities for performing inhalation studies are costly and limited in availability. Therefore, chronic inhalation studies have often been conducted by using daily exposures for a fraction of the lifetime (e.g., 6 to 18 months) and subsequently maintaining the animals for the duration of the lifetime. This protocol duplicates reasonably well exposures incurred by workers. Also, the track record of this protocol for revealing carcinogenic effects of a chemical is good, although few definitive studies comparing the results of daily administration for a lifetime versus a fraction of a lifetime are available.

A number of authors have demonstrated that toxicology studies exceeding 3 months in duration are usually unnecessary to reveal the potential chronic toxicity of chemicals (9–14). Weil and McCollister (15) compared the minimum dose causing a toxic effect and a maximum dose causing no effect for 33 chemicals in short-term (29 to 200 days) versus long-term studies. The ratios of the indicated dosage levels for short-term versus 2-year feeding studies were 2.0 or less for 50 percent of the chemicals. A ratio greater than 5 occurred for only three chemicals; the largest ratio for a maximum no-effect level was 12.0. These results suggest that in lieu of chronic toxicity data, selection of one-tenth the no-effect level in a subchronic study as a no-effect for chronic toxicity may be appropriate until a definitive study is conducted. Again judgment should be tempered by human experience when available.

As indicated earlier, subchronic toxicity studies often adequately reveal the potential chronic toxicity of a chemical. However, as more chemicals are studied, exceptions are becoming too common to permit overreliance on this generality. Therefore, subchronic toxicity studies should be considered only interim assessments of potential chronic toxicity.

As indicated previously, chronic studies by way of inhalation exposure constitute a major and costly undertaking. For many chemicals, only studies utilizing exposure by way of ingestion may be available. To estimate the maximum dose that will not cause an effect by way of inhalation, one may assume that a worker will inhale 10 m³ of air per 8-hr workday (16). Since not all the chemical inhaled is retained, a retention factor is suggested as follows (16, 17):

	Water Solubility (%)	Retention Factor (%)
Essentially insoluble	<0.1	10
Poorly soluble	0.1–5	30
Moderately soluble	5–50	50
Highly soluble	50–100	80

The estimated dose is calculated in accordance with the equation

$$\text{Estimated dose} = \frac{(10 \text{ m}^3) \, (\text{mg/m}^3) \, (\text{retention factor})}{\text{kg body weight}}$$

Estimation in accordance with the foregoing is frequently useful. However, it is not recommended that such an estimation be used to satisfy the need for more definitive studies (i.e., a chronic study using inhalation exposure or pharmacokinetic evaluation to characterize the fate of inhaled versus ingested chemicals).

Currently, assessment of the chronic toxicity potential of chemicals is considered the ultimate means of providing a data base to set tolerable chronic exposure levels for humans. Typically, in chronic studies, 50 to 100 animals per sex per species are exposed to each of three or four levels of the agent, together with an equal or greater number of controls. Two species are studied frequently. Unfortunately, tunnel vision has resulted too frequently in the use of such studies to study only carcinogenicity. Chronic toxicity studies should be aimed at assessing as best possible any manifestation of an untoward effect. Furthermore, interpretation of a carcinogenic response without assessment of other manifestations of toxicity is an unscientific exercise in futility, since secondarily induced carcinogenesis occurs (18) and since other manifestations of toxicity may be equally serious with respect to workers' health.

6 RELIABILITY OF ANIMAL STUDIES FOR PREDICTING TOXICITY IN HUMANS

There is no comprehensive, definitive means for assessing the reliability of predicting toxicity or lack thereof in humans with the use of toxicity data collected in animals. Although surprising initially, this apparent deficiency is understandable because such an assessment requires dose–response data for humans as well as animals. Doses of chemicals causing frank signs of toxicity cannot be administered intentionally to people. Even when human studies are conducted, reliance on symptoms such as nausea, headache, anxiety, depression, disorientation, weakness, or irritation as perceived by the subjects precludes, for the most part, a correlation with animal studies in which the parameters are limited to those that can be observed and measured by the investigator: signs versus symptoms.

Correlations of the adverse effects in humans with those in animals exist predominantly for drugs. Review of this extensive literature will cause considerable anxiety because it contains many instances of studies in animals that have revealed both false-positive and false-negative results with respect to humans (19, 20).

In the study of six unnamed drugs, Litchfield (21) found a significant relationship between the signs of toxicity in humans, rats, and dogs. However, 23 of 234 signs of toxicity were seen only in humans. Studies on rats and dogs did not predict symptoms unique to humans—headache, loss of libido, and so on. Dogs were found to be somewhat more useful in prediction of drug effects in humans than in rats.

Classical examples of drugs that induce false-positive results in animals are fluroxene, an anesthetic that has been used uneventfully in humans but kills dogs, cats, and rabbits; and penicillin, which produces lethality in guinea pigs in doses that would be therapeutic for humans (22).

Rall (23) demonstrated an excellent correlation between toxicity data for 18 anticancer agents in mice and humans when the dose was based on milligrams per square meter of body surface area (mg/m^2). The mouse was consistently twelvefold less susceptible than humans, when the data were examined on a milligram per kilogram body weight basis. In the original publication of the correlation of toxic responses in humans and animals for the antineoplastic agents, monkeys, dogs, and rats were also considered (24). Assessments in these and other animal species added little except to further substantiate the existence of good quantitative correlation when the dose is expressed as milligrams per square meter rather than milligrams per kilogram.

Administration of biologically active chemicals such as antineoplastic agents frequently induces equivalent responses when the dose is administered proportional to body surface area (25). This relationship is gaining recognition in estimating the risk incurred by humans from exposure to chemicals in the environment (26). In utilizing this relationship, however, one must recognize that the original relationship was developed for biologically active agents. Since metabolism and other physiological processes involved in detoxification are generally more active in smaller animals, the dose of a biologically active chemical per unit mass required to produce a given effect increases as the body mass decreases, whereas the dose per unit surface area remains relatively constant. For a chemical requiring activation to the biologically active toxic form, however, the opposite is likely to be generally true because the rate of transformation will be roughly proportional to the body surface area. For a chemical requiring conversion to the active toxicant, therefore, a greater dose of the active form may be received on a milligrams per kilogram body weight by a larger than smaller species of animal because the latter has a greater surface area:body weight ratio.

In reviewing animal experimental data for assessing the carcinogenicity of chemicals in humans, Wands and Broome (27) reported that studies in animals have revealed the carcinogenicity of all known human carcinogens except arsenic. Even for arsenic, an increased incidence of leukemia and malignant lymphomas in pregnant Swiss mice has been reported (28). Wands and Broome (27) also pointed out the existence of positive carcinogenic responses in animals for roughly 1000 compounds having no evidence for a carcinogenic response in humans.

With respect to the foregoing discussion of the apparent lack of reliability of data from animal experiments for predicting toxicity in humans, how can the usefulness of animal experimentation be advocated for this purpose? It must be recognized that exceptions to correlations rather than correlations per se tend to be reported. Existence of toxicity data in animals and application of precautions to avoid toxicity in humans for literally thousands of chemicals precluded visibility of a correlation.

Undoubtedly the false-positive finding of carcinogenic activity in animals for roughly 1000 chemicals is attributable to a number of factors such as the subjects' limited exposure to these agents. For many of these false-positive chemicals, the doses used to elicit a response were sufficient to overwhelm detoxication mechanisms and to cause prominent manifestation of toxicity other than carcinogenicity.

For drugs, it must be recognized that the apparent weakness, not absence, of a correlation for signs of toxicity is frequently qualitative. This is not surprising because administration of therapeutic or near therapeutic doses of biologically active agents to highly integrated biological systems may be expected to produce variable qualitative intra- and interspecies responses. Such differences in the qualitative manifestations of toxicity in response to toxic doses of a chemical should engender less concern than the absence of any manifestations of toxicity in animals given a dose of a chemical later found to be toxic in humans. Although of less concern, qualitative manifestations of toxicity in animal experimentation are important because they provide a basis for selecting parameters to be monitored in an exposed population of people.

For further conceptualization of the problems inherent in extrapolation of toxicity from animal experimentation for prediction of the hazard in humans, the dynamics of toxicity must be realized. The toxicity of a chemical is a function of the absorption of that chemical into the body, distribution, and biotransformation of the chemical. Once the chemical is absorbed in the body, interaction of the chemical per se or a biotransformation product with the biological receptor leads to the ultimate action, and excretion of the chemical itself or its biotransformation products. If each of these five processes is an independent variable and each has a correlation coefficient of 0.9 for humans versus animals, the overall correlation coefficient will be 0.6. In spite of the poor correlation predicted for equivalent toxicity in humans and animals, tolerable exposure levels based on animal data have effectively precluded manifestations of toxicity in humans. Undoubtedly, the safety factors used in establishing tolerable exposure levels have been largely responsible for this.

Regardless of the success experienced in using toxicity data from animal experimentation as a basis for recommending tolerable exposures for humans, future development of the science of toxicity must be directed at improving the reliability of this extrapolation. Consideration of the five factors influencing the toxicodynamics of chemicals immediately reveals how this can be done. First, it requires quantitative knowledge of the absorption, distribution, biotransformation, biological reaction, and excretion of a chemical or its biotransformation products in humans versus animals. This may be accomplished in pharmacokinetic evaluations of the chemical, which are becoming more common. Second, the reaction of the active chemical with the biological receptor or the mechanism of action must be elucidated. The mechanism of the toxicity of a chemical is not, however, as easily revealed as its pharmacokinetics. Modern technology and conceptualization are allowing mechanisms of toxicity to be elucidated for a few chemicals. In the meantime, elucidation of the pharmacokinetic parameters will dramatically improve the extrapolation process.

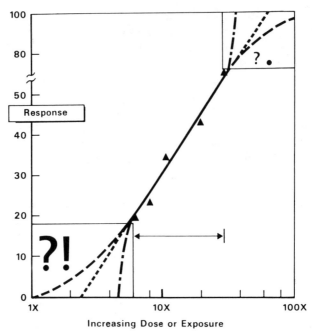

Figure 2.7 Simulated percentage of individuals responding adversely to the logarithm of selected doses. Measurable responses are represented by triangles. The sigmoid (dashed) curve represents a population described by normal distribution; in theory, the percentage responding never reaches zero on the low end or 100 on the high end. The other curves represent a threshold for the response. The boxed-in portions represent regions in which prediction of incidence depends on stochastic, statistical projection.

7 THRESHOLD

In a typical dose–response curve (Figure 2.7), the solid line represents an observable increase in the percentage of individuals responding to increasing doses. This type of dose–response curve occurs when the response of individuals within a population is distributed normally, which is the situation characteristic for most pharmacological and toxicological responses to chemicals. The concept of threshold has been accepted for most pharmacological and toxicologic responses. Biologically, the term "threshold" has been interpreted to mean that there is a dose below which no response will occur in a population of animals. Hence, when a dose is found that does not produce a toxic response in a reasonable number of subjects (laboratory animals or humans), it is assumed that the dose is subthreshold.

Threshold for an adverse response to a chemical differs with species and various physiological factors as well, such as age, sex, diet, and stress. In a population where these variables have been controlled, it is assumed that a large number of animals or people may be exposed to a subthreshold amount of the chemical without a response. Such an assumption renders relatively easy prediction of a "safe" exposure to a chemical for the population. After a given dose has been found to elicit no response in an experiment, a safety factor is applied to this dose to account for species, age, sex, diet, and other physiological differences. Subsequently, no experience of an adverse effect in a few individuals of a population is taken as confirmation of the appropriateness of the safety factor, and the existence of a subthreshold is assumed for the entire population.

A threshold concept is used to deem "safe" many events or materials in our lives other than chemicals. If a building is constructed to withstand a wind force of 100 mph, it is assumed that construction of millions of duplicates will not result in some that will be devastated by a 10-mph wind. In lieu of such a concept, a judgment of "safe" for any human endeavor is impossible.

For chemical carcinogenesis, mutagenesis, and to some degree teratogenesis, the threshold concept is not universally accepted. It is generally accepted that increasing doses of such agents will produce increased incidences of the response, represented by the solid line fitted to doses producing the discernible response shown in Figure 2.7. To establish existence of a threshold, however, statistically interpretable data would have to be acquired at doses below those that elicit experimentally discernible responses, represented by the boxed-in area containing the question mark and the exclamation point. The dilemma is whether the incidence of cancer, mutations, or terata goes quickly to zero as the dose is decreased, as in the case of the threshold phenomenon; whether the incidence will decrease as predicted by nonthreshold extrapolations of the experimentally discernible response function; or, indeed, whether the incidences may be greater than that predicted by such extrapolations. Because of the incidence of spontaneous tumors in animals, experimental resolution of this dilemma is precluded by the need for thousands to millions of animals per experiment.

The argument that no threshold exists for carcinogens or mutagens is assumed to be essentially as follows: cancer is an expression of a permanent, replicable defect providing for amplification of a defect initiated by reaction between a single molecule of a chemical carcinogen or mutagen and a critical receptor. Primary support for this argument is that exhaustive experiments on radiation-induced cancer have failed to reveal an absolute threshold within the realm of statistical reliability. Equating radiation-induced cancer with chemical-induced cancer is tenuous, however, since the latter involves absorption, distribution, biotransformation, and excretion, and the former does not. Even for radiation, Evans (29) has demonstrated clearly smaller incidences of cancer in individuals exposed to low doses of radium than were predicted by those exposed to large doses. Gehring and Blau (18) have enumerated evidence for the existence of a threshold for chemical carcinogens. In summary:

1. Carcinogenesis is a multistage process. Interference with any of the processes precludes cancer development.
2. There is evidence of mechanisms that suppress development of cancer even though cells programmed to become cancer exist in the body.

3. As the dose of carcinogens is reduced, the latent period for development of cancer increases. A latent period longer than the life of individuals in a population is for all practical purposes a threshold.

4. Chronic inflammation causes an increased incidence of cancer, as does stress. Therefore, it is not unreasonable to anticipate that administration of high, toxic doses of a chemical will also produce cancer secondarily.

5. There is a substantial and growing body of evidence that carcinogensis is subject to immunosurveillance, particularly cell-mediated immunity.

6. Mechanisms exist for repair of deoxyribonucleic acid (DNA) that has been reacted with a chemical in a manner that has programmed it for cancer induction. For example, a genetic deficiency of repair mechanisms in people having xeroderma pigmentosum renders them very susceptible to radiation-induced cancer (30).

7. Humans and animals are exposed to a sea of carcinogens, some artificial but most preexisting the contributions of our civilization. Overnutrition as well as overindulgence in exposure to chemicals increases the incidence of cancer. A much larger share of human cancer can be attributed to the former cause.

Significant advances in understanding the roles of genetic and nongenetic (epigenetic) mechanisms in the carcinogenic process have resulted in classification of chemical carcinogens into two broad classes (62). Genotoxic chemical carcinogens damage DNA directly, and nongenotoxic or epigenetic chemical carcinogens are those that exert other biological effects or indirectly (secondarily) affect DNA that can result in tumorigenesis. Genotoxic carcinogens are hypothesized to promote or enhance stages of the multistage carcinogenesis subsequent to the initial genetic event. The significance of this concept for risk assessment is that since promoters or enhancers of carcinogenesis do not act directly on DNA, characteristics such as reversibility and threshold should apply to those chemicals acting by means of nongenotoxic mechanisms. A more extensive discussion of one type of nongenotoxic mechanism (recurrent tissue injury) is addressed in the following section, citing examples.

In spite of the current advances, controversy will likely continue over the existence of thresholds or lack thereof for chemical carcinogens and mutagens (for teratogenesis it is generally acceptable that a threshold exists), since no totally definitive solution is possible in the foreseeable future. The same type of argument may be extended to almost all manifestations of chemical toxicity. Indeed, similar arguments can be used to label as unsafe almost all human activities since all constitute some degree of hazard. On the positive side, arguments for and against a threshold for chemical toxicity have provided great motivation to assess more definitively the risk of chemicals. The resulting technological advancement will undoubtedly benefit not only the science of toxicology, but other sciences devoted to human health and environment.

8 PHARMACOKINETICS AS AN AID IN INTER- AND INTRASPECIES EXTRAPOLATIONS OF TOXICITY DATA

As indicated previously, the dynamics of toxicity includes absorption of the chemical into the body, distribution, biotransformation of the chemical in the body, and excretion of the chemical itself or biotransformation products formed from the chemical. Toxic manifestations produced by a chemical are functions of the concentrations of the toxic entity at the target sites and the duration of exposure of these sites to the toxic entity. For assessment of the hazard of a chemical to humans as well as other species, therefore, it is essential to elucidate the kinetics for its absorption, distribution, biotransformation, and ultimate excretion—that is, its pharmacokinetics. Only with acquisition of such information can interspecies, intraspecies, and high-dose–low-dose extrapolations of potential toxicity be made definitively. Gehring et al. (31) should be referred to for a detailed discussion of the use of pharmacokinetics for assessment of the toxicologic and environmental hazards of chemicals.

8.1 Dose–Interspecies Response

The complex problem of extrapolating toxicology data may be envisioned in Figure 2.8a, which shows that chemical dose level (on a body weight basis), the species size (body weight), and the incidence (or severity) of the toxic response all increase in an outward direction from the coordinate intersection. The height of the surface above the dose–species plane in the absence of a chemical dose represents the normal background incidence of a given toxic response. The shaded area in the region of the higher dose levels in Figure 2.8a shows the experimentally observable response–dose relationship in smaller animal species. The heavy arrows on the surface of the plane indicate the directions of extrapolation of toxicology data across the toxicity surface in order to arrive at an estimate of the potential risk to humans at realistic exposure levels. When a toxic response is elicited by a chemical, the dose–response curve can be envisioned to lie in the vertical dose–response plane for a given species as shown in Figure 2.8b. As the dose level decreases, the toxic response also diminishes until it virtually vanishes into the background incidence.

At a given dose level (on a body weight basis) the toxic response may either increase or diminish as the species size increases, and this response can be envisioned to lie in the vertical species–response planes shown in Figures 2.8c and 2.8d, respectively. Since the basal metabolic rate of different mammalian species is proportional to the body surface area-to-volume ratio, and this ratio increases with decreasing body size, small animals will generally metabolize chemicals more rapidly on a body weight basis than larger animals. Therefore, whether a larger animal species will be more or less sensitive to a chemical than a smaller species is often dependent on whether metabolism of the parent chemical constitutes a detoxication or toxification process. Ultimately, the ratio

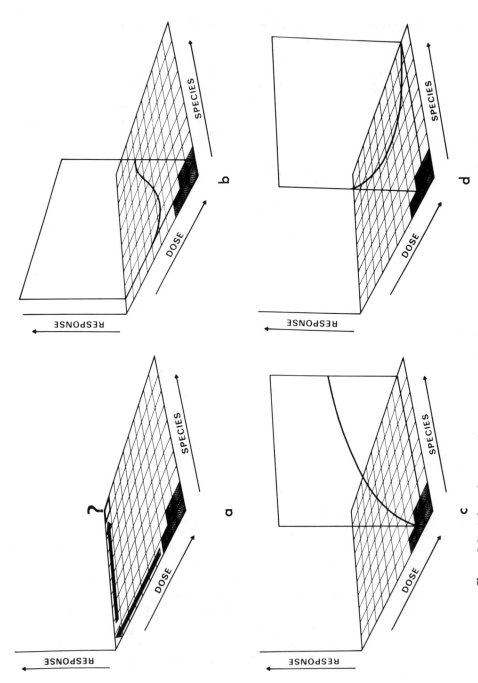

Figure 2.8 A three-dimensional representation of the dose–response–species interrelationship. All quantities increase in an outward direction from the coordinate intersection.

of the rates of activation and inactivation will determine the concentration of the biologically effective molecule and can be more complex than the simple size relationship as expressed above. The enhanced susceptibility of the rat versus the mouse for acetylaminofluorene-induced hepatocarcinogenesis has been related to the increased activity of the activating metabolic sulfotransferase activity in rats (32–34). However, as other species have been studied even this simple relationship has become more complex (35).

The foregoing considerations imply that the toxicity surface lying above the plane shown in Figure 2.8a may easily assume a complex shape between the extremes of dose level and species size and that the shape of this surface is quite likely different for different chemicals. Studies elucidating the predominant mechanisms of toxicity of a given chemical can reveal directional trends across this surface as both species and dose levels change. Likewise, a knowledge of the pharmacokinetic profile of a chemical can provide both qualitative and quantitative information concerning the expected toxic response with changing dose levels in different species. The complex nature of the surface emphasized the necessity of integrating the different disciplines of toxicology and utilizing all available information in a consistent manner in order to obtain the most realistic estimate of the potential risk to humans.

The following examples were selected to illustrate how knowledge of the pharmacokinetics, metabolism, and mechanism of actions of a chemical can influence the assessment of dose-related relationships and interspecies extrapolation.

8.2 2,4-Dinitrophenol (DNP): Linear Pharmacokinetics to Elucidate the Difference in Species Susceptibility to Cataractogenic Activity

2,4-Dinitrophenol was known for some time to be cataractogenic in fowl and humans but not in laboratory animals used routinely to evaluate toxicity. Gehring et al. (31) showed that rabbits less than 62 days of age, but not older, developed cataracts when treated with DNP.

To resolve the species and age difference in susceptibility to the cataractogenic activity of DNP, a pharmacokinetic study was undertaken to determine the concentration–time relationship for DNP in the serum, aqueous humor, vitreous humor, and lens of mature and immature rabbits and ducklings. The results appear in Figures 2.9, 2.10, and 2.11, respectively. The rate constants for the apparent first-order elimination of DNP from the plasma of mature and immature rabbits were 0.82 and 0.15 hr^{-1} ($t_{1/2}$ = 0.84 and 4.6 hr), respectively. For ducklings, elimination of DNP from plasma was biphasic and in acordance with a two-compartment model system; elimination rates for the two phases were 0.25 and 0.11 hr^{-1} ($t_{1/2}$ 2.77 and 6.3 hr), respectively. Thus susceptibility correlated with a slower rate of clearance of DNP from plasma.

Associated with the slower rate of clearance were the higher concentration of DNP attained in the aqueous humor, vitreous humor, and lens. Furthermore, mathematical analyses of the data for the concentration–time relationships for

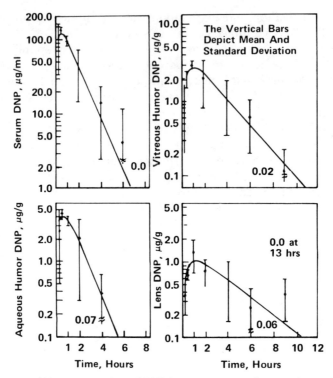

Figure 2.9 Mature rabbit: concentrations of DNP in serum, aqueous humor, vitreous humor, and lens as a function of time following intraperitoneal administration.

DNP in plasma and aqueous humor allowed determination of the ratio of the rate of movement of DNP from aqueous humor to plasma k_{ef} to the rate of movement into aqueous humor from plasma k_{in}. This ratio is, in essence, an expression of the blood–aqueous humor barrier; the higher the value of k_{ef}/k_{in}, the more substantial the barrier. For mature rabbits, immature rabbits, and ducklings, the values were 15.8, 6.3, and 3.5, respectively.

The results of these pharmacokinetic studies elucidate the reason for species and age differences in the cataractogenic activity of DNP. Demonstration that the blood–aqueous humor barrier served to decrease the cataractogenic activity, suggesting that conditions increasing the permeability of this barrier, such as trauma and infection, may increase susceptibility.

8.3 2,4,5-Trichlorophenoxyacetic Acid (2,4,5-T): Linear and Nonlinear Pharmacokinetics and Species Differences

2,4,5-Trichlorophenoxyacetic acid is a herbicide, the safety of which has been questioned because daily doses of 100 mg/kg or more during the period of organogenesis have been reported to be teratogenic, fetotoxic, and embryotoxic

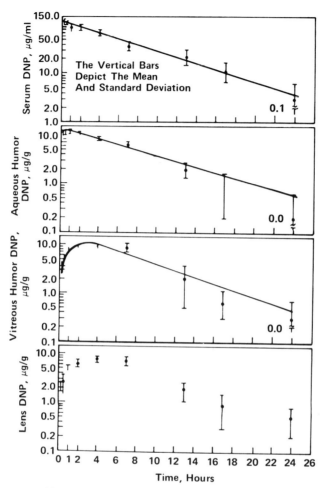

Figure 2.10 Immature rabbit: concentrations of DNP in serum, aqueous humor, vitreous humor, and lens as a function of time following intraperitoneal adminstration.

(36–40). In order to elucidate the potential hazard of 2,4,5-T, 5 mg/kg of ring-labeled 2,4,5-T was administered orally to rats and dogs (41). Figures 2.12 and 2.13 show elimination of ^{14}C activity from plasma and in urine, respectively. Plasma concentration–time curves were characterized by a one-compartment open model with a first-order absorption and clearance. The $t_{1/2}$ values for plasma clearance by rats and dogs were 4.7 and 77.0 hr, respectively. For elimination from the body, the urinary $t_{1/2}$ values were 13.6 and 86.6 hr. The more rapid elimination from the plasma of rats than in the urine suggested that the compound was concentrated in the kidney of rats prior to excretion. The slower rate of elimination by dogs than by rats correlates with the higher toxicity in dogs, for which the single oral LD_{50} is 100 mg/kg, compared to over

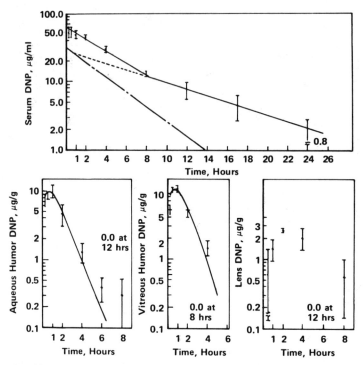

Figure 2.11 Duckling: concentrations of DNP in serum, aqueous humor, vitreous humor, and lens as a function of time following intraperitoneal administration. The lowest line depicting the disappearance of DNP from serum was determined by projecting the line representing the concentration of DNP in serum between 8 and 24 hr to the origin and subtracting the projected values from the experimentally determined values. Vertical bars depict means and standard deviations.

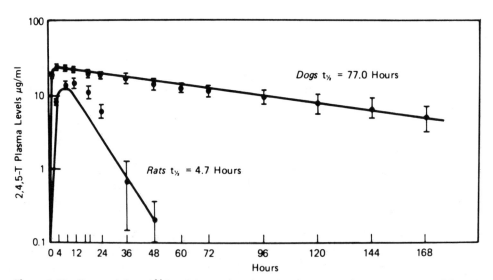

Figure 2.12 Concentration of ^{14}C activity expressed as microgram equivalents 2,4,5-T per milliliter of plasma in dogs and rats following a single oral dose of 5 mg/kg [^{14}C]-2,4,5-T. Each point represents a mean and SE.

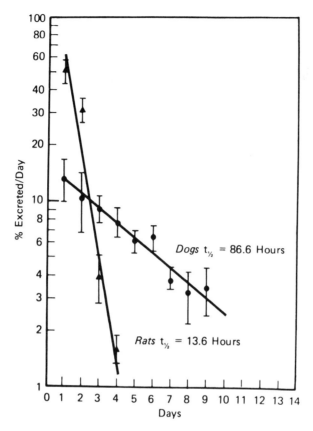

Figure 2.13 Percentage of administered ^{14}C activity excreted by dogs and rats during successive 24-hr intervals following a single oral dose of 5 mg/kg [^{14}C]-2,4,5-T. Each point represents a mean and SE.

300 mg/kg for rats (42, 43). Generally, the toxicity of an agent to an individual or species is related inversely to the ability of the individual or species to eliminate the compound.

Almost all of the 2,4,5-T excreted by rats was via the urine, whereas approximately 20% of that excreted by dogs was through the feces. Measurable amounts of breakdown products of 2,4,5-T were not produced by rats given 5 mg/kg, whereas 10% of the ^{14}C activity excreted in the urine of dogs was attributable to breakdown products. This illustrates a pertinent principle: if elimination by one route, in this case urine, is related, a greater fraction will be eliminated by other routes.

Since 2,4,5-T is an organic acid, it was hypothesized that the poorer ability of the dog than the rat to eliminate it was associated with the poor organic acid secretory process of the dog kidney. If an active secretory process of the kidney was the primary elimination process, elimination should be nonlinear, saturable

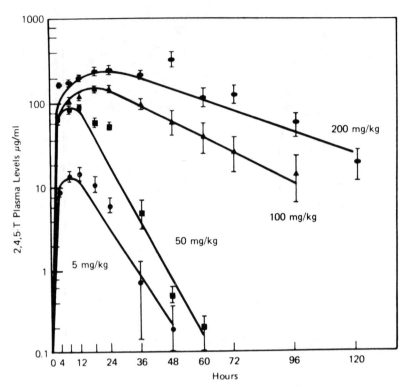

Figure 2.14 Concentration of ^{14}C activity expressed as microgram equivalents 2,4,5-T per milliliter plasma in rats following a single dose of 5, 50, 100, or 200 mg/kg [^{14}C]-2,4,5-T. Each point represents a mean and SE.

in the rat by administration of higher doses. Elimination from both the plasma and the body was shown to be nonlinear by the administration of 5, 50, 100, and 200 mg/kg (Figures 2.14 and 2.15). The previous suggestion that the rapid elimination from the plasma of rats given 5 mg/kg 2,4,5-T was due to uptake by the kidney was fortified by the finding that after doses of 100 or 200 mg/kg, plasma and body elimination rates were equivalent. The active uptake by the kidney had been saturated.

In addition to the lack of superposition of the concentration–time curves for the elimination of 2,4,5-T from plasma with increasing doses, another criterion for nonlinear pharmacokinetics was revealed. A larger percentage of the ^{14}C activity administered as [^{14}C] 2,4,5-T was excreted in the feces as the dose was increased. Furthermore, degradation products of 2,4,5-T were identified in the urine of rats given 100 or 200 mg/kg, whereas none were found in the urine of rats given 5 or 50 mg/kg.

In order to better characterize the nonlinear pharmacokinetics of 2,4,5-T, rats were given intravenous (IV) doses of 5 or 100 mg/kg (44). Elimination from the plasma of rats given 100 mg/kg (Figure 2.16) occurred in accordance

TOXICOLOGIC DATA IN CHEMICAL SAFETY EVALUATION

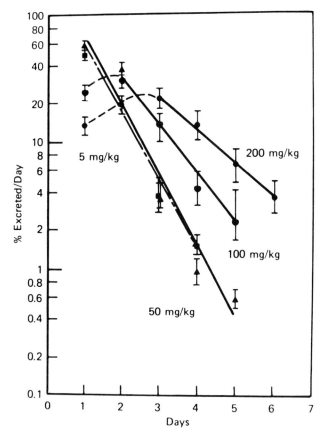

Figure 2.15 Percentage of administered ^{14}C activity excreted by rats during successive 24 hr intervals following a single oral dose of 5, 50, 100, or 200 mg/kg [^{14}C]-2,4,5-T. Each point represents a mean and SE.

with the Michaelis–Menten equation. The values for V_m and K_m were 16.6 ± 1.8 µg/g-hr and 127.6 ± 25.9 µg/g, respectively. During the linear phase of excretion, between 36 and 72 hr after administration of 100 mg/kg, the $t_{1/2}$ was 5.3 ± 1.2 hr. This value is not significantly different from that found for rats given 5 mg/kg.

The volume of distribution V_d increased from 190 ± 8 to 235 ± 10 ml/kg for rats receiving 5 and 100 mg/kg [^{14}C]-2,4,5-T, respectively. This increase in V_d indicates that as the dose is increased, a disproportionately larger fraction finds its way into tissue and cells. Undoubtedly, toxic effects will also increase disproportionately as a result.

As indicated previously, the discrepancy between the elimination rates of 2,4,5-T from the plasma and in the urine suggested that the kidney concentrated the compound prior to excretion. Table 2.1 depicts tissue:plasma ratios of ^{14}C

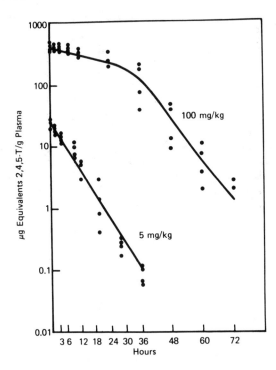

Figure 2.16 Concentration of ^{14}C activity expressed as microgram equivalents 2,4,5-T per gram of plasma of rats following a single intravenous dose of 5 or 100 mg/kg [^{14}C]-2,4,5-T. The curve for rats receiving the latter dose typifies dose-dependent elimination.

activity at various times after IV injection of 100 mg/kg [^{14}C] 2,4,5-T. The data show clearly that the kidney does indeed concentrate the compound and that its concentrating ability is saturable.

For further elucidation of the saturable active transport of 2,4,5-T by rat and dog kidney, *in vitro* studies were conducted with the use of renal slices (45). These studies demonstrated conclusively that (1) 2,4,5-T is actively transported by the organic acid excretory mechanism of the kidney, (2) kidney

Table 2.1 Concentration of ^{14}C Activity Expressed as Microgram Equivalents 2,4,5-T per Gram in Kidney and Plasma after Intravenous Administration of 100 mg/kg [^{14}C]-2,4,5-T

Time Interval (hr)	Concentration of 2,4,5-T (μg/g) in		Kidney / Plasma
	Plasma	Kidney	
4	415 ± 19	297 ± 11	0.71 ± 0.02
8	359 ± 20	281 ± 17	0.78 ± 0.03
16	315 ± 21	297 ± 29	0.94 ± 0.04
32	142 ± 21	198 ± 18	1.40 ± 0.11
64	25 ± 9	122 ± 45	4.95 ± 0.59

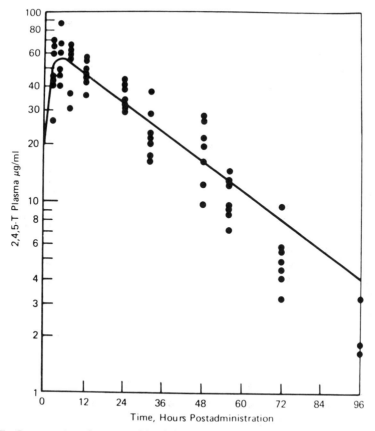

Figure 2.17 Concentration of 2,4,5-T in blood plasma of humans as a function of time following a single oral dose of 5 mg/kg. The points are values for seven different subjects.

tissue of dogs and newborn rats has less capacity to transport 2,4,5-T than does that of adult rats, and (3) large concentrations of 2,4,5-T overwhelm the transport system.

The fate of 2,4,5-T following oral doses of 5 mg/kg has also been investigated in humans (46). The elimination of 2,4,5-T from plasma and in the urine followed apparent first-order kinetics with a $t_{1/2}$ of 23.1 hr (Figures 2.17 and 2.18). Comparison of the elimination rates with those obtained for dogs and rats suggests that the toxicity of 2,4,5-T to humans would be greater than that to rats but less than that to dogs. The higher peak plasma levels attained with a dose of 5 mg/kg in humans than in either dogs or rats was associated with a greater degree of plasma protein binding in humans. In humans, the volume of distribution was only 80 ml/kg, attesting to the retention of 2,4,5-T in the vascular compartment.

Figure 2.19 illustrates simulated levels of 2,4,5-T in the plasma that would be attained in humans with repeated ingestion. If 0.25 mg/kg of 2,4,5-T were

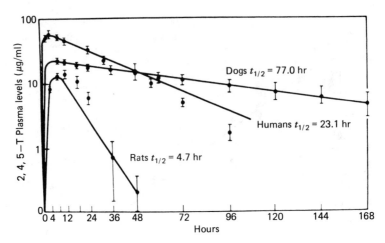

Figure 2.18 Concentration of 2,4,5-T in blood plasma of dogs, humans, and rats following a single oral dose of 5 mg/kg. Each point represents a mean and SD.

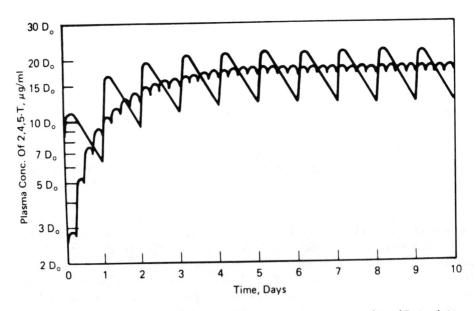

Figure 2.19 Predicted plasma concentrations of 2,4,5-T in humans ingesting a dose of D_0 (mg/kg) every 24 hr (curve with large excursions) or a dose of $\frac{1}{4}$ D_0 every 6 hr (curve with small excursions).

ingested daily, a level equaling that attained by ingesting a single dose of 5 mg/kg as in this study would never be attained. Finally, determination of renal clearance values revealed values exceeding glomerular filtration, as was expected from the results of the studies in animals described previously.

These pharmacokinetic studies of 2,4,5-T in humans and animals, together with the results of toxicological studies in animals, support the conclusion that the hazard from exposure to small amounts of 2,4,5-T encountered during its recommended use is nil.

9 CARCINOGENESIS

9.1 Role of Carcinogenic Mechanisms in Risk Assessment

Frequently, in the rodent carcinogenesis bioassay of a chemical, an excessive incidence of tumors is observed only after the prolonged administration of high dose levels of test substance. The use of these maximum tolerated dosages often results in considerable tissue toxicity but has been justified by some experimenters as being necessary to enhance the statistical sensitivity of bioassays. However, there is little chance for the use of these data for meaningful extrapolation of risk to humans unless interspecies pharmacokinetic data and data regarding resultant macromolecular events are also gathered on the chemical in question. It is important to integrate these three data bases (bioassay, pharmacokinetic, and macromolecular events) to more realistically assess the potential carcinogenic risk of a given chemical. Recently, studies have been directed toward understanding the different mechanisms of tumor formation and their implications for risk assessment. The potential role of cytotoxicity in the carcinogenic process is discussed in the next section.

9.2 Mechanisms of Carcinogenesis

Since the first report of chemically induced carcinogenesis by Yamagiwa and Ichikawa in 1918, numerous attempts have been made to mechanistically explain chemical carcinogenesis. Of these, the "somatic mutation theory," first proposed by Boveri (47), has gained wide acceptance. Simply stated, this theory dictates that mutations in critical sites of the genome of a somatic cell (i.e., DNA) as a result of chemical interaction (e.g., alkylation, intercalation) may result in the transformation of the cell to an anaplastic state. As envisioned by Comings (48), these critical site mutations would occur in regulator gene loci and their impact could be modified by the transcribable nature of corresponding structural genes capable of coding for cellular transforming factors. Mitigating factors in this progression are the various cellular DNA repair enzyme systems that may remove the chemically induced "lesion" before or after DNA replication (i.e., before the lesion is fixed as a mutation in the genome of the cell) (49, 50). However, it also appears that in some instances this repair itself may be the source of errors in DNA base sequences, possibly by an inducible error-prone DNA repair enzyme system analogous to SOS repair observed in bacteria (51–55).

As reviewed by Trosko and Chang (56), there are considerable experimental data to support the mutational origin of tumorigenesis as a result of chemical interaction. These data include findings regarding the clonal nature of tumors, the mutagenicity of many carcinogens, the correlation of high cellular mutation rates and skin cancer in individuals lacking normal DNA repair systems, and the involvement of mutation in the initiation phase of some polycyclic aromatic hydrocarbon-induced carcinogenesis. However, some experimental data do not support a mutational origin of carcinogenesis. These data have included observations that not all carcinogens appear to be mutagens (57), the induction of tumors by plastic or metal film implantation and hormone imbalance (58), and the apparent totipotency of some tumor cell genomes (59–61). Thus there is also substantial evidence of nonmutational mechanisms of carcinogenesis.

With the realization that chemicals may produce tumors through several different mechanisms of action, it has been possible to propose a classification scheme for chemical carcinogens even though the mechanistic details remain to be elucidated. An example of how chemical carcinogens may be classified based on their likely mechanisms of action has been presented by Weisburger and Williams (62). These authors have proposed that carcinogenic compounds be categorized under the general divisions of genotoxic (direct acting or primary carcinogens) and epigenetic carcinogens (solid-state carcinogens, hormones, immunosupressive agents, promotors, and cocarcinogens) based on their mutagenicity and ability to induce DNA damage and repair.

The distinction between chemicals causing tumors by means of a primarily genetic mechanism and of those causing tumors by a primarily epigenetic mechanism is of great significance in carcinogenic risk assessment. Genotoxic chemicals may cause tumors at nonmetabolically saturating dose levels, causing an irreversible initiation of target tissue cells at less than toxic dosages. Epigenetic carcinogens display characteristics of thresholds, and the cellular responses that they evoke are generally reversible. These responses are usually observed only on repeated administration of high dosages, which often result in nonlinear excretion pharmacokinetics or saturation of normal defense or repair processes in the body. Examples of epigenetic carcinogens are 1,4-dioxane and trichloroethylene.

Experimentally, it has been possible to separate chemicals into these general mechanistic classifications based on the ability of the chemical in question to interact directly with DNA. As used here, chemicals that alkylate or intercalate DNA are classified as those causing tumors by a genetic mechanism of action. Chemicals that do not alkylate DNA to a great extent are classified as epigenetic carcinogens. Possible mechanisms of action of epigenetic carcinogens include both mutational and nonmutational mechanisms.

9.3 Cytotoxic Mechanisms of Carcinogenesis

The involvement of cytotoxicity in a carcinogenic response to chemical exposure has been suggested for many years. Indeed, regenerative cellular proliferation in response to chemical cytotoxicity was the driving force of the "irritation theory of carcinogenesis" (63, 64). Although this theory was soon discounted as a comprehensive explanation of the mechanism of carcinogenesis, it still appeared to explain the promotion of "initiated" cells by some chemicals (e.g., phorbol esters) and the production of tumors in the absense of exogenous initiating compounds by repeated tissue injury (58). A logical question then arises as to whether this situation may be analogous to the tumors produced by the practice of bioassaying maximum tolerated, often cytotoxic, doses of chemicals in animals. Often in these bioassays, compounds are observed to cause tumors in animals only at very high dose levels, which result in severe recurrent cytotoxic responses or in animal strains predisposed to producing a high spontaneous incidence of the particular tumor in question. In understanding this relationship it is critical to explore the basis for "spontaneously produced" cellular mutations and the potential impact of cytotoxicity on this process.

Spontaneous mutations may originate in one of several ways: DNA thermodynamic degradation, natural genotoxic challenge, or replication errors. The DNA molecule and its maintenance do not represent a static situation *in vivo* even in the absence of measurable exogenous genotoxic challenge, but a dynamic structure with constant degradation and repair. Such naturally occurring genotoxic challenges as cosmic radiation, ultraviolet (UV) radiation, endogenously produced compounds (e.g., formaldehyde), and loss of nucleic acid bases due to thermodynamic instability require a constant DNA surveillance and repair process. The latter source of degradation alone has been estimated to account for over 11,000 depurinations and several hundred depyrimidinations and base deaminations of DNA molecules in a mammalian cell per day (65–67). Deamination of 5-methyl cytosine to form thymine may result in some loss of gene regulatory control associated with this base (68). Should these lesions be improperly repaired prior to replication, a base transversion or transition mutation may occur. Indeed, it has been demonstrated *in vivo* that DNA polymerases may copy past apurinic sites and that the fidelity of DNA synthesis by use of an apurinic DNA template was decreased proportionately to the degree of depurination (69). In addition, so-called proofreading repair enzyme(s) [exonuclease(s)] do not appear to recognize and remove the misincorporated bases at these apurinic sites (69). A low frequency of DNA base errors may also result during DNA replication or repair from base mispairing due to base keto–enol and amino–imino tautomerization, and anti–syn isomerization about the glycosyl bond (70). These base alterations may affect bond-forming capabilities and characteristics of the base pairs.

The replication process itself may also introduce errors into a growing DNA molecule or repair site. As reviewed by Loeb et al. (71) and Hartman (72), base mismatches may arise from DNA polymerase base selection errors and errors by proofreading enzymes as a result of lack of recognition or misincorporation. Estimates of *in vivo* misincorporation during DNA replication that escape proofreading repair range from 10^{-11} to 10^{-8} per base pair synthesized (73). With these sources of DNA base errors and of possible mutation associated with all other possible "natural" sources of mutation described previously, the end result in humans appears to be an observable mutation rate of approximately 10^{-6} to 10^{-5} mutations per gene per generation (74, 75). Indeed, it has been estimated that 10% of all human gametes contain at least one new mutation of their own as well as several inherited mutations (76).

Thus it appears that spontaneous mutations may occur in the absence of a measurable exogenous genotoxic challenge. Conceivably, some of these mutations may occur in critical regulatory sites, which may ultimately result in a spontaneous cellular transformation described by the somatic mutation theory of carcinogenesis. Tissue degenerative–regenerative changes in response to a cytotoxic dosage of a chemical are expected to have a profound impact on the incidence and rate at which this process may occur.

A shortened cell cycle in regenerating tissue would leave less time for DNA repair mechanisms for elimination of misincorporated bases, apurinic or apyrimidinic sites, and altered bases (pre- and postreplication). Berman et al. (77) have reported an increased mutation frequency (fivefold) in actively dividing versus nondividing rat liver epithelial cells exposed to the mutagenic chemical methyl methanesulfonate. Presumably, DNA repair mechanisms in the dividing cells had less time to repair altered DNA bases before the DNA replicated, resulting in a higher mutation rate. Work by Maher et al. (78) with the use of human fibroblasts has shown that cell survival is lower and mutation rates higher on UV irradiation in DNA excision repair incompetent cells than in normal cells. Normal cells were also observed to survive a usually lethal and mutagenic UV dose on being held in confluence (nondividing) for a period of time prior to assessment of survival and mutation rate. Presumably, this recovery was a result of the increased time given DNA repair mechanisms to remove thymine dimers prior to replication. Similar findings have also been reported by Lang et al. (79) with the use of benzo(*a*)pyrene, in which the actual number of DNA alkylation sites removed from DNA was observed to be directly related to cell survival.

The occurrence of abnormal DNA bases in response to hepatotoxicity has recently been reported by Barrows and Shank (80). Administration of high, cytotoxic doses of carbon tetrachloride, ethanol, or the noncarbonaceous chemical hydrazine to rats together with ^{3}H-labeled methionine was observed to cause detectable levels of O^6-methylguanine and *N*-7-methylguanine. These bases were previously believed to be formed only as a result of DNA alkylation by genotoxic compounds. O^6-Methylguanine in particular has been suggested as being very efficient at causing base transversion mutations, and its formation has been correlated with the carcinogenic potency of some chemicals (81–85). From data presented by these authors, levels of total abnormal guanine bases ranging from $44/10^6$ nucleotides for hydrazine to $118/10^6$ nucleotides for carbon tetrachloride may be calculated (based on 41 percent G + C content). These altered bases appeared to be the result of an enzymatic methylation of normal DNA bases utilizing S-adenosyl methionine as a methyl donor, suggesting that cytotoxicity in some way altered normal cellular DNA methylase or polymerase activity.

It is expected that chemical carcinogens will have a spectrum of activity ranging from those that cause tumors by means of a primarily genotoxic mechanism to those causing tumors by numerous epigenetic mechanisms, including recurrent cytotoxicity. In the former case, defense mechanisms such as DNA repair, detoxification of reactive chemicals, and the qualitative characteristics of DNA alkylation will be critical to the carcinogenic response for a given dose level. In the latter situation the prolonged exposure of an animal to high dosages of the compound will be required to elicit the multitude of possible cellular responses characterizing these mechanisms of action. It is also important not to rule out endogenous factors such as genetic predisposition working in concert with epigenetic mechanisms since many test species and strains of animals are predisposed to developing certain types of tumors unrelated to exogenous chemical exposure. The concept is also of paramount importance in assessing the relevance of extrapolating test results from sensitive animal species to humans.

9.4 Experimental Approaches: Some Examples

One experimental approach has been to develop methods for differentiating between genetic and epigenetic mechanisms of tumorigenesis. The direct interaction of chemicals with DNA (genotoxicity), *in vivo* alkylation of DNA by a radiolabelled compound and *in vivo* DNA repair, are indications of DNA–chemical interaction. Although there are inherent limitations to the use of total DNA alkylation data as an index of genetic interaction (i.e., lack of qualitative consideration), because of the paucity of information regarding critical sites of reaction on all DNA bases, total DNA alkylation is one of the means to ascertain a potential for *in vivo* genetic interaction with a chemical. Short-term *in vitro* assays of mutagenicity and *in vitro* DNA repair have also been useful as indicators of potential genotoxicity. In contrast, tissue cytotoxicity as observed in target tissues by measuring changes in organ size, DNA content, and DNA synthesis levels, as well as by the presence of histopathologic changes, is one indication of nonspecific cellular proliferation, a nongenotoxic endpoint. Chemicals to be discussed are sodium orthophenylphenol and 1,1,2-trichloroethylene. The genotoxic dimethylnitrosamine (DMN) was used as a positive reference compound for discussion purposes.

9.4.1 Sodium Orthophenyl Phenol

Hiraga and Fujii (86) reported that F344 rats consuming diets with high levels of sodium orthophenyl phenol (SOPP) developed tumors of the urinary tract. Several dose levels were studied, and a very sharp dose–response curve was obtained. Administration of diets containing 2 percent SOPP caused 19 of 21 animals to develop bladder tumors, whereas 0 of 21 animals consuming diets with 0.5 percent SOPP exhibited this lesion.

Reitz et al. (87) subsequently reported that administration of diets containing 2 percent SOPP to F344 rats produced increased cellular division in bladder epithelial cells after periods as short as 3 days. With continued administration of 2 percent SOPP, the lesion continued to increase in severity and was diagnosed as simple hyperplasia by light microscopy at 90 days.

Reitz et al. also studied SOPP metabolism in F344 rats. They found that when low doses of SOPP were administered, more than 90 percent of the material was eliminated from the animals in a two-step process involving conjugation of SOPP followed by excretion in the urine. Two major urinary metabolites were detected in the urine of animals receiving 50 mg/kg or less, and these were identified by gas chromatography/mass spectroscopy as sulfate and glucuronide conjugates of SOPP (Figure 2.20A, peaks I and II).

However, when 500 mg/kg of SOPP was administered to these rats, a third type of metabolite was detected. This metabolite (Figure 2.20B, peak III) was not detected in urine obtained from animals dosed with either 50 mg/kg or 5 mg/kg, suggesting that it is not produced in appreciable amounts until the primary metabolic pathways have been saturated (i.e., a pharmacokinetic threshold has been exceeded). Peak III was identified as an oxidation product of SOPP, the dihydroquinone of SOPP (DHQ–OPP) conjugated with either sulfate or glucuronic acid. Tunek et al. (88, 89) and Lutz and Schlatter (90) have described the production of similar intermediates during the metabolism of phenol and have suggested that these reactive intermediates may be partially responsible for the toxicity of phenol. The metabolic pathways for SOPP are summarized in Figure 2.21.

It was demonstrated that oxidation products of SOPP could be formed by the mixed function oxidase systems of rats by incubating ^{14}C–SOPP with purified rat-liver microsomes. More than half of the ^{14}C-SOPP was converted to a form that cochromatographed with an authentic sample of DHQ-OPP. During these incubations, ^{14}C material was covalently bound to microsomal protein, and this binding was dependent on the presence of active microsomes as well as a source of reduced cofactor (NADPH). Thus it appears that the alternate pathway of SOPP metabolism produces a reactive intermediate that can bind to macromolecules. To determine the dose dependency of this pathway *in vivo*, studies of macromolecular binding were conducted in rats. The results are summarized in Table 2.2. Binding is reported as a relative percentage of the administered dose that is bound to bladder tissue, normalized to the value obtained with 50 mg/kg. If all the processes involved in SOPP metabolism were

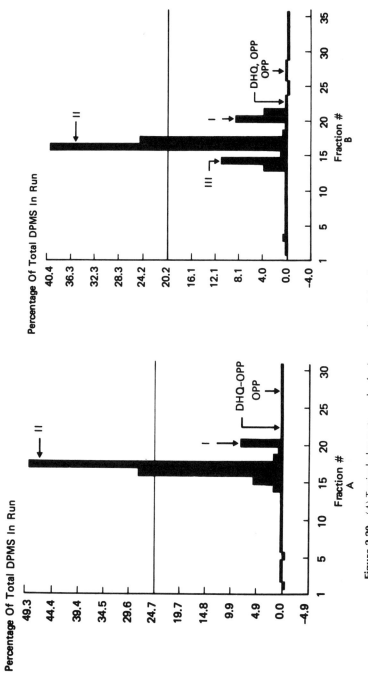

Figure 2.20 (A) Typical chromatograph of urinary radioactivity after gavage with 5 mg/kg SOPP. Urine collected 0 to 24 hr postgavage. Results are reported as percentage of the total radioactivity in the sample. (B) Mass spectra obtained with chromatographic fraction of rat urine. Urine was collected from 0 to 24 hr following gavage with 500 mg/kg ^{14}C–OPP. Spectrum 2A = metabolite I; spectrum 2B =, metabolite II; spectrum 2C = metabolite III.

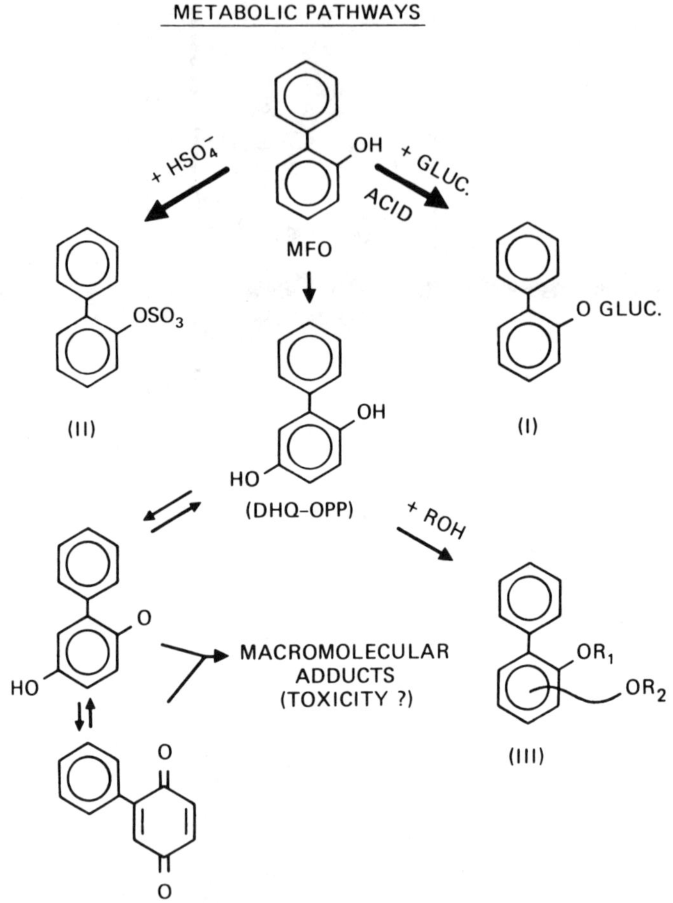

Figure 2.21 Hypothetical scheme for the metabolism of OPP or SOPP in rats, based on identification of metabolic products and dose–response studies in F344 male rats.

linear, the relative binding would equal 1.0 for each dose. Instead, it is clear that the primary metabolic pathways saturate somewhere between 200 and 500 mg/kg. There is a disproportionate increase in the macromolecular binding of SOPP to body tissues, particularly bladder tissue at the highest dose.

Increased macromolecular binding of SOPP appears to cause proliferation of bladder cells. DNA synthesis in bladder tissue of rats gavaged with 500 mg/kg of SOPP was elevated 2.2 ± 0.6-fold in one experiment and 2.5 ± 0.6-fold in another experiment (87). Because of the small amount of tissue available, DNA synthesis was measured with the DNA isolated from the entire bladder (containing both epithelial and muscular tissue). The stimulation of cellular regeneration was much higher when only epithelial cells of the bladder were examined by microautoradiography (Reitz et al., unpublished data). Thus it appears likely that some of the doses of SOPP administered by Hiraga and

Table 2.2 *In Vivo* Macromolecular Binding Following Oral Administration of ^{14}C–SOPP to Male Rats[a]

Sodium Orthophenylphenol (mg/kg)	Liver	Kidney	Bladder
50	1.0	1.0	1.0
100	1.2	0.9	0.6
200	1.8	1.8	2.9
500	12.5	4.0	19.8

[a] Values are normalized to 50 mg/kg dose for each tissue.

Fujii (86) saturated the primary metabolic pathways for elimination and produced a strong cytotoxic effect on bladder epithelial cells.

In contrast, there is little indication that SOPP is genotoxic, even at very high levels. Reitz et al. (87) failed to detect any covalent binding of radioactivity to purified bladder DNA after administration of 500 mg/kg of ^{14}C–SOPP to rats. The detection limit was less than one alkylation per million nucleotides in these experiments. Similarly, Reitz et al. (87) reported that SOPP failed to induce unscheduled DNA synthesis (UDS) when measured by the technique due to Williams (91) and did not increase the reverse mutation rate in Salmonella when assayed by the technique by Ames et al. (92).

The potential of SOPP (or its free phenol analogue, OPP) for producing mutations in bacteria has been studied by others (93–95), and only Hanada (93) concluded that SOPP might possess a mutagenic effect. Furthermore, OPP failed to induce cytogenetic effects (96, 97) and was reported as negative in dominant lethal tests (98) and teratogenicity tests (98, 99).

Consequently, it appears likely the high doses of SOPP administered by Hiraga and Fujii (86) saturated the primary pathways of SOPP detoxification (conjugation and urinary excretion) with the result that, above the pharmacokinetic threshold, reactive metabolites of SOPP were formed by an alternate metabolic pathway. These reactive intermediates have been demonstrated to bind to proteins and stimulate extensive cellular regeneration of bladder epithelial tissue. However, even the more reactive intermediate produced from SOPP by the alternate pathway did not produce genotoxic lesions when studied *in vivo* or *in vitro*. Thus an understanding of the molecular mechanisms involved in the toxicity of SOPP suggests that there is little risk of developing the (nongenotoxic) lesions observed in rat bladder tissue during exposure to levels of SOPP below the pharmacokinetic threshold.

9.4.2 1,1,2-Trichloroethylene (TRI)

The safety of 1,1,2-trichloroethylene has been questioned on the basis of the results of carcinogen bioassay conducted by the National Cancer Institute (NCI)

(100). In this study, B6C3F1 mice and Osborne–Mendel rats were repeatedly administered high gavage doses of TRI for 78 weeks and sacrificed 12 and 32 weeks later, respectively. An increased incidence of hepatocellular carcinoma was observed in male mice dosed with a time-weighted average daily dose of 1169 mg/kg (52 percent) and 2339 mg/kg (65 percent) relative to controls (5 percent). A lower incidence of tumors was observed in female mice and none in either sex of rats. A confounding factor in these results (100) was the use of TRI containing approximately 0.2 percent epichlorohydrin (101), which has previously been shown to cause nasal carcinomas in rats at toxic doses (102).

In a 3-week repeated TRI dosing experiment (Table 2.3), mice dosed with 0, 250, 500, 1200 or 2400 mg kg^{-1}/day^{-1} showed a clear dose-related hepatocellular hypertrophic response. A dose-related increase in liver weights (up to a maximum of 177 percent of control) and decrease in DNA content per gram of tissue (77 percent of control) were observed in mice repeatedly dosed with ≥500 mg/kg of TRI per day. Dose-related histopathological changes in hepatic tissue were observed at all dose levels. These ranged from slight increases in cytoplasmic eosinophilic staining of centrilobular hepatocytes (often associated with depleted glycogen or proliferation of smooth endoplasmic reticulum) in mice from the 250 and 500 mg kg^{-1}/day^{-1} TRI dose levels, to increased centrilobular hepatocellular swelling (1200 mg kg^{-1}/day^{-1} TRI group), to still more severe centrilobular hepatocyte swelling, giant cell inflammation, and mineralized cells (2400 mg kg^{-1}/day^{-1} TRI group). Although there was some indication of increased hepatic DNA synthesis at the higher dose levels, the biological variability precluded statistical significance, contrary to earlier observations.

Unlike the B6C3F1 mouse, only slight, nonsignificant changes were observed (Table 2.3) in hepatic tissue of Osborne–Mendel rats. In rats repeatedly dosed for 3 weeks (5 days/week) with 1100 mg/kg per day of TRI an increase in liver weights (118 percent of control) with no decrease in DNA content per gram of tissue, and no hepatic histopathology was observed. An increase in hepatic DNA synthesis (175 percent of control) in these animals was observed, however, indicating that the increased liver weight was due to an increase in the number of morphologically normal looking hepatocytes. Overall, the response of the Osborne–Mendel rats to administered TRI appeared to be considerably less than that of the B6C3F1 mice.

In Vivo **DNA Alkylation.** Investigation of the ability of TRI to alkylate hepatic DNA *in vivo* in mice indicated that TRI alkylated DNA to only a very small degree (Table 2.4). In mice dosed with a level previously associated with increased hepatic tumors (1200 mg/kg), a maximum estimate of the average DNA alkylation level of 0.62 (±0.42) alkylations/10^6 nucleotides was observed in three treated mice, whereas no ^{14}C-associated alkylated bases were detected in a fourth treated mouse. The possible erroneous contribution of trace contamination of the DNA isolate with alkylated proteins or of metabolic incorporation (C-1 pool) of TRI-derived radioactivity into normal DNA bases

Table 2.3 Liver: Body Weight Ratios, DNA Content, DNA Synthesis, and Histopathology Data from Male B6C3F1 Mice and Osborne–Mendel Rats Administered 1,1,2-Trichloroethylene by Gavage for 3 Weeks (5 days/wk)[a]

Dose Level	% $\dfrac{\text{Liver Wt}}{\text{Body Wt}}$	$\dfrac{\mu g \text{ DNA}}{g \text{ Hepatic Tissue}}$	DNA Synthesis $\dfrac{\text{dpm}^3\text{Hdt}[e]}{\mu g \text{ DNA}}$	$\dfrac{\text{Treated}}{\text{Control}}$	Histopathology[b]
B6C3F1 Mice					
Control (corn oil)	4.85 (0.28)	2772 (203)	128 (27.5)		—
Treated					
250 mg/kg	5.20 (0.53)	2517 (220)	108 (34.0)	0.84	+
500 mg/kg	5.46 (0.61)[c]	2391 (203)[c]	87.6 (15.4)[d]	0.68	+
1200 mg/kg	6.56 (0.42)[c]	2143 (122)[c]	140 (41.0)	1.09	++
2400 mg/kg	8.57 (0.46)[c]	2086 (356)[c]	156 (52.6)	1.22	+++
Osborne–Mendel Rats					
Control	3.77 (0.21)	3268 (396)	1.21 (0.61)		—
Treated					
1100 mg/kg	4.43 (0.21)[c]	3272 (262)	2.12 (0.96)[c]	1.75	

[a] Average of 10 to 12 mice and 4 rats ($\pm$ standard deviation).
[b] Histopathology: —, none observed; +++ > ++ > +, severity of observable histopathology.
[c] $p < .01$.
[d] $p < .05$.
[e] ^{3}Hdt—tritiated thymidine.

Table 2.4 *In Vivo* Alkylation of Hepatic DNA by 1,1,2-Trichloroethylene-UL-^{14}C in Male B6C3F1 Mice Dosed with 1200 mg/kg ^{14}C–TRI by Gavage

Animal	Detection Limit (Alkylations/10^6)	Non-C_1 DPM[a] >25 Percent Error	Maximum Alkylations 10^6 Nucleotides[b]	Maximum CBI[c]
1	0.12	18.0	0.50	0.055
2	0.52	ND[d]	—	—
3	0.15	7.1	0.27	0.030
4	0.18	26.6	1.1	0.120

[a] Disintegrations per minute not associated on HPLC analysis with normal DNA bases from normal metabolic incorporation and not attributable to counting error (i.e., mean background dpm + 2 standard deviations).

[b] "Maximum" alkylations possible due to the relatively large amount of dpm associated with protein binding and metabolic incorporation into normal bases.

[c] Covalent binding index—comparative hepatic CBI for dimethylnitrosamine, diethylnitrosamine, methylnitrosourea, and aflatoxin B-1 are approximately 5500, 125, 640, and 17,000, respectively:

$$\text{CBI} = \frac{\mu\text{mol adduct/mole nucleotides}}{\text{mmol/kg dose}}$$

[d] None detected.

to the alkylation value was eliminated to a large extent by high-performance liquid chromatographic (HPLC) analysis of isolated DNA digests. This is particularly necessary for compounds such as TRI that undergo considerable catabolism and protein binding *in vivo*.

Despite reports of a weak mutagenic response in a host-mediated assay with the use of yeast (103) and *in vitro* bacterial (104) mutagenicity assays, TRI (nonepoxide inhibited) was not observed to readily bind hepatic DNA of B6C3F1 mice *in vivo*. The observed maximum of only 0.65 (± 0.42) alkylated bases per 10^6 nucleotides contrasts sharply with the hundreds of alkylations observed when mice are dosed with powerful genotoxins such as dimethylnitrosamine or methylnitrosourea (105). Thus the genotoxic potential of TRI appears to be nil to very slight when the low potential for *in vivo* DNA alkylation coupled with numerous negative mutagenicity assays is considered.

The observed weak genotoxic potential and significant hepatotoxicity of TRI in B6C3F1 mice dosed with a tumorigenic dose level of TRI are consistent with a primarily epigenetic mechanism of tumorigenic action. The particular epigenetic mechanism of tumorigenicity is difficult to ascertain on the basis of the data available. The enhancement of the expression of preexisting oncogenic factors present in the B6C3F1 mouse is one plausible explanation, and it is supported by the high incidence of spontaneous hepatocellular carcinoma (16 percent) reported in this strain of mice (106). Another possible enhancement of this spontaneous tumor rate is related to the hypertrophic response of mouse hepatocytes to TRI similar to that observed for many inducers of the mixed-function oxygenase enzyme systems. (e.g., phenobarbital). Indeed, dosing of rats by repeated intraperitoneal administration of TRI (1400 mg/kg daily for 5 days) or inhalation of TRI (<50 ppm, 5 hr/day for 10 days) has been noted to induce hepatic aniline hydroxylase and aminopyrine hydroxylase activity (107–109).

In summary, B6C3F1 mice were observed to be more sensitive to the hepatotoxic effects of ingested TRI (maximum tolerated dose) than Osborne–Mendel rats. At a tumorigenic dose level in mice (1200 mg/kg), a maximum estimate of TRI-DNA interaction in the liver was very small, reflective of the reportedly weak positive or negative responses of TRI in microbial mutagenicity assays. These results suggest an epigenetic mechanism of tumorigenicity in the B6C3F1 mice due either to a repeated cytotoxicity mechanism or simply an enhancement of preexisting hepatic oncogenic factors in this mouse strain. Thus the tumorigenic response to TRI exposure observed in these mice could be explained by the repeated administration of high, cytotoxic dose levels of TRI.

The examples cited were selected to illustrate the value in understanding the overall pharmacokinetic–metabolic characteristics as well as potential mechanism of action in the assessment of risk when low-dose–high-dose and interspecies extrapolations are required.

10 CONCLUSIONS

Absolute safety is the goal of society in all endeavors. Absolute safety, however, is never achieved, whether in skyscraper construction or environmental management. An imperfect system is not to be condoned, but constantly improved through experience and research as rapidly as possible within limitations placed on the system by society itself.

In the evaluation of the safety of chemicals as in all other fields of human endeavor, some degree of risk must be considered acceptable to society. The alternative would be the needless prohibition of important benefits. The estimation of the safety (or hazard) of a chemical for a particular use can span an enormous range of complexity. Nevertheless, the objective of all safety testing is to ensure attainment of the desired benefits of use without incurring needless risks. There must, of course, be some balance between the benefit and the cost of assessment, just as there must be a balance between benefit and acceptable risk.

Investigations of "potential hazard" as applied to chemicals include a demonstration of the toxicologic properties of an agent by appropriate test procedures and a determination of an exposure level that does not produce detectable adverse effects by the same procedure. For humans, the margin of safety of an agent is evaluated by relating to the predicted human exposure the maximum exposure level in experimental animals that produces no detectable adverse effects. An agent with a small margin of safety has a high potential for producing adverse effects; conversely, an agent with a large margin of safety has a low potential for producing adverse effects.

For optimum assurance of safety within the existing limitations of capabilities and skills for evaluation, safety must be concentrated on environmental situations in which there is a reasonable expectation that exposure to chemicals may cause real hazards. The evaluation of toxicity data for judging the safety or hazard associated with the material can be made only in the light of the anticipated amount and circumstances of human exposure. If the anticipated human exposure is very close to, or perhaps greater than, the maximum dose that produces no observable effect in an adequately sized group of experimental animals, the type and extent of controls to be applied should be considered, in order to reduce human exposure to acceptable levels. The acceptable risk determination includes an evaluation of the benefits to the individual and to society of the proposed exposure and an evaluation of the nature and severity of the anticipated effects. To quote Goldberg: "Neither neglect nor panic is the answer (for safety evaluation). Somewhere between these two extremes lies the course of reasonable action appropriate to each chemical-contaminant. Given good science, good judgment, and above all, freedom from extraneous pressures, the right course can be found" (110).

REFERENCES

1. J. A. Zapp, *J. Toxicol. Environ. Health*, **2,** 1425, (1977).
2. C. S. Weil, *Toxicol. Appl. Pharmacol.*, **21,** 454 (1972).
3. J. T. Litchfield and F. Wilcoxon, *J. Pharmacol. Exp. Ther.*, **96,** 99 (1949).
4. G. L. Plaa, E. A. Evans, and C. H. Hine, *J. Pharmacol. Exp. Ther.*, **123,** 224 (1958).
5. P. J. Gehring, *Toxicol. Appl. Pharmacol.*, **13,** 287 (1968).
6. E. M. K. Geiling and P. R. Cannon, *JAMA*, **111,** 919 (1938).
7. E. B. Dewberry, *Food Poisoning*, 3rd ed., Leonard Hill Limited, London, 1950, pp. 205–217.
8. E. M. Adams, H. C. Spencer, V. K. Rowe, D. D. McCollister, and D. D. Irish, *AMA Arch. Ind. Hyg. Occup. Med.*, **6,** 50 (1952).
9. J. M. Barnes and I. A. Benz, *Pharmacol. Rev.*, **6,** 191 (1954).
10. G. E. Paget, *Proc. Eur. Soc. Drug Toxicol.*, **2,** 7 (1963).
11. H. J. Bein, *Proc. Eur. Soc. Drug Toxicol.*, **2,** 15 (1963).
12. J. P. Frawley, *Food Cosmet. Toxicol.* **5,** 293 (1967).
13. H. M. Peck, *Importance of Fundamental Principles in Drug Evaluation*, Raven Press, New York, 1968, pp. 449–471.
14. W. Hays, Jr., *Essays in Toxicology*, Vol 3, Academic Press, New York, 1972, pp. 65–77.
15. C. S. Weil and D. D. McCollister, *J. Agric. Food Chem.*, **11,** 486 (1963).
16. H. E. Stokinger and R. L. Woodward, *J. Am. Water Works Assoc.*, **50,** 519 (1958).
17. D. D. McCollister, W. H. Beamer, G. L. Atchison, and H. C. Spencer, *J. Pharmacol. Exp. Ther.*, **102,** 112 (1951).

18. P. J. Gehring and G. E. Blau, *J. Environ. Pathol. Toxicol.*, **1,** 163 (1977).
19. S. Baker, J. Tripod, and J. Jacob, *Proc. Eur. Soc. Drug Toxicol.*, **11,** 9 (1970).
20. S. Baker, *Proc. Eur. Soc. Drug Toxicol.*, **12,** 81 (1971).
21. J. T. Litchfield, *JAMA,* **177,** 34 (1961).
22. W. M. Wardell, *J. Anesthesiol.*, **38,** 309 (1973).
23. D. P. Rall, *Environ. Res.*, **2,** 360 (1969).
24. E. J. Freireich, E. A. Gehan, D. P. Rall, L. H. Schmidt, and H. E. Skipper, *Cancer Chemother. Rep.*, **50,** 219 (1966).
25. D. Pinkel, *Cancer Res.*, **18,** 853 (1958).
26. *Committee on Safe Drinking Water, Report, National Research Council*—National Academy of Sciences, Washington, D.C., 1977.
27. R. C. Wands and J. H. Broome, *in-Proceeding of the Fifth Annual Conference on Environmental Toxicology,* 1974. Page 237.
28. H. Osswald and K. Goertler, *Verh. Deut. Ges. Pathol.*, **55,** 289 (1971).
29. R. D. Evans, *Health Phys.*, **27,** 497 (1974).
30. J. E. Cleaver, *Genetics of Human Cancer,* Raven Press, New York, 1977, pp. 355–363.
31. P. J. Gehring, P. G. Watanabe, and G. E. Blau, *Advances in Modern Toxicology—New Concepts in Safety Evaluation,* Halstead Press, New york, pp. 195–270 (1976).
32. J. R. DeBaun, J. Y. Rowley, E. C. Miller, and J. A. Miller, *Proc. Soc. Exp. Biol. Med.*, **129,** 268 (1968).
33. E. C. Miller, J. A. Miller, and M. Enomoto, *Cancer Res.*, **24,** 2018 (1964).
34. J. A. Miller, *Cancer Res.*, **30,** 559 (1970).
35. P. H. Grantham, L. C. Mohan, and E. K. Weisburger, *J. Natl. Cancer Inst.*, **56,** 649 (1976).
36. D. K. Courtney, D. W. Gaylor, M. D. Hogan, H. L. Flak, R. R. Bates, and I. Mitchell, *Science,* **168,** 864 (1970).
37. D. K. Courtney and J. A. Moore, *Toxicol. Appl. Pharmacol.*, **20,** 396 (1971).
38. G. L. Sparschu, F. L. Dunn, R. W. Lisowe, and V. K. Rowe, *Food Cosmet. Toxicol.*, **9,** 527 (1971).
39. R. Roll, *Food Cosmet. Toxicol.,*, **9,** 671 (1971).
40. T. F. X. Collins and C. H. williams, *Bull. Environ. Contam. Toxicol.*, **6,** 559 (1971).
41. W. N. Piper, J. Q. Rose, M. L. Leng, and P. J. Gehring, *Toxicol. Appl. Pharmacol.*, **26,** 339 (1973).
42. V. A. Drill and T. Hiratzka, *Arch. Ind. Hyg. Occup. Med.*, **7,** 61 (1953).
43. V. K. Rowe and T. A. Hymas, *Am. J. Vet. Res.*, **15,** 622 (1954).
44. M. W. Sauerhoff, W. H. Braun, G. E. Blau, and P. J. Gehring, *Toxicol. Appl. Pharmacol.*, **36,** 491 (1976).
45. J. B. Hook, M. D. Bailie, J. T. Johnson, and P. J. Gehring, *Food Cosmet. Toxicol.*, **12,** 209 (1974).
46. P. J. Gehring, C. G. Kramer, B. A. Schwetz, J. Q. Rose, and V. K. Rowe, *Toxicol. Appl. Pharmacol,* **26,** 352 (1973).
47. T. Boveri, *The Origin of Malignant Tumors,* Williams and Wilkins, Baltimore, 1929.
48. D. E. Comings, *Proc. Natl. Acad. Sci.*, **70,** 3324 (1973).
49. P. C. Hanawalt, E. C. Friedberg, and C. F. Fox, *DNA Repair Mechanisms,* Academic Press, New York, 1978.
50. R. W. Hart, K. Y. Hall, and F. B. Daniel, *Photochem. Photobiol.*, **28,** 131 (1978).
51. S. Kondo, in: *Fundamentals in Prevention of Cancer,* University Park Press, Baltimore, 1976, pp 417–429.

52. S. Kondo, *Br. J. Cancer*, **35**, 595 (1977).
53. R. H. C. San and H. F. Stich, *Internatl. J. Cancer*, **16**, 284 (1975).
54. A. Sarasin and A. Benoit, *Mut. Res.*, **70**, 71 (1980).
55. E. M. Witkin, *Bacteriol. Ref.*, **40**, 869, (1976).
56. J. E. Trosko and C. C. Chang, *Photochem. Photobiol.*, **28**, 157 (1978).
57. A. Segaloff, *J. Steroid Biochem.*, **6**, 171 (1975).
58. I. Berenblum, *J. Natl. Cancer Inst.*, **60**, 723 (1978).
59. B. Mintz and K. Illmensee, *Proc. Natl. Acad. Sci.*, **9**, 3585 (1975).
60. R. G. McKinnell, B. A. Deggins, and D. D. Labat, *Science*, **165**, 394 (1969).
61. V. E. Papaioannou, M. W. McBurney, R. L. Gardner, and M. J. Evans, *Nature*, **258**, 70 (1975).
62. J. H. Weisburger and G. M. Williams, in: *Toxicology: Basic Science of Posions*, MacMillan, New York, 1980, pp. 84–138.
63. I. Berenblum, *Br. J. Exp. Pathol.*, **10**, 179 (1929).
64. I. Berenblum, *Arch. Pathol.*, **3**, 233 (1944).
65. T. Lindahl and B. Nyberg, *Biochemistry*, **11**, 3610 (1972).
66. T. Lindahl and B. Nyberg, *Biochemistry*, **13**, 3405 (1974).
67. T. Lindahl and O. Karlstrom, *Biochemistry*, **12**, 5151 (1973).
68. A. Razin and A. D. Riggs, *Science*, **210**, 604 (1980).
69. C. W. Shearman and L. A. Loeb, *J. Molec. Biol.*, **128**, 197 (1979).
70. M. D. Topal and J. R. Fresco, *Nature*, **263**, 285 (1976).
71. L. A. Loeb, L. A. Weymouth, T. A. Kunkel, K. P. Gopinathan, R. A. Beckman, and D. K. Dube, *Cold Spring Harbor Symp. Quant. Biol.*, **43**, 921 (1978).
72. P. E. Hartman, *Environ. Mut.*, **2**, 3 (1980).
73. J. W. Drake, *Nature*, **221**, 1132 (1969).
74. J. V. Neel, *Proc. Natl. Acad. Sci.*, **70**, 3311 (1973).
75. A. C. Stevenson and C. B. Kerr, *Mut. Res.*, **4**, 339 (1967).
76. J. W. Drake, in: *Advances in Modern Toxicology*, Hemisphere Publishing Corp., New York, 1978, pp. 9–26.
77. J. J. Berman, C. Tong, and G. M. Williams., *Cancer Lett.*, **4**, 277 (1978).
78. W. M. Maher, D. J. Dorney, A. L. Mendrala, B. Konze-Thomas, and J. J. McCormick, *Mut. Res.*, **62**, 311 (1979).
79. L. L. Lang, V. M. Maher, and J. J. McCormick, *Proc. Natl. Acad. Sci.*, **77**, 5933 (1980).
80. L. R. Barrows and R. Shank, *Toxicol. Appl. Pharmacol.*, **60**, 334 (1981).
81. R. Goth-Goldstein, *Nature*, **267**, 81 (1977).
82. P. D. Lawley, *Br. Med. Bull.*, **36**, 19 (1980).
83. R. F. Newbold, W. Warren, A. Medcalf, and J. Ames, *Nature*, **283**, 596 (1980).
84. A. E. Pegg, *Adv. Cancer Res.*, **25**, 195 (1977).
85. B. Singer, *J. Natl. Cancer Inst.*, **62**, 1329 (1979).
86. K. Hiraga and T. Fujii, *Food Cosmet. Toxicol.*, **19**, 303 (1981).
87. R. H. Reitz, T. R. Fox, J. F. Quast, E. A. Hermann, and P. G. Watanabe, *Chem. Biol. Interact.*, **43**, 99 (1983).
88. A. Tunek, K. L. Platt, M. Przylylski, and F. Oesch, *Molec. Pharmacol.*, **14**, 920 (1978).
89. A. Tunke, K. L. Platt, M. Przylylski, and F. Oesch, *Chem. Biol. Interact.*, **33**, 1 (1980).
90. W. K. Lutz and C. Schlatter, *Chem. Biol. Interact.*, **18**, 241 (1977).
91. G. M. Williams, *Cancer Lett.*, **1**, 231 (1976).
92. B. N. Ames, J. McCann, and E. Yamasaki, *Mut. Res.*, **31**, 347 (1975).

93. S. Hanada, *Eiyo To Shokuryo (Food Nutr.)*, **20,** 67 (1976).
94. D. Brusick, Kensington, Maryland 20795, as LBI Project No. 2547 (March 31, 1979).
95. A. Kojima and K. Hiraga, *Ann. Rep. Tokyo Metro. Res. Lab.*, **29,** 83 (1978).
96. S. Yoshida, M. Masubuchi, and K. Kiraga, *Pesti. Abstr.*, **80,** 2624 (1980).
97. S. Nawaii, S. Yoshida, J. Nakao, and K. Hiraga, *Ann. Rep. Tokyo Metro. Res. Lab.*, **30,** 51 (1979).
98. M. Kaneda, S. Teramoto, A. Shingu, and Y. Shirasu, *J. Pest. Sci.*, **3,** 365 (1978).
99. J. A. John, F. J. Murray, K. S. Rao, and B. A. Schwetz, *Fund. Appl. Toxicol.*, **1,** 282 (1981).
100. National Cancer Institute, *Carcinogensis Bioassay of Trichloroethylene*, CAS No 79-01-6, DHEW Publ. No. (NIH) 76-802, 1976.
101. D. Henschler, E. Eder, T. Neudecker, and M. Metzler, *Arch. Toxicol.*, **37,** 233 (1977).
102. S. Laskin, A. R. Sellakumar, M. Kuschner, N. Nelson, S. LaMendola, G. M. Rusch, G. V. Katz, N. C. Dulak, and R. E. Albert, *J. Natl. Cancer Inst.*, **65,** 751 (1980).
103. G. Bronzetti, E. Zeiger, and D. Frezaa, *J. Environ. Pathol. Toxicol.*, **1,** 411 (1978).
104. H. Greim, G. Bonse, Z. Radwan, D. Reichert, and D. Henschler, *Biochem. Pharmacol.*, **24,** 2013 (1975).
105. P. D. Lawley, in: *Screening Tests in Chemical Carcinogenesis*, IARC Scientific Publication No. 12, Lyon, 1976, pp. 181, 208.
106. R. M. Elashoff, D. L. Preston, and J. R. Fears, *J. Natl. Cancer Inst.*, **62,** 1209 (1979).
107. K. Norpoth, U. Witting, and M. Springorum, *Int. Arch. Arbeitsmed*, **33,** 315 (1974).
108. D. Pessayre, H. Allemand, J. C. Wandscheer, V. Descatorie, J.-Y. Artigou, and P.-P. Benhamou, *Toxicol. Appl. Pharmacol.*, **49,** 355 (1979).
109. H. Savolainen, P. Pfaffli, M. Tengen, and H. Vaino, *Arch. Toxicol.*, **38,** 229, (1977).
110. L. Goldberg, *Food Cosmet. Toxicol.*, **9,** 65 (1971).

CHAPTER THREE

Biological Indicators of Chemical Dosage and Burden

RICHARD S. WARITZ, Ph.D.

1 INTRODUCTION AND DEFINITIONS OF TERMS

"In analysis of biological specimens for solvents, however, it appears that each solvent represents a different problem, and some knowledge of the peculiarities of the metabolism and excretion of each is necessary, else one may be led astray in one's interpretation of the results" (1). Elkins's words of caution are as valid today as they were when he wrote them in 1954. Elkins defined "solvent" to include any liquid industrial chemical. The advice applies equally well to solids, gases, vapors, organic chemicals, and inorganic chemicals. As is discussed later, it also is important to know whether the exposure chemical or a metabolite is the toxicologically determining material and to be sure that the only source of the chemical being measured is the chemical whose exposure is being measured.

This chapter is not an in-depth review of particular biologic evaluations such as lead or mercury in blood or urine. Individual biological analyses usually have an extensive literature devoted to them and should be consulted by those interested in the particular analysis (2–4). Nor are analytical methodologies discussed in detail; they also are readily located in specialized texts and scientific articles. Biochemistry, enzyme kinetics, pharmacokinetics, and blood–protein binding of chemicals must be considered in the development of any biological assay, and an overview is presented. The reader is referred to relevant specialized texts and scientific papers for detailed guidance in these areas (5–8).

This chapter discusses (1) the general concept of biological analysis as a tool to measure workplace exposure to exogenous chemicals with some biological activity (xenobiotics), (2) possible utilities of the concept in the workplace, (3) the problems to be expected in developing a biological monitoring method for a particular chemical or element, (4) the problems involved in applying a method, and (5) the problems and uncertainties encountered in interpreting the results. Examples from the published literature are used for illustration.

The terms "biological analysis" and "biological monitoring" are used in this chapter to indicate analysis of exhaled air; analysis of some biological fluid, such as urine, blood, tears, or perspiration; or analysis of some body component, such as hair or nails for evaluation of past exposure to a chemical. The chemical of analytical interest is referred to as the "index" chemical.

2 RATIONALE FOR BIOLOGICAL EVALUATION OF EXPOSURE

Blood and urine analyses for certain components have long been used by the medical professions to facilitate differentiation between the normal and diseased states (9). Since most industrial chemicals that might cause systemic effects will be transported by the blood system, metabolized by enzymes, and excreted by one of the excretory systems, it seems a logical extension of the medical groundwork in these areas to attempt to measure exposure to industrial chemicals by biological analysis for the industrial chemical(s) of interest.

The objective of industrial hygiene is to prevent an effect level of an agent from reaching target organ(s) or tissue(s) of that agent in the worker. The agent may be physical or chemical. Physical agents such as sound or electromagnetic radiation or particulate radiation occur in only one form, and control of that one form will result in control of the worker's total exposure to that agent. However, workers may be exposed to chemicals in the form of gases, vapors, liquids, or solids, either singly or combined. Workers may be exposed to chemicals through inhaling them, absorbing them through the skin, or ingesting them. Exposure may be by several routes concurrently.

Within limits, the effect of a chemical on an organ or tissue is directly proportional to the amount of the chemical reaching or reacting with that organ or tissue. Ideally, then, to relate toxicity to dose, one should correlate the amount of chemical in the target organ(s) or tissue(s) with the effect seen. This would first be done in animal experiments and these results then extrapolated to humans. However, it is even more impractical to continually and routinely make such direct measurements on an organ in a human worker than in a laboratory test animal. Therefore, it would be desirable to have some secondary or tertiary measurements that could be made easily and routinely on a potentially exposed person—measurements that would reliably bear a known relationship to the appearance or absence of toxic effects. Ideally, the concentration of the index chemical found also would give some idea of what additional burden of the exposure chemical or its metabolites would be acceptable before reaching the threshold of a toxic effect.

Within limits, the amount of the chemical reaching the target organs or tissues usually will vary directly with the amount unbound (see Section 6.1.4) in the blood. Therefore, the amount reaching these target organs or tissues can be reasonably approximated by measuring blood levels. However, as with organ analysis, blood sampling is an invasive technique and is not suited to routine daily monitoring of workers.

In most cases, again within limits that vary for each chemical, the amount of the toxicologically determining chemical reaching the target organ(s) or tissue(s) will vary directly with the dose of the exposure chemical received by any or all routes. Historically, inhalation has been the major exposure route of concern for the industrial worker. Consequently, industrial hygienists have concentrated on keeping atmospheric levels of chemicals below effect levels. The Chemical Substances TLV Committee of the American Conference of Governmental Industrial Hygienists (ACGIH; TLV is a registered trademark of the ACGIH and is an acronym for threshold limit value.) has established acceptable industrial atmospheric levels for several hundred industrial chemicals. These are levels that they believe "nearly all workers may be repeatedly exposed (by inhalation) day after day without effects" (10). The TLVs for some of these chemicals also have been incorporated in the Occupational Safety and Health Act of 1970 (PL 91–596).

Examination of the TLV list shows that at least one member of most classes of organic compound listed has the notation "Skin," indicating that toxicologically significant amounts of that chemical in the liquid form or in solution can be absorbed through the skin. If, in addition to inhaling one of these chemicals, a worker spills the liquid form of one of these chemicals on skin, on shoes, or clothing, or if the worker's hands are immersed in this liquid, the target organs or tissues will receive the chemical through both inhalation and the skin. Obviously, in this case, atmospheric levels of the chemical do not give an accurate indication of the amount of chemical that might reach the target organ(s) or tissue(s). In fact, the amount reaching the target organ(s) or tissue(s) from inhaling the chemical may be only a small part of the total reaching them from both routes of entry (11–20). Since the amount of the liquid that might contact the skin in an industrial situation is variable, this concurrent route of exposure cannot be reliably assessed as an adjunct to an inhalation exposure of test animals and cannot be factored into deciding safe inhalation levels of the chemical.

Multiple routes of exposure also are possible when workers are exposed to particulate material. Inhaled dust particles may be cleared from the lungs through the trachea and thus be ingested. In these instances, exposure of workers by inhalation also constitutes an exposure by more than one route. If the absorption coefficient for gastrointestinal absorption is greater than that for pulmonary absorption, the second route of exposure again may be more important toxicologically than inhalation. The ratio of absorption by the two routes also may vary with particle size. However, in this situation, in contrast to the inhalation–skin absorption situation, inhalation exposure of test animals will give some measure of concurrent absorption by ingestion. If the particle size is different in the industrial situation than in the animal exposures, however, the amounts concurrently absorbed by the two routes may be more or less than expected from the inhalation studies in animals. Consequently, the amount of the chemical reaching the target organ(s) or tissue(s) may vary, and the animal inhalation exposures may not be quantitatively predictive of the effect on exposed humans.

Obviously, then, measurement of atmospheric concentrations of a chemical will not always give a reliable measure of the amount of the chemical that may reach target organ(s) or tissue(s) in exposed workers.

As already mentioned, the ideal measure would be the measure of the concentration in the target organ(s). This is impractical on a continuing basis as mentioned, although it may be practical for single, isolated analyses. Blood sampling, the next best measure of the amount in the target organ(s) or tissue(s), also is impractical on a continuing basis for two reasons: (1) hazards introduced in taking samples continuously and (2) worker resistance to the procedure. However, this technique also is suited for occasional isolated or repeated analyses.

The ideal method for measuring all the xenobiotic absorbed, and thus the total risk to the worker, is a noninvasive technique that accurately reflects the degree of toxic insult of the chemical to the worker and is not disruptive or repugnant to the worker.

3 HISTORY OF BIOLOGICAL MONITORING

As mentioned previously, biological monitoring has been used by the medical professions for decades. The formal application of this technique on a broad scale to industrial hygiene is probably most rightfully attributed to Hervey Elkins (1), although individual correlations of exposure levels, toxicity, and blood or excretory levels of industrial chemicals precede his broad-scale advocacy by many years (19, 21–25). Since his initial advocacy, Elkins has published several papers on the subject (26–28), and Dutkiewicz (29) has proposed an integrated index of absorption.

Elkins initially tried to correlate exposure levels, urinary levels, and toxicity. Since then, the concept of biological monitoring for xenobiotics has been extended to include monitoring of the biliary–fecal route, exhaled air, perspiration, tears, fingernails and toenails, hair, milk, and saliva.

Although analysis of feces, perspiration, tears, nails, hair, milk, or saliva has been used in special situations to give an estimation of body burden of a chemical, these substances are not suited for routine frequent monitoring of workers for determination of immediate past exposure. Analyses of exhaled air for volatile xenobiotics and/or their metabolites and of urine for water-soluble xenobiotics and/or their metabolites have received the greatest attention as measures of industrial exposure and body burden.

4 USEFULNESS OF BIOLOGICAL MONITORING

Since biological monitoring usually occurs after exposure, it cannot replace good industrial hygiene practices. It does, however, supplement a good industrial hygiene program and provides additional information of value in the overall worker protection program. Some specific utilities of biological monitoring are described in the following sections.

4.1 Indication of Unsuspected Employee Exposure

The unsuspected exposure may arise from skin absorption, as mentioned previously, or from unsuspected equipment leaks, which increase air levels of the exposure chemical.

4.2 Guidance to Physician in Deciding Whether to Administer Therapy that Also Carries Some Risk

Some antidotes, such as atropine or pyridine-2-carboxaldehyde monooxime (2-PAM) used for overexposure to certain organophosphates or carbamates, are used at dosages that might themselves be toxic if the patient had not received a prior excessive dosage of the organophosphate or carbamate.

4.3 Documentation of Overexposure or Acceptable Exposure

If a worker has received an excessive dose of an agent, good industrial hygiene practice dictates the worker be removed from that exposure. Biological monitoring can document the fact of overexposure or acceptable exposure. In the case of overexposure, it can give the degree of overexposure and indicate how long the worker must remain free of additional exposure.

4.4 Preemployment Screen for Metabolic Abnormalities

Some people are unusually sensitive to the adverse effects of chemicals (10). For their own protection, they should not be exposed to levels of a chemical that will not adversely affect the rest of the exposed population but that will affect them. In some cases, this unusual sensitivity is due to atypical metabolic routes or to atypical enzyme kinetics. Some of these cases could be evaluated in a preassignment physical examination by administering controlled, low doses of chemicals that the worker would be exposed to on the job. This would be followed by determining whether the worker's individual biological half-life values for these chemicals are in the normal range and whether the worker excretes the usual metabolites in the usual ratios.

A subnormal half-life would be acceptable, but a greater than normal half-life could predict bioaccumulation and possible hazard under exposure conditions that would not be harmful to most people (30). Abnormal metabolic products or ratios might or might not indicate greater hazard.

4.5 Validating TLVs Calculated for Concurrent Exposure to Chemicals Affecting the Same Target Organ(s)

When a worker is concurrently exposed to more than one chemical affecting the same target organ(s), exposure levels that were safe for the individual chemicals of exposure usually are not safe for the combined exposure. The TLV handbook contains a formula for calculating reduced atmospheric exposure levels for this situation (10). Direct biological monitoring for the index chemical will not document the safety of the calculated safe exposure levels. If the organ damage caused by overexposure results in elevated levels of some usual blood or excretory fluid component, however, normal postexposure levels of the component(s) would confirm the validity of the calculation. Conversely, abnormal values would indicate the inapplicability of the correction formula in this situation (see Section 5).

4.6 Validation of Area Monitoring

Personal monitoring, manual area monitoring, and automated area monitoring are the most common techniques now used to monitor worker exposure to atmospheric chemicals. All three offer opportunities for error. Regular biological monitoring may be useful for validation of the atmospheric monitoring. However, biological monitoring also offers opportunities for error as discussed in this chapter, and these must be considered in evaluating the biological monitoring results.

5 TYPES OF BIOLOGICAL MONITORING

Most monitoring of biological fluids (and, in some cases, hair, nails, and expired air) can utilize one of two approaches: direct or indirect.

Direct analysis is analysis for the index chemical per se.

Indirect analysis is the quantitation of second-generation effects that result from the action of the agent on some body system, organ, or tissue. For example, blood–urea–nitrogen or urinary glutamic–pyruvic transaminase (GPT) levels could be utilized as measures of effect on kidneys (9, 31). Changes in blood cell numbers per unit volume and distribution between types could be used as a measure of effect on the hemopoietic system (9). Changes in isoenzyme patterns could be used as a measure of effects, for example, on liver or heart (31), and changes in cholinesterase levels in whole blood or plasma could be used as an indirect measure of organophosphate exposure (32).

Indirect analysis is seldom a desirable approach for routine worker monitoring for at least the following reasons:

1. The measured effect usually occurs only after damage to the worker has occurred, and the thrust of industrial hygiene should be to prevent deleterious effects on workers from agents in the workplace.
2. Appearance of the measured effect may not occur for some time following exposure to the agent and thus might be difficult to attribute to a specific cause. Also, in this situation, there could have been protracted overexposure before the secondary effect was seen, and this is undesirable.
3. The effect measured may not vary directly with amount of exposure, making it difficult to determine the degree of exposure.
4. The effect may not be specific and may have several possible etiologies, not all of which occur in the workplace. This should not be interpreted to mitigate the health significance of the change for the individual worker, however.
5. If blood sampling is required, it is not suited for routine monitoring in the workplace on a frequent basis.

6 BIOCHEMICAL PROBLEMS ASSOCIATED WITH BIOLOGICAL MONITORING

As with most techniques, increased study has disclosed many possible sources of error, other than those associated with methodology. These apply not only to analyses carried out on most of the fluids or body components mentioned previously, but also to the interpretation of the results.

General factors which must be considered in developing, applying, and interpreting a biological analysis include:

1. Metabolism.
2. Changes in the ratio of bound to free chemical in the blood.
3. Special situations where excreted levels of the index chemicals do not indicate current exposure levels.
4. Concentration changes due to volume changes in the bioassay material.
5. Nonworkplace progenitors of the index chemical and resulting baseline variation in its concentration in the bioassay material.
6. Normal range of the index chemical to be expected in the bioassay material.
7. Time required for the index chemical to appear in the bioassay material.
8. Analytical methodology.
9. Route of exposure.

6.1 Metabolism

Xenobiotics can be handled by the body in many ways, including:

1. Metabolism to a component of a naturally occurring series of reactions. The component may then be oxidized to CO_2 and H_2O or else anabolized (biochemical synthesis) to a body component.

2. Excretion of the chemical unchanged.
3. Metabolism to a chemical that is more readily excreted.
4. Storage in an organ, bone, body structure, or organized element of some sort, such as body fat or brain.
5. Accumulation without storage (e.g., in the blood or the hepatobiliary–intestinal loop).

Many xenobiotics follow a combination of routes 2 and 3. For these, all the factors that can affect metabolism become important in determining how much of the original exposure chemical is excreted unchanged and by what route and how much of each possible metabolite is excreted and by what route.

Factors that can affect either or both the rate of excretion and the ratios of the xenobiotic and the various possible metabolites excreted obviously can affect the estimate of exposure based on analysis of excreted chemicals.

Anything that can increase or decrease the normal accumulation or the storage of xenobiotics handled by routes 4 or 5 will have the inverse effect on excreted materials. In these situations, biological analyses relied on to indicate level of exposure will indicate correspondingly high or low results.

Most of these factors apply to organic compounds. There are not as many opportunities for individual variation in biological monitoring of metal or inorganic metal salt exposures since, unlike organic compounds, these materials normally are not metabolized. Nevertheless, one should expect individual variations in route and rate of excretion of these materials. Also, to the extent that storage or excretion of these metals or salts may be hormonally controlled, concurrent exposure to medicinals or industrial chemicals that affect hormonal levels should be considered in evaluation of the body burden of metals or metal salts through analysis of excreted materials.

Inorganic metal salts normally are not absorbed through the skin. Thus monitoring of the workroom air usually will give a good estimation of worker exposures to such salts in contrast to organic chemicals that may be absorbed through the skin as well as inhaled. Organic salts, organometallic compounds, and organic complexes of metal ions or metals (e.g., butyl chromate, tetraethyl lead, tetramethyl lead, alkyl mercury, methylcyclopentadienyl manganese tricarbonyl, organic tins, and organic silanes) may be absorbed through the skin. The metabolism of the organic portion of these compounds, and consequently the storage and excretion of the metal component, may be affected by the factors mentioned earlier and discussed more fully in the paragraphs that follow.

Some organic industrial chemicals, such as polychlorinated or polybrominated biphenyls (33), and some metals, such as lead and mercury, are stored by the body. However, just as with foodstuffs, most industrial chemicals are metabolized and excreted (34, 35). Chemicals that are totally catabolized (biochemical breakdown) seldom yield characteristic end products and are not suitable for analysis other than in blood shortly after exposure. However, chemicals that are excreted either without modification, or after modification to metabolites

BIOLOGICAL INDICATORS OF CHEMICAL DOSAGE AND BURDEN

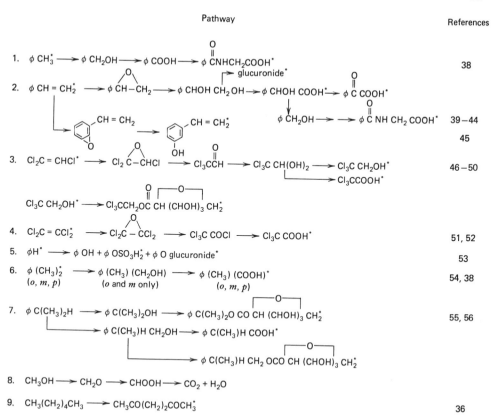

Figure 3.1 Principal metabolic pathways of some industrial chemicals. Other minor metabolites may be formed (38) and excreted, but they are not of value for biological monitoring. Symbols: φ = phenyl; * = identified in urine of humans or animals.

not occurring naturally, offer excellent opportunities for measuring worker exposure by measuring the amount excreted.

Most excretory systems of the body are aqueous. If the xenobiotic is not catabolized to a component of one of the many bodily metabolic sequences, the thrust of its metabolism will be toward increased water solubility. This means that, in general, a hydrocarbon xenobiotic will be hydroxylated or at least one carbon will be oxidized to a keto group. Frequently, alkyl side chains on aromatic compounds are oxidized to a carboxylic acid group or all the way to benzoic acid. An aliphatic alkane may have a terminal carbon oxidized to a carboxylic acid moiety or one or both β carbons oxidized to keto group(s) (36). Thus the chemical(s) excreted following exposure may not be the same as the exposure chemical. Examples of some of the metabolic routes that have led to metabolites in blood or to excretion products in liquid excreta are shown in Figure 3.1.

As can be seen, several metabolites of one chemical may be excreted in the urine. In addition, glucuronates, sulfates, mercapturic acids, or amides of metabolites may be excreted. These derivatives are called *conjugates*. Analyses of biological excretory liquids should measure conjugated, bound, and free forms of the index chemical since ratios may change.

Since the direction of metabolism usually is toward increased water solubility, metabolites generally will be less volatile than the chemical of exposure. Hence, metabolites generally will not be excreted in exhaled air.* Accordingly, some of the complications introduced by metabolism are correspondingly less important when biological monitoring can be based on analysis for the workplace chemical in exhaled air. For this latter monitoring, generally only the exposure chemical need be considered, regardless of the route by which it was absorbed.

Neither the kinetics of a particular metabolic pathway nor the relative kinetics of alternate metabolic pathways are constant in the same person at all times, or between individuals. Since a biological monitoring program relies on the rate of appearance or amount of the index chemical in the bioassay material, anything that affects this rate or amount will affect the estimate of the amount of workplace exposure.

Many factors can affect metabolic rates and thus excretion, singly or in combination. These include:

1. Individual variations in enzyme complement and enzyme kinetics.
2. Diet and dietary state.
3. Stimulation or inhibition of enzymes involved in the metabolic sequences.
4. Ratio of bound to free chemical in the blood.
5. Dose of the exposure chemical.
6. Competition for a necessary enzyme.
7. Obesity and fat:muscle ratio.
8. Age of the worker.
9. Disease.
10. Concurrent use of medicinals.
11. Sex of the worker.
12. Pregnancy.
13. Smoking.

Furthermore, any one or combination of many of these factors may be more or less important on any given day.

6.1.1 Individual Variations in Enzyme Complement and Kinetics

Individual variations in metabolic pathways and kinetics have been recognized in the field of biochemistry for years (57). As Hommes' book (57) indicates,

---

* There are some notable exceptions. For example, some metals, such as selenium and tellurium, are methylated by the body. The methylated compound is more volatile than the metal and is excreted in exhaled air (37).

these variations are known to be due in many cases to genetic variations in individuals. For years, they were regarded as only biochemical curiosities. The importance of these genetic variations in pharmaceutical metabolism was only recently appreciated as more effort was directed toward studying medicinal toxicity as a consequence of individual variations in metabolism and excretion. The study of genetic-based nonusual metabolic pathways and kinetics has been named *pharmacogenetics* by the pharmacologists. Pharmocogenetic-based perturbations of metabolism of xenobiotics is just as important to biological monitoring and metabolism/excretion of workplace chemicals as it is to metabolism/excretion and net dosage of medicinals.

If a worker is missing an enzyme or has an enzyme with specific activity different from that of most other workers and that enzyme is critical in the metabolism and excretion of a workplace chemical, analysis of his or her excretory products for the expected metabolite(s) will give a false picture of the worker's exposure. As discussed earlier, if, as a result of this aberrant enzyme, the elimination half-time ($t_{1/2}$) of a workplace chemical is increased, the error will be of the worst type; that is, it will indicate less exposure than the worker has received. As discussed earlier, increased $t_{1/2}$ for a chemical to which a worker will be exposed could lead to unexpected accumulation in the body with possible toxic consequences.

If metabolism by the worker leads to unexpected products, the fact and degree of the worker's exposure could be dangerously underestimated because of the absence of the expected excretory products.

Individual differences in the toxicity of sulfa drugs and the antitubercular agent isoniazid were found to be due to genetically controlled individual differences in the rate of acetylation and hence excretion of these medicinals (58, 59). It is believed that approximately 50 percent of the U.S. population can be classified as "slow" acetylators (60). Obviously, acetylation-dependent excretion of an industrial chemical would be expected to be slower in one half of the exposed U.S. workers. To further complicate the picture, Kalow (61) has reported that, of over 2000 Japanese individuals tested, only 12 percent were slow acetylators. This suggests that norms would have to be established or validated for each race for each biological monitoring procedure developed.

Sloan et al. have reported (62) that about 5 percent of a Caucasion population studied had reduced capability to hydroxylate aliphatic hydrocarbons. This reduced capability was of genetic origin. Genetically derived defective *O*-demethylation also has been reported (63).

Greim (Reference 60, p. 270) has tabulated other genetically controlled enzymatic aberrations of importance to pharmaceutical metabolism and excretion. He also reported their incidences in the population. To the extent that these enzymes are important in an individual's industrial chemical metabolism, a similar percentage of aberrations could be expected in workers.

Vessell and Passananti (64) have reported that the $t_{1/2}$ in plasma for dicumarol varied almost tenfold in fraternal twins. The $t_{1/2}$ in plasma for medicinals antipyrine and phenylbutazone varied threefold and sixfold, respectively, in these same twins. Variations at least as large are to be expected in the general population, as well as similar variations in the metabolism and excretion of industrial chemicals.

Tang and Friedman (65) examined the liver microsomal oxidase activity* from 10 humans varying in age from 3 days to 92 years. Six were between the ages of 22 and 57 years, inclusive. Both sexes were represented. Some had died violent deaths, some had died from disease, and one had died following surgery for an acute condition. Thus not all had died of disease and not all were aged, so Tang and Friedman's findings should be representative of what could be expected for a general worker population. Activity varied from "undetectable" to "highly active" as measured by mutations induced in *Salmonella typhimurium* TA-100 incubated with various aromatic amines known to require microsomal oxidase activation for mutagenic activity.

6.1.2 Diet and Dietary State

Many studies have been carried out showing the effects of diet on enzymatic activity and consequently on the metabolism of chemicals. For example, excessive (3 to 9 g/day) levels of niacinamide have been reported to cause elevation of serum glutamic–oxalic transaminase (SGOT) and alkaline phosphatase (APase) (66). Protein deficiency has been reported to cause reduced hepatic metabolism in rats (67) and could be expected to have the same effect in humans. High protein diet, on the other hand, has been shown to enhance the metabolism of chemicals in humans (68). A charcoal-broiled beef diet has been shown to increase microsomal oxidase (benzo[a]pyrene hydroxylase) activity in the liver of pregnant rats (69), suggesting a stimulation of the nonspecific microsomal oxidase systems. However, when humans were placed on a diet containing charcoal broiled beef, there was no change in the plasma half-life of the test chemical used to monitor changes in microsomal oxidase activity (70).

Brussels sprouts, cabbage, and cauliflower have been shown to contain chemicals that, when ingested by rats, increase liver microsomal oxidase activity as measured by oxidation of benzo[a]pyrene. Furthermore, specific activity was found to vary between cultivars of brussels sprouts and cabbage (71).

Diets high in corn oil and low in starch or sugar have also been reported to result in increased liver microsomal enzyme activity in rats (72).

It is thus probable that diet, particularly a diet unbalanced with regard to any component, could lead to changes in enzymatic activity followed by changes in metabolism and in excretion of workplace chemicals.

6.1.3 Stimulation of Inhibition of Enzymes that Metabolize the Chemical

As mentioned earlier, many xenobiotics are metabolized through the nonspecific microsomal enzymes that can catalyze at least four types of reaction: oxidation

---

* If a xenobiotic does not fit a specific enzyme, its metabolism generally will be affected by the nonspecific microsomal enzymes. These enzymes do not seem to require the enzymatic "fit" that characterizes most enzymatically catalyzed reactions. They are active in varying degrees on a variety of structures. They are known to carry out epoxidations, demethylations, reductions, and hydrolyses.

(epoxidation), demethylation, reduction, and hydrolysis. More than 200 chemicals are known to stimulate some or all of the known microsomal enzymes and thus increase the metabolism of many xenobiotics (73). This could result in an elevated concentration of the metabolites of workplace chemicals in excreted fluids of workers. If the evaluation of worker exposure is based on the normal concentration of excreted metabolites formed by means of the microsomal enzymes, this increased microsomal enzyme activity could result in an erroneously high estimate of the worker exposure. Conversely, if the analysis is based on unmetabolized xenobiotic, an erroneously low estimate of worker exposure would result.

Medicinal chemicals, pesticides, and industrial chemicals have been shown to stimulate the microsomal enzyme systems. These include phenobarbital [one of the most potent stimulators (73)], spironolactone (74), condensed polynuclear aromatic hydrocarbons, and polychlorinated or polybrominated biphenyls.

Cigarette smoke also has been found to stimulate the activity of liver microsomal enzymes in pregnant rats (70).

Pretreatment with phenobarbital has been shown to stimulate the oxidation of styrene in rats (43) and the oxidative step in the metabolism of 1,1,1-trichloroethylene (T_3CE) and 1,1,2,2-tetrachloroethylene (T_4CE) in rats and hamsters (30). In both cases, the oxidation rate was almost doubled. Other chemicals [e.g., morphine (74)] are known to inhibit the microsomal enzyme systems. Styrene and benzene epoxidation by microsomal enzymes also have been reported to be hindered in the presence of toluene (75). Obviously, these altered enzymatic rates could lead to increased or decreased appearance of metabolites in the excretory fluid analyzed. This, in turn, could lead to a high or low estimate, respectively, of workplace exposure if metabolite concentration in the excretory fluid were the basis of estimation of workplace exposure.

Regardless of whether the effect is stimulation or inhibition of the microsomal enzymes, the effect could alter the rate of appearance of metabolites in the excretory fluids only if the microsomal enzymes catalyzed rate-limiting steps in the metabolism of the xenobiotic.

Also, it has been known for many years that continued exposure to a chemical can stimulate the metabolism of that chemical (6, 76). Ikeda and Imamura (30) showed in addition that pretreatment with T_3CE can either stimulate or depress metabolism of subsequently administered T_3CE, depending on the pretreatment time and dose. Pretreatment with T_4CE, on the other hand, only stimulates the metabolism of subsequently administered T_4CE.

Ikeda et al. (77) also reported a subject who was addicted to sniffing T_3CE and whose $t_{1/2}$ for urinary excretion of metabolites was twice the usual $t_{1/2}$. It was not known whether the prolonged $t_{1/2}$ preceded or followed the addiction. Thus repeated exposure to a chemical may not always stimulate its metabolism.

6.1.4 Ratio of Bound to Free Chemical in the Blood

Many medicinal chemicals have been found to be transported in blood in the unbound state (free) as well as bound to proteins. Generally, the binding is to albumin in the plasma, but it also may be to other plasma proteins and/or proteins that are structural parts of formed elements in the blood (74). Since industrial chemicals basically differ from medicinal chemicals only in their application and in the relative magnitude of biological effect, there is no *a priori* reason why they should differ from medicinal chemicals in binding. Therefore, it is to be expected that industrial chemicals also would be transported in the blood in both the bound and free forms. Also, as with medicinal chemicals, the ratio of bound to unbound would be expected to vary from chemical to chemical.

The binding phenomenon in most cases probably is a simple complex formation, subject to all the mass action mathematics and dynamics of inorganic or other organic complexes. The stability constants also probably vary over several orders of magnitude, just as do those for classical complexes. Thus, bound and unbound fractions would be in dynamic equilibrium. This has important ramifications in biological monitoring, as discussed in the paragraphs that follow.

Total blood concentration (bound plus free) of the xenobiotic and its metabolites should reflect workplace exposure of chemicals that are absorbed. However, the amount excreted per unit time generally reflects only the unbound concentration in the blood. Again, as already mentioned, systemic toxicity generally will vary with the concentration of the unbound chemical. Situations can occur in which the normally bound fraction of a workplace chemical (the fraction that was bound when the correlations between dose, toxicity, and amount excreted were developed) is decreased. In these situations, the amount of index chemical excreted still will represent the unbound concentration in the blood and thus the hazard to the worker (78). However, it will not be as closely related to workplace exposure and will suggest recent exposures higher than those that occurred.

Changes in the ratio of bound to unbound chemical in the blood generally are caused by a decreased number of available binding sites. This is usually due to either a deficiency of binding protein or a concurrent administration of a chemical that is more tightly bound. Disease also has been shown to reduce binding of chemicals (78).

In some workers, a decreased amount of binding protein may be due to genetic factors. However, it most frequently is due to disease. Chronic conditions accompanied by lowered plasma levels of protein include chronic liver disease, rheumatoid arthritis, diabetes, and essential hypertension. Nephritis and gastrointestinal diseases such as peptic ulcer and colitis also can be accompanied by lowered plasma protein. Other effects of disease on the ratio of bound to unbound chemical are discussed in Section 6.1.9.

A decrease in the number of available binding sites accompanying a normal level of transport protein usually is caused by the presence of other chemicals that are more tightly bound. Phenylbutazone, a medicinal occasionally prescribed for arthritis and other aches and pains, is very tightly bound to plasma proteins. It is bound so tightly that it will displace most other bound chemicals from the blood transport proteins. If present in sufficient concentration in the blood, it could displace significant amounts of a bound workplace chemical. This would result in elevated blood levels of the free form of the latter and thus could result in an increased rate of metabolism and excretion. If workplace exposure were being estimated from excreted levels of the exposure chemicals and/or their metabolites, an erroneously high estimate of exposure could result. However, the excreted level would still reflect the *hazard* to the worker, since that would usually vary with the concentration of unbound chemical in the blood.

Another common medicinal that also is strongly bound is salicylic acid (78). Less common medicinals that could displace bound workplace chemicals include clofibrate, a cholesterol-lowering medicinal (78).

Since it is likely that workplace chemicals also are bound in varying degrees to blood proteins, as already mentioned, concurrent high exposure to either or both of two or more industrial chemicals could alter the binding ratio of the one with the weaker bond.

Among the diseases, cirrhosis has been shown to result in decreased binding of medicinals, with accompanying increased blood levels of the unbound form (78). There undoubtedly are other diseases that also reduce binding.

6.1.5 Effect of Dose of Exposure Chemical on Metabolism

It has been known for many years that the metabolism of chemicals can change with the dose administered, as mentioned in Section 6.1.3. This obviously can affect the dose–excretion curves for metabolites used to measure exposure. Quick (79), for example, found that when high doses of benzoic acid were administered to humans, the glucuronide became an important excretory product, as well as hippuric acid. Von Oettingen et al. (24) found that after about 8 hr of exposure to approximately 300 ppm of toluene, benzoic acid glucuronide also was excreted in addition to hippuric acid.

Tanaka and Ikeda (80) found the usual metabolic pathway for trichloroacetic acid (TCA) formation in humans from T_3CE became saturated after about 8 hr of inhalation exposure at 150 ppm. The excretion rate of TCA reached a plateau at that concentration.

More recently, Ikeda et al. (81) reported that the usual metabolic pathway for T_4CE was saturated after exposure at 50 ppm for several hours. They also reported that formation of phenylglyoxylic and mandelic acids in the rat plateaus at styrene exposure levels ≥ 100 ppm for 8 hr (82).

Götell et al. (83) reported that in man the urinary concentration of phenylglyoxylic acid not only reached a peak after exposure to styrene for approximately 8 hr at 150 ppm, but actually decreased at higher atmospheric levels.

Watanabe et al. (84) also have reported that the usual metabolic pathway for vinyl chloride monomer (VCM) is saturable in the rat. At low levels, VCM is principally metabolized and its metabolites are excreted in the urine. Only a small percentage is excreted unchanged in exhaled air. At high

VCM levels, the reverse is true and most of the VCM is excreted unchanged in exhaled air.

These studies all indicate the importance of not extrapolating biological monitoring equations or dose–response curves beyond the conditions used to derive the equations or reference curves.

6.1.6 Competition for a Necessary Enzyme

If a worker is concurrently exposed to two or more chemicals that require the same enzyme(s) for metabolism, the resulting pattern of metabolites and parent compound in blood and excreted fluids can be dramatically altered. The chemical that is preferentially metabolized by the enzyme(s) will have the most nearly normal pattern. This is elegantly illustrated by concurrent exposure to ethanol and methanol.

The first step in the metabolism of ethanol and methanol is a dehydrogenation catalyzed by the liver enzyme alcohol dehydrogenase (ADH) as shown in Reactions 1 and 2:

$$CH_3CH_2OH \xrightarrow{ADH} CH_3\overset{\overset{\displaystyle O}{\|}}{C}H \qquad (1)$$

$$CH_3OH \xrightarrow{ADH} H\overset{\overset{\displaystyle O}{\|}}{C}H \qquad (2)$$

ADH acts preferentially on ethanol. Therefore, in the presence of ethanol, only a small fraction of the methanol is dehydrogenated, and most of the methanol circulates unmetabolized in the blood. Consequently, blood, urinary, and exhaled methanol levels are elevated over what they would be if the same dose of methanol were not competing with ethanol for the available ADH (85). Under these conditions, biological monitoring would suggest higher workplace exposure to methanol than occurred. [In this situation, the elevated methanol levels reflect not increased hazard of eye damage, but rather decreased hazard, since the metabolite causing the eye damage appears to be formic acid, a metabolite (86).]

Enzyme competition also may explain the observation by Stewart et al. (87) that the urinary excretion of 1,1,1-trichloroethanol (TCE) was greater than expected and that of TCA was less than expected after T_3CE exposure, if the subject consumed alcohol. If T_3CE exposure were being measured by urinary TCE excretion, this also would result in higher estimates of exposure than occurred.

From a toxicological standpoint, the importance of all these possible metabolism variations depends on whether the most important toxicant is a metabolite of the exposure chemical or the exposure chemical itself. Obviously, anything that decreases the concentration of a toxicant decreases the hazard, and anything that increases the concentration will increase the hazard.

For biological monitoring purposes, these causes of variation indicate that the range of concentrations of a workplace chemical or its metabolites in blood, exhaled air, or excretory fluids following a given exposure normally will be broad. This also would be expected if one examines the range of blood and urine concentrations reported for chemicals normally occurring in these fluids in humans (9). Two- to threefold variations between the upper and lower "normal" values are usual. Furthermore, recognizing this normal range, physicians seldom rely solely on urine or blood levels of a chemical to diagnose a disease condition.

From a worker monitoring standpoint, one must be particularly careful about interpreting blood or excretory fluid levels of a chemical that is always present in this fluid and that also may arise from workplace exposure. In this situation, the concentration range to be expected from a particular exposure level may vary by four- to sixfold.

6.1.7 Obesity and Fat:Muscle Ratio

Obesity can decrease the rate of elimination of both water- and fat-soluble chemicals. A 50 to 100 percent increase in elimination half-life has been reported for both classes of medicinals (88). For the obese person and the fat-soluble medicinal, this was due to the increased amount of fat available for distribution and storage of the chemical. For the obese person and the water-soluble medicinal, this was due to an increased volume of distribution. Both of these factors tend to decrease blood concentrations of the exposure chemical and its metabolites and reduce their elimination rate.

If the fat:muscle ratio increases, the rate of elimination of a fat-soluble chemical could be expected to decrease for the reason already mentioned. The ratio of fat to muscle is usually higher in older workers. The ratio also is usually higher in women (see also Section 6.1.11).

Similar relationships would be expected to hold for industrial chemicals.

6.1.8 Age of the Worker

Blood level and excretion of medicinals following a given dose have been found to change with age. Since metabolism and excretion of industrial chemicals also utilize enzymes as well as the same excretion mechanisms as medicinals, similar changes could be expected for industrial chemicals. For example, in disease-free and pathology-free elderly subjects, Triggs et al. (89) found a 50 to 100 percent increase in the elimination half-life of two medicinals when compared with disease-free and pathology-free young subjects. Absorption of the medicinal by these groups did not differ significantly.

Glomerular filtration rate has been reported to decrease by about 35 percent in the elderly (90). This could result in delayed appearance of the index chemical in the urine, as well as a reduced slope of the exposure–urinary concentration curve if this curve had been determined in young workers.

Older workers also may have a higher proportion of fat or weigh more than younger workers of the same height. Both of these factors may cause decreased excretion rate of an index chemical (see Section 6.1.7).

Hepatic blood flow has been reported to decrease as much as 40 to 45 percent with age (98). This could result in a decrease in the rate of formation of the index chemical, if it is formed in the liver from the exposure chemical.

At the other end of the age spectrum, the enzymatic complement of the circumnate is not known for all enzymatically catalyzed reactions. For some reactions, it seems to be as complete and as active in the immediate prepartum and postpartum stages as it is in the adult. For other reactions, the activity may be totally missing or the specific activity of the enzyme may be severely decreased. Postpartum enzymatic activity of the circumnatal subject usually is of little concern industrially (see Section 11 for exceptions), but prepartum activity is. Too few investigations have been carried out at this time on which to base specific statements about various groups of enzymes.

Also, with few exceptions, we don't know all of the industrial chemicals and their metabolites that cross the human placenta. Therefore, biological monitoring has limited utility at present for estimation of fetal exposure or burden.

6.1.9 Disease

In addition to the effect of some diseases on the amount of binding proteins, diseases can affect the ratio of bound chemical to free chemical in other ways. This, in turn, may affect the rate of appearance of the index chemical in the biological assay material.

Any disease that affects metabolism and/or excretion also could be expected to alter the rate of appearance of the index chemical in the bioassay material. Decreased renal clearance, for example, could result in higher blood levels and lower urinary levels of the index chemical than would be expected from studies on subjects with normal renal clearance (78). As already mentioned, binding of xenobiotics to serum protein is decreased with cirrhosis. A biological monitoring program based on urinary levels of the index chemical would suggest lower exposure levels than occurred in this situation.

If the index chemical values were normalized by calculating the ratio to creatinine (see Section 6.4) and creatinine excretion also was correspondingly reduced, the error would be self-correcting. If the biological monitoring were based on blood, tears, perspiration, or saliva analysis, this situation would result in a higher than normal excretion through these fluids. This would result in higher estimates of workplace exposure than actually occurred.

6.1.10 Concurrent Use of Medicinals

In addition to the previous specific examples of the effects of concurrent administration of medicinals on any biological monitoring procedures, medicinals can cause erroneous results in other ways. For example, the elimination half-life of diazepam was increased in patients taking oral contraceptives with low estrogen content (92). Cimetidine has been reported to decrease liver blood flow and the elimination of propanolol (93). The possible effects of reduced blood flow on biological monitoring was discussed in Section 6.1.8.

These results indicate that the effect of concurrently administered common medicinals should be investigated for any biological monitoring procedure being developed. Such medicinals might include aspirin, acetaminophen, tranquilizers, phenytoin, barbiturates, antacids, and oral contraceptives.

6.1.11 Sex of the Worker

Chemicals are known to have different lethality in the two sexes. For example, many of the organophosphate insecticides are three to four times more acutely lethal to one sex than to the other (94), with the more sensitive sex varying with the chemical. This difference may be due, for example, to differences in absorption, metabolic route, or metabolic kinetics. Different exposure correlation factors would have to be used for each sex in this situation since the same level of index chemical in each sex of worker could indicate different exposures. Since these same factors would affect the appearance of other toxic signs or symptoms, it should not be surprising to find differences in these latter effects between sexes.

Daniel and Gage (95) found that, after dosing for up to 50 days, female rats had stored about 50 percent more orally administered butylated hydroxy toluene (BHT) in their fat than males at the dietary levels used. After single oral doses, females excreted about 50 percent more in the urine than did the males.

Ikeda and Imamura (30) found the urinary half-life of total trichloro compounds (TTC) from T_3CE or T_4CE exposure to be almost twice as great in female workers as in male workers. They reported similar findings for other index chemicals used to assess exposure to these two compounds. Conversely, Nomiyama and Nomiyama (96) reported a significantly greater ($p < .05$) urinary excretion of TCA in females than males in the first 24 hr following inhalation exposure to T_3CE. For the first 12 hr following exposure, males excreted twice as much TCE as females. In accordance with these findings, Nomiyama (97) found that it was necessary to use a larger factor for females than males when correlating exposure concentration with exhaled air concentration of T_3CE.

Nomiyama and Nomiyama (98) also found that female subjects absorbed approximately twice as much (41.6 percent) of a given inhalation dose of benzene as did males (20.3 percent) as mentioned previously.

Ikeda and Ohtsuji (99) found that the normal concentration of hippuric acid in about 30 female Japanese students was approximately twice that of about 36 male Japanese students (specific gravity and creatinine corrected). Conversely, the normal urinary level of TCA was twice as high in males as in females.

6.1.12 Pregnancy

It has been reported that the concentration of unbound medicinals in blood increased in pregnant women versus nonpregnant controls or the same women after delivery. This was true for weakly acidic medicinals, weakly basic medicinals, and a neutral steroid (100). The importance of decreased binding with regard to biological monitoring and worker hazard is discussed in Section 6.2. It can give a misleading evaluation of the exposure, but probably not of the hazard.

6.1.13 Smoking

Cigarette smoking has been reported to both stimulate (101) and have no effect (102) on xenobiotic metabolism. Metabolism of the same medicinal, chlordiazepoxide, has been reported to be both stimulated by smoking and to be unaffected by smoking (101, 102). This suggests that some unknown factor(s) may have been affecting these studies. Nevertheless, it would be prudent to try to evaluate the effect of smoking on any biological assay where the index chemical was a metabolite of the exposure chemical.

6.2 Changes in the Ratio of Bound to Free Chemical in the Blood

These changes have already been discussed in Section 6.1.4 as some of the factors that can alter metabolic rates. As indicated there, changes in the ratio also can affect the rate of appearance of the index chemical in the monitoring material and consequently the estimate of workplace exposure. Parenthetically, it still will correlate with the hazard to the worker.

6.3 Special Situations Where Execreted Levels of the Index Chemical Would Not Indicate Current Exposure Levels

Situations also exist where all the index chemical in the biological monitoring material may not arise from recent exposure or even from workplace exposure. Conversely, the index chemical may not be excreted in the monitoring material at the rate expected. Obviously, one should ensure that such factors are not operating in any particular assay on an individual.

It should be stressed here that, to the extent that the index chemical reflects the hazard to the worker, steps should be taken to reduce the hazard, if necessary, regardless of the etiology of the toxicant. If the index chemical did not arise largely from the workplace, however, attempts to reduce workplace exposure would not reduce the hazard to the worker (see also Section 6.5).

Competition for the same enzyme between concurrently administered methanol and ethanol is an example that already has been discussed. Leaf and Zatman (85) found that the urinary methanol levels following methanol exposure were up to 100 percent higher in the presence of ethanol than in its absence. Obviously correlations of urinary methanol and exposure, made in the absence of concurrently administered ethanol, would be invalid in this situation. Use of such correlations would result in great overestimation of the workplace exposure and the hazard.

Drug therapy for mobilization of stored or recycling chemicals can lead to misleading levels of a chemical in the biological monitoring materials. For example, the use of penicillamine for treatment of Wilson's disease (103) or schizophrenia (104) or the use of tetraethylthiuram disulfide for alcoholism therapy would lead to elevated excretory levels of copper that would be totally unrelated to current workplace exposure and the hazard.

Any medicinal that could release stored, recycling, or protein-bound index chemical or its progenitors could cause such erroneous results, regardless of the reason the medicinal was being given. As mentioned previously, it probably would reflect hazard to the worker but not workplace exposure (see also Section 6.8).

In the case of urinary measurement of the index chemical, decreased renal clearance, active lactation, excessive sweating, or excessive salivation all could cause decreases in urinary levels that would result in decreased estimates of workplace exposure.

Absorption of chemicals through the skin or by the lungs also can vary between individuals. If a person absorbed significantly more or less of the administered dose than did the subjects utilized to make the correlations of exposure and excretion, the individual analysis would indicate correspondingly greater or lower exposure than occurred. The hazard estimate, again, probably would be correct, however.

Individual absorption may vary by as much as 100 percent. In a study of oral absorption of ampicillin (115), the amount absorbed varied from 32 to 64 percent in nine healthy men aged 20 to 40 years. The absorption of benzene by inhalation has also been reported to vary by 100 percent between males and females, with females absorbing 41.6 percent of the dose and males 20.3 percent (98).

Assuming no metabolic differences, absorption differences between individuals thus could lead to a 100 percent difference in the level of index chemical in the biological monitoring material from two individuals with the same exposure.

6.4 Concentration Changes Due to Volume Changes in Biological Assay Liquids

The rate of appearance of the index chemical at the excretory organ generally will be independent of the rate of excretion of the biological monitoring material. Thus the concentration of the index chemical in the biomonitoring material will change with the volume of excretion of the latter. This is not a serious problem for hair or nail analyses. For exhaled air analyses, the subject generally is resting when samples are taken and will not be doing anything to greatly increase respiratory volume per unit time. Therefore, this also will not be an important factor in exhaled air analyses. However, the excretion rate of liquids can vary greatly. The rate of perspiration excretion during and closely following exertion versus resting, or the rate on a hot day versus a cool day, is obviously different. Similarly, the rate of excretion of tears in subjects who are crying is obviously different from that in those who are not crying. Salivation rate also can be altered by emotional, physical, and chemical factors. Since these biological fluids are seldom used for industrial biological monitoring, excretion rate changes of these fluids seldom are important. However, excretion rate is a variable that must be considered in any assay utilizing these fluids.

Urine volume normally varies by as much as 300 percent (600 to 2500 ml/24 hr) between individuals and in the same individual at different times (9). Factors that can cause volume variations include fluid intake, emotional state, medicines, pregnancy, disease, menstruation, ambient temperature, and body temperature.

Another dilution factor to be considered for urine samples is dilution by urine reaching the bladder before the index chemical reaches the kidneys. This will be most important in situations where (1) end-of-shift "spot" urine samples (see Section 7.3.4) are analyzed, (2) the index chemical has a relatively long excretion half-life, and (3) several hours elapse between the penultimate urine sample and the sample taken for analysis. Voiding at a set time prior to taking the urine sample for analysis might reduce the spot urine analysis variability due to this factor. This technique has been utilized by several investigators (18, 81, 99, 106).

If the content of index chemical in the total 24-hr volume of urine is being measured, the dilution effect will be minimized. However, if only spot samples of urine are being analyzed, dilution factors may be extremely important.

Recognizing that the same kinetics of presentation at the excretion organ apply to normal constituents of urine as well as to xenobiotics and their metabolites, investigators have used various properties or components of urine to normalize analytical values and thus compensate for concentration changes due to volume differences.

As mentioned previously, all other things being equal, the same ultimate insertion factors that control the concentration of the index chemical in urine generally will control the concentration of normal constituents in the urine. If it were possible to measure a colligative property of urine or the concentration of a usual urinary constituent that has a constant rate of presentation to the urine and rarely evolves from a xenobiotic, it should be possible to normalize urinary concentrations and thus correct for volume differences.

Specific gravity and osmolality are colligative properties of urine that are easy to measure. Specific gravity normally varies by only about tenfold, whereas osmolality varies by over twentyfold. Although the concentration of xenobiotic or its metabolites will make some contribution to specific gravity, it generally will be small compared to the effect due to normal constituents. The xenobiotic or its metabolites probably would make a greater contribution to changes in

Table 3.1 Effect of Two Normalizing Treatments on Urine Analysis for Hippuric Acid Following Toluene Exposure (107)

Exposure Sampling Period (hr):	0–4		4–8	
Toluene Exposure (ppm):	100	200	100	200
Hippuric acid concentration in urine, uncorrected (mg/liter)				
Mean	2.95	3.74	3.09	8.19
SD[a]	0.83	0.59	0.70	2.62
Specifity gravity, corrected (mg/liter)[b]				
Mean	2.58	3.71	2.81	5.85
SD	0.40	0.67	0.66	1.24
Rate (mg/min)				
Mean	2.09	3.13	3.10	4.61
SD	0.35	0.36	0.84	0.80

[a] SD = standard deviation.
[b] Corrected to 1.024.

osmolality. For whatever reasons, specific gravity has been used widely to normalize urine values, and osmolality has not. Of the possible reference chemicals for normalizing urinary analyses, creatinine has been used most extensively. Creatinine excretion normally varies less than twofold [1.1 to 1.6 g/24 hr or 15 to 25 mg/kg body weight/24 hr (9)].

Instead of specific gravity or creatinine content, Elkins normalized on the basis of sulfate excretion (1), using the same rationale discussed previously. However, this normalization has not been used extensively.

Ogata et al. (107) found that the "rate" of excretion of hippuric acid from 23 young adult human males following inhalation exposure to toluene is a more consistent measure than normalization by the use of specific gravity. To determine rate, they collected urine samples every 3 or 4 hr during a 3- or 7-hr exposure. The total amount of hippuric acid in the sample was then divided by the minutes of exposure. The resulting quotient was the rate in milligrams per minute. The rates for the 0- to 4-hr urine samples or the 4- to 8-hr samples were then plotted against the toluene concentration. The standard deviations were calculated for both the specific gravity and rate normalizing methods. At an airborne toluene concentration of 100 ppm, they were not greatly different for either time period as shown in Table 3.1. At toluene concentrations of 200 ppm, the standard deviations in the rates were less than the standard deviations of specific gravity-corrected concentrations.

As can be seen from Table 3.1, a direct dose–response relationship was found with both methods of normalizing. However, one standard deviation from the hippuric acid concentration value at 100 ppm overlapped one standard deviation at 200 ppm by either method of normalizing. Shorter sampling periods would be more desirable for validation of the concept. The approach would not be too practical in an industrial situation since it would require that the worker void urine a specific number of hours prior to voiding the urine sample for analysis. Levine and Fahy (108) found that calculation of rates did not result in reduced variability of lead analyses in the urine of exposed workers (109).

There has been some controversy as to whether 1.016 or 1.024 more nearly represents the average urinary specific gravity (110–112).

Normalizing will have the same effect on variability regardless of what number is used, because of the nature of the correction equation, as shown in Equations 3 and 4. Levine and Fahy (108) used Equation 3 to correct to specific gravity 1.024. Equation 4 is used by NIOSH (113) to correct to specific gravity 1.024.

$$\text{Corrected value} = (\text{observed value}) \times \frac{24}{(\text{specific gravity} - 1) \times 10^3} \quad (3)$$

$$\text{Corrected value} = (\text{observed value}) \times \frac{24}{\text{last two digits of specific gravity}} \quad (4)$$

It can be seen that Equation 3 reduces to Equation 4 in practice since (specific gravity $-$ 1) for a specific gravity of, say, 1.021 = 0.021. When this number is multiplied by 10^3, it becomes 21. These are the last two digits of the specific gravity. Obviously, Equation 3 or 4 could be used to normalize to any other specific gravity by substituting the last two digits of that specific gravity for 24 in the numerator of Equation 3 or 4.

The question of which specific gravity to use can be avoided by normalizing the urinary analyses on the basis of creatinine content, for the reasons already discussed. In the field of industrial hygiene, creatinine concentration was initially used by Hill (14) to normalize analyses of aniline in urine. Its use as a basis for normalizing has increased since then.

There have been mixed results reported from normalizing with either specific gravity or creatinine. As can be seen in the work of Ikeda or Ogata reported in Table 3.2, neither creatinine nor specific gravity corrections consistently gave significantly smaller standard deviations in comparison to each other or uncorrected values. This also has been the experience of other investigators. Seki et al. (114) found that specific gravity correction slightly reduced the scatter, whereas creatinine correction did not. Conversely, Pagnotto and Lieberman (115) found that creatinine correction reduced scatter over that found in uncorrected values. Ellis (116) found that neither correction significantly reduced scatter over that found in uncorrected samples.

Creatinine clearance also has been reported to decrease with age, especially in subjects over 60 years of age (90). No such change has been reported for specific gravity.

In practice, since both specific gravity and creatinine are usual urinary determinations, many authors now report values corrected by both methods as

Table 3.2 Effect of Corrections on Variations in Urine Analysis[a]

Exposure Chemical		Index Chemical Concentration in Urine[b]					
Name	Exposure Concentration (ppm)	Index Chemical Name	Uncorrected	Corrected to Specific Gravity 1.016	Corrected for Creatinine (g/g)	Number of Subjects	Reference
None	0	Phenylglyoxylic acid	0.017 (59–159)	0.011 (64–155)	0.013 (69–138)	35	82
None	0	Mandelic acid	0.057 (61–163)	0.036 (64–158)	0.043 (63–160)	35	82
None	0	Hippuric acid	0.35 (57–177)	0.29 (55–175)	0.24 (58–171)	31	106
Toluene	20	Hippuric acid	1.84 (66–152)[c]	1.18 (72–139)[c]	1.06 (76–132)[c]	10	106
	60	Hippuric acid	2.27 (59–171)[c]	1.21 (58–174)[c]	1.14 (66–151)[c]	10	106
None	—	Hippuric acid	0.30 (25–420)	0.29 (21–480)	0.23 (20–512)	36	99
None	—	Phenol	0.026 (32–312)	0.023 (31–316)	0.019 (32–318)	36	99
T_3CE	3	TTC	0.039 (67–149)	0.035 (75–133)	0.041 (74–135)	9	81
T_3CE	45	TTC	0.339 (72–138)	0.253 (77–130)	0.338 (79–126)	5	81
T_3CE	120	TTC	0.915 (85–118)	0.481 (86–116)	0.519 (68–146)	4	81

[a] Figures in parentheses are standard deviation ranges as a percentage of the geometric mean.
[b] Spot samples (see Section 7.3.4).
[c] Exposure values uncorrected for background level.

well as the uncorrected value. The data presently available do not indicate that either method of normalizing analyses reduces the standard deviation over that found in uncorrected values.

Reporting of the uncorrected value of the index chemical as well as the specific gravity and creatinine concentration is recommended, since this permits the readers to make their own corrections for comparison with other data or for other purposes.

6.5 Nonworkplace Progenitors of the Index Chemical and Resulting Baseline Variation in Its Concentration in the Biological Assay Material

If the index chemical also has a normal background concentration range in the biological assay material being analyzed, any biological monitoring procedure must determine (a) the normal background range, (b) nonoccupational factors that affect the normal range, and (c) whether or not a small fraction of the maximum safe workplace exposure to the index chemical or its workplace progenitor causes a significant elevation above the normal background concentration of the chemical. If one has a series of analyses on a worker over a period of time, interindividual variations become less important. However, if there is an elevation in one analysis, possible nonworkplace causes still should be investigated.

For example, extremely varied normal levels of various metals have been reported for human scalp hair. Topical contamination of the hair from such causes as cosmetics or hair dyes, which does not represent workplace exposure, may be the cause. Because of this great variation, hair levels that might be assumed to represent industrial exposure must be set quite high, unless one is regularly monitoring individual workers.

Examples of a similar situation with organic compounds include the use of (a) urinary hippuric acid levels for monitoring styrene, toluene, or certain *n*-alkyl benzene exposures or (b) urinary phenol levels as a measure of benzene or phenol exposure.

As mentioned earlier, styrene, toluene, and certain *n*-alkylbenzene compounds are metabolized to benzoic acid, which usually is condensed with glycine to form the excretory product, hippuric acid. Hippuric acid occurs normally in the urine, generally as a result of ingestion of benzoic acid, sodium benzoate, or quinic acid, a metabolic precursor of benzoic acid. Sodium benzoate is used as a food preservative. Coffee beans, prunes, plums, and cranberries are known to contain benzoic acid or quinic acid.

With widespread distribution of benzoic acid and its progenitors in foodstuffs, it is not surprising that its background level in urine should vary. As shown in Table 3.3, the normal urinary level of hippuric acid fluctuates threefold in U.S. subjects, up to thirteenfold in Japanese subjects, and almost fortyfold in Finnish subjects.

This variation essentially negates the use of hippuric acid as an index chemical for measuring styrene exposure to airborne concentrations up to four

Table 3.3 Normal Urinary Levels and Ranges (mg/liter) of Hippuric Acid as Reported by Various Authors[a]

Uncorrected	Corrected Value Based On Specific Gravity	Creatinine	Subjects: No., Sex, Country	Analytical Procedure	Reference
1583[a] (833–2583)	ND[b]	ND	9, M, U.S.	UV absorption	42
350[c] (199–616)	290[c,d] (160–510)	240[c]	31, M, Japanese	PC	106, 117
301[c]	290[c,d]	229[c]	36, M, Japanese		
398[c]	570[c,d]	449[c]	30, F, Japanese		
NR	1037[e] (126–4844)	739 (86–2340)	39, M, Finnish	GC	109
800 (400–1400)	ND	ND	NR, M, U.S.	UV absorption	115
NR	184[f] (35–444)	ND	NR, M, Japanese	PC	117
335	ND	ND	6, M, U.S.	Titration	24

[a] Calculated by RSW from data in reference. Assumed 1.2 l urine volume for 24 hr.
[b] ND = not determined; NR = not reported; M = male; F = female; PC = paper chromatography; GC = gas chromatography.
[c] Geometric mean.
[d] Corrected to specific gravity of 1.016.
[e] Corrected to specific gravity of 1.018.
[f] Corrected to specific gravity of 1.024.

times as high as the TLV for the following reasons. If one assumes for humans (a) 10 percent conversion of styrene to urinary hippuric acid (82), (b) a pulmonary absorption coefficient of 0.8 (118), (c) a 24-hr urine volume of 1.2 liters, and (d) an 8-hr exposure during which 10 m³ of air was breathed, an 8-hr exposure to 200 ppm of styrene would be necessary in order to raise the average level of urinary hippuric acid reported by Steward et al. (42) to the upper normal level they reported. In fact, Stewart et al. found that exposure to 100 ppm of styrene for 7 hr did not result in significant elevation of urinary hippuric acid concentration up to 48 hr postexposure.

In partial agreement with Stewart's findings, Ikeda et al. (82) found significantly elevated urinary hippuric acid levels in most subjects exposed to styrene for up to 160 minutes at concentrations of up to 200 ppm but no significant increase at concentrations of ≤60 ppm for 2 hr or less. Twenty-four urine samples were collected and specific gravity was corrected to 1.016. However, spot urine samples taken by these same investigators after approximately 6 hr of exposure of workers at levels of up to 30-ppm styrene did not show significant elevations of hippuric acid (119).

The insignificant urinary elevation of hippuric acid following these latter exposures might be explained by the delayed appearance of hippuric acid in the urine following styrene exposure. Ikeda et al. (82) found that hippuric acid did not appear in workers until about 24 hr postexposure. This would not explain Stewart's results since he collected total urine samples for over 24 hr postexposure.

Thus, the combination of small percentage conversion and high, variable, natural levels must be assumed to mitigate against the use of urinary hippuric acid as a marker for styrene exposure.

Hippuric acid in urine has been used successfully as a marker for toluene exposure. Since toluene is almost 70 percent metabolized to hippuric acid (107), the high background levels of hippuric acid become less important. Any exposure in excess of 50 ppm for 8 hr would be expected to elevate an average background hippuric acid level beyond Stewart's upper normal (42), assuming 79 percent absorption (107), 70 percent metabolism to hippuric acid, 10 m^3 of air respired, and a 24-hr urine volume of 1.2 liters.

In agreement with this, Engström et al. (109) found very poor correlation between blood levels of free toluene and urinary hippuric acid (creatinine corrected) in painters concurrently exposed for 8 hr to airborne levels of toluene less than 50 ppm and unspecified low concentrations of m- and p-xylene.

The use of urinary phenol levels as a measure of benzene exposure also could be frustrated at low benzene levels by the presence of phenol in urine from nonbenzene sources. For many of these sources, urinary phenol levels would correspond to much lower worker hazard than would be indicated if the phenol had arisen from benzene.

Phenol, as well as metabolic precursors of urinary phenol such as the amino acid tyrosine and the essential amino acid phenylalanine, occur in the diet. Whereas the normal urinary output usually is considered to be about 10 mg/day (35), the normal range can be much greater than this, as shown in Table 3.4. In three groups of Japanese subjects (in Japan) totaling 97 subjects, the geometric means of the three groups ranged from 18.2 mg/liter to 34.8 mg/liter. The 95 percent fiduciary limits varied from 7.3 to 123.8 mg/liter.

These values were for urine corrected to specific gravity 1.016. If the values are corrected to specific gravity 1.024 as recommended by NIOSH (113), they would be 50 percent higher. The use of geometric means by Ikeda et al. (99, 106) also suggests a distribution skewed toward the high values. Roush and Ott (121) also found preexposure urinary phenol values of up to 80 mg/liter (corrected to specific gravity 1.024). Fishbeck et al. (112) reported that the urinary phenol concentration of one person could vary by 100 percent (from 5.0 to 11.0 mg/liter) in a 6-week period. Because of such variations, Roush and Ott (121) have suggested that urinary phenol measurements do not reliably indicate benzene exposure if the exposure is less than 8 hr at 5 ppm. Van Haaften and Sie's data (110) support this contention. Thus urinary phenol measurements would not be suitable for routine monitoring of employee

Table 3.4 Normal Urinary Levels and Ranges (mg/liter) of Phenol as Reported by Various Authors

	Corrected[a]				
Uncorrected	Specific Gravity Adjustment	Creatinine Adjustment	Subjects: No., Sex, Country	Analytical Procedure	Reference
NR[b]	5.5[c]	NR	20, M/F, U.S.	GC	110
26.1[c] (8.3–81.5)[e]	23.3[c,d] (7.3–73.7)[e]	18.9[c] (6.0–60.1)[e]	36, M, Japanese	Colorimetric (120)	99
25.2[c] (8.4–75.5)[e]	34.8[c,d] (9.8–123.8)[e]	28.5[c] (11.0–71.1)[e]	30, F, Japanese	Colorimetric (120)	99
22.8[c] (8.9–58.3)[e]	8.2[c,d] (8.0–41.5)[e]	14.5[c] (7.1–29.9)[e]	31, F, Japanese	Colorimetric (120)	99
(1–80)			52, U.S.	GC	121
(4–5.5)	(5–11.0)[f]	NR	1, U.S.	GC	112
NR	(12–144)[d]	NR	109, British	Colorimetric (120)	111

[a] Only values obtained by use of the Gibbs colorimetric method (120) or a gas chromatographic method are reported here since the other common colorimetric method (122) is known to also give positive results with *p*-cresol (110). The *p*-cresol content in control urine may be 10 times greater than the phenol content (35, 110). Both Gibbs and the Theis and Benedict methods give reactions with *o*- and *m*-cresol, but these are usually present in urine in much lower concentrations than *p*-cresol.
[b] Abbreviations: NR = not reported; M = male; F = female; GC = gas chromatography.
[c] Corrected to specific gravity of 1.016.
[d] Geometric mean.
[e] 95 percent fiducial limit.
[f] Corrected to specific gravity of 1.024.

exposure to benzene under the proposed standard (123) for atmospheric benzene levels in the workplace. [An 8-hr time-weighted average (TWA) concentration of 1 ppm with no excursions exceeding 5 ppm averaged over 15 min.]

Another factor that also mitigates against the unqualified and sole use of some arbitrary urinary phenol level as an indicator of excessive benzene or phenol exposure is the presence of urinary phenol progenitors in many medicinals. Phenylsalicylate, phenol, and sodium phenate are active ingredients in common over-the-counter medicinals and prescription medicinals. Phenylsalicylate is used as an enteric coating in other common medicinals. Fishbeck et al. (112), for example, found that Pepto-Bismol® (a registered trademark of Procter & Gamble Co., Cincinnati, OH) or Chloraseptic® (a registered trademark of Procter & Gamble Co., Cincinnati, OH) cold lozenges, both of which contain phenylsalicylate, when taken as directed led to peak urinary phenol levels of 480 and 498 mg/liter, respectively (corrected to specific gravity of 1.024). Common medicinals containing phenol include "P&S" liquid or ointment, a shampoo suggested for use in cases of psoriasis and seborrheic dermatitis; Oraderm® (a registered trademark of R. Schattner Co., Washington, DC), a lip balm; and Campho-Phenique® (a registered trademark of Sterling Drug, Inc., New York, NY), a formulation suggested for cuts, burns, cold sores, and fever blisters.

Thus, before concluding that particular urinary phenol levels indicate particular benzene or phenol exposures, one should determine that the subject is not taking any over-the-counter or prescription pharmaceuticals that contain metabolic precursors of urinary phenol. It also would be desirable to have background levels of urinary phenol for the subject in addition to determining that there have been no changes in dietary habits that could increase urinary phenol levels.

6.6 Defining Normal Range of the Index Chemical to Be Expected in the Biological Assay Material

Section 6.5 discussed the wide concentration range of index chemical that can occur in the biological assay material if the index chemical also can arise in the assay material from nonwork sources. All the factors discussed in the preceding five subsections suggest that when the index chemical is a metabolite of the exposure chemical, a broad range of concentrations of the index chemical in the biological assay material is to be expected, even if the index chemical does not occur naturally. This range also would be expected from medical experience with urinary analyses. If measurements are repeatedly made on the same group of workers over a period of time, the ranges should be narrower and an excursion more readily seen and quantitated.

It recently has been reported that the amount of a nonspecific metabolite excreted in urine decreased after 1 month on the job even though worker exposure did not decrease. By 4 months, the concentration had increased to preexposure levels, and by 5 months the amount excreted exceeded preemployment values. Presumably the kinetics of excretion also varied. Measurements were made on 177 workers, so the results cannot be explained by the influence of one worker (124). This cannot be explained on the basis of adaptive enzyme formation and represents another possible variable that should be examined in developing any biological monitoring procedure.

In breath analyses, the index chemical usually is the exposure chemical, not a metabolite. Therefore, the concentration range of exhaled chemical for a particular exposure concentration would be expected to be narrower. Nevertheless, even here a range is to be expected.

Table 3.5 Effect of Conjugate Hydrolysis on Apparent Urinary Mandelic Acid Concentration in Humans (44)

	Urinary Mandelic Acid Concentration (mg/liter)		
Subject	Without Hydrolysis	With Hydrolysis	Increase Following Hydrolysis (%)
1	12	16	33
2	61	118	93
3	349	598	71

6.7 Time Required for the Index Chemical to Appear in the Biological Assay Material

The excretion half-life of chemicals varies from a few hours to days (see Section 7.3.4). This means that some xenobiotics will appear in biological assay materials in a few minutes, and others will not appear in significant concentrations for days following exposure. Obviously, in the latter situation, it would be useless to monitor at the end of the shift as a measure of that day's exposure to the chemical. There also may be differences in rate of appearance of the various possible biological assay materials. Thus, whereas many inorganic cations will appear reasonably promptly in urine, tears, perspiration, or saliva, they will not appear in the external part of the hair shaft or nails for weeks because of the slow generation of these assay materials.

Hippuric acid from styrene exposure, for example, does not appear in urine until about 24 hr after exposure (80). Similarly, TCA does not appear in significant amounts in the urine of exposed subjects until 24 to 36 hr after exposure to T_3CE (125).

6.8 Analytical Methodology

Another factor that must be considered in biological monitoring is the analytical procedure used. The analytical procedure potentially introduces two principal sources of variation: (1) conjugated versus free index chemical and (2) the analytical procedure itself.

As shown in Figure 3.1, many index chemicals are excreted, in urine at least, in both the conjugated and the free form. Also, as discussed in Section 6.1.4, chemicals may be transported in the blood in either the free or protein-bound form. Obviously, analyses of only the free chemical should not be compared with analyses of the total chemical: free + (conjugated or bound). The data of Slob (44) shown in Table 3.5 illustrate this very well. The three subjects were exposed to various unspecified concentrations of styrene in the workplace and the exposure increased from subject 1 to subject 3. The analyses are for mandelic acid. As can be seen, the apparent urinary concentration of mandelic acid almost doubled following hydrolysis of the mandelic acid conjugate that also was excreted.

The measuring methodology also can be an important source of variation. In general, two types of instrument methodology have been used: visible or UV spectrophotometry, and other instruments.

Many chemicals present in biological assay materials may have visible or UV absorption spectra similar to the index chemical, thus potentially introducing large and uncontrolled error. Light absorption values of colored derivatives measured at a particular wavelength also can be misleading, since most chromophoric reagents are nonspecific and may react with other chemicals having structures similar to the index chemical. For example, the Theis and Benedict reagent for phenol (122) also detects *p*-cresol. The concentration of *p*-cresol normally in the urine from unexposed subjects may be 10 times the concentration of phenol (110). The analytical procedure of Ikeda and Ohtsuji (126) for urinary TTC also detects dichloro compounds.

This lack of specificity can be overcome to some extent by removing interfering materials or isolating the index chemical. However, the isolation procedure may not be complete. For example, Pagnotto and Lieberman (115) and Walkley et al. (127) attempted to purify their urinary phenol by steam distillation, not knowing what interfering materials they were trying to remove. Unfortunately, *p*-cresol, the major analytical interference, should steam distill almost as well as phenol (128, 129). Therefore, their purification procedure did not remove the principal impurity that would give false-positive results with their colorimetric procedure. For this reason, the data this group obtained with this method has qualitative value but is of uncertain quantitative value. Either removal of interfering materials or isolation of index chemical frequently leads to losses of the index chemical, which creates further error if the losses are not regular. Even if the losses are regular, this will introduce discrepancies between the method requiring clean-up and other methods that do not. In addition, the removal of interfering substances may be so time-consuming that it is impractical for routine analysis of biological materials from workers.

Gas chromatography, high-performance liquid chromatography (HPLC), and atomic absorption spectrophotometry are three instrumental techniques that usually are readily amenable to separation of the index chemical without loss and give results that are characteristic and unique for the index chemical.

6.9 Route of Exposure

As mentioned earlier, one of the goals of biological monitoring is to measure total exposure to a chemical regardless of the route by which it is absorbed. This assumption must be shown to be true for each method developed. In at least two reported cases, this assumption apparently may not be true.

Dutkiewicz and Tyras (16) found that about 5 percent of ethylbenzene absorbed through the skin is converted to urinary mandelic acid, whereas Bardodej and Bardodejova (130) found 60 percent of inhaled ethylbenzene excreted as urinary mandelic acid. Dutkiewicz and Tyras thus concluded that urinary monitoring of mandelic acid as a measure of total ethylbenzene exposure would not be acceptable.

Yant and Schrenk (13) also reported different relative methanol levels in several organs and tissues when the methanol was administered by inhalation versus subcutaneous administration. This suggests that blood levels of unbound methanol vary depending on the administration route. If that is the case, excretion levels would be expected to vary with the two routes of administration.

7 URINE ANALYSIS FOR MEASUREMENT OF INDUSTRIAL EXPOSURE

7.1 Correlation of Exposure with Index Chemical Concentrations in Urine

Urine analysis for measurement of exposure usually relies on analysis for a metabolite. As already discussed, a great number of factors affect the concentration of the unbound exposure chemical and its metabolites in the blood and thus the amount excreted per unit of urine per unit time. Thus it should not be surprising that there are great variations reported between authors for the translation of urinary levels of the index chemical to exposure levels of the workplace chemical.

Despite the large uncertainty factor in translating urinary levels to body burden or exposure, there is no question but that for many (and probably all) chemicals a relationship does exist. There also is no question, as stated in the introduction, that we do not yet know all the factors that contribute to variability and thus do not know how to correct for all of them.

At the present stage of development, urine analyses for most organic compounds probably are best suited to (1) preemployment monitoring to detect obvious metabolic abnormalities that could increase individual hazard and (2) regular monitoring of an employee to develop a continuing baseline against which an overexposure would be obvious and could lead to remedial measures in the workplace. At the present stage of development for most industrial organic chemicals, a single urine analysis could be correlated only within a broad range of body burden or prior exposure.

The problem is compounded when the index chemical can arise from diet or other nonoccupational sources, as has been discussed for the use of urinary phenol to evaluate phenol or benzene exposure and hippuric acid to evaluate styrene or toluene exposure. In the latter three instances, natural levels are so high and so variable that they invalidate the technique for exposure levels of interest for routine worker monitoring. For example, as previously discussed, urinary levels of phenol from subjects without any workplace exposure to benzene or phenol can exceed 80 mg/liter (121). Roush and Ott (121), and also

Van Haaften and Sie (110), have published data suggesting that urinary phenol levels are not statistically valid measures of individual benzene exposure at levels of 5 ppm or below, because of normal variability. For these same reasons, Ikeda and Ohtsuji (99) suggested that urinary hippuric acid was not statistically valid for estimation of toluene concentrations below 130 ppm. As already mentioned, the data of Ogata et al. (107) suggest that it may not be valid below 200 ppm. Several publications suggest that urinary hippuric acid is not a usable index of styrene concentration at atmospheric levels of workplace interest because of this variability and high background level (82, 119).

Except for the situations mentioned earlier, where urinary phenol is elevated because of medication, it probably is a good index of workplace phenol exposure. The work of Ohtsuji and Ikeda (131) indicates that an 8-hr exposure to 5 ppm in the air would be expected to result in about 400 mg of phenol/liter of urine in a spot sample of urine taken at the end of the workday.

7.2 Biological Threshold Limit Values

Instead of using urinary levels of an index chemical to estimate the level of exposure, it has been proposed that excretion levels of an index chemical be used simply to indicate the presence or absence of overexposure. The term "biological threshold limit value" has been proposed by Elkins (28). The concept is not restricted to metabolites or to urine analyses but would apply to any index chemical. It would apply to blood and any liquid, solid, or gaseous biological excretion.

Nothing presently known about metabolism and excretion and their kinetics invalidates the concept. In general, the normal variation in excretion of chemicals, particularly metabolites, would have to be considered in setting such discriminator levels. The application of this concept would be further complicated where the index chemical occurs naturally in urine and thus has a high and variable background level. In this situation, two courses are possible: (1) the discriminator number could be either set high enough to eliminate all "normal" values or (2) it could be set at a lower fiducial or otherwise derived limit that could include many people that had not been overexposed to the workplace chemical. If the first alternative is utilized, the discriminator level might never be reached for some overexposed individuals.

The second alternative would require some alternate confirmation that the elevated levels really were due to workplace exposure. Urinary hippuric acid as an index of toluene or styrene exposure already has been discussed in this regard. The U.S. Department of Labor (123) has proposed that urinary phenol levels of >75 mg/liter, adjusted to specific gravity of 1.024, be considered indication of overexposure to benzene, leading to certain subsequent blood measurements. Yet, as discussed in Section 6.5, Roush and Ott (121) have shown that urinary phenol levels in unexposed individuals may exceed this value. Fishbeck et al. (112) also have shown that the recommended dosage of a common over-the-counter medicinal can lead to urinary phenol levels over five times this great. Even prescribed use of a common lozenge recommended for sore throats can cause a five- to sixfold elevation of urinary phenol. As discussed earlier, other common medicinal uses of metabolic precursors of urinary phenol also could cause levels of urinary phenol that could be interpreted as industrial overexposure to benzene.

In terms of worker hazard, the second course would be the safest, but it easily could be abused in situations where the index chemical has nonworkplace progenitors. However, as mentioned before, if urinary levels of the index chemical truly reflect hazard to the worker, the source of the chemical is irrelevant. Every effort should be made to find the cause of the elevated concentration and reduce the exposure. However, if the source is not the workplace, attempts to reduce workplace exposure will not result in reduced hazard to the worker.

For reasons discussed in the following paragraphs, the discriminator number for organic compounds also must be based on a urine sample collected a certain number of hours postexposure, or erroneously high or low values relevant to the discriminator number conditions will be obtained.

Some of the workplace exposure chemicals that have shown a urinary dose-response relationship are listed in Table 3.6. Also shown are the index chemicals, the exposure–excretion (dose–response) equations, the type of analysis, and the type of urine sample. In some cases the investigators did not calculate the dose response equation, and this was done by the present author. These instances have been noted. If the equation was based on analyses corrected by any of the methods discussed in Section 6.4, this also is noted. Expected metabolite concentrations based on the excretion equations also are shown for a particular atmospheric concentration of the workplace chemical (biologic TLVs).

Table 3.6 illustrates many of the problems of estimating total workplace exposure from the analysis of an index chemical in excretions. These are discussed in the following sections, using the data in the table for illustration.

Where more than one metabolite is excreted, as in the cases of T_3CE and styrene, it has been suggested that the total concentration of all of them may correlate better with exposure than the concentration of any one in particular (134).

Also, depending on the shape of the excretion curve, the equation giving the best fit may require use of the log of x and the log of y; for example, $\log y = a \log x + b$ (134).

7.3 Factors to Be Considered in Using the Methodology

7.3.1 Slope of the Dose-Response Curve

Methylchloroform metabolites (e.g., in urine) follow a normal dose–response relationship, at least to 50 ppm in air, the upper atmospheric exposure level studied. However, the standard deviation is approximately 40 percent of the

Table 3.6 Analytical Details for Analyses of Urine from Males Exposed to Various Chemicals

Exposure Chemical	Index Chemical	Equation	Correction	Analytical Method Hydrolyzes Conjugates
Methyl chloroform (1,1,1-trichloroethane)	TTC[a]	$Y = 0.27X + 0.54$ $= 0.18X + 0.29$ $= 0.28X + 0.16$	None Specific gravity = 1.016 mg/g creatinine	Yes
	TCE	$= 0.19X + 0.27$	None	
	TCA	$= 0.073X + 0.21$	None	
T_3CE	TTC	$Y = 8.37X + 17.12$	Specific gravity = 1.024	Yes
	TCE	$= 5.19X + 12.28$	None	
	TCA	$= 3.17X + 4.84$	None	
T_3CE	TTC	$Y = 7.25X + 5.5$ $= 4.97X + 1.9$ $= 5.50X + 6.2$	None Specific gravity = 1.016 mg/g creatinine	Yes
	TCE	$= 5.57X + 4.4$	None	
	TCA	$= 2.74X + 0.7$	None	
Benzene	Phenol	$Y = 19X + 30$[b]	Specific gravity = 1.016	Yes
Phenol	Phenol	$Y = 108X + 28$[b] $= 80X + 40$[b] $= 70X + 45$[b]	None Specific gravity = 1.016 mg/g creatinine	Yes
Toluene	Hippuric acid	$Y = 23X + 800$[b] $= 16X + 600$[b] $= 18X + 350$[b]	None Specific gravity = 1.016 mg/g creatinine	—
Toluene	Hippuric acid	$Y = 40X + 550$[b] $= 24X + 500$[b]	Specific gravity = 1.024 mg/g creatinine	—
Toluene	Hippuric acid	$Y = 9.8X + 400$[b]	None	—
Toluene	Hippuric acid	$Y = 31X + 400$[b] $= 25X + 300$[b]	None Specific gravity = 1.024	—
m-Xylene	m-Methyl–hippuric acid	$Y = 27X$[b] $= 26X$[b]	None Specific gravity = 1.024	—
p-Xylene	p-Methyl–hippuric acid	Not calculable Not calculable	None Specific gravity = 1.024	—
Styrene	Mandelic acid Mandelic acid Phenylglyoxylic acid Phenylglyoxylic acid	$Y = 18.4X + 149$[d] $= -14.0X + 6125$[e] $= 2.7X + 79$[d] $= -1.0X + 540$[e]	Specific gravity = 1.024 Specific gravity = 1.024 Specific gravity = 1.024 Specific gravity = 1.024	No

[a] Abbreviations: U = unknown; Y = concentration of index chemical in urine in mg/liter of urine or mg/g of creatinine; X = concentration of exposure chemical in ppm; NA = not available.
[b] Calculated by RSW from investigator's data.
[c] Calculated by investigators.
[d] Exposures ≤150 ppm.

Maximum Atmospheric Concentration Evaluated	Calculated Urinary Concentration (mg/liter) at Given Exposure ppm	Standard Deviation (ppm) at Given Exposure ppm	Urine Sample Type	Analytical Method	Reference
50 ppm × 8 hr × 3 days	14 @ 50 9 @ 50 14 @ 50 10 @ 50 4 @ 50	~20 @ 50[b] ~15 @ 50[b] ~15 @ 50[b] ~20 @ 50[b] ~25 @ 50[b]	Spot; after 3 days of exposure	Colorimetric (80)	114
40 ppm × 8 hr × 2.5 days	854 @ 100[e] 531 @ 100[e] 322 @ 100[e]	NA	Spot; after 2.5 days of exposure	Colorimetric (80)	125
175 ppm × 8 hr × 3 days	730 @ 100 500 @ 100 550 @ 100 575 @ 100 210 @ 100	~40 @ 100[b] ~20 @ 100[b] ~40 @ 100[b] ~40 @ 100[b] —	Spot; after 3 days of exposure	Colorimetric (80)	81
150 ppm × 8 hr	120 @ 50[b]	NA	Spot; at end of workday	Colorimetric (120)	111
3.5 ppm × 8 hr	568 @ 5[b] 440 @ 5[b] 395 @ 5[b]	~1 @ 3.5[b] ~1 @ 3.5[b] ~1 @ 3.5[b]	Spot; at end of workday	Colorimetric (132)	131
240 ppm × 6 hr	3100 @ 100[b] 2200 @ 100[b] 2150 @ 100[b]	~45 @ 100[b] ~45 @ 100[b] ~45 @ 100[b]	Spot; near end of workday	Colorimetric (99)	106
170 ppm × 8 hr	4550 @ 100[b] 2900 @ 100[b]	NA NA	Spot; at end of workday	Spectrophotometric[f]	115[g]
600 ppm × 8 hr 800 ppm × 6 hr	1380 @ 100[b]	NA NA	24-hr and end of day	Titration with 0.5N NaOH	24
200 ppm × 7 hr	3100 @ 100[b] 2500 @ 100[b]	~10 @ 100[b] ~15 @ 100[b]	Spot; at end of exposure	Colorimetric (133)	107
200 ppm × 7 hr	2700 @ 100[b] 2600 @ 100[b]	~55 @ 100[b] ~30 @ 100[b]	Spot; at end of exposure	Colorimetric (133)	107
100 ppm × 7 hr	1420 @ 100[b] 3090 @ 100[b]	NA NA	Spot; at end of exposure	Colorimetric (133)	107
290 ppm × 8 hr with occasional excursions of 1500 ppm × 10 min	1000 @ 50[e] 2000 @ 100[e] NA NA	NA	Spot; at end of workday	Colorimetric (119)	83

[e] Exposures >150 ppm.
[f] Toluic acid interferes.
[g] Includes female workers, but results for females were not separated by investigators.
[h] Observed values.

exposure value, and two standard deviations would be 80 percent. When this is coupled with the shallow slope of the curve (0.073 to 0.28, depending on the metabolite measured), it is doubtful that two standard deviations below the measured urinary metabolite level resulting from exposure to 350 ppm would be greater than two standard deviations above the zero exposure level of metabolites. The problem of the importance of the slope of the dose–response curve is treated in greater detail by Imamura and Ikeda (135). In general, because of the variability of metabolism and excretion, the dose–response curve should have a fairly steep slope. This is particularly important when the index chemical has nonworkplace progenitors.

7.3.2 Nonlinear Response

The equations found by Seki et al. (114) that relate urinary metabolite levels to inhalation exposure of methyl chloroform are straight lines, at least up to atmospheric concentrations of 50 ppm for 8 hr. However, they did not extend their observations to 350-ppm exposures, the present TLV of methyl chloroform. In some cases, as already discussed, the urinary metabolite concentration plateaus (82) or may even decrease (83) with increasing concentration. Thus one cannot extrapolate from lower to higher concentrations but must actually carry out measurements at the exposure levels of interest.

7.3.3 Different Analytical Responses for Conjugated and Unconjugated Forms of the Index Chemical

The importance of hydrolysis of conjugates prior to analysis already has been discussed in Section 6.8. As shown in Table 3.6, most analytical procedures include a procedure for conjugate hydrolysis.

7.3.4 Spot Urine Samples versus 24-hr Sample

In his analyses, Seki et al. (114) used so-called spot samples of urine. These are samples taken at a particular time, in contrast to 24-hr samples in which all urine voided over a 24-hr period or longer is collected. Elkins (27) has raised the question of whether spot samples of 24-hr samples are better. The present author believes that this is not a matter of real concern, but rather that it is more important that:

1. The sample be taken after there has been sufficient time for adequate metabolism and excretion to minimize variation due to analysis of small amounts and to reach a more or less steady state of excretion versus intake. This will vary with the urinary elimination half-life, which may vary from less than 4 hr for phenol to almost a week for T_4CE. Representative elimination half-life values for some industrial chemicals are shown in Table 3.7.

Table 3.7 Human Urinary Excretion Half-Life Values for Some Industrial Chemicals

Exposure Chemical	Index Chemical	Excretion Half-Life (hr)	Urine Samples	Reference
Styrene	Phenylglyoxylic acid	8.5	24 hr	82
	Mandelic acid	7.8	24 hr	
Styrene	Phenylglyoxylic acid	10	Not given	136
	Mandelic acid	10		
Toluene	Hippuric acid	~4 (α phase)[a]	24 hr	107
	Hippuric acid	~12 (β phase)[a]	24 hr	
Toluene	Hippuric acid	6.3^b	24 hr	107
T_3CE	TTC	41	Spot	30
T_3CE	Trichloroethanol	~7 (α phase)[a]	Spot	124
		~76 (β phase)[a]	Spot	
Phenol	Phenol	3.4^b	Every 2 hr	18
Xylene	Methyl hippuric acid	3.8^b	24 hr	107
Xylene	Methyl hippuric acid	~1 (α phase)	24 hr	137, 138
		~20 (β phase)	24 hr	
T_4CE	TTC	144^b	24 hr	30
1,1,1-Trichloroethane	TTC	8.7	Spot	114

[a] Estimated by RSW from graphs in publication: α phase is the initial, rapid decay; β phase is the later, slower decay.
[b] Calculated by Ikeda and Immamura (30) from data in publication.

2. The sample always be taken at the same time relative to exposure. This applies not only to the original investigator, but to others who want either to compare results or to use the original investigator's results. This is necessary because the kinetics of absorption, distribution, and excretion may give different dose–response equations for metabolite levels at different times after exposure or after a single exposure versus multiple exposures. This is particularly true if the chemical has a long elimination half-life.

7.3.5 Equivalency of Analytical Methods

Seki et al. (114) used a colorimetric method of analysis. The problems associated with various methods of analysis already have been discussed. If comparisons are to be made between investigators, all should either use the same analytical procedure or the equivalency of the different procedure should be demonstrated. If a discriminator number to indicate overexposure is to be used, the same requirements should be met.

When the same analytical procedure and comparable urine sampling strategy are used, comparable results are possible as shown by the equations developed by Ogata et al. (125) and Ikeda et al. (81) for urinary TTC, TCE, and TCA following T_3CE exposure. Accordingly, the urinary TTC, TCE, and TCA following T_3CE exposure at 100 ppm T_3CE are very close (Table 3.6).

Despite the similarity of the equations developed by Ogata et al. and Ikeda et al., when the urinary TCA data and TCE data reported by Stewart et al. (139) are substituted in these equations, the exposures reported by Stewart et al. are underestimated by approximately one-third on the basis of TCE content and approximately one-half on the basis of TCA content. Stewart et al. analyzed 24-hr urine samples after three $7\frac{1}{2}$ hr. exposures to 100 ppm of T_3CE, in contrast to Ogata et al. and Ikeda et al., who analyzed spot urine samples after two $\frac{1}{2}$-hr or three 8-hr exposures to maximum concentrations of 40 or 175 ppm, respectively, of T_3CE. This probably is not a major procedural difference. The major difference appears to be in analytical procedures. Ogata et al. and Ikeda et al. used the same colorimetric procedures for TTC, TCE, and TCA. Stewart et al. used a GC procedure. The data suggest that the slope of the standard curve for the colorimetric procedure is greater than the slope of the standard curve for the GC procedure.

The deviations to be expected with different analytical procedures also are illustrated in Table 3.6 by the equations developed by various investigators for urinary hippuric acid levels following toluene exposure. In this case, the situation is further complicated by the high variable background level of hippuric acid, as already mentioned. However, the investigators carried out the exposures at atmospheric levels that would minimize this interference. Thus these differences probably also principally reflect analytical differences, for the most part. Von Oettingen (24), for instance, whose equation indicates one of the lowest background levels of hippuric acid, precipitated the hippuric acid from urine and then titrated it. The precipitation almost certainly was not quantitative. Ogata (107), whose equation indicates the next lowest background level of hippuric acid, extracted the hippuric acid before forming a derivative that had a characteristic light absorption spectrum. It is possible that the initial extraction was not complete. The method of Pagnotto and Lieberman (115) also depends on an extraction, followed by determination of light absorption at a specific wavelength. Ikeda and Ohtsuji (106) separated the hippuric acid by paper chromatography followed by reaction to form a colored derivative, extraction of the derivative from the paper, and determination of its light absorption at a characteristic wavelength. Ogata et al. (133) later showed that the color depended on the filter paper used.

7.4 Analysis for Inorganic Ions

Urine analysis also can be used to measure exposure to metallic ions. Some of the metal ions excreted in urine include beryllium, cadmium, chromium, copper, iron, lead, lithium, magnesium, manganese, mercury, selenium, and zinc. The factors to be considered in developing and applying the methodology and interpreting the results do not differ from those already discussed. Dose–response relationships exist and can be utilized. As already mentioned, it is unlikely that metabolic factors will be as important for inorganic salt excretion as they are for organic chemical excretion. Therefore, the normal range of excretion accompanying a given exposure may be narrower.

Metals are unlike organic chemicals in that they cannot have nonworkplace precursors that are less toxic than the industrial exposure chemical. Consequently, metal ion excretion levels will represent the true hazard to the worker, except in the few instances discussed. Thus, biologic TLVs for metals can be set with greater assurance that excursions above the level represent greater hazard to the worker. However, it should be remembered that exposure to metals can also occur outside the workplace. Regardless of where they occurred, excretion levels generally indicate the true hazard and exposure should be reduced, whatever the source.

Elkins (28) has reviewed urine levels for several metals. On the basis of his experience, he feels that 0.2 mg lead/liter (corrected to specific gravity of 1.024) represents significant absorption but not necessarily a toxic level. Similarly, he reported that 0.25 mg mercury/liter would represent significant absorption of mercury but not necessarily a toxic effect.

The 1972 NIOSH review (140) of the data correlating urinary lead concentration with biological effects concurs with Elkins's assessment.

Urine analyses also have been used to measure exposure to inorganic anions. Extensive work has been reported, for example, on the correlation of urinary fluoride and exposure to inorganic fluoride in the aluminum industry. NIOSH (141) has suggested that the work of Kaltreider et al. (142) and Derryberry et al. (143) indicates that postshift urinary values of ≤ 7 mg fluoride/liter urine (corrected to specific gravity of 1.024) indicate exposure to inorganic fluoride levels that would not be expected to cause osteofluorosis.

However, because of the ubiquity of fluorine atoms, for example, inorganic fluoride in drinking water, tea, and cereal grains, and organic fluoride in refrigerants, degreasing solvents, and other substances, urinary fluoride levels may not always reflect solely occupational exposure to inorganic fluoride. This urinary level also cannot be taken to indicate safe exposure to inorganic fluorides such as oxygen difluoride, nitrogren trifluoride, sulfur pentafluoride, sulfur tetrafluoride, tellurium hexafluoride, or any other inorganic fluoride with innate toxicity significantly greater than the inorganic fluorides used to set this standard.

Organic fluoride levels may indicate more or less hazard than corresponding inorganic fluoride levels, depending on the comparative toxicity of the organofluorine compound and any of its metabolites.

7.5 Optimum Conditions for Use of Urine Analyses

The studies to date indicate that urine analyses for a particular organic index chemical will be most reliable if:

1. The index chemical has no nonworkplace progenitors.
2. The slope of the dose–response curve is fairly steep (≥ 0.5).
3. The elimination half-life is no greater than 8 hr and preferably no greater than 4 hr.
4. The analytical method is specific for the index chemical. This requirement tends to eliminate colorimetric procedures. If GC procedures are used, the peak identity should be confirmed with a second column, or a GC/mass spectrometer combination should be used in developing the method.
5. The method is validated in humans at the highest exposure level of interest. The dose–response curves should not be extrapolated beyond the highest level experimentally validated.
6. Urine collection times are consistent and consider excretion half-life.
7. Urine samples are analyzed shortly after collection. If they cannot be analyzed promptly, they should be frozen or at least refrigerated (28, 111).
8. The method and equations are first validated for the group of workers of interest before routine application of the correlation equations.
9. The worker is not on a diet, suffering from a disease, or taking a medicine that could alter the relevant kinetics of the reactions of interest or any normalizing procedures used.
10. The worker is not being exposed off the job to the index chemical or another progenitor of the index chemical.
11. The urinary level of index chemical is relatable to the amount of exposure chemical absorbed by all routes.
12. The dose–response equations have been shown to apply to both men and women, or separate equations are developed for (and applied to) each sex.
13. The workday of the group of interest and of the group used to develop the equations and/or discriminator number are the same, if the half-life of elimination is much greater than 8 hr.

8 EXHALED AIR ANALYSIS FOR MEASUREMENT OF INDUSTRIAL EXPOSURE

8.1 Background and Advantages of Methodology

It has been known for decades that many industrial chemicals that are inhaled and enter the vascular system of the human body are later excreted to some degree in exhaled air. Metabolites of some also may be exhaled. A representative listing is shown in Table 3.8.

It has been observed in the past few decades that the concentration of many of these chemicals in exhaled air decreases regularly with time (145). It also has been observed that some chemicals absorbed through the skin are excreted, unchanged, in exhaled air. The concentration of these chemicals in the exhaled air also has been found to decrease regularly with time (15). If the concentration of the exposure chemical in exhaled air varies in some regular fashion with body burden, regardless of the route of absorption, a very desirable method

Table 3.8 Some Industrial Chemicals Detected in Human Exhaled Air Following Exposure

Chemical	Reference
Benzene	143
Carbon tetrachloride	15, 145
Diethyl ether	146
Methanol	1
Methyl acetate	1
Methylene chloride	15, 147
Styrene	42, 83
Toluene	24
1,1,1-Trichloroethane	15
T_3CE	15, 96
T_4CE	148–150
Vinyl chloride monomer	151
Xylene	137

for measuring industrial exposure would be provided. It would have at least the following advantages over various other methods:

1. Metabolism usually would not be involved, so all the metabolic factors that can affect the rate of appearance of the index chemical would not be involved. In most cases, physical or physicochemical factors would predominate, and these should be fairly constant between individuals.

2. The index chemical appears rapidly in the exhaled air. It is not necessary to wait hours or weeks for the index chemical to appear in the biological assay material as in many other biological assays.

3. The analysis would be amenable to GC techniques. These can be made quite specific, thus eliminating analytical interference from nonindex chemicals. Gas chromatographic techniques are able to quantitate small amounts and can be used for simultaneous analysis of several chemicals. Thus, concurrent exposure to several chemicals could be quantitated fairly rapidly and inexpensively.

4. Several samples can be taken in rapid succession, so the assessment of exposure can be based on either a kinetic analysis or substitution in an already derived equation.

5. In many cases the subject could be observed while providing the sample to assure that the instructions are being followed.

6. Very few nonworkplace progenitors exist for the index chemical. Thus, the excreted material more likely represents workplace exposure than in many analyses of other excretion materials.

7. The technique is noninvasive.

8. The technique measures individual exposure without the bother of a personal monitor or the uncertainty of area monitoring.

8.2 Factors Affecting Exhaled Air Levels of Index Chemicals

Assuming that transport through the lungs into the vascular system and back into the lungs is simple diffusion rather than "active transport" (Reference 2, p. 113), several factors can be proposed that possibly will affect the postexposure concentrations of index chemical in exhaled air.

1. *Concentration of index chemical in inhaled air.* If movement of the chemical into the bloodstream is a simple diffusion process, the amount entering the blood will vary with the rate of diffusion through the alveolar wall and the partial pressure (concentration) of the chemical in the workplace atmosphere. It should not vary with the duration of exposure, once equilibrium between blood and the atmosphere has been established.

2. *Blood concentration.* If movement in and out of the blood is a simple diffusion process, the concentration of the index chemical in exhaled air will vary directly with the concentration of that chemical in the blood. There will be, of course, some lag due to the rate of diffusion. The important blood concentration usually will be the concentration of unbound chemical. (Factors affecting binding already have been discussed in Section 6.1.4.) The rate of metabolism of the chemical also will affect the unbound concentrations in the blood. (Factors affecting this were discussed in Section 6.1.)

In addition to the general effects of disease (discussed in Section 6.1.9), emphysema can uniquely affect the interpretation of the results. If the effective lung surface is decreased, the total diffusion rate of the index chemical from the blood into the atmospheric side of the lungs will be decreased, resulting in a lowered concentration in the exhaled air in comparision to the nondiseased person. This would decrease the calculated exposure. Since the absorption coefficient of this emphysematous person also would be decreased in comparison to the normal person, the exhaled air concentration of index chemical still would be a measure of that person's body burden, but the equation relating the exhaled air concentration to exposure concentration would be different.

3. *Solubility in tissues or fat and binding to them.* Some chemicals will be more soluble in, or more strongly bound to, these materials than will others. Except in unusual situations where there is a high degree of solubility or binding, this should not have a significant effect on exhaled air concentrations of the index chemical at the concentrations and time frame of interest.

4. *Dilution in exhaled air.* This is the pulmonary counterpart of urinary dilution discussed in Section 6.4. Every postexposure breath dilutes the index chemical in the alveoli with air that does not contain the index chemical. This lowers the concentration of index chemical in the exhaled air, which usually will decrease the precision of the analysis. This dilution can be overcome to some extent by analyzing "end-tidal" or "alveolar" air. To do this, the subject

inhales and exhales normally through the collecting apparatus two or three times. At the end of the last breath, the final few milliliters are either collected in the sampler for later analysis (123) or diverted directly into the analytical instrument (150). According to DiVincenzo et al. (147), the alveolar air may be 50 percent richer in index chemical than the usual exhaled air. This technique is possible because instruments such as the gas chromatograph can routinely analyze a few hundredths of a milliliter of gas with high precision and accuracy.

Factors that have been shown to affect index chemical concentration in exhaled air include:

1. *Nonworkplace progenitors of the index chemical.* Although the instrumentation generally used for exhaled air sampling is quite specific, it cannot indicate whether the chemical it is detecting appeared in the breath sample from the bloodstream or the mouth. It cannot tell if the chemical got into the bloodstream from prior inhalation or from ingestion. For example, phenol from a lozenge for sore throats (see Section 6.5) will invalidate exhaled air analyses for phenol. Acetone from severe untreated diabetes will similarly confound exhaled air analyses for acetone from the workplace. Depending on the chemical of interest, lozenges, candy, chewing gum, tobacco, mouthwash, or toothpaste also could be sources of interference.

2. *Respiratory rate.* Until the blood and extracellular fluid compartment are saturated, respiratory rate can affect the rate of uptake. It similarly can affect the rate of desorption as shown by DiVincenzo et al. (147). This suggests that correlation equations derived from subjects at rest should not be used on subjects that have been exerting themselves, and vice versa.

3. *Sex.* As already mentioned, the absorption coefficients for some vapors have been reported to vary for the two sexes (18, 98). This will affect both the time to saturation and, probably, the rate of desorption. Nomiyama (97) also has reported an absolute difference between sexes in the concentration of T_3CE in the exhaled air in the β phase (see footnote *a* in Table 3.7) of respiratory elimination. The female concentration is lower. However, the slopes of the two β-phase curves are parallel. The difference is due to a prolonged α phase in women. Stewart et al. (148), on the other hand, found no sexual difference in the decay curves for T_3CE in the exhaled air of men and women following concurrent exposure of both sexes to T_3CE. If there are sexual differences, the use of equations derived for one sex on analyses of exhaled air from the other sex would give erroneous estimates of workplace exposure. Another factor that will affect the rate of saturation of the blood and extracellular fluid compartment is the difference in blood volume and, probably, total extracellular fluid. For instance, the male blood volume is 75 ml/kg body weight, whereas the female volume is only 90 percent of that, or 67 ml/kg body weight (9). Since there easily can be a 100 percent difference in body weights between the sexes, there could be large differences in the volume of this compartment and thus the amount of the index chemical in the compartment after a given

exposure. Again, this could result in the use of an inappropriate correlation equation for estimation of workplace exposure.

Obesity and the fat:muscle ratio will alter the amount of fat-soluble exposure chemicals that are stored and, consequently, the rate of excretion, as already discussed in Section 6.1.7.

4. *Skin absorption of the exposure chemical.* In addition to these factors that could lead to erroneous correlations between exposure levels and index chemical concentration in exhaled air, Stewart and Dodd (15) have shown that the alveolar decay kinetics for some solvents may be different for skin absorption exposure than for inhalation exposure. If this is true for other solvents, exhaled air concentration may not be a reliable index for evaluating total body burden when the burden arises from various routes of exposure.

Stewart and Dodd studied the skin absorption of carbon tetrachloride, methylene chloride, T_3CE, T_4CE, and methylchloroform (1,1,1-trichloroethane). They found that, for thumb immersion or nonoccluded topical exposure, the postexposure alveolar air decay was linear, whereas for total hand immersion the decay was exponential. Thus, the breath excretion kinetics may vary with the size of the dose for noninhalation administration. The decay curves developed to date for excretion in breath following inhalation administration have all been exponential decays.

As the technique is further studied, additional possible causes of variation undoubtedly will be found.

Obviously, just as with urine analyses, it will have to be demonstrated for each chemical that the kinetics of its excretion in breath are the same no matter what the route of exposure. If the kinetics change, multiple routes of exposure will invalidate the methodology for measuring body burden and exposure.

8.3 Shortcomings of the Methodology

Studies to date indicate that the technique has some shortcomings that must be recognized:

1. It does not appear to be widely usable for samples taken within, variously, 0–2 hr postexposure. The decay curves for various concentrations of xenobiotic and times of exposure are frequently indistinguishable in this time period. If a chemical is excreted so rapidly in exhaled air that none remains the next morning, samples would have to be taken off the job by the worker and returned for analysis the next morning. DiVincenzo et al. (147) have found that workers may not exactly follow instructions for collection of alveolar air samples.

2. Although the body tends to integrate the exposure, thus giving exhaled breath concentrations representative of the average exposure, some data suggest that samples taken shortly after exposure will principally reflect the latest exposure level (148).

8.4 Experimental Studies

For inhalation administration, the results published to date have been very encouraging for all the compounds studied. The methodology has given good correlation between concentration decrease of the index chemical in postexposure exhaled air and prior exposure, at least for the first 3–5 hr postexposure.

As mentioned previously, the decay curve is exponential and generally fits Equation 5:

$$\frac{C_t}{C_0} = K_A e^{-\alpha t} + K_B e^{-\beta t} + \cdots + K_N e^{-nt} \qquad (5)$$

where C_t = concentration of the index chemical in exhaled air at time t
C_0 = exposure concentration
α, β, n = rate constants for respective decay periods
K_A, K_B, K_N = zero time coefficient for that curve segment

Except for protracted studies over many hours or for chemicals where the decay curve seems to have more than two segments, all terms beyond the first two usually are unnecessary. The terms K_A, α, K_B, β, K, and n all can be obtained graphically from the semilogarithmic plot of C_t/C_0 versus t (7). The value of C_t is determined experimentally at time t. The equation can then be applied to unknown exposure situations and C_0 can be calculated.

The data of Fernandez et al. (150) suggests that the time of exposure, at least for T_4CE, determines the decay curve. They studied alveolar air concentrations up to 4 hr postexposure. They found that the same equation described the postexposure decays after 2 hr of exposure to 100 or 200 ppm, if the ratio of C_t/C_0 was plotted against postexposure time. A similar result was found for 4-hr exposures to 100, 150, or 200 ppm of T_4CE and for 8-hr exposures to these same three concentrations. The curves for the three time periods were different. Their equations are shown in Table 3.9. Baretta et al. (151) found that the decay curves for VCM in human alveolar air following 8-hr exposures to 50, 100, 250, or 500 ppm also were a family for the decay period studied. Calculation of C_t/C_0 for all curves at various common reported postexposure times yielded a common curve, within experimental limits, for the 20-hr decay period reported (see footnote d in Table 3.9). Because neither the y intercept(s) for these curves nor the data points for the first hour postexposure are shown in the paper by Baretta et al., it is not possible to determine whether the first term in the equation describing the decay curves will be the same for all. Nevertheless, the data are strongly suggestive, and the time period reported would be suitable for calculating exposure levels. Decay curves for shorter exposure times were not reported, so it is not possible to determine whether those curves would be different from those for the 8-hr exposure.

Within experimental limits, Stewart et al. (148) also found that human alveolar air decay curves for the first 20 hr following 7.5 hr of exposure to 20,

Table 3.9 Postexposure Decay Equations for Xenobiotics in Human Exhaled Air

Chemical	Exposure Period (hr)	Exposure Concentration (ppm)	Postexposure Study Period (hr)	Air Sample	Subjects (No., Sex)	K_A	K_B	α	β	Reference
Methylene chloride	2	100, 200	3	E^b	11, M^c	1.8×10^{-2}	10^{-1}	1.1×10^{-1}	10^{-2}	147
Styrene	8	25, 115, 260	5	E	15, M	1.5×10^{-2}	5×10^{-3}	1.01	1.3×10^{-1d}	83
	7	99	6	A	6, M	1.5×10^{-2}	5×10^{-3}	1.01	1.3×10^{-1d}	109
T$_4$CE	7	101	110	A	16, M	3.5×10^{-2}	3.5×10^{-2}	8.3×10^{-2}	8.8×10^{-3d}	87
	2	100, 150, 200	4	A	23, M; 1, F	2.9×10^{-1}	1.6×10^{-1}	1.2×10^{-2}	8.0×10^{-3}	150
			4			2.1×10^{-1}	2.0×10^{-1}	9.9×10^{-2}	7.5×10^{-3}	
			4			10^{-1}	2.2×10^{-1}	4.9×10^{-2}	4.4×10^{-3}	
T$_3$CE	1	20, 100, 20	22	A	3–9 (M, F)	4.5×10^{-2}	4.8×10^{-2}	4.6×10^{-1}	5.1×10^{-2d}	148
	3					$5 \times 10^{-?}$	1.3×10^{-2}	1.6	8.1×10^{-2d}	
	7½					$9 \times 10^{-?}$	2.6×10^{-2}	1.8	6.9×10^{-2d}	

[a] Coefficients and Exponents ($Y = K_A e^{-\alpha t} + K_B e^{-\beta t}$)

[a] Y = ratio of concentration in exhaled air at time t (hr) to exposure concentration (i.e., C_t/C_0).
[b] E = normal exhaled air; A = alveolar air.
[c] M = male; F = female.
[d] These calculations were made by R. S. Waritz, using data obtained from graphs presented in the author's original scientific paper and without access to the original raw data. Therefore, they are subject to errors introduced by the original translation to graph form and by printing. They should be considered illustrative only.

100, or 200 ppm T₃CE also were a family. For this family, as with the T₄CE decay curves given by Fernandez et al. (150), the curves expressing the ratio of C_t/C_0 versus postexposure time were coincident within experimental limits and could be represented by the one equation shown in Table 3.9. Likewise, the curves for alveolar decay ratios following 1- or 3-hr exposures at 100 and 200 ppm were identical for both exposures at each time period. The coefficients and exponents for the curves from all three exposure times (1, 3, and 7.5 hr) were not identical. The decays following the 1- and 3-hr exposures at 20 ppm were not followed by Stewart et al. for a long enough time to provide enough data points for comparisons.

The data of DiVincenzo et al. (147) also suggest that the equations for the decay curves of the C_t/C_0 ratio following 2-hr human exposures to either 100 or 200 ppm of methylene chloride also will be the same for both concentrations. This equation is shown in Table 3.9.

Both Stewart et al. (42) and Götell et al. (83) studied the decay of styrene in exhaled air following 8-hr exposures of humans. Stewart's subjects were exposed in a chamber to 99 ppm of styrene. Götell's subjects were workers exposed to a time-weighted average of 89–139 ppm in the workplace. Stewart collected alveolar air samples for analysis. Götell did not specify the samples collected.

Despite these inconsistencies, the agreement between the two decay curves was remarkable. Both decay curves when transformed to C_t/C_0 versus time (in hours) could be expressed as the same equation over the time period studied. This is shown in Table 3.9 and is in contrast to the findings of Fernandez et al. (150) and Stewart et al. (87) for T₄CE breath decay. In this case, the former group found that the β phase did not start until about 1.5 to 2 hr following an 8-hr exposure to 100 ppm. Stewart's group did not find the β phase starting until approximately 45 hr following a 7-hr exposure to 101 ppm.

There seems to be no question that in order to calculate exposure from exhaled air decay curves, it will be necessary to know the exposure time.

Although this aspect of the use of exhaled air decay curves to calculate exposure is very encouraging, the variability reported for some solvents is discouraging. For example, DiVincenzo et al. (147) found that with 11 subjects the breath decay curves for 100 and 200 ppm of methylene chloride were within two standard deviations of each other. Similarly, Götell's (83) data for styrene indicate his decay curves would not reliably distinguish between 25 and 115 ppm or 115 and 260 ppm.

Conversely, Baretta et al. (151) found that exhaled air decay curves for VCM could readily distinguish between 50 and 100 ppm or 100 and 250 ppm with only four to six subjects. Unfortunately, not enough studies have been published with such comparisons to indicate the probable general situation.

The technique appeared to demonstrate worker accumulation of T₄CE after four and five 7-hr exposures to about 100 ppm (87).

Overall, exhaled air analyses seem to hold good promise as a way of determining previous exposure to certain industrial chemicals. As can be seen in Table 3.9, not all chemicals are excreted at the same rate, and calibration curves will have to be developed for each chemical. It is probable that curves also will have to be developed for each group of workers, but not enough data have yet been developed to judge. The data strongly suggest that in order to apply the technique quantitatively, knowledge of the time of exposure will be necessary. However, Stewart et al. (148) have published data suggesting that, within certain concentration and time limits, the decay curve is determined by the product of concentration and time.

For optimum utilization of the technique, it may be necessary to develop individual decay data over a period of time. The technique, with or without concurrent urine analysis, should be usable to screen a worker for abnormal metabolism and excretion of many workplace chemicals prior to assignment to an area where the worker may be exposed to these chemicals.

9 BLOOD ANALYSIS AS A MEASURE OF INDUSTRIAL EXPOSURE

Blood analysis is an invasive technique that carries some resultant risk to the worker. Most workers also find it objectionable and would probably object strongly to daily or even weekly samples being drawn. Generally, the order of worker preference for sampling for biological monitoring would be expected to be exhaled air, urine, and, last of all, blood.

Blood analysis has not been used extensively in industrial hygiene for measuring exposure, and comparatively few papers have been published in this area.

Blood analyses are used extensively in pharmacology and in clinical trials for new medicinals and are a valuable tool in these areas. They have provided great insight into the metabolism and excretion of xenobiotics and in the individual variation in these body processes. Pharmaceutical chemists and biochemists have been responsible for the greatest developments in these areas. The reader is referred to texts in this specialized field for in-depth information on its utility and drawbacks (8, 60, 74).

There is no question that total blood levels of xenobiotics and metabolites reflect dosage. There also is no question that the factors discussed in Section 6.1 play a great role in individual variations. Thus all the caveats presented in that section must be considered in interpreting individual analyses for organic and inorganic industrial chemicals.

The correlations of inorganic lead exposure, biological effects, and blood levels for inorganic lead probably have been more extensively studied than for any other industrial compound. The early work by Kehoe and others has been reviewed by Kehoe (152). This work and more recent studies also have been reviewed by NIOSH (140). The reader is referred to these reviews for excellent summaries and discussions of the work that has led to the present proposals for biological monitoring of inorganic lead exposure.

Kehoe (152) has suggested that the maximum acceptable blood level for inorganic lead in adult workers is <80 µg/100 g of blood; NIOSH concurred

(140), whereas OSHA (153) initially proposed a level of >60 μg/100 g as the level that would dictate worker removal from exposure areas. The OSHA lead standard that finally issued has an action level of 40 μg/100 g of whole blood. It also has a sliding scale of maximum blood levels, initially coupled with maximum airborne exposure levels of lead. Both decrease with the age of the standard. Allowable blood levels start at 80 μg/100 g of whole blood, decreasing to 50 μg/100 g of whole blood in the fifth year.

In their proposed National Ambient Air Quality Guide, EPA (155) has suggested that mean blood lead levels in excess of 15 μg/100 ml of blood in children aged 1 to 5 years could be accompanied by biological effects.

As discussed in Section 5, blood analyses also are routinely used in the medical field to detect organ damage by measuring transferase enzyme levels, blood urea nitrogen, and so on. However, the increase is not specific for any causative agent. Anything damaging the organ could cause an increase in the blood level of these enzymes. The causative agent could be disease, medicinals, or industrial chemicals. Thus such measurements are nonspecific measures of organ damage; therefore, this approach is deficient in two respects:

1. It is not specific to a particular industrial chemical or even to industrial chemicals.
2. It measures an effect that occurs after injury, instead of measuring a leading effect that could be used to forestall injury, which is the goal of industrial hygiene.

In summary, there is no question that blood analyses can be developed to measure exposure to workplace chemicals. There also is no question that blood levels can be used to set acceptable exposure levels for industrial chemicals. However, they will show individual variability already seen in the medical profession for naturally occurring blood chemicals and medicinals. The causes of the expected variability already have been discussed in Section 6 and 7. Because of worker objections to the technique and a slight risk to the worker, blood analyses probably will not be used if urine analyses or exhaled air analyses can provide equivalent reliability in measurements of worker exposure.

10 HAIR ANALYSIS TO MEASURE INDUSTRIAL EXPOSURE

It has been known for many years that hair contains metals. As discussed earlier, it may be considered an excretory mechanism for these metals, since the metals appear to have no functional role in the hair protein and, for at least a few of the metals, their content in hair seems to vary with exposure to the metal (156–159).

Some of the metals reported in hair to date are aluminum, arsenic, beryllium, cadmium, calcium, copper, iron, mercury, lead, manganese, molybdenum, potassium, selenium, silicon, thallium, titanium, and zinc (146, 156, 159–169).

In addition, the nonmetals, chlorine and phosphorus, also have been reported (159).

Historically, hair has not been considered suitable for measurement of exposure to organic compounds. Since hair is protein with a small amount of colorant, it was not believed capable of carrying organic compounds from systemic circulation. In 1974 Harrison et al. (170) reported the presence of ^{14}C in the hair of guinea pigs that had been dosed intraperitoneally with ^{14}C–labeled amphetamine. Since then, heroin and morphine metabolites have been reported in human scalp hair of admitted users of these drugs (171). The amount and location of the drug along the hair shaft correlated directly with the use history. The analytical procedure used was radioimmune assay. Analyses were confirmed by urine analysis. Using the same methodology, these authors later reported the presence of phencyclidine (172) and cocaine (173) in scalp hair of admitted users. Similar correlations were found. Again, radioimmune assay was used, and drug use was confirmed by urine analysis.

The presence of organomercurials in the scalp hair of humans after eating fish from lakes contaminated with organomercurials and inorganic mercury also has been reported (174).

Hair is not suitable as a dynamic system for evaluating immediate past exposure to chemicals since the individual hair shafts do not have access throughout their length to any fluid transport system. Thus, the xenobiotic content reflects that available at the time any particular portion of the shaft was being synthesized. Since hair grows at the rate of about 1 cm in 30 days, clippings would be expected to reflect a historical exposure at best. How many months in history would be indicated by the length of the clipping and the length of the remaining proximal hair shaft, in centimeters, as already mentioned.

However, if the range of index chemical content of hair normally is sufficiently narrow in a population and the hair concentration varies regularly with blood concentration, suitably timed postexposure analyses could be used to confirm or refute suspected overexposure or continuing acceptable exposure. Unfortunately, as with most evolving areas of science, there are conflicting reports in the literature on the utility of hair analyses. Also, in addition to some of the already discussed factors that can lead to aberrant exposure estimates based on urinary concentration of chemicals, hair analyses have unique factors that may lead to aberrant conclusions. These must be considered in development and application of any procedure for correlating hair levels with exposure levels.

One of these additional unique problems of hair analysis is that of suitable cleansing of the hair to remove adsorbed contaminant prior to analysis. Hair normally develops an oily coating, and this oil can trap exogenous metal or organic material. Since this adsorbed material does not represent body burden, it must be removed prior to analysis.

Several preanalysis hair cleaning procedures have been reported. The simplest is washing with detergent followed by distilled deionized water rinses and oven drying (157, 161, 171, 175). Additional washings have included acetone (163, 164), methanol (171), nitric acid (156, 161), ether (160, 163), and trisodium ethylenediamine tetraacetic acid (161). Most investigators assume that detergent washes, followed by (a) distilled water washes to remove detergent and (b) organic solvent washes, remove all adsorbed materials. In some cases, the initial wash has been with the organic solvent (178). Nishiyama and Nordberg (161) found that washing procedures that removed all adsorbed cadmium also removed the endogenously derived cadmium in the hair. Experimental results similar to those of Nishiyama have been repored for hair analyses of other metals (165).

Petering et al. (163) suggested that ionic detergents were more appropriate than nonionic ones for washing metals from hair because the former could complex the metal ion or form salts with it, thus aiding its removal from the exterior of the hair. Obviously, if the detergent charge could be a factor in metal ion removal, anionic detergents would be more suitable than cationic detergents, unless the metal is present as a negatively charged complex or radical.

Renshaw et al. (160) reported that the concentration of lead in single hairs from one woman increased with the distance from the scalp. Since they had cleansed the hair by diethyl ether reflux in a Soxhlet extractor prior to analysis, they assumed that all adsorbed external lead had been removed, and this distal increase represented lead that had deposited on the hair from external sources and then diffused into the body of the hairs. No details of the work history, residence, or cosmetics use of the woman were given. The increased lead content at the distal portion also could be surface lead that had not been completely removed by the ether wash. Certainly, the mass of the evidence suggests that hair concentrations of metals, and apparently organic compounds, do bear a relationship to body burden (156, 167, 179) if the hair is adequately cleansed. The uncertainty regarding contamination by airborne material presumably could be removed by using body, pubic, or axillary hair instead of scalp hair. However, hair from these sites may grow at different rates from scalp or vertex hair. Axillary hair grows at about two-thirds the rate of scalp hair and pubic hair at about one-half the rate. Body hair grows at approximately the same rate (169). Rate differences obviously would affect time correlations.

The same factors discussed in Sections 6 and 7 can control the appearance of the index chemical in hair. For example, many of the metals are transported in blood predominantly in the bound form or are stored in the body. Anything that caused their release, with a resulting increase of the free metal in liquid transport systems accessing the hair root, could result in a short or extended shaft section with elevated concentration. This could be interpreted as a short- or long-term overexposure. Medicinals or industrial chemicals that were more tightly bound to transport or storage proteins could replace, and thus cause the release of, bound metals. This would be followed by increased hair uptake of the released metal ion. Wasting diseases that liberate stored metals could have a similar effect.

Therefore, interpretation of isolated hair analyses for judgments against workplace exposure should be coupled with a careful and complete medical history to assure that elevated local concentrations along the hair shaft due to nonwork causes are not attributed to work exposure.

Diet also can be expected to affect the level of metals found in hair. Green plants are notorious scavengers of metals from the ground in which they are grown. Thus, hair levels of metals could be due not only to the direct ingestion of fruits, vegetables, and cereals but also to the ingestion of meat from animals grazed or fed hay or cereal grains. Conversely, high-fiber diets apparently lower uptake of metals from the intestine (178) and would be expected to eventually result in lowered metal content of hair from nonindustrial exposure.

Cosmetics and hair dyes also may contain various metal salts or complexes and may contribute to hair levels of metals either through (a) absorption of the metal salt or complex followed by uptake by the hair root, or (b) by adsorption on the hair shaft. Various salts or complexes of metal ions are permitted in the coloring agents of hair dyes, dye formulations, or cosmetics. These include aluminum, arsenic, barium, chromium (+3), cobalt, copper, iron, lead, mercury, titanium, and zinc (179).

Thus, many of the metals found in hair also appear in cosmetics or hair dyes.

Obviously, the cosmetic and hair-dye use history of the worker also must be determined before attempts are made to correlate metal content of a worker's hair with workplace exposure to a metal.

In order to use metal content of hair as an index of occupational exposure, sample preparation procedures (e.g., washing) and the analytical method must be validated. In addition, a normal baseline must be established. From the medical experience with urinalyses and the industrial experience with urinalyses mentioned earlier, this would be expected to be a range, rather than a line. Variations with sex also might be expected. The literature data shown in Tables 3.10 and 3.11 indicate this to be the case. For some metals, there also were variations due to age. Therefore, before judging isolated hair levels of metals with regard to occupational exposure, one must judge the hair analyses in these cases against not only medical and dietary background but also cosmetic and hair-dye use and against control ranges for sex and age or prior analyses of the same employee's hair.

As can be seen from Table 3.10, in general, women with no known industrial exposure have higher hair levels of lead than do their male counterparts in the studies. The exception was the group of middle-class urban white females from Cincinnati, Ohio, studied by Petering et al. (163). Petering also found that the lead level in male hair decreased with age from 2 to 88 years. In his study, the lead level increased rapidly in women from age 14 to 30 and then decreased rapidly from age 30 to 84. For both males and females, several values of about 35 to 40 µg/g were observed. Other studies attempted to correlate lead content of hair with age. This would appear from the data of Petering et al. to be necessary although the slope of the line for men was not great and the upper

Table 3.10 Hair Lead Concentrations in Control Populations[a]

Arithmetic Mean Value (μg/g) and [Range]			Significantly Different?	Analytical Method	Reference
Male	Female	Sex Unknown			
14.7 (A)	19.2 (A)		Yes; $p < .001$	Atomic absorption	175, 180
9.9 (A)	14.6 (A)		ND	Unknown	167
17.8 (A)	19.0 (A)		No; $p > .05$	Atomic absorption	166
		12 (U), [4–25]	—	Atomic absorption	160
6.1 (T)		9.4 (U), [3–26]	—	Dithizone (149)	156
24.5 (AA), [1.1–52.1]	34.6 (AA), [8.7–78.7]		Yes; $p < .001$	Atomic absorption	158
22 (P)	—		ND	Atomic absorption	176
17 (T)	6 (T)			Atomic absorption	164
14 (W)	11 (W)				
11 (R)	10 (R)				
		14.5 (P)	—	Atomic absorption	157
4.1 (W)			—	Unknown	169

[a] Symbols are as follows: A = adult; AA = all ages, urban and rural; ND = not determined; P = preschool; R = >60 years; T = 6–20 years; U = age unknown; W = 20–60 years.

Table 3.11 Hair Content of Various Metals in Control Populations[a]

Metal	Range of Concentrations (μg/g)			Significantly Different?	Analytical Method	Reference
	Male	Female	Sex Unknown			
Aluminum	1.6–7.8 (T)		2–9 (AA)	ND	Neutron activation	165
	1.2–9.2 (T,P)			—	Unknown	169
	4.4–5.5 (W)					
Antimony	0.1–1.4 (T)		0.5–4 (AA)	ND	Neutron activation	165
	0–4.4 (T)			—	Unknown	169
	0.07–0.2 (W)					
Arsenic	0.4–7.9 (T)		0.18 (P)	—	Atomic absorption	157
	0.7–5.3 (T)			—	Neutron activation	165
Cadmium	0.08 (mean; A)	0.15 (mean; A)		Yes	Neutron activation	181
	1–2 (P)			ND	Atomic absorption	164
	2 (T)	1–1.3 (T)				
	1.5–2 (W, R)	1.3–2 (W)				
		2–1.5 (R)				
			30–530 (AA)	—	Atomic absorption	175
			1.06 (P)	—	Atomic absorption	157
			>1000 (U)	—	Unknown	161
	2.76 (AA)	1.77 (AA)		•No ($p > .05$)	Atomic absorption	166
	0.47 (W)			—	Unknown	169

Chromium	0.46 (mean; A)	0.34 (mean; A)	No	Neutron activation	181
Copper	.13–30 (P)	20–25 (AA)	ND	Atomic absorption	163
	30 (T)				
	30–15 (W)				
	15–10 (R)				
	7–93 (T)	7.8–234 (AA)	—	Neutron activation	165
	8–150 (T)				
	16.1 (AA)	55.6 (AA)	$p < .001$	Atomic absorption	166
	15–17 (W)		—	Unknown	169
	15.7 (mean; A)	31 (mean; A)	Yes	Neutron activation	181
Mercury	0.3–34 (T)	0.1–33 (AA)	—	Neutron activation	165
	0.5–53 (T)				
	1.7–1.9 (W)		—	Unknown	169
Selenium	1.30 (mean; A)	1.10 (mean; A)	No	Neutron activation	181
Zinc	100–110 (P)	200 (P)	No	Atomic absorption	163
	110–140 (T)	200–180 (T)	No		
	140–125 (W, R)	180–150 (W)			
	101–186 (T)	51–602 (AA)	—	Neutron activation	165
	85–166 (T)				
	167 (AA)	172 (AA)	No ($p > .05$)	Atomic absorption	166
	150–190 (W)		—	Unknown	169
	140 (mean; A)	160 (mean; A)	Yes	Neutron activation	181

[a] Symbols are as follows: A = adult; AA = all ages, urban and rural; ND = not determined; P = preschool; R = >60 years; T = 6–20 years; U = age unknown; W = 20–60 years.

95 percent fiducial limit at 60 years was within the fiducial limits at 20 years. The changes with age were dramatic for the women in Petering's study, and age matching for women of working ages definitely would appear to be necessary. If individual historical controls were used, allowance would have to be made for the changes that accompany age.

Klevay (176), however, reported a relationship between age and hair lead content for only males in Panama. In his population, this relationship appeared to reach a nadir between 11 to 20 years. He reported ranges of 1.1 to 51.2 µg of lead per gram of hair for males and 8.7 to 78.7 µg of lead per gram of hair for females.

It also is obvious that the normal control adult average lead concentration varies by a factor of about 2 for males and about 1.5 for females. Furthermore, values varying by factors of 10 to 50 between control individuals have been reported (175).

Suzuki et al. (167) suggested that 30 µg of lead per gram of hair be considered the upper normal level. El-Dakhakhny and El-Sadik (156), on the basis of correlation of hair lead levels and clinical signs and/or symptoms, suggested that hair lead content greater than 30 µg/g be considered indicative of excessive lead exposure. They found that 30 µg of lead per gram of hair corresponded to approximately 90 µg of lead per 100 g of blood.

However, the blood and hair analyses were carried out concurrently, and no mention was made of the residual hair length or the length of the hair sample. Therefore, unless the work exposure to lead had not changed over the number of months represented by the distance of the hair sample from the scalp in centimenters, the comparison is not valid.

Klevay's data (176) indicate that 30 µg/g may be too low, at least for women. He proposes an upper normal value of 35 to 40 µg/g, and some of his hair samples from presumably unexposed persons even exceeded this value, as mentioned previously. However, some of these high values came from urban, nonindustrial areas and may reflect unique urban exposures. In Klevay's study, values greater than 35 µg/g were seen in male hair only from subjects less than 10 years of age. They were seen in female subjects of all ages. Petering et al. (164) also reported urban male and female values of approximatly 40 µg/g from subjects with no apparent industrial exposures. Neither Petering nor Klevay reported any clinical signs or symptoms of plumbism in their subjects, but there is no indication that they looked for them. Thus, the data suggest that even 40 µg/g may not be the upper normal level for lead content in hair.

The data shown in Table 3.11 show age and sex variations reported for normal hair content of several metals, including cadmium, copper, and zinc. In some cases, the variations appear to be significant and in others, not. For instance, the data of Petering et al. (163) show an age-related decrease in zinc content for both males and females, but the slopes of the lines are so shallow and the equations of the lines so similar that very little allowance need be made for age or sex. Copper content of hair peaked at about 10 years for males in the study of Petering et al. but showed only a gradual increase with age for females. The 95 percent fiducial limits for the two sexes overlapped for about 20 years. Thus, their data suggest that hair analyses for copper need to be evaluated against controls matched for age and sex.

Arunachalam et al. (181) reported significant sex differences for arsenic and copper but not for chromium and selenium.

The available data suggest that because of the variability of individual hair content of metals, it would be difficult to set trigger numbers or action numbers for most metals that, taken by themselves, would indicate excessive exposure. Also, since appearance of the metal in the external hair shaft would follow exposure by possibly a month or so, such analyses would not provide as early warning of overexposure as would urine analyses.

Because of the individual variations reported, hair analyses would be of most value if individual histories of metal content could be developed over a period of years. As with urine and exhaled air analyses, comparisons between populations is difficult, particularly if different washing procedures and analytical techniques are used.

Also, as with other biological assays, high metal content of the hair can come from nonworkplace sources or may not represent recent past workplace exposure for the other reasons discussed. Obviously in these cases workplace exposure may be trivial, and attempts to reduce exposure will not reduce the worker's hazard. If an undesirably high level of metal in the hair is found, the source should be conscientiously sought so that possible hazard to the worker may be reduced.

11 MILK ANALYSIS FOR MEASUREMENT OF INDUSTRIAL EXPOSURE

Historically, human breast milk analysis has not been used to monitor industrial exposure to chemicals. It has been realized for over 90 years that milk could be used by the body as an excretory mechanism and that even a mouse mammary cancer virus could be transmitted through the mother mouse's milk. However, it is only within the past two decades that its importance as an excretory mechanism has been appreciated. Because of the small number of lactating women in the workplace, milk analysis has very restricted utility for monitoring workplace exposure.

Nursing infants could have body weights as small as one-twentieth the body weight of the mother. Thus, milk levels and corresponding blood or tissue levels that would not be an effect level for the mother could be an effect level, or could accumulate to an effect level, in the infant. In addition, the neonatal enzyme system for metabolizing the zenobiotic could be incomplete, leading to toxic accumulations.

These considerations, rather than industrial exposure monitoring, have been principally responsible for the recent interest in monitoring mother's milk. It is unlikely that such analyses will achieve much use for industrial monitoring, but a nursing industrial employee could seek assurance that she did not have a possible effect level of workplace chemicals in her milk.

Since the concern has been with principally nonindustrial exposure, very few studies have related human milk levels of industrial chemicals to blood or plasma levels or to storage levels.

Pesticide levels reported in mother's milk have been summarized (183). Among the pesticides reported in mother's milk are β-benzene hexachloride, "benzenehexachloride", DDT and its metabolites, dieldrin, heptachlor epoxide, "hexachlorobenzene", and polychlorinated biphenyls (PCBs). Strassman and Kutz (184) also reported finding oxychlordane and *trans*-nonachlor, metabolites of chlordane and heptachlor, in mother's milk. Curley et al. (185) also have reported finding α-benzene hexachloride, endrin, aldrin, and mirex in mother's milk. Kepone also has been reported in mother's milk (186).

Mercury and lead (187), molybdenum (168), and iron (188) also have been reported in mother's milk.

Many medicinals also are excreted in mother's milk. These have been reviewed in several papers (186, 189–192). Medicinals reported in mother's milk include barbiturates, sulfonamides, some hormones, oral contraceptives, lithium salts, narcotics, ergotamine, some hypoglycaemic agents, acetylsalicylic acid (aspirin), antibiotics, opiates, caffeine, sulfanilamide, penicillin, and erythromycin.

Subsequent to exposure to styrene, alkyl benzenes, toluene, or trichloroethylene, their metabolites have been found in mother's milk (189). Halothane has been reported in the milk of a nursing anesthesiologist (193).

Although few analyses for industrial chemicals have been carried out on mother's milk following industrial exposure, their appearance should be expected since the factors governing the appearance of pesticides or medicinals also would operate for industrial chemicals. Although few studies have been carried out to elucidate the factors, there appear to be many. Thus, a simple spot analysis of an individual mother's milk may be indicative of a range of possible plasma concentrations rather than a specific concentration.

Because milk has an aqueous and a lipid phase, concentrations of excreted chemicals might be expected to be related to the unbound blood concentration of the chemical (189) and also to levels stored in fatty depots. This is in contrast to urine, saliva, perspiration, and tears that have no, or a much smaller, lipid component. The pH of mother's milk is slightly lower (ca, 7.0) than that of plasma (ca. 7.4). Therefore, weak, unbound bases might be expected to partition preferentially to milk. PCBs appear to partition into mother's milk (183), as do polybrominated biphenyls (PBBs) (194). The milk levels of the latter were reported to be 100 times the maternal serum levels, which would lead to a very unusual standard curve in a biological assay for PBBs with the use of mother's milk.

Other factors that appear to affect the concentration of xenobiotics in mother's milk appear to include age, weight, number of previous pregnancies (183), sampling time of day, time since last sample, and fat content of the milk

(195). These latter authors also reported that the fat content was higher in mother's milk at the end of a nursing period than at the beginning. Thus, the level of fat-soluble materials in the milk would be expected to be higher at the end of a nursing period than at the beginning.

The factors discussed in Section 6 also would affect the level of a xenobiotic or an index chemical in mother's milk (196). A comprehensive review of mechanisms, factors, and kinetics of excretion of xenobiotics in mother's milk has recently been published (197).

Impairment of an alternate excretory mechanism also would be expected to increase milk levels of the index chemical. Conversely, lactation could be expected to alter the usual urinary ratios and concentrations for a nonbound, fat-soluble xenobiotic and its metabolites. This would lead to erroneous calculated exposure levels of the xenobiotic in the workplace if only the urine were analyzed and the fact of lactation were ignored. Furthermore, the resulting error would be underestimation of workplace exposure. Similar considerations would hold for analyses of sweat and tears from lactating women.

Increased photo period has been shown to increase milk output 15 percent in Holstein cattle (198). The effect on the concentrations of chemical components was not studied. It is not known whether photoperiod would affect human milk output or the concentrations of xenobiotics and metabolites. In any case, it is likely that the fiducial limits of the normal value would be considerably greater than ±15 percent, thus, any effect would be insignificant in determining whether the measured concentration reflected a safe or unsafe exposure level for the mother.

Xenobiotics and their metabolites would be expected to appear in milk at least as rapidly as in urine. Thus, milk analysis would have an advantage over hair and nail analyses in that it would give a measure of current or immediate past exposures rather than historical exposures.

In summary, mother's milk is a possible excretory fluid for assessment of workplace exposure to chemicals. However, because of the limited occurrence of lactating women in the workplace, development of correlations with plasma levels and toxic effect levels in the mother is difficult to justify in preference to similar analytical development for urine and exhaled air.

12 SALIVA, TEARS, PERSPIRATION, AND NAIL ANALYSES AS MEASURES OF INDUSTRIAL EXPOSURE

Saliva, tears, and perspiration, although not usually considered excretory fluids, do contain chemicals transferred from blood. Xenobiotics can partition between these fluids and blood. Saliva and perspiration levels of chemicals seem to reflect unbound concentrations of the chemicals in blood (199–201). It is likely that tear levels also will reflect blood concentrations.

Very little work has been reported on the use of these excretion fluids for biological monitoring. It is probable that all the factors affecting xenobiotic levels discussed in Sections 6 and 7 will apply to these fluids. There are obvious problems collecting samples of tears and perspiration.

Saliva analysis has demonstrated some utility for following blood levels of drugs such as the antiepileptic agent phenytoin (199, 200).

Metals or metallic salts reported in saliva following industrial exposure include cadmium (202, 203) and mercury (204). Like PBB in mother's milk discussed previously, cadmium concentrations were higher in saliva than in plasma (203).

Saliva analyses might overcome one of the problems of urine analysis: variable dilution of the xenobiotic or its metabolites by urine already in the bladder. Samples could be collected easily at the end of the workday and could provide a viable alternative to urine or blood analyses for nonvolatile workplace chemicals.

Salicylic acid and urea are among the industrially important organic chemicals reported in saliva and tears, respectively (188).

Some of the chemicals of industrial importance that have been reported in perspiration include lead (200), arsenic, mercury, iron, manganese, zinc, magnesium, copper, ethanol, benzoic acid, salicylic acid, urea, and phenol (188). Arsenic appears in perspiration very quickly after administration (169).

Although metals have been detected in human fingernails or toenails (169, 205, 206), analysis of human nails has not been used as a technique for monitoring industrial exposure. Some of the metals reported in human nails are copper (206) and zinc (205). It is very likely that the same considerations discussed above relative to hair analysis will apply to nail analysis, and they are not discussed further here.

In relating nail levels of a metal to previous exposure, one should note that human fingernails grow approximately 100 μm/day and that human toenails grow only approximately 25 μm/day (169).

REFERENCES

1. H. P. Elkins, *AMA Arch. Ind. Hyg. Occup. Med.*, **9**, 212 (1954).
2. J. K. Piotrowski, *Exposure Tests for Organic Compounds in Industrial Toxicology*, National Technical Information Service Publication PB-274 767, 1977.
3. R. R. Lauwerys, *Industrial Chemical Exposure: Guidelines for Biological Monitoring*, Biomedical, Davis, CA, 1983.
4. R. C. Baselt, *Biological Monitoring Methods for Industrial Chemicals*, Biomedical, Davis, CA, 1983.
5. E. S. West and W. R. Todd, *Textbook of Biochemistry*, 4th ed., Macmillan, New York, 1966.
6. J. B. Neilands and P. K. Stumpf, *Outlines of Enzyme Chemistry*, 2nd ed., Wiley, New York, 1958, pp. 379–381.
7. M. Gibaldi and D. Ferrier, *Pharmacokinetics*, 1st ed., Marcel Dekker, New York, 1975, pp. 284–287.
8. B. N. LaDu, H. G. Mandel, and E. L. May, *Fundamentals of Drug Metabolism and Drug Disposition*, 1st ed., Williams and Wilkins, Baltimore, 1971.
9. J. Wallach, *Interpretation of Diagnostic Tests*, 2nd ed., Little, Brown and Co., Boston, 1974.

10. American Conference of Governmental Industrial Hygienists, *TLV's. Threshold Limit Values for Chemical Substances and Physical Agents in the Workroom Environment and Intended Changes for 1983*, Cincinnati, OH, 1983.
11. W. A. Eldridge, *Report 29, Chemical Warfare Service*, (1924), through B. R. Allen, M. R. Moore, and J. A. A. Hunter, *Br. J. Dermatol.*, **92**, 715 (1975).
12. C. P. McCord, *Am. J. Pub. Health*, **24**, 677 (1934).
13. W. P. Yant and H. H. Schrenk, *J. Ind. Hyg. Toxicol.*, **19**, 337 (1937).
14. D. L. Hill, *AMA Arch. Ind. Hyg. Occup. Med.*, **8**, 347 (1953).
15. R. D. Stewart and H. C. Dodd, *Ind. Hyg. J.*, **25**, 439 (1964).
16. T. Dutkiewicz and H. Tyras, *Br. J. Ind. Med.*, **24**, 330 (1967).
17. T. Dutkiewicz and H. Tyras, *Br. J. Ind. Med.*, **25**, 243 (1968).
18. J. K. Piotrowski, *Br. J. Ind. Med.*, **28**, 172 (1971).
19. C. P. McCord, *Ind. Eng. Chem.*, **23**, 931 (1931).
20. O. Tada, K. Nakaaki, S. Fukabori, and J. Yonemoto, *J. Sci. Labour Part 2 (Rodo Kagaku)*, **51**, 143 (1975); through *Chem. Abstr.*, **83**, 54178w (1973).
21. E. M. P. Widmark, *Biochem. Z.*, **259**, 285 (1933).
22. M. Neymark, *Skand. Arch. Physiol.*, **73**, 227 (1936); through *Chem. Abstr.*, **30**, 4930^2 (1936).
23. H. H. Schrenk, W. P. Yant, S. J. Pearce, F. A. Patty, and R. R. Sayers, *J. Industr. Hyg. Toxicol.*, **23**, 20 (1941).
24. W. F. von Oettingen, P. A. Neal, and D. D. Donahue, *J. Am. Med. Assoc.*, **118**, 579 (1942).
25. R. R. Sayers, W. P. Yant, H. H. Schrenk, J. Chornyak, S. J. Pearce, F. A. Patty, and J. G. Linn, "Methanol Poisoning: I. Esposure of Dogs to 450–500 ppm. Methanol Vapor in Air," R. I. 3617, U.S. Department of the Interior, Bureau of Mines, 1942; through *Chem. Abstr.*, **36**, 4596^2 (1942).
26. H. P. Elkins, *Pure Appl. Chem.*, **3**, 269 (1961).
27. H. P. Elkins, *J. Am. Ind. Hyg. Assoc.*, **26**, 456 (1965).
28. H. P. Elkins, *J. Am. Ind. Hyg. Assoc.*, **28**, 305 (1967).
29. T. Dutkiewicz, *Bromatol. Chem. Toksykol.*, **4**, 39 (1971); through *Chem. Abstr.*, **76**, 10786u (1972).
30. M. Ikeda and T. Imamura, *Int. Arch. Arbeitsmed.*, **31**, 209 (1973).
31. H. H. Cornish, *Crit. Rev. Toxicol.*, **1**, 1 (1971).
32. K. R. Long, *Int. Arch. Occup. Environ. Health*, **36**, 75 (1975); through *Chem. Abstr.*, **85**, 67416p (1976).
33. R. S. Waritz, J. G. Aftosmis, R. Culik, O. L. Dashiell, M. M. Faunce, F. D. Griffith, C. S. Hornberger, K. P. Lee, H. Sherman, and F. O. Tayfun, *J. Am. Ind. Hyg. Assoc.*, **38**, 307 (1977).
34. R. T. Williams, *Detoxication Mechanisms*, 2nd ed., Wiley, New York, 1959.
35. D. V. Parke, *The Biochemistry of Foreign Compounds*, 1st ed., Pergamon Press, New York, 1968, p. 146.
36. L. Perbellini, F. Brugnone, and I. Pavan, *Toxicol. Appl. Pharmacol.*, **53**, 220 (1980).
37. K. P. McConnell and O. W. Portman, *J. Biol. Chem.*, **195**, 277 (1952).
38. O. M. Bakke and R. R. Scheline, *Toxicol. Appl. Pharmacol.*, **16**, 691 (1970).
39. C. P. Carpenter, C. B. Shaffer, C. S. Weil, and H. F. Smyth, Jr., *J. Ind. Hyg. Toxicol.*, **26**, 69, (1944).
40. I. Danishefsky and M. Willhite, *J. Biol. Chem.*, **211**, 549 (1954).
41. A. M. El Masri, J. N. Smith, and R. T. Williams, *Biochem. J.*, **68**, 199 (1958).
42. R. D. Stewart, H. C. Dodd, E. D. Baretta, and E. D. Schaffer, *Arch. Environ. Health*, **16**, 656 (1968)

43. H. Ohtsuji and M. Ikeda, *Toxicol. Appl. Pharmacol.*, **18,** 321 (1971).
44. A. Slob, *Br. J. Ind. Med.*, **30,** 390 (1973).
45. P. Pfäffli, A. Hesso, H. Vainio, and M. Hyvönen, *Toxicol. Appl. Pharmacol.*, **60,** 85 (1981).
46. T. C. Butler, *J. Pharmacol. Exp. Ther.*, **97,** 84 (1949).
47. J. R. Cooper and P. J. Friedman, *Biochem. Pharmacol.*, **1,** 76 (1958).
48. B. Souček and D. Vlachová, *Br. J. Ind. Med.*, **17,** 60 (1960).
49. K. H. Byington and K. C. Liebman, *Molec. Pharmacol.*, **1,** 247 (1965).
50. W. J. Cole, R. G. Mitchell, and R. F. Salamonsen, *J. Pharm. Pharmacol.*, **27,** 167 (1975).
51. S. Yllner, *Nature*, **191,** 820 (1961).
52. J. W. Daniel, *Biochem. Pharmacol.*, **12,** 795 (1963).
53. J. W. Porteous and R. T. Williams, *Biochem. J.*, **44,** 46 (1949).
54. H. G. Bray, B. G. Humphris, and W. V. Thorpe, *Biochem. J.*, **45,** 241 (1949).
55. D. Robinson, J. N. Smith, and R. T. Williams, *Biochem. J.*, **59,** 153 (1955).
56. W. Seńczuk and B. Litewka, *Br. J. Ind. Med.*, **33,** 100 (1976).
57. F. A. Hommes, Ed., *Inborn Errors of Metabolism*, Academic Press, New York, 1973.
58. D. A. P. Evans and T. A. White, *J. Lab. Clin. Med.*, **63,** 394 (1964).
59. J. W. Jenne, *J. Clin. Invest.*, **44,** 1992 (1965).
60. H. A. Greim, in: P. Jenner and B. Testa, Eds., *Drugs and Pharmaceutical Sciences*, Vol. 10, *Concepts in Drug Metabolism*, Part B, Marcel Dekker, New York, 1981, p. 270.
61. W. Kalow, *Trends Pharmacol. Sci.*, **1,** 403 (1980).
62. T. P. Sloan, A. Mahgoub, R. Lancaster, J. R. Idle, and R. L. Smith, *Br. Med. J.*, **2,** 655 (1978).
63. I. Kitchen, J. Tremblay, J. Andres, L. G. Dring, J. R. Idle, R. L. Smith, and R. T. Williams, *Xenobiotica*, **9,** 397 (1979).
64. E. S. Vessell and G. T. Passananti, *Clin. Chem.*, **17,** 851 (1971).
65. T. Tang and M. A. Friedman, *Mut. Res.*, **46,** 387 (1977).
66. S. L. Winter and J. L. Boyer, *N. Engl. J. Med.*, **289,** 1180 (1973).
67. P. M. Newberne, *Lab. Animal*, **4,** (7), 20 (1975).
68. A. P. Alvares, K. E. Anderson, A. H. Conney, and A. H. Kappas, *Proc. Natl. Acad. Sci. U.S.A.*, **73,** 2501 (1976).
69. Y. E. Harrison and W. L. West, *Biochem. Pharmacol.*, **20,** 2105 (1971).
70. A. H. Conney, E. J. Pantuck, K. C. Hsiao, R. Kuntzman, A. P. Alvares, and A. Kappas, *Fed. Proc., Fed. Am. Soc. Exp. Biol.*, **36,** 1647 (1977).
71. W. D. Loub, L. W. Wattenberg, and D. W. Davis, *J. Natl. Cancer Inst.*, **54,** 985 (1975).
72. S. Yaffe, B. Sonawane, H. Lau, P. Coates, and O. Koldovsky, *Fed. Proc., Fed. Am. Soc. Exp. Biol.*, **39,** (3), 751 (1980).
73. A. H. Conney, *Pharmacol. Rev.*, **19,** 317 (1967).
74. R. J. Gillette, *Ann. NY Acad. Sci.*, **179,** 43 (1971).
75. M. Ikeda, H. Ohtsuji, and T. Imamura, *Xenobiotica*, **2,** 101 (1972).
76. W. E. Knox and A. H. Mehler, *Science*, **113,** 237 (1951).
77. M. Ikeda, H. Ohtsuji, H. Kawai, and M. Kuniyoshi, *Br. J. Ind. Med.*, **28,** 203 (1971).
78. M. M. Reidenburg, *Med. Clin. North Am.*, **58,** 1103 (1974).
79. A. J. Quick, *J. Biol. Chem.*, **92,** 65 (1931).
80. S. Tanaka and M. Ikeda, *Br. J. Ind. Med.*, **25,** 214 (1968).
81. M. Ikeda, H. Ohtsuji, T. Imamura, and Y. Komoike, *Br. J. Ind. Med.*, **29,** 328 (1972).
82. M. Ikeda, T. Imamura, M. Hayashi, T. Tabuchi, and I. Hara, *Int. Arch. Arbeitsmed.*, **32,** 93 (1974).

83. P. Götell, O. Axelson, and B. Lindelof, *Work Environ. Health* **9,** 76 (1972).
84. P. G. Watanabe, G. R. McGowan, and P. J. Gehring, *Toxicol. Appl. Pharmacol.*, **36,** 339 (1976).
85. G. Leaf and L. J. Zatman, *Br. J. Ind. Med.*, **9,** 19 (1952).
86. T. R. Tephly, *Fed. Proc., Fed. Am. Soc. Exp. Biol.*, **36,** 1627 (1977).
87. R. D. Stewart, E. D. Baretta, H. C. Dodd, and T. R. Torkelson, *Arch. Environ. Health*, **20,** 224 (1970).
88. D. R. Abernethy, D. J. Greenblatt, M. Divoll, J. S. Harmatz, and R. I. Shader, *J. Pharmacol. Exp. Ther.*, **217,** 681 (1981).
89. E. J. Tiggs, R. L. Nation, A. Long, and J. J. Ashley, *Eur. J. Clin. Pharmacol.*, **8,** 55 (1975).
90. J. W. Rowe, R. Andres, J. D. Tobin, A. H Norris, and N. W. Shock, *Ann. Intern. Med.*, **84,** 567 (1976).
91. M. C. Geokas and B. J. Haverback, *Am. J. Surg.*, **117,** 881 (1969).
92. D. R. Abernethy, D. J. Greenblatt, M. Divoll, R. Arendt, H. R. Ochs, and R. I. Schrader, *N. Engl. J. Med.*, **306,** 791 (1982).
93. J. Feely, G. R. Wilkinson, and A. J. J. Wood, *N. Engl. J. Med.*, **304,** 692 (1981).
94. W. J. Hayes, Jr., *Clinical Handbook on Economic Poisons*, Government Printing Office, Washington, DC, 1963, p. 13.
95. J. W. Daniel and J. C. Gage, *Food Cosmet. Toxicol.*, **3,** 405 (1965).
96. K. Nomiyama and H. Nomiyama, *Int. Arch. Arbeitsmed.*, **28,** 37 (1971).
97. K. Nomiyama, *Int. Arch. Arbeitsmed.*, **27,** 281 (1971).
98. K. Nomiyama and H. Nomiyama, *Ind. Health (Kawasaki, Jap.)* **7,** 86 (1969); through *Chem. Abstr.*, **72,** 82685a (1970).
99. M. Ikeda and H. Ohtsuji, *Br. J. Ind. Med.*, **26,** 162 (1969).
100. M. Dean, B. Stock, R. J. Patterson, and G. Levy, *Clin. Pharmacol. Ther.*, **28,** 253 (1980).
101. H. Lick, *Med. Clin. North. Am.*, **58,** 1143 (1974).
102. P. V. Desmond, R. K. Roberts, G. R. Wilkinson, and S. Schenker, *N. Engl. J. Med.*, **300,** 199 (1979).
103. J. M. Walshe, *Am. J. Med.*, **21,** 487 (1956).
104. G. A. Nicolson, A. C. Greiner, W. J. G. McFarlane, and R. A. Baker, *Lancet*, 344 (February 12, 1966).
105. G. MacLeod, H. Rabin, R. Ogilvie, J. Ruedy, M. Caron, D. Zarowny, and R. O. Davies, *Can. Med. Assoc. J.*, **111,** 341 (1974).
106. M. Ikeda and H. Ohtsuji, *Br. J. Ind. Med.*, **26,** 244 (1969).
107. M. Ogata, K. Tomokuni, and Y. Takatsuka, *Br. J. Ind. Med.*, **27,** 43 (1970).
108. L. Levine and J. P. Fahy, *J. Ind. Hyg. Toxicol.*, **27,** 217 (1945).
109. K. Engström, K. Husman, and J. Rantanen, *Int. Arch. Occup. Environ. Health*, **36,** 153 (1976).
110. A. B. Van Haaften and S. T. Sie, *J. Am. Ind. Hyg. Assoc.*, **26,** 52 (1965).
111. S. G. Rainsford and T. A. L. Davies, *Br. J. Ind. Med.*, **22,** 21 (1965).
112. W. A. Fishbeck, R. R. Langner, and R. J. Kociba, *J. Am. Ind. Hyg. Assoc.*, **36,** 820 (1975).
113. U.S. Dept. Health, Education and Welfare, *Criteria for a Recommended Standard ... Occupational Exposure to Benzene*, Cincinnati, OH, 1974, p. 112.
114. Y. Seki, Y. Urashima, H. Aikawa, H. Matsumura, Y. Ichikawa, F. Hiratsuka, Y. Yoshioka, S. Shimbo, and M. Ikeda, *Int. Arch. Arbeitsmed.*, **34,** 39 (1975).
115. L. D. Pagnotto and L. M. Lieberman, *J. Am. Ind. Hyg. Assoc.*, **28,** 129 (1967).
116. R. W. Ellis, *Br. J. Ind. Med.*, **23,** 263 (1963).
117. M. Ogata, K. Sugiyama, and H. Moriyasu, *Acta Med. Okayama*, **16,** 283 (1962): through References 84 and 94.

118. W. V. Lorimer, R. Lilis, W. J. Nicholson, H. Anderson, A. Fischbein, S. Daum, W. Rom, C. Rice, and I. J. Selikoff, *Environ. Health Perspect.*, **17**, 171 (1976).
119. H. Ohtsuji and M. Ikeda, *Br. J. Ind. Med.*, **27**, 150 (1970).
120. H. D. Gibbs, *J. Biol. Chem.*, **72**, 649 (1927).
121. G. J. Roush and M. G. Ott, *J. Am. Ind. Hyg. Assoc.*, **38**, 67 (1977).
122. R. C. Theis and S. R. Benedict, *J. Biol. Chem.*, **61**, 67 (1924).
123. U.S. Dept. of Labor, *Fed. Reg.*, **43**, 5917 (1978).
124. I. Kilpikari and H. Savolainen, *Br. J. Ind. Med.*, **39**, 401 (1982).
125. M. Ogata, Y. Takatsuka, and K. Tomokuni, *Br. J. Ind. Med.*, **28**, 386 (1971).
126. M. Ikeda and H. Ohtsuji, *Br. J. Ind. Med.*, **29**, 99 (1972).
127. J. E. Walkley, L. D. Pagnotto, and H. B. Elkins, *Ind. Hyg. J.*, **22**, 362 (1961).
128. D. R. Stull, *Ind. Eng. Chem.*, **39**, 517 (1947).
129. S. Glasstone, *Elements of Physical Chemistry*, 1st ed., Van Nostrand, New York, 1949, pp. 362–363.
130. Z. Bardodej and E. Bardodejova, *Cesk. Hyg.*, **6**, 537 (1961); through *Chem. Abstr.*, **65**, 6086g (1966).
131. H. Ohtsuji and M. Ikeda, *Br. J. Ind. Med.*, **29**, 70 (1972).
132. M. Ikeda, (J. Biochem. (Tokyo), **55**, 231 (1964); through *Chem. Abstr.*, **60**, 16407b (1966).
133. M. Ogata, K. Tomokuni, and Y. Takatsuka, *Br. J. Ind. Med.*, **26**, 330 (1969).
134. V. J. Elia, L. A. Anderson, T. J. MacDonald, A. Carson, C. R. Buncher, and S. M. Brooks, *J. Am. Ind. Hyg. Assoc.*, **41**, 922 (1980).
135. T. Imamura and M. Ikeda, *Br. J. Ind. Med.*, **30**, 289 (1973).
136. M. S. Wolff, R. Lilis, W. V. Lorimer, and I. J. Selikoff, *Scand. J. Work Environ. Health*, **4** (Suppl. 2), 114 (1978).
137. J. Riihimäki, P. Pfäffli, K. Savolainen, and K. Pekari, *Scand. J. Work Environ. Health*, **5**, 217 (1979).
138. K. Engström, K. Husman, and V. Riihimäki, *Int. Arch. Occup. Environ. Health*, **39**, 181 (1977).
139. R. D. Stewart, C. L. Hake, A. J. Lebrun, J. E. Peterson, et al., *Biologic Standards for the Industrial Worker by Breath Analysis: Trichloroethylene*, U.S. Department of Health, Education, and Welfare, Cincinnati, OH, 1974, p. 96.
140. U.S. Dept. Health, Education and Welfare, *Criteria for a Recommended Standard ... Occupational Exposure to Inorganic Lead*, Cincinnati, OH, 1972.
141. U.S. Dept. Health, Education and Welfare, *Criteria for a Recommended Standard ... Occupational Exposure to Inorganic Fluorides*, Cincinnati, OH, 1975.
142. N. L. Kaltreider, M. J. Elder, L. V. Cralley, and M. O. Colwell, *J. Occup. Med.*, **14**, 531 (1972).
143. O. M. Derryberry, M. D. Bartholomew, and R. B. L. Fleming, *Arch. Environ. Health*, **6**, 503 (1963).
144. J. Teisinger, V. Bergerová-Fišerová, and J. Kudrna, *Pracouní Lékařství*, **4**, 175 (1952); through *Chem. Abstr.*, **49**, 4181i (1955).
145. R. D. Stewart, H. H. Gay, D. S. Erley, C. L. Hake, and J. E. Peterson, *J. Occup. Med.*, **3**, 586 (1961).
146. H. W. Haggard, *J. Biol. Chem.*, **59**, 737 (1924).
147. G. D. DiVincenzo, P. F. Yanno, and B. D. Astill, *J. Am. Ind. Hyg. Assoc.*, **33**, 125 (1972).
148. R. D. Stewart, C. L. Hake, and J. E. Peterson, *Arch. Environ. Health*, **29**, 6 (1974).
149. E. Guberan and J. Fernandez, *Br. J. Ind. Med.*, **31**, 159 (1974).
150. J. Fernandez, E. Guberan, and J. Caperos, *J. Am. Ind. Hyg. Assoc.*, **37**, 143 (1976).
151. E. D. Baretta, R. D. Stewart, and J. E. Mutchler, *J. Am. Ind. Hyg. Assoc.*, **30**, 537 (1969).

BIOLOGICAL INDICATORS OF CHEMICAL DOSAGE AND BURDEN 141

152. R. A. Kehoe, in: F. A. Patty, Ed., *Industrial Hygiene and Toxicology*, 2nd rev. ed., Vol. II, Wiley, New York, 1963, p. 941.
153. U.S. Department of Labor, Occupational Safety and Health Administration, *Fed. Reg.*, **40,** 45934 (1975).
154. 29CFR 1910.1025.
155. U.S. Environmental Protection Agency, *Fed. Reg.*, **42,** 63076 (1977).
156. A. El-Dakhakhny and Y. M. El-Sdaik, *J. Am. Industr. Hyg. Assoc.*, **33,** 31 (1972).
157. A. M. Yoakum, *Natl. Tech. Inform. Serv. Report* EPA-600/1-76-029 (1976).
158. D. I. Hammer, J. F. Finklea, R. H. Hendricks, C. M. Shy, and R. J. N. Norton, *Air Pollut. Contr. Off. (U.S.) Publ.*, **AP91,** 125 (1972).
159. E. C. Henley, M. E. Kassouny, and J. W. Nelson, *Science*, **197,** 277 (1977).
160. G. D. Renshaw, C. A. Pounds, and E. F. Pearson, *Nature*, **238,** 162 (1972).
161. K. Nishiyama and G. F. Nordberg, *Arch. Environ. Health*, **25,** 92 (1972).
162. J. A. Hurlburt, *Natl. Tech. Inform. Serv. Report TID-4500-R64* (1976).
163. H. G. Petering, D. W. Yeager, and S. O. Witherup, *Arch. Environ. Health*, **23,** 202 (1971).
164. H. G. Petering, D. W. Yeager, and S. O. Witherup, *Arch. Environ. Health*, **27,** 327 (1973).
165. L. C. Bate and F. F. Dyer, *Nucleonics*, **23,** 74 (1965).
166. H. A. Schroeder and A. P. Nason, *J. Invest. Dermatol.*, **53,** 71 (1969).
167. Y. Suzuki, K. Nishiyama, and Y. Matsuka, *Tokushima J. Exp. Med.*, **5,** 111 (1958); through L. M. Klevay, *Arch. Environ. Health*, **26,** 169 (1973).
168. M. Anke, A. Hennig, M. Diettrich, G. Hoffman, G. Wicke, and D. Pflug, *Arch. Tierernaehr*, **21,** 205 (1971).
169. H. C. Hopps, *Trace Subst. Environ. Health*, **8,** 59 (1974).
170. W. H. Harrison, R. M. Gray, and L. M. Solomon, *Br. J. Dermatol.*, **91,** 415 (1974).
171. H. M. Baumgartner, P. F. Jones, W. A. Baumgartner, and C. T. Black, *J. Nucl. Med.*, **20,** 748 (1979).
172. A. M. Baumgartner, P. F. Jones, and C. T. Black, *J. Forensic Sci.*, **26,** 576 (1981).
173. W. A. Baumgartner, C. T. Black, P. F. Jones, and W. H. Blahd, *J. Nucl. Med.*, **23,** 790 (1982).
174. R. W. Phelps, T. W. Clarkson, T. G. Kershaw, and B. Wheatley, *Arch. Environ. Health*, **35,** 161 (1980).
175. V. G. Oleru, *J. Am. Ind. Hyg. Assoc.*, **37,** 617 (1976).
176. L. M. Klevay, *Arch. Environ. Health*, **26,** 169 (1973).
177. Z. S. Jaworoski, *Atompraxis*, **11,** 271 (1965); through L. M. Klevay, *Arch. Environ. Health*, **26,** 169 (1973).
178. Anon., *Lancet*, 337 (August 13, 1977).
179. Commerce Clearing House, Inc., *Food Drug Cosmetic Law Reporter*, Chicago, 1978.
180. H. Kraut and M. Weber, *Biochem. Z.*, **317,** 133 (1944); through L. M. Klevay, *Arch. Environ. Health*, **26,** 169 (1973).
181. J. Arunachalam, S. Gangadharan, and S. Yegnasubramanian, *Nucl. Act. Tech. Life Sci., Proc. Int. Symp.*, 499 (1979).
182. R. G. Keenan, D. H. Byers, B. E. Salzman, and F. L. Hyslop, *J. Am. Ind. Hyg. Assoc.*, **24,** 481 (1963).
183. Z. W. Polishuk, M. Ron, M. Wasserman, S. Cucos, D. Wasserman, and C. Lemeson, *Pest. Monit. J.*, **10,** 121 (1977).
184. S. S. Strassman and F. W. Kutz, *Pest. Monit. J.*, **10,** 130, (1977).
185. A. Curley, M. F. Copeland, and R. D. Kimbrough, *Arch. Environ. Health*, **19,** 628 (1969).
186. G. P. Giacoia and C. S. Catz, *Clin. Perinatol.*, **6,** 181 (1979).

187. J. A. Knowles, *Clin., Toxicol.*, **7,** 69 (1974).
188. P. Lanzkowsky, *N. Engl. J. Med.*, **298,** 343 (1978).
189. C. M. Stowe and G. L. Plaa, *Ann. Rev. Pharmacol.*, **8,** 337 (1956).
190. R. L. Savage, *Adverse Drug Reaction Bulletin*, p. 212, 1976.
191. G. J. White and M. K. White, *Vet. Hum. Toxicol.*, **22,** (Suppl. 1), 1 (1980).
192. E. J. Lien, *J. Clin. Pharmacol.*, **4,** 133 (1979).
193. C. J. Cote, S. B. Reed, and G. E. Strobel, *Br. J. Anaesth.*, **48,** 451 (1976).
194. J. T. Eyster, H. E. B. Humphrey, and R. D. Kimbrough, *Arch. Environ. Health*, **38,** 47 (1983).
195. J. Mes and D. J. Davies, *Chemosphere*, **7,** 699 (1976).
196. E. C. Shepherd, L. W. Robertson, N. D. Heidelbaugh, S H. Safe, and T. D. Phillips, *The Toxicologist*, **3,** 6 (1983).
197. J. T. Wilson, R. D. Brown, D. R. Cherek, J. W. Dailey, B. Hilman, P. C. Jobe, B. R. Manno, J. E. Manno, H. M. Redetzki, and J. J. Stewart, *Clin. Pharmacokinet.*, **5,** 1 (1980).
198. R. R. Peters, L. T. Chapin, K. B. Leining, and H. A. Tucker, *Science*, **199,** 911 (1978).
199. D. Schmidt and H. J. Kupferberg, *Epilepsia*, **16,** 735 (1975).
200. D. Schmidt, J. W. Paxton, B. Whiting, F. J. Rowell, J. G. Ratcliff, and K. W. Stephen, *Lancet*, 639 (September 18, 1976).
201. B. R. Allen, M. R. Moore, and J. A. A. Hunter, *Br. J. Dermatol.*, **92,** 715 (1975).
202. S. Dreizen, B. M. Levy, W. Niedermeier, and J. H. Griggs, *Arch. Oral Biol.*, **15,** 179 (1970).
203. L. Gervais, Y. Lacasse, J. Brodeur, and A. P'an, *Toxicol. Lett.*, **8,** 63 (1981).
204. M. M. Joselow, R. Ruiz, and L. J. Goldwater, *Arch. Environ. Health*, **17,** 35 (1968).
205. R. W. Goldblum, S. Derby, and A. B. Lerner, *J. Invest. Dermatol.*, **20,** 13 (1953).
206. W. B. Barnett, *Clin. Chem.*, **18,** 923 (1972).

CHAPTER FOUR

Body Defense Mechanisms to Toxicant Exposure

JAMES S. BUS, Ph.D., and
JAMES E. GIBSON, Ph.D.

1 INTRODUCTION

On any one day the human body is exposed to an incalculably large number of events capable of causing adverse responses in the body. Although many of these potentially damaging reactions can be attributed to voluntary or involuntary exposure to synthetic substances, it is also true that many of the events are linked to endogenous biological reactions or to agents in the natural environment and hence are totally inescapable. It is obvious, however, that the vast majority of these potential reactions do not result in tissue damage; otherwise the expected lifespan would be short, indeed. The reason that the human body, or for that matter, all organisms, is able to sustain life in the face of such a continuous onslaught of toxic reactions lies in the ability of the body to defend itself against these reactions by a wide variety of mechanisms.

Although the number of defense mechanisms is exceedingly diverse, they can be divided into three general classifications. Physical or anatomic defenses primarily function as barriers to entry of an agent either into the body or into a tissue or organ once entry to the body has occurred. Physiological defenses, which also have as their primary function the prevention of absorption of agents, have the additional important capacity of specific response to a variety of toxic insults. The third classification of defense mechanisms, biochemical defenses, act at the cellular level to prevent or repair damaging reactions induced by toxicants.

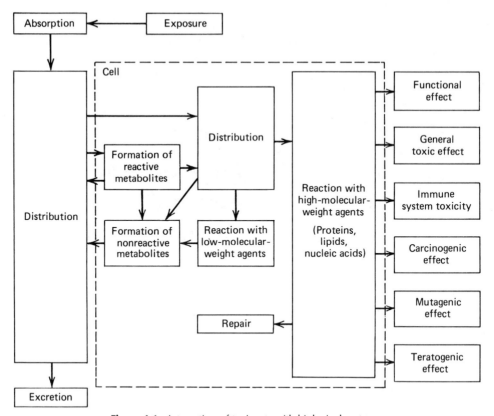

Figure 4.1 Interaction of toxicants with biological systems.

An important feature of all defense systems is the fact that, despite their diversity, they frequently function as an integrated unit to diminish the consequences of toxicant exposure. This is readily seen by an examination of Figure 4.1, which depicts the potential interactions of chemicals with biological systems. Before exogenously administered compounds interact with important intracellular macromolecules, they must first penetrate a series of defense mechanisms. The first barrier is to absorption itself, and if this is obtained, additional barriers may prevent distribution of the agent to the various organs. Should a toxicant gain entry to the cell it still faces further barriers to intracellular distribution and in addition is subject to a host of biochemical transformations capable of rendering it less toxic and enhancing its susceptibility to excretion. Finally, even if a toxicant should manage to survive these defenses and interact with critical cell macromolecules, the potential remains for the damaged macromolecules to be repaired, thus preventing or diminishing expression of a toxic response.

The purpose of this chapter is to provide a general overview of some of the important defense mechanisms that modulate the toxicity of foreign compounds.

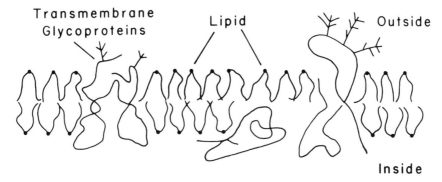

Figure 4.2 Conceptual depiction of a cell membrane. Redrawn from Bus and Gibson (44).

It must be pointed out that, for the sake of brevity, not all proposed or demonstrated detoxication mechanisms are discussed.

2 MEMBRANE DEFENSES TO ABSORPTION AND DISTRIBUTION

For a toxicant to exert its activity at the cellular level, it first must pass through a series of membranes. The membranes encountered include those at the site of absorption, primarily skin, lungs, and gastrointestinal tract, and also those at the tissue level such as the cell, mitochondrial, and nuclear membranes. The ability of membranes to limit passage of toxicants can be considered the body's first important line of defense to toxic agents.

The composition of cell membranes, although not constant from tissue to tissue, nonetheless possesses sufficient commonality to permit construction of a unified model describing the ability of chemicals to pass through them. Before consideration of the factors that affect the ability of a chemical to cross cell membranes, however, it is necessary to describe the current understanding of the structure of biological membranes.

2.1 Factors Influencing Membrane Transfer

In general, membranes consist of two layers of lipid molecules, with each layer spatially oriented perpendicular to the membrane surface (Figure 4.2). The hydrophobic inner core of the membrane is composed primarily of fatty acid hydrocarbon side chains of phospholipids. Varying amounts of cholesterol also are located in the hydrophobic inner core. Each surface of the membrane is composed of the hydrophilic ends of the phospholipid molecules, which in turn are interspersed with various proteins either partially embedded in or completely transversing the membrane (1).

The majority of agents cross membranes by the process of simple diffusion. Although low-molecular-weight hydrophilic compounds can diffuse through aqueous channels in the membrane, the passage of higher-molecular-weight compounds (≥ 100 mass units) is determined primarily by their lipid solubility. Because of the nature of the body membranes, lipophilic agents will diffuse across more readily than agents with greater hydrophilic properties. Chemicals may cross membranes by mechanisms other than simple diffusion. These mechanisms, active transport and facilitated diffusion, seldom contribute to the defensive nature of the membrane since they increase the rate of transport across the membrane. In fact, active transport can lead to selective accumulation of toxicants in tissues and subsequent overwhelming of the local tissue defense mechanisms (Section 5.4.2). In certain instances, however, transport mechanisms may facilitate elimination of toxicants from the body (Section 4.1).

Passive diffusion of toxicants through biological membranes is governed by three basic factors: existence of a concentration gradient; lipid solubility, characterized by the oil–water partition coefficient; and degree of ionization (2). Since most toxicants are rapidly distributed away from the inner surface of the membrane on absorption, the latter two factors represent the primary rate-limiting steps for xenobiotic absorption. For many toxicants lipid solubility is reflected by the degree of ionization, which in turn is determined frequently by the pH of the surrounding media. Agents existing in nonionized form will generally have greater lipid solubility than their ionized form and thus will penetrate membranes more readily.

Ionization of an organic acid or base is a direct function of the pH of the solution in which it is dissolved and the pK_a of the agent. The pK_a of a chemical is the negative logarithm of the acidic dissociation constant and represents the pH at which a chemical is 50 percent ionized. The degree of ionization of organic acids or bases at any selected pH is given by the Henderson–Hasselbalch equations:

$$\text{For acids: } pK_a - pH = \log \frac{\text{(nonionized)}}{\text{(ionized)}} \qquad (1)$$

$$\text{For bases: } pK_a - pH = \log \frac{\text{(ionized)}}{\text{(nonionized)}} \qquad (2)$$

It should be clear from these equations that the extent of ionization of an organic acid, for example, benzoic acid, will decrease with decreasing pH, whereas that for an organic base, such as aniline, will increase with decreasing pH. Thus organic acids will cross membranes more easily when present in a low-pH environment and organic bases, when in a high-pH solution.

The absorption of toxicants through membranes is governed by the principles outlined in this section. However, each primary site of entry of chemicals into the body has unique characteristics that influence the absorption rate from that site.

2.2 Skin

Skin, which probably has the greatest amount of exposure to environmental agents, serves as a relatively impermeable barrier to a large number of potential toxicants. It must be pointed out, however, that skin, like all lipid-containing membranes, will readily permit passage of highly lipid-soluble materials. For example, many cases of human poisoning associated with the agricultural use of the organophosphate insecticides can be directly attributed to the high lipid solubility of these agents and their subsequent ability to readily pass through the skin (3).

For a chemical to move through skin, it must pass the membranes of a large number of cells. Skin is composed of two multicellular layers, the external epidermis and the underlying dermis (4). The epidermis consists of an outer 8- to 16-cell-thick layer of flattened keratinized cells termed the *stratum corneum*. Beneath the stratum corneum are several layers of living cells that undergo continual differentiation, ultimately ending up as the dead, keratinized cells of the stratum corneum. It is the stratum corneum, however, that offers the primary resistance to passage of chemicals through the epidermis. Removal of this layer of cells by abrasion or other means such as chemical irritation greatly enhances the movement of most chemicals through the skin to the systemic circulation. Once a chemical gains access to the dermis, rapid and complete absorption is usually assured.

Passage of toxicants through the intact epidermis is influenced by several factors (4). The stratum corneum normally contains only small amounts of water, approximately 5 to 15 percent by weight. If the skin is occluded, which increases hydration by preventing escape of perspiration, the stratum corneum may contain up to 50 percent water. Hydration of the skin can increase permeability up to four- to fivefold. Absorption of chemicals may also be altered by various organic solvents. Dimethyl sulfoxide, for example, has been shown to facilitate movement of chemicals through the skin. Percutaneous absorption rates may also vary with the anatomical location of the skin. The skin of the palm and heel may be 100 to 400 times thicker as that of the scrotum. This difference in thickness can significantly alter the amount of chemical absorbed. The ability of nonionized toxicants to pass through various areas of the skin has been estimated as follows (3): scrotal > forehead > axilla = scalp > back = abdomen > palm.

2.3 Gastrointestinal Tract and Lung

The epithelial cells lining the gastrointestinal tract and lung present significantly less a barrier to absorption than the skin. Unlike skin, there is essentially only a single layer of columnar cells between the lumen of the gastrointestinal tract and the membranes of the systemic capillary beds. In the lung, the cells lining the alveoli also are very thin and are intimately associated with underlying capillaries. Thus, these thin layers of cells offer little resistance to the penetration of lipid-soluble molecules.

The pH of the gastrointestinal tract varies from 1 to 3 in stomach to approximately 6 in the intestines (3). As might be expected, this difference in pH significantly affects the absorption of ionizable compounds. Absorption of organic acids occurs more readily from the stomach since the unionized state is preferred in the low-pH environment of the stomach. In contrast, organic bases are more likely to be absorbed from the intestine, where the higher pH favors the nonionized state.

Although the epithelial cells lining the respiratory tract afford little resistance to the systemic absorption of toxicants, the respiratory tract possesses a number of important anatomic and physiological defenses that decrease the potential of inhaled gases, vapors, and particulates to be absorbed. These specialized defense mechanisms are described in Sections 3.1.1 to 3.1.5.

2.4 Membrane Barriers to Distribution in the Body

The absorption of a toxicant from the external environment to the systemic circulation does not ensure its even distribution throughout the body. For entry into the various tissues from the blood, additional membranes must be crossed. The anatomic environment and physicochemical composition of the capillary membranes serving the various tissues is not uniform, however, and this can affect the ability of a chemical to penetrate any given tissue. As with other membranes, lipid solubility and ionization state are the primary rate-limiting factors affecting movement across tissue membranes. Thus highly lipophilic agents and ionizable substances that exist primarily in the nonionized state at physiological pH will readily enter tissues.

The distribution of chemicals to the central nervous system, testis, and embyro–fetus complex are examples of nonuniform distribution within an organism. In the central nervous system (CNS) the capillary beds are largely encircled with glial cells, which create additional layers of membranes to be penetrated before access to tissue is obtained (5). A similar situation also exists in the placenta and testis, in which multiple cell layers are interposed between the capillary and the ultimate target cells, the embryo and germ cells respectively (5, 6). Because of the restricted passage of toxicants across these membrane systems, these systems are termed the *blood–brain*, *blood–testis*, and *placental barriers*. It must be pointed out, however, that the term "barrier" does not imply an absolute prevention of transfer, but only that entry of toxicants to these sites may be more limited when compared to other tissues.

3 NONMEMBRANE DEFENSES TO ABSORPTION AND DISTRIBUTION

3.1 Respiratory Tract

The primary function of the lung is that of gas exchange. It is not surprising, therefore, that the design of the respiratory tract is ideally suited to accomplish this purpose. Inspired air is rapidly humidified and temperature controlled during passage through the upper portions of the respiratory tract. This conditioned air is then delivered to the lung, which, with its vast surface area and only thin separation between the air and the capillary circulation, readily absorbs the oxygen necessary for sustenance of life. Unfortunately, the air that is breathed all too frequently contains numerous potential toxicants, whether in gaseous, vapor, or aerosol form. In the absence of defense mechanisms, these toxicants would have access to all areas of the respiratory tract, where they not only could damage the delicate membranes of gas exchange but also could undergo systemic absorption.

Fortunately, the respiratory tract has multiple anatomical and physiological defenses mechanisms that prevent or moderate the interaction of inhaled toxicants with the exposed tissue. These mechanisms can be divided into four general classes: (1) anatomical barriers, represented by the convoluted pathways that inhaled material must pass through before reaching deep portions of the lung; (2) mucociliary clearance mechanisms, which transport toxicants from the respiratory tract following their deposition there; (3) alveolar macrophages, which remove inhaled debris from the alveoli; and (4) respiratory reflex mechanisms, which reduce the amount of toxicant inhaled. The physicochemical properties of the inhaled toxicant, such as its physical state (aerosol, gas, or vapor) and lipid solubility, and tissue irritant properties determine which, if any, of these defense mechanisms will affect respiratory absorption of the toxicant.

3.1.1 Anatomical Defenses to Particle Deposition

The respiratory tract can be divided into three anatomic regions: the nasopharyngeal, tracheobronchial, and pulmonary regions (7). The nasopharyngeal region begins at the nose and continues to the larynx. This region is a complex series of passages in which the minimum channel width may approach 2 mm in humans (8, 9). The tracheobronchial region extends from the larynx to the terminal bronchioles. The trachea bifurcates into the main bronchus to each lung. Each bronchus may have up to 20 divisions before ending in the terminal bronchiole, which has an approximate width of 0.7 mm. The pulmonary region consists of the respiratory bronchioles, alveolar ducts, atria, alveolar sacs, and alveoli. The primary function of this region is gas exchange. The width of a typical alveolus is 0.15 mm.

The complex anatomy of the respiratory tract is a key defense to the deposition of liquid or solid aerosols in the lung. Deposition of particles in various regions of the respiratory tract is controlled by four mechanisms: inertial impaction, gravitational settling, interception, and diffusion (9). Inertial impaction is the process whereby particles carried in an airstream impact into the wall of passage at a bend in the airstream. Since this phenomenon is governed by the inertia of the particle, impaction is a function of particle density, size, and velocity. Gravitational settling, or sedimentation, occurs in areas of low air velocity, primarily the small bronchi, bronchioles, and alveoli. Settling of particles begins when gravitational force on the particles equals the sum of the buoyancy and air resistance, at which time settling will continue at what is

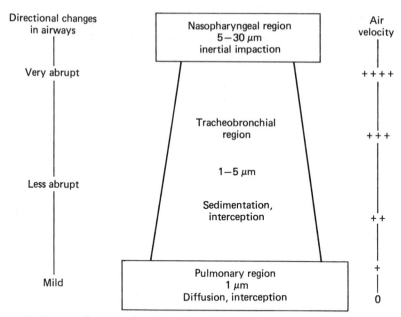

Figure 4.3 Factors influencing deposition of particles in various regions of the respiratory tract.

described as the terminal settling velocity. Interception is the phenomenon in which particles traveling near the side of a passageway come into contact with the surface. Interception is particularly important for particles of irregular shape such as fibers. Fibers that have small diameters and long lengths may escape impaction in the upper portions of the respiratory tract because of their low density. Fibers encountering the small-diameter openings in the terminal bronchioles and below may not be able to pass, whereas particles of equivalent density but more regular shape pass readily. Diffusion is limited to particles of 0.5 μm or less in size and is due to the impact of gas molecules on the particle. This deposition mechanism is important for particles reaching the small airways.

The net effect of the particle deposition mechanisms, acting in concert with the respiratory anatomy, is that the delicate membranes of the alveoli are well protected from deposition of toxic particulates. Relatively large particles of 5 to 30 μm, which enter the nose at high velocity and rapidly encounter sharp changes in airflow direction, are deposited by impaction in the nasopharyngeal region. Particles of lower density and smaller diameter (1 to 5 μm) are trapped primarily by sedimentation in the tracheobronchial region, where velocity of the airstream is reduced and less abrupt changes in airflow occur. Only particles of less than 1 μm in size penetrate to the alveoli, where deposition occurs primarily by diffusion (Figure 4.3).

3.1.2 Mucociliary Clearance

Despite the effectiveness of the anatomical barriers in reducing deposition of particles in the lower and upper respiratory tracts, it should be obvious that

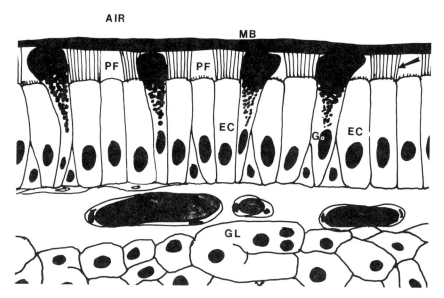

Figure 4.4 Diagram of the components of the mucociliary apparatus. The high-viscosity mucus blanket (MB) floats on the less viscous periciliary fluid (PF) and is propelled by the action of cilia (arrow) extending from the epithelial cells (EC). The mucus is produced by goblet cells (GC) in the epithelium and glands (GL) in the underlying connective tissue.

significant particle deposition is still likely to occur. Particles deposited in the nasopharyngeal and tracheobronchial regions of the respiratory tract are removed by an additional defense mechanism, the mucociliary clearance system (7, 10), whereas those deposited in the pulmonary region are cleared by alveolar macrophages (7, 11). These systems not only prevent accumulation of toxic particulates in the respiratory tract but also diminish or prevent contact of particles with the membranes of cells lining the lumen of the airways.

The nasopharyngeal and tracheobronchial regions of the respiratory tract are lined with both ciliated and mucus-secreting cells (10, 12). There are two sources of mucus, epithelial goblet cells and submucosal glands. Mucus produced from these sources provides a thin cover for the epithelial cells of the airways. The mucus layer is thought to be biphasic, consisting of a high-viscosity mucus sheet overlaying a layer of low-viscosity serous fluid. The cilia of the ciliated cells extend primarily into the serous layer of the mucus, with only the cilia tips contacting the high-viscosity outer layer (Figure 4.4). Cilia beat rapidly (approximately 400 to 1000 beats/min) in a coordinated pendular fashion, creating a flow of mucus from the tracheobronchial and nasopharyngeal regions that is ultimately swallowed to the stomach. Mucociliary clearance of particles from the respiratory tract varies with location. Mucus flow of 0.4 to 0.6 mm/min has been reported in the distal areas of the tracheobronchial region and up to 10 mm/min in portions of the trachea (7, 12).

3.1.3 Alveolar Macrophages

Particles deposited in the pulmonary region of the respiratory tract are removed by alveolar macrophages (11). These cells, which originate from bone marrow and migrate into the alveolar spaces from the pulmonary interstitium, rapidly phagocytize small particles deposited on the epithelial cells lining the alveoli. Macrophages containing engulfed particles move to the respiratory bronchioles, at which point they are transported out of the respiratory tract by the mucociliary clearance system. Some macrophages also are cleared from the alveoli by migration into the interstitial spaces and lymph systems of the lung.

Ingestion of particles by macrophages does not always result in clearance from the alveoli. Several types of particles, particularly silica and asbestos, cause cell dysfunction and death on ingestion by the macrophages. The toxicity of these types of particles to the macrophages results in the release of toxic constituents, such as lysosomal proteases, into the alveolar spaces. Thus the development of lung toxicity as a result of exposure to silica and asbestos appears to be the direct result of an initial insult to a respiratory defense mechanism, the alveolar macrophage (13).

3.1.4 Mucus as a Protective Barrier

The mucus layer has an additional protective function above that associated with the mucociliary clearance system. Mucus is composed of approximately 95 percent water, with the remainder consisting of varying amounts of glycoproteins, free proteins, salts, and other materials (14). Consequently, highly water soluble toxicants such as sulfur dioxide, chlorine, and formaldehyde are completely scrubbed by the mucus layer in the upper portion of the respiratory tract and do not penetrate to the sensitive, unprotected alveolar membranes of the deep lung. Less water soluble agents such as ozone are not effectively scrubbed during passage through the upper respiratory tract and may cause significant toxic injury to the pulmonary region of the lung (7).

Respiratory tract mucus also serves as a physicochemical barrier to the direct contact of inhaled toxicants with epithelial cells lining the airways. Thus, water-soluble toxicants trapped by mucus do not necessarily diffuse quantitatively through to the epithelial cells, where potential toxic injury may occur. Toxicants absorbed by mucus may react with or otherwise be neutralized by the various components of mucus, effectively reducing the amounts of toxicant delivered to the underlying epithelial cells (14). Since mucus is constantly removed from the respiratory tract and is freshly renewed by mucus-secreting cells, exposure to low concentrations of toxicants may be insufficient to saturate the neutralization capacity of the mucous layer. There is some evidence that such a protective mechanism may be functioning with exposure to the irritant gas formaldehyde, which produces nasal tumors in rats exposed to high concentrations (15 ppm) of the gas for 2 years (15). Exposure of rats to high concentrations of formaldehyde resulted in both mucostasis and ciliastasis and was associated with injury to the nasal epithelial cells. Exposure to lower concentrations of formaldehyde (0.5 ppm), however, did not alter mucociliary function and was not associated with epithelial cell injury. The ability of formaldehyde to penetrate the mucus barrier, therefore, may have been an important factor in the dose-dependent development of epithelial cell damage following inhalation exposure to formaldehyde.

3.1.5 Respiratory Tract Reflexes

Reflex responses to exposure to inhaled toxicants include (a) coughing or sneezing, which rapidly expel the toxicant; (b) reduction in minute ventilation, which minimizes further uptake of toxicant; and (c) alterations in bronchomotor, cardiovascular, and mucus-secreting activities (12). The precise mechanism of the initiation of the reflexes is not known, although they are thought to be mediated through stimulation of a variety of nerve receptors in the upper respiratory tract (16).

The receptors in the nasal cavity are associated with the trigeminal nerve and are readily stimulated by a variety of irritant agents. The characteristic response seen when these receptors are stimulated by an inhaled irritant is a rapid decrease in the minute ventilation, which effectively reduces the subsequent dose of the irritant. The reflex inhibition of respiration varies with species, and it is known, for example, that the reflex response of mice to an irritant agent is far greater than that of rats (16). This difference in response appears to have a profound influence on expression of the ultimate toxic effect in the respiratory tract.

A comparison of the differential response of mice and rats to formaldehyde is an example of the role of respiratory reflexes in modulating the toxic effects of exposure to an irritant agent. As mentioned in Section 3.1.4, chronic exposure of rats to 15-ppm formaldehyde produced a 50 percent incidence of tumors in the nasal cavity (15). Mice similarly exposed to formaldehyde exhibited only a 3.3 percent tumor incidence. At a lower (6 ppm) exposure to formaldehyde, a 0 and 1 percent incidence of nasal tumors was found in mice and rats, respectively. An evaluation of respiratory reflex response of mice and rats to acute exposures of formaldehyde suggested that this defense mechanism may explain the dose-dependent and species-specific toxicity of formaldehyde (17). Rats exposed to 15-ppm formaldehyde had only a slight reflex-mediated reduction in minute ventilation, whereas respiration at this concentration in mice was reduced approximately 70 percent (Figure 4.5a). Since almost all of the inhaled formaldehyde is retained by the nose, the total dose of formaldehyde delivered to the nasal epithelium of each species can be calculated by normalizing the amount of inhaled irritant per unit time over the surface area of the nasal cavity. The amount of inhaled formaldehyde, of course, is dependent on the minute ventilation of the animal, and the surface area of the nasal cavity can be estimated by morphometric techniques. When these calculations are compared for mice and rats exposed to 15-ppm formaldehyde, it can be seen that

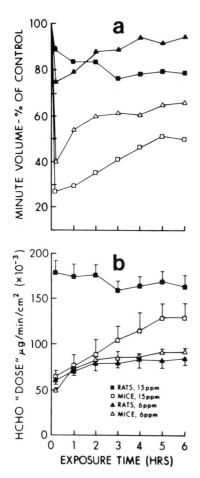

Figure 4.5 (a) Time–response curves for minute volume from rats and mice exposed to 6 or 15 ppm of formaldehyde for 6 hr. (b) Time-weighted averages of the theoretical formaldehyde dose available for deposition on the nasal passages of rats and mice during a 6-hr exposure to 6 or 15 ppm. Reprinted with permission from Swenberg et al. (15).

mice received a substantially lower dose of the irritant than did rats (Figure 4.5b). In fact, because of the marked reduction in minute ventilation in mice exposed to 15-ppm formaldehyde, the actual delivered dose to the nasal cavity at this concentration was comparable to that of rats exposed to 6-ppm formaldehyde. Since these equivalent delivered doses resulted in nonsignificant tumor incidences in both mice and rats, these data suggest that the tumorigenic activity of high doses of formaldehyde in rats is a direct reflection of the inability of rats to reduce their exposure by reflex inhibition of respiration.

3.2 Protein Binding as a Barrier to Toxicant Distribution

Binding of toxicants to high-molecular-weight proteins may be an important factor in controlling subsequent distribution to critical target macromolecules within cells. Many toxicants, on absorption into the systemic circulation, bind with varying affinity to plasma proteins. The binding to proteins, consisting of hydrogen, van der Waal's and ionic bonds, is reversible (3, 5). Despite the weak nature of the binding, it creates a high-molecular-weight polar complex incapable of crossing cell membranes. Only the fraction of free, unbound toxicant is available for diffusion across membranes, as influenced by the factors described in Section 2.1.

The binding of most toxicants to plasma proteins is nonspecific; consequently, agents with similar physicochemical characteristics may displace a bound toxicant from the proteins. Such displacement increases the amount of unbound toxicant available to diffuse into cells, often resulting in enhanced toxicity. An example of this phenomenon is the displacement of unconjugated bilirubin from plasma albumin by sulfonamide drugs. In newborn infants, sulfonamide-induced release of bilirubin from plasma protein binding sites may increase the risk of bilirubin encephalopathy (18). Similarly, sulfonamide drugs also have been shown to displace antidiabetic drugs from plasma proteins, resulting in induction of hypoglycemic coma (5).

There are two examples in which binding of toxicants to blood protein have been used as antidotal therapy for the toxicant. The toxicity of both cyanide and hydrogen sulfide is mediated by their complexation with the ferric heme moeity of cytochrome oxidase, resulting in inhibition of oxidative respiration. The toxicity of these agents can be ameliorated, however, by induction of methemoglobinemia. The mechanism of this protective effect is attributed to the fact that methemoglobin contains a ferric heme group capable of competing effectively with the heme moeity of cytochrome oxidase for cyanide and sulfide binding and thus provides an innocuous binding site for these toxicants (19).

Intracellular protein binding is an important detoxication mechanism for metals such as cadmium and mercury. Many different tissues synthesize a metal binding protein termed *metallothionein*. This unique protein has a high cysteine content, providing abundant thiol binding sites for these metals. Thus the binding of metals to metallothionein prevents their further distribution and binding to other functionally important intracellular sites (20). An interesting aspect of metallothionein function is that its synthesis is rapidly induced following low-level exposure to cadmium. Although the synthesis of additional binding protein offers protection against further exposure to higher doses of cadmium, several investigators have suggested that metallothionein does not function primarily as a defense against cadmium (21, 22). Rather, this protein may play an important role in the homeostatic control of tissue zinc and copper concentrations. The protection afforded by metallothionein against cadmium toxicity, therefore, may represent only a fortuitous interaction of a toxic metal with an endogenous control mechanism.

4 EXCRETORY DEFENSE MECHANISMS

Absorbed toxicants are eliminated from the body by a variety of mechanisms. Most toxicants and their metabolites (see Sections 5.1 and 5.2) are excreted by the kidney or liver, although excretion from the lung, gastrointestinal tract, milk, sweat, and saliva can also occur (5).

4.1 Renal Excretion

Passive glomerular filtration represents the predominate mechanism by which toxicants are eliminated by the kidney (5). Compounds with molecular weights of less than 60,000 will readily pass into the urine through pores in the glomerular membrane. Of course, binding of toxicants to plasma proteins will restrict passage into the urine because of the high molecular weight of the toxicant–protein complex. Elimination of toxicants in the urine after filtration through the glomerulus is directly related to lipid solubility of the individual agents. Compounds with high lipid solubility will passively diffuse back into the systemic circulation through the renal tubule cells. Metabolism of toxicants favors elimination in urine, however, since the products of these reactions are generally more water soluble than the corresponding parent compounds (Sections 5.1 and 5.2).

A second important mechanism by which toxicants are excreted into urine is by active transport through renal tubule cells. There are two types of transport systems, one for organic anions and another for organic cations. The transport systems are not necessarily specific for any given substrate, however, and in certain circumstances elimination of a toxicant may be hindered when multiple substrates are presented to the transport mechanism. For example, the ability of exogenous organic acids to precipitate attacks of gout is due to their ability to compete with endogenous uric acid for active secretion into the urine.

4.2 Liver (Biliary) Excretion

Toxicants (or their metabolites) with molecular weights of approximately 325 or greater may be excreted into the bile (5). Although the mechanisms by which toxicants are excreted into bile are not fully understood, transport mechanisms for organic acids, bases, and neutral compounds have been identified.

The ability of a toxicant to be excreted into bile may profoundly influence its toxicity. In newborn animals the hepatic excretory system is not fully functional. Thus, the fortyfold increased toxicity of ouabain in newborn rats compared to adult animals has been attributed to an inability of the newborn to eliminate ouabain in the bile.

5 BIOCHEMICAL DEFENSES

The absorption of most toxicants into the body is directly influenced by the lipophilicity of the agent; thus the more lipid soluble an agent is, the more likely it is to be absorbed (Section 2.1). This same property, however, decreases the ability of a toxicant to be excreted. This is particularly true for the primary

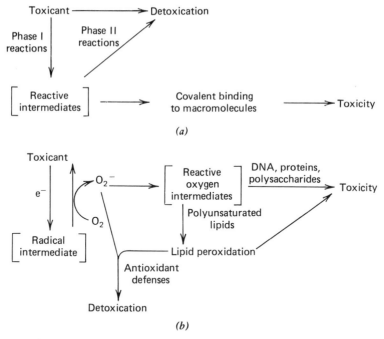

Figure 4.6 Metabolism of toxicants to reactive intermediates capable of (a) covalently binding to cell macromolecules or (b) generating toxic forms of activated oxygen. Because of the intervention of detoxication mechanisms, production of either type of reactive intermediate may not result in toxic responses within the cell.

organ of excretion, the kidney, where the aqueous environment of the urine favors elimination of hydrophilic compounds. Fortunately, the body possesses a tremendous capacity to transform foreign compounds into more water soluble forms, rendering them more susceptible to excretion.

The biotransformation of toxicants to water-soluble metabolites is generally catalyzed by a broad spectrum of cell enzymes. These enzymatic reactions have been divided into two phases: Phase I reactions, consisting of oxidation, reduction, and hydrolysis; and Phase II reactions, mediating conjugation and synthesis (23, 24). Phase I reactions involve the insertion of a polar reactive group into the toxicant molecule and usually prepare the toxicant for subsequent Phase II reactions.

It must be pointed out, however, that Phase I and II metabolic transformations do not always result in detoxication. In many cases intermediate metabolites having greater biological reactivity than the parent compound are formed. Although generation of reactive intermediates usually results in enhanced toxicity, the ability of such metabolites to attack important biological macromolecules is nonetheless modulated by further Phase I or II reactions (Figure 4.6).

In addition to formation of reactive metabolites capable of covalently binding to tissue macromolecules, Phase I reduction reactions also may catalyze electron-transfer reactions between endogenous electron sources, toxicants, and molecular oxygen. The product of this type of reaction, the superoxide radical (O_2^-), potentially may form extremely toxic oxidant species such as the hydroxyl radical ($OH^.$) (Figure 4.6). Generation of O_2^- does not always result in toxic injury to the cell, however, since several defense mechanisms have been identified that detoxify these reactive oxygen products.

5.1 Phase I Detoxication Reactions

5.1.1 Microsomal Mixed-Function Monooxygenases

The majority of Phase I transformations of toxicants is carried out by a nonspecific multienzyme system frequently termed the *mixed-function monooxygenase system*. This metabolic system is located in the endoplasmic reticulum (microsomal fraction) of cells and is composed of two component enzymes, NADPH cytochrome P-450 reductase and cytochrome P-450 (24). Both molecular oxygen and reducing equivalents in the form of NADPH are required for catalysis of the oxidation reactions. Multiple forms of cytochrome P-450 have been identified, and each form has some overlap in substrate specificity with the other forms (25).

The cytochrome P-450 monooxygenases catalyze a broad spectrum of oxidation reactions (Table 4.1). The products of the oxidation reactions are more water soluble than the parent compound and therefore have taken the first of perhaps several steps leading to detoxication. Several products of Phase I metabolism are significantly more reactive than the corresponding parent compounds, however, and formation of such products may be responsible for expression of toxicity. An example of this type of reaction is the formation of epoxides (reaction 4, Table 4.1). Epoxides result from addition of an oxygen atom across an aromatic or olefinic double bond and are sufficiently electrophilic to form a covalently bound adduct with biological macromolecules. Fortunately, epoxides, along with most other reactive intermediates formed in Phase I reactions, are subject to further Phase II reactions. These reactions generally result in formation of readily excretable, detoxified metabolites. Sections 5.2 and 5.4 describe biochemical defense mechanisms frequently associated with detoxication of products resulting from mixed-function monooxygenase metabolism. It should be noted, however, that if the parent compound contains a reactive structural element similar to that resulting from cytochrome P-450 metabolism, it, too, will be subject to the same detoxication reactions.

In addition to catalyzing oxidation reactions, the mixed function monooxygenase system enzymes also catalyze several reduction reactions (Table 4.1). Single-electron reductions of aromatic nitro groups, bipyridylium compounds, and halogenated hydrocarbons may be carried out by either NADPH cytochrome P-450 reductase or cytochrome P-450. The products of these

Table 4.1 Phase I Metabolic Transformations Catalyzed by the Microsomal Mixed-Function Oxidase System Enzymes[a]

Reaction	Example
Oxidation reactions	
1. Aromatic hydroxylation	R—C$_6$H$_5$ → R—C$_6$H$_4$—OH
2. Aliphatic hydroxylation	R—CH$_2$—CH$_2$—CH$_3$ → R—CH$_2$—CHOH—CH$_3$
3. N-, O-, or S-dealkylation	R—(N, O, S)—CH$_3$ → R—(NH$_2$, OH, SH) + CH$_2$O (H on heteroatom of product)
4. Epoxidation	R—CH=CH—R′ → R—CH(—O—)CH—R′ (epoxide)
5. Desulfuration	R$_1$R$_2$P(=S)—X → R$_1$R$_2$P(=O)—X + S
6. Sulfoxidation	R—S—R → R—S(=O)—R′
7. N-hydroxylation	R—NH—C(=O)—CH$_3$ → R—NOH—C(=O)—CH$_3$
Reduction reactions	
8. Halogenated hydrocarbon reduction	CCl$_4$ → CCl$_3\cdot$
9. Nitro reduction	R—NO$_2$ → R—NO$_2\cdot$
10. Bipyridylium reduction	CH$_3$—$^+$N(C$_5$H$_4$)—(C$_5$H$_4$)N$^+$—CH$_3$ → CH$_3$—$^+$N(C$_5$H$_4$)—$\cdot$(C$_5$H$_4$)N$^+$—CH$_3$

[a] Modified from Neal (24).

reduction reactions are highly reactive free radicals and cause cell toxicity by either direct reaction with cell macromolecules or reoxidation by molecular oxygen with resultant formation of toxic oxygen species (26, 27). The potentially toxic consequences associated with generation of these reactive intermediates within the cell do not go unchecked since several biochemical defenses directed to controlling free radical-initiated oxidant reactions have been identified (Section 5.4).

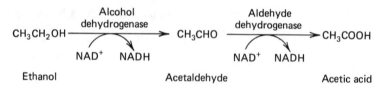

Figure 4.7 Detoxication of ethanol by oxidation to acetic acid (acetate).

5.1.2 Other Microsomal Phase I Metabolism

Although the cytochrome P-450 monooxygenase system is responsible for the majority of Phase I microsomal xenobiotic metabolism, two additional metabolic systems, flavin-containing monooxygenase (28) and epoxide hydrolase (29), have been identified in microsomes. Like cytochrome P-450-mediated metabolism, the flavin-containing monooxygenase requires molecular oxygen and NADPH. The substrate specificity of this oxidizing system is more limited than that of cytochrome P-450, however, and is restricted to oxidation of amine- and sulfur-containing compounds.

Epoxides formed from cytochrome P-450 oxidation of aromatic or olefinic double bonds are potentially reactive with critical cell macromolecules. One mechanism by which reactive epoxides are detoxified is through hydration to dihydrodiols, a reaction catalyzed by epoxide hydrolase. The observation that both epoxide hydrolase and the epoxide generating enzyme, cytochrome P-450, are in close association within the endoplasmic reticulum suggests the importance of this the enzyme as an important defense mechanism.

5.1.3 Alcohol and Aldehyde Dehydrogenases

The toxicity of a variety of primary alcohols is modulated by the metabolic action of alcohol and aldehyde dehydrogenases (Figure 4.7). The majority of alcohol dehydrogenase activity is in the cytosolic fraction of tissues, whereas aldehyde dehydrogenase activity is found in cytosol, mitochondria, and microsomal fractions (30, 32). The multiple isozymes of both enzymes require NAD^+ as a cofactor. It is important that these two defense enzymes act in sequence since both alcohols and their corresponding aldehydes are potentially toxic; alcohols frequently cause narcotic effects, and aldehydes readily form Schiff base adducts with cell macromolecules (24). Inhibition of aldehyde dehydrogenase by disulfiram, for example, has been used in the treatment of alcoholism. Thus, the discomfort caused by accumulation of toxic levels of acetaldehyde when alcohol is consumed by disulfiram-treated individuals acts as a deterrent to further alcohol consumption (32).

An additional aldehyde dehydrogenase, formaldehyde dehydrogenase, appears to have as its primary function the specific detoxication of formaldehyde (31). Formaldehyde is formed not only from the oxidation of methanol but also from N-demethylation of endogenous and exogenous compounds (see reaction 3, Table 4.1). Unlike the aldehyde dehydrogenases described previously, formaldehyde dehydrogenase uses both NAD^+ and reduced glutathione as cofactors.

5.1.4 Hydrolysis Reactions

A wide variety of toxicants contain ester or amide linkages capable of being hydrolyzed by a group of nonspecific esterases and amidases. The products of the hydrolysis reactions, carboxylic acids, alcohols, and free amines, are subject to further Phase I and II detoxication reactions (33).

The importance of enzymatic hydrolysis as a detoxication mechanism is readily appreciated when individuals bearing genetically mediated deficiencies in this enzyme activity are exposed to toxicants deactivated by this mechanism. For example, the duration of action of the neuromuscular blocking agent succinylcholine in humans is usually extremely brief because of its rapid hydrolysis by plasma and liver esterase to less active products. In a few individuals, however, prolonged apnea results from succinylcholine treatment. The aberrant response in these individuals has been attributed to the existence of an atypical pseudocholinesterase incapable of hydrolyzing succinylcholine (34).

Variations among species in the ability to detoxify agents by hydrolysis reactions has been exploited in the development of pesticides. Malathion, a widely used organophosphate insecticide, requires metabolic activation for expression of toxicity in both insects and humans. Although both species rapidly form the toxic intermediate, humans are far more capable than insects of subsequently hydrolyzing the intermediate to nontoxic products. This differential ability to catalyze hydrolysis of reactive intermediates is in part responsible for the species-specific toxicity of malathion and other organophosphate insecticides (35).

5.2 Phase II Detoxication Reactions

5.2.1 Glucuronic Acid Conjugation

Glucuronidation is a major Phase II reaction involved in the detoxication of aliphatic and aromatic alcohols, carboxylic acids, and certain types of amine- and thiol-containing compounds (36, 37). This reaction is characterized by conjugation of glucuronic acid with the toxicant or its metabolite and is catalyzed by a family of closely related enzymes, the glucuronyl transferases. The majority of the body enzyme activity is associated with the endoplasmic reticulum of the liver, although activity also has been found in the kidney, intestine, skin, brain, and spleen.

Conjugation with glucuronic acid does not occur directly but is mediated by formation of an active intermediate, uridine diphosphate glucuronic acid (UDPGA). The UDPGA substrate of the glucuronyl transferases is in the α

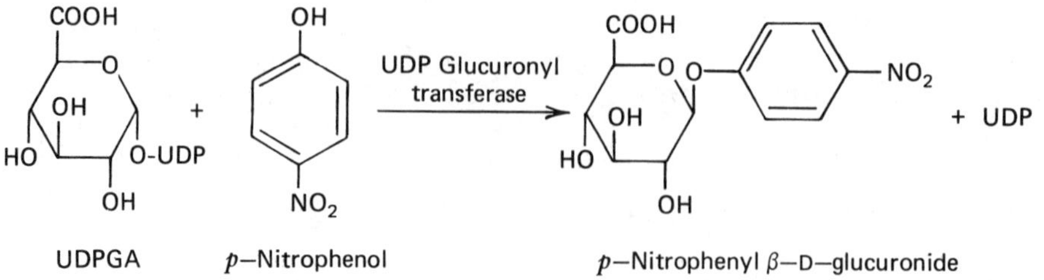

Figure 4.8 Conjugation of p-nitrophenol with glucuronic acid.

configuration, whereas the glucuronide conjugate is inverted to the β configuration (37) (Figure 4.8). Glucuronide conjugates are excreted in both urine and bile.

5.2.2 Sulfate Conjugation

Conjugation of toxicants with sulfate is catalyzed by a group of sulfotransferases found in several tissues (38). The sulfate ester products are highly water soluble and readily excreted by the kidney. Like glucuronic acid conjugation, sulfate conjugation proceeds by means of an activated sulfate intermediate, 3′-phosphoadenosine-5′-phosphosulfate (PAPS). The primary substrates for sulfate conjugation are primary, secondary, and tertiary alcohols, phenols, and to a lesser extent, aromatic amines. In most species, however, conjugation of these substrates with glucuronic acid predominates over sulfate conjugation. The preference for glucuronic acid conjugation has been attributed to the limited pool of PAPS available relative to that of UDPGA (24).

5.2.3 Glutathione Conjugation

The toxicity of many compounds is mediated through an electrophilic attack on nucleophilic sites of cellular macromolecules. The resultant covalently bound adducts may cause cell dysfunction or even death. Exposure to electrophilic agents is inescapable, however, since they are ubiquitous in the environment and also are formed in many endogenous metabolic reactions. Conjugation of these reactive agents with glutathione represents the primary mechanism by which the body prevents their attack on critical cellular macromolecules.

Glutathione is a tripeptide consisting of the sequence γ-glutamate–cysteine–glycine and is found in all tissues. Its concentration ranges up to 10 mM in organs such as the liver, which is highly active in forming reactive metabolites from toxicants. The thiol group of the cysteine residue is the functionally important element in detoxication reactions, serving as a readily available nucleophile for electrophilic attack.

Although electrophilic agents spontaneously react with glutathione to varying degrees, this reaction is greatly facilitated in cells by a family of cytosolic enzymes, the glutathione-S-transferases. These enzymes, which have broad substrate specificity, are particularly important in catalyzing the conjugation of hydrophobic, electrophilic toxicants with glutathione (24, 39).

Conjugation with glutathione enhances elimination of the toxicant from the body in two ways. First, to be excreted into the bile, toxicants must have a molecular weight of 325 or greater (37). Since glutathione has a molecular weight of 307, conjugation of even the simplest electrophilic molecules, when it occurs in the liver, frequently results in extensive biliary excretion. Excretion of glutathione conjugates into bile results in their subsequent elimination in the feces. In certain cases, however, glutathione conjugates excreted into bile may be metabolized by bacteria of the intestinal microflora, resulting in formation of lipophilic toxicants that may be reabsorbed (5). This cyclical process has been termed *enterohepatic circulation* and may result in a significant prolongation of the time required for elimination of certain toxicants from the body.

A second mechanism by which glutathione conjugates are eliminated from the body is by metabolism of the conjugates to mercapturic acids (37). Mercapturic acids are formed by sequential removal of glutamate and glycine from the glutathione conjugate, followed by acetylation of the α-amino group of the remaining cysteine conjugate. The enzymes responsible for catalyzing these reactions are found primarily in liver and kidney. Mercapturic acids are very water soluble and consequently are readily excreted by the kidney.

An examination of the dose-dependent hepatotoxicity of acetaminophen provides a good example of the importance of glutathione as a detoxication mechanism (40). Acetaminophen, a commonly used analgesic agent, produces severe liver necrosis after high doses in both humans and animals. The liver toxicity has been attributed to microsomal mixed-function monooxygenase metabolism of acetaminophen to an electrophilic intermediate capable of forming covalent adducts with critical cell macromolecules. The data in Figure 4.9, obtained from hamsters treated with increasing doses of acetaminophen, illustrate the protective role of glutathione in preventing arylation of cell macromolecules and associated liver necrosis. As the dose of acetaminophen was increased, the concentration of glutathione in the liver progressively declined as a result of conjugation of the reactive intermediate with hepatic stores of glutathione. It was not until glutathione concentrations were reduced to approximately 30 percent of control values, however, that any marked increase in covalent binding, liver necrosis, and mortality was observed. These data clearly suggested that toxicity only resulted when inadequate amounts of glutathione were available to trap the reactive metabolite, thereby freeing it to react with critical cellular macromolecules.

The role of glutathione in modulating acetaminophen toxicity was further demonstrated in a study in which hamsters were treated with diethyl maleate, an agent that reduces hepatic glutathione, prior to administration of acetaminophen (40). In diethyl maleate-pretreated animals, doses of acetaminophen that were minimally hepatotoxic to naive animals now produced extensive liver

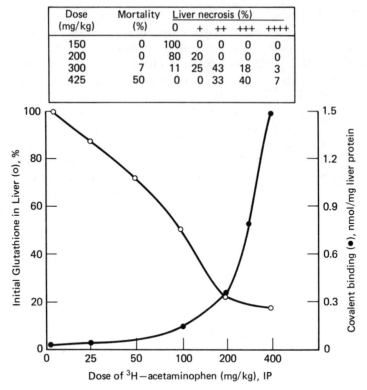

Figure 4.9 Relationship of liver glutathione concentrations and covalent binding to development of acetaminophen toxicity in hamsters. Modified from Potter et al. (40).

necrosis. The enhanced toxicity was presumed to be due to decreased availability of glutathione to trap the reactive intermediate of acetaminophen.

The results obtained with acetaminophen represent only one example of many in which the toxicity of electrophilic agents is clearly modulated by availability of tissue glutathione. An important implication from all of these studies is that low doses of an electrophilic toxicant may not result in any direct expression of toxicity and that toxicity may result only when the dose is sufficient to overwhelm the glutathione defenses.

5.2.4 Acetylation

Aromatic primary amines may be rendered less toxic by formation of N-acetyl conjugates. This conjugation reaction is catalyzed by a group of cytosolic enzymes, the N-acetyltransferases, which utilize acetyl–coenzyme A as the acetyl donor cofactor (24, 37, 41). Although the N-acetyltransferases catalyze acetylation of hydroxy and thiol groups of endogenous substrates, amines appear to be the only exogenous substrates for these enzymes.

BODY DEFENSE MECHANISMS TO TOXICANT EXPOSURE

1. Activation of toxicant:

$$\text{CoASH} + \text{RCOOH} \xrightarrow{\text{ATP}} \text{RCOSCoA}$$

Coenzyme A Acid-containing Acyl-coenzyme A
 toxicant intermediate

2. Conjugation of acyl-CoA intermediate with amino acid:

$$\text{RCOSCoA} + \text{H}_2\text{N CH}_2\text{ COOH} \xrightarrow{N\text{-acyl-transferase}} \text{RCONHCH}_2\text{COOH}$$

Acyl-coenzyme A Glycine Glycine conjugate
intermediate

Figure 4.10 Multistep conjugation of a toxicant containing a carboxyl group with the amino acid glycine.

An interesting feature of the N-acetyltransferases is that in humans and rabbits a genetically determined polymorphism exists in the ability to acetylate exogenous amines. Thus one population of individuals is capable of rapidly acetylating toxicants, whereas another conducts this conjugation slowly. This difference in the ability to acetylate makes a profound difference in the suceptibility of these populations to toxicity of amine-containing compounds. For example, individuals who are slow acetylators are more susceptible to the neurotoxic effects of the drug isoniazid, which is eliminated from the body after acetylation. Individuals who are fast acetylators do not accumulate toxic concentrations of isoniazid in their tissues as do the slow acetylators.

5.2.5 Amino Acid Conjugation

Toxicants containing aromatic carboxylic acids may be detoxified by conjugation with glucuronic acid (Section 5.2.1) or amino acids (24, 37). Like glucuronic acid conjugation, amino acid conjugation is a multistep process and involves formation of an activated intermediate (Figure 4.10). The first step of the reaction requires formation of an acyl-S-coenzyme A derivative of the toxicant. This reaction is carried out in mitochondria by enzymes thought to be identical with the fatty acid activating system. The activated intermediate is subject to nucleophilic attack by α-amino groups of several amino acids, primarily glycine and glutamate, and is catalyzed by mitochondrial and cytosolic N-acyltransferases specific for each amino acid substrate. Conjugation with amino acids not only reduces toxicity of the agent but also permits rapid renal excretion.

5.2.6 Methylation Reactions

Amine-, phenol-, and thiol-containing toxicants may form N-, O-, and S-methyl conjugates in reactions catalyzed by a variety of methyl transferases. The methyl donor for these reactions is S-adenosylmethionine (37, 42). Methylation of N- and O-containing toxicants is seldom regarded as a significant detoxication reaction, however, since the products formed are less water soluble, which may delay elimination of the compound.

There is some indication that S-methylation may represent an important detoxication pathway for hydrogen sulfide produced as a result of bacterial metabolism of dietary protein in the intestines (42). Hydrogen sulfide generated from methionine metabolism in the gut is sequentially metabolized to methanethiol and demethylsulfide by methyltransferases located in the mucosal lining cells of the intestine and liver. Since both methanethiol, and particularly demethylsulfide, are significantly less toxic than hydrogen sulfide, formation of these conjugates may protect organs such as the brain from the toxic effects of endogenously produced hydrogen sulfide. In fact, coma induced in humans with diminished liver function has been attributed to an accumulation of methanethiol in the blood resulting from an inability to fully methylate this endogenous toxicant. There is no evidence to date, however, that suggests that methylation represents a significant detoxication route for exogenously administered hydrogen sulfide (43).

5.3 Detoxication of Cyanide

The extreme toxicity of cyanide is due to its ability to inhibit oxidative respiration by complexation with cytochrome oxidase. This toxicity can be ameliorated in part by induction of methemoglobinemia, which provides a biologically inert binding site for cyanide (Section 3.2). Although cyanide is tightly bound to methemoglobin, additional steps must be taken to detoxify the small amounts of cyanide that dissociate from the complex. This is accomplished by intravenous administration of thiosulfate, which serves as a sulfur donor for the enzyme rhodanese. Rhodanese converts cyanide to thiocyanate, which is much less toxic than cyanide and is rapidly excreted by the kidney (24).

5.4 Oxidant Defense Mechanisms

A number of toxicants that cause oxidant injury in tissues have been identified. These toxicants include environmental air pollutants such as ozone and nitrogen dioxide, chlorinated hydrocarbons such as carbon tetrachloride, and agents capable of generating reactive forms of molecular oxygen such as the herbicide paraquat (26, 44).

An important mechanism by which oxidants damage tissue is through the membrane damaging process of lipid peroxidation (44) (Figure 4.11). Lipid peroxidation is a chain reaction process in which lipid radicals are formed by abstraction of allylic hydrogens of polyunsaturated lipids. Once formed, lipid radicals may abstract hydrogens from nearby polyunsaturated lipids, resulting in additional radical formation. Since polyunsaturated lipids are a major component of biological membranes, stimulation of lipid peroxidation may lead to cell dysfunction or even death.

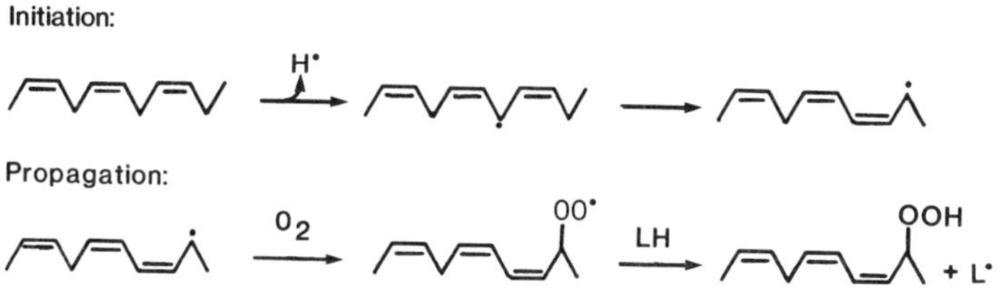

BODY DEFENSE MECHANISMS TO TOXICANT EXPOSURE

Initiation:

/=\/=\/=\/ —H·→ /=\/=\/=\/ ⟶ /=\/=\=\·

Propagation:

/=\/=\=\· —O₂→ /=\/=\=\(OO·) —LH→ /=\/=\=\(OOH) + L·

Termination:

L· + L·	→	nonradical products
L· + LO₂·	→	nonradical products
LO₂· + LO₂·	→	nonradical products

LH = polyunsaturated lipids; L· = lipid radical; LO₂· = lipid peroxy radical

Figure 4.11 Reactions associated with toxicant-initiated lipid peroxidation. Hydrogen abstraction may be mediated by organic radicals, hydroxyl radicals, or other lipid radicals formed in propagation reactions of lipid peroxidation. The chain reaction process of lipid peroxidation is quenched by a series of termination reactions. Antioxidants such as vitamin E inhibit lipid peroxidation by preferentially undergoing lipid radical-mediated hydrogen abstractions, resulting in formation of nonradical products.

Toxicants stimulate lipid peroxidation by a variety of mechanisms (44). Both nitrogen dioxide and ozone react directly with membrane lipids, resulting in lipid radical generation. Carbon tetrachloride is reductively metabolized by microsomal cytochrome P-450 to the trichloromethyl radical, which is capable of directly abstracting hydrogens from unsaturated lipids. A large number of toxicants appear to stimulate lipid peroxidation through generation of reactive forms of molecular oxygen. The primary mechanism by which toxicants stimulate reactive oxygen formation in tissues is through an enzyme-catalyzed single-electron reduction–oxidation reaction (26). In this type of reaction the parent toxicant molecule undergoes a single-electron reduction catalyzed by cellular reductases such as NADPH cytochrome P-450 reductase. The radical products of the reduction reaction are frequently unstable in the oxygen environment of the cell and are immediately reoxidized to the parent compound. An additional product of the reoxidation reaction is the superoxide radical (O_2^-) (Figure 4.6). Although O_2^- has weak oxidant activity itself, it is likely that O_2^- per se is not responsible for the oxidant damage associated with O_2^- generation in tissues. Superoxide apparently undergoes an iron-catalyzed Haber–Weiss reaction (Equation 3), however, in which O_2^- and hydrogen peroxide (H_2O_2) react to form the extremely potent oxidant, the hydroxyl radical (OH·). This radical is readily capable of abstracting hydrogens from

$$O_2^- + H_2O_2 \rightarrow OH\cdot + OH^- + O_2 \qquad (3)$$

unsaturated lipids and consequently stimulating lipid peroxidation reactions.

Toxicants containing bipyridylium, quinone, or nitro structural elements have been recognized as frequent participants in redox cycling reactions resulting in O_2^- production (26).

Generation of O_2^- has additional toxic consequences beyond the stimulation of lipid peroxidation (26). *In vitro* studies have indicated that hydroxyl radicals may produce DNA base and sugar phosphate damage and also may cause DNA strand scission. This type of damage to the genetic material may result in both mutagenic or carcinogenic responses. Hydroxyl radicals also react with proteins and polysaccharides, resulting in inactivation of other important functional macromolecules.

It should be clear that, if unchecked, toxicant-mediated oxidant reactions have tremendous potential for producing toxic injury within the cell. In fact, an adequate defense against oxidant attack is required of cells even in the absence of exposure to exogenous toxicants. In aerobic biological systems O_2^- is a product of many endogenous metabolic reactions (26). For example, O_2^- is formed from autoxidation of endogenous substrates such as hydroquinones, catecholamines, and thiols and is also produced as a product of the catalytic activity of enzymes such as xanthine oxidase and aldehyde oxidase. In aerobic systems, therefore, the ability to combat oxidant stress is an essential element of survival. In view of the extremely detrimental consequences of O_2^- generation within cells, it is not surprising that a multitiered defense system exists to combat oxidant stress originating not only from endogenous metabolic reactions but also from toxicant exposure (Figure 4.12).

5.4.1 Superoxide Dismutase, Catalase, and Peroxidase System Enzymes

The first line of defense against O_2^- generation is the enzyme superoxide dismutase, which catalyzes the dismutation of O_2^- to hydrogen peroxide and molecular oxygen. Three distinct forms of this metalloenzyme have been identified: iron- or managanese-containing enzymes are commonly found in prokaryotes, whereas a copper–zinc enzyme is found in the cystosol of eukaryotes (45). Mitochondria of eukaryotes, however, contain only the manganoenzyme.

The importance of superoxide dismutase as a biological defense against oxidant toxicity has been elegantly demonstrated by Fridovich and co-workers in a series of experiments in bacteria (45). The ability of *Escherichia coli* to survive in hyberbaric oxygen was found to be directly correlated to the activity of superoxide dismutase, that is, bacteria containing high levels of superoxide dismutase activity tolerated elevated oxygen tensions far better than did those with low enzyme activity. In addition, superoxide dismutase increased in *E. coli* grown under hyperbaric oxygen, which was presumed to represent an adaptive response to oxidant stress. A similar response was observed when bacteria were treated with paraquat, a toxicant that stimulates O_2^- production by redox cycling reactions. An important implication of these findings was that superoxide dismutase does not merely offer a static defense against increased oxidant

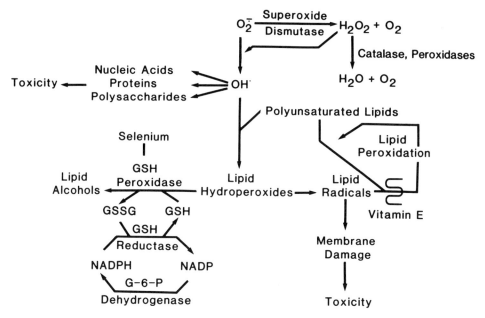

Figure 4.12 Bichemical defense mechanisms that modulate toxic reactions initiated by superoxide radicals. Modified from Bus and Gibson (26).

fluxes but also has the capacity to respond with increased synthesis of the enzyme.

Mammalian systems also have been shown to increase tissue superoxide dismutase activity in response to oxidant exposure. Rats exposed to 85 percent oxygen for 7 days have a marked increase in pulmonary superoxide dismutase activity (46). These rats subsequently exhibited prolonged survival when exposed to an oxidant environment (100 percent oxygen) that produces complete lethality in nonadapted animals. Return of rats exposed to 85 percent oxygen to room air resulted in a gradual decline of the elevated superoxide dismutase activity to normal levels, which was associated with a corresponding loss of tolerance to 100 percent oxygen exposure.

For effective function of superoxide dismutase as a defense mechanism, the product of the dismutation reaction, H_2O_2, must also be detoxified. Hydrogen peroxide not only has intrinsic oxidant activity but is also a necessary participant with O_2^- in the Haber–Weiss reaction in the production of toxic hydroxyl radicals. Two mechanisms participate in detoxication of H_2O_2: catalase and the glutathione peroxidase system enzymes, which consist of glutathione peroxidase, glutathione reductase, and glucose-6-phosphate dehydrogenase. The three enzymes of the glutathione peroxidase system act in concert, with glucose-6-phosphate supplying reducing equivalents needed for glutathione reductase activity, which in turn maintains adequate concentrations of glutathione required for glutathione peroxidase activity. The consequences of an inadequate ability to detoxify H_2O_2 are evidenced in humans who have a genetically determined deficiency in erythrocyte glucose-6-phosphate dehydrogenase activity. Extensive erythrocyte hemolysis occurs when these individuals are exposed to drugs or toxicants that generate increased concentrations of H_2O_2 within the erythrocyte, presumably because adequate concentrations of reduced glutathione cannot be maintained in the face of diminished glutathione peroxidase system activity (19).

5.4.2 Antioxidant Protection Against Lipid Peroxidation

A second line of defense exists to combat the deleterious effects of lipid peroxidation initiated by oxygen radicals, organic radicals, or gaseous toxicants such as ozone and nitrogen dioxide (Figure 4.12). The lipid-soluble antioxidant vitamin E serves to terminate the chain reaction process of lipid peroxidation, whereas glutathione peroxidase consumes lipid hydroperoxides that otherwise are a source of additional lipid radicals (44).

The significance of these defense mechanisms in controlling the toxic effects of oxidant agents can be seen from the example of paraquat, which stimulates O_2^- generation in cells. Mice with decreased glutathione peroxidase activity or vitamin E concentrations, obtained by respective administration of diets deficient in selenium or vitamin E, were markedly more susceptible to paraquat-induced lethality (26). It is interesting to note that paraquat is primarily a lung toxicant, which is due in large part to the fact that paraquat is preferentially taken up into this organ by an active transport mechanism. However, in mice on selenium-deficient diets that had significantly reduced glutathione peroxidase in both liver and lung, paraquat also produced extensive liver toxicity (47). These data suggest that paraquat causes lung toxicity in normal animals because accumulation and retention of toxicant overtaxes the oxidant defense mechanisms in that organ. Toxicity in liver, which does not retain paraquat, occurs only when a defense mechanism is compromised.

The glutathione peroxidase system enzymes respond to oxidant stress in the same manner as superoxide dismutase (44). Thus, rats exposed to low concentrations of ozone had significantly increased activity of all three enzymes of this defensive system. This increased oxidant defense capacity appeared to be in part responsible for the development of tolerance to subsequent exposure to high concentrations of ozone.

5.5 DNA Repair

Many toxicants are capable of reacting with nucleic acids, forming lesions that may result in mutagenic or carcinogenic responses. Not all toxicant–DNA interactions lead to these responses, however, since cells have the capacity to repair many types of DNA damage. DNA repair is accomplished by two general mechanisms, excision repair and postreplication repair (48).

DNA excision repair in mammalian cells involves several differing processes. DNA damage resulting from exposure to UV radiation or electrophilic toxicants with large ring systems (e.g., polycyclic aromatic hydrocarbons) is repaired by what has been termed a "long-patch" mechanism. This mechanism involves removal and subsequent replacement of up to 100 adjacent damaged nucleotides. DNA damage caused by exposure to methylating or ethylating agents or by ionizing radiation is repaired by a short-patch mechanism. Short-patch repair involves removal and replacement of short segments of apurinic and apyrimidinic sites in DNA resulting from both nonenymatic and enzymatic removal of alkylated DNA bases. Although both excision repair processes involve a complex series of enzymatic reactions, excision repair is regarded as a relatively error free process.

In contrast to the excision repair mechanism, postreplication DNA repair is regarded as being an error-prone process. As its name implies, postreplication repair occurs only during the replication, or S phase, of the cell cycle. DNA damaged by both chemical toxicants or radiation is subject to repair by this mechanism.

The potential importance of DNA repair mechanisms in controlling the onset of cancer can be illustrated from two examples. The human disease xeroderma pigmentosum is characterized by a genetically mediated deficiency in the DNA repair enzymes responsible for removing pyrimidine dimers induced in skin cells as a result of exposure to UV radiation (49). Individuals suffering from this disease exhibit a predisposition for skin cancer as a result of exposure to sunlight. In animal studies the organotropic, or organ-specific, carcinogenicity of methylating- or ethylating-type carcinogens was found to directly correlate with a persistence of alkylated DNA bases in the affected tissue (50). For example, following administration of N-ethyl-N-nitrosourea to rats, O^6-ethylguanine was removed at a far slower rate from brain, which is a target tissue for this carcinogen, than from liver, which is a nontarget tissue. In both of these examples, therefore, the susceptibility to environmentally induced carcinogenesis directly correlated with the ability of the affected tissue to repair promutagenic lesions in DNA.

6 IMMUNOLOGICAL DEFENSE MECHANISMS

The immune system has been recognized in recent years to play a potentially important role in control of the development of cancer (51). The incidence of cancer is markedly higher, for example, in humans chronically treated with immunosuppressive agents or afflicted with primary immunodeficiency diseases. In addition, studies in animals have demonstrated that chemically induced tumors produce strong antitumor immune responses that may ultimately result in regression or elimination of the tumor. These observations suggest the possibility that, even if a toxicant should penetrate the host of defense mechanisms protecting the genetic material of a cell and initiate a malignant transformation, intervention by the immune system may prevent development of a tumor.

7 CONCLUSIONS

Information regarding the interaction of specific chemical agents with the body defense mechanisms has important practical application in the assessment of the potential hazard associated with exposure to these agents. The toxicity of any foreign compound is necessarily dependent on the concentration of the toxicant at its receptor site(s) within cells or tissues. Thus, in part through intervention of the body defense mechanisms, the dose ingested is seldom equivalent to the dose delivered to the target site. Toxic responses associated with exposure to many agents exhibit a dose threshold or a dose below which no response is likely to occur. Since intervention of the body defense mechanisms may be a primary determinant underlying the existence of many thresholds, a comparative evaluation of defense mechanism function among species may be extremely useful in extrapolating animal toxicity data to humans.

REFERENCES

1. J. D. Robertson, *Arch. Intern Med.*, **129**, 202 (1972).
2. J. A. Timbrell, *Principles of Biochemical Toxicology*, Taylor and Francis, London, 1982, pp. 21–47.
3. F. E. Guthrie, "Absorption and Distribution," in: E. Hodgson and F. E. Guthrie, Eds., *Introduction to Biochemical Toxicology*, Elsevier, New York, 1980, p. 10.
4. B. Idson, *J. Pharm. Sci.*, **64**, 901 (1975).
5. C. D. Klaassen, "Absorption, Distribution, and Excretion of Toxicants," in: J. D. Doull, C. D. Klaassen, and M. O. Amdur, Eds., *Toxicology: The Basic Science of Poisons*, 2nd ed., Macmillan, New York, 1980, p. 28.
6. R. L. Dixon, "Toxic Responses of the Reproductive System," in: J. D. Doull, C. D. Klaassen, and M. O. Amdur, Eds., *Toxicology: The Basic Science of Poisons*, 2nd ed., Macmillan, New York, 1980, p. 332.
7. D. B. Menzel and R. O. McClellan, "Toxic Responses of the Respiratory System," in: J. D. Doull, C. D. Klaassen, and M. O. Amdur, Eds., *Toxicology: The Basic Science of Poisons*, 2nd ed., Macmillan, New York, 1980, p. 246.
8. D. L. Swift, and D. F. Proctor, "Access of Air to the Respiratory Tract," in: J. D. Brain, D. F. Proctor, and L. M. Reid, Eds., *Respiratory Defense Mechanisms*, Part 1, Marcel Dekker, New York, 1977, p. 63.
9. R. F. Hounam and A. Morgan, "Particle Deposition," in: J. D. Brain, D. F. Proctor, and L. M. Reid, Eds., *Respiratory Defense Mechanisms*, Part 1, Marcel Dekker, New York, 1977, p. 125.
10. D. F. Proctor, "The Mucociliary Clearance System," in: D. F. Proctor and I. Andersen, Eds., *The Nose: Upper Airway Physiology and the Atmospheric Environment*, Elsevier, New York, 1982, p. 245.
11. G. M. Green, G. J. Jakab, R. B. Low, and G. S. Davis, *Am. Rev. Resp. Dis.*, **115**, 479 (1977).
12. J. G. Widdicombe, "Defense Mechanisms of the Respiratory Tract and Lungs," in: J. G. Widdicombe, Ed., *Respiratory Physiology II*, Vol. 14, University Park Press, Baltimore, 1977, p. 291.

13. A. C. Allison, "Mechanisms of Macrophage Damage in Relation to the Pathogenesis of Some Lung Diseases," in: J. D. Brain, D. F. Proctor, and L. M. Reid, Eds., *Respiratory Defense Mechanisms*, Part 2, Marcel Dekker, New York, 1977, p. 1075.
14. J. M. Creeth, *Br. Med. Bull.*, **34**, 17 (1978).
15. J. A. Swenberg, C. S. Barrow, C. J. Boreiko, H. d'A. Heck, R. J. Levine, K. T. Morgan, and T. B. Starr, *Carcinogenesis*, **4**, 945 (1983).
16. Y. Alarie, *C. R. C. Crit. Rev. Toxicol.*, **2**, 299 (1973).
17. J. C. F. Chang, E. A. Gross, J. A. Swenberg, and C. S. Barrow, *Toxicol. Appl. Pharmacol.*, **68**, 161 (1983).
18. S. E. Mayer, K. L. Melmon, and A. G. Gilman, "Introduction; the Dynamics of Drug Absorption, Distribution, and Elimination," in: A. G. Goodman, L. S. Goodman, and A. Goodman, Eds., *The Pharmacological Basis of Therapeutics*, 6th ed., Macmillan, New York, 1980, p. 1.
19. R. P. Smith, "Toxic Responses of the Blood," in J. Doull, C. D. Klaassen, and M. O. Amdur, Eds., *Toxicology: The Basic Science of Poisons*, 2nd ed., Macmillan, New York, 1980, p. 311.
20. P. B. Hammond and R. P. Beliles, "Metals," in J. Doull, C. D. Klaassen, and M. O. Amdur, Eds., *Toxicology: The Basic Science of Posions*, 2nd ed., Macmillan, New York, 1980, p. 409.
21. M. G. Cherian and M. Nordberg, *Toxicology*, **28**, 1 (1983).
22. M. Webb and K. Cain, *Biochem. Pharmacol.*, **31**, 137 (1982).
23. R. T. Williams, *Detoxication Mechanisms*, 2nd ed., Chapman and Hall, London, 1959.
24. R. A. Neal, "Metabolism of Toxic Substances," in: J. Doull, C. D. Klaassen, and M. O. Amdur, Eds., *Toxicology: The Basic Science of Poisons*, 2nd ed., Macmillan, 1980, p. 56.
25. M. J. Coon and A. V. Persson, "Microsomal Cytochrome P-450: A Central Catalyst in Detoxication Reactions," in: W. B. Jakoby, Ed., *Enzymatic Basis of Detoxication*, Vol. 1, Academic Press, New York, 1980, p. 117.
26. J. S. Bus and J. E. Gibson, "Role of Activated Oxygen in Chemical Toxicity," in: J. R. Mitchell and M. G. Horning, Eds., *Drug Metabolism and Drug Toxicity*, Raven Press, New York, 1984, p. 21.
27. M. W. Anders, "Biotransformation of Halogenated Hydrocarbons," in: J. R. Mitchell and M. G. Horning, Eds., *Drug Metabolism and Drug Toxicity*, Raven Press, New York, 1984, p. 55.
28. D. M. Ziegler, "Microsomal Flavin-containing Monooxygenases: Oxygenation of Nucleophilic Nitrogen and Sulfur Compounds," in: W. B. Jakoby, Ed., *Enzymatic Basis of Detoxication*, Vol. 1, Academic Press, New York, 1980, p. 201.
29. F. Oesch, "Microsomal Epoxide Hydrolase," in: W. B. Jakoby, Ed., *Enzymatic Basis of Detoxication*, Vol. 2, Academic Press, New York, 1980, p. 271.
30. W. F. Borson and T.-K. Li, "Alcohol Dehydrogenase," in: W. B. Jakoby, Ed., *Enzymatic Basis of Detoxication*, Vol. 1, Academic Press, New York, 1980, p. 231.
31. H. Weiner, "Aldehyde Oxidizing Enzymes," in: W. B. Jakoby, Ed., *Enzymatic Basis of Detoxication*, Vol. 1, Academic Press, New York, 1980, p. 261.
32. J. M. Ritchie, "The Aliphatic Alcohols," in: A. G. Gilman, L. S. Goodman, and A. Gilman, Eds., *The Pharmacological Basis of Therapeutics*, 6th ed., Macmillan, New York, 1980, p. 376.
33. E. Heymann, "Carboxylesterases and Amidases," in: W. B. Jakoby, Ed., *Enzymatic Basis of Detoxication*, Vol. 2, Academic Press, New York, 1980, p. 291.
34. D. W. Nebert, "Human Genetic Variation in the Enzymes of Detoxication," in: W. B. Jakoby, Ed., *Enzymatic Basis of Detoxication*, Vol. 1, Academic Press, New York, 1980, p. 25.
35. A. P. Kulkarni and E. Hodgson, "Comparative Toxicity," in: E. Hodgson and F. E. Guthrie, Eds., *Introduction to Biochemical Toxicology*, Elsevier, New York, 1980, p. 106.
36. C. B. Kasper and D. Henton, "Glucuronidation," in: W. B. Jakoby, Ed., *Enzymatic Basis of Detoxication*, Vol. 2, Academic Press, New York, 1980, p. 3.

37. W. C. Dauterman, "Metabolism of Toxicants: Phase II Reactions," in: E. Hodgson and F. E. Guthrie, Eds., *Introduction to Biochemical Toxicology*, Elsevier, New York, 1980, p. 92.
38. W. B. Jakoby, R. D. Sekura, E. S. Lyon, C. J. Marcus, and J.-L. Wang, "Sulfotransferases," in: W. B. Jakoby, Ed., *Enzymatic Basis of Detoxication*, Vol. 2, Academic Press, New York, 1980, p. 199.
39. W. B. Jakoby and W. H. Habig, "Glutathione Transferases," in: W. B. Jakoby, ed., *Enzymatic Basis of Detoxication*, Vol. 2, Academic Press, New York, 1980, p. 63.
40. W. Z. Potter, S. S. Thorgeirsson, D. J. Jollow, and J. R. Mitchell, *Pharmacology*, **12,** 129 (1974).
41. W. E. Weber and I. B. Glowinski, "Acetylation," in: W. B. Jakoby, Ed., *Enzymatic Basis of Detoxication*, Vol. 2, Academic Press, New York, 1980, p. 169.
42. R. A. Weisiger and W. B. Jakoby, "S-Methylation: Thiol S-Methyltransferase," in: W. B. Jakoby, Ed., *Enzymatic Basis of Detoxication*, Vol. 2, Academic Press, New York, 1980, p. 131.
43. R. O. Beauchamp, Jr., J. S. Bus, J. A. Popp, C. J. Boreiko, and D. A. Andjelkovich, *C. R. C. Crit. Rev. Toxicol.*, **13,** 25 (1984).
44. J. S. Bus and J. E. Gibson, *Rev. Biochem. Toxicol.*, **1,** 125 (1979).
45. I. Fridovich, *Science*, **201,** 875 (1978).
46. J. P. Crapo and D. F. Tierney, *Am. J. Physiol.*, **226,** 1401 (1974).
47. S. Z. Cagen and J. E. Gibson, *Toxicol. Appl. Pharmacol.*, **40,** 193 (1977).
48. D. J. Holbrook, Jr. "Chemical Carcinogenisis," in: E. Hodgson and F. E. Guthrie, Eds., *Introduction to Biochemical Toxicology*, Elsevier, New York, 1980, p. 310.
49. J. E. Cleaver, *Nature*, **218,** 652 (1968).
50. P. Kleihues, "The Role of DNA Alkylation and Repair in the Toxic and Carcinogenic Effects of Alkylnitrosoureas," in: G. L. Plaa and W. A. M. Duncan, Eds., *Proceedings of the First International Congress on Toxicology: Toxicology as a Predictive Science*, Academic Press, New York, 1978, p. 191.
51. J. H. Dean, M. J. Murray, and E. C. Ward, "Toxic Modifications of the Immune System," in: J. Doull, C. D. Klaassen, and M. O. Amdur, Eds., *Toxicology: The Basic Science of Poisons*, 3rd ed., Macmillan, New York, in press.

CHAPTER FIVE

Biological Rhythms, Shiftwork, and Occupational Health

M. H. SMOLENSKY, Ph.D.,
D. J. PAUSTENBACH, Ph.D., and
L. E. SCHEVING, Ph.D.

1 INTRODUCTION

Human beings are classified as a diurnally active species. This means that a majority of people prefer a daily routine of daytime work alternating with nighttime sleep. In spite of strong preference for diurnal work schedules, an increasing and significant percentage of persons in the United States and other industrialized nations are engaged in shift schedules requiring nighttime work (1–4). Even in developing countries, shiftwork is gaining popularity (5–11). Even though shiftwork schedules are common in many industries, we now know that *not* all individuals are able to work these schedules without suffering some adverse effects.

Basic research and epidemiologic studies have demonstrated that some employees are able to tolerate shiftwork without complaint, even for several decades, whereas others can be biologically or psychologically intolerant of it even for relatively short spans of time—6 months or so. In the past, complaints of shiftwork intolerance were often viewed by management as organized labor's attempt to justify concessions rather than as legitimate difficulties of shiftwork itself. In an effort to resolve this issue, researchers have, over the last 30 years, investigated the potential effects of shiftwork on the health and safety of workers. Finally, after many years of painstaking research and the publication of over 5000 manuscripts, the biological and psychosocial bases of the problems of shiftworkers are better understood.

We now know that human functions undergo periodic alterations during each 24 hr so as to prepare the various systems (e.g., cardiovascular, endocrine) of the body for the predictable variation in human activity and rest that occurs each day. These 24-hr bioperiodicities, called *circadian* (ca. 24-hr) *rhythms*, represent an important aspect of our genetic inheritance. The timing of the circadian peaks and troughs of the various biological functions is determined by the daily scheduling of the sleep and activity periods. For example, persons who work only during the day and sleep during the night exhibit a timing of biological rhythms with regard to clock hour that differs from that in permanent nightworkers who are routinely active during the nighttime and who sleep during the daytime. Sometimes, people need to work irregular schedules and these can tax the flexibility of some biological systems. For example, some work schedules require workers to rotate every few days between day and night shifts. As a result, some circadian rhythms can become disorganzied. This lack of organization, termed desynchronization, can produce a group of symptoms such as upset stomach, insomnia, and fatigue, which are common among many rotating shiftworkers.

At one time, health professionals were concerned that employment on rotating shiftwork schedules could cause increased morbidity and mortality among these workers. Fortunately, the results of numerous biological rhythm and epidemiologic studies indicate that shiftwork, in itself, *does not* directly induce disease. However, we also know that for many persons, shiftwork and especially rotating shifts that require periodic night duty can affect the overall well-being of employees by increasing the likelihood of developing certain minor discomforts such as gastrointestinal and sleep disorders. Although these disorders may be classified as minor with respect to their ability to jeopardize survival, such annoyances can produce very real difficulties that can seriously affect the workers' lives.

Researchers are actively attempting to develop practical approaches to lessening the potential adverse health hazards or other difficulties associated with shiftwork. Ways to minimize the adverse physiological consequences of rotating between day and night shifts by altering the starting and stopping times have been proposed (13–20). So that persons engaged in the usual rotating 8-hr as well as the so-called unusual work schedules (e.g., 10- to 12-hr) will have the same degree of protection against exposure to air contaminants as persons working normal, 8-hr, daytime-only schedules, mathematical models for adjusting workplace exposure limits for these substances have been proposed (21–29). Other research efforts involve ways to minimize the effect of shiftwork on employee moral, job productivity, accident occurrence, and family life (19, 30–32). As much as possible, research in these areas is coordinated among the groups so that a complete picture of the cause of difficulties will result. Thus far, those closest to the efforts are optimistic that the psychosocial and physiological problems often associated with shiftwork, although complex, will be solvable. The results of these multidisciplinary investigations have produced a number of recommendations for minimizing certain shiftwork-induced disorders; these have been reviewed by Rutenfranz (12) and Reinberg (15) and are discussed in this chapter.

In order to more fully comprehend and alleviate the difficulties experienced by shiftworkers, additional multidisciplinary research efforts are needed. One discipline that provides useful solutions is *chronobiology*—that branch of science that investigates biological rhythms. During the past 40 years, most persons trained in the biological sciences were taught that human biological systems were best described by the homeostatic theory, a concept that was first put forward by Claude Bernard in the nineteenth century and later amplified by Walter Bradford Cannon during the early years of this century. According to the homeostatic theory, a set of regulatory systems maintains the constituents of the environment surrounding the cells of the body in relative constancy. Although this theory was extremely useful for understanding the mechanisms of human physiological function during the first part of this century, findings from chronobiologic research conducted during the past three to four decades have provided and broadened insight into the functioning of biological processes. We now know that the physiologic and metabolic activities of all living organisms, including human beings, are rhythmic over time in a predictable manner rather than homeostatic. In short, the functions of even the simplest cells are structured not only in space, as in anatomy, but also in time.

Now that the existence and role of biological rhythms has been identified, occupational health professionals have an opportunity to understand how and why human beings who are involved in shiftwork can respond to various types of chemical challenges, medicines, and work schedules different than persons on regular shifts. As in most research, our understanding of biological rhythms also has been useful for answering questions beyond basic chronobiology. For example, our knowledge of biological rhythms has allowed us to improve the practice of clinical medicine (32–43) and to lessen the risks and discomforts associated with shiftwork (12–15, 17–19, 44).

The topic of shiftwork is of such importance that it has been the subject of several international conferences, the more recent of which were held in Paris, France (18) and Kyoto, Japan (17). These were organized as a part of the activity of the Scientific Committee on Shift Work of the Permanent Commission and International Association on Occupational Health (PCIAOH). Numerous papers presented at these international symposia and conferences have been discussed elsewhere (45). The objectives of this chapter are to address all aspects of shiftwork and chronobiology as they relate to the physical and emotional healthfulness of workers. This task was particularly difficult since the intent of much rhythm research was not to answer questions surrounding shiftwork. The nature and popularity of shiftwork schedules and human biological rhythms are reviewed. The significance of biological rhythms as they pertain to the predictable difference in employees' biological tolerance to various shiftwork schedules as well as the effects of shiftwork on the physiology, health, and well being of employees is emphasized. This chapter is unique in that it attempts to critically review the major research efforts that have not previously been evaluated for their usefulness in minimizing the adverse effects of shiftwork.

2 WORK SCHEDULES

Work schedules are the time patterns that employees follow on the job. They represent an attempt to match the production requirements of industry to the needs, wishes, and availability of workers. For persons who fill jobs that involve night duty, some degree of physiological and psychosocial adjustments will often be required. Unfortunately, not all persons are capable of making all the adjustments necessary for permanent work on these jobs. In general, persons who find shiftwork intolerable are sincerely interested in the work but because of genetic or environmental factors cannot adjust. Perhaps the primary reason night shifts are rarely well received by workers is the natural (biological) human preference for diurnal activity and nocturnal sleep. Chronobiologists prefer to use the term "diurnal" when referring to daytime activity and the term "nocturnal" to nighttime activity. When speaking of a 24-hr biological rhythm, the term "circadian" (circa = about; dian = day) is most appropriate.

As noted by Taylor (46), a celebrated shiftwork researcher who conducted extensive epidemiologic investigations of morbidity and mortality patterns in dayworkers and shiftworkers, "man is a diurnal animal and society is designed for the day worker." We now know that human preference for diurnal activity and nocturnal rest relates to both psychosocial and biological factors. With regard to the first of these factors, most people desire a work schedule that allows for sufficient meaningful leisure time for family and friends and for attending cultural, sporting, or other events of interest. With regard to the latter factor, most if not all biological functions are circadian rhythmic in support of the alternating work–rest routine, wherein persons are typically active during the day and asleep at night.

Nearly all of the available data indicate that in order for a person to routinely work a night shift schedule, adjustment of nearly all circadian rhythms is required. In a majority of persons, this biological adjustment will require several days. During this time many unpleasant symptoms, like those of "jet lag," may be experienced. The symptoms of jet lag and shiftwork, both caused by a change in the clock hours of sleep and activity, commonly include fatigue, reduced mental and physical efficiency, and impaired performance as well as sleepiness and hunger at inappropriate times. In short, the physiological and emotional conflicts that can be caused by shiftwork include the disorganization of one's biological rhythms and the disruption of patterns of socialization. The result is that people who may otherwise find their jobs agreeable can become less productive, more fatigued, less happy, and perhaps more vulnerable to the adverse effects of exposure to chemical and physical agents in the workplace when they convert from fixed day work to shiftwork.

Traditionally, the fixed work schedule adopted by most firms consists of five 8-hr workdays weekly with each starting at 8 A.M. and ending at 5 P.M. The need to operate certain manufacturing processes 24 hr/day, however, has made it necessary for more than one 8-hr crew of employees to be present. This has led to the development of numerous shift schedules, the most popular being the three-team rotating shift system, which includes nighttime hours. The use of three crews, each working 8 hr, enables uninterrupted production throughout the 24-hr period every day of the year so that a process never need be shut down because of a worker shortage. Frequently additional overtime hours are necessary in order to accommodate routine requirements to maintain equipment, fill in for absent workers, and meet seasonal loads. In certain situations, the overtime hours may increase the hazards associated with shiftwork.

Many continuous process operations, such as those found in oil refineries, chemical operations, steel and aluminium mills, pharmaceutical manufacturing, glass plants, and paper mills, cannot be shut down without causing serious production problems and thus require "round-the-clock" monitoring. Consequently, the inability to halt certain chemical and physical processes coupled with economic factors, such as the high cost of equipment and increased foreign competition, have forced these capital-intensive industries to make shiftwork a necessary part of manufacturing.

From a sociological perspective, both the standard rotating and, to a lesser extent, unusual work schedules have been shown to affect the worker's family and community life. This is not surprising because the nature of rotating shiftwork schedules requires the time patterns of recreation and socializing undergo reorganization weekly, sometimes even more often. Commonly, when shiftworkers are at home, their friends and family may be at work, school, or asleep. Another frequent shortcoming is that shiftworkers may have only one full weekend per month during which they *are not* at work. As a result, shiftworkers and their family members are frequently disappointed with the quantity and quality of time that they spend together. Years of research have shown that shiftwork is likely to affect the well-being and health of employees. Often, the adverse effects include difficulties in sleeping, excessive fatigue, and irritability (12, 15, 44–50). Other problems that are frequently reported include constipation, gastritis, duodenal and peptic ulcers, high absenteeism, and lessened productivity (12, 51–54). Because of these and other difficulties, numerous kinds of modified work schedules have been developed and implemented by companies seeking an alternative to the standard rotating 8-hr, 40-hr/week work schedule. Modified or unusual work schedules are different as they do not involve an 8-hr/day, 5-day workweek. Generally, unusual schedules consist of shortened workweeks but lengthened workdays.

So-called *unusual workshifts* have been devised in an attempt to alleviate, or at least minimize, some of the difficulties caused by the standard three-team rotating 8-hr shiftwork schedule. These atypical shifts generally involve work durations much longer than 8 hr and, in some instances, require rapid changes between night and daywork. Often, these unusual shifts involve work periods of 10, 12, 14, and 18 hrs per day. To evaluate the potential adverse health effects of shiftwork, occupational health professionals must draw on knowledge of many different disciplines, including sociology, psychology, physiology, toxicology, pharmacology, and chronobiology, as well as their understanding of the rationale on which occupational exposure limits have been established.

3 REVIEW OF POPULAR WORK SCHEDULES IN INDUSTRY

3.1 The Standard 5-day, 40-hr Workweek

Clearly, the most common work schedule is the standard 5-day, 40-hr week. Typically, morning arrival to work is at 8 A.M. and evening departure is at 5 P.M. Among all nonfarm workers, more than 80 percent work 5 days a week. By far the largest percentage of employees (41 percent) work exactly 40 hr weekly (see Figure 5.1). Nearly half work between 35 and 40 hr. For example, the average duration of the workweek for all nonfarm wage earners was 38.6 hr in 1979; it was less than 40 hr because of inclusion of part-time employment in the estimate. It should be noted that there is considerable variation in the length of the workweek depending on the industry and the age and sex of the employees. In May 1978, for men aged 25 to 44 years, it averaged 44 hr, whereas for women of the same age it was 36 hr. On the average, wage and salaried persons engaged in manufacturing worked 40.9 hr in 1979, whereas service industry workers averaged 36 hr (1). Because of the wide range of

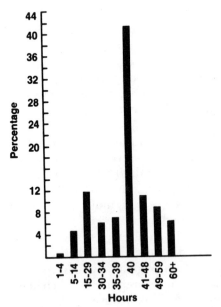

Figure 5.1 Distribution of weekly work hours. From Employment and Training Report to the President. U.S. Government Printing Office, Washington, D.C., 1979.

Table 5.1 Categories of Shiftwork Systems[a]

I. Permanent shift systems
   A. Permanent morning shift
   B. Permanent afternoon shift
   C. Permanent night shift
   D. Split shifts with each consistently timed for a given worker
II. Rotating shift systems
   A. Systems without night shifts
      1. Discontinuous
         a. Nonoverlapping (e.g., crew 1, 6 A.M. to 2 P.M.; crew 2, 2 P.M. to 10 P.M.)
         b. Overlapping (e.g., crew 1, 6 A.M. to 2 P.M.; crew 2, 1:30 P.M. to 9:30 P.M.)
      2. Continuous
         a. Nonoverlapping (e.g., crew 1, 6 A.M. to 2 P.M.; crew 2, 2 P.M. to 10 P.M.)
         b. Overlapping (e.g., crew 1, 6 A.M. to 2 P.M.; crew 2, 1:30 P.M. to 9:30 P.M.)
   B. Systems with night shift
      1. Discontinuous
         a. Two-team (e.g., each crew works 12-hr shifts)
         b. Three-team (e.g., each crew works 8-hr shifts)
      2. Continuous (regular)
         a. Two-team (e.g., each crew works 12-hr shifts)
         b. Three-team (e.g., each crew works 8-hr shifts)
      3. Continuous (irregular)
         a. Varying numbers of teams
         b. Varying numbers of cycle lengths

[a] Modified from Knauth and Rutenfranz (55).

schedules available to industry and business, no one of them achieves a majority of usage; nonetheless, the standard 5-day, 40-hr workweek outdistances all the others by a wide margin (1).

3.2 Shiftwork

"Shiftwork" generally refers to employment that entails the presence of employees in the workplace between 4 P.M. and 8 A.M. A great variety of shift schedules are utilized in industry. We have modified the Knauth–Rutenfranz (55) classification (Table 5.1) for the purpose of discussing the different shiftwork patterns.

Category I refers to permanent shift systems. These are rather common in the United States. With these systems the start and end times of work are always the same for a given employee. This is in contrast to the situation for rotating schedules in which a crew systematically varies its work and offtimes at specific, for example weekly, intervals. Permanent systems are those in which a given crew consistently works the same shift, whether in the morning, evening, or night. Permanent systems also include the nonvarying split-shift pattern of the merchant navy in which the work "day" is divided into two portions with a significant duration of offtime between each. For example, one crew can be assigned to duty from midnight to 4 A.M. as well as from noon to 4 P.M., a second from 4 A.M. to 8 A.M. and again from 4 P.M. to 8 P.M., and a third from 8 A.M. to noon and again from 8 A.M. to midnight.

Category II pertains to rotating shiftwork systems. These may be continuous or discontinuous in form. Discontinuous shiftwork systems are defined as work patterns that require the presence of employees Monday through Friday or Saturday with at least one or two days offtime, usually over the weekend. These are quite common in industrial and commercial plants where continuous operation is not necessary. "Continuous shift systems," on the other hand, refer to those schedules that include weekend work. These predominate in the continuous process industries and in the public service sectors, such as fire, ambulance, and police departments. Rotating shift systems do not necessarily include nighttime hours. This is the case for both the so-called two-team "double-day" pattern that typically consists of morning and afternoon shifts as well as the three-shift (morning, afternoon, and evening) systems. The evening shift of the latter commonly involves part-time workers. However, many industries require rotating shiftwork with regularly scheduled nighttime duty. When this is the case, different crews, usually two (each working 12 hr) or three (each working 8 hr), rotate their hours of work at set intervals. In certain industries the speed of crew rotation between shifts is slow, every 7 days or more; in others it may be much faster, every 2 or 3 days; this is termed *rapid rotation*. In most cases the order of rotation between shifts is in a forward direction of morning to afternoon (or evening) to nighttime work schedules; however, for some persons the order of rotation is in a backward direction from nighttime to afternoon (or evening) to morning work schedules. In most large industries a variety of shift schedules are likely to be in use simultaneously to meet the needs of different operations.

The number of employees or percentage of the workforce engaged in shiftwork in the United States is not trivial (1–3). In general, rotating shiftwork schedules are especially prevalent in industries requiring uninterrupted service, manufacturing operations that cannot be routinely halted, and those in which the monetary investment is so great that the noncontinuous operation of equipment is economically prohibitive. Shiftwork is very common, for example, in the automotive, electronics, health, petrochemical, and pharmaceutical industries.

Dependence on shiftwork schedules in the United States has increased steadily since World War II (4). As expected, the bulk of shiftwork occurs in industries located in metropolitan centers, as opposed to rural regions. In 1978 4.9 million persons were working evening shifts, whereas an additional 2.1

million were working night shifts (2). Overall, during 1978 it was estimated that 16 percent of all employees were shiftworkers (2).

3.3 Flexible and Staggered Work Hours

A schedule of flexible work hours, or "flexitime" as it is termed, is one in which employees choose their starting and quitting times within limits set by management. In general, flexitime schedules differ with regard to (a) daily versus periodic (e.g., weekly or monthly) differences in the starting and quitting times, (b) variable versus constant length workday (whether credit and debit hours are allowed), and (c) core time—the hours of the day when all employees are required to be present (1).

In 1980 it was estimated that about 11 percent of all organizations and 9 percent of all workers, or 7 to 9 million people, were using flexitime in the United States (1). If professionals, managers, salespeople, and self-employed persons, who have long set their own hours without calling it flexitime, were to be included, the usage rate might be as great as 81 percent. Flexitime has been used by all major businesses and industries, but with a somewhat heavier concentration in financial and insurance companies and governmental agencies than in manufacturing facilities. There has been a roughly equal popularity of the three major flexitime models with perhaps a slightly greater use of "gliding time," which involves flexibility not only in the times of reporting to and leaving work but also in the number of hours worked daily. At present, flexitime and gliding time are more common in Europe than in the United States. For example, in Germany and Switzerland more than one-third of the workforce has flexible hours (2, 57).

4 UNUSUAL WORK SHIFTS

An unusual work shift is different from the standard 8-hr one, which generally begins at 8 A.M. and ends at around 5 P.M. By definition, the unusual work shift may be longer or shorter in duration with regard to the hours worked per day. They are often shorter (compressed) with regard to the number of days worked per week. Over the past 15 years, "other-than-standard" work schedules, which have been variously termed or classified, such as unusual, modified, altered, abnormal, exceptional, abbreviated, nonnormal, novel, extraordinary, odd, nontraditional, special, compressed, and even weird, have received increased attention by both labor and management. A permanent committee of the American Industrial Hygiene Association (AIHA) was convened in 1981 to review and promote research on the occupational health aspects as well as recommend approaches to adjusting occupational exposure. This committee decided that the term "unusual work schedule" seemed to best describe those schedules that consist of workdays that are either longer or shorter than 8 hr. Thus, this term is used to describe any schedule that is markedly different in length (either longer or shorter) than the "standard" 8 hr/24 hr, generally 5-day workweek, pattern.

A *compressed* workweek schedule is one type of unusual work shift that has been used in many nonmanufacturing and manufacturing settings. It refers to full-time employment accomplished in less than 5 days/week. Many compressed schedules are used but the most common are (1) workweeks of four separate 10-hr shifts, (2) workweeks of three separate 12-hr shifts, (3) workweeks of four separate 9-hr shifts plus one 4-hr span (usually on Fridays), and (4) the 5/4–9 plan of alternating 5-day (or night) with 4-day (or night) workweeks, with each shift 9 hr in duration (1).

In 1980, 2.2 percent of all full-time nonfarm wage and salaried employees or 1.7 million people were on unusual (compressed) workweeks. Of this number, two-thirds were working 4-day weeks. Compressed workweeks were used more in some industries than in others. Initially, their heaviest use was in certain city service departments and in small manufacturing firms (Table 5.2).

Unusual work shifts do not always involve compressed workweeks. They may be in the form of a great variety of fairly complicated schedules that have been implemented in manufacturing facilities over the past two decades. Examples of some types of unusual work schedule and an indication of an industry that has used each include (a) four 10-hr workdays per week (chemical); (b) a 6-week cycle of three 12-hr workdays for 3 weeks followed by four 12-hr workdays for 3 weeks (pharmaceutical); (c) a 6-hr/day, 6-day workweek (rubber); (d) a 56/21 schedule involving 56 continuous days of work, each of 8 hr duration followed by 21 days off (petroleum); (e) a 14/7 schedule involving 14 continuous days of work, each of 8 to 12 hr in duration, followed by 7 days off (petroleum); (f) a 3/4 schedule involving only three 12-hr workdays in one week followed by a week of four 12-hr workdays (pharmaceutical); (g) four 12-hr workdays followed by three 10-hr workdays, followed by five 8-hr workdays, then 4 days off (petroleum); (h) five 8-hr workdays, followed by two 12-hr days, then five more 8-hr workdays followed by 5 days off work (military); (i) a 24-hr day followed by two days off (maritime) and numerous other variations of these (23, 24, 58).

Unusual (including compressed) workweeks began to gain widespread popularity first in Canada and the United States in the early 1970s. Later they gained the attention of the petrochemical industry in Europe. The use of the compressed workweek grew rapidly from 1970, when very few persons worked these types of schedules, to 1.3 million workers, or 2.2 percent of the labor force, in 1975. Until 1981 the usage rate has remained steady (Figure 5.2). In 1981, as a result of a desire to conserve energy, more companies began utilizing compressed workweeks. The statistics on these types of work schedules can be deceiving. Although it is true that the total number of employees (mostly office personnel) working unusual shifts has lessened or leveled off, the number of employees in the chemical and other manufacturing industries being placed on unusual shifts is still growing (1, 21, 23, 58).

Table 5.2 Use of the Compressed Workweek in the United States According to Industry and Occupation as of 1980[a,b]

	Employees on <5-day Workweek	
	Number (1000s)	Percent
Industry		
Mining	17	2.0
Construction	148	3.4
Manufacturing	422	2.2
Transportation, public utilities	143	2.7
Wholesale, retail trade	267	2.4
Finance, insurance, real estate	76	1.8
Professional services	382	2.8
Other services	148	3.6
Federal public administration, except postal service	34	2.1
State public administration	31	3.6
Local public administration	180	10.8
Occupation		
Professional and technical personnel	303	2.5
Managers and administrators	83	1.1
Sales workers	63	1.9
Clerical workers	225	1.7
Craft workers	244	2.4
Operatives	290	3.3
Laborers	73	2.5
Service workers	439	6.7

[a] *Source:* U.S. Bureau of Labor Statistics.
[b] All figures refer to nonfarm wage and salary workers who usually work full time.

Many unusual work schedules make use of a rapid shift rotation between two 10- or 12-hr shifts to minimize the number of nights that employees must be present on the job. In these schedules, persons transfer between day and night work such that the number of consecutive nights worked is limited to two or three before changing to daywork or offtime (15, 58, 59). One type, for example, involves three daytime 12-hr shifts followed by 4 days off, then three 12-hr night shifts followed by 3 days off. This rotation usually takes 13 weeks to complete and provides for five weekends off (58).

Not all employees find the rapid rotation between night and daywork to be physically or emotionally agreeable; specifically, some persons simply find it too stressful and taxing for long-term use. Research is being directed at evaluating the psychological and physiological effects of the rapid rotating shifts, especially with regard to the potential effects of these schedules on

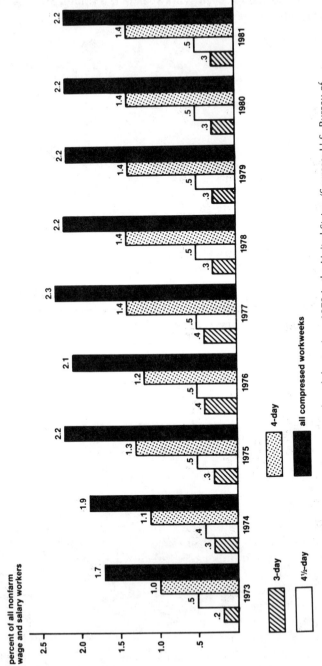

Figure 5.2 Usage of compressed worksheets since 1973 in the United States. (Sources: U.S. Bureau of Labor Statistics, 1982; Hedges, 1975).

circadian rhythms (15, 59–61) and methods for predetermining which persons will be most appropriate for this type of shiftwork (15).

In summary, unusual shifts have been devised and are now being used as one means to decrease some of the undesirable aspects of shiftwork. These schedules consist of shift durations that are generally longer than 8 hr and workweeks that are shorter than or equal to 4 days. Many of these schedules involve rotation between night and day shifts on alternating weeks. The two most common types of unusual schedules used in larger manufacturing facilities are the 10-hr, 4-day workweek and the 12-hr, 3- or 4-day workweek (1, 58).

4.1 A History of Unusual Schedules and Compressed Workweeks

In the early 1960s the so-called compressed workweek was initiated in various companies that were primarily involved in light manufacturing or clerical services. This schedule has become very popular among office workers, although it has not been as well received by manufacturing facilities because of the difficulties involved in covering continuous processes. In 1970, the petrochemical industry began using the 12-hr workshift. Soon after initiation, it was apparent that this schedule had many advantages. Today, the 12-hr/day schedule is the most popular of the modified work schedules in those industries requiring continuous processing (23). It has the advantage of providing complete manpower coverage over the 24 hr while providing more leisure and off time than the rotating 8-hr shift. Importantly, in comparison to the typical 8-hr shift, the 12-hr one allows more time for family and friends as well as one's favorite activities. Numerous other advantages also have been reported (58–60). Complex in its application, the 12-hr shift has nonetheless been very positively received by those industrial plants adopting it (58).

By and large, the pressures on corporations to develop a more acceptable schedule of work than the 8-hr rotating one were sufficiently great that modified work schedules were initially adopted with primarily only four considerations in mind: (1) cost, (2) safety, (3) productivity, and (4) morale (58). Interestingly, in most firms, it appears that potential impact on the health of the employees was *not* considered until *after* the adoption of the unusual work schedule. A listing of the constraints and incentives surrounding unusual shifts is provided in Table 5.3.

4.2 Why the 12-hr Work Shift Became Popular

The earliest major use of the 12-hr schedule is reported to have occurred in 1955 at a manufacturing facility of Eli Lilly and Company (58). The plant was involved in producing pharmaceuticals and had previously operated 7 days each week, but for no more than 16 hr daily. When a change in the chemical processing required that the facility be operated continuously for 24 hr, 365 days per year, a 12-hr work shift was adopted in an attempt to minimize hardship on the employees. Although initially there was concern that some persons would find the schedule fatiguing, the shift was well received by both the operators on the line and the management (58). As anticipated, the employees were pleased to have a greater opportunity to be involved in social activities during the daytime, which by means of the new work schedule amounted to continuous blocks of time equalling up to 3 to 4 days weekly.

In spite of the generally favorable acceptance of this schedule, the 12-hr/day schedule was not used widely in the United States until the 1970s. However, in 1971, when Eli Lilly and Company started a new facility in Clinton, Indiana, it was decided to place all plant employees (except senior management) on the 12-hr schedule. Based on its prior experience, management felt that such a work pattern would entice top-quality employees to join the firm and also help maintain a smoother 24-hr operation. In an effort to confirm that the workers' attitudes were favorable toward the 12-hr schedule, the company conducted a number of employee surveys. These revealed a strong preference for the 12-hr/day rapid rotation over the standard weekly rotation of 8 hr/day. Fueled by the resulting increased productivity and efficiency, as well as high employee morale experienced in Clinton, Eli Lilly and Company adopted this work pattern at a number of its other production facilities. In part as a result of this company's success with this unusual shift, the 12-hr workshift has been adopted by several other pharmaceutical manufacturers. Today, over 20,000 workers in this industry are involved in 12-hr shifts (23).

At about the same time, but not in conjunction with the pharmaceutical manufacturers, the petrochemical industry began experimenting with the 12-hr shift (58). In 1971, Imperial Oil, Ltd., Exxon's Canadian affiliate, initiated it, and their efforts were responsible for bringing about the eventual widespread adoption of this work pattern in the petroleum industry. The Canadian company had experienced numerous complaints from their 8-hr/day shift employees regarding discontent with their restricted social and family lives. In response, the company began a trial of the 12-hr/day work schedule, evaluating it over a 9-month span for productivity, safety, administrative costs, and employee morale. Following the trial, the new system was unanimously approved by both the workers and management. It was soon adopted by five other Imperial Oil facilities. Within the subsequent 2 years, many companies within the United States also began experimenting with the 12-hr shift. Facilities within Exxon, Dow, Ciba-Geigy, Shell, Chevron, Pfizer, Goodyear, Monsanto, E. I. DuPont, Kodak, Owens-Corning, and numerous other firms have since implemented or experimented with this schedule (23, 58).

It is estimated that during 1984 at least 60,000 persons in the United States were employed in chemical industries utilizing 12-hr shifts (21). In general, use of this shift has been restricted to the chemical, petrochemical, rubber, electronic, transportation, and pharmaceutical industries, which all have a high level of automation. Industries requiring more strenuous physical labor such as steel, foundry, and automotive manufacturing, all of which require manual material handling, have shown less interest since physiological fatigue is a comparatively greater potential problem during extended workdays.

As noted before, some firms have implemented a rapid shift rotation along with the longer workday so as to minimize the number of consecutive nights during which workers must be on nighttime duty each week. Rapid rotation also allows for longer blocks of offtime. A myriad of rapid rotation schedules are possible (15, 59) but they are all characterized by changes within short intervals between day and night work.

According to Reinberg et al. (62), a secondary benefit of rapid rotation is reduction in significant alterations of circadian rhythms, such as those observed in regularly rotating shiftworkers having night duty (15, 59). This is important since alteration in the temporal staging of circadian rhythmic processes due to the disruption of one's usual activity–rest routine when working several (usually three or more) night shifts in succession can result in symptoms resembling those of jet lag. Most employees on rapid rotation, when working nights, *do not* appear to undergo significant alteration in the staging of their 24-hr biological rhythms. Thus, they have a greater chance of being exempt from many of the unpleasant symptoms and complaints typically found in non-rapid-rotation 8-hr shiftworkers. Research now in process (15, 62, 63) is aimed at developing sensitive screening tests to identify those persons who are likely to be highly tolerant to the different types of rotating shiftwork schedule.

4.3 Advantages and Disadvantages of Unusual Schedules

The industrial interest in unusual work shifts and the compressed workweek is understandable. Many firms that use 12-hr/day shifts find that their 24-hr continuous processes demonstrate better consistency of operation since two rather than three different crews of employees are involved. Occasionally, fewer persons may be needed to complete the same set of tasks. Workers, especially younger ones, usually desire the longer work hours per shift because of the higher annual income resulting from the scheduled overtime hours, more full days off from work, and less time spent per week commuting.

As with any work schedule, however, there are some drawbacks of the 12-hr shift (Table 5.3). It has been reported that older employees who have become accustomed to 8-hr work durations and who are more susceptible to fatigue are often less than enthusiastic about schedules involving longer shift durations (59, 60). Some companies have found that younger workers should be available and trained to fill the jobs of older ones so that they can be transferred as needed to "normal" shiftwork schedules. Indeed, some firms, including Eli Lilly and Company, have recognized that an aging workforce may not prefer the 12-hr work schedule and have kept a fraction of their production lines on 8-hr shifts to provide these employees an opportunity to work a schedule that best suits their needs.

Eventually, federal and/or state legislation concerned with protecting worker health may make the 12-hr shift less attractive since stricter occupational health standards might be recommended to protect these persons (58). For example, the American Conference of Governmental Industrial Hygienists (ACGIH) has

Table 5.3 Main Factors Affecting Usage of Compressed Workweeks and Unusual Work Shifts

Constraints	Incentives
Stringent work technology requirements may disrupt production operations; interface and coverage problems	Can smooth out production operations in some cases; output and productivity can increase
Potential fatigue of workers could cause productivity losses and increase risk of injury through accidents	Utility costs are reduced if buildings are shut down during off days
Supervision must often be stretched over a longer day; management more difficult	Unpleasant aspects of shift work can be partially alleviated with the compressed or unusual shift schedules
Labor law requires premium pay for hours worked in excess of 8 per day for workers on government contract or under some collective agreements	Morale, absenteeism, and recruiting often improved; income is usually greater each year because of scheduled overtime, shift differential, and unscheduled overtime
Many employees, especially women, older workers, and parents with young children, do not like them	
Family life and weekday social and civic activities often are disrupted	Commuting trips reduced, saving time and money for workers
Sometimes no improvement in employee–employer relationship is observed; does not always improve quality of work life	Leisure time redistributed into longer blocks, which most workers like
High failure rate among one-time user companies and the need to maintain a fairly automated process discourages many firms	Rapid rotation schedules can eliminate a lifetime of night shiftwork

considered adjusting the Threshold Limit Value (TLV) for the occupational exposure to noise from the present 85 dBA for 8 hr of work to something less than this for those on the 12-hr shift (58). Several groups, including the Occupational Safety and Health Administration (OSHA), have suggested that the time-weighted average (TWA) exposure to airborne toxicants should be reduced for persons working unusual work schedules (24–29, 64, 65). These approaches are discussed in detail in another chapter of this volume (21). In addition, overtime during work on the 12-hr shift poses specific problems, since physical and emotional fatigue is likely when persons work longer than 12 hr per shift for several consecutive days.

The Wharton Business School at the University of Pennsylvania has conducted a comprehensive evaluation of the 12-hr shift, the most popular of the extraordinary work shifts (58). This evaluation involved the polling of 50 plant sites that had used or were using this shift schedule. The investigation examined the costs of implementation, impact on productivity, maintenance of safety standards, and compliance to the Fair Labor Standards and Walsh-Healey Acts.

Overall, the investigators concluded that in many industries, such as the highly automated and clerical ones, there were distinct advantages to adopting the 12-hr schedule. For others, such as heavy manufacturing and research, the 12-hr shift system was less advantageous or inappropriate (1, 58).

5 EMPLOYEE DISSATISFACTION WITH SHIFTWORK

Wilson and Rose (58) as well as many other researchers (59, 60) have reported that one of the major reasons for the move toward the 12-hr and other unusual work shifts was concern about the sociological problems caused by standard 8-hr rotating shiftwork. In general, the study by Wilson and Rose (58) revealed that the complaints of workers centered around three major problem areas:

1. The necessity to periodically work unusual hours with few weekends off.
2. The imposition of a work schedule that requires performance of activities at a time that is contrary to the optimum functioning of bodily processes with respect to the organization of circadian rhythms.
3. The decay of family and social life because shiftworkers, although averaging approximately 40-hr weekly, find their leisure time occurring during hours when friends are at work, children are at school, or most other people are at sleep.

Moreover, in addition to these problems, shiftworkers commonly express concerns about difficulty with sleep, fatigue, digestion, anxiety, and feelings of social isolation (12, 49, 55, 66).

5.1 Psychosocial Problems of Shiftworkers

Workers who are employed nights, either permanently or periodically, because of employment on a rotational schedule, experience a life-style that differs markedly from their daytime-working friends and also from their diurnally active family members. Human beings as a species are diurnally active; that is, they are active during the daytime and sleep during the nighttime. This means that a majority of businesses, recreational facilities, cultural events, and various other types of activity are scheduled mainly to meet the needs of diurnally active persons. Even though a certain number of businesses in most large metropolitan centers have evening hours, with some remaining open round-the-clock, the choices for rotating shiftworkers are less than those of permanent dayworkers. Although it is difficult to generalize for all workers and shift systems, a variety of studies have revealed that work schedules involving nighttime duty results in at least some degree of difficulty (67–75).

The drawbacks of shiftwork are exemplified by the findings of a large study of British steel mill shiftworkers (75). Complaints centered around not having weekends free or not having enough time for pursuing a full social life, watching

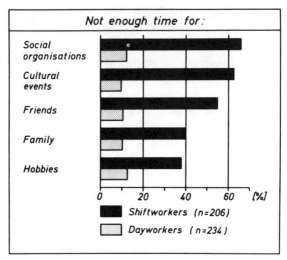

Figure 5.3 Frequency distribution of reports of insufficient time for designated social activities in a group of 206 rotating shiftworkers and in a group of 234 permanent dayworkers. From Nachreiner and Rutenfranz (73).

sports, attending social functions, planning social engagements, and following a regular television series (75). Nachreiner and Rutenfranz (73) found vast disparities between shiftworkers and dayworkers with regard to the type and number of complaints about not having sufficient time for pursuing social and leisure time desires. Figure 5.3 reveals that four to six times as many shiftworkers as compared to dayworkers in the chemical industry complained of not having enough time for social activities, including friends and family. On the other hand, it is noteworthy that the British shiftworkers studied by Wedderburn (75) felt they were, relative to dayworkers, better off with regard to pay, freedom during the daytime, time for one's ownself, free time, and variety of working hours.

Shiftwork has significant impact not only on the employees themselves, but also on members of their families (12, 15, 49, 69, 70). Time budget studies clearly point out that wives of shiftworkers have a more difficult time trying to keep the family together. Among other things, they find it necessary to change the timings of meals to fit the husband's shift schedule, which may change weekly. Moreover, in families in which one spouse is a rotating shiftworker and the other a dayworker, great strain is placed on the marriage. Communication between husband and wife, as well as children, becomes restricted or curtailed. In addition, the shiftworking parent finds it difficult or impossible to attend important school and/or other functions. In certain situations, children become resentful toward the shiftworking parent. Spouses also can become disenchanted; they may feel alone and neglected and marital discord may result. Thus, many researchers believe that rotating shiftwork can increase the likelihood of divorce and contribute to the development of personality disorders in children. Moreover, when it is the wife who is a rotating shiftworker, off-the-job demands and responsibilities, such as tending to the children and home, further add to the stresses of the rotating shiftwork pattern. These responsibilities usually take priority over making up sleep deficits resulting from working nights. This contributes to the high level of fatigue experienced by most women shiftworkers who have families (69).

5.2 Sleep Disruption and Fatigue

Shiftworkers often complain about the inadequacy of the quantity and quality of their sleep. This subject has been studied by many investigators (47, 61, 76–95). The vast research on this topic has recently been reviewed by Rutenfranz and his colleagues (12, 55, 66). The need for sleep, independent of the type and schedule of shiftwork, varies greatly between individuals and also according to age. However, the duration required is definitely more than that achievable with night work as demonstrated through studies on locomotive engineers by Rutenfranz et al. (94). These investigators used self-assessment questionnaires to sample a group of 329 shiftworking locomotive engineers. Figure 5.4 shows the results of their survey. When the responses to the inquiry pertaining to the duration of sleep needed are plotted in the form of a frequency distribution, it is clear that a majority of the workers (ca. 85 percent) expressed a preference for 7 to 9 hr. When the same persons were queried as to the amount of sleep attained when working nights, the distribution of sleep durations was very different. Compared to the former curve, the one of attained sleep was skewed to the left, meaning that there was a rather large percentage of the workers who obtained less than the quantity of sleep that was required. Since sleep loss and fatigue are known risk factors of industrial accidents, the matter of sleep quantity and quality as it relates to accidents and sickness is a major interest of shiftwork researchers.

Workers frequently complain of problems concerning a lack of sleep, particularly during assignment to the night rotation. Results of various surveys indicate that as many as 90 percent complain about their sleep when working nights. In contrast, only 5 to 20 percent of dayworkers and shiftworkers not having to work nights complain of sleep problems. It is important to point out that there are a number of factors that can contribute to sleep difficulties in shiftworkers. These include housing conditions, times of going to bed, age, noise, marital status, the presence and age distribution of children in one's household, and the regularity of shift rotation, for example. Noise, in particular, and the lack of access to a quiet place to take daytime sleep make sleep difficult when working nights.

Figure 5.5, from Knauth and Rutenfranz (66), provides factual information about sleep duration in relation to the type of work schedule based on a thorough study of almost 5000 shiftworkers engaged in various occupations. For the purpose of classifying the different work schedules, those commencing

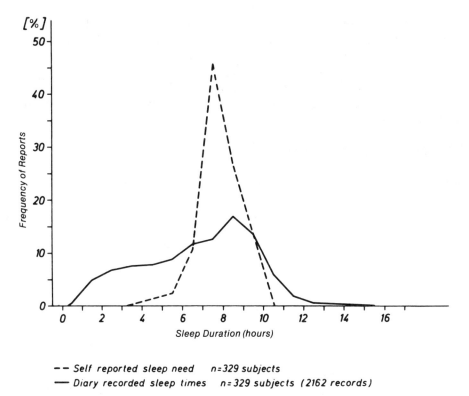

- - Self reported sleep need   n = 329 subjects
— Diary recorded sleep times   n = 329 subjects (2162 records)

Figure 5.4 Frequency distribution of sleep duration reported to be required by 329 rotating shiftworkers employed as locomotive engineers compared to the frequency distribution of their sleep durations (based on 2162 sleep records) between two consecutive shifts. Sleep duration data were obtained by self-assessment diaries. From Rutenfranz et al. (94).

between 5 A.M. and 9:59 A.M. were termed *morning shifts*; between 10 A.M. and 1:59 P.M., *midday shifts*; between 2 P.M. and 6 P.M., *evening shifts*; and between 10 P.M. and 3:30 A.M., *night shifts*. The data in the upper portion of the figure suggest that shiftworkers either did not choose to sleep during the daytime unless it was absolutely necessary, or if they did, to take less than that typical of nighttime sleep. In particular, day sleep during a day off or after returning home from working a morning shift was taken only by about 10 percent of the rotating shiftworkers, apparently as a nap averaging less than an hour. This is not unexpected; shiftworkers prefer to partake in activities with friends and family who are diurnally active. Only when persons worked the night shift was sleep more commonly taken during the day. In this instance more than two-thirds took sleep during the daytime prior to the first night duty and especially after the last night shift when day sleep averaged about 4 hrs. Between successive night shifts, sleep had to be scheduled during the daytime; however, the duration was found to average only about 6 hr. The duration of nighttime

BIOLOGICAL RHYTHMS, SHIFTWORK, AND OCCUPATIONAL HEALTH

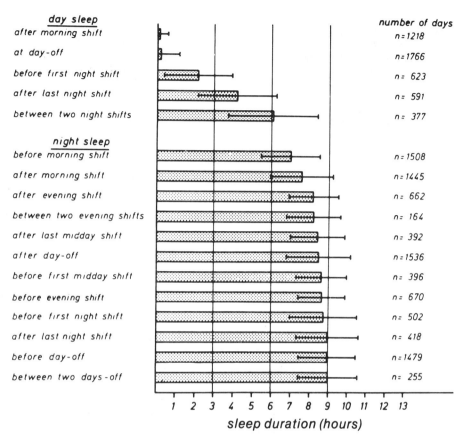

Figure 5.5 Average (shown with the standard deviation) duration of sleep as a function of the shift schedule worked and the timing of rest during either daytime or nighttime. The data summarized in the figure represent the findings from sleep diary records from 1230 persons encompassing in total of 9840 24-hr spans. See text and Knauth and Rutenfranz (95) for details.

sleep when working morning, midday, and evening shifts does not vary greatly. Only as a consequence of working the morning shift was the average duration of nightly sleep somewhat shorter than 8 hr. Since human beings tend to be rather inflexible about the time (clock hours) when they go to bed (55), reduction in sleep duration is inevitable when the morning shift commences early.

The inability to obtain sufficient sleep during the daytime when one is working the night shift results from at least two types of problems. First, it is difficult to obtain restful sleep during the daytime because of noises emanating from diurnally organized activities (66, 94). Workers most often complain that their daytime sleep is disrupted because of noises (Figure 5.6) made by children, neighbors, or road traffic (66). Second, for biological reasons, sleep taken

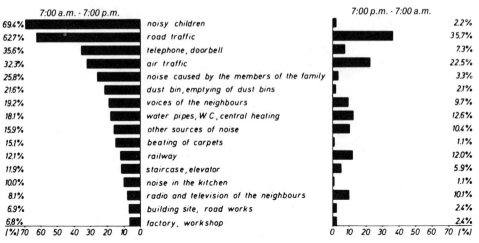

Figure 5.6 Relative frequency of sources of noise resulting in interrupted daytime (7 A.M. to 7 P.M.) or nighttime (7 P.M. to 7 A.M.) sleep. Data from 808 European shiftworkers who complained about frequent noise-induced sleep disruptions. From Rutenfranz et al. (66).

during the daytime, especially around midday (Figure 5.7), tends to be of shorter duration than nighttime sleep (76, 78, 95). When taken during the daytime, sleep is forced to occur at a biological time that is inappropriate and incompatible with the staging of circadian rhythms. The adjustment of circadian rhythms to nighttime duty in a majority of shiftworkers requires at least a few days. Until this adjustment is complete, the capability of sleeping during the daytime is low. The subject of biological rhythm adjustments in relation to shiftwork is addressed in great detail in Section 8, especially Section 8.2.2, of this chapter.

In their review, Rutenfranz et al. (66) found that the frequency of complaints about sleep disturbances varied with the type of task performed and the shift schedule worked (Figure 5.8). Examination of the data from a study of 5766 employees revealed that sleep disturbances were not uncommon. They were a complaint even for as many as 10 to 20 percent of the dayworkers. In comparison, as many as between 70 and 80 percent of those employees who rotated between day and night work experienced sleep disturbances. An interesting and important finding is that the incidence of sleep disturbances in former shiftworkers while engaged in rotating shiftwork could be as great as 90 percent. Although it was found that transfer to permanent daytime work reduced the incidence of complaints, sleep disturbance was still a significant problem for more than 15 percent.

In summary, it may be concluded that permanent daywork and shiftwork schedules that exclude nighttime duty do not ordinarily cause problems with sleep. On the other hand, rotating shiftwork schedules that incorporate a nighttime rotation or involve permanent nightwork, necessitating the displace-

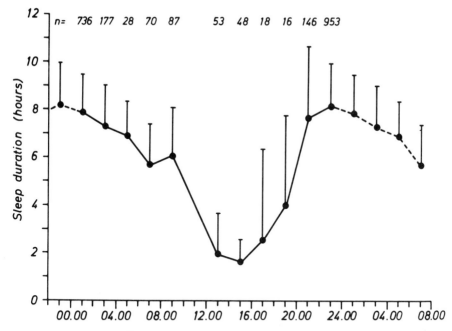

Figure 5.7 Sleep duration ($\bar{X}$ and SD) based on data from 304 shiftworkers as a function of the time during the day and night when it began. Indicated across the top is the number of reports for sleep onset according to clock hour from which the means and standard deviations were calculated. The revealed rhythmic trend found by Knauth and Rutenfranz (95) is similar to that shown by field studies on train drivers by Foret and Lantin (78) and through laboratory studies by Akerstedt and Gillberg (76). From Knauth and Rutenfranz (95).

ment of sleep to times other than the night, are more likely to cause sleep problems in some persons. The major concerns about sleep disturbance and reduction when work occurs at night (during the usual span of sleep) involve the worker's perception of lowered physical well-being, potential increased risk of illness (66), and increased fatigue with loss of vigor (Figure 5.9) (47). The loss of vigor coupled with insufficient sleep has been shown to contribute to a reduction in worker performance and productivity as well as an elevated risk of accidents in the workplace (30, 95, 96) (see Section 9.6).

5.3 Shiftwork as a Potentiator of Illness

The relationship between shiftwork and ill health is controversial. Although it has been speculated that illness is more common in shiftworkers in comparison to dayworkers, the findings of epidemiologic investigations on the morbidity and mortality of such employees have been inconclusive (97–101). Some have reported increased morbidity or mortality in shiftworkers, whereas others have reported the opposite or no difference. From an epidemiologic perspective,

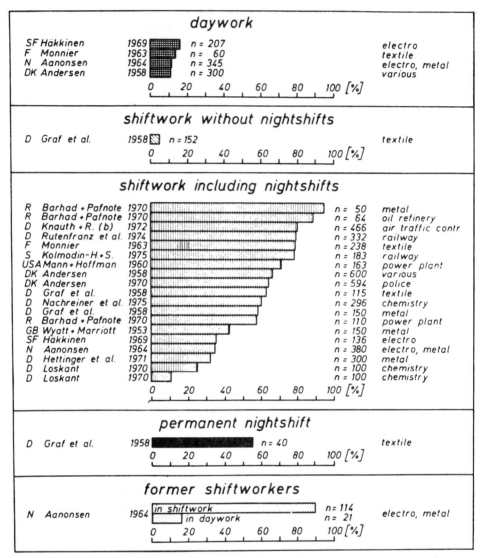

Figure 5.8 Frequency of complaints about sleep disturbances according to shift system, industry, and investigator. From Rutenfranz et al. (66).

failure to select appropriate control groups has weakened the conclusions of most studies. A common belief of some epidemiologists is that shiftworkers represent a type of survival population, since those employees who are incapable of tolerating shiftwork, for one reason or another, eventually withdraw to daywork schedules. As a result, the "dropouts" from rotating shiftwork schedules become part of the daytime controls that are utilized in typical case comparison

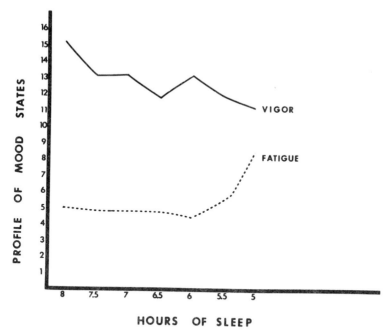

Figure 5.9 Effect of sleep reduction on self-assessed feelings of vigor and fatigue. The figure indicates a definite elevation in the self-perceived level of fatigue when sleep duration is reduced to less than 6 hr. The curve of vigor exhibits a trend of decline as sleep becomes reduced in length from 8 hr. From Johnson (47).

investigations by epidemiologists. Some investigators, believing that shiftwork dropouts who have been switched to daywork might represent a special group, have separately compared the morbidity patterns of these to other dayworkers who have never attempted shiftwork, as well as to rotating shiftworkers (93, 102).

Specifically, former rotating shiftworkers who have transferred to permanent day work in general reveal a greater fragility to illness than do either other dayworkers without previous history of shiftwork or rotating shiftworkers who exhibit good tolerance to this type of work schedule (93, 102). Consequently, it is now thought that dropouts from shiftwork exhibit a greater susceptibility to disease than do most day and tolerant shiftworkers. Some researchers believe that for certain types of persons, shiftwork poses a special risk in that it may promote the development of certain diseases. Thus, studies that compare separately the health profiles of dropouts from shiftwork to those of tolerant shiftworkers and permanent dayworkers are preferable (93, 102). However, even though better designed epidemiologic studies are now being conducted, the earlier findings, perhaps compromised because of methodological drawbacks, cloud our understanding of whether shiftwork contributes to or promotes the development of illness.

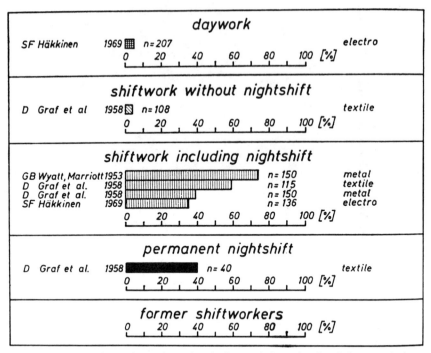

Figure 5.10 Frequency of complaints about disturbed eating habits related to shift system, industry, and investigator. From Rutenfranz et al. (66).

Today it is generally accepted that shiftwork can be a risk factor in exacerbating or potentiating certain ailments or disorders (66). Changes in the timing of work, meals, and sleep affect appetite during nightwork and may be related to the observation of an elevated incidence of dyspepsia and certain other gastrointestinal disturbances or ailments of shiftworkers (66, 93, 98, 103, 104). Figure 5.10 (12, 66) summarizes the frequency of complaints of altered appetite in relationship to shift system and industry. Disturbance of eating habits seems to be relatively infrequent in dayworkers and shiftworkers without nighttime duty. In comparison, it occurs in about 35 to 75 percent of the workers who rotate with a night shift, depending on the industry and study, and in about 40 percent of those engaged in permanent night work.

It is of interest that small but statistically significant differences in the intake of nutrients and calories have been detected in rotating shiftworkers by Reinberg and co-workers (105). In particular, a slightly greater quantity of calories was consumed during the morning shift (from 6 A.M. to 1 P.M.) as compared to the night shift (from 9 P.M. to 6 A.M.), when it was the lowest. Total protein, lipid, and carbohydrate consumption appeared to be increased slightly during the morning shift. However, none of these shift-related differences in the intake of nutrients was found to be statistically significant. A surprising finding of the

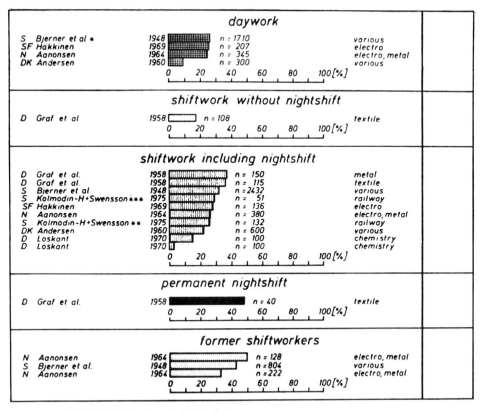

Figure 5.11 Frequency of complaints about gastrointestinal disturbances related to shift system, industry, and investigator. From Rutenfranz et al. (66).

studies due to Reinberg et al. (105) was that rotating shiftworkers, when working nights, exhibited a pattern of frequent (as many as seven) nocturnal snackings of carbohydrate-rich foods. Nonetheless, this nibbling did not result in an overall increase in daily carbohydrate intake during nightshifts in comparison to working the other shifts. It is presumed that this nibbling behavior was related to the odd hours of sleep and wakefulness during night rotation.

Perhaps related to the alteration of eating habits, meal timing and disruption of biological rhythms caused by shiftwork is the observed increased incidence of dyspepsia. Figure 5.11 illustrates the frequency of gastrointestinal disturbances in relation to shift system and industry (12, 66). Overall, gastrointestinal complaints were found in 10 to 25 percent of the dayworkers and about 17 percent of the shift employees not assigned to night work. Such complaints were found in from 5 to 35 percent of the employees engaged in rotating shiftwork involving a night rotation, depending on the industry, and in about 50 percent of those permanently working nights. Surprisingly, gastrointestinal complaints were a problem for 30 to 50 percent of those persons who had switched from rotating shiftwork to permanent day work. Apparent from these findings is the wide range in the incidence of complaints between groups of day and shiftworkers having to rotate between day and night shifts. Nonetheless, it is striking that former shiftworkers exhibited such a high incidence of gastrointestinal disturbances.

Many researchers have reported that the incidence of ulcers is higher in shiftworkers than in dayworkers. Figure 5.12 summarizes the findings of studies pertaining to the incidence of ulcers according to shift system and industry (12, 66). Gastric ulcers were reported among 0.3 to 7.0 percent of the dayworkers, about 5 percent of the shiftworkers not having a night shift, and approximately 2.5 to 15 percent of those having to rotate to a night shift. Gastric ulcers were reported to occur in 10 to 30 percent of former shiftworkers who had changed to day hours. The findings reveal a rather wide overlap in the incidence of this health problem. It has been suggested that differences in findings between studies on comparable groups might represent methodological biases of the various investigations (12). On the basis of large differences in incidence of this disorder between the various groups studied, Rutenfranz and his colleagues (12) question whether rotating shiftwork involving nighttime duty represents a true risk factor. Nonetheless, it is the opinion of a majority of researchers that ulcers are a risk of certain susceptible shiftworkers.

In summary, it appears that shiftworkers, in comparison to permanent day employees who have never worked nights, are at increased risk of experiencing sleep irregularities as well as gastrointestinal difficulties, with the latter manifest as dyspepsia, appetite disturbances, and/or duodenal or peptic ulcers. It is of interest that as a group former rotating shiftworkers, even after being transferred to permanent daytime employment with nighttime rest, exhibit the highest incidences of these health-related complaints. This finding is consistent with a view held by some that tolerant shiftworkers constitute a survivor-type population who as a group exhibit elevated resistance to a variety of common illnesses in comparison to those who display an intolerance to shiftwork involving a nighttime rotation. In short, the major difficulties associated with rotating shiftwork are sleep alteration and an increased risk of gastrointestinal disorders. Although surveys of other diseases such as cardiovascular, neurologic, and psychiatric illnesses in shiftworkers have been conducted, no effects of rotating shiftwork have been demonstrated (101).

5.4 Tolerance to Shiftwork

It is apparent that even at a young age not everyone is equally tolerant of the rigors of shiftwork. Typically, dropout is particularly common during the first 6 to 12 months; however, it is not unusual for biological intolerance to develop suddenly after 10, 25, or even 30 years of shiftwork, around the age of 50

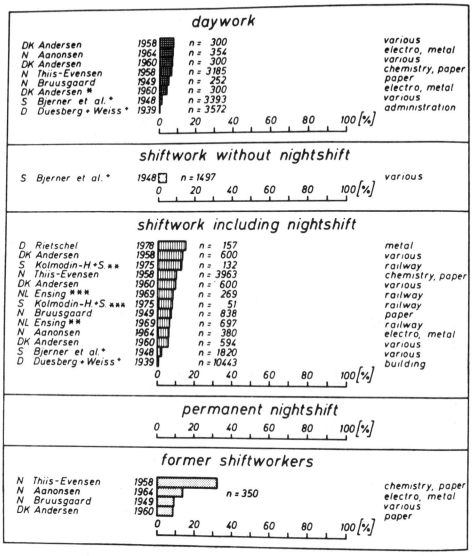

Figure 5.12 Frequency of gastric and duodenal ulcers related to shift system, industry, and investigator. From Rutenfranz et al. (66).

years or after. In the seasoned worker who has successfully handled the demands of shiftwork for many years, intolerance can develop for no apparent reason. The employee may suddenly find it difficult to sleep, especially during the daytime, and may even resort to using sleep-inducing medications, often without relief and sometimes causing harm. The intolerant shiftworker typically complains of excessive fatigue due to the inability to experience sufficiently restful sleep. The only sensible solution for such a person is to request transfer to permanent daywork.

Why persons suddenly develop a biological intolerance to shiftwork is not completely known. Reinberg and his colleagues (106) have conducted field studies of both tolerant and intolerant petrochemical shiftworkers and now believe that intolerance may be related, at least in part, to a disorganization of the circadian temporal structure in susceptible persons. The problem of shiftworker intolerance is again considered in detail from a chronobiologic (biological rhythm) point of view in later sections of this chapter (Sections 9.1 to 9.4).

6 CHRONOBIOLOGIC ASPECTS OF SHIFTWORK

Pertinent to understanding the nature of the biological adjustments to shiftwork is a fundamental understanding of chronobiology—biological rhythms or biological time structure. Biological adjustments to shifts involving an alteration in the sleep–wake pattern with respect to clock hour entail changes in the timing of the peaks and troughs of circadian rhythms and possibly rhythms of other frequencies. In this section, the topic of chronobiology is reviewed in preparation for the discussion of the chronobiologic aspects of shiftwork.

6.1 Introduction to Chronobiology

Chronobiology is the branch of science that explores mechanisms of biological time structure, including the important rhythmic manifestations of life. Although it is a comparatively young investigative science, ancient civilizations were well aware of rhythmic processes, for example, in the seasonal breedings of wild and domestic animals and the planting and harvesting of agricultural or wild plant foods. In addition, early writers, such as the poets, were fascinated with rhythmic events, particularly as they pertained to the leaf and petal movements of plants. Nonetheless, advances in our understanding of rhythms were rather slow, especially in higher animals and humans, until the middle of this century; since then great strides have been made, particularly for human beings.

Rhythms of many frequencies have been demonstrated at all levels of animal life as well as biological organization by many investigators (14, 34, 107–117). As a result of a multitude of studies, rhythmicity has been firmly established as a fundamental property of all living things such that today chronobiology represents a new and powerful field of scientific endeavor. A contemporary chronobiologist and geneticist, Charles Ehret (118), has made the following observation: "The newest of the integrating disciplines is circadian regulation. The other integrating disciplines have been evolution, genetics, and developmental biology. Chronobiology cuts across the conventional strata of levels of organization from the molecular to the levels of complex systems of social structure and deals with biological time structure of relatively long duration." The regularity of these rhythms has caused some to refer to them as biological or physiological clocks (119, 120).

The range of frequencies that has been found in living systems extends from cycles of less than 1 sec to cycles of 1 year or more (109, 110). It is noteworthy that many, but not all, biological rhythmicities clearly correspond to physical environmental frequencies, such as the 24-hr and 365.25-day natural light–dark cycles. There is strong evidence that many rhythms are adaptive and serve to adjust organisms to predictable and regularly occurring environmental changes. It is safe to say that the rhythmicity of life has been imprinted throughout the eons on living organisms by continuous exposure to this periodic environment during evolution. There is no question that the day–night and lunar cycles, as well as the seasons of our planet, were in existence before life began; thus, from the beginning, life was subjected to these and many associated periodic events. Therefore, in order to survive, animals throughout their evolutionary history had to be biologically prepared for rather drastic differences in the environment occurring over 24 hr, 28 days, and the year. In addition to the numerous examples of annual cycles in reproductive behavior in animals, hibernation represents one obvious example of a rhythm of 1-year duration in the level of activity and metabolism in certain species. Biological rhythmicity, representing predictable temporal changes in bodily functions and processes, constitutes a significant adaptation to expected cyclic alterations in the geophysical environment in which one lives.

6.2 Biorhythm Theory

We wish to emphasize that the science of chronobiology is in no way related to or connected with the popularized fad of biorhythm forecasting (121, 122), although it has been used unwittingly by some industries as part of a safety and accident prevention program. The biorhythm theory claims that it is possible to predict the occurrence of one's best or worst physical and intellectual capability as well as efficiency and performance. The theory postulates the existence of a 23-day physical cycle governing strength and endurance, a 28-day cycle regulating changes in sensitivity and emotionality, and a 33-day intellectual cycle affecting intelligence, alertness, and awareness. Proponents of biorhythm theory claim to be able to predict human behavior in terms of "good" or "bad" days at any time throughout one's lifetime. A mathematical model based on a number of assumed premises is used. These premises are:

1. Each cycle is described by a sine curve having a positive and negative phase with two so-called crossover points. The days corresponding to the crossovers of each cycle are called "critical" days. When the curves of two of the cycles cross the medium line on the same day, criticality is doubled. When all three cross simultaneously, criticality is tripled. "Critical days" supposedly are those that are most dangerous in terms of accident proneness and vulnerability.
2. The cycling of the three postulated component biorhythms commences at the moment of birth, each one always initiated with the positive phase.
3. In every person, at all times, each of the 23- 28-, and 33-day cycles occurs with a precise immutable length set by the date and time of birth.
4. The positive phase of all cycles corresponds to best performance, the negative phase to poorer performance. Proponents of biorhythm theory claim that performance is reduced, vulnerability increased, and accident proneness elevated on critical days (especially on those days when one or more cycles attain their critical days on the same date and when the others are in a negative phase or on days when all cycles are in negative phase).

The possibility of predicting human performance capacity was eagerly regarded by the management of certain labor-intensive industries as a means to lower accident rates and/or increase worker efficiency and productivity. Biorhythm consulting companies sprang up so quickly that the biorhythm concept became commercialized before it had been objectively tested for either its biological or mathematical merits. As stated earlier by Schonholzer et al. (123) in 1972, a time when the popularity of biorhythms had peaked, "Besides some recent expert statements which have not been published and, therefore, are not available, there are an endless number of press-releases, propaganda brochures, mass-media commentaries, courtesy certificates, etc. The very few regularly published papers which allow a scientific review more or less belong to the past. New sound and scientific publications do not exist. Therefore, there is no really controllable base." During the 1970s the biorhythm theory was investigated independently in different scientific laboratories mainly by comparison of actual events such as accidents, deaths, and athletic records, with those expected to occur at random. A second research method made use of computed correlations between the actual performance of a laboratory task with the phasing of the postulated biorhythm cycles. Klein and Wegmann (122), in their comprehensive review of the papers dealing with biorhythm theory in 1979, found *no substantiation* for its merits. The failure to verify biorhythm theory through application of scientific methods contrasts with those claims of its successful application in reducing accident rates in industry (121,124, 125). These latter findings suggest the possibility of a potential suggestive power, as a placebo effect, from the application of the biorhythm theory to employees who have confidence in its purported predictive ability. It is perhaps because of this placebo effect that accident rates might be reduced in some workers under certain circumstances and not because of an existence of biorhythm

CHRONOBIOLOGY

ILLUSTRATIVE SPECTRUM OF BIOLOGICAL RHYTHMS

DOMAIN	HIGH FREQUENCY $T<0.5h$	MEDIAL FREQUENCY $0.5h<T<6d$	LOW FREQUENCY $T>6d$
MAJOR RHYTHMIC COMPONENTS	$T\sim 0.1s$ $T\sim 1s$ et cetera	ULTRADIAN ($0.5<T<20h$) CIRCADIAN ($20<T<28h$) INFRADIAN ($28<T<6d$)	CIRCASEPTAN ($T\sim 7d$) CIRCAMENSUAL ($T\sim 30d$) CIRCANNUAL ($T\sim 1yr$)
EXAMPLES Rhythms in	Electroencephalogram Electrocardiogram Respiration	Rest-Activity Sleep-wakefulness Responses to drugs Blood constituents Urinary variables Metabolic processes, generally	Menstruation 17-Ketosteroid excretion with spectral components in all regions indicated above and in other domains

Domains and regions [named according to frequency (f) criteria] delineated according to reciprocal f, i.e., period (T) of function approximating rhythm. s = second, h = hour, d = day.

Several variables examined thus far exhibit statistically significant components in several spectral domains.

Figure 5.13 Illustrative spectrum of biological rhythms.

phenomena per se. We believe that the biorhythm concept is without scientific merit. It has no place in the management of industrial safety nor shiftwork schedules.

6.3 Naturally Occurring Biological Rhythms

Living beings are precisely organized in space as an anatomy; they are precisely organized in time as well, as biological rhythms that demonstrate predictable variation over specific time domains. Biological processes exhibit several different types of bioperiodicity. These are categorized according to their duration, that is, by the amount of time required for a single repetition.

In general, rhythms are divided into three broad classes (Figure 5.13):

1. Ultradian rhythms are those that exhibit frequencies of less than 1 cycle in 20 hr. Examples of ultradian rhythms include the basal activity levels of neurons that typically exhibit one cycle per 0.1 second or so, the pulsatile secretion of hormones, for example, from the adrenal cortex, and the repetition of sleep stages such as rapid eye movement (REM) during nightly rest at about 90-min intervals, to mention just a few.

2. Infradian rhythms categorize those bioperiodicities that are greater in duration than 28 hr. Examples of infradian cycles include the circaseptan (about 7-day), circamensual (about monthly, also termed *circatrigintan* or about 30-day), and circannual (about 1-year) rhythms.

3. In between are rhythms that have bioperiodicities in the range of 20 to 28 hr. Rhythms having these durations are termed *circadian* (meaning about 1

day or 24 hours in length) after F. Halberg (108, 109). Many of these rhythms have important medical significance (34).

Circadian rhythms in human beings and other animals have received more attention by researchers than any of the others, and thus a great deal is known regarding the 24-hr temporal organization of biological processes. It is known that circadian rhythms exist at all levels of biological organization from that of the single cell to complete organ systems. All life forms, whether plant, animal, or human, are made up of a multitude of interrelated circadian physiological, metabolic, and behavioral rhythms. All of these act in harmony in the healthy organism. Thus, chronobiologists commonly refer to these 24-hr periodicities collectively as a circadian system, much in the same way in which biologists speak of the nervous, circulatory, or digestive systems. It must be kept in mind that rhythms of many other frequencies are intermingled and may even modulate or be modulated by the circadian frequency.

The biological consequences of rotating shiftwork schedules and jet lag (referred to as *desynchronosis*) on industrial performance and accidents among others have been studied by chronobiologists with particular emphasis on the role of the human circadian system. In rodents, the circadian system has been examined for the occurrence of susceptibility–resistance rhythms in the effects of hazardous chemical, physical, and microbial substances as a means of modeling predictable changes over the 24 hr in human vulnerability. The findings of these studies are critical to those involved in medicine, toxicology, and occupational health. The findings as well as their significance are reviewed in subsequent sections (Sections 7.3 and 7.4) of this chapter, especially in relation to our understanding of the potential challenges of shiftwork.

6.4 Examples of Circadian Rhythms

Almost all biological processes and functions thus far studied have been shown to exhibit circadian variation. The examples that are discussed here demonstrate that circadian changes can be found at various levels of hierarchical organization and that the prominence of rhythms, defined by the peak-to-trough difference, although significant, is not the same for each.

One of the earliest documented and most extensively studied hormonal rhythms in mammals is the serum corticosteroid rhythm. This rhythm, plotted in Figure 5.14 for both the rat and human, is used to illustrate some of the basic properties of rhythms and the special terminology used to describe their behavior (126).

In the diurnally active human, the serum corticosteroids, such as cortisol, begin to be secreted from the cortex of the adrenal gland during sleep and usually reach a peak just after the daytime activity begins. It is noteworthy that this temporal pattern and most other rhythms persist even if one remains awake through the 24 hr; moreover, the occurrence of this rhythm, but not all rhythms, is independent of meal timing (127). In the night-active rodent, the

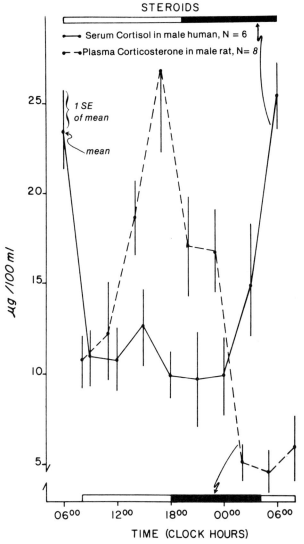

Figure 5.14 Prominent circadian fluctuation of the predominant serum steroids of rats and humans. For 2 weeks prior to study rats were standardized to a synchronizer schedule of light for 14 hr and darkness for 10 hr (darkness from 1800 or 6 P.M. to 0400 or 4 A.M.) with free access to food and water. For humans, mealtimes daily were at 7 A.M., 12:45 P.M., and 4:45 P.M. with rest or sleep between 9 P.M. and 6 A.M., except during study, when the participants were awakened at midnight and 3 A.M. for sampling. Note that the timing of peak serum corticosteroid levels of the two species differ by 12 hr clock time; yet when the peaks are referenced with respect to the rest–activity span for each species, the circadian rhythms are found to be identically timed—the peak occurs just prior to the commencement of the respective activity spans. Note that clock hour is shown in military fashion: 1800 = 6 P.M.; 0000 = midnight. Darkened portion of upper horizontal axis denotes the timing and duration of nightly rest alternating with activity (designated by an absence of shading) in the human subjects. Darkened portion of horizontal axis at the bottom of the graph denotes the hours of nocturnal activity in the studied rodents. From Scheving (37).

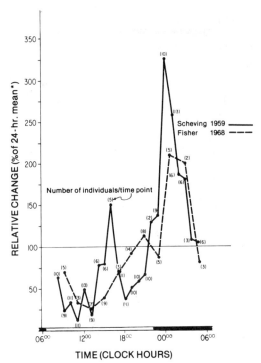

Figure 5.15 Reproducibility of the mitotic–index rhythm in adult human epidermis. Standard errors have been omitted to avoid a cluttered graph. Note: 1800 = 6 P.M.; 0000 = midnight. Span of sleep or rest indicated by darkened portion along the lower horizontal axis. From Scheving and colleagues (129, 265).

peak occurs shortly before the nocturnal activity period begins. The fourfold or greater variation seen along the 24-hr time scale clearly shows that such variation is simply not a minor fluctuation around a 24-hr mean that could be attributed to errors in experimental design or analysis. Generally, the rhythms of the nocturnally active rodent and diurnally active human differ in the timing of peak values by about 12 hr, with reference to clock time. However, it is important to point out that this is not the case for all rhythmic variables; serum prolactin and melatonin, for example, are known exceptions (128). It should be noted that rhythms with higher than circadian (ultradian) and even lower frequencies (infradian ones such as circaseptan, circamensual, and circannual) also characterize serum corticosteroid secretion as well as many other variables (39, 110). Other examples of high-amplitude circadian rhythms include cell proliferation (129) (Figure 5.15), airway resistance in asthmatics (130) (Figure 5.16), and venous lymphocytes (131) (Figure 5.17). Examples of relatively small amplitude circadian rhythms include heart rate, body temperature, eye–hand coordination, and others shown (132) in Figure 5.18.

In regard to blood pressure it is noteworthy that a National Center for Health Statistics survey (133) done in 1964 reports that, on the average, human blood pressure varies during the day by no more than 3 to 4 mm Hg. Although

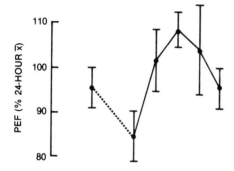
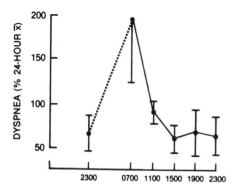

Figure 5.16 Circadian variation of peak expiratory flow (PEF) (a direct measure of airway patency) and dyspnea self-measured by 13 diurnally active (sleep from 11 P.M. to 7 A.M.) untreated asthmatics at designated clock hours daily. The crest in dyspnea was self-rated as greatest on arising at 7 A.M. and least at 1500 (3 P.M.). The 24-hr pattern of PEF was as expected 12 hr out of phase with that of dyspnea. PEF was lowest at 7 A.M. and greatest at 1500 (3 P.M.). The data of each graph are plotted as percentages (means and standard errors) from the group 24-hr average set equal to 100 percent. Smolensky (130).

undoubtedly the statement can be made for a particular set of averaged data, the variation for individuals is artificially reduced. Such statistics can be very misleading, especially when it can be demonstrated that diastolic blood pressure in a presumably healthy resting subject can vary as much as 10 times the amount mentioned in the survey report.

Averaging of data, as illustrated in Figure 5.18 and other figures, also may be misleading simply because any of the estimated parameters may vary greatly from individual to individual; this is especially true for the amplitude and 24-hr mean. For example, the average range of change for oral temperature during the 24-hr period was about 2°F (1.9°C) for the subjects whose data are shown in Figure 18. For pulse, the range of change averaged 30 percent for the group, but the change was as great as 81 percent and as small as 21 percent for single individuals within the group. The range of change for systolic blood pressure was 15 percent for the group, but for single individuals it was as great as 57 percent and as small as 10 percent. The change in diastolic blood pressure for the group averaged 14 percent, but it was as great as 78 percent for one individual and as small as 13 percent for another. At one phase of the circadian cycle, 2 of 13 presumably healthy young men had blood pressure readings that

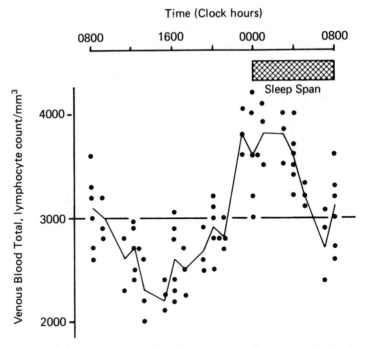

Figure 5.17 A circadian rhythm in the total lymphocyte count/mm^3 of venous blood substantiated in a sample of 12 healthy young adults (6 men, 20 to 28 years of age, and 6 women, 23 to 24 years of age) during December 1974 in Paris, France. Subjects were synchronized with diurnal activity from 0800 (8 A.M.) to 0000 (midnight) alternating with nocturnal rest (sleep). Blood samples were obtained every 4 hr at fixed clock hours with the exact times of sampling differing between three subgroups of four subjects each. Raw data (filled circles) are displayed as a function of time (clock hours). The arithmetic mean of each of the time points (thin line) when connected appears to resemble a sine wave with small swings. Redrawn from Reinberg and Smolensky (34).

might suggest at least diastolic elevation (150/108 and 148/96); the lowest readings recorded for these same men were 122/68 and 120/70, respectively. Figure 5.19 illustrates the extremes of variation in urine 17-ketosteroids and potassium in two individuals and compares them with the group rhythm. All subjects consumed an identical diet and adhered to a similar routine during study (134–136).

From our studies on several different populations, including the one on young individuals discussed previously, the elderly (137), and patients with leprosy (138), it seems evident that individuals fall into one of three categories: (1) those having a high range of change over the 24 hr for a particular variable, (2) those having a low range of change, and (3) those (possibly the largest group) who fall in between these extremes. What, if anything, do these differences imply as far as health is concerned? It is conceivable that the amplitude of a rhythm may be important in evaluating human physiology, and

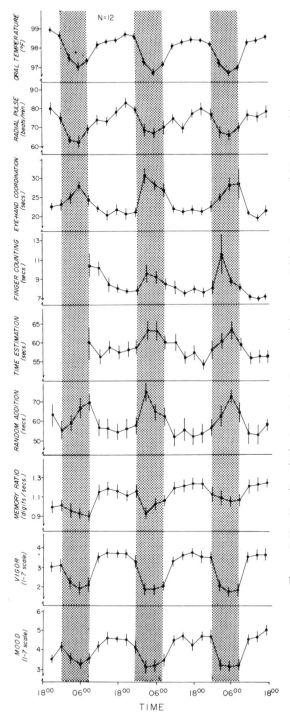

Figure 5.18 Circadian rhythmic variation in a set of diverse variables in a group of 12 presumably healthy young men studied over a 72-hr span with measurements done every 3 hr. Meal times were at 6:15 A.M., 12:15 P.M., and 4:30 P.M. with rest between 9 P.M. and 6 A.M. except when subjects were awakened for sampling at midnight and 3 A.M. Note that circadian differences occur in the three investigated performance variables: eye–hand coordination, finger counting, and random addition speed. Clock hour time designated in military fashion where 1800 = 6 P.M. and 0600 = 6 A.M. From Scheving (132).

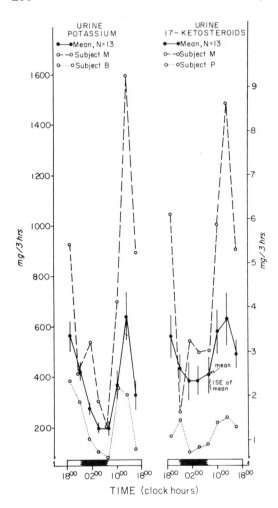

Figure 5.19 Individual and group circadian variation in the two illustrated variables of urinary potassium and ketosteroid concentration in young, presumably healthy soldiers consuming identical diets and adhering to the same rest–activity routine. Time as clock hours is designated in military fashion where 1800 = 6 P.M. and 0200 = 2 A.M. The hours of rest are indicated by the darkened portion of the horizontal axis. From Scheving et al. (134).

this is discussed further in Sections 9.3 and 9.4. Some variables in presumably healthy individuals do indeed have so-called unusual values at certain times of the circadian cycle, whereas at others they are well within conventional, "normal" ranges. This raises such questions as: Will the younger individual who shows high blood pressure readings (near the currently postulated limits of hypertension) only during the peak of the rhythm become hypertensive at all phases of the circadian cycle with increasing age? Or are these "high" readings within the individual's normal range? Will a person with high intraocular pressure at one stage of the circadian system become a glaucoma victim? How does the physiology of cholesterol metabolism differ between one who has a daily fluctuation of 6 percent and another whose daily cholesterol level fluctuates as much as 100 percent? Similar questions may be asked for almost any variable measured (139).

6.5 Synchronizers of Biological Rhythms

A discussion of the concept of *synchronization* is essential to appreciating the temporal integration of the circadian system and for understanding the chronobiologic aspects of shiftwork. In short, a synchronizer is a cue or signal from one's environment that serves to lock circadian and other bioperiodicities into fixed frequencies. For example, many animals in nature are synchronized to the natural light–dark cycle, whereas those in the laboratory are usually synchronized to an artificial 24-hr light–dark cycle (140). The serum corticosteroid rhythm illustrated in Figure 5.14 is such a synchronized rhythm. The synchronizer (108, 141) sometimes is called a *Zeitgeber* (142), an entraining agent (119), clue, or cue; all are used synonymously. It is important to recognize that synchronizers do not cause or are not the source of rhythmicity. They are capable only of influencing certain of the characteristics, most prominently the timing of the peak and trough of a given bioperiodicity.

6.5.1 Rodents

The best evidence that light is the dominant synchronizer of rodent rhythms comes from inversion studies. If one inverts the ambient light–dark cycle by 12 hr clock time, the circadian system will eventually, but seldom immediately, invert to the same degree. This means that the peak of each circadian process shifts in time by an amount equivalent to the shift in the synchronizer. Figure 5.20 illustrates the inverted pattern for both the serum corticosterone level and the mitotic index of the corneal epithelium (126). The rhythms of some variables invert rapidly following alteration of the synchronizer schedule, whereas others do so slowly and at differential speeds. The significance of this is referred to later in the discussion of the chronobiologic adjustments of employees to rotating shiftwork. The results of many studies on rodents demonstrate the dominance of the ecological or artificial light–dark cycle as a synchronizer of their rhythms. There is much evidence indicating that human physiology and behavior, as well, is influenced by synchronizers.

6.5.2 Human Beings

The circadian rhythms of humans, in comparison to those of rodents, are less dependent on the environmental light–dark cycle. The human circadian system appears to be more strongly influenced by the social cycle, that is, the timing according to clock hour of work and other activities alternating with sleep (143, 144). For example, in permanent day workers, the circadian peak of the rhythm in serum cortisol occurs during the *morning* around the time of awakening, 7 or 8 A.M. In permanent night workers, the circadian rhythm of cortisol resembles that of day workers; however, the occurrence of the circadian crest does not occur at the same clock hour. Following adjustment of the circadian system to the atypical activity–sleep routine, it generally occurs in the evening or at night

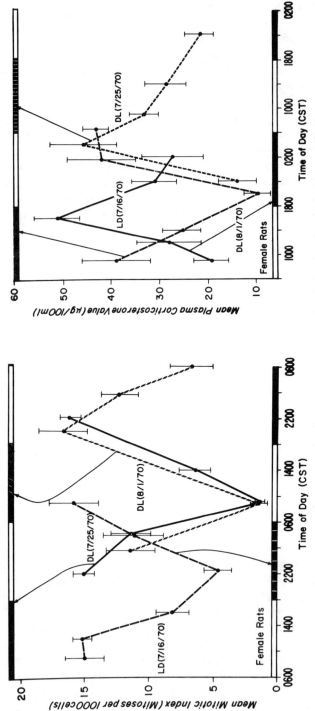

Figure 5.20 In these two graphs, the curves LD (7/16/70) identify the circadian rhythmic pattern for the mitotic index of corneal epithelium (left) and serum corticosteroid (right). (Note the inverse phasing of these two circadian rhythms.) At this time the synchronizer schedule was darkness from 6 P.M. (1800) to 6 A.M. (0600). The curves identified as DL (7/25/70) represent the pattern seen for both variables 7 days after a 12-hr reversal of the light (L) and dark (D) cycles. Complete phase shifting of these variables had occurred after 7 days with the persistence of the same inverse relationship in staging between the two rhythms. The curves labeled DL (8/1/70) represent the twelfth day after the reversal of the LD cycle, indicating that the reversed rhythm for both variables is "locked in" to the changed LD cycle. Arrows from each curve in each figure point to the existing LD synchronizer schedule according to the dated curves. From Scheving et al. (236).

since such persons are at rest during the day and work at night; that is, they have a vastly different synchronizer schedule. In persons who always work the night shift, the circadian peak appears to be very different from that of day workers when referenced to clock time, that is, 7 P.M. rather than 7 A.M. However, when the crest of this rhythm is referenced, respectively, to the middle of the day worker's and night worker's sleep span, the circadian peak is similarly timed, occurring about 3 to 4 hr after this marker of the synchronizer schedule. It is clear, then, that the timing of a work shift as it affects the timing of sleep and activity can, and does, represent the primary synchronizer for the human circadian system. Also, it can be appreciated that rotating shiftwork schedules that include nighttime duty play havoc with the body's circadian functions since this represents a regularly occurring change in the synchronizer pattern. Establishment of a consistent pattern or schedule for such important activity as eating, elimination, sleep, or even relaxing, becomes difficult or for some impossible. Thus such commonly reported complaints—fatigue, gastrointestinal disorders, loss of appetite, constipation, and disorientation—are not surprising.

6.6 The Significance of Synchronization

For organisms (plants or animals) in the synchronized state, rhythmic variables normally have a fixed time relationship. For example, body temperature for both humans and the rodents is typically highest during the active or awake stage of the sleep–rest cycle. This means that the temperature and activity rhythms are internally synchronized, as their timing or staging under normal conditions is nearly the same. The temporal staging of other variables in the normal synchronized state, however, could be very different, as much as 12 hr out of phase. This is the case, for example, for the phase relationship between the human circadian rhythms of serum cortisol concentration and the white cell numbers in the blood. The circadian peak of the former occurs around the commencement of the daily activity span, whereas that of the latter occurs later around bedtime. The different staging of the various circadian functions, which are, for example, 4, 6, or 12 hr out of phase with one another, is attributable to the temporal organization of biological processes acquired through an adaptation to predictable variations in the geophysical environment that occurred during evolution. This temporal organization represents an important aspect of the body's efficiency in conserving and optimizing its capabilities.

When organisms are synchronized and the synchronizer schedule is known, it is possible to determine and then predict rather accurately on subsequent days the staging of a great many circadian rhythms. For example, Figure 5.21 presents certain aspects of the circadian system of rodents synchronized to light from 6 A.M. to 6 P.M. alternating with 12 hr of darkness (110). Figure 5.22 presents selected aspects of the circadian system of diurnally active human beings (110). In both these figures the circadian peak (termed *acrophase*) of each rhythm is represented by a filled circle; the extensions to the right and

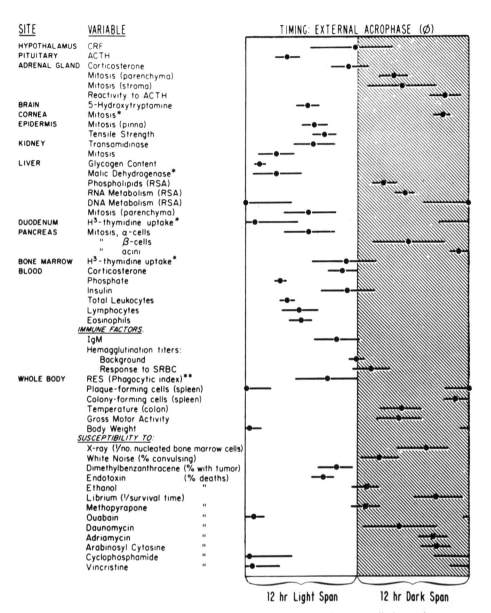

Figure 5.21 The circadian temporal structure of the mouse is shown by a so-called acrophase map—a display by graphic means of the circadian acrophase or crest (designated by ∅ for a phase reference corresponding to the middle of light—animal rest—span) of different biological functions. The acrophase, black dot, and the 95 percent confidence limits, lateral extensions, depict the timing with respect to the 24-hr domain of highest expected values for each designated function. By knowing the light–dark synchronizer schedule, one can determine with confidence the biological time structure at a given clock hour of sampling or experimentation. From Halberg and Nelson (110).

left of the symbol indicate graphically the 95 percent confidence range of each acrophase. These so-called acrophase charts provide a temporal mapping of the circadian crests of various functions. Just as an anatomical atlas is useful for relating the functional integration of biological structure in space, so is it that acrophase charts, such as those in Figures 5.21 and 5.22, are useful for relating and knowing the functional integration of circadian (or other period) processes in time.

The two acrophase charts reveal there is a definite temporal staging for each variable with regard to the light–dark schedule in rodents and the activity–rest routine of human beings. Moreover, the information in these charts defines the fixed temporal relationships between biological variables. It is apparent that not all the circadian rhythms attain their peak (acrophase) or trough values at the same identical times. Acrophase charts such as these have been very helpful in revealing many rhythms in humans and are useful in biomedical studies, including shiftwork research. Similar types of acrophase charts have been constructed for circamensual (145, 146) and circannual (34, 146) systems of mankind. With respect to the circadian system, if one has knowledge of the synchronizer schedule, it is possible to *predict* rather accurately, for a given clock hour when a medical test or research procedure is conducted, the temporal staging of a great many rhythmic variables of the circadian system. This type of information can be critical for the proper interpretation of biological data obtained from a single clinical test on patients or single timepoint biological samplings, such as for monitoring workplace exposures to hazardous substances, taken from persons studied in industrial settings. Today, clinical laboratories of several large hospitals have developed so-called time-qualified reference values so as to take into consideration circadian rhythms and synchronizer schedule in relation to the clock hour of blood or urine sampling. The topic of time-qualified reference values is discussed in greater detail in Section 10.1.

The phasing of circadian and other rhythms in human beings is internally organized in time to anticipate the different metabolic requirements of activity and rest when adhering to a fixed life routine (147). However, when *inversion* of the synchronizer schedule takes place, such as when having to work nights and sleep days, as is commonly the situation with rotating shiftwork schedules, or following rapid geographic displacement by jet aircraft entailing the sudden alteration of the clock hours of sleep and activity, the acrophase of each circadian function eventually shifts in an amount equal to that of the change in the synchronizer schedule (148). Generally this adjustment requires several to many days. Some persons seem to be able to adjust rather quickly and others slowly (62). It is this type of biological variability among individuals that makes the estimation of the time needed to "invert" difficult. Some persons are biologically or emotionally intolerant and incapable of rotating shiftwork (106, 149–151). Consequently, it is important that occupational health professionals understand the scientific bases for shiftwork intolerance so that wise selection of shiftwork candidates is made and employees and management comprehend

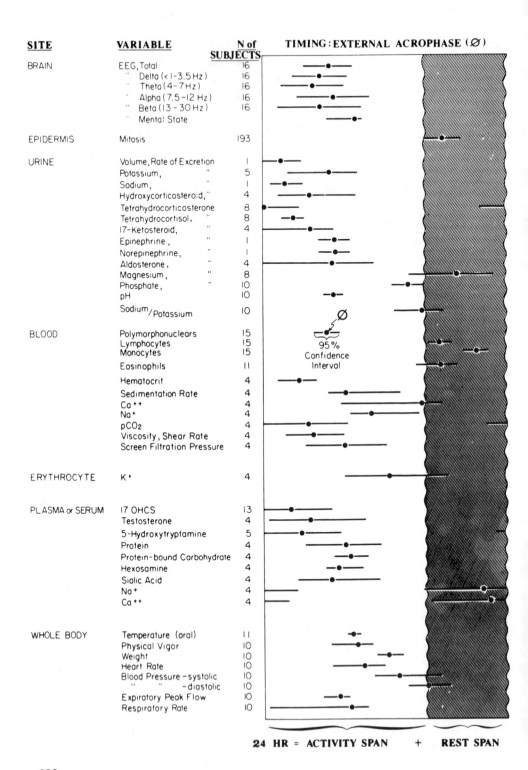

BIOLOGICAL RHYTHMS, SHIFTWORK, AND OCCUPATIONAL HEALTH

the nature of the difficulties inherent in adhering to rotating shiftwork schedules requiring nighttime duty.

During the span of the chronobiologic adjustment to a new work–rest routine, the rate at which each circadian process shifts differs (95, 96, 148, 152, 153). For a few days, alterations in the usual phase relationships between a large number of circadian rhythms occur. The state of (usually transient) altered phase relationships within the circadian system is termed *internal desynchronization* (95, 96, 148, 152). This transient state of internal desynchronization between circadian functions results in a less than optimal temporal organization such that biological efficiency can be decreased (30, 95, 96, 148, 153, 155). This period of transition of less than optimal biologic efficiency is responsible for a specific type of fatigue, which is different from that due to sleep loss, itself, commonly experienced by travelers and shiftworkers during the first few days following a sudden change in their sleep–activity routine.

6.7 Free-Running Rhythms

Free-running rhythms are defined as bioperiodicities that in the absence of synchronizers exhibit durations other than expected, in the case of circadian rhythms period durations shorter or longer than exactly 24 hr. Scientists once thought that biological rhythms represented nothing more than direct responses to periodicities in the geophysical environment, such as in the light–dark cycle, to temperature and relative humidity, or to nutrient uptake. As long ago as 1729, DeMarian (156), the mathematician and astronomer, was fascinated with the daily periodic movement of plant leaves. To his surprise, he found that the leaves of plants would continue their periodic movements when kept in a darkened cave protected from the sunlight or open air. This finding was troublesome to the scientists of that time since it suggested that some endogenous mechanism within the plant was responsible for this behavior.

Interestingly, we now know that when a rodent or other animal is removed from the influence of the synchronizer, that is, the light–dark cycle, which can be accomplished by blinding or subjecting the animal to continuous light or darkness, circadian rhythms continue to persist (157–161). This is the case, for example, for the circadian rhythms of body temperature, serum corticosterone, cell proliferation, feeding, and rest and activity, among others. It should be noted, however, it is rare that the frequencies of these rhythms will average precisely 24 hr. For humans they are usually greater than 24 hr. For example, in human beings dwelling in experimental settings without time cues, the sleep–

←

Figure 5.22 The circadian temporal structure of humans is shown by a so-called acrophase map—a display by graphic means of the circadian acrophase or crest (designated by Ø for a phase reference corresponding to the middle of the sleep span) of different biological functions. The acrophase, filled circle, and the 95 percent confidence limits, lateral extensions, depict the timing with respect to the 24-hr domain of highest expected values for each designated function. By knowing the rest–activity synchronizer schedule, one can determine with confidence the biological time structure at a given clock hour of sampling or experimentation. From Halberg and Nelson (110).

wake pattern elongated to almost 25 hr (161). It is of interest that women who remained in such settings for several months exhibited elongated menstrual cycles as well as longer circadian periodicities (162, 163). Under constant conditions, each bioperiodicity exhibits a cycle duration that differs from the others. Because the duration of these cycles is not exactly 24 hr, the clock time of the peaks and troughs of individual circadian functions differs from one day to the next. When this is the situation, clock time can no longer be used to interpret the staging of specific circadian functions. These circadian, non-24-hr, bioperiodicities are termed *free-running rhythms*. This phenomenon has been observed in both plant and animal life following the removal of synchronizers (111, 159–161, 163–168). These findings, among others, constitute strong evidence for the innate and endogenous nature of biological rhythms, that is, for their genetic origin.

The frequency of a free-running rhythm can be influenced by several factors. In animals other than human beings these include the intensity of light and the magnetic field (164). These factors have been shown to be capable, at least in certain organisms, of delaying or advancing the stage of free-running rhythms. The period length of free-running rhythms is remarkably resistant to chemical perturbation. If the free-running period has a fixed duration of say 25 hr and 4 min, for example, it is difficult to change. However, there is a class of chemicals that are used in psychiatry, such as lithium, imipramine, and clorgyline (169–174), that can alter it; it is perhaps important to note that these chemicals also are known to influence cell membrane function.

Since the early 1960s a number of investigations have focused on the free-running rhythms of humans. Many of the early studies were conducted in caves. Frequently, the subject was a speleologist who would perform some type of exploration or endurance record in the depths of caves, whereas chronobiologists would monitor and interpret certain rhythmic behavior (162, 163, 166, 175). The late J. N. Mills (176), a well-known English chronobiologist, has reviewed many of the cave studies, and these make for interesting reading as their results have been very important to our understanding of biological rhythms. In recent years, chronobiologists have used specially constructed isolation chambers where humans can be studied in comparative comfort but separated from all social and recognizable environmental cues (161, 177, 178). As recently as 1983, chronobiologists have submitted themselves to isolation in such; one of the authors (LES), a hypertensive, recently did so for 45 days, demonstrating, among other things, that systolic and diastolic blood pressure and heart rate free run with a period greater than 24 hr (179).

It has been demonstrated that even when humans are isolated from all time cues, certain rhythmic variables, even though they are free running, remain entirely internally synchronized; that is, they retain the same phase relationship that was demonstrated under the normal synchronized state. On the other hand, some bioperiodicities such as the sleep–activity cycle and body temperature rhythm may become internally desynchronized in certain individuals, that is, they exhibit an atypical phase relationship, in experiments performed under constant conditions. This may occur at the beginning or spontaneously at any time during isolation. This means that the stagings of certain circadian functions are more strongly integrated than are others. Such a phenomenon almost certainly implicates the existence of more than one control center for biological rhythms (114, 161, 180, 181).

Interestingly, it has been shown that in some persons the free-running period such as the sleep–wake rhythm may be different from that of the body temperature rhythm. Consequently, the body temperature may on occasion reach its peak during sleep, and only periodically do the temperature and activity rhythms obtain the same relationship that characterized them in the normal synchronized state. Simply stated, the time relationship of these two rhythmic variables is continuously changing in relation to one another. R. Wever, who has been an active investigator of this phenomenon, has reviewed many aspects of free-running rhythms as they relate specifically to humans (161). With respect to shiftwork, certain susceptible persons appear to exhibit free-running rhythms; this condition is believed to contribute to the biological intolerance of some persons to rotating shiftwork schedules involving nightwork (106, 151). The occurrence of internal desynchronization in synchronized persons has implications in medicine; some investigators believe that it explains in certain patients the periodic exacerbation of mental disorders (106, 171, 182).

Ambient temperature cycles may serve as a dominant synchronizer for a number of cold-blooded animals, or even sometimes for warm-blooded animals in the absence of the usual dominant synchronizer, such as the day–night cycle (183). However, ambient temperature probably does not play a significant role in synchronizing humans. It is interesting to note that free-running rhythms are relatively independent of temperature; that is, one can increase the ambient temperature within a wide range, but the period of free running will change very little, if at all; however, the 24-hr average or the rhythm amplitude may be altered. It was this unexpected observation that stimulated in the early 1950s great interest in chronobiology among scientists (37).

In summary, it is clear that the mechanism(s) of free running and synchronization is ideally designed for permitting an organism in nature to successfully cope with a cyclically changing environment. Life without such a mechanism would be difficult. Yet, this same mechanism controls the rate of adjustment of human biological rhythms following a rapid change in the human synchronizer schedule. Consequently, a chronobiologic adjustment to a shift in the synchronizer schedule typically involves a span of transient internal desynchronization as a result of acrophase shifts occurring at individualized, but varying rates, in circadian rhythms following a drastic change in the clock hours of rest and activity, as is the case after rapid travel through several time zones or after commencement of the nightshift rotation.

The need to consider the chronobiologic aspects of adjustment to different activity–rest routines such as in shiftwork is gaining increasing attention in industry. Chronobiologists are becoming more involved in assisting companies that wish to adapt both standard rotating 8-hr as well as unusual workshifts. Thus far, chronobiologists have been useful in devising methods to identify those persons who are likely to be intolerant of shiftwork and in suggesting ways to minimize the consequences of rotating time patterns of work (15, 55, 62).

6.8 The Genetic Basis of Biological Rhythms

The finding that the length of a free-running rhythm may be characteristic for each individual and for each biological function constitutes strong evidence that rhythmicity has a genetic basis. The results of recent genetic studies on plants (184) and more recently the fruit fly, *drosophila*, supports this view. For *drosophila*, it was found that a small region on the X chromosome is responsible for the circadian rhythm in both activity and emergence as well as the length of the natural free-running period (185). Additional studies on the unicellular algae *Chlamydomonas reinhardi* (186) and the fungus *Neurospura crassa* (187) also support a genetic basis for rhythms [see reviews by Feldman and Dunlap (188) and Edmunds (120).]

Studies on human beings also reveal that there is a genetic basis for rhythms. For example, the characteristics of circadian rhythms in monozygotic twins are more nearly alike than those of dizygotic twins. Similarly, the rhythm characteristics of monozygotic and dizygotic twins are much more alike than those of nontwin siblings (39, 189).

6.9 Mechanisms of Biological Rhythms

Many attempts have been made through experiments on mammals to elucidate the origin and mechanisms controlling rhythmicity. These studies employed the classic endocrinological approaches of adrenalectomy, hypophysectomy, cerebral ablation, and pinealectomy. Recently, a great deal of interest has been given to the possible role of the suprachiasmatic nucleus, located in the brain, as a rhythm generator (114, 180, 181). At this time the results of these studies indicate that no single regulator can account for the control of all rhythmic variables (147).

Halberg has suggested that hormonal secretions from the adrenal cortex deserve serious consideration as one principal mechanism that possibly underly human biological adaptation to the daily activity–rest cycle. Incidentally, it has been shown that the adrenal as well as the pineal gland will continue to secrete and respond circadianly for several cycles, even when isolated in *in vitro* preparations. Halberg (39) has recently reviewed the role of the adrenal gland and other possible regulators in mammals. Although there have been numerous attempts to locate a single central regulator of circadian phenomena at the unicellular level, thus far these have been unsuccesful. Several researchers feel that the secret to understanding temporal rhythms lies in the communication between cell membranes. Some (180, 190–192) have postulated that the circadian oscillation may be generated and synchronization effected by the temporal variation within cell membranes and that regulation of all systems starts with the cycles of the organelles within the cells since mitochondria and the endoplasmic reticulum have both been shown to exhibit rhythms in many processes (193). Edmunds (120) has recently reviewed the many diverse mechanisms that have been proposed as regulators. These include a number of strictly molecular, feedback loop (network), transcriptional (tape-reading) models, as well as membrane models mentioned previously.

In spite of the numerous models that have been proposed and the large volume of data that have been gathered, it is still unclear whether bioperiodicities are driven by a single master oscillator or by a population of noncircadian biochemical oscillators (114). Even though the majority of chronobiologists accept the endogenous nature of rhythms, one school postulates that circadian rhythms are generated by an interaction of several external forces, such as light, temperature, electromagnetic variation, and possibly yet unknown and subtle geophysical forces (194). In this chapter we cannot adequately comment on this interesting concept, but we do wish to call attention to it for the interested reader. Clearly, evidence exists for geophysical forces affecting life.

As soon as the mechanisms underlying biological rhythms are identified and well understood, it will be possible to better comprehend and hopefully alleviate the health-related problems of rotating shiftworkers. It may even be possible to hasten the chronobiologic adjustments of rotating shiftworkers and travelers to changes in their sleep–activity routine through the use of special medications called *chronobiotics*, as has been attempted earlier by Simpson and his colleagues (195, 196). Unfortunately, up to this time, studies on human beings using candidate medications have not been fruitful. Nonetheless, investigators are continuing their search for a useful chronobiotic.

7 MEDICAL IMPLICATIONS OF BIOLOGICAL RHYTHMS

A great deal of research on rodents and human beings indicates that biological rhythms are so significant they must be taken into account in clinical medicine (34–36, 39, 107, 170, 172, 173, 197, 198). In the diagnosis of allergy with the use of intradermal injections of antigens, for example, the timing of tests, that is, morning, afternoon, or night, appears to be almost as important as what is tested! On the average, the cutaneous sensitivity of patients to several common allergens, such as house dust or mixed pollens, evidences a 3.5-fold difference depending on whether the timing of the diagnostic test is done in the morning or evening. Some patients show as great as a 7- or even 11-fold variation (199–201) (Figure 5.23). Moreover, in certain ailments the occurrence or exacerbation of symptoms is predictable over time as biological rhythms (130, 202, 203). This is the case for allergic, cardiovascular, neurologic, and inflammatory diseases, among others. In addition, the pharmacokinetics and pharmacodynamics of a large number of medications vary significantly according to their

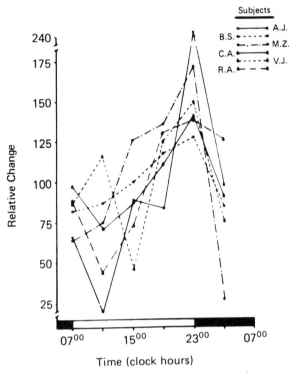

Figure 5.23 Individual and mean chronograms (time plots) for the circadian susceptibility rhythm of the skin studied by intradermal injection of house dust extract at different times of the day and night in six house-dust-allergic adult patients. The temporal changes are shown as percentage deviations from each person's 24-hr average response. There exists large variation in cutaneous reactivity with respect to clock hour of testing as well as between persons. Clock hour time is given in military fashion: 1500 = 3 p.m., 2300 = 11 p.m., and 0700 = 7 a.m. The patients were synchronized with sleep (shown as shading along horizontal axis) from 11 p.m. to 7 a.m. From Reinberg and colleagues (200, 201).

timing with respect to the circadian system (33, 34, 39, 109, 204, 205). For some medications, such as those used to treat arthritis, synthetic corticosteroids, or a variety of nonsteroid anti-inflammatory preparations (206–209), the timing of the medication is critical. The side effects of synthetic corticosteroids in day-active persons may be so great if these medications are taken before bedtime that the patient may have to give up their use. Yet when taken in the morning there is little or only minor side effect, depending on drug form and dose (206). Similarly, many arthritic patients do not tolerate nonsteroid anti-inflammatory medications if taken in the morning, but do so with good relief when taken before bedtime (207–209). We wish to add that biological rhythms in the pharmacokinetics and effects of most drugs are not dependent on day–night differences in posture nor meal timing and composition.

7.1 Biological Rhythms and Human Illness

The significance of biological rhythms with regard to the occurrence or intensification of symptoms of human disease was recognized even in biblical times, when it was noted that the illness we now call *asthma* tended to worsen or not occur at all until nightfall. Today, a division of chronobiology, chronopathology, is concerned with understanding the relationships between alterations of rhythmic processes and the development of human illness as well as the predictable, rhythmic exacerbation of symptoms. A chapter in a recently published book by one of us (130) addresses the issue of chronopathology in great detail. Just a few examples have been selected to illustrate the concept.

Because of the high incidence of cardiac problems and the assumption that they occur primarily in the evening, these ailments were some of the first to be studied. The onset of symptoms (as opposed to the admission time to the hospital emergency room) of myocardial and cerebral infarction, cerebral hemorrhage, and angina pectoris is strongly circadian rhythmic. For example, in a group of more than 1200 diurnally active patients, myocardial infarction (Figure 5.24) was found to be most frequent between 8 A.M. and 10 A.M. (130). There was a secondary peak 12 hr later. However, for cerebral infarction (Figure 5.25), the phasing was different; the occurrence was very much more frequent during sleep with a peak at 3 A.M. (210). Also, it was found that spontaneous intracerebral hemorrhage (130) (Figure 5.26) was primarily an evening event with a peak at around 7 P.M. The clinical signs of Prinzmetal's variant angina as evidenced by ST-segment elevations of the electrocardiogram (Figure 5.27) were more common during sleep between 2 and 4 A.M. and quite uncommon around midday and during the afternoon (211).

When biological rhythms are found in the occurrence or exacerbation of human illnesses, the question frequently arises as to whether they result from cycles in ambient conditions, rest–activity, and/or rhythms in related biological processes. For example, asthmatic patients exhibit a high-amplitude circadian rhythm of airway patency and dyspnea. The data in Figure 5.16 show that the breathing of asthmatic persons tends to be considerably easier during midday than at night or on awakening. With regard to peak expiratory flow, the rhythm represents a 25 percent variation in airway function over 24 hr (212). The acrophase chart in Figure 5.28 also confirms that the dyspnea of asthma in day-active patients is a nocturnal event, with the symptoms becoming worse between 11 P.M. and 5 A.M. This acrophase chart (203) reveals that the staging of several circadian rhythms that are known to influence airway caliber contribute to the nocturnal predilection of asthma. In particular, the dyspnea of asthma worsens during the nighttime when the urine and blood levels of anti-inflammatory hormones—adrenal corticorsteroids (serum cortisol and urinary 17-hydroxycorticosteroids) and catecholamines [noradrenalin, adrenaline, and vanillylmandelic acid (VMA)]—are at their circadian minimum. The nocturnal worsening of the dyspnea of asthma corresponds in time also to

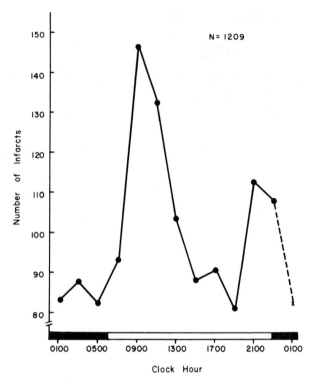

Figure 5.24 The temporal distribution of myocardial infarction (MI) is nonrandom; MIs are more common in occurrence around 0900 and 2100 (9 P.M.) than 0500 and also between 1500 (3 P.M.) and 1900 (7 P.M.). The shaded portion of the horizontal axis shows the presumed sleep span for the sample of patients. From Smolensky (130).

poorest airway function [as determined by the 1-sec forced expiratory volume ($FEV_{1.0}$), vital capacity (VC), and peak expiratory flow (PEF)]. It corresponds as well to greatest susceptibility of the airways and skin to antigens, such as house dust, and to the nonspecific chemical irritants of histamine and acetylcholine. Acrophase charts such as the one for asthma help to explain why these patients become worse at night. This type of information also indicates that temporal variations in one's physiology may be equally or even more important than those of the ambient environment in explaining the cyclic nature of certain human illnesses and also in defining their etiology.

Evidence for the importance of biological rhythms in human illness was obtained in an exemplatory study by Halberg and Howard (213) on a man exhibiting prominent circadian variation in the occurrence of his overt epileptic seizures (Figure 5.29). During the years when he was active between 6 A.M. and 9 P.M., seizures were most commonly experienced after awakening, especially between 7:30 A.M. and 1:30 P.M. After he altered his sleep–activity routine to accommodate nighttime work, the clock hours during which his symptoms were

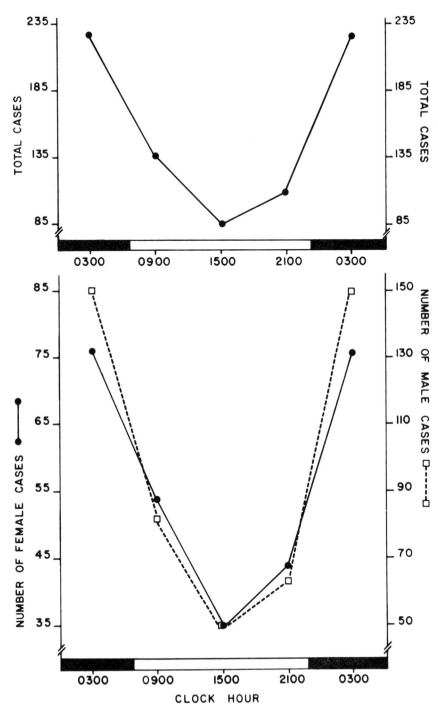

Figure 5.25 Circadian rhythm in the occurrence of cerebral infarction (morbidity). In presumably diurnally active males and females cerebral infarction was considerably more common at night around 0300 (3 A.M.) than during the afternoon at 1500 (3 P.M.). Shaded portion at bottom shows the presumed usual sleep span. Data from Marshall (210).

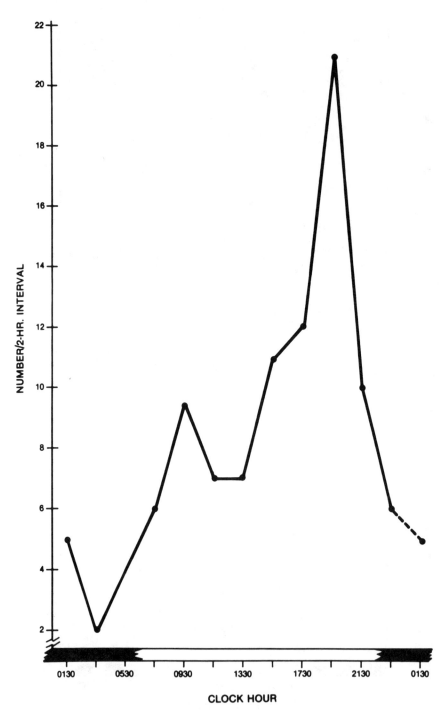

Figure 5.26 The occurrence of spontaneous intracerebral hemorrhage (SICH) is not evenly distributed over the 24 hr. In a sample of 100 cases for which the time of the event was known. the susceptibility was considerably greater between 1730 (5:30 P.M.) and 2130 (9:30 P.M.) than between 2330 (11:30 P.M.) and 1330 (1:30 P.M.). Shaded portion at bottom depicts the presumed sleep span. From Smolensky (130).

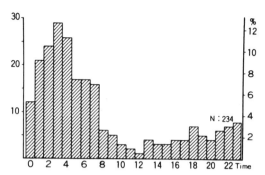

Figure 5.27 Circadian variation in the occurrence of 234 episodes of ST-segment elevation, a characteristic sign of Prinzmetal's variant angina, as detectable from the continuous recording of the electrical activity of the heart muscle in a sample of diurnally active patients suffering from this form of heart disease. The number of episodes of ST-segment anomaly per hour is greatest during the sleep span with the peak around 0300 (3 A.M.). Few episodes of ST-segment anomaly are detected throughout the daytime span of activity. From Kuroiwa (211).

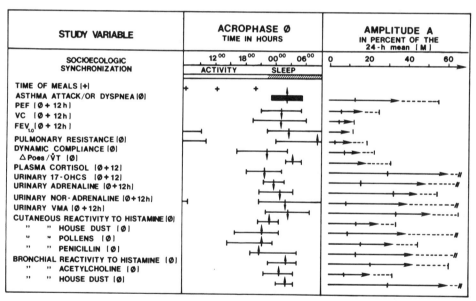

Figure 5.28 Temporal occurrence in the peak or trough of circadian functions affecting the 24-hr pattern of airway function in diurnally active asthmatic patients. The arrows in the middle column pointing upward indicate the timing of the circadian peaks (termed *acrophases*), whereas those arrows pointing downward indicate the timing of the circadian troughs. A horizontal line extending to the right and left of the acrophase or trough marker represents the 95 percent confidence limit of each. The amplitude, a measure of the 24-hr rhythmic variability, shown in the column to the right (the length of the arrow is proportional to the circadian rhythmic variation), is expressed as a percentage of the 24-hr time series mean, termed the *mesor*. The timing of the peak and trough values of these and other biological rhythms is believed to contribute to the heightened susceptibility of patients to asthma nocturnally. Reproduced from Smolensky et al. (202).

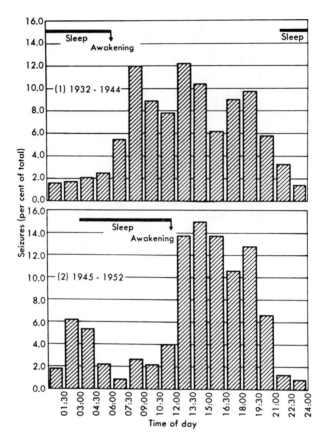

Figure 5.29 Circadian pattern in the initiation of epileptic seizures in one patient studied longitudinally over several years while residing in an institution. The temporal distribution with regard to the clock hour of the seizures changed following transfer to a different work–rest pattern when the sleep schedule was delayed from one of 9 P.M. to 6 A.M. to one of 3 A.M. to 11 A.M. Clock time along horizontal axis is given in military fashion, for example, 1500 = 3 P.M.; 2100 = 9 P.M. From Halberg and Howard (213).

most frequent had changed to between 12 P.M. and 7 P.M. However, with reference to the sleep–activity schedule of this individual, the occurrence of the majority of his seizures was comparable under both sleep–activity routines; the occurrence was always most frequent during the several hours immediately following the termination of sleep.

Information pertaining to circadian and other rhythms in symptoms of disease is important for three reasons: (a) occupational as well as clinical practitioners can improve their treatment of certain ailments through an understanding of chronopathology, (b) appropriate steps may be taken to minimize or prevent the occurrence of disease or exacerbation of existing conditions, and (c) such information can be utilized by physicians to determined whether a disease might be occupationally induced. For example, a frequent

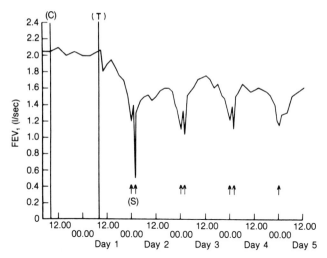

Figure 5.30 Temporal variation in airway patency of a presumably diurnally active asthmatic patient following a single pulmonary challenge on the test day (numbered 1) with an antigen to which hypersensitivity had developed (budgerigar serum) through occupational exposure. Although the $FEV_{1.0}$, a measure of airway patency, was close to normal during the day, it was extremely low during the night, giving rise to severe episodes of dyspnea over several consecutive nights. From Taylor et al. (215).

difficulty for some workers employed in certain industries is occupationally induced recurrent nocturnal asthma (214–217). Patients suffering from this disorder quite often exhibit marked circadian patterns in the status of their airway function. Figure 5.30 presents one example of circadian changes of airway patency in an individual who became sensitized to budgerigar serum due to exposures arising from breeding parakeets. It is readily apparent that in this day-active person a brief exposure to the offending antigen during the morning around 10 A.M. resulted initially in only a small effect on airway patency as indicated by the pulmonary function $FEV_{1.0}$ measurements. However, around midnight well after the single brief morning exposure the effect was fully manifest. Moreover, it was only at night, for several subsequent 24-hr spans, that the symptoms and effects of the asthmatic condition were worsened even though no subsequent exposure to the antigen occurred.

Occupationally induced recurrent nocturnal asthma is not uncommon and is often observed in employees of several different industries, including those who work with wood, plastics, resins, grains, and pharmaceuticals (214). The prior example illustrates how susceptible workers who are exposed during daytime work need *not* develop significant symptoms until much later in the evening, only to recur on several consecutive nights even without further exposure to the offending antigen. The recurrent nocturnal exacerbation of the asthma is believed to represent, at least in part, the circadian organization of processes that affect airway patency. This example serves to illustrate that epidemiologic studies of occupationally induced disease should take into account

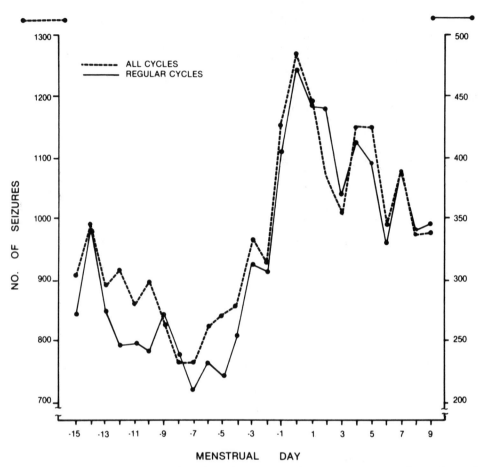

Figure 5.31 In this sample of 50 epileptic but otherwise normal menstruating women, the incidence of overt epileptic seizures varied with the phase of the menstrual cycle. The incidence just before (day −1) and on the first day (0) of menstruation was nearly twice that 7 days before. Similar patterns were exhibited independent of the regularity or irregularity in the menstrual cycle duration. From Laidlaw (218).

rhythmic variation in the symptoms or occurrences of disease. Quite often, patients and physicians do not associate the etiology of certain diseases with employment when complaints occur many hours after leaving the workplace. This appears to be especially true for occupationally induced asthma, for which the interval between daytime exposure at work and the manifestation of the nighttime illness at home can be quite lengthy.

It is important to remember that chronopathologies can show frequencies different from the circadian one. For example, the occurrence of asthma and epilepsy in women can be circamensual (218) (Figure 5.31) as well as circadian rhythmic, and the occurrence of asthma and cardiovascular disorders is known to be circannual as well as circadian rhythmic (130).

7.2 Biological Rhythms and Medications

Chronopharmacology is the study of the rhythmic aspects of the absorption, distribution, metabolism, excretion, and action of therapeutic chemical agents in animals, including humans. Research conducted primarily during the last three decades reveals that the behavior or fate of medications and other xenobiotics can vary rather dramatically as a function of the time of their administration with respect to the scale of 24 hr and, for some, the time of year (33, 204, 205, 219–222). The emergence of new findings from chronopharmacologic research has resulted in a different perspective about the metabolism and response of not only therapeutic medications but toxic chemical substances as well (223–225). These findings have, in turn, given rise to a set of new concepts and terms, such as *chronokinetics*, *chronesthesy*, and *chronergy* (33, 222) for describing the chronopharmacologic aspects of medications or chronotoxicologic properties of other chemical agents. In the subsequent sections the concepts of chronopharmacology are reviewed in preparation for discussing the topic of chronotoxicology and shiftwork.

7.2.1 Chronokinetics of Medications

Many medications exhibit important differences in the time course and pharmacokinetics of their concentration in the blood or urine, depending on *when* treatment is given during the day or night. For example, the time to achieve the peak blood concentration, magnitude of peak plasma concentration, half-life of elimination, and the area under the time–concentration curve for a given drug can vary predictably in the same person depending when in the course of a day (and perhaps year) it is administered as a result of the influence of bioperiodicities. Rhythms in pharmacokinetic phenomena due to the effect of naturally occurring bioperiodicities on chemical substances is termed *chronokinetics*. When chronokinetic findings were first detected and reported, most clinical pharmacologists were skeptical. Most believed the findings were the result of improper research methods, effects of posture, or the influence of meal timings and composition. However, as an increasing number of findings have been published, it has become clear that for a majority of the medications the circadian differences represent the result of endogenous rhythmic processes.

A wide range of drugs is sensitive to the timing, with regard to rhythms, of administration (33, 219–222). For example, as shown in Figure 5.32, the chronokinetics of indomethacin, a nonsteroid anti-inflammatory drug used to treat arthritis and other tissue inflammatory disorders, can be seen to differ according to its administration time when studied under carefully controlled conditions (226). When this drug is given to diurnally active patients in the morning at either 7 or 11 A.M., the peak plasma concentration is achieved much more rapidly that at 3, 7, or 11 P.M. Moreover, the peak concentration varies also as a function of its timing. It is greater for the 7 and 11 A.M. dosings than it is for the others. These findings show that indomethacin when taken in the

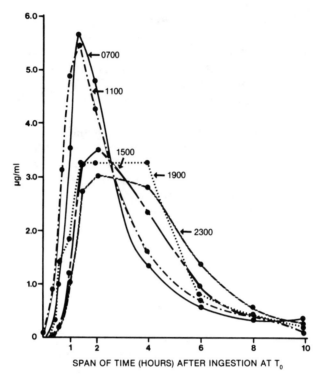

Figure 5.32 In random order and at weekly intervals, nine healthy subjects 19 to 29 years of age, synchronized with activity from 7 P.M. to 0000, received a single oral dose (100 mg) of indomethacin at fixed hours: 0700 (7 A.M.), 1100 (11 A.M.), 1500 (3 P.M.), 1900 (7 P.M.), and 2300 (11 P.M.). Venous blood (sampled at 0, 0.33, 0.67, 1.0, 1.5, 2.0, 4.0, 6.0, 8.0, and 10.0 hr postingestion) was obtained for plasma drug determinations. Ingestion at 1900 and/or 2300 led to the smallest peak height and longest time to attain the peak concentration, whereas ingestion at 0700 and/or 1100 led to the largest peak height, shortest time to attain the peak height, greatest area under the time–concentration curve, and fastest disappearance rate from the blood. A circadian rhythm of both peak height and time to peak was statistically validated. From Clench et al. (226).

morning quickly reaches very high plasma concentrations. It is pertinent that patients' complaints of indomethacin intolerance are much more frequent when the drug is taken in the morning, when the peak height is greatest, than when taken midday or in the evenings, when the peak height is more moderate.

The disposition and efficacy of propranolol, a widely used medication for treating certain cardiovascular ailments, also have been shown to be sensitive to human circadian rhythms (227). Figure 5.33 shows that the time to reach peak plasma concentration, the concentration at the peak, the elimination rate, and the area under the plasma–time concentration curve all vary significantly according to when propranolol is given. Predictable circadian differences in the pharmacokinetics of many commonly used medications, such as aspirin, ethanol, theophylline, and antihistamines, as well as analgesics and carcinostatics, among others, are now known (34, 222).

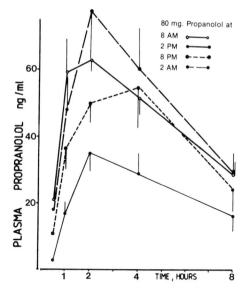

Figure 5.33 Mean and standard error of plasma propranolol concentrations after acute single administrations of 80 mg at selected clock hours. The same six diurnally active healthy persons were studied at each timepoint. Note the large difference in the pharmacokinetics of propranolol as a function of the treatment time. Statistically significant differences were detected in the peak height concentration, time to peak, and the area under the curve (AUC) pharmacokinetic indices. From Markiewicz et al. (227).

The urinary kinetics of medications also exhibit important differences dependent on treatment time. This is the case for aspirin (228) in human beings. The shape of the time–concentration curve for the excretion of salicylates, the major constituent of aspirin, differs, for example, for treatment at 7 A.M. versus 7 P.M. (Figure 5.34). The time to reach the peak and the peak concentration of salicylates in the urine are dependent on administration time, as is their elimination from the body. The shortest time to achieve highest urinary concentration, the greatest peak height, and the most rapid elimination occur when 1 g of pure aspirin is taken at 7 or 11 P.M., in comparison to when taken at 7 or 11 A.M.

The results of recent studies on mequitazine (222), a new antihistamine, as well as β-methyldigoxin (229), a medicine used to treat heart disease, indicate the the shape of the time–blood or time–urinary concentration curve may be so radically different for the same group of persons treated at different times, during the morning or around midday in comparison to during the evening or midnight, that the data must be dealt with in a different manner. For example, some drugs when studied under rigorously controlled conditions may exhibit patterns that can be best described by a one-compartment model if given in the morning but by a two-compartment model if administered in the afternoon or evening. This phenomenon has been shown to exist for the time–urinary excretion curve of mequitazine (Figure 5.35). In a carefully controlled experiment, including the randomization of subjects for the sequence of treatment times, a group of persons were given a single dose at 7 A.M. on one occasion and again at 7 P.M. on another a week later (222). The shape of the time–urinary excretion curve following the 7 A.M. treatment suggested a two-compartment model, whereas the shape of the curve following the 7 P.M. one

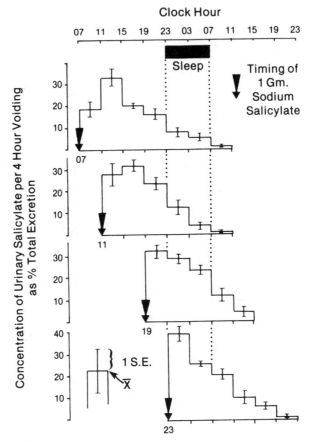

Figure 5.34 Circadian differences in the urinary chronopharmacokinetics of aspirin (sodium salicylate). Six subjects in good health synchronized to sleep from 11 P.M. to 7 A.M., meal timing and composition controlled. For a single 1-g oral administration of aspirin the time to achieve the peak urinary concentration is fastest for a 1900 (7 P.M.) ingestion; the greatest peak height concentration coincides with ingestion at 2300 (11 P.M.). Slowest elimination of salicylate follows ingestion at 0700 (7 A.M.); fastest elimination follows ingestion at 1900 (7 P.M.). From Reinberg et al. (228).

suggested a simple one-compartment model. Although there are many possible explanations for this difference in behavior, it is clear that circadian differences in one or more of the following processes—absorption, metabolism, distribution, and elimination—are involved.

7.2.2 Chronesthesy

It frequently has been assumed that in the same person the effect produced by a given dose of a drug will be approximately the same independent of the time when it is administered. Studies by many investigators have shown that

BIOLOGICAL RHYTHMS, SHIFTWORK, AND OCCUPATIONAL HEALTH 239

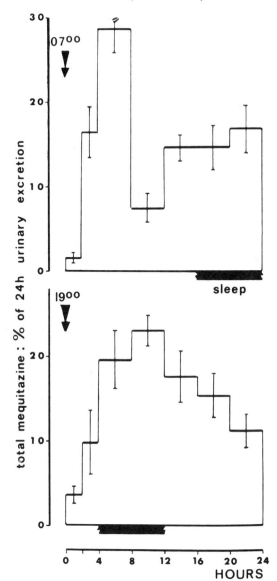

Figure 5.35 Circadian time dependence of pharmacokinetic models. A single 5-mg dose of mequitazine (an antihistaminic agent) was given orally to six healthy subjects at 0700 (7 A.M.) and 1900 (7 P.M.), 1 week apart. Participants were synchronized with diurnal activity from approximately 7 A.M. to about 11 P.M., alternating with nocturnal rest. Determinations of mequitazine were made from urine voidings collected at first every 2 hr (twice) and thereafter at 4-hr intervals during 24 hr. The results are graphed as the percentage of mequitazine excreted per urinary collection interval with reference to the entire 24-hr collection span. The kinetics were characterized by two peaks (a two-compartment model) when the antihistamine was ingested at 0700 and by 1 peak (a one-compartment model) only when given at 1900. Values for mequitazine and the occurrence of sleep are shown with respect to time posttreatment at 0700 (top) or 1900 (bottom) in hours. From Reinberg (222).

the quantitative and qualitative effects of a given medication, however, can differ greatly according to the timing of treatment. The differential effect of a medication or toxicity of a chemical contaminant as a function of its timing is termed *chronesthesy*. Chronesthesy is defined as circadian and other rhythmic susceptibilities to chemical agents arising from bioperiodicities primarily related to receptor sites, enzyme activity, and metabolic processes.

Lidocaine-induced analgesia is one response for which a circadian chronesthesy has been demonstrated (Figure 5.36). The data show that a dose of

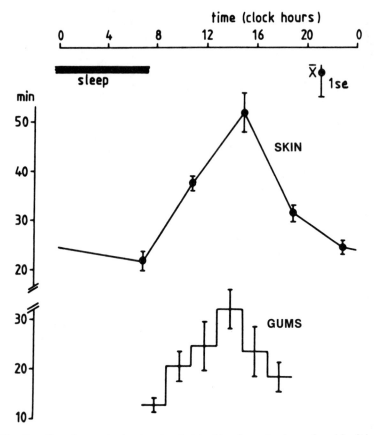

Figure 5.36 Circadian changes in the duration (min) of local anesthesia produced by lidocaine. Top curve: In six apparently healthy adults synchronized with diurnal activity from 7 A.M. to midnight and nocturnal rest, 0.1 ml of a 2 percent lidocaine solution was injected intradermally every 4 hr during 24 hr, at specified clock hours. The flexor surface of both forearms was used exclusively. The duration of anesthesia was determined by measuring the time in minutes from injection to the recovery of cutaneous sensitivity. The mean duration was only about 20 min at 0700 (7 A.M.): it was 52 min at 1500 (3 P.M.) and about 25 min at 2300 (11 P.M.). Differences between the longest and shortest durations are statistically significant ($p < .0005$). Bottom curve: for rigorous standardization, the study was restricted to selected patients suffering from decay (dentin caries) in a living, single-rooted upper front tooth. Lidocaine (2 ml 2 percent solution) was injected in the para-apical region. A stopwatch was started at the end of the injection. Thereafter, the tooth was drilled to remove decay, but the cavity was left unfilled until the return of sensitivity as determined by a set of tests. A group of 35 subjects (apparently healthy apart from their tooth decay) was investigated. Each patient was treated only once. The timing of treatment was randomized between 7 A.M. and 7 P.M. There were six subgroups of five to seven patients studied at 2-hr intervals. All subjects had diurnal activity and nocturnal rest. The duration of local anesthesia was about 12 min for the test interval from 7 A.M. to 9 A.M, about 32 min for that of 10 A.M. to 2 P.M. ($p < .005$), and about 19 min for that of 3 to 7 P.M. ($p < .025$). For both of these experiments illustrating the chronesthesy of lidocaine, differences between the longest and shortest effect are statistically significant. From Reinberg and Reinberg (230).

lidocaine when injected in either the forearms or the apical region of the gum above a tooth with decay at midday is twice as effective as it is when given around 8 A.M. or 4 P.M. (230). Other medications that exhibit circadian chronesthesys are antihistamines, analgesics, synthetic hormones, hypnotics, and stimulants (33, 208, 221, 222).

Another type of chemically induced chronesthesy produced by a xenobiotic is illustrated by the response of the airways to histamine aerosols when briefly inhaled by asthmatic, emphysemic, or bronchitic persons at various times during the day or night. In these types of patients the airways are known to exhibit a hyperreactivity to a variety of chemical agents, including histamine, causing an increase in the resistance of airflow in the lungs. The results of studies by DeVries and his colleagues (231) (Figure 5.37) indicate that there is large variation in the threshold concentration required to provoke a 15 percent decrease of airway patency with the use of this nonspecific irritant chemical. The airways are least susceptible; that is, they exhibit the highest threshold around noontime and are most susceptible around midnight—that is, they exhibit the lowest threshold at that time. The difference in the threshold concentration between the times of greatest and lowest reactivity can be as great as 100 percent or more (231, 232). These findings suggest that persons with small or large airway disease who are also involved in shiftwork and are exposed to irritant chemicals may experience more airway reactivity and provocation during one shift (the evening or night shift) in comparison to another. In other words, a given concentration of a chemical irritant that may be rather well tolerated during the day shift may not be tolerated at all when encountered during the night shift.

7.2.3 Chronergy

Chronergy is defined as rhythmic variations in either the desirable or undesirable effect of medications or other chemical agents. The *chronergy* of a chemical substance is dependent on both its *chronokinetics* and its *chronesthesys*. Homeostatically designed, single-timepoint, pharmacology and toxicology research generally has revealed that the effects of a chemical agent are related directly to its concentration in the serum or tissue. However, chronobiologic studies reveal that this need not be the case; in fact, it is rare that the chronesthesy of a medication coincides with its chronokinetics. This is exemplified in Figures 5.38 and 5.39, which review the results of a chronopharmacologic study of ethanol (233, 234).

In this study, it was shown that ethanol (0.67 g/kg body weight) ingested by diurnally active healthy persons demonstrated significant chronopharmacokinetic differences. Ethanol, when taken at 7 A.M., resulted in the greatest serum peak height, shortest span of time to reach this peak, and fastest disappearance rate (Figure 5.38) (234). In comparison, ethanol ingestion during the evening, at 7 or 11 P.M., resulted in the lowest peak ethanolemia, longest span of time to reach this peak, and slowest disappearance rate. It is of interest that the

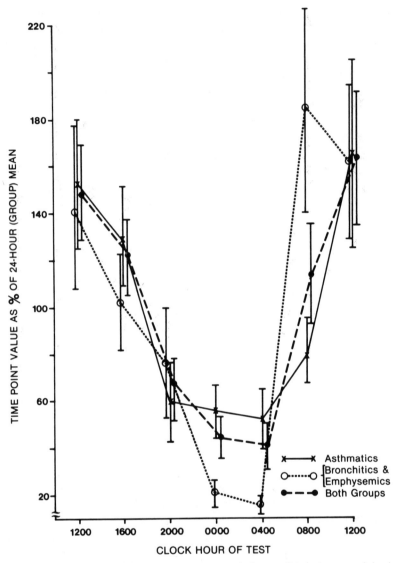

Figure 5.37 The circadian difference in hyperreactivity to inhalation of histamine aerosols by the airways of asthmatic, bronchitic, and emphysemic patients is great. At midnight (0000) or 4 A.M. (0400), times of heightened responsiveness, the concentration required to induce a 15 percent decrease in airway patency relative to the time-specific control baselines, is from 100 to 160 percent less than that required 12 hr later, at noon or 1600 (4 P.M.), when the airways are least reactive. Note that the data of each curve are plotted as deviations from the given group's 24-hr mean reactivity, which for the purpose of graphing has been set equal to 100 percent. From Smolensky et al. (202).

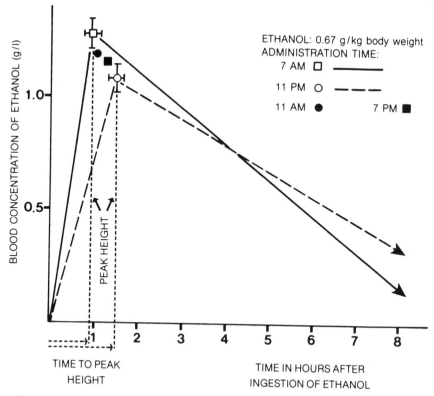

Figure 5.38 In diurnally active persons, the greatest peak height, shortest span to reach peak height, and fastest elimination for ethanol occur when this substance is given at 7 A.M. In comparison, ingestion at night at 7 or 11 P.M. by the same persons is associated with lowest peak height, longest time to reach this peak, and slowest elimination. Redrawn from Reinberg et al. (234).

chronesthesy of ethanol defined as the level of ebriety self-estimated at 60 min following its consumption also differed as a function of the time of intake (222, 233, 234). In short, ebriety was greatest following an 11 P.M. ingestion and least following the 7 or 11 A.M. ingestions. This study shows that although ethanol was always consumed in the same dosage, both the pharmacokinetics and effect varied with the time of ingestion. The effects on the brain were less at 7 A.M., when the serum peak height concentration was greatest and the disappearance rate from the blood fastest, than at 7 or 11 P.M., when the peak height was lowest and the disappearance rate was slowest (222, 234).

Chronergic effects of ethanol on the oral temperature circadian rhythm in humans have been examined by Reinberg et al. (233, 234). The findings are related here as a further example of how kinetic data may not always be accurately predictive of systemic effects. In these studies ethanol in the same dosage (0.67 g/kg body weight) was given at either 7 A.M., 11 A.M., 7 P.M., or 11 P.M., with each study time being done in a random order and on different

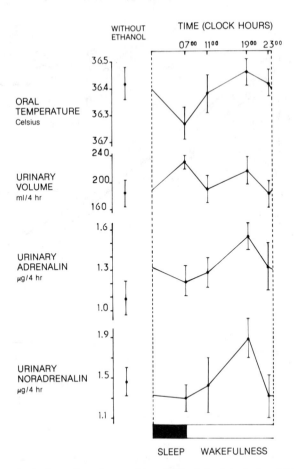

Figure 5.39 The 24-hr rhythm-adjusted mean (M ± 1 SE) of physiological variables after the ingestion of ethanol (0.67 g/kg body weight) at different test times [0700 (7 A.M.), 1100 (11 A.M.), 1900 (7 P.M.), and 2300 (11 P.M.)]. Control data were collected over 24 hr without ethanol ingestion. Timed ethanol ingestions were performed at least 1 week apart, in random order. Subjects were six healthy young adult males synchronized with activity from about 0700 to 0000 alternating with nocturnal rest. Subjects had fasted about 7 hr before and 7 hr after each ethanol ingestion. Measurements and integrated urine samples were gathered at 4-hr intervals. Relative to the respective control values, ethanol ingested at 0700 induced a decrease of the temperature 24-hr mean—M, but no change in the catecholamine M values; ethanol ingested at 1900 induced a rise of the catecholamine M values but no change of the temperature M. The temperature decrease due to ethanol at 0700 is presumably related to the fact that the ethanol-induced peripheral vasodilation is not counteracted at this time by a change in catecholamines. On the contrary, ethanol ingestion at 1900 is followed by higher levels of catecholamines that presumably resulted in peripheral vasoconstriction compensating for the ethanol-induced vasodilation with no change of body temperature. From Reinberg et al. (234).

days. A placebo liquid was also given to obtain baseline data on several biological rhythms, including that of oral temperature. Figure 5.39 summarizes the circadian-stage dependent effect of ethanol on oral temperature (222, 234). Shown are the 24-hr average oral temperature values calculated using data collected following each of the separately timed ethanol ingestions. The results indicate that only when ethanol was ingested at 7 A.M. was a statistically significant posttreatment hypothermia induced relative to control conditions. When ethanol was ingested at the other test times, no hypothermia occurred. The hypothermia that followed the ingestion at 7 A.M. could have been related to the fact that a drug-induced peripheral vasodilation was not counteracted at this time by an appropriate catecholamine (adrenalin and noradrenalin) secretion as suggested by the data in Figure 5.39. On the other hand, ethanol ingestion at the other test times was followed by an elevation of catecholamine secretion presumably causing peripheral vasoconstriction and compensation for the ethanol-induced vasodilation so that there was no net change in the body temperature.

Although the examples put forward to describe chronergys do not have direct occupational relevance, the concept that blood levels need *not* always be predictive of the same level of biological effect because of chronobiologic factors is very important. A potentially harmful concentration of a toxicant may or may not lead to symptoms of toxicity, depending on when, during the 24 hr with respect to the circadian organization of biological functioning, it occurs. Liver enzymes at one circadian time may exhibit greater efficiency in detoxification than at others. Similarly, target tissues may be less susceptible to toxicants at one circadian time than another. This is discussed in greater detail in the subsequent sections of this chapter.

In summary, an awareness of the potential impact of chronokinetics, chronesthesy, and chronergy on the human response to medications is useful in understanding and improving the treatment of human illnesses, including ones experienced by shiftworkers. More pertinent to the focus of this chapter, these phenomena have relevance to predicting temporal differences in the pharmacokinetics and adverse effects of exposure to contaminants in the workplace. The implications to the field of industrial hygiene, toxicology, and occupational medicine are abundant, especially with respect to evaluation of the potential adverse effects of shiftwork on human health. Persons may exhibit different degrees of risk from exposure to chemical or physical agents depending on the shift worked.

7.3 Effect of Biological Rhythms on Toxic Responses

Chronobiologic studies on rodents conducted during the late 1950s and 1960s revealed that the acute toxicity of many chemical agents differs in intensity depending on the time of exposure with respect to circadian and perhaps circannual rhythms. F. Halberg (108), an eminent chronobiologist, initially referred to this as the "hours of diminished resistance" (Figure 5.40); he

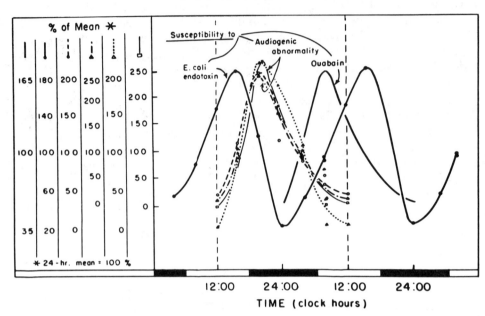

Figure 5.40 Chronotoxicology was first considered as the hours of diminished resistance. Experimental results reported by Halberg demonstrated the existence of circadian rhythms in the susceptibility of mice exposed to various potentially noxious agents, for example, a fixed "dose" of E. coli endotoxin, ouabain, or white noise. The endpoint of response in the studies on mice herein summarized was the number of deaths per treatment group per time of testing. From Halberg (108).

pioneered the concept and research of susceptibility–resistance rhythms in rodents as well as human beings.

7.3.1 Toxicology Evaluation and Susceptibility–Resistance Rhythms in Rodents

Following many years of investigation, it has become quite clear that the acute and chronic toxicity of a chemical is definitely influenced by the time when it is administered, as shown by many investigators (108, 127, 225, 235–244). Numerous studies have shown, for example, that the most common and fundamental test for toxicity evaluation, the LD_{50}, can be dramatically affected by rhythmic processes. The inconsistencies between laboratory reports for LD_{50} values could result from differences in the timing of studies and/or variations in the synchronization of the test animals. In certain instances, the effect of rhythms may be so great that the LD_{50} may vary by one or two orders of magnitude (127, 235). In addition, more subtle indicators of toxicity such as the concentrations of serum liver enzymes (see Section 7.3.2) as well as tests of behavioral toxicity also may vary depending on the timing of dosing (224, 225, 233, 235). We wish to point out that other types of toxicity tests can be affected by biological rhythms as well. These include ones pertaining to teratogenicity, target organ toxicity, mutagenicity, and carcinogenicity as reviewed by Scheving (127, 235). On the basis of the fundamentally rhythmic nature of biological function at the cell and higher levels of organization such as exhibited by the endocrine, nervous, and cardiovascular systems, among others, the timing with regard at least to circadian stage of toxicity evaluations is of critical importance.

Chronobiologists believe that toxicologists should more carefully conduct their studies of chemical agents. Since the results of toxicologic studies have tremendous impact on the ability of a firm to register, manufacture, and market chemicals, it is surprising that industrial and research laboratories as well as regulatory agencies have not been sensitive to the methods and findings of chronobiology.

Representative examples of circadian susceptibility–resistance rhythms in rodents are presented in Figures 5.41–5.43. As shown in these figures, the acute toxicity of many agents such as nicotine, strychnine, amphetamine, and ethanol, as well as urethane, paraquot, malathion, and mercury chloride, is characterized by large-amplitude circadian patterns. The mortality induced by exposure to X-ray irradiation, bacterial endotoxin, and 100 percent oxygen also are strongly circadian rhythmic.

Rhythms also have been shown to influence tests used to determine the teratogenic potential of chemicals (248, 251). Sauerbier (251) reported that in pregnant mice injected once with cyclophosphamide (20 mg/kg) or Th-R (N-mustard, 2 mg/kg) at one of four different circadian stages (7 A.M., 1 P.M., 7 P.M., or 1 A.M.) on day 12 of gestation at precisely the same number of hours from impregnation at each timepoint, there was a strong circadian-stage dependence. The highest incidence of malformations due to maternal treatment with cyclophosphamide was found to be associated with the dark-to-light transition (7 A.M.), whereas the lowest occurred at 1 A.M. In contrast, the teratogenic action of Th-R was strongest at the onset of darkness (7 P.M.) and the lowest at 7 A.M. Interestingly, the mean embryotoxicity of both compounds was reported to be subjected to seasonal modifications, being highest during the spring and summer and lowest during the winter. The authors conclude in the evaluation of teratogenic potentials of drugs that rhythms cannot be neglected.

Chaudhry and Halberg (252) and Halberg (253) reported that the induction of submandibular gland sarcomas in hamsters with the use of dimethylbenzanthracene was shown to be remarkably circadian-stage dependent (Figure 5.44). More recently, Iversen and Kauffman (254), with only two-timepoint sampling and using the chemical carcinogens methylnitrosourea (MNU) and β-propiolactone (BPL) on hairless mice, concluded that the skin was more sensitive to these two quick-acting carcinogens at 1200 than at 0000. The differences were statistically significant. A more extensive review of the chronobiology of carcinogens may be found in one of our earlier publications (236).

The influence of circadian rhythms on the acute toxicity of various medications used to treat human cancers is summarized at the bottom of the

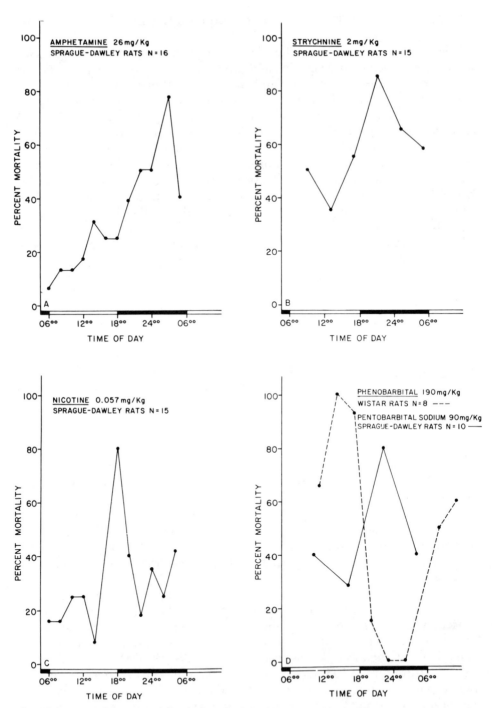

Figure 5.41 Circadian-susceptibility rhythm of rodents in response to each of four agents, using mortality as the endpoint. [For details, see Scheving et al. (237) for amphetamine data, Tsai et al. (242) for strychnine and nicotine data, and Müller (249) for phenobarbital sodium data.] Note in the case of the long-acting barbiturate (phenobarbital) that at one circadian stage 100 percent of the rats survived, whereas at another they all died. In all studies animals were synchronized with 12 hr of light alternating with 12 hr of darkness, the timing of the latter indicated by shaded portion of horizontal axis.

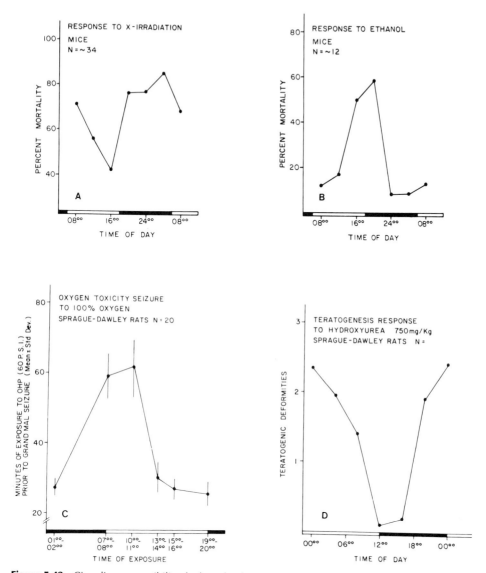

Figure 5.42 Circadian-susceptibility rhythm of rodents in response to each of four agents: (A) mortality from x-irradiation [Haus et al. (246)]; B) mortality from ethanol [Haus (247)]; (C) seizures due to oxygen toxicity [Hof et al. (245)]; and (D) teratogenesis from hydroxyurea [Clayton et al. (248)]. The response to each agent using the indicated end points is markedly circadian rhythmic. Shaded portion of horizontal axis indicates the timing of darkness and activity for the nocturnally active rodents.

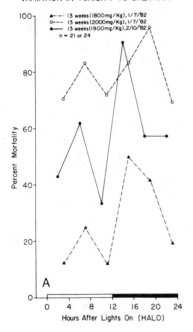
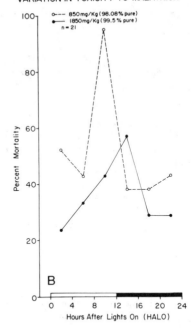
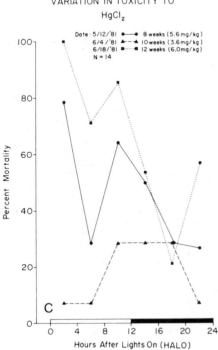
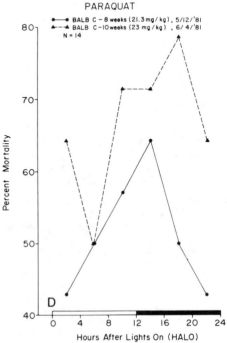

BIOLOGICAL RHYTHMS, SHIFTWORK, AND OCCUPATIONAL HEALTH

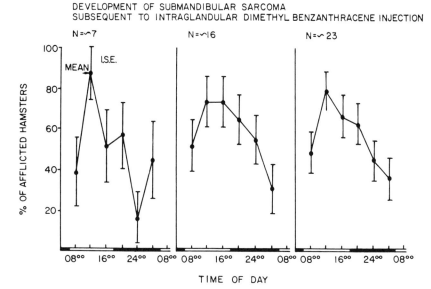

Figure 5.44 Circadian variation in tumor development from a carcinogen (dimethylbenzanthracene) injected into the salivary glands of hamsters. Note that the highest incidence of tumors was recorded in animals injected during the light period rather than the dark period; the latter is indicated by shading along the horizontal axis. Note clock hour given in military designation with 1600 = 4 P.M.; 2400 = midnight and 0800 = 8 A.M.). From Scheving (235).

acrophase map shown in Figure 5.21. This is discussed further in Section 7.3.3. Clearly, circadian changes within the cells of the target organ and/or in the metabolism of the chemical agent give rise to the observed circadian differences. For example, the duration of hexobarbital sodium-induced sleep in rodents (Figure 5.45) is markedly influenced by circadian rhythms. Nair (255) found that sleep was longest in rodents treated at 2 P.M., a time corresponding to the middle of the animals' rest span, and was shortest in rats treated at 10 P.M. The time of longest induced sleep duration coincided with the trough of the circadian rhythm of hepatic hexobarbital oxidase, the enzyme responsible for the metabolism of hexobarbital sodium. It has been shown in a number of studies that many subcellular organelles that are responsible for the manufacture or elaboration of enzymes involved in metabolism are circadian rhythmic (244).

←

Figure 5.43 Circadian-susceptibility rhythm in response (mortality) of rodents to (A) urethane, three dosages (Tsai et al., unpublished); (B) malathion, two dosages (Tsai et al., unpublished); (C) mercuric chloride, three dosages [Tsai et al. (243)]; and (D) paraquat, two dosages [Tsai et al. (243)]. The synchronizer schedule was 12 hr of light alternating with 12 hr of darkness (shown by shading on the horizontal axis). For A to D, the data are plotted with reference to the hours after lights on (HALO) in the animal colonies.

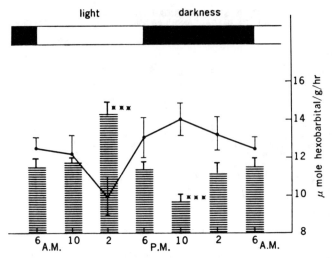

Figure 5.45 Circadian variation in the duration of hexobarbital-induced sleep (vertical bars) compared to the circadian changes in hepatic hexobarbital oxidase activity (solid line). The duration of induced sleep at each clock hour of study represents the average of 6 to 10 animals. Hexobarbital sodium (150 mg/kg) was given intraperitoneally. The enzyme activity is expressed as micromoles of hexobarbital metabolized per gram of tissue per hour; each datum represents the mean of six to eight animals. The shorter vertical lines represent the standard errors; the triple asterisks (***) = $p < .001$. The timing of highest and lowest hexobarbital oxidase activity at 10 and 2 P.M. coincides with shortest and longest hexobarbital-induced sleep. Redrawn from Nair (255).

7.3.2 Circadian Rhythms in Liver and Brain Enzymes

Critical to understanding and evaluating the potential harmful effects of human exposures to chemical and physical agents when engaged in rotating shiftwork is a recognition of circadian differences in the levels of metabolizing enzymes. A review of the findings from rodent studies conducted in the laboratory of one of the authors (256) is summarized by the acrophase chart found in Figure 5.46. Circadian differences in the activity level of enzymes of both the liver and brain are well known. The acrophase chart denotes statistically significant ($p \leq .05$) circadian rhythmicity if the 95 percent confidence limits shown by extension to the right and left of the acrophase symbol (a triangle) are shown. It can be seen that circadian rhythms are depicted for 4 of the 6 listed brain enzymes and 11 of the 13 listed liver enzymes. It should be mentioned that subsequent investigations have revealed statistically significant circadian differences in the brain enzymes of pyruvate kinase and malate dehydrogenase as well as the liver enzymes of fatty acid synthetase and aldolase fructose-1-phosphate. Almost all the liver enzymes reveal greatest activity during either the beginning, middle, or end of the nocturnal hours of activity. The brain enzymes have a tendency to peak somewhat later than those of the liver. It is significant that these rhythms of enzyme activity in both the liver and brain are susceptible to

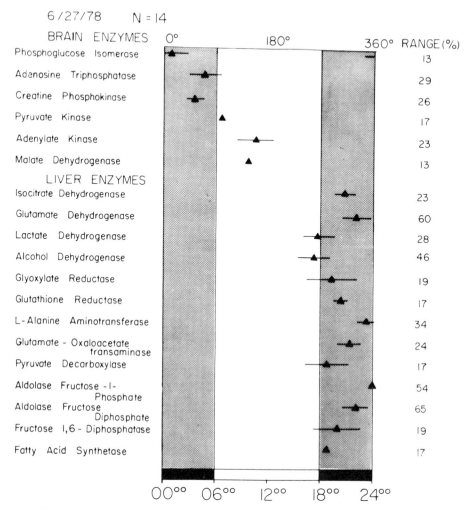

Figure 5.46 Acrophase mapping of selected enzymes of the brain and liver in CD2F, male mice that were standardized to alternating 12-hr durations of light and darkness (from midnight to 6 A.M.). There were two replicate studies done within 30 days with the findings exhibiting remarkable reproducibility. From North et al. (256).

inversion. The acrophases of these circadian rhythms undergo phase shifts of 12 hr, but at variable rates with some of the enzyme rhythms taking as long as 2 weeks after the light–dark synchronizer in the animal colony is altered by 12 hr (257).

The existence of circadian changes in enzymes such as these very likely contributes to the temporal differences in the susceptibility–resistance rhythms of rodents and presumably to chemical agents to which persons are exposed during rotating 8- or 12-hr shifts. One early toxicology study on rodents that quantitatively evaluated the changes in acute histopathology and clinical chemistry due to circadian rhythms was conducted by Craft (225). His research was designed to evaluate the circadian variation in the toxic response of the male albino rat to CCl_4. He measured changes in the susceptibility to CCl_4 following a shift in the lighting regimen to which the experimental animals had been previously adapted.

Craft (225) found that the susceptibility of the albino rat to a given dose of CCl_4, as measured by histopathological evaluation of liver tissue and changes in certain enzymes—serum glutamic–oxaloacetic transaminase (SGOT), and serum lactic dehydrogenase (LDH)—varied significantly depending on the time of day that the dose was received. Different subgroups of animals all living under conditions standardized for periodicity analysis were injected (IP) at 4-hr intervals throughout a 24-hr period with 1.0-ml CCl_4 per kilogram body weight. This dose produced only small effects at one time of day but resulted in extensive hepatic necrosis, marked serum enzyme elevations, and even death when administered at another time of day. Histopathologic review confirmed the liver damage suggested by the increased levels of the liver enzymes. The reported differences were much greater than those that could be attributed to the increased blood flow and ventilation rate that is associated with increased physical activity during the night in these nocturnally active animals.

In Craft's study (225), the mean serum LDH, SGOT, calcium, and potassium levels of normal, untreated albino rats were shown to vary in a circadian manner. However, the phase and magnitude of the rhythms in the untreated animals were different from those of animals treated with CCl_4. The isoenzyme patterns of LDH revealed that the liver is the organ principally affected by a dose level of 1.0 ml/kg and that there is no apparent variation with time of day in the response of other tissues. Interestingly, the serum enzyme results revealed an increased hepatic susceptibility on the day that the animals' synchronizer, lighting, had been changed. Then, for a period of 6 days after the shift, there was a progressively diminished susceptibility response. Within 10 days after the lighting shift, the rhythm was reversed from its preshift phase relationship with reference to the clock hour. These findings suggest that shiftwork transitions in themselves may alter hepatic function, resulting in the increased risk to hepatotoxicity in those exposed to chemical hazards. These findings await confirmation in prospective animal and human investigations.

More recently, the laboratory of one of the authors (LES) has demonstrated that the response in activity of certain enzymes in the liver and brain of mice to certain peptides known to have significant roles in metabolism is circadian-stage dependent. For example, a single IP injection of insulin increased by 23 percent the response in activity of the pivitol enzyme, hepatic pyruvate kinase (PK), when administered toward the end of the dark span and determined 4 hr later (258). No statistically significant effect on enzyme activity was seen if the identical dose of insulin was injected toward the end of the animals' rest span and determined either 4, 8, or 12 hr later. When glucagon was administered toward the end of the light span, there was a 17 percent increase in the enzyme

BIOLOGICAL RHYTHMS, SHIFTWORK, AND OCCUPATIONAL HEALTH 255

activity 12 hr after treatment, but when administered toward the end of the dark span no statistically significant response was seen 4, 8, or 12 hr later. When epidermal growth factor (EGF) was administered toward the end of the dark span, there was a statistically significant increase of 23 percent in PK activity when determined 12 hr later; no such response was seen when EGF was administered at the end of the light span. Similar findings were found for many other enzymes. In fact, some enzymes showed a positive response at one circadian stage, a negative one at another, or no response at still another. One such example was recorded in the response of hepatic malic enzyme to glucagon; when glucagon was administered at one time, the response was a statistically significant increase, whereas at another time it was a statistically significant decrease.

In summary, on the basis of results of chronobiologic studies, circadian difference in the effects of toxic and tumor-inducing chemicals are to be expected. Differences in toxic effect over the 24-hr period appear to be dependent on circadian variations in pharmacokinetic phenomena and in enzyme activity in the liver and target tissues. Circadian rhythms in enzyme activity help to explain temporal patterns in the toxic effects of chemicals and seemingly are important in helping to explain 24-hr patterns of metabolism in general.

7.3.3 Significance of Circadian Chronotoxicity for Medicine and Occupational Health

A logical question regarding the many rhythms in susceptibility–resistance that we have shown above is—can they be exploited in any way? The answer is "yes," and we will cite one example, among many.

Schering and others have carried out a series of studies that document the fact that susceptibility to several anticancer agents that are cell-cycle specific is strongly circadian rhythmic (Figure 5.47). One of our interests has been how the administration of these treatments can be optimized in experimental cancer chemotherapy (41, 127, 259–266). Figure 5.48 presents one such example. In this particular case, cyclophosphamide (100 mg/kg) and adriamycin (5 mg/kg) were found to be synergistic in treating mice that had been inoculated with 1 $\times$ 10^5 L1210 leukemia cells 4 days prior to treatment. With only one course of treatment, there was a dramatic circadian variation in response as monitored by mean survival time and cure rate. The variation in cure rate (mice alive and apparently free of disease 75 days posttumor inoculation) as a function of treatment timing ranged from 8 percent to 68 percent in male animals standardized to 12 hr of light alternating with 12 hr of darkness. Similarly, in female mice standardized to 8 hr of light alternating with 16 hr of darkness, the cure rate ranged from 0 to 56 percent, depending on when the drugs were injected during the 24-hr span. No cures were obtained with either drug alone (127, 266). The maximum cure rate was recorded when the two drugs were administered in the early part of the dark portion of the light–dark cycle

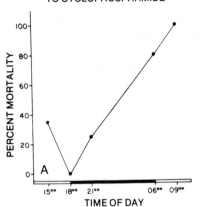
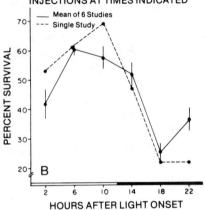
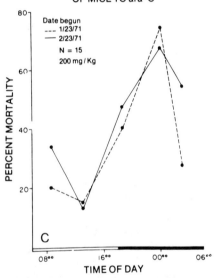

Figure 5.47 Circadian variation in the toxicity (measured by mortality) from three different carcinostatic agents: (A) cyclophosphamide [Haus et al. (261) and Cardoso et al. (264)], (B) adriamycin [Lévi et al. (260)], and (C) arabinosylcytosine [Kühl et al. (262) and Scheving et al. (263)] among several that have been studied [Levi et al. (260)]. The timing of darkness is indicated by shadings along each horizontal axis. The peak susceptibility of each of the agents varies, occurring between the end of the dark and the beginning of the light span for cyclophosphamide, 6 to 10 hr after light onset for adriamycin and during the middle of the dark span for ara-C.

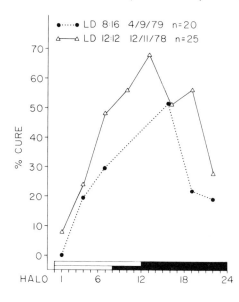

Figure 5.48 Variation in cure rate of "leukemic" mice depending on the timing of combined treatment with cyclophosphamide (CTX) and adriamycin (ADR). Study 1 (solid line) on male mice synchronized to 12 hours of light alternating with 12 hours of darkness; study 2 (dotted line) on female mice synchronized to 8 hours of light alternating with 16 hours of darkness. HALO = hours after lights on. Reproduced from Scheving and colleagues (127, 265).

(whether the animals were synchronized to 12 hr of light and 12 hr of darkness or 8 hr of light and 16 hr of darkness); maximum mortality occurred following treatment in the light span. The data also documented that maximal therapeutic advantage was obtained when the two drugs were separated by 2- or 3-hr intervals and that this effect of drug sequencing was strongly circadian-stage dependent (127, 265, 266).

The data from experimental models demonstrate clearly that one should not ignore circadian variation in tolerance when dealing with chemotherapy or sequencing studies. The evidence is compelling that the temporal organization in general carries with it significant implications, not only for cancer chemotherapy and radiotherapy, but equally important for basic research on normal as well as abnormal growth (127, 235, 246, 261). Perhaps we can come to better grips with the mechanism of such diseases as cancer if we consider in the first place the rhythmic nature of cell proliferation. We have just reached the stage where most scientists accept without question the existence and the potential importance of such rhythms, even though some may still ignore them in experimental design. On the basis of the facts, we must abandon the erroneous concept that somehow sampling at the same time of day takes care of the rhythm problem (127).

With regard to shiftwork it can be expected that exposure to chemical agents may result in different degrees of toxicity or irritation, depending on the timing of the exposure relative to the circadian system and the dose. On the basis of fundamental research of the late H. von Mayersbach (244), it is now well understood that the functional and structural organization of the hepatocyte, a primary site of chemical biotransformation, differs tremendously throughout the 24 hr (Figure 5.49). Microscopic studies have shown that the ultrastructure

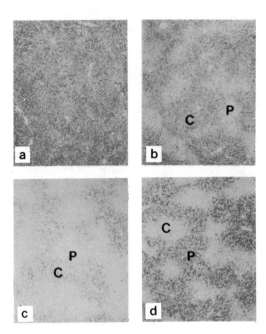

Figure 5.49 Histological demonstration of circadian variation in the glycogen content in the hepatocyte of rats synchronized to 12 hr of light (6 A.M. to 6 P.M.) alternating with 12 hr of darkness: (a) when the liver exhibits maximal glycogen content, at 0600, just after the end of the activity span; (b) when glycogen stores are in the process of depletion, at 1200, while the animals are resting; (c) when glycogen content is at its minimum, at 6 P.M., just before the start of nocturnal activity; (d) when glycogen is being replenished, at 2 A.M., while the animals are still active. Note that the spatial pattern of depletion and replenishment is from the periportal (P) to centrolobular (C) region. Circadian variation in hepatocyte protein (enzyme) synthesis also occurs with a phase difference from the rhythm of glycogen content by 12 hr. From von Mayersbach (244).

of the liver cell during the middle to later portion of the activity span is organized mainly to support glycogen deposition, whereas 12 hr later the ultrastructure appears remarkably different in support of enzyme synthesis. If it is acceptable to assume the existence of a similar circadian rhythm in hepatocyte function in human beings, then for day workers during the early to middle hours of daytime work, the hepatocyte should be expected to be involved primarily in protein and enzyme synthesis, whereas during the late evening it is involved primarily in glycogen deposition. The findings of von Mayersbach (244) on circadian rhythms in liver ultrastructure and function are consistent with the results of studies showing circadian variation in the potency of hepatotoxins. Von Mayerbach's findings support the supposition that rhythms are a very important component of biological adaptability, representing a readied capacity during activity to react to challenges and changes within our geophysical environment. Chronobiologically, organisms differ in their tolerance and reactivity to various types of challenge; we are *not* the same,

BIOLOGICAL RHYTHMS, SHIFTWORK, AND OCCUPATIONAL HEALTH 259

biochemically and physiologically, in the morning in comparison to the evening or night.

7.3.4 Methods for Conducting Chronotoxicology Evaluations in Rodents

Conducting chronobiological studies on animals in the past has required sampling round-the-clock, thus creating logistical problems and an enormous demand on the stamina of the personnel involved (267). This demanding effort more than anything else has impeded research along this line. Earlier we discussed the fact that the circadian system of the experimental animal could be synchronized to the light–dark cycle, and we believe that an explanation of how to exploit this in carying out toxicologic research and evaluations in industry warrants some comment in this chapter.

The ability to synchronize a rhythm to the light–dark cycle permits an investigator to carry out 24-hr rhythmic studies during the span of an 8-hr working day. We have spent a great deal of time exploring the reliability of the various techniques used to accomplish this. One that has been successful is to subject half of the animals to be studied to the conventional light–dark cycle (assume, for the purpose of explanation, that the light is on from 0600 to 1800 and that darkness is from 1800 to 0600) while the other half is subjected to the opposite schedule (with light on from 1800 to 0600 and darkness from 0600 to 1800). Thus, for sampling at six circadian stages, animals need to be taken from both environments at only three different clock hours. Figure 5.50*A* (235) shows a model illustrating the laborious conventional approach of "staying up" around the clock to sample at six different circadian stages, and Figure 5.50*B* represents the inverted light–dark technique. The same type of sampling as illustrated in Figure 5.50*A* can be obtained from the two different environmental light–dark schedules shown in Figure 5.50*B* by conducting studies on the animals at approximately the beginning of light or darkness, in the middle of light or darkness, and toward the end of light or darkness.

It *must* be kept in mind that standardization of animals in this manner does take time, and one should *not* attempt shortcuts. In animals subjected to an inverted light–dark cycle, some rhythmic variables may shift completely in 1 week, whereas others may take as long as 3 or 4 weeks. While the phase shifting is going on, the circadian system is in a state of transient internal desynchronization. Therefore, to avoid the effects of such desynchronization, it has been routine practice in our laboratory to standardize animals to the respective light–dark synchronizer schedules for at least 3 weeks prior to each experiment. Data are collected only after we are convinced that the rhythmic variables to be studied are likely to have phase shifted; this is done in a preliminary study using this time span. Essentially, if the data are not already available in the literature, one must determine just how long it takes to invert the variable under question. Also, it is *very* important that the animals, such as the rodent, are not subjected to any light perturbation during the dark span; this could, under the right circumstances, cause a phase shift or at least a disturbance in

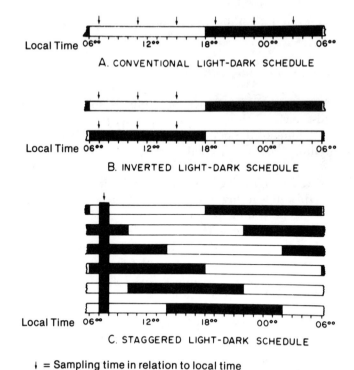

Figure 5.50 Models of three different ways of sampling animals for studies of circadian rhythms. Arrows indicate the times of designated sampling for each design. Note that sampling done by a conventional design (A, top of figure) requires round-the-clock work. The inverted (B, middle of figure) or staggered (C, bottom of figure) designs enable the completion of toxicity studies within the regular 8-hr workday (see text for details). From Scheving (235).

the rhythm that could be seen on subsequent days. If it is necessary to view or attend to the animals during their dark span, a dim red light should be used; we routinely use approximately 0.5 lux at the level of the mouse eye.

Figure 5.50C represents still another way of sampling all six circadian stages, by subjecting six separate groups of animals to six different light–dark schedules; such a technique is commonly referred to as a "staggered" light–dark schedule. With such a technique, one must allow sufficient time for all rhythms to phase shift within the various groups, all on different light–dark schedules. The phase shifting of various rhythms may be quick or slow, but the complexity must be recognized, and the tendency is for it to be greater in the staggered light–dark scheme. In addition, how fast a rhythm phase shifts will depend on whether the process involves a phase advance or phase delay (see Sections 8.1 to 8.2.2). Certainly, if enough time is allowed, the staggered light technique can be effective in synchronizing the circadian system. Nonetheless, from our experience, we prefer the simple inverted light–dark cycle as illustrated in 5.50B when carrying out circadian rhythm studies.

Some investigators have argued that keeping animals in continuous light each 24 hr is one way to avoid the "nuisance" of biological rhythms. We stress that there are many reasons why animals should not be kept in continuous light, and these were previously discussed in detail (158, 267). It is clearly documented that continuous light has a deleterious effect. Even a short exposure to such an environment may significantly alter a variable in comparison to one measured in animals synchronized to a light–dark schedule or even in animals housed in continuous darkness. For example, animals drink and eat less in continuous light (268). The not uncommon practice of keeping rodents in continuous light should be abandoned. We suspect that some of the controversy that arose in the first place about the so-called lack of rhythmicity in certain studies resulted from this practice. Editors have not always required that the conditions under which the animals were standardized be documented. Such should be a *sine qua non* for the reporting of data in the future.

We manage to implement the different models shown in Figure 5.50 by keeping subgroups of animals in simple lighttight boxes such as seen in Figure 5.51. These are sound attenuated "boxes" having their own ventilation system as well as a mechanism for programming the light–dark cycle. They are relatively inexpensive to construct. Others use separate rooms; we prefer the boxes because there is less total disturbance. For example, animals used at the end of the sampling span need not be disturbed in any way by removal of animals from the room at the beginning of the study; this eliminates a considerable amount of experimental "noise" (235, 265).

7.4 Susceptibility–Resistance Rhythms of Human Beings

Susceptibility–resistance rhythms in human beings, although less studied because of ethical reasons, have been shown to exist. Much of our understanding is based primarily on studies of patients allergic to substances such as dust and pollen or of responses to chemical irritants (200, 201, 269, 270). Circadian rhythms have been identified for the susceptibility–resistance of the airways of asthmatic patients sensitive to histamine (231, 232) (Figure 5.37) and acetylcholine (271) as well as house dust, in house-dust-sensitive asthmatic patients (272). Inhalation of an allergen by asthmatic patients who are specifically sensitized to it typically induces an allergic response of the airways characterized by an elevation in the resistance to airflow. The degree to which airway resistance is increased as the result of a standardized house-dust provocation was found to be dependent on when inhalation occurred as shown in Figure 5.52. It is clear that a given concentration of house dust extract when inhaled in aerosol form around 3 P.M. was well tolerated since there was only an 8 percent transient increase in airway resistance. In contrast, the same persons when exposed in the identical manner to the exact same concentration of the antigen at 11 P.M. experienced a much more severe reaction; the increase in airway resistance was nearly 23 percent. At this time the patients exhibited chest rales and dyspnea that persisted for several hours. In other words, exposure to house dust in

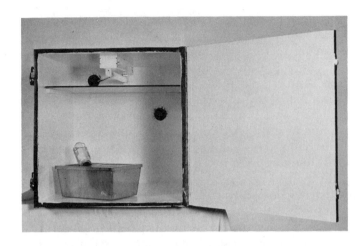

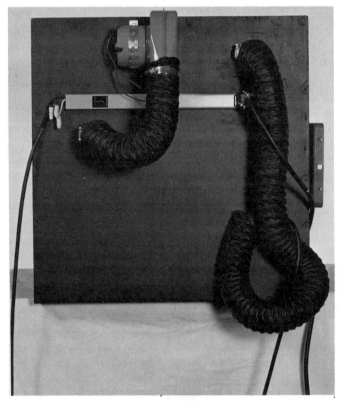

Figure 5.51 The upper photograph shows the interior of the box with the air inlet below and the outlet above; a glass partition separates the two to prevent accumulation of heat from the light. Each box can accommodate five cages such as the one shown. The lower photograph is of the same box showing the back and how the equipment is attached to it. An automatic timer is attached to the side of each box to control the lighting schedule. In 1984 the approximate cost for all materials and equipment for each easy-to-construct box was about $125.00.

BIOLOGICAL RHYTHMS, SHIFTWORK, AND OCCUPATIONAL HEALTH

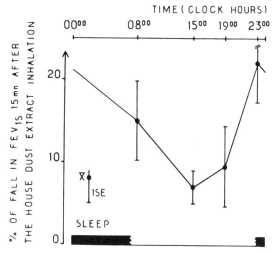

Figure 5.52 Circadian changes in the susceptibility of the airways to an aerosol of house dust in diurnally active house-dust-sensitive asthmatic patients. The pre- to post-test difference in the $FEV_{1.0}$ at 1500 (3 P.M.) was relatively minor compared to that at 2300 (11 P.M.) when the airways exhibited marked hyperreactivity. From Gervais et al. (272).

house-dust-sensitive asthmatic patients was well tolerated as long as it was encountered during the middle of the daytime activity span. When exposure to the same concentration occurred late at night, it was poorly tolerated.

The presence of rhythms is sometimes so obvious to the affected persons that scientific research involving quantitative data gathering is not necessary. For example, some house-dust-sensitive asthmatic patients often recognize a circadian pattern in their tolerance to moderate concentrations of dust. Some persons set up schedules for dusting and vacuuming their homes during the middle of the day because they have found it difficult or troublesome to breath as a consequence of conducting these activities at night. Occasionally, workers in certain jobs within the cotton, pharmaceutical, fiberglass, woodworking, grain, and foundry industries recognize that there are periods of the day during which they are better able to tolerate chemical or dusty environments and thus try to schedule "dirty" tasks for these times of day.

8 THE CHRONOBIOLOGY OF SHIFTWORK

To appreciate the biological changes that workers experience when they alter the timing of their sleep and activity patterns to accommodate rotating shiftwork, it is necessary to understand certain basic aspects of chronobiology. The temporal organization of biological processes beginning at the level of the cell to the level of the organ system must be precise and synchronized if the functional capacity of the organism is to be optional. At the cellular level (244), structural rearrangements occur predictably over the span of 24 hr to accommodate predictable cycles in rest and activity. At higher levels of organization, such as the nervous, cardiovascular, or endocrine system, the integration of rhythmic functions is equally prominent (110).

The endocrine system is one that shows strong rhythmic behavior. For example, if one studies the circadian variation in the secretion of the adrenocortical hormone cortisol, the integration of biological rhythms becomes readily apparent. In day-active human beings, the secretion of cortisol from the adrenal cortex commences to increase during the mid to latter portion of the sleep span in expectation of the onset of activity during the ensuing hours. The circadian rhythm in serum cortisol concentration is of very high amplitude; serum values may be as great as 20 μg/liter around the start of daily activity or as low as 1 to 2 μg/liter from the early to middle of the sleep span. This rhythm is influenced by a circadian rhythm in adrenocorticotrophic hormone (ACTH), which is produced in the anterior lobe of the pituitary gland of the brain. The release of ACTH from the pituitary is, in turn, affected by the circadian rhythm of another hormone from the hypothalamus, corticotrophin-releasing factor (CRF). The secretion of CRF and ACTH are circadian rhythmic, as is the secretion of cortisol from the adrenal cortex (273).

It is of interest that the CRF and ACTH rhythms exhibit different circadian acrophases (273). The acrophase for the rhythm of serum ACTH precedes that of serum cortisol by about 6 hr; the acrophase of the circadian rhythm in CRF precedes that of pituitary ACTH secretion by about 18 hr (see top portion of Figure 5.21). Thus, the circadian rhythm in serum cortisol, having an acrophase coinciding with the commencement of the usual activity span, is dependent on circadian rhythmic processes having different acrophases and occurring at anatomically different locations, specifically, the brain and the adrenal cortex. It is of interest that the adrenal cortex also exhibits a circadian susceptibility in its capacity to respond to ACTH. This has been demonstrated in both laboratory rodents (273, 274) and human beings (273, 275, 276).

The rhythm of serum cortisol itself is a very important bioperiodicity since it serves as a circadian pacemaker or modulator for a great many bodily processes that are cortisol-dependent or influenced. For instance, immune surveillance, energy metabolism, enzyme activity, cell proliferation, electrolyte balance, and many other diverse biological rhythms are phase-related to the 24-hr variation in the secretion of this hormone. In other words, cortisol and perhaps the other hormones secreted from the adrenal cortex may be responsible for regulating a multitude of circadian functions in various tissues located at anatomically distinct sites throughout the body as reviewed by Halberg (39).

8.1 Shiftwork, Sleep–Wake Schedules, and Circadian-Rhythm Adjustments

As discussed earlier in this chapter, persons who are employed on rotating shift schedules that include nighttime work must periodically change their sleep–

activity pattern in order to accommodate their jobs. Although the number of hours worked per week may remain unchanged, the clock time of sleep and activity changes. Similarly, travelers and airline crews who are rapidly transferred across several time zones face the same situation—a resulting alteration in the sleep–activity pattern relative to clock time at home. A major alteration in the clock hours of the sleep–activity pattern represents a change in the synchronizer schedule. Such a change, in turn, requires compensatory adjustment of the circadian system, in case the synchronizer schedule remains altered for some critical duration, usually considered to be 2 or 3 days for a typical person. It is important to note that not all persons can tolerate rotating between nights and days, thereby giving rise to intolerance to shiftwork.

Under normal circumstances the biological rhythms of most persons do not immediately adjust following an alteration in the sleep–activity synchronizer schedule. The jet-lag syndrome experienced by vacationers after being rapidly transported across several time zones and the dull lackluster feeling of lassitude and fatigue experienced by shiftworkers for a few days after the start of night duty represent in large part a reduction in biological efficiency resulting from a disturbed circadian system. This is not surprising since as a minimum at least several days are usually required for circadian and other bioperiodicities to undergo complete acrophase adjustment to a new synchronizer schedule (148). The exact duration of time needed, however, varies with the biological function, the individual, and extent of alteration in the synchronizer schedule. It also depends on whether the initial alteration occurs by a delay in the timing of sleep and activity (as would be the case if rotating in sequence and direction from an afternoon to a nighttime shift or for someone traveling from Europe to the United States) or by an advance in the timing of the sleep–activity schedule (as would be the case in rotating in sequence and direction from a morning (e.g., working from 6 A.M. to 2 P.M.) to a nighttime shift (e.g., working from 11 P.M. to 6 A.M.) or for one traveling from the United States to Europe). The chronobiologic factors of shiftwork are considered in detail in the sections that follow.

8.2 Phase Shifting of Biological Rhythms

The primary synchronizer of circadian rhythms in plants, animals, and insects is the environmental light–dark cycle (see Sections 6.5.1 and 6.5.2). When the synchronizer is changed with regard to clock time, the circadian system eventually compensates in that a multitude of interrelated rhythmic variables undergo acrophase shifts equal in amount to the alteration of the synchronizer. In controlled investigations of acrophase shifts, nonhuman organisms, plants, insects, or rodents are initially standardized to an environment in which the hours of lights on and darkness are regulated. For example, light in the case of rodent studies has traditionally been from 6 A.M. to 6 P.M., alternating with 12 hr of darkness. Thereafter, either the light or the dark span may be

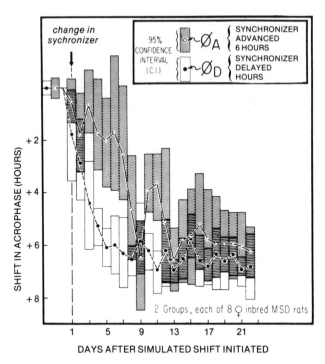

Figure 5.53 Rate of acrophase adjustment ($\Delta\emptyset$) by the circadian rhythm in body temperature subsequent to a synchronizer alteration in the form of (1) an advance ($\emptyset_A$) by 6 hr, an earlier timing of the light and dark cycle by 6 hr causing the rest and activity cycle to occur earlier by 6 hr as compared to the reference synchronizer schedule and (2) a delay ($\emptyset_D$), a later timing of the light and dark cycle by 6 hr, causing the activity and rest cycle to be delayed by 6 hr as compared to the reference synchronizer schedule. Note that the acrophase shift of 6 hr was completed and "locked" in much quicker, within 5 days, following a delay as compared to an advance, which is quite slow, taking 9 to 13 days. Adapted from Halberg et al. (167).

lengthened or shortened by several hours to study the acrophase adjustments of the circadian system to the alteration of the synchronizer.

8.2.1 Phase Shifting of Biological Rhythms: Rodent Models

Much data on phase shifting comes from studies on rodents. If the light–dark synchronizer schedule is altered in an animal colony, the acrophase of each rhythmic function ultimately adjusts to the new schedule by the same amount. This is exemplified by the variable of body temperature continuously monitored in groups of rodents (148) (Figure 5.53). When, after 12 days of exposure to a standardized 12-hr light–12-hr dark regimen, a delay or advance in the synchronizer light–dark schedule is initiated, such as by either prolonging or shortening the light schedule by 6 hr, respectively, the clock hours of the synchronizer become displaced. In the first case, activity and rest are now

Table 5.4 Rates of Acrophase Shift (min/day) in Selected Circadian Functions Following Either Westbound or Eastbound Flights Corresponding to Delays and Advances of Synchronizer Schedule[a]

Study Variable	Westbound (Delay)	Eastbound (Advance)
Urinary adrenalin	90	60
Urinary noradrenalin	180	120
Urinary 17-OHCS	47	32
Psychomotor performance	52	38
Reaction time	150	74
Heart rate	90	60
Body temperature	60	39

[a] From Klein and Wegman (96, 97).

initiated each day 6 hr later than before; in the second case they are now initated 6 hr earlier than before. Under usual conditions, when animals are housed under light from 6 A.M. to 6 P.M. alternating with darkness, the acrophase of the circadian rhythm in body temperature occurs at around midnight. As shown in Figure 5.53, when the synchronizer is changed by 6 hrs, the acrophase of the body temperature rhythm changes with regard to clock time by this same amount; however, this adjustment requires several days. It took only 3 to 5 days when the synchronizer was delayed; it required two to three times as long when the synchronizer was advanced (148).

The speed at which circadian rhythms adjusts varies with the direction of the change in the synchronizer schedule, that is, advancing (comparable to beginning rest or going to sleep earlier) versus delaying (comparable to beginning rest or going to sleep later) the clock times of rest and activity. This difference in the rate of adjustment by the circadian system is termed *polarity*. It represents an asymmetry in the rate of circadian processes to adjust their staging as a function of the direction of the alteration in the synchronizer schedule. The polarity of circadian rhythms can exhibit differences between functions in the same individual, among individuals, and also among species (95, 96, 148).

8.2.2 Phase Shifting of Biological Rhythms: Human Beings

Thus far, all the data indicate that human beings react to an alteration in their sleep–activity patterns much like the rodent (148). Although there are individual differences, human beings tend to adjust more rapidly to a delay, that is, a later onset of sleep and activity, than to an advance, that is, an earlier onset of sleep and activity, in their schedules. This point is illustrated by the data in Table 5.4. Klein and Wegmann (96, 97), through a large number of investigations dealing with jet lag, have quantified the average rates of which certain biological functions of human beings resynchronize following a delay or an advance in the sleep–activity schedule. Westbound flight constitutes a delay in the synchronizer schedule in that retiring to sleep and also awakening occur later relative to the local clock time before travel. Conversely, eastbound flight constitutes an advance in the synchronizer schedule in that retiring to bed for sleep and also awakening occur earlier relative to the local clock time before travel.

The findings presented in Table 5.4 substantiate in human beings two phenomena of biological rhythms: (a) the rate of adjustment of circadian processes such as those listed vary such that it is faster when the sleep–wake cycle is delayed than when it is advanced; and (b) individual circadian functions can differ in their rate of adjustment following change in the hours of sleep and activity. For example, after either a westbound or eastbound flight, the circadian rhythms of psychomotor performance, body temperature, and adrenocortical hormone secretion adjust rather slowly whereas those of urinary noradrenalin and reaction time adjust at relatively rapid rates. Because interdependent rhythmic variables shift at differential rates, a transient state of internal desynchronization, characterized by atypical and changing phase relationships between circadian processes from one day to the next until adjustment is complete, results. During the span when circadian rhythms are in the process of adjustment, the transient state of internal desynchronization, along with sleep disruption, leads to increased feelings of mental and physical fatigue, the elevated likelihood of disrupted performance, and perhaps susceptibility to illness (225).

The endogenous polarity in the adjustment of the circadian system favors the utilization in traditional 8-hr rotating shiftwork systems involving periodic nighttime duty of work sequences that represent a delay rather than an advance of the synchronizer schedule (81, 277). Thus, traditional 8-hr rotating shiftwork schedules that incorporate a later timing of duty hours, that is, a forward rotation, from one shift to the next tend to be better biologically tolerated than those that incorporate an earlier timing, that is, a backward rotation. It must be emphasized that because of the possibility of individual differences between persons with regard to the polarity of their circadian systems there will be exceptions to any generalization. Moreover, certain workers may prefer, for nonbiological reasons, to rotate in a backward rather than a forward direction.

Knauth and Rutenfranz (81) recommend forward rotation, especially for the discontinuous three-shift systems having a 5-day work and 2-day off pattern. With typical shift changes of 6 A.M., 2 P.M., and 10 P.M., there is a significant difference in the total amount of time off between successive shifts when rotation is in a forward in comparison to when it is in a backward sequence. A forward rotation sequence allows 72 hr offtime between morning and afternoon shifts, 72 hr off between afternoon and night shifts, and 48 hr between night and morning shifts (without a postnight shift day sleep). A backward rotation, in contrast, allows only 56 hr between night and afternoon shifts (without a postnight shift day sleep), 56 hr off between afternoon and morning shifts, and 80 hr off between morning and night shifts. With the backward rotation pattern, the immediate transfer from an afternoon to a morning shift or from a night to an afternoon shift results in between-shift intervals that are too short. Since forward rotation results in only one short between-shift interval (48 hr between the night and morning shifts), whereas the backward rotation results in two (56 hr between the night and afternoon shifts as well as between the afternoon and morning shifts), backward rotation is not recommended.

In summary, results of chronobiologic studies on travelers and shiftworkers have been useful in gaining insight as to how to better design rotating shiftwork schedules. By optimizing the clock hours when shifts begin and end, the interval of time between shift rotations, and the direction of shift sequences, some of the undesirable aspects of shiftwork have been moderated. Nonetheless, chronobiologists continue their research to devise ways to minimize and alleviate continuing shiftwork problems as well as develop selective criteria specific for predicting long-term tolerance to rotating shiftwork. In this regard the results of field studies on shiftworkers, as described in the following sections, are most exciting.

8.3 Chronobiologic Studies of Shiftworkers

To a large extent, our knowledge about the chronobiologic adjustments to 8-hr and other rotating shiftwork schedules comes from investigations on volunteers studied under simulated shiftwork conditions. Although the findings from these are helpful, their usefulness for deriving general principles is limited. Frequently, the volunteers of such investigations have been studied under atypical conditions. Although they are subjected to simulated synchronizer shifts, they may not be required to perform tasks similar to those found in an industrial setting. Furthermore, the social and experimental milieu of a simulated study can be very different from that of shiftwork. For these reasons, actual field studies of shiftworkers are necessary.

Field studies of the chronobiologic adjustments of rotating shiftworkers have been conducted on selected groups of volunteering employees (15, 59, 60, 76). The results of these studies are discussed in Sections 9.1 through 9.4. For investigation of the biological rhythms of workers, several measurements separated by more or less equal intervals over the activity span are necessary. Recent developments have made available a variety of portable, light-weight instruments for conducting self-measurements on a variety of health-related functions. Moreover, numerous questionnaires useful for studying aspects of shiftwork have been developed, field tested, and validated.

The utilization of medical instrumentation to self-monitor biological functions repeatedly over time for the purpose of detecting and quantitatively describing biological rhythms is termed *autorhythmometry* (278). Autorhythmometry has been used to study and screen blood pressure in children and adults, evaluate airway patency in asthmatic persons, quantitatively assay the irritant effects of air pollutants, and evaluate the biological rhythm-dependent effects of medications (148, 203, 212, 272· 279· 280). Autorhythmometry has also been used to intensively study various chronobiologic aspects of shiftwork in employees of certain industries (15, 131). It has also been used for the purpose of teaching certain principles of health education as discussed by Glasgow et al. (280).

8.4 Criteria for Conducting Chronobiologically Oriented Field Studies on Shiftworkers

When conducting chronobiologic field studies on shiftworkers, three major types of difficulties must be overcome. These usually can be placed in one of the following categories: social, methodological, or analytical (281).

8.4.1 Social Obstacles to Shiftwork Research

The first problem to solve when conducting field studies on shiftworkers involves social obstacles. For many reasons, few companies or unions wish to pursue research on shiftworkers. Therefore, in the design of research protocols, both the aims and benefits that can be expected from the investigation must be stated clearly and publicly to management, union, and participating employees. The priorities relative to the aims and methods of the study should be noted. If the study could pose any risks, these, too, must be defined. Usually, the development of a workable protocol requires several meetings between the interested parties if they are to achieve close cooperation and trust. At the conclusion of such field studies, it is the responsibility of the investigators to report pertinent and unbiased information to management, organized labor, and select others with the objective of improving the health and well-being of workers while maximizing safety and productivity.

8.4.2 Methodological Difficulties of Shiftwork Research

The methodological problems of chronobiologic field studies are mainly those related to data gathering. The simultaneous investigation of a set of behavioral physiologic and metabolic variables usually requires the participation of a large number of trained investigators and the use of rather large and cumbersome medical instrumentation. Meeting these experimental requirements is difficult. The appropriate equipment is not only expensive but generally is available only in well-equipped but all too often remote laboratories. One reasonable solution to the problems of data gathering, namely, cost, manpower, and instrumentation, is to select a set of easily portable, moderately priced, yet medically and socially acceptable instruments and to educate participants in their proper use so that they can accurately self-measure certain important biological functions. In principle, such an approach is workable, but at the present time in the United States and certain other countries it is abundant with practical problems. To succeed, dedicated investigators are required to devote a great deal of time and effort to assure that each participant is properly trained as well as motivated so that the research procedures are carried out correctly.

In doing field studies, according to Reinberg and his colleagues (281), who have conducted several successful large-scale investigations on shiftworkers in European industries, selection of biological measurements and instrumentation should be based on the following criteria: (a) measurements should be biologically meaningful; (b) instruments should be easy to use, calibrate, and transport as well as reasonably priced; and (c) measurements should not require too much time to conduct. Reinberg, one of the most respected researchers in the field of shiftwork chronobiology, in a series of studies with his colleagues on more than 100 shiftworkers obtained data by autorhythmometry on the following biological and psychological variables: mood and activity level, time estimation, heart rate, oral temperature, card sorting (a measure of eye–hand skill), peak expiratory flow (a measure of airway patency), and grip strength of the left and right hands (using a dynamometer) (15, 131, 281). In addition, urine samles were collected so that electrolytes as well as adrenocorticoid and catecholamine hormones could be determined. Self-measurements and self-assessments as well as the collection of urine samples were done at 4-hr intervals or less, only during the waking hours, for several consecutive days in some cases, or on selected days of the week in others. Some of the long-term studies involved the sampling of rhythms on only the first and last days of each shift rotation. Information pertaining to the onset and end times as well as the qualitative aspects of sleep also were obtained by use of specifically designed and field-tested questionnaires.

In Reinberg's studies (15) selected participants also were invited to record what, how much, and when meals and snacks were consumed in order to learn of possible differences in the intake of carbohydrates, proteins, lipids, and calories according to shift. Such data provide basic and important information on the eating habits of shiftworkers. This information is relevant to exploring the etiology of the elevated incidence of digestive complaints and also peptic and duodenal ulcers found in rotating shiftworkers. In general, it is the experience of Reinberg and his co-workers that persons who volunteer to participate in shiftwork field studies are dependable and accurate in carrying out their self-assessments.

In summary, autorhythmometric methods can be useful under careful supervision for investigating the chronobiologic adjustments of shiftworkers to alteration in the activity–sleep routine. Self-assessments also can be evaluated for alterations in function, and these are likely to give insight into the type and degree of biological adaptations that occur in shiftworkers. Autorhymometry also can be used to conduct biological monitoring as discussed in Section 10.2 as a means to evaluate the health status of employees at special risk to workplace contaminants.

8.4.3 Methods for Biological Rhythm Detection and Description

As chronobiology developed, it became increasingly evident to some investigators that the classic statistical procedures, such as analysis of variance (ANOVA) and Student *t*-test, were not adequate because they did not always provide the desired quantitative numerical endpoints. Thus a number of mathematical models for evaluating and quantifying time series were developed.

Periodogram. The adaptation of the periodogram was introduced by Schuster (282) for the detection and analysis of geophysical time series. A periodogram provides quantitative estimates for both the period and the amplitude of a given periodicity. It can be an appropriate technique for the detection of bioperiodicity when there exists a sufficiently long series of data with a consistent rhythm overlain by noise (random events that distort the detection of periodic functions), but its indiscriminant use in the study of biological time-series data may lead to the description of spurious as well as real periods, and real periods sometimes may not be revealed. For a more detailed treatment of the computational aspects of this method and its usefulness and limitations for describing biological time-series data, see Koehler and others (283).

Spectral Analyses. Generalized harmonic or power spectral analyses also have been used for the study of periodic phenomena. Spectral analysis is based on the assumption that a time series represents the sum of infinite sinusoidal functions, so-called harmonics with different periods. The variance spectrum or so-called power spectrum, which may be obtained by electronic computer methods, shows the prominence of different periodicities in terms of the variance accountable by each. This method was made practical for application to limited, noisy time series found in biological material through the development by Tukey (284) of methods for determining the statistical reliability of spectral estimates. The variance spectrum method was adopted for use on biological time series by Halberg and Panofsky (285) and Halberg, et al. (286). This method detects different frequencies in a noisy time series and evaluates their statistical significance; it is particularly useful if a biological variable exhibits changes in the form of several frequencies, some of which may be ill-defined. Nonetheless, the method of spectral analysis has its limitations. It does not give information on phasing (the occurrence of peak values) of the rhythms detected, nor does it provide a numerical endpoint for the amplitude. Another shortcoming, especially for conducting chronobiologic studies on human beings, is that the technique requires sampling at regular intervals over a relatively long time span.

There are other ways of analyzing data, including the multiple complex demodulation technique (287), which we do not discuss at this time, but that might be considered as the method of choice in some instances, although even this analytical tool has limitations.

Cosinor Analysis. Since the data from field studies are collected only during the hours of activity, so as not to disturb the sleep span, the time series consists of measurements that are derived at unequal intervals. Special statistical methods must be used to analyze these since most conventional so-called time series analyses require that the collected data be spaced at equal intervals. One method that is very useful is the *cosinor*. This technique frequently is employed by chronobiologists; it has been the method chosen to analyze some data included herein and is widely encountered in the literature. One should be familiar with the method if the data analyzed by it are to be fully appreciated. The cosinor method was originally described by Halberg, et al. (141) and modified by Halberg and others (278). A brief explanation of what the cosinor is and what it does follows.

Cosinor analyses consist of the least-squares approximations of one or more cosine curves to time series data in order to objectively detect and quantify rhythms. The quantitative endpoints used to describe rhythms include the period (the duration of time needed to complete one cycle of a rhythmic variation), acrophase (the crest time relative to a defined reference such as some aspect of the synchronizer schedule, e.g., the middle of the rest span), amplitude (a measure of the prominence of the rhythm and equivalent numerically to one-half the extent of rhythmic change between the peak and trough), and mesor (the rhythm-determined average, e.g., in the case of a single cosine approximation, the value midway between the peak and trough). The cosinor indicates the existence of statistically significant rhythmicity if the amplitude is found by F-test to be significantly different from zero (141). Cosinor analyses have been used to evaluate the chronobiologic adjustments of various biological rhythmic parameters in humans due to shiftwork as well as jet travel across several time zones. They have also been used to substantiate circadian patterns of industrial performance and accidents as well as the response to chemical substances. A more recently developed analysis (151) enables one to determine the exact period length for a given rhythm when the data of the time series are of unequal interval. This represents an important step forward in that it allows one to determine, for example, whether an employee's intolerance to shiftwork is related to an internal desynchronization of circadian processes (Section 9.4).

9 SELECTION CRITERIA FOR SHIFTWORKERS

Currently, there is no reliable method for predicting the long-term biological capability of employees for shiftwork. Although the individual differences in the manner in which persons adjust to shiftwork have been linked to personality characteristics (73, 288–290), the significance of these have not been appropriately validated through controlled field studies. At this time, therefore, direct on-the-job experience continues to be the best indicator of one's tolerance to shiftwork. Reinberg et al. (15), through analyses of a series of field studies on both tolerant and intolerant shiftworkers, however, has detected chronobiologic differences between the two types of shiftworkers. Reinberg and his colleagues believe these have relevance to improving the selection of employees who are likely to have a high chronobiologic tolerance for specific rotating shiftwork schedules.

9.1 Intolerance to Shiftwork

Most occupational health professionals are aware that not all persons are capable of tolerating shiftwork and that there are various differences between those who do shiftwork and those who find it intolerable. Some persons can adhere to rotating shift schedules that include periodic nighttime duty throughout their working life, 35 or more years, without health problems or complaints of any kind. On the other hand, some young healthy workers may exhibit intolerance to the rigors of shiftwork even at a very early age.

Both young and older workers may develop intolerance to rotating shiftwork schedules. Older workers in particular, for as yet unknown reasons, after being successfully engaged in rotating shiftwork for several decades can suddenly exhibit biological intolerance to shiftwork at around the age of 40 to 50 years. This intolerance often is manifested by the onset or exacerbation of the typical shiftwork-associated medical complaints (see Sections 5.1 to 5.3). Sleep alterations such as shortened durations, frequent awakenings, and insomnia are commonly reported. Persisting fatigue, not always the result of these sleep difficulties, is another. The intolerant shiftworker often complains of tiredness immediately after awakening, during weekends away from work, and during vacations. This type of persisting fatigue is different from the physiological fatigue resulting from physical and/or mental effort on the job since the latter nearly always dissipates after an adequate night's rest. Intolerant shiftworkers may also exhibit changed behavior. This can be manifest as an increase in irritability and excitability with malaise and lowered performance. Digestive complaints such as dyspepsia, epigastric pain, and peptic ulcer are also common difficulties of the intolerant shiftworker.

9.2 Major Factors Associated with Intolerance to Shiftwork

Over the years, methods to evaluate and minimize intolerance to shiftwork have been addressed by many investigators. Some have identified or postulated biological attributes that might be responsible for intolerance, whereas others have focused on nonbiological factors. Figure 5.54 presents a summary of the various factors either known or thought to affect one's capacity for long-term tolerance to rotating shiftwork schedules (291). This schematic is intended to introduce the subsequent sections on shiftwork tolerance as well as place in perspective those variables that have been identified or proposed as affecting shiftwork tolerance.

For the purpose of discussion, the factors that can affect tolerance have been divided into two categories: endogenous, that is, biological in nature; and exogenous, that is, nonbiological in nature. With regard to the endogenous factors, age, health status, sleep variables, physiologic tolerance to night work,

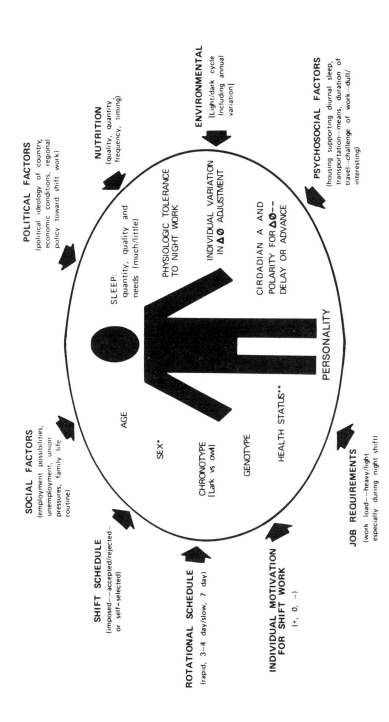

Figure 5.54 Major factors underlying tolerance to shiftwork.

*In certain countries such as France, night and shift work is forbidden by law, except for nurses, airline stewardesses and aircraft personnel. A difference between males and females for tolerance to shift work has not been documented experimentally.

**Personal and/or family history for certain diseases such as peptic ulcer, epilepsy, depressive illness, diabetes, etc.

genotype, and personality have been discussed in the literature (12, 55). At this time there is no convincing evidence indicating a differential capability between men and women for rotating shiftwork schedules. However, it has been shown that women with children tend to find shiftwork very demanding. In particular, married women usually have less available time to catch up on lost sleep, especially when working the nightshift, because of having major responsibility for organizing the everyday activities of the family (31, 55, 292). Individual variation in the $\Delta\varnothing$, a symbol that designates the rate of adjustment of the circadian system following alteration in the sleep–work pattern, is addressed in Section 9.3. The significance of polarity with regard to the rapidity of acrophase adjustment, such as found in studies of travelers, has already been discussed (Section 8.2.2). "Chronotype" as used in this diagram refers to proposed slight differences in the temporal staging of biological rhythms between persons. It has been postulated that some persons actually prefer to be active late into the night and sleep later into the day; they have been termed "owls" by scientists. It is postulated as well that others prefer to retire early to bed in favor of arising very early in the morning; they have been termed "larks." The question of whether there is a true biological difference in tolerance for rotating shiftwork between so-called owls, as opposed to so-called larks, has been suggested but not resolved with certainty (73, 288–290). Obviously, if there are measurable and quantifiable biologic differences between these persons, it might be helpful in screening those persons who would be able to tolerate or even benefit from shiftwork or unusual shift schedules. However, we wish to point out that only about 10 to 20 percent of the population can be classified into either one of these two (owl or lark) categories. Most persons cannot be classified into either.

9.3 Differences Between Workers in the Rate of Biological Rhythm Adjustments When on the Night Shift

Aschoff (293) and Wever (161) have conducted many studies on volunteers living in isolation under controlled constant conditions that provide no time cues or synchronizers. Their research was significant in that their findings suggested that persons exhibiting a high-amplitude oral temperature circadian rhythm were less disturbed chronobiologically when exposed to conditions lacking synchronizers. These findings led Reinberg to reexamined his data from field studies on shiftworkers to determine whether the rate of change in the acrophase (designated as $\Delta\varnothing$) for a set of circadian rhythms varied with the magnitude of the respective rhythm amplitude. Reinberg et al. (62, 149) found for the circadian rhythms of oral temperature and urinary 17-hydroxy-corticosteroids that the $\Delta\varnothing$ was negatively related to the respective circadian amplitude. Specifically, the greater the amplitude of the rhythm exhibited by a worker, the slower the rate of change in that person's acrophase following a shift in the sleep–activity synchronizer schedule. Figures 5.55 and 5.56 show the association between the circadian amplitude and the rate of acrophase shift for these two rhythms.

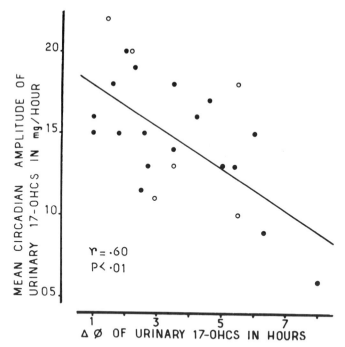

Figure 5.55 Correlation between individual mean amplitude A (in milligrams per hour) and acrophase shift $\Delta\emptyset$ (in hours) of the urinary 17-hydroxycorticosteroid (17-OHCS) circadian rhythm; $\Delta\emptyset$ corresponds to the difference of acrophase location on the 24-hr scale between control days and the first night shift. The graph shows that the greater a person's circadian amplitude, the slower the acrophase shift of the 17-OHCS circadian rhythm. Symbols: ●, 18 shiftworkers; Reichstett study; ○, 6 shiftworkers, Petit–Couronne study. From Reinberg et al. (62).

Data from field studies of tolerant and nontolerant shiftwork employees in the steel, chemical, oil, and coal industries suggest that individuals who exhibit a large-amplitude oral temperature circadian rhythm possess a better capability for the long-term biological tolerance of rotating shiftwork that includes nighttime duty (62, 63, 149–151, 294). As shown in Table 5.5 shiftworkers with oral temperature circadian rhythms greater than 0.30°C were found to display good tolerance to rotating shiftwork without medical problems. On the other hand, employees who were found to exhibit a small-amplitude oral temperature circadian rhythm (≤0.23°C) were those who experienced biological intolerance to rotating shiftwork schedules (151).

The overall findings of chronobiologic field studies on tolerant and nontolerant shiftworkers, according to Reinberg and his co-workers (151), suggest the following conclusions: (a) workers differ chronobiologically in their ability to do rotating shiftwork; and (b) persons who have demonstrated a good tolerance for rotating shift schedules that require night work, for perhaps as long as 15 to 30 years, are likely to possess a high-amplitude oral temperature circadian rhythm. This trait indicates that they adjust their circadian system

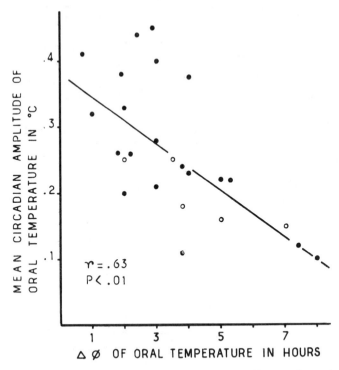

Figure 5.56 Correlation between mean amplitude A (in degrees Celsius) and acrophase shift Δø (in hours) of the oral temperature circadian rhythm; Δø corresponds to the difference of the acrophase locations on the 24-hr scale between control days and the first nightshift. Symbols: ●, 20 shiftworkers, Reichstett study; ○, 5 shiftworkers, Petit–Couronne study. From Reinberg et al. (62).

slowly following alteration in their sleep–activity patterns such as when working nights. In short, these shiftworkers have a strong (high-amplitude) circadian temporal structure that is evidenced by the fact that their circadian system seems to show little ill effect by repeated short-term alterations in the sleep–activity schedule during many years of rotating shiftwork duty. In addition, persons who exhibit a small-amplitude oral temperature circadian rhythm and who rapidly adjust their circadian system following alteration in the sleep–activity pattern appear to have a weaker circadian temporal organization. Individuals of this type tend to be at risk of becoming shiftwork intolerant because their rhythms so quickly invert during change from a day to night schedule of work. These rapid "inverters" are likely to be at greater risk to suffering from those medical maladies typical of shiftwork intolerance and at an early age.

If these findings are validated in prospective field studies on large groups of shiftworkers, they have the following implications. Employees exhibiting large-amplitude circadian rhythms, such as in the rhythms of oral temperature and urinary 17-hydroxycorticosteroid concentration, appear to be best suited

Table 5.5 Circadian Rhythm in Oral Temperature of Subjects with Good, Adequate, or Poor Tolerance to Shiftwork: Single Cosinor Summary[a,b]

Group Mean Age (Years)	Tolerance to Shiftwork (Number of Subjects)		Rhythm Detection p Value	Mesor (24-hr Rhythm Adjusted Mean ± SE) (°C)	Circadian Amplitude (95 Percent Confidence Limit)	Acrophase φ in hr and min, Referenced to Midnight (95 Percent Confidence Limit)
I 25.3	Good	(6)	<0.005	36.53 ± 0.08	0.37 (0.29–0.45)	1549 (1438–1700)
II 50.0	Good	(10)	<0.005	36.51 ± 0.07	0.35 (0.30–0.40)	1534 (1438–1627)
III 50.2	Adequate	(6)	<0.005	36.53 ± 0.11	0.30 (0.24–0.36)	1657 (1541–1905)
IV 47.4	Poor	(7)	<0.005	36.52 ± 0.12	0.23 (0.17–0.29)	1711 (1607–1814)

[a] From Reinberg et al. (106).
[b] All subjects were shiftworkers at the same oil refinery. Groups differed mainly by a good (e.g., group II) versus a poor (group IV) tolerance to shiftwork as well as by a large (group II) versus a small (group IV) amplitude. Group II (good tolerance) involved 10 senior operators (no history of shiftwork difficulty; mean age = 50 years, range 44 to 57 years; and mean shiftwork duration = 25.1 years, range 15 to 32 years). Group IV involved seven senior operators (who were to be discharged from shiftwork due to nontolerance; mean age = 47.4 years, range 30 to 56 years; mean shiftwork duration = 22.9 years, range 9 to 29). In any group, amplitude differs from zero with $p < .005$; mesor, no difference between groups; amplitude (one-half of total peak to trough variability), large in groups I and II (good tolerance), small in group IV (poor tolerance to shiftwork); acrophase φ (crest time), no difference between groups.

for rapid, 2- to 3-day duration, rotating shiftwork schedules. This is because those having such high-amplitude rhythms tend to adjust the acrophases of their circadian system so slowly that by the last "day" of the nighttime shift, day 2 or 3, the alteration of the circadian acrophase is minimal. During rotation back to a pattern of nighttime sleep alternating with daytime work, therefore, the chronobiologic organization of such employees needs to undergo only slight readjustment, if any. Individuals with this type of temporal organization, according to Reinberg and his co-workers (62, 63) seem to be most appropriate for rapidly rotating shiftwork systems since their circadian rhythms are not likely to be much affected by a short-term change in their sleep–wakefulness schedule. On the other hand, persons exhibiting low-amplitude circadian rhythms, such as in oral temperature and 17-hydroxycorticosteroid concentration, are likely to undergo comparatively rapid shifts in circadian system following a change to nighttime work requiring daytime sleep. For these, chronobiologic adjustment is likely to be complete by the end of the first or second day of the nighttime shift. For this type of person, therefore, rapid, 2- to 3-day, rotational shiftwork is less appropriate than are 7-day or longer rotations.

9.4 Internal Desynchronization of Circadian Rhythms and Intolerance to Shiftwork

Internal desynchronization of biological rhythms is defined as an alteration in the circadian period or phase relationship between interdependent rhythmic functions. Generally, circadian (and other) rhythms are synchronized such that a fixed phase relationship exists between interdependent bioperiodicities of a given frequency (see Figures 5.21 and 5.22). Transient internal desynchronization of circadian rhythms, in terms of an alteration of the phase relationship of biological functions, occurs during the chronobiologic adjustment of shiftworkers after changing to nighttime duty involving daytime sleep. Why it is that only certain, perhaps prone, employees develop intolerance to a rotating shiftwork schedule remains to be explained. It has been proposed by Reinberg et al. (106, 151) that some persons may be predisposed genetically and thus may be susceptible to rapid internal desynchronization when subjected to recurring, periodic alteration in sleep–activity schedule.

Using newly developed analytical techniques, Reinberg et al. (151) have found in their sample of biologically intolerant shiftworkers, defined clinically by the characteristic clinical symptoms as discussed in Sections 5.2 and 5.3, evidence of a persisting internal desynchronization of the circadian system. The period duration of at least certain rhythmic variables are longer than 24.0 hr. For example, Table 5.6 shows the findings for studies of the oral temperature circadian rhythm in workers who at the time of study reported good (subjects 1 to 11) or poor (subjects 12 to 15) biological tolerance for rotating shiftwork involving the periodic alteration of the sleep and activity pattern. With the exception of subject 1, the circadian period of the oral temperature circadian rhythm in tolerant shiftwork employees was 24.0 hr. Moreover, the amplitude of this rhythm in the tolerant workers was large, at least 0.21°C. In contrast, for the intolerant workers of this sample, the circadian period for the oral temperature rhythm was longer, between 24.9 and 25.7 hr. [In some cases, shiftwork intolerance was associated with circadian periods shorter than 24 hr (Reinberg, personal communication).] Also, the circadian amplitudes were very small, between 0.04 and 0.13°C. It is of interest that since these findings were published, it was necessary to transfer subject 1 from rotating shiftwork to permanent day work (Reinberg, personal communication) because of a developed biological intolerance to shiftwork. Thus, it appears that one important component of shiftwork tolerance relates to the strength of the circadian-rhythm temporal organization, as best gauged at present by the magnitude of the amplitude of the oral temperature circadian rhythm. According to Reinberg and his colleagues (151), shiftworkers with large-amplitude rhythms appear to

BIOLOGICAL RHYTHMS, SHIFTWORK, AND OCCUPATIONAL HEALTH

Table 5.6 Characteristics of Circadian Rhythm of Body Temperature in Tolerant and Nontolerant Shiftworkers[a]

Subject No. (Age in Years)		History of Rotating Shiftwork	Dominant Circadian Period[b] (hr)	Circadian Amplitude[c] (°C)
		Tolerant		
1	(20)[d]	1 month	25.1	0.21
2	(22)	2.5 years	24.0	0.23
3	(22)	2.5 months	24.0	0.27
4	(25)	3 years	24.0	0.31
5	(25)	4 years	24.0	0.25
6	(27)	1 month	24.0	0.35
7	(27)	3 months	24.0	0.19
8	(31)	1.5 years	24.0	0.21
9	(36)	2.5 years	24.0	0.27
10	(35)	4 years	24.0	0.36
11	(43)	2 years	24.0	0.42
		Nontolerant		
12	(20)	1 month	25.7	0.04
13	(28)	3 months	25.1	0.04
14	(30)	3 months	24.9	0.13
15	(31)	3 years	25.2	0.09

[a] From Reinberg et al. (151).
[b] Estimated by power spectrum.
[c] Estimated by the single cosinor method.
[d] Following publication of these data, subject 1 had to be transferred from rotating shiftwork to permanent daywork because of a developed intolerance to the former.

be less prone to developing an internal desynchronization of circadian functions and intolerance to shiftwork.

The phenomenon of internal desynchronization in an intolerant as compared to a tolerant shiftworker is clearly apparent in Figure 5.57 (151). The upper left-hand portion of the figure shows the clock hour timing of the circadian acrophase (indicated by closed circles) of the oral temperature rhythm for one subject (subject 3) who exhibited good biological tolerance to shiftwork. The timing of the acrophase of this rhythm was quite stable from day to day, despite the regular rotation to night work (shown by the asterisked horizontal bars) during the weekdays. The bottom left-hand portion of Figure 5.57 shows the so-called power spectrum of the oral temperature data for this worker. This method was used by Reinberg et al. to indicate through a partitioning of the variance in the data set of oral temperature the existence of prominent bioperiodicities in each of the various shiftworkers studied. It can be seen for subject 3 that the rhythm of oral temperature was exactly 24.0 hr in duration.

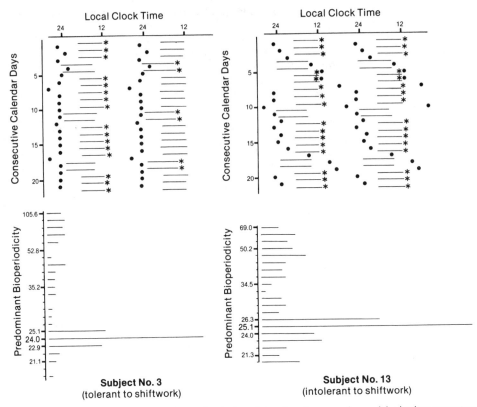

Figure 5.57 The plots at the top indicate the timing of the circadian acrophase of the body temperature rhythm of a person exhibiting either good tolerance (subject 3, left side) or poor tolerance (subject 13, right side) to rotating shiftwork. The acrophase of the rhythm is consistently timed from one day to the next in subject 3 when sleeping during the day (shown by asterisks) for 5 of 7 days each week. The acrophases for subject 13 exhibit a systemic delay with each consecutive day. This latter phenomenon suggests a period length for this rhythm greater than 24 hr. This is documented statistically by spectral analysis (bottom right). The most prevalent period for the body temperature of subject 13 was shown to be 25.1 rather than 24 hr as shown for subject 3 in the lower graph to left. From Reinberg et al. (151).

The upper right-hand portion of Figure 5.57 gives an example of a desynchronized circadian rhythm of an intolerant subject (subject 13). It can be seen that the timing of the circadian acrophase (closed circles) of the oral temperature rhythm occurred progressively later from one day to the next. Such a finding would be expected if the circadian period of his oral temperature rhythm was longer than 24 hr. In this nontolerant shiftworker, the period of this rhythm as shown by the power spectrum analyses (lower right-hand portion of Figure 5.57) was 25.1, rather than the expected 24.0, hr in duration. In this biologically intolerant shiftworker, therefore, the later timing of the circadian acrophase resulted from a prolongation of the period length of at least the oral temperature circadian rhythm.

BIOLOGICAL RHYTHMS, SHIFTWORK, AND OCCUPATIONAL HEALTH 283

In this and other presumably susceptible persons, internal desynchronization apparently results from recurring changes in the hours of sleep and activity associated with shiftwork schedules that include a rotation of work during the nighttime and sleep during the daytime. Although reports dealing only with desynchronization of the oral temperature circadian rhythm have been published, other circadian functions appear to become internally desynchronized in nontolerant shiftworkers (Reinberg, personal communication). Presumably, alteration in the temporal organization, manifest as an internal desynchronization of the circadian system, constitutes an important aspect of the biological nontolerance to shiftwork.

It has been suggested that in predisposed persons, an internal desynchronization is associated with a set of symptoms that are common to both clinical depression and shiftwork intolerance (106). In connection with this, Michel-Briand et al. (295) reported a greater incidence of depression and affective illness in *retired shiftworkers* than in *retired dayworkers* in whom cardiovascular and locomotor illnesses and disorders were the more common. In patients suffering from depressive illness, effective treatment with specific medications such as lithium and other antidepressant drugs, which are known to influence the circadian period of certain rhythmic variables, moderates or corrects the internal desynchronization. Often nontolerant shiftworkers resort to using medications of another type—sleep-inducing aids, such as barbiturates, tranquilizers, diazepam derivatives, and antihistamines. These medications, which lack the ability to affect the period of circadian functions, frequently are overused, leading to addiction, although typically without effective resolution of the employee's sleep disorders or persisting fatigue.

It is important to note that persons intolerant to rotating shiftwork do recover from their untoward symptoms when they are returned to a regular pattern of activity and sleep. Permanent daywork with nighttime sleep leads to a synchronization of biological rhythms and usually alleviates or modulates the associated clinical symptoms of intolerance to shiftwork. However, complete regression of the symptoms may require in certain cases as long as 6 months or so. In this regard, the decision to discharge an employee from shift to permanent daytime work must be done early when the signs and symptoms of biological intolerance first become apparent. This is important in order to increase the likelihood that the biological effects of shiftwork intolerance will be minimal and that recovery will be rapid.

9.5 Proposed Countermeasures to Hasten Chronobiologic Adjustments to Shift Changes

Medications or procedures to facilitate the chronobiologic adjustments to change in the sleep–activity routine have long been sought (153, 195, 196). Ehret (296–301) and his co-workers, on the basis of results from a series of investigations on rodents involving the careful manipulation of several known synchronizers for these animals, including food and drugs, were able to accelerate phase shifts of selected circadian functions. On the basis of their findings on animal investigations, Ehret and his co-workers have now proposed a diet plan to hasten the chronobiologic adjustments of travelers and shiftworkers (302). The Argonne anti-jet-lag diet (Figure 5.58), as it is called, is founded on three premises: (a) methylated xanthines, such as those found in caffeinated beverages (coffee and tea) as well as the medication theophylline, cannot only serve as a stimulant but can induce or aid phase shifts in certain rhythms when properly timed in rodents; (b) timed food intakes in rodents can be a potent synchronizer of biological rhythms; and (c) the reported tendency that a high protein meal stimulates the synthesis of catecholamine hormones—adrenalin and noradrenalin—which promotes alertness, whereas a meal high in carbohydrates stimulates the synthesis of serotonin, which promotes sleep.

The underlying concept of the diet plan is to (a) maximize the synchronizing effects of meal timing by the careful scheduling of "feasting" and "fasting" on alternating days preceding a flight across several time zones (or a shift in the sleep–activity schedule such as necessitated by shiftwork); (b) restrict consumption of the methylated xanthines to a biologically appropriate time in the circadian cycle on the day of flight departure (or before a shift change), for example, in the morning if traveling west; and (c) on arrival (or transfer to the night shift) to one's destination to vary the protein: carbohydrate content of the meals according to the appropriate phase of the rest–activity cycle for the new time zone (or work shift). The latter consideration is based on the assumption that meals high in protein taken in the morning and at lunch on the day of arrival following displacement over several time zones (or during the nighttime of a nightshift) will facilitate a rise in brain catecholamine synthesis, which is usually the situation for the active phase of the work–rest routine. Conversely, it is assumed that a high carbohydrate dinner ingested at a time in synchrony with the population of the destination (or at a time before diurnal sleep in workers on the night rotation) will facilitate an increase in brain serotonin synthesis, which typically precedes sleep.

The anti-jet-lag diet is intended to hasten the chronobiologic adjustments of vacationers and shiftworkers. It has been used by travelers and business representatives and has been reported to have been tried even by one of the presidents of the United States. It has thus far received great attention by the popular press, including newspapers, magazines, and a book (302). As a result, the unions and managements of certain industries and businesses have become aware of the claim that the proposed diet plan can expedite the adjustment of biological rhythms of shiftworkers.

The Argonne anti-jet-lag diet has been derived from research mainly on rodents. The objective and scientific substantiation of the diet plan in human beings, as of this writing, is inadequate. Outside of several subjective evaluations or volunteered testimonials, little properly controlled human research on this diet plan has been conducted. The work of Graeber and his co-workers (152, 154) apparently represents the only published study on human beings that purportedly has evaluated the diet. In their research, several synchronizers—

THE ARGONNE ANTI-JET-LAG DIET

The Argonne Anti-Jet-Lag Diet is helping travelers quickly adjust their bodies' internal clocks to new time zones. It is also being used to speed the adjustment of shiftworkers, such as power plant operators, to periodically rotating work hours. The diet was developed by Dr. Charles F. Ehret of Argonne's Division of Biological and Medical Research as an application of his fundamental studies of the daily biological rhythms of animals. Argonne National Laboratory is one of the U. S. Department of Energy's major centers of research in energy and the fundamental sciences. Argonne National Laboratory, 9700 South Cass Avenue, Argonne, Illinois 60439

How to avoid jet lag:

1. **DETERMINE BREAKFAST TIME** at destination on day of arrival.
2. **FEAST-FAST-FEAST-FAST** — Start four days before breakfast time in step 1. On day one, FEAST; eat heartily with high-protein breakfast and lunch and a high-carbohydrate dinner. No coffee except between 3 and 5 p.m. On day two, FAST on light meals of salads, light soups, fruits and juices. Again, no coffee except between 3 and 5 p.m. On day three, FEAST again. On day four, FAST; if you drink caffeinated beverages, take them in morning when traveling west, or between 6 and 11 p.m. when traveling east.
3. **BREAK THE FINAL FAST** at destination breakfast time. No alcohol on the plane. If the flight is long enough, sleep until normal breakfast time at destination, *but no later*. Wake up and FEAST on a high-protein breakfast. Stay awake and active. Continue the day's meals according to mealtimes at the destination.

FEAST on high protein breakfasts and lunches to stimulate the body's active cycle. Suitable meals include steak, eggs, hamburgers, high-protein cereals, green beans.

FEAST on high-carbohydrate suppers to stimulate sleep. They include spaghetti and other pastas (but no meatballs), crepes (but no meat filling), potatoes, other starchy vegetables, and sweet desserts.

FAST days help deplete the liver's store of carbohydrates and prepare the body's clock for resetting. Suitable foods include fruit, light soups, broths, skimpy salads, unbuttered toast, half pieces of bread. Keep calories and carbohydrates to a minimum.

COUNTDOWN

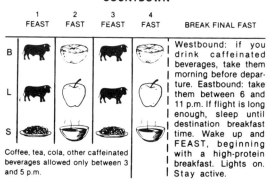

Figure 5.58

including the use of the Argonne anti-jet-lag diet, methyl xanthines (caffeine), and a sleep-inducing medication—were used to enhance the speed of adjustment of circadian functions of military personnel undergoing rapid deployment across many time zones. In this study, made complicated because of several logistic problems, the countermeasures relying on the use of multiple synchronizers (*diet as well as methyl xanthines and hypnotics*) seemed to be advantageous, for example, in reducing fatigue and the need for sleep in comparison to controls who were not provided with treatment. According to Graeber (personal communication), it is uncertain whether the diet plan alone was the sole reason for a better tolerance of the experimental group of soldiers to their rapid deployment since in this study each member of the experimental group received a 100-mg dose of dimenhydrinate, a hypnotic, to induce sleep at an appropriate time in addition to adhering to the anti-jet-lag diet. Moreover, more recent chronobiologic studies suggest that with the exception of only a few circadian variables, meal timing is only a *weak synchronizer* of 24-hr rhythms in *human beings* (303). In rodents the situation is somewhat different in that meal timing has been shown to be capable of strongly affecting certain circadian functions; on the other hand, certain others may be affected only to a slight degree or not at all (262, 303). The subject of meal timing and its role as a synchronizer in rodents and human beings has been reviewed by Reinberg and his colleagues (105, 303), with the conclusion that it is of relatively minor significance as a synchronizer in human beings.

In summary, the countermeasure involving the anti-jet-lag diet that has been proposed for hastening the chronobiologic adjustments of human beings following alteration of the sleep–activity pattern, such as is the case when traveling across time zones or when engaged in shiftwork entailing nighttime hours, has not been rigorously tested as of yet. Thus, it is premature at this time to recommend the anti-jet-lag diet for shiftworkers or travelers as a validated countermeasure to facilitate chronobiologic adjustment.

9.6 Circadian Rhythms, Shiftwork, and Human Performance

Although laboratory investigations of circadian differences in human mental and physical performance are relatively common (19, 30, 96, 97, 155, 304–306), during the last four decades only the findings of a small number of field studies have been published on temporal changes in human performance or accidents over a 24-hr period. Results of these investigations indicate that there is a problem of impaired efficiency in night shiftwork. Recently Folkard (155) reviewed the findings of six field studies; these are summarized in Figure 5.59. The findings of each study are presented in this figure from top to bottom in the order of publication date. In this figure arbitrary scales have been selected to make the amplitudes of the six curves approximately equal. For each curve, the lower the reading, the poorer the performance.

In apparently the earliest study, Browne (307) examined the speed with which shiftworking switchboard operators answered calls at different times of

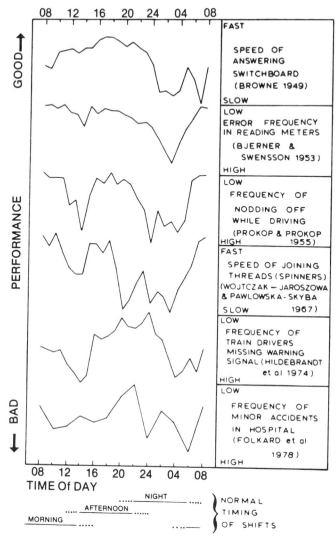

Figure 5.59 Temporal patterns in performance, vigilance, and accidents over 24 hr as indicated from a review of six studies. Time is shown in military fashion: 1600 = 4 P.M., 2000 = 8 P.M., 2400 = midnight. Indicated at the bottom is the timings of the night, afternoon, and morning shifts. In general, performance and vigilance are reduced during the nighttime as well as for a short duration after midday. From Fokard (155).

the day or night (top panel of the figure). Browne corrected the data to account for the number of calls at any given time of the day. performance speed improved in a fairly linear manner from 8 A.M. to 6 P.M. and dropped sharply after about 10 P.M. such that it was slower during the night than at any other time.

Bjerner et al. (51) studied the 24-hr variation of committing errors by shiftworking meter readers (second panel from the top of Figure 5.59). They found that performance decreased only slightly over most of the normal working day with evidence of a slight "postlunch dip" (19, 305). There was a fairly sharp drop in accuracy after about 10 P.M. with a minimum during the nighttime hours. Prior to the publication of this study the existence of a postlunch dip in performance generally was not recognized.

The third panel from the top presents the frequency with which professional truck drivers reported falling asleep ("nodding off") while driving at different times during the day or night (308). In this data set the postlunch dip is very apparent, with the frequency of nodding off almost as high around 2 to 3 P.M. as during the night period. The reason for this is uncertain, although it has been suggested that the postlunch dip is more marked in people who are relatively sleep deprived (309). Again, however, there was clear evidence of an impairment during the night hours.

Wojtczak-Jaroszowa and Pawlowska-Skyba (20) examined the speed with which five spinners joined broken threads (middle panel, Figure 5.59). This study was a particularly detailed one, with about 5000 measurements taken. The studied employees had all been on shiftwork for at least 10 years and were highly proficient at their tasks. They were studied on the morning (5:30 A.M. to 1:30 P.M.), afternoon (1:30 P.M. to 9:30 P.M.), and night (9:30 P.M. to 5:30 A.M.) shifts, although it was not reported how rapidly they rotated. In Figure 5.59 the data from these three shifts have been combined as a continuous 24-hr curve. As the authors pointed out, performance speed was about 10 percent slower when the employees worked the night than during either the morning or afternoon shifts.

The results of another detailed study are shown in the next-to-bottom panel. In this study by Hildebrandt et al. (309), automatic recording devices were fitted into the cabs of 10 train locomotives. Approximately every 20 min a warning light appeared for 2.5 secs, followed by an auditory warning signal for an additional 2.5 secs. If neither of these signals was heeded, there was a 30-sec sounding of a loud horn, compelling the engineer to operate the safety gear to avoid automatic braking of the locomative. Hildebrandt et al. (309) recorded a total of 2238 occurrences of signal sounding, an indication that neither of the two warning signals had been heeded. Despite the fact that the warning light was more visible at night, more warning signals were missed; that is, the signal sounded more often during the night than during the day. Also apparent was a decrease in vigilance during the period immediately after midday.

The bottom portion of Figure 5.59 shows the findings of Folkard, et al. (310), who studied the frequency with which patients incurred minor accidents while hospitalized. In this study, the records of a large modern hospital were examined for a 5-year period. A total of 1854 "unusual incidents" occurred during this period, of which 1576 were minor accidents involving individual patients, and for which there was a clear indication of the time of occurrence. Only 30 percent of these accidents resulted in injury to patients, and in the majority of these cases (80 percent) only minor scratches or bruises were sustained. The frequency of the incidents tended to decrease throughout the normal waking period and increase during the night, despite the fact that the patients were asleep during most of the nighttime. The two sharp peaks at 10 P.M. and 8 A.M. apparently are associated with the patients' activity related to the need to urinate before going to sleep at night and on awakening in the morning.

Taken together, the results shown in Figure 5.59 suggest that there is a problem of impaired efficiency and increased accident proneness during the nighttime. This conclusion is supported by the results of other studies that have examined performance efficiency on different shifts but have been unable to extract relatively continuous data. Thus, for example, accidents on the nightshift have been found to be more serious, although less frequent than on the morning or afternoon shifts (311), whereas production quality has been found to be poorer at night (312). A problem in many of these studies is that usually there are a large number of confounding factors. For example, Meers (312) not only reports lowered production quality at night in a sugar refinery and wire pulling factory but also notes that maintenance of the machinery at night was of lesser quality than during the daytime. It is, therefore, unclear as to whether the lowered quality was due to the impaired efficiency of the shiftworkers themselves or to the poorer job of maintaining the machinery during the night shift, which in itself might be indicative of circadian variation in performance of maintenance crews, or to both.

In summary, a variety of investigations reveal circadian variation in performance and accidents. A majority of studies reveal decreased efficiency during the night shift as well as during the afternoon. Knowledge of these predictable differences is fundamental not only to the development of an effective industrial safety program but also to maintaining quality control and productivity from one shift to the next.

10 CHRONOBIOLOGIC ASPECTS OF BIOLOGICAL MONITORING

Since all human functions are rhythmic with respect to the 24-hr span and perhaps the year, the protocol for conducting biologic sampling should take this into account. In certain instances, the results of biologic monitoring may reflect more the influence of bioperiodicities than the result of exposure to workplace contaminants. This confounding variable can be an especially serious problem in those cases where the parameter being monitored exhibits a large circadian amplitude such as liver enzymes and blood cholesterol. Indeed, in certain clinical situations circadian variation in some biological function can be so large that failure to take it into account may result in misleading findings and, perhaps, inappropriate interpretations of laboratory tests (197, 198). In order to more effectively interpret findings and better conduct biological sampling, chronobiologists recommend the use of so-called time-qualified, with regard to biological rhythms, reference values as well as other approaches such as repeated self-assessments over time (autorhythmometry).

10.1 Time-Qualified Reference Values

Physicians and clinical pathologists have begun to realize that for certain functions the time *when* blood or urine samplings are obtained or *when* biological measurements are carried out can dramatically influence the results of tests. In recognition of this, Haus and his colleagues (197, 198) have devised a large set of circadian time-qualified references for evaluating and interpreting hematologic and other parameters used by clinical pathologists and others working in hospital laboratories and occupational settings.

One example of the effect of rhythms on biologic monitoring is shown in Figure 5.60. This figure presents the time-qualified reference system, with respect to the 24-hr span, for the hormone, serum cortisol. The graph shows the 5 to 95 percentile ranges for cortisol as a function of the timing of blood samplings from a large group of healthy men and women who work during the day and sleep at night. It illustrates how the time-qualified ranges can differ in upper and lower limits as a function of the time along the 24-hr scale. In particular, it can be seen that a cortisol level of 23 µg/dl is well within the normal range of values for the blood of both healthy men and women when obtained at the beginning of the activity span, at 8 A.M. On the other hand, if this were to be the value for a sample drawn at 8 P.M. (labeled 20:00 in Figure 5.60), it would be considered abnormal since it is well beyond the time-qualified range of normal values. Similarly, a cortisol level of 3 to 5 µg/dl for a blood sample drawn at 8 P.M. or midnight is well within the respective time-qualified ranges; however, it would be considered abnormal if it were found for a blood sample drawn at 8 A.M. or noon. Currently, circadian rhythm time-qualified reference values are available for more than 25 hematologic parameters, including serum electrolytes, red and white blood cell populations, and hormones. These are being used in many clinical laboratories around the world to assist with the interpretation of test results.

Time-qualified references are very useful in clinical medicine. They are potentially important in occupational medicine as well. Our understanding of the potential importance of "timing" of biologic sampling now makes it necessary for occupational medicine departments in various firms to generate their own "diurnal" time-qualified reference values. This could, in part, be accomplished by the careful scheduling of preplacement or other physical examinations of employees at different clock hours. The time-qualified references obtained in this manner or, if necessary, from other sources would be useful for the subsequent interpretation of data obtained from periodic medical reexaminations of employees. Also, these values would be useful for the purpose of medical screenings and monitorings, such as when checking the status of white and red cell levels in workers exposed to the chemical or physical agents known to affect these.

Circadian Variation of Usual Range* in Plasma Cortisol**

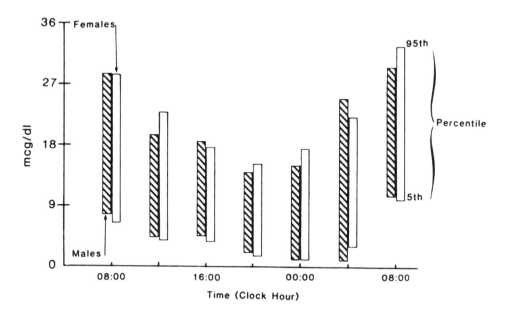

*expressed as 5th to 95th percentile

**158 Clinically Healthy Subjects (71 Females, 87 Males)

Figure 5.60 An example of a time-qualified reference system—the variable of plasma cortisol—for diurnally active healthy persons. Other such time-qualified references have been obtained for many other meaningful variables. A value of 3 μg/dl from a sample of blood drawn at 0800 (8 A.M.) from a diurnally active worker is outside the appropriate time-qualified range of normal values at this time. The same value, however, found for a blood sample drawn at 8 P.M.) (2000) is within this (8 P.M.) time-qualified normal range. From Haus et al. (197, 198).

Use of time-qualified references requires knowledge of only two critical pieces of information: (a) the synchronizing sleep–activity schedule of the person and (b) the timing, with regard to clock hours, of when the blood or other biological sample or measurement was obtained. In the case of a urine sample, information on the time of its collection as well as the time of the previous voiding is needed in order to determine the duration over which the sampling is representative. All this information is easily obtainable by simply questioning the worker, for example, about the usual time of retiring at night and time of awakening in the morning. With respect to urine samples, persons generally must be reminded to record the time of the previous voiding on the day of scheduled monitorings.

Time-qualified references are useful for studies involving permanent day as well as night workers, since their rhythms are not periodically subjected to acrophase shifts. For example, in the case of permanent night workers, time-qualified references should be used in the following manner. Instead of referencing the time-qualified ranges according to clock hour time, they can be referenced to the hours after the end of the workers' usual sleep span. In the example shown in Figure 5.60 involving diurnally active persons, 8 A.M. corresponds approximately to 2-hr postsleep; 4 P.M. corresponds to 10-hr postsleep, and so forth. Replacement of clock hour with another system of identification, such as hours postsleep, makes the time-qualified references applicable for both permanent day and night workers. Since the rhythms of shiftworkers undergo alteration in their circadian acrophases during nighttime duty, time-qualified references are not always useful in interpreting their biological data unless tests are performed on the last day of a given shift. For employees working slow rotation schedules, the circadian rhythms of shiftworkers are more likely to have adjusted completely by the last shift 'day'.

Time-qualified references are being used at the present time in a number of aerospace research programs, among others, in an attempt to better assess the effects of space flight on the blood chemistry of astronauts. Early on, researchers recognized that unless these values were evaluated with regard to circadian variation, it would not be possible to know whether changes were the result of space flight. During the Space 1 Laboratory flight (November 1983), baseline, time-qualified reference values were obtained for selected astronauts during three 28-hr study spans. Specifically, blood samples were collected every 4 hr, even during the usual sleep span, from astronauts who adhered to a specific routine of activity by day with sleep at night. After their flight, single blood samples were obtained and the time of day was recorded in relationship to the activity–sleep routine. The effect of the conditions of space flight on selected hematological variables was then assessed according to the appropriate time-qualified reference for each astronaut. It is of interest that on this particular space mission, astronauts worked 12-hr shifts; some worked nights (relative to the time in Houston, Texas), whereas others worked days. Thus, for the night-active astronauts, the time-qualified reference system was adjusted for time-after sleep termination.

10.2 Biological Monitoring and Autorhythmometry

For a number of years, biological monitoring has been a controversial topic for industry and organized labor. Physicians have disagreed about its usefulness. The myriad of difficulties that arise while attempting to accurately collect and assay the samples as well as the problems of proper intervention make it a less than straightforward way to assess either the amount of toxicant in the workroom air or the amount of toxicant contained in certain tissues of the worker. The influence of the pharmacokinetics of the chemical and the possibility of skin absorption have also made it difficult to evaluate the meaning of the results when matching the biological results with air sampling data. Nonetheless, the advantages of biological monitoring tend to outweigh the difficulties in inter-

Table 5.7 Autorhythmometry Method: Changes in PEF Mesor Related to CrO_3 Exposure of a Chromium-Plating Worker[a,b]

Activity (Situation)	Autorhythmometry Month (No. of Days)	PEF Circadian Mesor in liters/min ± 1SE	p Value (Vacation vs. Work Each Span)
Work (1)	July (5)	394 ± 20	<.02
Work (2)	July (5)	396 ± 22	<.05
Vacation (3)	August (9)	464 ± 15	—
Work (4)	September (7)	394 ± 18	<.01

[a] From Gervais et al. (279).
[b] A male (38 years old) working in a chromium-plating industry for 4 years suffering from dyspnea and/or asthma for 3 years. Skin tests and bronchial challenge to chromium were inconclusive. Peak expiratory flow (PEF)—an indication of airways patency-self-measurments were done at fixed clock hours (8 A.M., 11 A.M., 3 P.M., 7 P.M., and 11 P.M.) for 5 to 9 days, both when working during the day (situations 1, 2, and 4) and during vacation (situation 3) in a rural area. With reference to values obtained in the latter situation, a statistically significant fall in the PEF circadian mesor (24-hr mean) occurred when this man was exposed to CrO_3.

pretation since the results will reflect the degree of body burden and subsequent risk of injury such as the case for blood cholinesterase inhibition in pesticide workers. Consequently, the approach is still recommended for those situations wherein the risk to health necessitates quantification.

Autorhythmometry, the self-measurement of pertinent biological functions at regular intervals during waking for several consecutive days with the data thus obtained analyzed and quantified for rhythms (278), is useful for purposes other than study of chronobiologic adjustments to rotating shiftwork. It can be used to investigate or monitor specific health functions or indicators on a more or less continuous basis or at set intervals, such as throughout the first and last days of each shift rotation. The instrumentation and methods for autorhythmometry, with regard to studying the chronobiologic adjustments of shiftworkers, were discussed in Sections 8.4 to 8.4.3. This methodology has been quite useful in the past. For example, autorhythmometry has been used as a means to screen persons for hypertension and to investigate the relationship between air quality and airway function in asthmatic patients (148, 203, 279, 280). In these studies, portable light-weight Wright or Hildebrant peak flow instruments appropriate for measuring the patency of the larger airways of the lungs were provided and instructions given for their correct use. In occupational medicine autorhythmometry has been utilized in industrial surveys of occupationally induced asthma in employees exposed to fumes from soldering fluxes and toluene diisocyanate (214, 279).

Gervais et al. (279) has used autorhythmometry to detect the occupationally induced airway hypersensitivity of one worker to CrO_3. Table 5.7 shows the findings for this worker, who monitored his own peak expiratory flow many times (at least four or five times) daily for several consecutive days during several spans of work and vacation. It is clear that values for the measurements made during the work weeks were different from those made on vacation. In particular, the 24-hr mean levels of airway patency, which were consistent in value from one studied work span to another, were reduced in comparison to those obtained away from work during vacation. The peak flow is a direct measurement of airway patency; the lower the flow, the lower the patency. The data of this study indicate that for this employee, exposure to CrO_3 is associated with a statistically significant impairment of airway function. Pepys (personal communication) and others (203, 279) have recommended that workers employed in tasks associated with an elevated risk of pulmonary disorders, in addition to undergoing timely laboratory workups, should self-monitor their airway functions periodically by use of autorhythmometric methods as a means of detecting early signs of impaired airway function. We concur, but advise as well that other relevant endpoints used in the biological monitoring of employees and which may be influenced by rhythms be self-assessed using autorhythmometric methods.

10.3 Chronotoxicity and Rapid Rotation Shift Schedules

The rapid rotation schedules have been useful in minimizing the adverse effects of night shifts. For employees engaged in rapid rotational shifts of 2 to 3 days' duration, the avoidance of jet-lag-like shiftwork symptoms is possible since the acrophase of most circadian processes undergo only slight change in timing, if any, by the last night shift of the rota. On the other hand, occupational physicians have expressed concern about the methodology and interpretation of biological monitorings and samplings of such workers. The issue is whether exposure to workplace contaminants at biologically atypical times in persons who work during the nighttime, on the rapid rotation schedule, in the absence of significant acrophase shifts should be compared to those exposures which occur during the daytime and which coincide with a different staging of the circadian system. Based on modeling studies on rodents, concern is justified. For example, rodent investigations (see Figures 5.41–5.43) clearly document the existence of greatly differing levels of resistance-susceptibility to a variety of chemical and physical agents as a function of the biological timing of exposure.

Most physicians and toxicologists suggest that the interpretation of biological monitorings and other health-related indices of response need not be adjusted for circadian rhythm effects because the already promulgated threshold limit values (TLV) or permissible exposure limits (PEL) are sufficiently conservative to protect even persons who work on rotating shifts. Unfortunately, appropriate studies on workers while on the night as well as the day shift have yet to be conducted. Such studies are required to validate the TLV or PEL, especially for certain chemical exposures.

10.4 Chronotoxicity and Unusual Shift Schedules

Unusual work shifts pose at least two special kinds of problems. The first problem relates to the necessity of adjusting TLVs for unusual work schedules to ensure that employees will not be at greater risk than those on regular shifts. The second problem involves the circumstance that the promulgation of all workplace standards has been guided by the biological concept of homeostasis. This concept infers that all bodily functions and processes are maintained in relative constancy with variations occurring at random or resulting from environmental causes. This concept infers also that the biological response to toxic chemicals in individuals does not vary over time—over 24 hrs, menstrual cycle, and year. Homeostasis does not take into account the existence of biological rhythmicity which, as has been discussed, can be quite significant, especially to occupational health concerns. It is only in recent years that investigations have begun to model in rodents workplace exposures, taking into consideration the role circadian rhythms play in the metabolism and effects of workplace and other toxicants (127, 223–225, 291). In connection with this, one of the objectives of chronobiologists working with the Unusual Work Shifts Subcommittee of the American Industrial Hygiene Association is to evaluate the importance of biological rhythms on the toxicity of workplace chemicals in persons employed on unusual shifts.

With regard to unusual as with most shift schedules, emphasis is placed upon the peak body burden of the chemicals to which the worker is exposed by the end of the last day, for example, of 10- or 12-hr in duration shifts. Biological monitorings, when conducted, are done typically on shiftworkers when they are on daytime duty, when the occupational health department is well staffed to carry out such samplings. Inferences are then drawn, assuming homeostatic principles, about the safety of the workplace environment for employees when they work the night shift. When from week-to-week employees alternate between day and night work, chronobiologic factors can be significant. In the case of unusual work schedules, with work periods of 10 to 12 hr or more in length, exposures of workers to chemicals and other potentially hazardous substances not only occur for longer durations than they do during the standard 8-hr shift, but they occur as well during different circadian stages. The concern of chronobiologists is that circadian susceptibility–resistance phenomena are important considerations for workers on unusual shift schedules. Moreover, from a chronobiologic perspective, when chemical exposures are involved, biological rhythms in their disposition have to be considered. This entails the chronokinetics of chemicals (see Section 7.2.1), that is, circadian or other rhythmic influences on the time to reach peak concentration, peak height, clearance, and area under the time–concentration curve. As is the case for special chemical agents such as medications, these chronokinetic parameters can be vastly different, depending on the biological time of exposure (33, 34, 221, 222). Although at this time sufficient data do not exist either from animal or human beings to assess the significance of circadian chronokinetic phenomena with respect to the need for adjusting TLVs for unusual work shifts, current research efforts are directed at evaluating for which types of exposures circadian chronokinetic changes are critical.

The biological effects of xenobiotic exposures in the case of both usual and unusual work schedules are dependent not only on their chronokinetics but, in addition, on their chronesthesy (see Section 7.2.2). Exposures to substances either during differently timed 8-hr shifts or 10- to 12-hr (or longer) unusual work shifts may result in different biological effects due to circadian differences in the chemical susceptibility of target tissues. As in the case of medications, circadian chronesthesys may represent among others 24-hr variations in the number of receptor sites and/or the capacity of various biochemical pathways to metabolize, transform, or incorporate chemical agents. Taking into consideration the findings from chronopharmacologic investigations of medications (Sections 7.2.1–7.2.3) as well as knowledge about the chronotoxicology of chemicals (Sections 7.3–7.3.4), circadian rhythms in liver and brain enzymes (Section 7.3.2), and known circadian resistance–susceptibility rhythms of human beings (Section 7.4), the possibility that the biological response to or effect of the peak body burden of chemical agents when working, for example, the daytime in comparison to the nighttime shift being different, especially for workers on rapid rotation shiftwork schedules, cannot be ignored.

Currently, except for medications, circadian chronesthesys of chemical substances in human beings have yet to be demonstrated. This is due primarily to the fact so few investigations of this type have been attempted. Nonetheless, they are known for substances used in the diagnosis of human diseases, for example, the acetylcholine, histamine, and house-dust airway challenge tests, which serve to evaluate the hyperreactivity of airways in patients exhibiting obstructive airways disease. In rodents, additional data are available, for example, for the effect of carbon tetrachloride (CCl_4) on liver enzyme activity and toxicity. (21–23, 224, 225, 313). Moreover, other findings from studies on rodents reveal circadian differences in the effects of a large variety of toxic substances as exemplified in Figures 5.41–5.43. Since circadian variation occurs at all levels of biological organization, circadian chronesthesys are possible at each. Keeping in mind that clinically meaningful chronesthesys are common with medications, it behooves occupational health researchers and physicians to evaluate findings from laboratory models and medical records for temporal patterns of susceptibility and toxicity to chemicals found in the workplace to determine the need for altering exposure limits for unusual, as well as standard 8-hr, rotational shiftwork schedules.

11 SUMMARY

As of 1985, the facts clearly demonstrate that practically every physiological or behavioral variable amenable to measurement is characterized by being rhythmic, at least in the healthy organism. The spectrum of rhythms is broad.

The one frequency that has received the greatest attention in the past few years is that which approximates the frequency of the rotation of the earth, or the circadian. It is this frequency, primarily in the mammalian organism, upon which we have focused. Biological rhythms in most physiological variables are not apparent to us in the same sense that the respiratory or menstrual rhythm is; they only become overt when they are properly measured at frequent intervals over the 24-hr time scale. Because of their somewhat "invisible" nature there is a tendency on the part of some investigators to slight or to ignore them. In spite of all that is known they simply have not been accorded the attention they deserve; this undoubtedly is in large part due to the fact that the science of chronobiology is relatively young. Perhaps this explains why many occupational health professionals are not aware of practical developments in this emerging science. On the other hand, many health professionals are confused about the significance of biological rhythms as reflected by their thoughts about them.

Some rationalize that biological rhythms represent no more than minor fluctuations around a daily mean and consequently do not warrant the additional amount of work and expense required to properly explore for their significance. Those who believe this simply are ignoring the scientific facts available to them. Admittedly, oral or rectal temperature may change only one or two degrees centigrade over the 24-hr time scale; there are, however, few physiologists who would minimize the effect that this small-amplitude rhythm has on metabolic or physiological activity. On the other hand, the corticosterone concentration in the plasma of human beings may vary three- or fourfold; the same applies to many enzymes in the liver and plasma of rodents and probably human beings as well. When 5-hydroxytryptamine levels in the whole brain of the rodent are measured at frequent intervals over a 24-hr period, the difference between the lowest and highest recorded mean may represent only an 18% change, but the same substance in the pineal may vary as much as 900%. When similarly compared, the cell proliferation in a number of mitotically labile tissues, such as the cornea, may represent over a 1200% increase (265). Generally, fluctuations in man are equally as great as the daily change in rodents; the evidence to support the above statements is extensively documented. Because the biological system is continually changing, the organism is a different biochemical entity at different stages of its circadian system. If an organism such as the rodent is subjected to a potentially noxious agent at one time, it may succumb, whereas at another phase of the circadian system it may evidence very little harm. Circadian differences in the response of human beings, although less studied, are well documented also. This circadian differential in response has led to the concept of the *Hours of Changing Resistance*. The overwhelming evidence that supports this concept necessitates total rejection of the idea that circadian time structure represents no more than minor and, hence, unimportant fluctuations around a daily mean. Furthermore, the facts render completely untenable the concept of homeostatic balance as is presently taught in many freshman medical school physiology courses as well as industrial toxicology and hygiene curricula.

Other health professionals have expressed the opinion that circadian rhythms represent nothing more than day–night changes, or that they are some kind of direct response to the ingestion of a meal, and that if one simply samples once during the day and once during the night in some vague way one will have taken care of this nuisance variable. We wish to point out that sampling only twice per 24 hrs does not account for the possible influence of circadian temporal structure. One may happen to sample at times which approximate the 24-hr mean level, the mesor, even though the times of sampling are separated by 12 hrs. Also, sampling by clock time only, without regard to synchronizer schedule, especially when shiftwork is concerned, is meaningless. This attitude towad rhythmicity simply cannot be defended, yet it still is held by most investigators! Although facts demonstrate that bioperiodicities are responsive to changes in the environment, they represent endogenous and inherited attributes. Teleologically, their properties provide the mechanisms that enable the organism to adjust not only to predictable geophysical changes over time but, in particular, to shiftwork.

A large number of investigators (especially toxicologists) conduct research or carry out biological samplings each day at the same time, ostensibly to avoid or minimize dealing with circadian rhythms. In view of what has been learned about their properties, such as their endogenous nature and in rodents their ability to phase shift with alteration in the environmental light–dark synchronizer schedule or in man with changes in the sleep and activity pattern and their tendency to freerun in continuous light or darkness, it is this widespread practice that is most difficult for the chronobiologist to accept as the means for avoiding the effects of rhythms. All that this practice of conducting procedures at the "same time of day" does is to assure an investigator that he is sampling at one particular clock hour, perhaps indicative of a single circadian stage of one or more rhythms—in the trough, on the incline, at the peak or on the decline. However, he can only be sure of this if he has first gone to the trouble of mapping the rhythm(s) under controlled conditions, that is, if for rodents the establishment of a carefully controlled synchronizing cycle such as that of light and darkness on a fixed alternating schedule; even then the circadian system phase may systematically shift with the changes in season so that sampling at 0900 June 1, may not be the same in terms of biological time as sampling at 0900 January 1. This practice of sampling once daily always at the same time of day can lead the investigator into pitfalls.

There are many excellent toxicologists who have at least an intuitive appreciation of the importance of rhythms, for example in industrial toxicology, and would explore their effects in depth but find that proper control of the animal quarters simply cannot be obtained. Many animal facilities are geared more to housekeeping activities and the work schedules of animal caretakers than to careful control of the light–dark synchrony. We believe this is a major reason for staying away from this type of investigation; it certainly is the most frequent excuse given.

No longer is it necessary for one to have to "stay up round the clock" to get a handle on circadian variation and we have pointed out how this can be done in Section 9.5. If the determination of an LD_{50}, and so on, is important it should be reliable. We are at a stage where an effort must be made to come to grips with biological oscillation and we offer the following suggestions as to how this might be done. The leadership in the biological scientific community, including industrial toxicologists and occupational health physicians, must critically examine the body of knowledge upon which chronobiology presently is founded. We assume that if this were done it would lead one to enthusiastically recognize the potential that this science has to contribute to the better design of shiftwork schedules and protection of shiftworkers from hazardous exposures, keeping in mind the existence of circadian differences in resistance–susceptibility.

With regard to improving shiftwork conditions in industry, Knauth and Rutenfranz (55), reviewing a large number of investigations, including chronobiologic ones, suggest a number of recommendations. For the good of the worker, (a) a shift system should have few successive night shifts to minimize jet-lag-like shiftwork symptoms and sleep problems with resulting fatigue; (b) the morning shift should not commence too early, relative to clock hour, keeping in mind the sleep schedules of workers and the local traffic patterns, for example; (c) the shift change times should allow individuals some degree of choice or flexibility, for example, by being allowed to participate in the planning of shift timing changes, and so on; (d) the length of the shift should depend on the mental and physical demands of the job and when possible the duration of the night shift should be shorter than the morning and afternoon shifts; (e) short intervals of time (e.g., 7 to 12 hr) between two successive shifts, which is often common to irregular shift systems, should be avoided since they result in too much fatigue; (f) continuous shift systems should include some free weekends with a minimum of two successive full days off to enable socialization with friends who also are free at this time; (g) in the case of continuous shift systems, a forward rotation rather than a backward rotation is preferred for chronobiologic as well as other reasons discussed in Section 8.2.2; (h) the duration of the shift cycle (the pattern of shifts and days off until the sequence repeats on the same day of the week) should be reasonable in length and not too long in order to favor planning for leisure activity; (i) shift rotas should be regular to facilitate the planning of activities away for work. The aforementioned suggestions are offered with the understanding that complying with all of them is difficult since many factors must be considered simultaneously. Nonetheless, these recommendations are offered as a means for improving the conditions of many shiftwork schedules now in operation in industrial plants worldwide.

In recent years rapid progress has been realized in the chronobiology of industrial hygiene and occupational health; yet, much more remains to be done. Chronobiologists, as do other health professionals, look forward to continued collaboration with management, labor, and scientists of various disciplines to alleviate the many still troublesome problems associated with rotating shiftwork schedules.

ACKNOWLEDGMENT

Some of the work reported herein was supported by grant OH-00952 from the National Institute for Occupational Safety and health to Professor L. E. Scheving. Dr. Smolensky and Dr. Paustenbach thank their colleagues in the fields of toxicology, clinical medicine, and industrial hygiene who encouraged them to assemble and evaluate this information.

REFERENCES

1. S. D. Nollen, *Ind. Eng.*, 58–63 (1981).
2. S. D. Nollen, "Work Schedules," in: *Handbook of Industrial Engineering*, G. Salvendy, Ed., Wiley-Interscience, New York, 1983.
3. M. Colligan, "Methodological and Practical Issues Related to Shiftwork Research," in: *Biological Rhythms, Sleep and Shift Work*, L. C. Johnson, D. I. Tepas, W. P. Colquhoun, and M. Colligan, Eds., SP Scientific and Medical Books, New York, 1981, pp. 197–204.
4. W. B. Webb, "Work/Rest Schedules: Economic, Health and Social Implication", in: L. C. Johnson, D. I., Tepas, W. P. Colquhoun, and M. Colligan, Eds., SP Medical and Scientific Publications, New York, 1981, pp. 1–10.
5. R. Mahathevan, *J. Human Ergol.*, **11** (Suppl.), 139–145 (1982).
6. A. Manuba, *J. Human Ergol.*, **11** (Suppl.), 147–153 (1982).
7. A. Khaleque and A. Rahman, *J. Human Ergol.*, **11** (Suppl.), 155–164 (1982).
8. F. M. Fischer, *J. Human Ergol.*, **11** (Suppl.), 177–193 (1982).
9. S. E. G. Perera, *J. Human Ergol.*, **11** (Suppl.), 201–208 (1982).
10. C. N. Ong and B. T. Hoong, *J. Human Ergol.*, **11** (Suppl.), 209–216 (1982).
11. M. Wongphanich, H. Saito, K. Kogi, and Y. Temmyo, *J. Human Ergol.*, **11** (Suppl.), 165–175 (1982).
12. J. Rutenfranz, *J. Human Ergol.*, **11** (Suppl.), 67–86 (1982).
13. A. N. Nicholson, Ed., *Sleep, Wakefulness and Circadian Rhythm*, AGARD Lecture Series **105**, 1979.
14. L. E. Scheving and F. Halberg, Eds., *Chronobiology: Principles and Applications to Shifts in Schedules*, NATO Advanced Study Institutes Series D, Behavioural and Social Sciences, No. 3, Sijthoff and Noordhoff, The Netherlands, 1980.
15. A. Reinberg, Ed., Chronobiological Field Studies of Oil Refinery Shift Workers, *Chronobiologia*, **6** (Suppl.) (1979).
16. L. C. Johnson, D. I. Tepas, W. P. Colquhoun, and M. Colligan, "Preface," in: *Biological Rhythms, Sleep and Shift Work*, L. C. Johnson, D. I. Tepas, W. P. Colquhoun, and M. Colligan, Eds., SP Scientific and Medical Books, New York, 1981.
17. K. Kogi, T. Miura, and H. Saito, Eds., *Shiftwork: Its Practice and Improvement*, Center for Academic Publications, Tokyo, 1982.
18. A. Reinberg, N. Vieux, and P. Andlauer, Eds., *Night and Shift Work: Biological and Social Aspects*, Pergamon Press, Oxford, 1981.

19. W. P. Colquhoun, Ed., *Biological Rhythms and Human Performance*, Academic Press, London, 1971.
20. J. Wojtczak-Jaroszowa and K. Pawlowska-Skyba, *Medycyna Pracy*, **18,** 1 (1967).
21. D. J. Paustenbach, "Occupational Exposure Limits, Pharmacokinetics and Unusual Work Schedules," in: *Patty's Industrial Hygiene and Toxicology, IIIA, Industrial Hygiene Aspects*, 2nd ed., Wiley-Interscience, New York, in press.
22. D. J. Paustenbach, G. S. Born, G. P. Carlson, and J. E. Christian, "A Comparative Study of the Pharmacokinetic and Toxic Effects of Exposing Rats to Carbon Tetrachloride Vapor During a Standard or Novel Workweek,' in: *Proceedings of the American Industrial Hygiene Conference*, Portland, Oregon, 1981.
23. D. J. Paustenbach, "The Effect of the Twelve-Hour Workshift on the Toxicology, Disposition and Pharmacokinetics of Carbon Tetrachloride in the Rat," Doctoral Dissertation, Purdue University, West Lafayette, Indiana, 1982.
24. R. S. Brief and R. A. Scala, *Am. Ind. Hyg. Assoc. J.*, **36,** 467–471 (1975).
25. J. W. Mason and H. Dershin, *J. Occup. Med.*, **18,** 603–607 (1976).
26. J. W. Mason and J. Hughes, *Scand. J. Work Env. Health* (in press).
27. J. L. S. Hickey, *Am. Ind. Hyg. Assoc. J.*, **41,** 261–263 (1980).
28. J. L. S. Hickey, *Am. Ind. Hyg. Assoc. J.*, **38,** 613–621 (1977).
29. J. L. S. Hickey and P. C. Reist, *Am. Ind. Hyg. Assoc. J.*, **40,** 727–734.
30. S. Folkard, "Shiftwork and Performance," in: L. C. Johnson, D. I. Tepas, W. P. Colquhoun, and M. Colligan, Eds., *Biological Rhythms, Sleep and Shift Work*, SP Medical and Scientific Publications, New York, 1981, pp. 283–306.
31. B. Kolmodin-Hedman, *J. Human Ergol.*, **11** (Suppl.), 447–456 (1982).
32. F. Brown and R. C. Graeber, Eds., *Rhythmic Aspects of Behavior*, LEA, New Jersey, 1982.
33. A. Reinberg and M. H. Smolensky, *J. Clin. Pharmacokinetics*, **7,** 401–420, 1982.
34. A. Reinberg and M. H. Smolensky, *Biological Rhythms and Medicine*, Springer Verlag, New York, 1983.
35. J. P. McGovern, M. H. Smolensky, and A. Reinberg, Eds., *Chronobiology in Allergy and Immunology*, Thomas, Springfield, IL, 1977.
36. M. H. Smolensky, A. Reinberg, and J. P. McGovern, Eds., *Recent Advances in the Chronobiology of Allergy and Immunology*, Pergamon Press, 1980.
37. L. E. Scheving, *Endeavour*, **35,** 66–72 (1976).
38. L. E. Scheving, *Trends Pharmacol. Sci.*, 303–307 (1980).
39. F. Halberg, *Am. J. Anat.*, **168,** 543–594 (1983).
40. F. Halberg, E. Haus, S. S. Cardoso, L. E. Scheving, J. F. W. Kühl, R. Shiotsuka, G. Rosene, J. P. Pauly, W. Runge, J. F. Spalding, J. K. Lee, and R. A. Good, *Experientia*, **29,** 909–1044 (1973).
41. E. Haus, F. Halberg, L. E. Scheving, and H. Simpson, *Int. J. Chronobiol.*, **6,** 67–107 (1979).
42. W. J. M. Hrushesky, *Am. J. Ant.*, **168,** 519–542 (1983).
43. W. J. M. Hrushesky, "Chemotherapy Timing: An Important Variable in Treatment and Response," in: *Toxicity of Chemotherapy*, M. Perry and J. Yarbo, Eds., Grune and Stratton, Chapter 18, in Press, Orlando, Florida.
44. L. C. Johnson, D. I. Tepas, W. P. Colquhoun, and M. Colligan, Eds., *Biological Rhythms, Sleep and Shift Work*, SP Medical and Scientific Publications, New York, 1982.
45. J. Rutenfranz and W. P. Colquhoun, "The Scientific Committee on Shiftwork: A Short Account of Its History, Aims and Achievements," in: *Night and Shift Work: Biological and Social Aspects*, A. Reinberg, N. Vieux, and P. Andlauer, Eds., Pergamon Press, Oxford, 1981, pp. 11–12.

46. P. J. Taylor, "The Problems of Shift Work," in: *Proceedings of an International Symposium on Night and Shiftwork*, Oslo, Sweden, 1969.
47. L. C. Johnson, "On Varying Work/Sleep Schedules: Issues and Perspectives as Seen by a Sleep Researcher," in: *Biological Rhythms, Sleep and Shift Work*, L. C. Johnson, D. I. Tepas, W. P. Colquhoun, and M. J. Colligan, Eds., SP Medical and Scientific Books, New York, 1981, pp. 335–346.
48. P. G. Rentos and R. D. Shepard, Eds., *Shift Work and Health—A Symposium*, U. S. HEW PHS, National Institute of Occupational Safety and Health, Washington, DC, 1976.
49. J. Rutenfranz, W. P. Colquhoun, P. Knauth, and J. N. Ghata, *Scand. J. Work, Environ., Health*, **3**, 165–192 (1977).
50. J. Wojtczak-Jaroszowa, *Physiological and Psychological Aspects of Night and Shift Work*, National Institute of Occupational Safety and Health, Cincinnati, OH, 1977.
51. B. Bjerner, A. Holm, and A. Swensson, *Br. J. Ind. Med.*, **12**, 103–110 (1955).
52. B. Bjerner, A. Holm, and A. Swensson, "Studies on Night and Shiftwork," in: *Shiftwork and Health*, A. Aanonsen, Ed., Scandinavian University Books, Oslo, 1964.
53. E. Thiis-Evensen, *Ind. Med. Surg.*, **27**, 493–513 (1958).
54. D. L. Tasto, M. J. Colligan, E. W. Skjei, and S. J. Polly, *Health Consequences of Shift Work*, U.S. DHEW PHS, National Institute of Safety and Health, Washington, DC, 1978.
55. P. Knauth and J. Rutenfranz, *J. Human Ergol.*, **11** (Suppl.), 337–367 (1982).
56. U.S. Census, USBLS, May 1975, Series P, No. 10.
57. H. Allenspach, *Flexible Working Hours*, International Labor Office, World Health Organization (WHO), Geneva, Switzerland, 1975.
58. J. T. Wilson and K. M. Rose, *The Twelve-Hour Shift in the Petroleum and Chemical Industries of the United States and Canada: A Study of Current Experience*, Wharton Business School, University of Pennsylvania, Philadelphia, PA, 1978.
59. G. D. Botzum and R. L. Lucas, "Slide Shift Evaluation—a Practical Look at Rapid Rotation Theory," in: *Proceedings of Human Factors Society*, 1981, pp. 207–211.
60. T. A. Yoder and G. D. Botzum, "The Long-Day Short-Week in Shift Work—A Human Factors Study," in: *Proceedings, Human Factors Society*, 1971.
61. E. D. Weitzmann and D. F. Kripke, "Experimental 12-Hour Shifts of the Sleep-Wake Cycle in Man: Effects on Sleep and Physiologic Rhythms," in: *Biological Rhythms, Sleep and Shift Work*," L. C. Johnson, D. I. Tepas, W. P. Colquhoun, and M. J. Colligan, Eds., SP Medical and Scientific Books, New York, 1982, pp. 93–110.
62. A. Reinberg, N. Vieux, J. Ghata, A.-J. Chaumont, and A. Laporte, "Consideration of the Circadian Amplitude in Relationship to the Ability to Phase Shift Circadian Rhythms of Shift Workers," in: *Chronobiological Field Studies of Oil Refinery Shift Workers*, A. Reinberg, Ed.; *Chronobiologia*, **6** (Suppl.), 57–63 (1979).
63. A. Reinberg, N. Vieux, P. Andlauer, P. Guillet, and A. Nicolai, "Tolerance to Shift-Work, Amplitude of Circadian Rhythms and Aging," in: *Night and Shift-Work Studies: Biological and Social Aspects*, A. Reinberg, N. Vieux, and P. Andlauer, Eds., Pergamon Press, Oxford, 1981, pp. 341–354.
64. S. A. Roach, *Am. Ind. Hyg. Assoc. J.*, **39**, 345–364 (1978).
65. Occupational Safety and Health Administration, *Compliance Officers: Field Manual*, Department of Labor, Washington, DC, 1979.
66. J. Rutenfranz, P. Knauth, and D. Angerbach, "Shift Work Research Issues," in: *Biological Rhythms, Sleep and Shift Work*, L. C. Johnson, D. I. Tepas, W. P. Colquhoun, and M. J. Colligan, Eds., SP Medical and Scientific Books, New York, 1981, pp. 165–196.
67. B. Barhard and M. Pafnotte, *Le Travail Humain*, **33**, 1–19 (1970).
68. D. L. Bosworth and P. J. Dawkins, "Private and Social Costs and Benefits of Shift and Night Work," in *Night and Shift Work: Biological and Social Aspects*, A. Reinberg, N. Vieux, and P. Andlauer, Eds., Pergamon Press, Oxford, 1981, pp. 207–214.

69. C. Gadbois, "Women on Night Shift: Interdependence of Sleep and Off-the-Job Activities," in: *Night and Shift Work: Biological and Social Aspects*, A. Reinberg, N. Vieux, and P. Andlauer, Eds., Pergamon Press, Oxford, 1981, pp. 223–228.
70. M. Kundi, M. Koller, R. Cervinka, and M. Hiader, "Job Satisfaction in Shift Workers and Its Relation to Family Situation and Health," in: *Night and Shift Work: Biological and Social Aspects*, A. Reinberg, N. Vieux, and P. Andlauer, Eds., Pergamon Press, Oxford, 1981, pp. 237–245.
71. M. Maurice, "Shiftwork: Economic Advantages and Social Costs," International Labor Office, WHO, Geneva, 1975.
72. P. E. Mott, F. C. Mann, Q. McLoughlin, and D. P. Warwick, *Shift Work: The Social, Psychological and Physical Consequences*, The University of Michigan Press, Ann Arbor, 1965.
73. F. Nachreiner and J. Rutenfranz, "Sozialpsychologische, Arbeitspsychologische und Medizinische Erhebungen in der Chemischen Industrie," in: *Schichtarbeit bei kontinuierlicher Produktion*, F. Nachreiner et al., Eds., Wirtschaftsverlag Nordwest GmbH, Withelmshaven, 1975, pp. 83–117.
74. A. A. I. Wedderburn, *Occup. Psychol.*, **41**, 85–107 (1967).
75. A. A. I. Wedderburn, "How Important Are the Social Effects of Shiftwork?" in: *Biological Rhythms, Sleep and Shift Work*, L. C. Johnson, D. I. Tepas, W. P. Colquhoun, and M. Colligan, Eds., SP Medical and Scientific Books, New York, 1981, pp. 257–269.
76. T. Akerstedt and M. Gillberg, "The Circadian Pattern of Unrestricted Sleep and its Relation to Body Temperature, Hormones and Alertness," in: *Biological Rhythms, Sleep and Shift Work*, L. C. Johnson, D. I. Tepas, W. P. Colquhoun, and M. Colligan, Eds., SP Medical and Scientific Books, New York, 1981, pp. 481–498.
77. J. Foret and O. Benoit, in: *Chronobiological Field Studies of Oil Refinery Shift Workers*, A. Reinberg, Ed., *Chronobiologia*, **6** (Suppl), 45–53 (1979).
78. J. Foret and G. Lantin, "The Sleep of Train Drivers: An Example of the Effects of Irregular Work Schedules on Sleep," in: *Aspects of Human Efficiency: Diurnal Rhythms and Loss of Sleep*, W. P. Colquhoun, Ed., The English University Press Ltd., London, 1972, pp. 273–282.
79. H. Fukuda, S. Endo, T. Yamamoto, Y. Saito, and K. Nishihara, *J. Human Ergol.*, **11** (Suppl.), 245–257 (1982).
80. N. Kleitman, *Sleep and Wakefulness*, University of Chicago Press, Chicago, 1963.
81. P. Knauth and J. Rutenfranz, "The Effects of Noise on the Sleep of Nightworkers," in: *Studies of Shiftwork*, W. P. Colquhoun and J. Rutenfranz, Eds., Taylor and Francis Ltd., London, 1980, pp. 111–120.
82. K. Kogi, *J. Human Ergol.*, **11** (Suppl.), 217–231 (1982).
83. D. F. Kripke, B. Cook, and O. F. Lewis, *Psychophysiology*, **7**, 377–384 (1971).
84. F. Lille, *Le Travail Humain*, **30**, 85 (1967).
85. D. Minors and J. M. Waterhouse, "Anchor Sleep as a Synchronizer of Rhythms on Abnormal Routines," in: *Biological Rhythms, Sleep and Shift Work*, L. C. Johnson, D. I. Tepas, W. P. Colquhoun, and M. Colligan Eds., SP Medical and Scientific Books, New York, 1981, pp. 399–414.
86. D. I. Tepas, *J. Human Ergol.*, **11** (Suppl.), 325–336 (1982).
87. D. I. Tepas, J. Walsh, and D. Armstrong, "Comprehensive Study of the Sleep of Shift Workers," in: *Biological Rhythms, Sleep and Shift Work*, L. C. Johnson, D. I. Tepas, W. P. Colquhoun, and M. Colligan, Eds., SP Medical and Scientific Books, New York, 1981, pp. 347–356.
88. D. I. Tepas, *J. Human Ergol.*, **11** (Suppl.), 1–12 (1982).
89. S. Torii, N. Okudaria, H. Fukuda, H. Kanamoto, Y. Yamashiro, M. Akiya, K. Nomoto, N. Katayama, M. Hasegawa, M. Sato, M. Hatano, and H. Hemoto, *J. Human Ergol.*, **11** (Suppl.), 233–244 (1982).

90. W. J. Price and D. C. Holley, *J. Human Ergol.*, **11** (Suppl.), 291–301 (1982).
91. J. Walsh, D. I. Tepas, and P. Moss, "The EEG Sleep of Night and Rotating Shift Workers," in: *Biological Rhythms, Sleep and Shift Work*, L. C. Johnson, D. I. Tepas, W. P. Colquhoun, and M. Colligan, Eds., SP Medical and Scientific Books, New York, 1981, pp. 371–382.
92. W. B. Webb and H. W. Agnew, Jr., *Aviation, Space Envir. Med.*, **49**, 384–389 (1978).
93. G. S. Tune, *Br. J. Industr. Med.*, **26**, 54–58 (1969).
94. J. Rutenfranz, P. Knauth, G. Hildebrandt, and W. Rohmert, *Int. Arch. Arbeitsmed.*, **32**, 243–259 (1974).
95. P. Knauth and J. Rutenfranz, "Duration of Sleep Related to the Type of Shift Work," in: *Night and Shift Work: Biological and Social Aspects*, A. Reinberg, N. Vieux, and P. Andlauer, Eds., Pergamon Press, Oxford, 1981, pp. 161–167.
96. K. E. Klein and H. M. Wegmann, "The Effects of Transmeridian and Transequatorial Air Travel on Psychological Well-Being and Performance," in: *Chronobiology: Principles and Applications to Shifts in Schedules*, NATO Advanced Study Institutes Series D: Behavioural and Social Sciences, No. 3, Sijthoff and Noordhoff, The Netherlands, 1980, pp. 339–352.
97. K. E. Klein and H. M. Wegmann, in: *Sleep, Wakefulness and Circadian Rhythm*, A. N. Nicholson, Ed., AGARD Lecture Series No. 105, NATO, 1979, pp. 2:1–2:9.
98. T. Akerstedt and L. Torsvall, *Ergonomics*, **21**, 849–856 (1978).
99. P. J. Taylor, *Br. J. Industr. Med.*, **24**, 93–102 (1967).
100. P. J. Taylor and S. J. Pocock, *Br. J. Industr. Med.*, **29**, 201–207 (1972).
101. J. M. Harrington, *Shift Work and Health: A Critical Review of the Literature*, Her Majesty's Stationery Office, London, 1978.
102. M. Koller, M. Kundi, and R. Cervinka, *Ergonomics*, **21**, 835–847 (1978).
103. D. Angersbach, P. Knauth, H. Loskant, M. J. Karvonen, K. Undeutsch, and J. Rutenfranz, *Int. Arch. Occup. Environ. Health*, **45**, 127–140 (1980).
104. R. Doll and F. A. Jones, "Occupational Factors in the Aetiology of Gastric and Duodenal Ulcers," Medical Research Council, Special Report Series No. 276, His Majesty's Stationery Office, London, 1951.
105. A. Reinberg, C. Migraine, M. Apfelbaum, L. Brigant, J. Ghata, N. Vieux, A. Laporte, and A. Nicolai, in: *Chronobiological Field Studies of Oil Refinery Shift Workers*, A. Reinberg, Ed., *Chronogiologia*, **6** (Suppl.), 89–102 (1979).
106. A. Reinberg, N. Vieux, P. Andlauer, and M. Smolensky, *Adv. Biol. Psychiatr.*, **11**, 35–47 (1983).
107. L. E. Scheving and J. E. Pauly, "Chronopharmacology: Its Implication for Clinical Medicine," *Ann. Rep. Med. Chem.*, **11**, 251–260 (1976).
108. F. Halberg, *Cold Spring Harbor Symp. Quant. Biol.*, **25**, 289–310 (1960).
109. F. Halberg, *Ann. Rev. Physiol.*, **31**, 675–725 (1969).
110. F. Halberg and W. Nelson, "Chronobiologic Optimization of Aging," in: *Aging and Biological Rhythms*, H. V. Samis, jr. and S. Capobianco, Eds., Plenum Press, New York, 1978, pp. 5–56.
111. F. Halberg, M. Engeli, C. Hamburger, and D. Hillman, *Acta Endocr.*, **103** (Suppl.), 5–54 (1965).
112. L. E. Scheving, F. Halberg, and J. E. Pauly, Eds., *Chronobiology*, Igaku Shoin Ltd., Tokyo, 1974.
113. A. Reinberg and J. Ghata, *Biological Rhythms*, Walker, New York, 1964.
114. M. C. Moore-Ede, F. M. Sulzman, and C. A. Fuller, *The Clocks That Time Us*, Harvard University Press, Cambridge, 1982.
115. I. Assenmacher and D. S. Farner, Eds., *Environmental Endocrinology*, Springer-Verlag, Berlin, 1978.
116. G. Luce, *Biological Rhythms in Psychiatry and Medicine*, U.S. DHEW Publ. No. (ADM) 78-247, 1970.

117. E. T. Pengelley, Ed., *Circannual Clocks*, Academic Press, New York, 1974.
118. C. F. Ehret, "The Importance of Fundamental Research in Chronobiology to Human Health," Report to National Conferences on Health-Related Principles, DHEW (NIH) 79-1892, Bethesda, Md., Vol. 1, Appendix C, 1979, pp. 91–95.
119. C. S. Pittendrigh, *Cold Spring Harbor Symp. Quant. Biol.*, **25**, 159–184 (1960).
120. L. N. Edmunds, Jr., *Am. J. Anat.*, **168**, 389–431 (1983).
121. G. Thommen, *Is This Your Day?*, Universal Publishing and Distributing Corp., New York, 1969.
122. K. E. Klein and H. M. Wegmann, "Appendix: Circadian Rhythms of Human Performance and Resistance: Operational Aspects," in: *Sleep, Wakefulness and Circadian Rhythm*, A. N. Nicholson, Ed., AGARD Lecture Series No. 105, 1979, pp. 2:10–2:17.
123. G. Schönholzer, G. Schilling, and H. Müller, *Schwerz Z. Sportmed.*, **1**, 7 (1972).
124. H. R. Willis, "Biorhythm and Its Relationship to Human Error," in: *Proceedings of the 16th Annual Meeting, Human Factors Society*, Beverly Hills, CA, 1973, pp. 274–282.
125. H. R. Willis, "The Effect of Biorhythm Cycles. Implication for Industry." American Industrial Hygiene Association Conference, Miami Beach, FL, 1974; cited in: T. M. Khalil and Ch. N. Kurveg, *Ergonomics*, **20**, 397 (1977).
126. L. E. Scheving and J. E. Pauly, *Chronobiologia*, **2**, 3–21 (1974).
127. L. E. Scheving, "Chronotoxicology in General and Experimental Chronotherapeutics of Cancer," in: *Chronobiology: Principles and Applications to Shifts in Schedules*, L. E. Scheving and F. Halberg, Eds., NATO Advanced Study Institutes Series D: Behavioral and Social Sciences, No. 3, Sijthoff and Noordhoff, The Netherlands, 1980, pp. 455–480.
128. L. E. Scheving and J. D. Dunn, "The Cyclic Nature of Prolactin in Mammals," in: *Chronobiological Aspects of Endocrinology*, J. Aschoff, F. Ceresa, and F. Halberg, Eds., Schattauer Verlag, Stuttgard; *Symposia Medica Holchst.*, **9**, 193–201, 1974.
129. L. E. Scheving, *Anat. Rec.*, **135**, 7–20 (1959).
130. M. H. Smolensky, "Aspects of Human Chronopathology," in: *Biological Rhythms and Medicine*, A. Reinberg and M. H. Smolensky, Eds., Springer-Verlag, New York, 1983, pp. 131–209.
131. A. Reinberg and M. H. Smolensky, "Investigative Methodology for Chronobiology," in: *Biological Rhythms and Medicine*, A. Reinberg and M. H. Smolensky, Eds., Springer-Verlag, New York, 1983, pp. 23–46.
132. L. E. Scheving, "Chronobiology, a New Perspective for Biology and Medicine," in: *Proceedings of the 11th Collegium Internationale Neuro-Psychopharmacologicum (CINP) Congress* (Vienna), B. Salolu, Ed., Pergamon Press, Oxford, 1978, pp. 629–642.
133. National Center for Health Statistics, Division of Health Examination Statistics, "Blood Pressure of Adults by Race and Area, 1960–1962," U.S. Government Printing Office, Washington, DC, 1964.
134. L. E. Scheving, F. Halberg, and E. L. Kanabrocki, "Circadian Rhythmometry on 42 Variables of Thirteen Presumably Healthy Young Men," in: *12th International Conference Proceedings* (International Society of Chronobiology), II Ponte, Milan, 1977, pp. 47–71.
135. E. L. Kanabrocki, L. E. Scheving, F. Halberg, R. Brewer, and T. Bird, *Space Life Sci.*, **4**, 258–270 (1973).
136. E. L. Kanabrocki, L. E. Scheving, F. Halberg, R. Brewer, and T. Bird, "Circadian Variation in Presumably Healthy Young Soldiers," *National Technical Information Service*, U.S. Dept. Commerce, Document No. PB22847, 1974, 56 pp.
137. L. E. Scheving, C. Roig III, R. Halberg, J. E. Pauly, and E. A. Hand, "Circadian Variations in Residents of a Senior Citizine's Home," in: *Chronobiology*, L. E. Scheving, F. Halberg, and J. E. Pauly, Eds., Igaku Shoin Ltd., Tokyo, 1974, pp. 353–357.
138. L. E. Scheving, C. C. Enna, F. Halberg, R. R. Jacobson, A. Mather, and J. E. Pauly, *Int. J. Leprosy*, **43**, 364–377 (1975).

139. L. E. Scheving, "Temporal Variation in Man," in: *Rhythmische Funktionen in Biologischen Systemen*, G. Lassman and F. Seitelberger, Eds., Facultas-Verlag, Vienna, 1977, pp. 49–74.
140. F. Halberg, *Zeit f. Vitamin–Hormone–und Fermentforschung*, **10**, 225–296 (1959).
141. F. Halberg, Y. L. Tong, and E. A. Johnson, "Circadian System Phase, An Aspect of Temporal Morphology: Procedures and Illustrative Examples," in: *The Cellular Aspects of Biorhythms*, H. von Mayersbach, Ed., Springer-Verlag, Berlin, 1967, pp. 20–48.
142. J. Aschoff, *Naturwissenschaften*, **41**, 49–56 (1954).
143. F. Halberg, E. Halberg, C. P. Barnum, and J. J. Bittner, "Physiologic 24-Hour Periodicity in Human Beings and Mice, the Lighting Regimen and Daily Routine," in: *Photoperiodism and Related Phenomena in Plants and Animals*, R. B. Withrow, Ed., American Association for the Advancement of Science Publication No. 55, Washington, DC, 1959, pp. 803–878.
144. F. Halberg and H. Simpson, *Hum. Biol.*, **39**, 405–413 (1967).
145. A. Reinberg and M. H. Smolensky, "Secondary Rhythms Related to Hormonal Changes in the Menstrual Cycle: General Considerations," in: *Biorhythms and Human Reproduction*, M. Ferin, F. Halberg, R. M. Richart, and R. L. Vande Wiele, Eds., Wiley, New York, 1974, pp. 241–258.
146. F. Halberg, M. Lagoguey, and A. Reinberg, *Int. J. Chronobiol.*, **8**, 225–268 (1983).
147. L. E. Scheving, T. H. Tsai, and L. A. Scheving, *Am. J. Anat.*, **168**, 433–465 (1983).
148. F. Halberg, A. Reinberg, and A. Reinberg, *Waking and Sleeping*, **1**, 259–279 (1977).
149. A. Reinberg, N. Vieux, J. Ghata, A. J. Chaumont, and A. Laporte, *Ergonomics*, **21**, 763–766 (1978).
150. A. Reinberg, N. Vieux, P. Andlauer, P. Guillet, A. Laporte, and A. Nicolai, "Oral Temperature Circadian Rhythm Amplitude, Aging and Tolerance to Shift-Work (Study 3)," in: *Chronobiological Field Studies of Oil Refinery Shift Workers*, A. Reinberg, Ed.; *Chronobiologia*, **6** (Suppl.), 67–85 (1979).
151. A. Reinberg, P. Andlauer, P. Teinturier, J. DePrins, W. Malbecq, and J. Dupont, *C. R. Acad. Sci.* (Paris), **296**, 267–269 (1983).
152. C. R. Graeber, "Alterations in Performance Following Rapid Transmeridian Flight," in: *Rhythmic Aspects of Behavior*, F. M. Brown and R. C. Graeber, Eds., LEA, New Jersey, 1982, pp. 173–212.
153. G. A. Christie and M. Moore-Robinson, *Clin. Trials J.*, **7**, 45 (1970).
154. R. C. Graeber, H. C. Sing, and B. N. Cuthbert, "The Impact of Transmeridian Flight on Deploying Soldiers," in: *Biological Rhythms, Sleep and Shift Work*, L. C. Johnson, D. I. Tepas, W. P. Colquhoun, and M. J. Colligan, Eds., SP Medical and Scientific Books, New York, 1981, pp. 513–537.
155. S. Folkard, "Circadian Rhythms and Human Memory," in: *Rhythmic Aspects of Behavior*, F. Brown and R. C. Graeber, Eds., LEA Publishers, New Jersey, 1982, pp. 241–272.
156. DeMarian, J., "Observations Botaniques," *Hist. Acad. Roy. Sci.* (Paris), 35–36 (1729).
157. L. E. Scheving, J. E. Pauly, H. von Mayersbach, and J. D. Dunn, *Acta Anat.*, **88**, 411–423 (1974).
158. L. E. Scheving, G. Sohal, C. Enna, and J. E. Pauly, *Anat. Rec.*, **175**, 1–6 (1973).
159. E. Haus, D. Lakatua, and F. Halberg, *Exp. Med. Surg.*, **25**, 7–45 (1967).
160. J. Ghata, F. Halberg, A. Reinberg, and M. Siffre, *Ann. Endocr.* (Paris), **30**, 245–260 (1969).
161. R. Wever, *The Circadian System of Man. Results of Experiments Under Temporal Isolation*, Springer-Verlag, New York, 1979.
162. A. Reinberg, "Eclairment et cycle menstruel de la femme," in: *La Photorégulation chez les Oiseaux et les Mammifères*, J. Benoit and I. Assenmacher, Eds., Coll. International CNRS Publication No. 172, Paris, 1970, pp. 529–546.
163. A. Reinberg, F. Halberg, J. Ghata, and M. Siffre, *Compt. Rend. Acad. Sci.*, **262**, 782–785 (1966).

164. J. Aschoff, *Cold Spring Harbor Sym. Quant. Biol.*, **25**, 11–28 (1960).
165. A. P. de Candolle, *Physiologie Végétale*, Béchet Jeune, Paris, 1832.
166. M. Siffre, *Hors du Temps*, Tulliurd, Paris, 1963.
167. F. Halberg, A. Reinberg, E. Haus, J. Ghata and M. Siffre, *Bull. Nat. Speleol. Soc.*, **32**, 89–115 (1970).
168. M. Siffre, A. Reinberg, F. Halberg, J. Ghata, G. Perdriel, and R. Slind, *Presse Med.*, **74**, 915–919 (1966).
169. W. Engelmann, *Z. Naturf.*, **28c**, 733–736 (1973).
170. D. F. Kripke, D. J. Mullaney, M. Atkinson, and S. Wolf, *Biol. Psychiatry*, **13**, 335–351 (1978).
171. D. F. Kripke, "Phase Advance Theories for Affective Illnesses," in: *Circadian Rhythms in Psychiatry*, F. K. Gookwin and T. A. Wehr, Eds., Boxwood Press, Los Angeles, 1982.
172. T. A. Wehr, A. Wirz-Justice, F. K. Goodwin, W. Duncan, and J. C. Gillin, *Science*, **206**, 710 (1979).
173. T. A. Wehr, A. Wirz-Justice and F. K. Goodwin, "Advanced Circadian Rhythms and a Sleep-Sensitive Switch Mechanism in Depression," in: *Circadian Rhythms in Psychiatry*, F. K. Goodwin and T. A. Wehr, Eds., Boxwood Press, Los Angeles, 1982.
174. A. Wirz-Justice, M. S. Kalka, D. Naber, and T. A. Wehr, *Life Sci.*, **27**, 341–347 (1980).
175. F. Halberg, M. Siffre, M. Engeli, D. Hillman, and A. Reinberg, *Compt. Rend. Acad. Sci.*, **260**, 1259–1262 (1965).
176. J. N. Mills, *Trans. Brit. Cave Res. Assoc.*, **2**, 95 (1975).
177. C. A. Czeisler, "Human Circadian Physiology: Internal Organization of Temperature, Sleep–Wake and Neuroendocrine Rhythms Monitored in an Environment Free from Time Cues," Ph.D. Dissertation, Stanford University, Stanford, CA, 1978.
178. C. A. Czeisler, G. S. Richardson, J. C. Zimmerman, M. C. Moore-Ede, and E. D. Weitzman, *Photochem. Photobiol.*, **34**, 239–247 (1981).
179. L. E. Scheving and W. S. Kals, *Time and You*, Doubleday, New York, in press.
180. H.-G. Schweiger and M. Schweiger, *Int. Rev. Cytol.*, **51**, 315–342 (1977).
181. W. J. Reitveld and G. A. Gross, "The Role of the Suprachiasmatic Nucleus: Afferents in the Central Regulation of Circadian Rhythms," in: *Biological Rhythms in Structure and Function*, H. von Mayersbach, L. E. Scheving, and J. E. Pauly, Eds., A. R. Liss, New York, 1981, pp. 205–211.
182. D. F. Kripke, *Chronobiologia*, **10**, 137 (1983).
183. K. Hoffmann, *Z. Vergl. Physiol.*, **37**, 253 (1955).
184. E. Bünning, *The Physiological Clock*, Springer-Verlag, New York, 1968.
185. R. J. Konopka and S. Beuzer, *Proc. Natl. Acad. Sci.*, **58**, 2112–2116 (1971).
186. V. G. Bruce, *Genetics*, **70**, 537–548 (1972).
187. J. F. Feldman and M. H. Hoyle, *Genetics*, **75**, 605–613 (1973).
188. J. F. Feldman and J. C. Dunlap, *Photochem. Photobiol. Rev.*, **7**, 319–368 (1983).
189. R. Barcal, J. Sova, M. Krizanovska, J. Levy, and J. Matousek, *Nature*, **220**, 1128–1131 (1968).
190. B. M. Sweeney, *Int. J. Chronobiol.*, **2**, 25–33 (1974).
191. D. Njus, F. M. Sulzman, and J. W. Hastings, *Nature*, **248**, 116–120 (1974).
192. R. D. Burgoyne, *Fed. Exp. Biol. Soc. Lett.*, **94**, 17–19 (1978).
193. H. von Mayersbach, *Arzneim.-Forsch.*, **28**, 1824–1836 (1978).
194. F. A. Brown, "The Exogenous Nature of Rhythms," in: *Chronobiology: Principles and Applications to Shifts in Schedules*, L. E. Scheving and F. Halberg, Eds., NATO Advanced Studies Institutes Series D: Behavioural and Social Sciences, No. 3, Sijthoff and Noordhoff, The Netherlands, 1980, pp. 127–135.
195. H. M. Simpson, N. Bellamy, J. Bohlen, and F. Halberg, *Int. J. Chronobiol.*, **1**, 287–311 (1973).

196. H. W. Simpson, "Chronobiotics: Selected Agents of Potential Values in Jet Lag and other Dyschronisms," in: *Chronobiology: Principles and Applications to Shifts in Schedules*, L. E. Scheving and F. Halberg, Eds., NATO Advanced Study Institutes Series D: Behavioural and Social Sciences, No. 3, Sijthoff and Noordhoff, The Netherlands, 1980, pp. 433–446.
197. E. Haus, D. J. Lakatua, J. Swoyer, and L. Sackett-Lundeen, *Am. J. Anat.*, **168,** 469–517 (1983).
198. E. Haus, D. J. Lakatua, L. Sackett-Lundeen, and J. Swoyer, "Chronobiology in Laboratory Medicine," in: *Clinical Aspects of Chronobiology*, W. Reitveld, Ed., Madition in Buaen, 1984, in press.
199. R. E. Lee, M. H. Smolensky, C. S. Leach and J. P. McGovern, *Ann. Allergy*, **38,** 231–236 (1977).
200. A. Reinberg, E. Sidi, and J. Ghata, *J. Allergy*, **36,** 273–283 (1965).
201. A. Reinberg, *Perspect. Biol. Med.*, **11,** 111–128 (1968).
202. M. H. Smolensky, A. Reinberg, and J. Queng, *Ann. Allergy*, **47,** 234–252 (1981).
203. M. H. Smolensky, A. Reinberg, R. J. Prevost, J. P. McGovern, and P. Gervais, "The Application of Chronobiological Findings and Methods to the Epidemiological Investigations of the Health Effects of Air Pollutants on Sentinel Patients," in: *Recent Advances in the Chronobiology of Allergy and Immunology*, M. H. Smolensky, A. Reinberg, and J. P. McGovern, Eds., Pergamon Press, New York, 1980, pp. 211–236.
204. A. Reinberg and F. Halberg, *Ann. Rev. Pharmacol.*, **11,** 455–492 (1971).
205. R. Takahashi, F. Halberg, and C. A. Walker, Eds., *Toward Chronopharmacology*, Pergamon Press, New York, 1982.
206. M. H. Smolensky and A. Reinberg, *Nurs. Clin. N. Am.*, **11,** 609–620 (1976).
207. I. C. Kowanko, R. Pownall, M. S. Knapp, A. J. Swannell, and P. G. C. Mahoney, *Br. J. Clin. Pharmacol.*, **11,** 477–484 (1981).
208. F. Lévi, C. LeLouarn, and A. Reinberg, *Ann. Rev. Chronopharmacol.*, **1,** 345–348, 1984.
209. V. Rejholec, V. Vitulova, and J. Vachtenheim, *Ann. Rev. Chronopharmacol.*, **1,** 357–360, 1984.
210. J. Marshall, *Stroke*, **8,** 230–231 (1977).
211. A. Kuroiwa, *Jpn. Circ. J.*, **42,** 459–476 (1978).
212. A. Reinberg, P. Guillet, P. Gervais, J. Ghata, D. Vignaud, and C. Abulker, *Chronobiologia*, **4,** 295–312 (1977).
213. F. Halberg and R. B. Howard, *Postgrad. Med.*, **24,** 349–358 (1958).
214. J. Pepys and R. J. Davies, "Occupational Asthma," in: *Allergy, Principles and Practice*, E. Middleton, C. E. Reed, and E. F. Ellis, Eds., Mosby, New York, 1978, pp. 812–842.
215. A. N. Taylor, R. J. Davies, D. J. Hendrick, and J. Pepys, *Clin. Allergy*, **9,** 213–219 (1979).
216. B. Gandevia and J. Milne, *Br. J. Ind. Med.*, **27,** 235–244 (1970).
217. A. Siracusa, F. Curradi, and G. Abbritti, *Clin. Allergy*, **8,** 195–201 (1978).
218. J. Laidlaw, *Lancet*, **2,** 1235–1237 (1956).
219. B. Lemmer, "Chronopharmacokinetics," in: *Topics in Pharmaceutical Sciences*, D. O. Breimer and Speiser, Eds., Elsevier/North-Holland Biomedicine Press, Amsterdam, 1981, pp. 49–68.
220. B. Lemmer, *Chronopharmakologie. Tagesrhythmen und Arzneimittel-Wirkung*, Wissenschaft Verlagsgesellschft. MbH, Stuttgart, 1983.
221. A. Reinberg, M. H. Smolensky, and G. Labrecque, Eds., *Annual Review of Chronopharmacology*, Pergamon Press, New York, 1984.
222. A. Reinberg, "Clinical Chronopharmacology: An Experimental Basis for Chronotherapy," in: *Biological Rhythms and Medicine*, A. Reinberg and M. H. Smolensky, Eds., Springer-Verlag, New York, 1983, pp. 211–263.
223. J. V. Bruckner, R. Luthra, G. M. Kyle, S. Muralidhara, R. Ramanathan, and D. Acosta, *Ann. Rev. Chronopharmacol.*, **1,** 373–376, 1984.

224. J. G. Lavigne, P. M. Belanger, F. Dore, and G. Labrecque, *Toxicology*, **26**, 267–273 (1983).
225. B. J. Craft, "The Effects of Circadian Rhythms on the Toxicological Response of Rats to Xenobiotics," Ph. D. Dissertation, University of Michigan, Ann Arbor, 1970.
226. J. Clench, A. Reinberg, Z. Dziewanowska, J. Ghata, and M. H. Smolensky, *Eur. J. Clin. Pharmacol.*, **20**, 359–369 (1981).
227. A. Markiewicz, K. Semenowicz, J. Korczynska, and H. Boldys, "Temporal Variations in the Response of Ventilatory and Circulatory Functions to Propranolol in Healthy Man," in: *Recent Advances in the Chronobiology of Allergy and Immunology*, M. H. Smolensky, A. Reinberg, and J. P. McGovern, Eds., Pergamon Press, New York, 1980, pp. 185–193.
228. A. Reinberg, J. Clench, J. Ghata, F. Halberg, C. Abulker, J. Dupont, and Z. Zagula-Mally, *C. R. Acad. Sci.*, **280**, 1697–1700 (1975).
229. L. Carosella, P. DiNardo, R. Bernabei, A. Cocchi, and P. Carbonin, "Chronopharmacokinetics of Digitalis: Circadian Variations of Beta-Methyl-Digoxin Serum Levels After Oral Administration," in: *Chronopharmacology*, A. Reinberg and F. Halberg, Eds., Pergamon Press, New York, 1980, pp. 125–134.
230. A. Reinberg and M. Reinberg, *Naunyn Schmiedebergs Arch. Pharmacol.*, **297**, 149–159 (1977).
231. K. DeVries, J. T. Goei, H. Booy-Noord, and N. G. M. Orie, *Int. Arch. Allergy*, **20**, 93–101 (1962).
232. G. J. Tammeling, K. DeVries, and E. W. Kruyt, "Circadian Pattern of Bronchial Reactivity to Histamine in Healthy Subjects and in Patients with Obstructive Lung Disease," in: *Chronobiology in Allergy and Immunology*, J. P. McGovern, M. H. Smolensky, and A. Reinberg, Eds., C. Thomas, Springfield, IL, 1977, pp. 139–149.
233. A. Reinberg, J. Clench, N. Aymard, M. Gaillot, R. Bourdon, P. Gervais, C. Abulker, and J. Dupont, *C. R. Acad. Sci.*, **278**, 1503–1505 (1974).
234. A. Reinberg, J. Clench, N. Aymard, M. Gaillot, R. Bourdon, P. Gervais, C. Abulker, and J. Dupont, *J. Physiol.* (Paris), **70**, 435–456 (1975).
235. L. E. Scheving, "Circadian Rhythms in Cell Proliferation: Their Importance when Investigating the Basic Mechanism of Normal Versus Abnormal Growth," in: *Biological Rhythms in Structure and Function*, H. von Mayersbach, L. E. Scheving, and J. Pauly, Eds., A. R. Liss, New York, 1981, pp. 39–79.
236. L. E. Scheving, H. von Mayersbach, and J. E. Pauly, *Eur. J. Toxicol.*, **7**, 203–227 (1974).
237. L. E. Scheving, D. F. Vedral, and J. E. Pauly, *Nature*, **219**, 612–622 (1968).
238. J. Fisch, A. Yonovitz, and M. H. Smolensky, *Ann. Rev. Pharmacol.*, **1**, 385–388, 1984.
239. H. Bafitis, M. H. Smolensky, B. Hsi, S. Mahoney, and H. Kresse, *J. Pharmacol. Toxicol.*, **11**, 251–258 (1978).
240. N. K. Synder, M. H. Smolensky, and B. P. Hsi, *Chronobiologia*, **8**, 33–44 (1980).
241. G. M. Kyle, M. H. Smolensky, and J. P. McGovern, "Circadian Variation in the Susceptibility of Rodents to the Toxicity Effects of Theophylline," in: *Chronopharmacology*, A. Reinberg and F. Halberg, Eds., Pergamon Press, Oxford, 1979, pp. 239–244.
242. T. H. Tsai, L. E. Scheving, and J. E. Pauly, *Jpn. J. Physiol.*, **20**, 12–29 (1970).
243. T. H. Tsai, L. E. Scheving, and J. E. Pauly, "Circadian Variation in Host Susceptibility to Mercuric Chloride and Paraquat in Balb/Cann Female Mice," in: *Toward Chronopharmacology*, R. Takahashi, F. Halberg, and C. A. Walker, Eds., Pergamon Press, New York, 1982, pp. 249–255.
244. H. von Mayersbach, "An Overview of the Chronobiology of Cellular Morphology," in: *Biological Rhythms and Medicine*, A. Reinberg and M. H. Smolensky, Eds., Springer-Verlag, New York, 1983, pp. 47–78.
245. D. G. Hof, J. D. Dexter, and C. E. Mengel, *Aerospace Med.*, **42**, 1293–1296 (1971).
246. E. Haus, F. Halberg, M. K. Loken, and Y. S. Kim, "Circadian Rhythmometry of Mammalian Radiosensitivity," in: *Space Radiation Biology and Related Topics*, C. A. Tobias and P. Todd, Eds., Academic Press, New York, 1974, pp. 435–474.

247. E. Haus, "Biological Aspects of a Chronopathology," Ph.D. Dissertation, University of Minnesota College of Medicine, 1970.
248. D. L. Clayton, A. W. McMullen, and C. C. Barnett, *Chronobiologia*, **2**, 210–217 (1975).
249. O. Müller, "Circadian Rhythmicity and Response to Barbituates," in: *Chronobiology*, L. E. Scheving, F. Halberg, and J. E. Pauly, Eds., Igaku Shoin, Tokyo, 1974, pp. 187–190.
250. J. E. Pauly and L. E. Scheving, *Int. J. Neuropharmacol.*, **3**, 651–658 (1965).
251. I. Sauerbier, "Circadian System and Teratogenicity," in: *Progress in Clinical and Biomedical Research*, H. von Mayersbach, L. E. Scheving, and J. E. Pauly, Eds., A. R. Liss, New York, 1980, pp. 143–149.
252. A. P. Chaudhry and F. Halberg, *J. Dent. Res.*, **39**, 704 (1960).
253. F. Halberg, *Mkurse. Arztl. Forbild.*, **14**, 67 (1964).
254. O. H. Iversen and S. L. Kauffman, *Int. J. Chronobiol.*, **8**, 95–104 (1982).
255. V. Nair, "Circadian Rhythm in Drug Action; A Pharmacologic, Biochemical and Electromicroscopic Study," in: *Chronobiology*, L. E. Scheving, F. Halberg, and J. E. Pauly, Eds., Igaku Shoin, Tokyo, 1974, pp. 182–186.
256. C. North, R. J. Feuers, L. E. Scheving, J. E. Pauly, T. H. Tsai, and D. A. Casciano, *Am. J. Anat.*, **162**, 183–199 (1981).
257. R. J. Feuers, L. A. Scheving, R. R. Delongchamp, T. H. Tsai, D. A. Casciano, J. E. Pauly, and L. E. Scheving, *Chronobiologia*, **10**, 125–126 (1983).
258. L. A. Scheving, R. J. Feuers, L. E. Scheving, R. R. DeLongchamp, T. H. Tsai, D. A. Casciano, and J. E. Pauly, *Chronobiologia*, **10**, 155–156 (1983).
259. E. Haus, F. Halberg, L. E. Scheving, S. Cardoso, A. Kühl, R. Sothern, R. Shiotsuka, D. S. Hwang, and J. E. Pauly, *Science*, **177**, 80–82 (1972).
260. F. Lévi, W. Hrushesky, E. Haus, F. Halberg, L. E. Scheving, and B. J. Kennedy, "Experimental Chrono-oncology," in: *Chronobiology: Principles and Applications to Shifts in Schedules*, L. E. Scheving and F. Halberg, Eds., NATO Advanced Study Institutes Series D: Behavioral and Social Sciences, No. 3, Sijthoff and Noordhoff, The Netherlands, 1980, pp. 481–512.
261. E. Haus, G. Fernandes, J. Kühl, E. J. Yunis, J. K. Lee, and F. Halberg, *Chronobiologia*, **3**, 270–277 (1974).
262. J. F. K. Kühl, E. Haus, F. Halberg, L. E. Scheving, J. E. Pauly, S. Cardoso, and G. Rosene, *Chronobiologia*, **1**, 316–317 (1974).
263. L. E. Scheving, S. S. Cardoso, J. E. Pauly, F. Halberg, and E. Haus, "Variations in Susceptibility of Mice to the Carcinostatic Agent Arabinosyl Cytosine," in: *Chronobiology*, L. E. Scheving, F. Halberg, and J. E. Pauly, Eds., Igaku Shoin Ltd., Tokyo, 1974, pp. 213–217.
264. S. S. Cardoso, T. Avery, J. M. Venditti, and A. Goldin, *Eur. J. Cancer*, **14**, 949–954 (1978).
265. L. E. Scheving, J. E. Pauly, T. H. Tsai, and L. A. Scheving, "Chronobiology of Cellular Proliferation: Implications for Cancer Chemotherapy," in: *Biological Rhythms and Medicine*, A. Reinberg and M. H. Smolensky, Springer-Verlag, New York, 1983, pp. 79–130.
266. L. E. Scheving, E. R. Burns, J. E. Pauly, and F. Halberg, *Cancer Res.*, **40**, 1511 (1980).
267. L. E. Scheving and J. E. Pauly, "Several Problems Associated with the Conduct of Chronobiological Research," in: *Die Zeit und das Leben*, J. H. Scharf and H. von Mayersbach, Eds., *Nova Acta Leopoldina*, **46**, 237–258 (1977).
268. I. Zucker, *Physiol. Behav.*, **6**, 115–126 (1971).
269. J. P. McGovern, M. H. Smolensky, and A. Reinberg, "Circadian and Circamensual Rhythmicity in Cutaneous Reactivity to Histamine and Allergenic Extracts," in: *Chronobiology in Allergy and Immunology*, J. P. McGovern, M. H. Smolensky, and A. Reinberg, Eds., C. Thomas, Springfield, IL, 1977, pp. 76–116.
270. S. Flannigan, "Cutaneous Reactivity to Contact Irritants," Master's Thesis, The University of Texas School of Public Health, 1981.

271. A. Reinberg, P. Gervais, M. Morin, and C. Abulker, *C. R. Acad. Sci.*, **272**, 1879–1881 (1971).
272. P. Gervais, A. Reinberg, C. Gervais, M. H. Smolensky, and O. DeFrance, *J. Allergy Clin. Immunol.*, **59**, 207–213 (1977).
273. E. Haus and F. Halberg, "Endocrine Rhythms," in: *Chronobiology: Principles and Applications to Shifts in Schedules*, L. E. Scheving and F. Halberg, Eds., NATO Advanced Study Institutes Series D: Behavioral and Social Sciences, No. 3, Sijthoff and Noordhoff, The Netherlands, 1980, pp. 137–188.
274. F. Ungar and F. Halberg, *Science*, **137**, 1058 (1962).
275. A. Reinberg, W. Dupont, Y. Touitou, M. Lagoguey, P. Bourgeois, C. Touitou, G. Murianx, D. Przyrowsky, S. Guillemant, J. Guillemant, L. Briere, and B. Zean, *Chronobiologia*, **8**, 11–31 (1981).
276. A. Reinberg, S. Guillemant, N. J. Ghata, J. Guillemant, Y. Touitou, W. Dupont, M. Lagoguey, P. Bourgeois, L. Briere, G. Fraboulet, and P. Guillet, *Chronobiologia*, **7**, 513–523 (1980).
277. C. A. Czeisler, M. Moore-Ede, and R. M. Coleman, *Science*, **217**, 460–463 (1982).
278. F. Halberg, E. A. Johnson, W. Nelson, W. Runge, and R. Sothern, *Physiol. Teacher*, **1**, 1–11 (1972).
279. P. Gervais, A. Reinberg, C. Fraboulet, C. Abulker, O. Vignaud, and M. E. R. Delcourt, "Circadian Changes in Peak Expiratory Flow of Subjects Suffering from Allergic Asthma Documented Both in Areas of High and Low Air Pollution," in: *Chronopharmacology*, A. Reinberg and F. Halberg, Eds., Pergamon Press, Oxford, 1979, pp. 203–212.
280. D. R. Glasgow, L. E. Scheving, J. E. Pauly, and J. A. Bruce, *J. Arkansas Med. Soc.*, **79**, 81–91 (1982).
281. A. Reinberg, N. Vieux, A.-J. Chaumont, A. Laporte, M. Smolensky, A. Nicolai, C. Abulker, and J. Dupont, "Aims and Conditions of Shift Work Studies," in: *Chronobiological Field Studies of Oil Refinery Shift Workers*, A. Reinberg, Ed.; *Chronobiologia*, **6**, (Suppl. 1), 7–23 (1979).
282. A. Schuster, *Trans. Cambridge Phil. Soc.*, **18**, 107–135 (1900).
283. F. Koehler, F. K. Okano, L. R. Elbeback, F. Halberg, and J. J. Bittner, *Exp. Med. Surg.*, **14**, 5–30 (1956).
284. J. W. Tukey, "The Sampling Theory of Power Spectrum Estimates," in: *Proceedings of Symposium on Applications of Autocorrelation Analysis to Physical Problems*, Woods Hole, Massachusetts, Office of Naval Research, Washington, DC, 1949, pp. 46–67.
285. F. Halberg and H. Panofsky, *Exp. Med. Surg.*, **19**, 284–309 (1961).
286. F. Halberg, H. Panofsky, and H. Mantis, *Ann. NY Acad. Sci.*, **117**, 254–270 (1964).
287. J. DePrins and G. Cornelissen, "Methods Workshop," in: *Chronobiology: Principles and Applications to Shifts in Schedules*, L. E. Scheving and F. Halberg, Eds., NATO Advanced Study Institutes Series D: Behavioral and Social Sciences, No. 3, The Netherlands, Sijthoff and Noordhoff, 1980, pp. 249–260.
288. M. J. K. Blake, "Treatment and Time of Day," in: *Biological Rhythms and Human Performance*, W. P. Colquhoun, Ed., Academic Press, London, 1971, pp. 109–148.
289. W. P. Colquhoun and S. Folkard, *Ergonomics*, **21**, 811–817 (1978).
290. O. Oestberg, *Ergonomics*, **16**, 203–209 (1973).
291. M. H. Smolensky, "The Conceptual Implication of Chronotoxicology and Chronopathology for Occupational Health and Shift Work," in: *Chronobiology; Principles and Applications to Shifts in Schedules*, L. E. Scheving and F. Halberg, Eds., NATO Advanced Study Institutes Series D: Behavioural and Social Sciences, No. 3, Sijthoff and Noordhoff, The Netherlands, 1980, pp. 325–337.
292. D. Brown, *J. Human Ergol.*, **11** (Suppl.), 475–482 (1982).
293. J. Aschoff, *Ergonomics*, **39**, 739–754 (1978).

294. R. Leonard, "Amplitude of the Temperature Circadian Rhythm and Tolerance to Shift Work," in: *Night and Shiftwork Studies: Biological and Social Aspects*, A. Reinberg, N. Vieux, and P. Andlauer, Eds., Pergamon Press, Oxford, 1981, pp. 323–329.
295. C. Michel-Briand, J. L. Chopard, A. Guiot, M. Paulmeier, and G. Struder, "The Pathological Consequences of Shift-Work in Retired Workers," in: *Night and Shiftwork Studies: Biological and Social Aspects*, A. Reinberg, N. Vieux, and P. Andlauer, Eds., Pergamon Press, Oxford, 1981, pp. 399–407.
296. C. F. Ehret, K. R. Groh, and J. C. Meinert, "Circadian Dyschronism and Chronotypic Ecophilia as Factors in Aging and Longevity," in: *Aging and Biological Rhythms*, H. V. Samis and S. Capobianco, Eds., Plenum Press, New York, 1978, pp. 185–214.
297. C. F. Ehret, V. R. Potter, and K. W. Dobra, *Science*, **188**, 1212–1215 (1975).
298. C. F. Ehret and V. R. Potter, *Int. J. Chronobiol.*, **2**, 321–325 (1974).
299. A. L. Cahill and C. F. Ehret, *J. Neurochemistry*, **37**, 1109–1115 (1981).
300. A. L. Cahill and C. F. Ehret, *Am. J. Physiol.*, **243**, R218 (1982).
301. N. D. Horseman and C. F. Ehret, *Am. J. Physiol.*, **243**, R373 (1982).
302. C. F. Ehret and L. W. Scanlon, *Overcoming Jet Lag*, Berkley Books, New York, 1983.
303. A. Reinberg, "Chronobiology and Nutrition," in: *Biological Rhythms and Medicine*, A. Reinberg and M. H. Smolensky, Eds., Springer-Verlag, New York, 1983, pp. 266–300.
304. R. T. Wilkson, "The Relationship between Body Temperature and Performance Across Circadian Phase Shifts," in: *Rhythmic Aspects of Behavior*, F. Brown and R. C. Graeber, Eds., LEA Publications, New Jersey, 1982, pp. 213–240.
305. W. P. Colquhoun, "Circadian Variations in Mental Efficiency," in: *Biological Rhythms and Human Performance*, W. P. Colquhoun, Ed., Academic Press, London, 1971, pp. 39–107.
306. M. I. Härma, J. Ilmarinen, and I. Yletyinen, *J. Human Ergol.*, **11** (Suppl.), 33–46 (1982).
307. R. C. Browne, *Occup. Psychol.* **21**, 121 (1949).
308. O. Prokop and L. Prokop, *D. Zeitschrift Gesamte Gerichtlichen Medizin*, **44**, 343 (1955).
309. G. Hildebrandt, W. Rohmert, and J. Rutenfranz, *Int. J. Chronobiol.*, **2**, 175–180 (1974).
310. S. Folkard, T. H. Monk, and M. C. Lobban, *Ergonomics*, **21**, 785–799 (1978).
311. P. Andlauer and B. Metz, "Le travail en equipes alternantes," in: *Physiologie du Travail—Ergonomie II*, J. Scherrer, Ed., Masson, Paris, 1967, pp. 272–281.
312. A. Meers, "Performance on Turns of Duty within a Three-Shift System and Its Relation to Body Temperature—Two Field Studies," in: *Experimental Studies of Shiftwork*, W. P. Colquhoun, S. Folkard, P. Knauth, and J. Rutenfranz, Eds., Westdeutscher-Verlag, Opladen, West Germany, 1975, pp. 188–205.
313. D. J. Paustenbach, G. P. Carlson, J. E. Christian and G. S. Born. *Fund. Appl. Toxicol.* (Accepted for publication in 1986).

CHAPTER SIX

Work Costs and Work Measurements

STEVEN M. HORVATH, Ph. D.

Human performance needs to be evaluated on three general levels: (1) determination of the maximum capability of humans to perform, (2) evaluation of the percentage of the maximal human capacity required to perform in a particular work situation, and (3) determination of human ability to perform for sustained periods of time at the submaximal levels required in the workplace. It would also be important to determine whether years of performing at certain high levels of work have a detrimental influence on the worker's health. Unfortunately, this latter factor has been the least studied and the most difficult to evaluate since long-term studies on workers are few in number and subject to the lack of critical evaluation and basic background history. A study conducted retrospectively on longshoremen does indicate that the type, level, and duration of job-related physical activity is associated with the presence of cardiovascular disease. Additional long-term studies are essential for clarification of this issue. The data on the longshore workers suggest that short, intensive bouts of activity continued over the years may reduce the incidence of cardiovascular disease. In order to substantiate the observations of this epidemiological observation, it is essential that studies on the capacity and the degree to which human capability is utilized in the actual workplace be initiated and available for retrospective analysis. The material presented in this chapter deals with methods designed to measure performance in relation to capability and provide illustrative examples of studies that have been conducted in the workplace. Unfortunately, most of the available information in the workplace has been obtained on men, and relatively little is known as to how women perform in the work situation.

Although sporadic attempts to evaluate the physical fitness of various populations have been conducted for many years, serious interest in methods designed to evaluate endurance fitness arose only during world War II. Since that time an everincreasing number of tests to evaluate criteria for fitness have been developed. There are two favored procedures: the first designed to directly measure the maximum aerobic power by measuring maximal oxygen uptake and the second utilizing measurements of heart rate and oxygen uptake during submaximal levels of exercise to predict the maximal capacity of the individual. The basis for all tests of endurance fitness lies in the concept that the overall determinant of endurance is the ability of the body to transport oxygen from the atmosphere to the working cells and incidentally to remove the end products of the biochemical reactions in the cells away from the active cells. However, this cannot be accomplished completely and the body can go into an oxygen debt, which can be repaid after the work is completed. These procedures are ideally designed to measure cardiovascular fitness while other tests are available to determine strength and flexibility.

1 PHYSICAL FITNESS TESTS

1.1 Relative Work Loads

The type of exercise performed (dynamic or static) determines to some degree the cardiorespiratory effects produced. Dynamic exercise (walking, running, swimming, etc.) results in a certain amount of external work, whereas no mechanical work is produced by static exercise. The respiratory responses are quite different. In dynamic work ventilation increases until a steady state is achieved, whereas in static work ventilation may be temporarily suppressed or does not exceed resting levels and the hyperpnea may be delayed until after the work ceases.

Physical activity during dynamic muscular activity can be roughly defined as rest, light work, moderate work, heavy work, and very heavy work (Table 6.1). This grading is more applicable to young subjects if absolute work loads are used to define these terms, but they are usable as approximations for individuals of different ages if considered in terms of relative loads in respect to maximum aerobic capacity. In young adult males the relative work load can also be related to the heart rate accompanying work requiring short periods of time. In the case of prolonged work periods (up to and above 2 to 4 hrs) heart rates tend to increase with duration of work. Heart rates (beats/min) for light work are generally less than 100, whereas for moderate work there is a range of 100 to 124, for heavy work the range is from 124 to 150, and for very heavy work heart rates exceed 150. A more precise method is to express the work load cost (in terms of oxygen uptake per minute) as a percentage of the maximum oxygen uptake. This relative work index provides some correction for and an appreciation of the differences due to age, sex, and work capacity. Another frequently utilized index is to express work loads as metabolic equivalents, that

WORK COSTS AND WORK MEASUREMENTS

Table 6.1 Classification of Work Levels and Associated Physiological Costs

	Work Load					
	Very Light	Light	Moderately Heavy	Heavy	Very Heavy	Extremely Heavy
Oxygen uptake (liters/min)	<0.5	0.5–1.0	1.0–1.5	1.5–2.0	2.0–2.5	>2.5
kcal/min	<2.5	2.5–5.0	5.0–7.5	7.5–10.0	10.0–12.5	>12.5
Heart rate (beats/min)	—	75–100	100–125	125–150	150–175	>175
Rectal temperature (°C)	37.0	37.0	37.5–38.0	38.0–38.5	38.5–39.0	>39.0
Sweating rate (ml/hr) (average 8-hr workday)	—	—	200–400	400–600	600–800	>800

is, levels above resting metabolic requirements. The resting level is expressed as 1 Met and activity levels as 2, 3, 4, ..., (n) Met.

Static work (isometric contractions) such as lifting and holding heavy weights or exerting a constant tension induce physiological reactions very different from those observed during dynamic activity. Depending on the level of the static effort, blood flow in the contracting muscles is impeded by mechanical compression induced by the muscle activity. If muscle contractions are sustained or exceed some 30 percent of these muscles' maximum contraction, blood flow to the muscles may completely cease, resulting in a diminished ability to maintain muscular force. Oxygen supplies to these tissues are reduced, and end products of biochemical reactions, such as lactates, increase and effort is discontinued with the accompaniment of extreme pain. A cardiovascular reflex response (1, 2) occurs, resulting in large increases in systolic and diastolic blood pressure and moderate increases in heart rates and cardiac output. Static work occurs most frequently with upper body (arm work) and in addition to its cardiovascular effects induces changes in the respiratory system. Classification of static work is not precise, although for certain kinds of effort such as in arm work it has been expressed as a percentage of maximal grip strength. Systemic arterial pressure rises with dynamic exercise. Brachial pressure is raised during exercise in linear proportion to the increase in oxygen cost of the work load. Systolic pressure rises to as high as 200 torr, but since diastolic pressure changes only slightly, the mean blood pressure is only slightly elevated.

The effect of exercise on blood pressure is markedly influenced by the type of exercise performed, as is evidenced by the marked increase observed in individuals performing sustained contractions even if the isometric work being performed is as low as 20 percent of the maximum. These latter effects appear to be induced by a reflex. It should also be noted that this reflex also induces relatively marked heart rate increases considering the small mass of muscle involved and the total energy produced in comparison to the heart rate responses at equivalent energy expenditures during performance of dynamic work. It is clear that isometric work such as lifting heavy objects induces a greater cardiovascular stress. This suggests that individuals with cardiovascular diseases who are required to perform isometric work may be at some additional risk. Åstrand et al. (3) demonstrated this response when they compared the blood pressure responses when men were hammering nails into a ceiling with the pressures observed when they were hammering nails into wood at bench heights.

1.1.1 Indirect Measurement of Maximal Aerobic Power

Direct measurements of maximum oxygen uptake (undoubtedly the best if properly conducted) cannot be utilized indiscriminately on all population groups. Individuals with various cardiorespiratory diseases need to be carefully medically screened prior to any exercise testing and probably should even then be approached with caution if direct maximal aerobic power tests are to be employed. Because of these and factors related to proper surveillance of the test, the use of indirect estimates of maximum oxygen uptake are more commonly employed. It is also more difficult to determine whether a maximum effort has occurred. In most cases, the subject's self-determination of the inability to continue has been the criterion employed. More precise attempts to determine the attainment of this maximum have utilized certain physiological criteria such as (a) a respiratory gas exchange ratio exceeding 1.0, (b) the attainment of a plateau in oxygen uptake, that is, where the last few minutes of oxygen uptake prior to the termination of the test is within 100 to 200 ml/min or 2 ml/kg body weight, (c) a decrease in the oxygen uptake in the last minute of the test, (d) sudden decreases in blood pressure, and (e) a high blood lactate level. In individuals suffering from cardiovascular disease no real determination of maximum values can be obtained since in these subjects alterations in the electrocardiogram (ECG) result in termination of the test. However, such changes give some indication of this individual's maximum capacity in the sense that occurrence of ECG changes reflect the limits of performance.

Most indirect measurements of maximum oxygen uptake are based on several assumptions: (a) that submaximal oxygen uptakes are linear up to maximum levels; (b) that heart rate is linearly related to oxygen uptake over a wide range of work loads (Figure 6.1); and (c) that there is a constant maximum heart rate for any particular population. [For example, the most common formula used is maximum heart rate = 200 − age (years), and so for an individual of 20 years, the maximum heart rate is 180 beats/min whereas for a 60-year-old

WORK COSTS AND WORK MEASUREMENTS

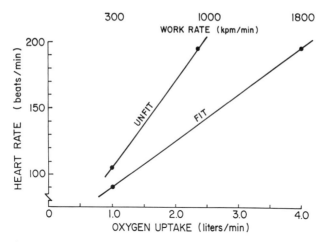

Figure 6.1 The relationship between heart rate and oxygen uptake (level of work) observed in fit and unfit individuals.

individual the maximum rate would be 140 beats/min.] This assumption as to maximum heart rate does not allow for individual variation.

1.1.2 Measurement of Energy Cost

Evaluation using Time–Motion Studies. This technique has been employed to determine the individual's total daily expenditure. It is dependent on two factors: (a) knowledge of the energy expenditure for any task, which is frequently not known but is estimated on the basis of data such as presented in Table 6.2 (there is considerable individual variation in these estimates); and (b) accurate recording of the time the subject devotes to each activity. This method is tedious and can be further complicated by the number and variety of the tasks being performed. Further complications are related to the reliability of the observer and whether the observed subject is performing the tasks in the usual manner. An additional factor complicating the final results relates to the number of individuals studied and the frequency with which a subject is observed. Ambient environmental conditions may also vary from one observation period to another. Despite all these caveats, it is possible that with trained observers and frequent repetitions, energy expenditures can be consistently recorded within 10 percent of the actual cost. A less accurate procedure is to request the worker to keep a personal activity record and use this to estimate energy expenditures. A procedure that has been employed with success in studies on Alcoa smelter workers is to follow individual workers for successive days (7 to 14) and record not only time devoted to a task but simultaneously measure heart rates and body temperatures (5). Simultaneously with these observations oxygen uptake measurements were made, and consequently more precise information as to energy expenditure was obtained (see Table 6.3).

Table 6.2 Duration of Work (mins) at Various Levels of Maximal Aerobic Capacity by Men Workers Having Different Levels of Physical Fitness

	Trained[a]	Average[a]	Untrained[a]
Duration of effort (mins)	480	480	480
Percentage of maximal aerobic power	50	35	25
Duration of effort (mins)	240	240	240
Percentage of maximal aerobic power	70	40	30
Duration of effort (mins)	120	120	120
Percentage of maximal aerobic power	80	50	35
Duration of effort (mins)	60	60	60
Percentage of maximal aerobic power	85	65	50
Duration of effort (mins)	30	30	30
Percentage of maximal aerobic power	95	80	65
Duration of effort (mins)	10	10	10
Percentage of maximal aerobic power	100	95	90
Duration of effort (mins)	6	3	1
Percentage of maximal aerobic power	100	100	100

[a] Maximal aerobic capacity (trained, 58; average, 43; untrained, 32) in ml $O_2/kg \cdot min^{-1}$.

Oxygen Uptake Procedures. Two excellent presentations providing information as to the basic procedures to measure the maximal aerobic capacity (power) of humans are available (6, 7). Maximum aerobic power has been assessed by measurement of the highest oxygen uptake obtained in dynamic muscular exercise. Direct measurement is based on performance of muscular exercise with increasing intensity until a work rate is established beyond which a further increase in work output does not induce any additional increase (or in fact a decrease in oxygen uptake). If one determines the individual's maximum aerobic power, that is, potential for oxygen transport, one can predict human potential for performing various intensities of muscular work. These measures of maximum capacity do not guarantee good performance on a specific job. Technique, strength, motivation, and experience may be important modifying factors, although the limits of performance capability are definable.

There are a number of suggested test procedures available for determining the maximum aerobic power. In general, they employ either of two devices to induce increased work effort, that is, treadmills or bicycle ergometers. Treadmill test procedures utilize either walking or running and are either continuous at a fixed speed and elevation or are conducted at fixed speeds and progressively increasing elevations. The following describes some of the treadmill tests employed for normal, healthy individuals. It should be noted that modifications of these tests have been made in order to evaluate individuals suffering from various disease states. Measurements made on subjects tested vary from simple to complex. The simplest tests measure only heart rates, whereas the more complex and complete tests determine oxygen uptakes (ml/min), carbon dioxide

Table 6.3 Summary of Energy Expenditures[a] of Aluminum Workers in Smelting Plant Expressed as a Multiple of Basal Metabolic Rate[b]

Job Description	Plant 1	Range	Plant 2	Range
Prebake Plant				
Walking	3.0	(2.5–3.5)	2.5	(2.0–3.0)
Sitting			1.5	(1.1–1.6)
Using 14-lb hammer to break carbon butts	5.5			
Crowbar breaking crust	6.5	(6.0–7.5)	5.5	
Crowbar on carbon	8.0	(7.5–8.5)		
Picking up and throwing carbon butt	8.5			
Rowelling pot	5.5		4.5	
Setting siphon	3.4		4.9	
Working around crucible	4.9		4.0	
Skimming metal in crucible	6.0	(5.5–6.5)		
Setting carbons	3.5	(3.1–8.0)	3.0	
Working carbons into alignment	4.0	(3.0–8.0)	3.6	
Ored 2 carbons	4.5	(4.1–5.1)		
Ored 3 carbons	5.5	(4.5–6.9)		
Pneumatic crust breaker	4.0	(3.0–5.1)		
Remove radiant shield			5.5	
Unhook carbons			3.0	
Loosen bolts on carbons with air gun			4.0	
Cleaned 1½ carbon butts with jackhammer			5.0	
Using jackhammer to clean crucible			4.8	(2.8–6.0)
Riding tricycle at inspection speed			2.5	
Soderborg Plant				
Pneumatic crust breaker			3.7	(2.6–3.8)
Walking, sweeping, and oreing			3.1	(2.9–3.4)
Crust breaking with hand jackhammer			4.5	(4.3–4.6)
Standing recovery from breaking crust			2.5	

[a] Average energy expenditure = 35 percent maximum aerobic capacity.
[b] Adapted from References 4 and 5.

production, minute ventilation, multiple electrocardiograms, heart rates, blood pressures, blood gases, cardiac output, and a variety of blood biochemical substances. There are a number of significant changes in the cardiovascular system as the level of work increases. Figure 6.2 illustrates some of the changes observed as work loads progressively increase from rest to near-maximal effort. Although not discussed in this chapter, these changes are important factors in determining human ability to perform in hot and cold environments.

The development of treadmill exercise testing at the Harvard Fatigue Laboratory to determine maximum aerobic power has been described by Horvath and Horvath (8). This test consisted of a sitting rest period prior to a

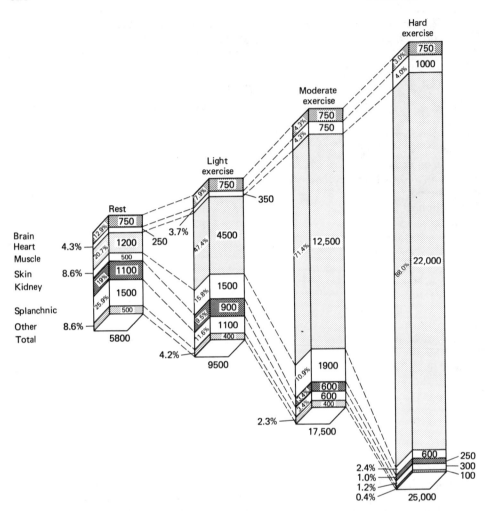

Figure 6.2 Cardiovascular changes and regional (oxygen) blood flow as observed at various intensities of exercise. Note the changes in muscle, hepatic, and renal blood flow as exercise level increases.

15-min walk (a submaximal effort) on a motor-driven treadmill at 3.5 miles per hour (mph) and followed by running at 7.5 mph and an 8.6 percent grade for 5 min or to the point of exhaustion. The duration of the run was important. Oxygen uptakes were determined continuously. In general, if the individual had actually attained maximum level, the last oxygen uptake value would be less or equal to the previous one.

A more precise criterion for the maximum was developed by Taylor et al. (9), who suggested that oxygen uptake determinations in consecutive order separated by a grade of 2.5 percent should be within 150 ml of each other. They also modified the test procedure in that the speed was maintained at 7 mph. The grade was raised from zero 2.5 percent every 3 min. In view of the difficulty of having all subjects run at these various speeds, and especially to permit studies to be conducted on children and aging adults, a walking test [modified Balke (10)] has been utilized at the Institute of Environmental Stress to study individuals into their ninetieth year. This test requires subjects to walk at a fixed speed (2 to 3 mph) and the treadmill elevation is rasied one percent every minute. The main disadvantage of this test is that very fit subjects could walk up to 30 min and the final steep grades induced considerable muscle problems.

The second exercising device utilized bicycle ergometers. These devices were in common use in Scandinavian laboratories prior to the availability of treadmills. They became a favorite tool (possibly related to cost) after the results of studies by Åstrand and Rhyming (11) became more widely known. The procedure followed in their testing was as follows. A practice session (or several) to acquaint the subjects with the bicycle and other measuring devices was undertaken. Heart rates, minute ventilations, oxygen uptakes, and carbon dioxide productions were monitored, usually at 30- or 60-sec intervals. A warm-up period of low-intensity work (00 kpm/min) preceded the minute-by-minute increments in load of 150 kpm and continuing to attainment of maximum oxygen uptakes. The pedaling frequency was 50 revolutions per minute (rpm). A modification of this progressive test (11) conducted at various loads (900, 1200, 1500, 1800, up to 2400 kpm/min) estimated to produce exhaustion within 4 to 8 min. The definitive test accompanied by all metabolite measurements was carried out on another subsequent day. Other modifications of tests utilizing the bicycle ergometer employed fixed work loads for periods of 4 to 6 min, a period of no work followed by another 4 to 6 min bout at a higher work load. This sequence may be repeated as required to attain maximum values.

Since subjects are required to pedal against a fixed external load or resistance, individuals with smaller leg muscle mass are at some disadvantage. Peripheral muscle fatigue may occur before cardiorespiratory limitations are reached. It should be further noted that in general maximal oxygen uptake determined on bicycle ergometers are some 10 percent lower than those measured utilizing treadmills. However, there are advantages to using bicycle ergometers. They are not costly, can be easily transported to test sites, and, most importantly, total work performed can be precisely calculated.

On some occasions these more complex but complete tests can be modified so as to provide an estimate of maximum aerobic capacity. Typically the bicycle ergometer can be used with the subject working at two submaximal levels while either or both heart rates and oxygen uptakes are measured. Åstrand and Rhyming (11) have developed nomograms that can be utilized to predict maximum capacity. A modification of the nomogram is available to convert for age differences. The predictions may be incorrect by ±10 percent in some instances.

In view of the concern that age and sex may not be adequately addressed by the Åstrand and Rhyming procedures, several modifications of the test procedure have been suggested. The most recent and probably the best of these modifications has been presented by Siconolfi et al. (1982) (12). Their procedure provides for, at least for the general population, adjustments in load for men and women and age. In the test for men under 35 years, the initial exercise load (bicycle ergometer at 50 rpm) began at 49.0 W (300 kpm/min), whereas for men 35 years and older and all women that initial load began at 24.5 W (150 kpm/min). After 2 min of exercise the work load was respectively increased by 49.0 W or 24.5 W for the younger men and older men and women. The work time was again 2 min in duration. The work load continued to be increased by these loads until the heart rate attained a threshold level of 70 percent of the predicted level assumed to be 200 beats minus age was achieved. The $\dot{V}O_{2max}$ was then estimated from the Åstrand–Rhyming nomogram by use of the mean steady-state heart rate and the final exercise heart rate. The multiple-regression equations developed by Siconolfi et al. (12) for men are

$$Y = 0.348(X_1) - 0.035(X_2) + 3.011$$

and for women

$$Y = 0.302(X_1) - 0.019(X_2) + 1.593$$

where Y = max $\dot{V}O_2$ (liters/min), X = $\dot{V}O_2$ (liters/min) from the nomogram and X = age.

These predictive equations had R values of 0.86 and 0.97 with standard errors of estimate of 0.359 liters/min and 0.199 liters/min, respectively.

Another approximate test that provides an indication of relative physical fitness was developed at the Harvard Fatigue Laboratory (8). It involves simple measurements of heart rates and time while subjects step up and down a fixed bench (height of bench determined by subject sex). There is a fixed cadence to the on-and-off stepping. This test has been employed frequently by investigators in industrial settings to provide a rough index of workers fitness.

Maximum aerobic power is markedly influenced by the age of the worker. Figure 6.3 illustrates the decline in maximum oxygen uptake with advancing age. Peak levels occur between 20 and 30 years and decrease linearly with age. However, not all individuals show this typical pattern, as some age more slowly and some more rapidly regardless of their chronological age. It is of some concern that over the past quarter century maximum capacity has shown a tendency to decrease. Women have smaller maximum powers, with maximum oxygen values some 20 percent lower than men of equivalent ages. However, as women have become more interested in active exercise programs, the differential is decreasing and many women not only have higher maximum oxygen uptakes than earlier reported but appear to remain at a higher level well into their middle fifties.

As noted in Figure 6.1, there is an approximately linear relationship between heart rates and oxygen uptake. This rough relationship holds for individuals

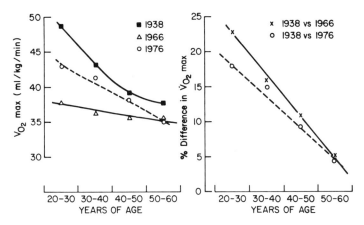

Figure 6.3 Maximal aerobic capacity (ml kg^{-1}min^{-1}) in men measured at three different time periods. A significant decrease in maximum capacity was observed following a 28-year period.

of differing levels of physical fitness and thus provides some problems in estimating maximum aerobic power if the fitness level is not actually determined by the various tests available. However, as shown in Figure 6.4, which also shows the decline in maximum oxygen uptake with age, improvement in maximum capacity does occur if individuals, regardless of age, maintain their physical activity levels by engaging in training programs. It is also apparent that the magnitude of the improvement consequent to a training program is greater in individuals with the lesser potential and smaller in those individuals who have the larger $\dot{V}O_{2max}$. The value of a sustained and maintained physical activity at any age is quite clear.

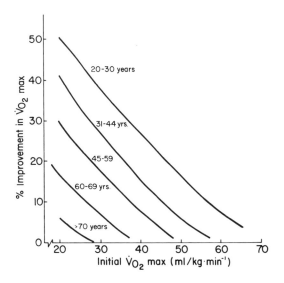

Figure 6.4 There is a significant decrease in maximal aerobic capacity as men become older. The potential improvement in maximum capacity is shown to be related to both age and the maximum capacity. The more highly trained individuals show less relative improvement regardless of age and initial capacity, whereas those with smaller capacities tend to improve more if they engage in a serious training program.

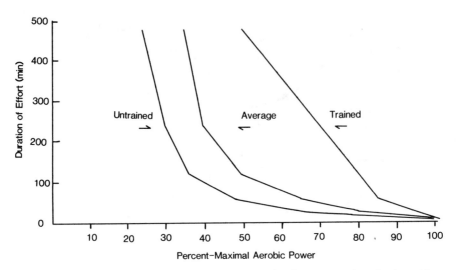

Figure 6.5 The duration of work time (minutes) at various levels (percent) of maximal aerobic power (capacity) in men having different degrees of physical fitness.

It is also apparent that one's maximum aerobic power will influence one's ability to perform submaximal levels of work. Table 6.2 and Figure 6.5 present data illustrating the differences in time that work can be performed and the relative cost to the individual in doing such work. For example, trained individuals can work at 85 percent of their maximum $\dot{V}O_2$ for 1 hr, whereas untrained persons can work at only 50 percent of their capacity for 1 hr. It is also apparent that when people are required to work at their highest capacity, trained people can do such work for up to 6 min whereas the average person is limited to some 3 min.

Since not all workers are employed in sea-level industries, it is important to consider the effects of working at various higher altitudes. Decrements in maximum aerobic power are minimal at altitudes up to 5000 ft but become more evident at higher altitudes (Figure 6.6). The effects of altitude on maximum $\dot{V}O_2$ are similar in men, women, and aged men. Such decrements are of considerable significance in one's ability to perform work at altitude.

Energy Costs in the Industrial Situation. Procedures similar to those employed to determine maximum aerobic power have been utilized to measure the energy requirements for various industrial jobs. Measurements of heart rate during work activities and relating them to oxygen uptakes have been most frequently utilized. However, this approach is fraught with some problems, such as relative fitness of the workers. The most precise approach is to actually measure the oxygen uptake during each particular operation. The measurements made include minute ventilation, oxygen uptake, and carbon dioxide production. Expired gases can be collected in bags. The use of bags, mouthpieces, and nose

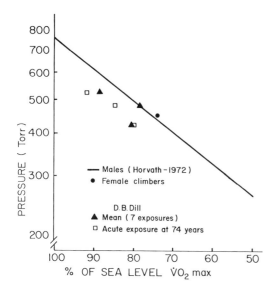

Figure 6.6 Work capacity is markedly influenced if maximum work is undertaken at altitudes ranging from sea level to 450 torr. The data indicate that the magnitude of decreased performance is similar in young men, young women, and old individuals.

clips are potentially cumbersome but new techniques have eliminated some of the problems. Unless several bags are hooked up in series, measurements can be made for only limited time periods. The availability of the Kofrangi–Michaelis respirometer (13) has made it easier to make determinations over a longer time. This unit weighs 3.6 kg and consists of a back harness, a dry gas meter, and an aliquoting device. This latter device continuously removes a fraction of each breath of expired air, which can then be analyzed for its oxygen and carbon dioxide content. It has a limitation in that it is not effective if minute ventilations exceed some 60 to 80 liters. Despite its obvious advantages, it has had limited use in this country.

There are available a considerable number of experimental studies on industrial workers. It would be difficult in the limitations of this chapter to include all the available reports. Some indication of the efforts expended to evaluate work costs has been presented in Table 6.1 and various references. Table 6.3 provides an insight into the research conducted in metal reduction plants (4, 5). Åstrand (14) has presented information as to the excellent relationship between the individual physical work capacity as measured in the laboratory and the actual work output of men engaged in the building industry. Her data indicate that the occupational workload level spontaneously chosen by the individual had a definite positive relationship to aerobic work capacity. This level corresponded to approximately 40 percent of the individual maximal capacity. However, individuals with a large capacity are probably more productive than those with smaller capacities. Åstrand et al. (3) evaluated the circulatory responses to arm exercise by nailing at bench level, into wall at head level, and into ceiling 10 cm above the head. Heart rate, blood pressure, and blood lactate concentrations during arm exercise were higher for nailing into

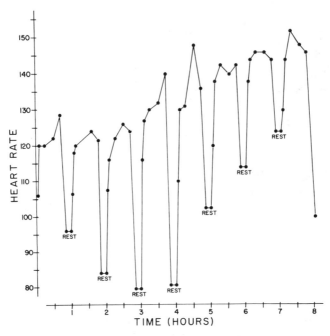

Figure 6.7 The heart rate response to repetitive equivalent work loads. Note that as the day's work progresses there are two major changes in recorded heart rate: (1) gradual increases in peak rates with repetitions during the work day and (2) that heart rates following cessation of effort show a failure to recover as rapidly as the work day continues, suggesting the development of work fatigue.

ceiling than for nailing into wall or bench. In comparison to bench nailing, bicycle exercise at equivalent oxygen uptakes induced lower heart rates and blood pressure elevations.

Investigators who have studied various operations of the Eastman Kodak Company (15) present information that effectively facilitates the optimum design of jobs, workplaces, and equipment. Additional similar studies in other industries would provide information of value to both workers and management. Miller et al. (16) performed studies on repetitive lifting. They concluded that relationships among energy expenditure rate, pulmonary ventilation, heart rate, and work load were essentially linear through the range of lifting work loads from 1 to 50 W in men ranging in age from 21 to 50 years. A general review of work physiology has been presented by Miller and Horvath (17). An interesting and suggestive observation was made by Wojtczak-Jaroszowa and Banaszkiewicz (18), who reported that maximum aerobic power was lower during the night than during the day. Although this observation has not been confirmed by others, it has significant implications for night workers. Figure 6.7 presents some data on the heart rate responses of a worker performing repetitive tasks during an 8-hr workday. It indicates some of the problems associated with only utilizing heart rates to evaluate the energy costs of a task.

The data clearly indicate that fatigue may modify the heart rate response, as not only is the heart rate during the task higher as the day progresses but recovery from the task is slower.

These few references to studies in the workplace do not represent the efforts expended to evaluate workers' performance. They have been presented in order to stimulate other investigations. Hopefully, as more information is accumulated, our understanding of and appreciation for the capacity of men and women workers to accomplish their occupational requirements without compromising their health status will be on a more firm scientific basis.

1.1.3 Work Output During Typical Workday

Assessment of the physiological strain resulting from the job activity and the associated environment can be obtained from a determination of the energy expenditure of the working person. Not only does such a determination serve to classify job requirements but knowledge of the cost of the effort can be useful in appropriate placement of workers according to their capacity and the demand of the job. The maximum level of energy expenditure over a full day's work could be assessed directly for each person by evaluating individual oxygen uptake and consequently the energy expenditure by making measurements over the 8-hr workday. It has been proposed that a level of energy expenditures equivalent to 33 percent of the maximum oxygen uptake be set as the standard of reference. This level agrees well with the findings of Michael et al. (19), whose subjects could walk or ride a bicycle ergometer for 8 hr without undue fatigue if the energy cost did not exceed 35 percent of the maximum oxygen uptake. A similar acceptable work load (35 percent) has been suggested to be applicable to Indian workers (20). They estimated that this level of energy expenditure would be accomplished with a heart rate of 110 beats/min. They also emphasized that it would be important to further evaluate these criteria levels on the basis of physical fitness level of the workers.

However, the physiology of prolonged work has not received adequate attention and there remain many unresolved problems, especially in determining the role of various limiting factors. In particular, it is not clear that workers can maintain their performance level without undue fatigue if they continue to work for successive days at 35 percent of their capacity. Some indirect evidence suggests that the level may be closer to 25 percent on a long-term basis.

Applied physiologists have long held the view that the maximum 24-hr energy expenditure should not exceed 4800 kcal (21). Rates of energy expenditure in excess of that value are thought to be associated with feelings of chronic fatigue and increased susceptibility to many physiologic dysfunctions. It is of interest that the pre-referenced studies on energy expenditure during the workday suggest that most workers expend some 1300 to 1800 kcal, which leaves a considerable excess energy potential to be utilized in other activities. Numerous studies have evaluated the energy costs of particular work proce-

Table 6.4 Average Values of Energy Expenditures (kcal/min)

	Young Adult Males		Young Adult Females
Rest			
Sleeping	1.1		0.9
Sitting	1.4		1.1
Standing	1.7		1.4
Work			
Truck driving	1.6	Office work	1.6
Office work	1.8	Retail trade	1.6
Printing	2.3	Floor sweeping	2.0
Light industry	2.4–4.1	Dusting	2.0
Bricklaying	2.5	Walking on level	2.2
Walking, slow, level	2.6	Domestic work	2.4
Tailoring	2.9	Machine tooling	2.5
Shoemaking	3.0	Bookbinding	2.5
Average industrial	2.5–3.0	Chemical industry	2.7
Driving locomotives	3.8	Bed making	3.4
Walking, 10 kg	4.0	Shopping	3.4
Carpentry	4.1	Light industry	3.4
Sawing, power saw	4.8	Window cleaning	3.4
Pushing wheelbarrow	5.0	Walking, carrying 10 kg	3.4
Ploughing (horsedrawn)	5.0	Baking	3.4
Construction work (laborers)	6.0	Scrubbing floors	4.0
Farming, mechanical	6.0	Farming, traditional	4.8
Farming, nonmechanical	6.5	Beating carpets	5.0
Mining	6.8		
Shoveling	6.2–10.0		
Mowing oats	7.4		
	7.6		
Pulling carts	6.8–10.8		
Tending furnace (steel)	8.2		
	8.2		
Sledgehammering	6.0–9.0		
Digging (spade)	8.4		
Sawing	9.0		
Felling trees	9.8		
Slag removal (steel)	14.6		

dures. Some examples of these observations are given in Table 6.4. Additional information on other tasks is available (22, 23). Figure 6.8 presents information on both oxygen uptake and energy expenditure for some common tasks as well as for walking and running, A practical illustration of energy expenditures in a single industry is given in Table 6.3. It illustrates the techniques employed

WORK COSTS AND WORK MEASUREMENTS

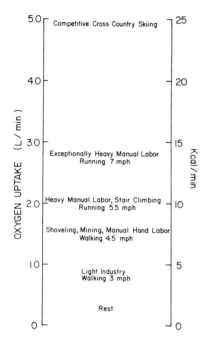

Figure 6.8 Oxygen uptake and energy cost of performing various levels of work.

and to some extent the variable energy expenditure seen even in the performance of a single task. The lack of more complete and detailed information on energy costs of jobs in industry prevents a complete evaluation of the load placed on the working population. It is important that additional studies be conducted, partially in view of the development of more automated devices that reduce physical loads but may induce more psychological loads and partially to consider the cost of performing physically demanding tasks in workers of underdeveloped countries. We also need additional information as to the influence that nutritional status, in particular undernutrition, affect work performance and work efficiency.

1.2 Secondary Factors Influencing Performance

Workers, as do others, occasionally suffer from various infectious diseases. In general because of an individual's economic drive to return to work and the pressure from the industrial organization not to lose production, many return to the workplace as soon as the febrile state of disease has subsided. Although studies on the influence of infectious disease on the work capacity of the worker are limited, sufficient information is available to suggest that work performance is influenced by the prior disease. Generalized malaise accompanied by myologia, negative nitrogen balance, breakdown of muscle protein, and enforced bed rest is often associated with infectious diseases. These changes undoubtedly influence the ability of man to maintain a normal level of performance. Friman (24) reported that immediately after illness maximal muscular isometric strength decreased significantly (2 to 15 percent). Bed rest in these individuals (over the same time period) had no effect on muscular strength. Friman (25) also found that isometric endurance capacity at 67 percent of maximal isometric strength was reduced some 13 to 17 percent. It should be noted that not all muscles exhibit this deterioration, although in general arm, upper torso, and leg extensor appear to be the most seriously affected. Psychological factors may be involved in these altered responses, but their roles have not been elucidated.

There is also suggestive information that both maximal aerobic capacity and the physiologic responses to submaximal work are influenced by prior infection. The most definitive study was conducted by Henschel et al. (26), who studied the effect of experimentally induced malaria on performance. The maximal aerobic capacity decreased 19 percent during the febrile period, and although the malaria was appropriately terminated, maximum $\dot{V}O_2$ had not returned to preinfective levels even after some 28 days. A complicating factor was the marked decrease in hemoglobin from 15.1 to 11.8 g/100 ml, which may have sufficiently reduced the oxygen-carrying capacity to be at least partly responsible for the decrements observed. However, Friman (27) also found reduced maximal capacities in patients hospitalized for various infectious diseases. Only a small fraction of this loss could be attributed to the associated bed rest.

Submaximal work is also influenced by infectious diseases. Henschel et al. (26) reported that the heart rate response to a fixed submaximal work (60 min in duration) was some 30 beats higher when this work was performed during the febrile phase of malaria. This increased heart rate persisted for some 20 days following the febrile state of the disease. Similar greater physiological cost for submaximal work was reported by Friman (27) and Grimby (28).

It is apparent that not only is there a decrease in maximum capacity and a loss of muscular strength as a consequence of an infective process, but that submaximal work (the usual level of activity of workers) is accomplished with greater physiological cost. Probably of even greater importance to the worker is the long period (up to several months) required for recovery to peak capacity. Unfortunately, the data on these effects of infective diseases are minimal, yet there are sufficient anecdotal references to such impairment that careful and assiduous attention should be paid to the effects on the workers and on production if they return to their job situation too early. Bed rest per se not associated with febrile diseases also alters the capacity to perform maximal work. Friman (29) found that after 1 week of bed rest the maximal oxygen uptake was reduced to 94 percent of the person's initial value. Bed rest for longer periods up to 20 days results in further decline in maximum oxygen uptake to as much as 75 percent of the pre-bed-rest values (30). These decreases are a reflection of the associated reduction in maximum cardiac output due primarily to a decrease of up to 30 percent in stroke output. Subjects with the higher initial aerobic power appear to have more marked decrements than those with smaller maximum capacities. These observations on bed rest may have some relationship to the ability of individuals with various previous diseases to perform work.

2 SUMMARY

There is no doubt that there are appropriate tools available to evaluate human performance in the workplace. Sufficient information is present to suggest the need for extension of studies to more fully categorize human potential for work and the actual work cost of jobs and the impact of environmental factors. Missing at this time are data for evaluation of the long-term effects on workers of their job-related activities. We further lack information on the interrelationships between the workplace and the other activities that engage the attention of the workers and subsequently influence their performance.

REFERENCES

1. M. Alam and F. H. Smirk, *J. Physiol.*, **89**, 372 (1937).
2. W. W. Tuttle and S. M. Horvath, *J. Appl. Physiol.*, **10**, 294 (1957).
3. I. Åstrand, A. Guharay, and J. Wahren, *J. Appl. Physiol.*, **25**, 528, (1968).
4. S. M. Horvath, M. O. Colwell, and P. B. Raven, *Arch. Environ. Health*, **25**, 323 (1972).
5. P. B. Raven, M. O. Colwell, B. L. Drinkwater, and S. M. Horvath, *J. Occup. Med.*, **15**, 894 (1973).
6. K. L. Andersen, P. J. Shephard, H. Denolin, E. Varnauskas, and R. Maseroni, *Fundamentals of Exercise Testing*, World Health Organization, Geneva, Switzerland (1971).
7. C. F. Consolazio, R. E. Johnson, and L. J. Pecora, *Physiological Measurements of Metabolic Functions in Man*, McGraw-Hill, New York, 1963.
8. S. M. Horvath and E. C. Horvath, *The Harvard Fatigue Laboratory—Its History and Contributions*, Prentice-Hall, Englewood Cliffs, NJ, 1973.
9. H. L. Taylor, E. Buskirk, and A. Henschel, *J. Appl. Physiol.*, **8**, 73 (1955).
10. B. Balke, G. P. Grillo, E. B. Konecci, and U. C. Luft, *J. Appl. Physiol.*, **7**, 231 (1954).
11. P.-O. Åstrand and I. Ryhming, *J. Appl. Physiol.*, **7**, 218 (1954).
12. S. F. Siconolfi, E. M. Cullinane, R. A. Carleton, and P. D. Thompson, *Med. Sci. Sports Exercise*, **14**, 335 (1983).
13. E. Kofrangi and H. F. Michaelis, *Arbeitsphysiologie*, **11**, 148 (1940).
14. I. Åstrand, *Ergonomics*, **10**, 293 (1967).
15. H. L. Davis, T. W. Faulkner, and C. I. Miller, *Human Factors*, **11**, 157 (1969).
16. J. C. Miller, D. E. Farlow, and M. L. Seltzer, *Aviat. Space Environ. Med.*, **48**, 984 (1977).
17. J. C. Miller and S. M. Horvath, "Work Physiology," in: *Human Factors/Ergonomics for Civil Engineers and Construction Managers*, M. Helander, Ed., Wiley, New York, 1981, pp. 109–139.
18. J. Wojtczak-Jaroszowa and A. Banaszkiewicz, *Ergonomics*, **17**, 193 (1974).
19. E. D. Michael, Jr., K. E. Hutton, and S. M. Horvath, *J. Appl. Physiol.*, **16**, 997 (1961).
20. P. N. Saha, S. R. Datta, P. K. Barenjie, and G. G. Narayne, *Ergonomics*, **22**, 1059 (1979).
21. G. Lehman, *Ergonomics*, **1**, 328 (1958).
22. R. Passmore and I. V. G. A. Durnin, *Physiol. Rev.*, **35**, 801 (1955).

23. AIHA Ergonomics Committee, *Am. Ind. Hyg. Assoc. J.*, **32,** 560 (1971).
24. G. Friman, *Scand. J. Clin. Lab. Invest.*, **37,** 303 (1977).
25. G. Friman, *Uppsala J. Med. Sci.*, **83,** 105 (1978).
26. A. Henschel, H. L. Taylor, and A. Keys, *J. Clin. Invest.*, **29,** 52 (1950).
27. G. Friman, *Acta Med. Scand. Suppl.*, **592** (1976).
28. G. Grimby, *Scand. J. Clin. Lab. Invest.*, **14,** (Suppl.), 67 (1962).
29. G. Friman, *Acta Med. Scand.*, **205,** 389 (1979).
30. J. E. Greenleaf and S. Kozlowski, *Med. Sci. Sports Exercise*, **14,** 477 (1982).

CHAPTER SEVEN

Interpreting Exposure Levels to Chemical Agents

CHARLES H. POWELL, Sc.D.

1 INTRODUCTION

The position of an industrial hygienist as the cental professional in the evaluation of occupational health hazards is still almost unique to the United States. However, to properly evaluate the environment, the industrial hygienist must call on members of other scientific and professional disciplines, such as occupational physicians, toxicologists, chemists, control engineers, nurses, and production personnel. The industrial hygienist may take a position of leadership, but the evaluation and control of the chemical and physical workplace environment is a team effort requiring the input and knowledge of many disciplines; thus this position is one of coordination and recognition of the significance of the contributions made by the other members of the team. In addition to the professional assistance, the worker who may be exposed can make a significant contribution to the evaluation of the workplace. It must always be remembered that the plant management may know how it should be done, but the worker knows *how it is done*. The inclusion of both management and the worker as part of the evaluation team also improves the possibility of effective communication between management,the workers, and the health professionals. In many cases this type of communication is necessary to fulfill both the ethical responsibilities of the industrial hygienist and the physician, and the requirements of the Occupational Safety and Health Act of 1970 (1).

Most workers are employed in industrial establishments that are small in total employment, and as a result it is not possible for the management to have a staff of health professionals with the diversity of background necessary to make a valid, comprehensive evaluation of possible exposures to chemical agents. Thus it is often necessary to make such an evaluation through the use of consultants, or the individuals available must have some knowledge of the other disciplines and the necessary contributions of these disciplines. This can be achieved through study of professional publications, training, or exchanges of ideas at professional meetings. It is important that the individual making the evaluation recognize that the first responsibility is to the worker and that recommendations and evaluations must extend only to the limit of professional competency.

To be able to make a comprehensive, scientific evaluation of a chemical workplace hazard, the individual charged with this responsibility must have at least a working knowledge of the chemistry of the compound, its toxicologic effect on workers, the routes of entry and elimination, the meaning of specific medical tests, the related signs and symptoms, the industrial process in which possible exposure could occur, and the methods to control exposure. Whatever the professional training and experience of the individual making the evaluation, some knowledge in the other professional disciplines is required.

2 OBJECTIVES OF DATA ACQUISITION

The general principles for evaluation of an occupational environment remain the same regardless of the specific objective for which the information is being collected. The final evaluation is dependent on the sampling strategy and planning and familiarization with the process prior to the collection of any data. The basic principles of evaluation have been outlined by Hosey and Kusnetz (2). These principles include recognition, evaluation, and control. It is important to consider that the recognition of a possible health hazard, an understanding of the industrial process and how it may result in a significant employee exposure, and the biologic action that may occur are all part of the preplanning necessary to determine the extent of exposure. To plan for an effective evaluation program, the final objective must be determined prior to sampling, so that the sampling and the study can be planned in conjunction with other available information to answer the specific evaluation objective.

In the preplanning stage, consideration must be given to the following:

1. *What to sample.* The workroom environment and/or biologic samples from the worker.
2. *Where to sample.* The breathing zone of the worker, the general room air, specific samples of workroom air collected while machinery is in operation, biological specimens collected from the worker, raw material, settled particulate samples, and wipe samples of contaminated surfaces.
3. *Whom to sample.* The selection of the workers exposed, the number to be tested, and the duration of sampling.

4. *When to sample.* On all shifts, only during the day, on the weekend, with doors and windows open or closed, for seasonal variations, during "normal" operations, or during maintenance or shutdown.

Once all the data have been collected, they must be considered in light of the specific objectives. Normally the objective of data collection falls into one of four categories, discussed further in Sections 2.1 to 2.4.

2.1 Compliance with Governmental Regulations

The procedures to be followed for compliance with government regulations may be somewhat more comprehensive and restrictive than those for almost any other type of evaluation except for research. This restriction takes the form of the limitation of scientific judgment of the industrial hygienist and substitutes for that individual scientific judgment specific government regulations. To accomplish this objective, it is necessary to be familiar with the government regulations and the intent of the regulations and the law from which they are derived. Many times the intent of the regulations and the specific action to be taken have not been included in the regulation, and it may be necessary to contact the local or higher government office to determine precisely what is expected. Many government regulations are derived from administrative laws and as a result are little help in determining requirements, except on a broad philosophical basis. As an example, most of the present health standards of the Occupational Safety and Health Administration (OSHA) (3) were derived from the 1968 American Conference of Governmental Industrial Hygienists, (ACGIH) threshold limit values (TLVs) (4). Since analytical methods are not included in the TLVs, the method being used by OSHA must be determined and either the same or similar analytical procedures selected to ensure comparability of the two methods. At a minimum, the collection of data for compliance should include the following materials, steps, and procedures.

2.1.1 Sampling Equipment

Selection of sampling equipment and procedures that are the same as, or similar to, the equipment or procedures used by the regulating agency.

2.1.2 Calibration of Sampling Equipment

Calibration is as important in sampling for government compliance as it is in the collection of data from any other industrial hygiene evaluation effort.

2.1.3 Analytical Methods

Procedures must be similar to, or the same as, those used by the governmental agency, to ensure comparability of results. In addition, it is advisable to have the analytical work accomplished by a laboratory that is accredited by the American Industrial Hygiene Association (AIHA). This accreditation will add credence to the assessment of the data.

2.1.4 Sampling Strategy

Consideration must be given to the selection of the environmental sampling procedures to be used, based on the type of standard that has been promulgated (time-weighted average, ceiling value, etc.). In some cases it is desirable to collect samples of several different types even though the regulatory agencies may not require them. For example, the industrial hygienist might decide to collect data for both respirable and total dust when the exposure substance in question is free silica and only one type of sample, the respirable, is required by the regulatory agency. The total dust samples can be used to develop specifications for engineering control. The number of samples that must be collected is seldom delineated in the standard. However, the National Institute for Occupational Safety and Health (NIOSH) (5) has developed a statistical method to determine how many samples should be collected for the determination of noncompliance with occupational health standards.

2.1.5 Quality Control

It is very important in the collection data for the evaluation of compliance with government regulations that every effort be made to assure that the sampling, analysis, and evaluation are not only of the highest quality, but the sequence of events can be well documented and a well-designed quality control system has been included. The use of internal "check samples" for analytical procedures and cooperative cross-check of samples with other laboratories, and utilization of the Proficiency Analytical Testing (PAT) program of NIOSH and the AIHA Laboratory Accreditation Program are recommended. In addition, good quality control on preventive maintenance, repair, and calibration of sampling equipment is essential.

2.1.6 Recordkeeping

Data are only as good as the records that are maintained. The development of a "log" or record of the chain of events that have occurred in any system for the collection and analysis of data is important under any circumstances, but particularly if the data are to be used as exhibits to establish exposure levels in a legal sense, to meet the recordkeeping requirements of a regulatory agency, to save for use at some future date in epidemiologic studies, or to evaluate environmental levels where clinical symptoms have occurred. Detailed recordkeeping may be less critical if the data will be used for short-term evaluations only, such as use of general room or source sampling to determine the effectiveness of new control methods. Even under these circumstances it may be advisable to maintain records for future use, to show good faith to a regulatory agency.

Recordkeeping is also required for accreditation of laboratories by AIHA. Such a system need not be complicated, but the records must be maintained in an orderly manner and must be easily retrievable. Comprehensive records are important in trying to develop retrospectively a picture of exposure. It is interesting to note that such records have not been maintained in many industrial operations, and as a result it is frequently impossible to develop a long-term exposure record.

When a recordkeeping system is being developed, one must plan to include environmental, medical, and personnel data that can be used to establish a dose–response relationship. Many industrial concerns are presently developing computer-based systems for long-term use for clinical and epidemiologic studies as well as short-term use for regulatory requirements and for management information systems to determine the present status of the health of the worker and environmental levels, as well as trends in environmental levels over an extended period (see Chapter 7, Volume 3A).

Quality control and maintenance of records is particularly important in any data acquisition for compliance evaluation, and a sequential chain of records is valuable to assure acceptability of data by a regulatory agency.

2.2 Surveillance

Surveillance data acquisition serves several purposes and can be acquired in such a manner as to answer several objectives at the same time. These data can be used not only to characterize the individual worker exposure but, also, at least in part, to characterize a specific work area or job as well as to establish trends and exposure levels for any given operation. It is likewise possible to use the information collected in this manner for long-term research studies and to measure effectiveness of control procedures. The specifics for this type of atmospheric measurement have been discussed in detail in Chapter 11, Volume 3A. Several considerations must be kept in mind when collecting data for surveillance. These are outlined next.

2.2.1 Sampling and Analytical Procedures

The sampling and analytical procedures must be selected so that the data can be related, either directly or by some meaningful comparison factor, to data that have been collected previously and to similar investigations that have been reported in the published literature. It is important that the data collected be usable in establishing trends and that they be standardized sufficiently to be usable at a later date for epidemiologic studies.

2.2.2 Position Descriptions

Position descriptions to assure proper identification of duty assignments for the workers are necessary. These must include all job elements of any exposures that have been determined to be related to the specific job. It may be that the position description used by the personnel office does not "fit" the actual duty of the worker and the worker's exposure. As a result, some companies have found it advisable to prepare an Occupational Health Position Description that more adequately describes the exposure conditions of a given job. These "occupational health" position descriptions are usually maintained in a separate health file and do not become part of the personnel file.

The position descriptions must be constantly updated and changed as the worker moves from one position to another. This information is very significant in any future use of the data for a retrospective epidemiologic study.

2.2.3 Chronological Log

During the collection of the data, a chronological log of events that have occurred is necessary for the proper evaluation of the data. This is true for both short-term "ceiling values" and other types of sampling as well as for assessment of 8-hr time-weighted average exposures. This information should also indicate whether the exposures are continuous or sporadic, and whether the equipment is operating properly; any unusual occurrences should be described. Evaluation of the data necessitates the recording of any unusual activities, the condition of housekeeping, the use of provided ventilation, any recent engineering control changes, and the size of the work force available on the day of sampling. The housekeeping may be poor on Monday morning, for example, if the maintenance crew has not worked over the weekend. This type of housekeeping could result in higher exposures on Mondays than would normally be the case throughout the rest of the work week. The condition and operation of any local exhaust ventilation system can have a similar effect. Times of vacations or high absenteeism can result in increased work demands for the individual worker. These conditions need to be noted in relation to the data being evaluated.

2.2.4 Personal Protective Devices

The use of personal protective devices by the worker can also have an effect on the evaluation of surveillance data. For example, the difference between the levels of exposure of a worker to a toxic chemical when a respirator is worn as opposed to when such protection is not used is as significant, as is the information that the respirator worn is the right type for the hazard, is effective at the level of exposure, and is approved by NIOSH/MSHA (the U.S. Mine Safety and Health Administration) for the specific hazard. The same type of information in relation to skin absorption, protective clothing, smoking, and eating habits in the work area must also be considered and recorded.

2.2.5 Weather Conditions

Weather conditions can have a major impact on the evaluation of the industrial hygiene surveillance data. The differences that closed or open doors and windows can make on the effectiveness of general room or dilution ventilation are well documented and can exert a major effect on the environmental levels. This can also be true, usually to a lesser extent, for local exhaust systems. The effects of rain in reducing the dust levels in any outside operation are easily recognized. A change in humidity can also influence environmental levels of some contaminants. Although these are not items that can be controlled, they do represent information that is valuable in the evaluation of environmental surveillance data.

2.2.6 Supplemental Data

Supplemental data that do not relate directly to the environmental data obtained, but must be considered as a major modifier to any evaluation of environmental data, are the effectiveness of engineering controls and the attitude and information available to the employee.

2.2.7 Engineering Controls

The effectiveness and the availability of engineering contnrols have an effect on the evaluation of the environment. For example, if it is found that the local exhaust ventilation system is normally turned off to conserve energy when the exhaust system is not in use, the evaluation must consider whether the local exhaust system, when operational, is sufficient to meet the requirements for environmental control, and whether employee training is sufficient to assure that the system would be turned on when necessary for hazard control.

Efficiency of control measures must be considered in relation to the maintenance of equipment as well as design characteristics. The evaluation of surveillance data cannot proceed without knowledge of the methods used to assure that preventive maintenance programs are effective and the time from submission of a work order until the work is completed; there must also be assurance that all work orders are handled on a priority basis to meet health needs.

Housekeeping is an important yardstick of plant attitude and concern about safety and health. For example, the accumulation of dust in the general work area, on machinery, rafters, and floors not only create a secondary health hazard but in some cases may make it impossible to evaluate the effectiveness of the existing ventilation system until the dust level is reduced by cleanup of the secondary dust sources.

Any ventilation system that is installed must be evaluated to assure that the actual capacity is sufficient to control the hazard. Frequently additional exhausts incorporated into an already existing and well-designed ventilation system have served to reduce the effectiveness of the system to an unsatisfactory level; furthermore, the system may have become ineffective because of changes made during maintenance, such as reversing the fan direction, or lack of maintenance leading to plugging of the ductwork may have resulted in reduced airflow.

If the general room or dilution ventilation is in use, consideration must be given to the seasonal effectivenss of this type of ventilation, since the patterns of air movement may change when windows and doors are open during warmer periods of the year. The use of dilution ventilation must also be evaluated in conjunction with the toxicity, size, and number of sources of contaminants in the work area.

Isolation of the worker from hazardous materials as a method of engineering control must also be taken into account. The isolation of workers from the hazardous material by the use of respiratory protection or other personal protective devices, or the use of isolated hazard-free work areas, such as control booths, can be effective, as well as isolating the source from the general work area and the worker. If the source of hazard has been enclosed as a means of control, the enclosure should be checked to determine whether there is any leakage, perhaps because the enclosure is under positive pressure.

A respirator program is difficult to assess but must be considered to determine whether the proper respirator protection is available for the type and concentration of the contaminant and whether the respirator is approved by a federal agency such as NIOSH/MSHA. It also must be noted whether the use of respirators is required or voluntary, whether there is a record of environmental data to support the type of respirator chosen and used, the type of respirator "fit" program that is used to assure that each person who must wear a respirator has one that gives a good fit, and that there is a program for care and cleaning of the respirators, and so on.

The effectiveness of other controls is also relevant in an evaluation of worker protection. "Other controls" include length of the workday, administrative procedures to reduce the period of exposure, the substitution of less toxic substances for a more hazardous material, and the availability of adequate (for the hazard) shower room facilities, changes of work clothes, and so on.

2.2.8 Worker Training and Attitude

It is vital that the attitude of the employees and their understanding of work procedures designed to assure their ability to work safely are understood. The Occupational Safety and Health Act speaks of informing the employee of the exposure hazards; of even more importance to the evaluation of surveillance activities, however, is the question, Does the worker understand these hazards and the proper work procedures? The answer usually must be arrived at subjectively, for few companies have testing programs for such health work rules. The evaluation of worker attitude is always subjective. The attitude of the workers, many times, is similar to the attitude of management, or a reflection of normal management relationships. However the attitude of a single worker under certain circumstances may reflect a negative attitude toward the health professionals collecting the industrial hygiene data. This is usually exhibited by "salted" samples, pumps out of calibration, and similar misadventures designed to confuse the health professional making the survey.

2.3 Research on Characterization of Chemical Agents

A number of different strategies must be employed in the acquisition of data for use in occupational health research. For the most part, the approach that gives the most definitive results will be similar to the strategies for the collection and analysis of workroom environments that are used for other types of evaluation, including surveillance. The main difference between the other types of evaluation and those for research is the additional consideration given to the design of the research program to assure that the results will be statistically significant.

The research effort usually involves the determination of methods to characterize the environment for epidemiology, animal studies, or clinical cases. The basic objectives of such research could also be to evaluate government health requirements, new products or by-products, and to reevaluate old processes or exposures.

2.3.1 Characterization Methods

The characterization of exposure patterns many times involves the development of a sampling protocol for the in-depth evaluation of a specific operation or industry using presently available methods, or it may require the development of a totally new means of sampling and analysis. For example, exposures to dust containing silica have been evaluated by existing methods and through the development of new procedures. In an extensive report in 1929 on "The Health of the Worker in the Dusty Trades, II. Exposure to Silicatious Dust in the Granite Industry" (6), the impinger was used as the means of collection. The particles counted were divided into two groups; those above and those below 10 µm in size. At the same time samples were collected and weighed and reported on the basis of total dust, even though a modification of this technique was not generally accepted until many years later.

With increased knowledge in respirable diseases and sampling techniques, it was later concluded that measurement of the total mass weight of dust and the analysis of the actual percentage of free silica in airborne dust would give a better measure of silica exposure. In addition, the range of particle size that had the highest biologic activity had been narrowed to the 1- to 5-µm range. This knowledge resulted in size-selective respirable mass collection of particulates, which has been used in both research and surveillance (7), including a reevaluation of the silica exposures in the granite industry (7, 8). The analytical procedures for analysis of free silica have also been changed to take advantage of new and improved instrumentation. Although the colorimetric analytical method of Talvitie (9–11) is the most widely used procedure for the determination of free silica, the use of both X-ray diffraction and infrared (IR) instrumentation is gaining in acceptance because of the stability of the samples, the ability to select specific silica polymorphs and to differentiate from high background material, and the lowered level of operator dependence.

In any research evaluation the development of an understanding of the pattern of possible exposure throughout the normal work cycle is necessary to determine the significance of exposures and the biologic effect. The biologic effects may be best measured for a short-term high exposure, or for a time-weighted average exposure throughout the normal work shift, or in some cases a combination of several different types of exposure patterns. The evaluation of work patterns and job cycles is probably more important for this type of research characterization than any other type of evaluation. This characterization, coupled with some expectation of the route of entry into the body and the biologic action, will determine to a great extent the type of sampling and analysis procedures to be utilized.

In some cases, when the exact chemical configuration of the hazardous material is not known, it may be necessary to predict possible contaminants and to base the design of the sampling strategy on these predictions. An understanding of the chemistry and reactions in the process will improve the accuracy of the predictions, which are usually not as comprehensive and as sensitive as would be desirable. Consequently minor yet possibly highly toxic contaminants may be overlooked in the development of a prediction model. For example, predictions of the decomposition products of certain missile fuels were based on the decomposition of the fuel and the oxidizer, and consideration was given to the primary constituents only. As a consequence, hydrogen cyanide, a highly toxic material that was later found to be a decomposition product in the parts per million range, was not considered in the prediction model.

One example of an effort to develop new methods for both sampling and analysis, necessary to characterize the chemical content of an operation and to attempt to predict compounds that may be hazardous agents, is the research underway to characterize the emissions from aluminum reduction operations, to predict what chemicals may be in the mixture, and if possible to select a compound or compounds that could be used as an "index" of exposure.

In this case it is necessary to select the proper solvent to remove the combination of particulates and vapors from a collecting medium and to quantitate the compounds that may be considered to be carcinogens or cocarcinogens. Thus the possible contaminants that may be in the emissions must be predicted and measured separately. For these studies measurements are being made of chemical compounds with three to six organic rings because of the suspected carcinogenic potency of this class of organic compounds. Although the compounds have not been precisely identified, nor is the percentage of each compound in the emissions known, this information, coupled with the data on the carcinogenic action of each compound, is necessary to develop a dose–response relationship. Predicting the dose–response relationship of individual compounds is difficult. The problem is far more complex with chemical mixtures, and even worse than the case described when the composition of the mixture and the percentage of chemicals in it are subject to change with minor adjustments in the process operations.

2.3.2 Epidemiology

Evaluation of environmental data for the purposes of epidemiologic studies may present a significant challenge, since it is usually necessary to collect data over an extended period of time and/or to attempt to reevaluate data collected earlier in relation to the methods used at the present time.

Chapter 5, Volume 1, covers the epidemiologic approach and the evaluation of disease in specific populations in detail, and this material is not repeated here. It is essential, however, to discuss the importance of environmental data that must be made available for any epidemiologic study designed to produce dose–response data.

If the epidemiologic method chosen is, for example, a prospective study, it is possible to collect the necessary data for fulfilling the objective of the epidemiologic study without assembling or evaluating environmental data. The basic question in this type of study is, Is there a difference in the incidence of disease between the exposed and nonexposed groups? Many studies, such as those involving death certificates, worker's compensation records, can be classified as this type of investigation. Even in these studies, however, there are several reasons for considering the availability of environmental data. If the results are positive (there is a difference in disease patterns between the exposed and nonexposed groups), the next and logical question is, What is the dose–response relationship between the disease and the chemical agent? In other investigations, when a causal relationship has been shown to exist between a disease and a job or industrial operation and the specific chemical agent or agents have not been identified, efforts to characterize the environment and to develop a dose–response relationship have been intensified. In some cases a prospective study is run in parallel with efforts to develop field sampling and analytical methods for the characterization of the occupational environment.

Another method of difference is the retrospective study, which measures the relative risk by difference in exposure between "disease" and "no disease" groups. This type of observational research does not necessarily require collection of environmental data. Milham (12) compared "disease" and "no disease" groups by classification of specific industrial process without measurement of exposure; his work is an example of this type of investigation.

The method of concomitant variation poses the question, Does the disease vary as exposure varies? This approach requires at least some attempt to measure or estimate the environmental exposure. When the latent period between onset of exposure and disease production is long, the best estimate may be only one of generalized exposure levels, such as high, medium, and low. This environmental evaluation is difficult and may be totally dependent on "opinions" of individuals who have been familiar with the operation over an extended time. If this is the only information available, the conclusions may not be entirely valid.

In addition, the method of intervention could be used: here the suspected causitive agent is removed or controlled, and the decline or absence of disease is measured. This type of epidemiologic study may be done as a follow-up to previous studies and after engineering controls have been instituted.

These epidemiologic methods may not require environmental data initially, but in almost all cases the need exists for a follow-up study to establish a dose–response relationship. This is of ever-increasing importance, even when the results of the epidemiologic study are negative. As more emphasis is placed on mandatory health standards, there is an increasing demand to establish the lower limit of biologic response. This requires quantitating negative disease data with environmental levels of exposure.

The development of research protocol must be such that the use of available environmental data is maximized, as well as the assessment of the disease. This means that the environmental data developed delineate the exposure conditions in the greatest possible detail as outlined earlier in this chapter (i.e., duration of exposure, changes in process, housekeeping, and job requirements). This information is valuable in the evaluation and in any follow-up studies.

Changes in sampling methods and analytical procedures can have a major impact on the establishment of any dose–response information and can result in an attempt to develop a relationship between the old and the new methods. This many times presents more new questions than it answers, since it is not possible to hold static other parameters in an industrial process. The efforts to relate old analytical and sampling methods to new procedures have been addressed in epidemiologic studies in the past by Reno (8) for silica exposures and by Ayer et al. (13) in the investigation of asbestos exposures. These problems are primarily of concern in retrospective evaluation of the environment, but can also present difficulties if new techniques are developed during the epidemiologic studies.

Development of retrospective environmental data is the most difficult of tasks and must be addressed with skepticism about the value of any old data as well as some data developed in other countries, where it may not be possible to define with certainty the conditions under which the data were collected. Detailed information is needed to relate analytical methods, field sampling techniques, reproducibility of analytical results, calibration procedures for sampling equipment, duration of sampling, and the development of time-weighted averages. The method of sampling (e.g., general room or breathing zone), length of workday, and any change in the process must not be overlooked.

It is desirable to be able to estimate the effectiveness of control methods and to know whether they were operational during previous studies, particularly when the control is designed to isolate the worker from the hazardous condition. If it is felt that the door to the isolation booth, for example, was never closed, or that the respiratory protection provided was not generally in use, these types of control should be discounted when evaluating environmental levels and biological reaction to exposure.

The exact nature of the job as related to exposure must be ascertained: personnel job descriptions may not give a true picture of exposure. Any interferences resulting from exposure to multiple contaminants must be evaluated, particularly if they may have an adverse effect on the same organ system or if they use the same route of entry. Changes in process may result in exposure to chemical intermediates and/or by-products that could cause a biologic variation. The question of an interaction of two or more chemicals and a possible synergistic or antagonistic or carcinogenic cofactor must be evaluated. For some chemicals the possibility of two or more routes of absorbtion must also be considered. For example, the route of entry of ethylene dibromide (EDB) may be percutaneous absorption of the liquid or inhalation of the vapor (14).

The possibility must be closely scrutinized that the signs and symptoms that furnish the corroborative evidence of intoxication may be the result of or intensified by nonoccupational disease, or may emanate from other exposures.

2.3.3 Animal Studies

In recent years there has been increasing recognition that the development of research protocols and the interpretation of laboratory research information requires evaluation by physicians, industrial hygienists, and others in addition to the evaluation made by the toxicologist. This expertise and differing professional background is required in the assessment of animal studies that must be related finally to an assessment of the effect of the workroom environment on man and compared with data collected in epidemiologic investigations, clinical cases, chemical tests, and in some cases similarity of molecular structure (see Chapter 2).

The passage on October 11, 1976, of the Toxic Substances Control Act (PL 94-469) coupled with increasing demands by OSHA for definitive data for the development of a scientific base for federal regulatory standards, may result in the loss of some latitude of the individual researcher in the manner and procedures that will be used for the collection and interpretation of research data related to exposures in the workplace. There is recognition today that the interpretation of a single animal study may be the sole basis for an enforceable federal standard. This is not the objective for much of the laboratory animal research that is done, and it puts an increasing burden on the researcher. In the development of a protocol for an animal study, many more parameters of the experiment can be controlled than in epidemiologic or clinical studies, such as duration of exposure and exposures at a specific concentration. However failure to control even one of these many variables can result in a difference in response in the test animal by 100- to 1000-fold, according to OSHA in the proposed cancer policy and standard (15).

The acquisition of meaningful data from animal studies on which evaluation of exposure of the worker to chemical agents can be based, requires that careful consideration be given to the design of the study, so that the response expected or seen in the experimental animal can be extrapolated to humans and to the exposure parameters that can be expected to occur either on the job or in the community environment. For many animal studies the route of entry that is representative of worker exposure is by inhalation, and this requires long-term exposure studies that can be related to chronic inhalation exposures in man. In this type of investigation one must take account of the methods for generation of atmospheric concentrations of the chemical and for evaluation of the environmental data from the animal exposure. Inhalation studies may be planned after some suggestive toxic effect has been determined by acute and/ or subacute investigations. These studies may not require the generation and assessment of environmental exposures, but they do produce valuable data that can be used to give some estimate of the type of response that can be expected, the environmental concentrations that would be most appropriate, and reasonable time periods for exposure of the animal.

The generation of environmental data for the purpose of animal inhalation studies presents some interesting problems not only in the generation of known concentrations of a specific chemical, but in the ability to maintain this concentration in equilibrium with the experimental parameters for a long time. For example, the design and construction of animal exposure chambers may result in condensation of the chemical being tested on the walls of the exposure chamber. In addition, the equilibrium of the chemical concentration in the exposure chamber may be disturbed by condensation of the chemical on the skin or hair of the experimental animal. This condensation will increase as the area of the skin surface of the animal is increased in relation to the total volume of the exposure chamber and the concentration of the chemical. In some instances these difficulties with generation and maintenance of specific exposure conditions throughout the exposure chamber necessitate measurement and evaluation of the exposure levels in the breathing zone of the animals.

The choice of the sampling and analytical methods, as with other research and surveillance efforts, can have a significant impact on the evaluation of the data collected. The sampling strategy, of course, must reflect consideration of the exposure parameters of the study, both as in any other sampling procedure and with respect to sampling in a small closed system and relating these results to the occupational environment. These methods must meet the "test" of being applicable to both the laboratory and the workplace environments. The need for quality control of analytical results is, of course, critical in the evaluation of environmental data from animal inhalation studies. Additional information in these subject areas is contained in Chapters 2 and 3.

2.3.4 Clinical Cases

Epidemiologic and clinical cases have the advantage that it is not necessary to extrapolate animal data to humans. It is seldom that a human population can be controlled in such a manner as to reduce the effect of outside influences, such as smoking, dietary considerations, and intake of drugs or alcohol. Even when it is possible to control such influencing factors by clinical experimentation, the number of subjects is greatly reduced, and the differences in individual susceptibility must be considered to be a major factor in any recorded variation.

The controlled clinical studies can be of great value, as illustrated by the study of lead absorption and excretion pursued for a number of years at the Kettering Laboratory, University of Cincinnati (16). The clinical studies such as those carried out at the University of Cincinnati represent controlled clinical experiments where most of the parameters of the experiment are closely controlled and are evaluated with confidence in the results. The environmental concentrations are also well controlled even over extended periods. These controlled experiments are usually developed around the use of recognized sampling and analytical procedures. The evaluations of chemical exposures in clinical experiments are, as in the lead studies, necessarily of low levels of exposure to evaluate preclinical signs or symptoms.

The retrospective reconstruction of work exposure conditions that resulted in a clinical manifestation of disease presents a difficult exposure evaluation. In these clinical cases there is no opportunity to plan "beforehand" the parameters of the study. Rather, the investigator must attempt to develop a retrospective exposure profile in much the same way as is done in epidemiologic studies when the question of dose/response must be answered. Because of the limited number of people normally exposed in the clinical cases, it is necessary to probe in more depth for other possible causes of the signs and symptoms than in the morbidity studies. Although it is possible to estimate exposures that are reasonably accurate for clinical cases of acute exposure, the longer the latent period, the more difficult it is to make a reasonable estimate of past exposures. Even in acute exposure cases, it must be recognized that any sample collection and analysis may or may not represent the same environmental conditions that resulted in the clinical response.

In investigations of recent acute clinical cases any retrospective environmental sampling and analysis can at least be accomplished with methods that are presently available and generally accepted. In any attempt to evaluate data from clinical cases that occurred in the past, it must be recognized that it may not be possible to verify the analytical results and to develop a retrospective profile of environmental conditions, let alone relate these experiences to present conditions. Under these circumstances the environmental exposure data may be grouped with confidence only into categories of high, medium, and low exposure levels.

These limitations in the use of clinical cases have been recognized for a number of years. For example, in 1928 Bloomfield and Blum (17), when reporting on a study of workers in chrome-plating operations, made an attempt to approximate the extent of exposure of 17 of 19 employees who had symptoms of chromic acid exposure. They indicated that the estimate of exposure could be considered only an approximation except for a few of the workers who had been employed during the period that the ventilation system had been in operation. The estimate of exposure had been based on the occupational histories and the results of 39 air samples taken in six plants where the 19 workers were employed. The value of the data from these grouped clinical cases may be somewhat questionable, but they do provide some information of the cause and effect.

Lacking any more definitive information, many times data collected in clinical cases can be used with caution to develop exposure guidelines to assure some degree of worker protection that otherwise would not be possible. The study by Bloomfield et al. (17) has been used as partial justification by the ACGIH to establish a TLV for chromic acid (18).

The development of a retrospective environmental profile necessitates the careful consideration of all factors that may influence the conclusions to be made from the clinical studies. The possibility of interference and/or synergism or antagonism from other exposures must be investigated to assure that the cause is well defined when it is related to an effect. An example of the effect of exposure to a second chemical, particularly an off-the-job exposure, is the interaction between inhalation of trichlorethylene and the ingestion of ethyl alcohol, which can result in a marked increase in the effect of the trichloroethylene on the body (19, 20). In clinical cases the possibility of reaction from medication that was being given at the time of exposure to the industrial chemical must be examined. Exposure from a second job and from hobbies are other possibilities that must be investigated as having been either the primary cause or an aggrevation of the industrial chemical exposure. The need to develop a complete environmental profile of exposure is necessary to assure that a true cause-and-effect relationship is established when attempting to reconstruct the circumstances that led to a medical aberration in the worker.

2.4 Adequacy and Performance of Control Program

Acquisition of data on the adequacy and performance of engineering controls, and the effectiveness of administrative controls, may be considered to be integral parts of the data collection for routine surveillance activities. However, under some conditions the collection of performance data for engineering design specifications and/or the determination of the effectiveness of administrative or operational controls may dictate specific data acquisition strategies.

The development of specifications for engineering control calls for the measurement of the environmental levels of the chemical to be controlled, and the measurement of the airflow in any existing ventilation systems. The assessment of the effectiveness of the new engineering controls requires repetition of the environmental measurements for comparison with previous environmental levels. If a new ventilation system is added or any existing system is modified, airflow measurements should be taken.

The selection of specific methods of control is discussed in Chapter 12, Volume 3A, and the design specifications of a number of hoods and local exhaust ventilation systems are developed in detail in the *Industrial Ventilation Manual* (21).

2.4.1 Testing Procedures

The requirement for environmental testing to develop engineering specifications depends on the type of engineering control under consideration and may include process sampling or sampling at the source of generation of the contaminant, general room sampling, personal monitoring, and short-term monitoring.

Environmental sampling at the source of generation or the release of a contaminant into the work area is usually done with stationary samplers. The selection of specific sampling equipment and the procedures for acquisition of these data depend on the conditions of release of the contaminant. Source sampling gives data on the amount of material being released into the work area. If this information is not available by material balance or rate of consumption calculations, this type of data acquisition is essential when total enclosure of the source of the chemical contaminant may be necessary or when the use of local exhaust ventilation to control the hazardous material is being considered.

The collection of environmental data is only one element in the design of engineering controls. Evaluation must be made of the effect of air movements caused by motion of machinery, the workers, the material being processed, general room air currents, and thermal air movements from any hot processes or the heating and cooling equipment, as well as the capture velocity of any local exhaust ventilation system in the general area. Sampling at source is usually not applicable when dilution ventilation is being considered, since the consumption or evaporation rate of the chemical contaminant in a work area normally can be calculated. General room sampling or personnel monitoring will give a more definitive assessment of the exposure and of the dilution rate necessary to reduce exposures to below the TLV if dilution ventilation is the method of control under consideration.

The method for the calculation of airflow required to maintain the levels below the TLV is covered in the *Industrial Ventilation Manual* (21). The limitations for effective use of dilution ventilation must be considered before this approach is selected as the control method. Normally it can be used to successfully control exposures only when the toxicity of the material is low (high TLV), the rate of consumption and the rate of evaporation are low, the workers do not work directly at the source of the contaminant, and the hazardous material is released at a constant rate throughout the workday. The use of general room and personnel monitoring are acceptable methods for the evaluation of the effectiveness of dilution ventilation.

Continuous monitoring can be a very effective method of evaluating the engineering control devices used for highly toxic and life-threatening acute hazards. For example, air samples have been collected on continuous tape and analyzed directly for beryllium in aerospace operations, and permanent fixed monitors for hydrogen sulfide have been used extensively in the petroleum refining industry and has been recommended by NIOSH as a means of monitoring for hydrogen sulfide under certain conditions (22).

When it is necessary to analyze any samples collected to evaluate the adequacy of the performance of controls, the choice of analytical procedure would normally be the one that best fits the requirement for selectivity, precision, and accuracy for the levels of the chemical contaminant expected to be collected. The standard method for compliance or surveillance purposes is not necessary if the results will only compare "before" and "after" the installation of controls; however, if the results are used to compare worker exposure before and after installation of controls, or for compliance purposes, the same analytical method that is used for compliance and routine surveillance would be the procedure of choice.

2.4.2 Evaluation of Local Exhaust Ventilation

Airflow measurements are the best method to evaluate the effectiveness of any existing exhaust system and to determine whether any new ventilation system is meeting its design criteria. Airflow measurements have a number of applications and are necessary to assure proper evaluation of any ventilation system.

These measurements can be used to evaluate any new exhaust system to determine whether it meets the design criteria, to balance the airflow if the exhaust system is designed for the use of blast gates, to obtain additional data for similar installations that may be required in the future, and to determine whether the delivered airflow meets the requirements of any regulatory or voluntary code.

In any exhaust systems that are already installed, airflow measurements can be used to determine the system's capability to deliver the necessary airflow to meet governmental and trade association codes, to collect design data for similar installations, to determine whether the present system has sufficient capacity for additional exhaust outlets, and for repeated evaluation of the airflow to determine whether repair, maintenance, or readjustment of the airflow is necessary.

When an evaluation is made of either an existing or new installation, the results of all air measurements, the location in the exhaust system where the measurements were obtained, and the comparison made with any previous collected air measurements should be retained just as data from environmental measurements are retained in the industrial hygiene files.

Air measurements on any new system should include velocity and static pressure measurements in the main, and in all branches of the ductwork. Static pressure measures should be obtained at each exhaust opening, static and total pressure measurements should be made at the inlet and outlet fan openings, and the differential pressure should be determined at the inlet and outlet of the collection equipment.

These testing procedures need not be repeated at regular intervals, since there is usually little change in any existing exhaust system unless the process is changed, or the exhaust system is altered, or unless routine maintenance and repair is inadequate. For routine evaluation of any existing ventilation system, the measurement of hood suction is the usual and acceptable procedure. The criteria for selection of measurement devices for assessment of ventilation systems are discussed in detail in the *Industrial Ventilation Manual* (21).

2.4.3 Administrative Control

Administrative controls have been widely used where engineering controls of exposure are inadequate—for example, to reduce the hazard of working with risks through training programs, to inform the employees of the hazards, and to remove an employee from exposure to allow sufficient time to recover from the effects of exposure. "Recovery time" has many uses in industry: as a cooling-off period for workers who have been exposed to hot environments or for those who need to recover from the effects of abnormally high or low air pressure (e.g., deep sea divers, caisson workers, tunnelers, and airplane crew members). Chemical exposures have also resulted in the need for recovery time while the worker metabolizes or excretes the hazardous chemical and the body burden is reduced to acceptable levels. During the recovery time the worker is placed in a nonhazardous job or excused from work. These procedures have been used extensively in response to exposures to heavy metals, particularly lead and mercury, and to a lesser extent to organophosphorus compounds, among others. These administrative procedures require the ability to monitor the biologic fluids of the worker as the basis for deciding when the individual should be removed from, and returned to, work. this approach usually is an indication of lack of sufficient engineering controls or that the chemical is extremely hazardous and employees must be monitored on a continuing basis to be assured that they are in fact being protected from the hazardous material. The use of biologic monitoring is covered later in this chapter.

The administrative controls that require a detailed evaluation of the environment are those that should be used only until the proper engineering controls can effectively protect the workers. These administrative controls require that the workers spend only part of their time during the normal work shift in a highly hazardous area (above the TLV) and that the remainder be in a low or nonhazardous area to assure that the combined exposure is below TLV.

This type of administrative control would normally be used only for chemicals whose exposure standard is based on a time-weighted average and no adverse acute effect from high exposures, and the exposure is reasonably consistent throughout the work shift. Silica exposure in quarry operations, for example, could be controlled by this approach. Benzene exposure should not be controlled by administrative reduction of exposure time because an acute high exposure can cause systemic poisoning, can act as a primary irritant, and is absorbed through the skin. In addition, chronic exposure by inhalation can cause blood changes, and benzene may be a carcinogen.

Prior to use of administrative control, a sampling strategy that will result in a complete understanding of the exposure characteristics of the job must be selected. Any cyclicity of the exposure, of work procedures, and exposure times must be estimated. If the preliminary evaluation suggests that the exposure would be reasonably constant throughout the work shift, it may be possible to collect data over the entire work period, at least as a preliminary analysis of exposure; later, detailed time and motion studies must be made of the work process, with specific exposure values for each operation and total time-weighted average exposure for the work shift.

The assessment of these data forms the basis for determining whether reduction of exposure is possible by administratively reducing exposure time. It must be remembered that such administrative control usually means that some other worker or workers must assume part of the toxicologic burden of the employee who had worked on this specific high hazard job. If the time spent off the high hazard job has no exposure, the high hazard exposure time is the total exposure. For the worker moved to a low exposure for the remainder of the workday, however, the total exposure will be the time-weighted average exposure of both the high and the low exposure jobs. This total exposure assessment must also be made for any employee who helps "fill in" on the high hazard job.

The use of such control demands close and continuing monitoring of these exposures both in the high and low or no exposure positions to assure that no overexposure occurs. The increase in the number of workers who are exposed on the high hazard operation is not an acceptable long-term alternative to engineering control. Even for a short period the impact of the increase in the number of workers exposed may result in additional medical examinations, specific medical procedures related to the exposure, informing the employee of the hazards, possible fitting with respirators, and increased environmental monitoring.

The analytical method as well as the sampling procedures must be the acceptable method used for routine surveillance and regulatory sampling to assure that the results can be compared to other data collected.

3 IMPORTANCE OF MODES OF ENTRY

An understanding of the physical and chemical characteristics of a workroom contaminant, its route of entry, and the effect of the hazardous material in workers is essential in the development of sampling strategies that will satisfy evaluation objectives. Several methods are used for classification of the modes of entry and the biologic reaction of the hazardous material in humans. The specific and detailed discussion of these classification systems are covered in Chapter 6, Volume I. These classification systems present a convenient method for the development of evaluation strategies.

Three classification methods—physiological, physical, and chemical—should be considered individually and collectively: these allow for an organized method to answer the basic questions for the development of an evaluation strategy. How does the toxic agent enter and react in the body (physiological classification)? What are the physical characteristics of the contaminant (physical classification)? What is its chemical structure (chemical classification)? This section deals with these classification systems and how they are used in the development of sampling strategies.

3.1 Inhalation

The inhalation of airborne hazardous material into the respiratory system is the most common route of entry for the contaminants found in the industrial workplace. Most of the occupational health standards and the TLVs of the ACGIH are based primarily on this route of entry. A few standards are based on other modes of entry, but for the most part they are considered to be an additional and/or complicating route of entry. A group of hazardous chemicals that may use one or several of the routes of entry into the body are those that result in a carcinogenic reaction in humans. The standards for this biologic effect group may not be based on an evaluation of an airborne dose–response relationship. The basic industrial hygiene approach to the evaluation of occupational hazards has been the collection of environmental data and the relating of that information to an adverse effect in workers. This approach relies on the assumption that the primary, and usually only, route of entry is the inhalation of the airborne contaminant into the respiratory system from the workroom air. The effect of exposure by other routes is less well defined.

The methods of collection of airborne contaminants are based primarily on their physical characteristics, and the method of analysis is based on their chemical characteristics. The selection of a method for field collection and/or chemical analysis may have to be modified as a function of other considerations. For example, the "analysis" for asbestos has been based on its physical characteristics, not its chemical classification. In this case any fibers exhibiting a defined physical characteristic are considered to be asbestos. This evaluation technique is not able to differentiate between fibers that have different chemical identities, such as tremolite asbestos and fibrous talc dust, which cause entirely different diseases in humans. This suggests the need for a more complex chemical analysis to differentiate between the several types of fibers. Asbestos is found in many commercial talcs (23).

As a group, particulates, including dust, fibers, fumes, and mist, have been difficult to assess. Generally the collection and analysis of dust and fibers have been based on physical characteristics and modified in relation to the reaction of the respiratory system to the contaminant; for example, the sampling method for dust has been modified based on the deposition in the lung of sphericallike particulates of a specific aerodynamic size. The lung retention models used for aerodynamic sphericallike particulates that have been used for silica, lead, and other particulates were found to be inapplicable to asbestos because of its fiber configuration. This resulted in the development of the concept of counting only fibers with a certain aspect ratio between fiber length and diameter.

Normally fumes and mist, though having been formed by different physical conditions, are collected by the same methods employed for the other particulates, since they all exhibit similar deposition characteristics in the respiratory tract. For example, NIOSH recommends that for breathing zone sampling for lead dust and fumes (24), chromic acid dust or mist (25), asbestos fibers (26), and silica dust (27), all should be collected on some type of filter media. These four chemical particulates all require some type of evaluation after collection, since direct reading instrumentation is not normally used and three of these (silica, lead, and chromic acid) require chemical analysis. Only the asbestos assessment is based on physical characteristics alone.

The collection and analysis of environmental samples of gases and vapors can be considered together, since the need to differentiate between the two physical states is primarily in the evaluation of the hazard potential of a contaminant, not in the consideration of how the substance is collected and analyzed. In the case of gases and vapors the collection method may be based on the contaminant's physical or chemical characteristics. The basic collection methods may be physical entrapment on one hand, or chemical reaction of the contaminated air with an absorbing or adsorbing medium. Physical entrapment is usually accomplished by collecting the air to be analyzed in an evacuated container, such as a bottle or a plastic bag. The methods based on the chemical characteristics of the airborne contaminant and of the collection medium rely on the collecting medium to remove the contaminant from the airstream as it moves through or over the medium, which may be a solvent or a sorbent, such as silica gel or activated carbon.

Under certain conditions gases or vapors are deposited on particulates while still in the air and may be inhaled as particulates. Where both particulates and gases and vapors may be contaminants, such as in aluminum reduction and coke oven operations, consideration must be given to the method of collection and analysis of both the particulates and the gases and vapors.

3.2 Ingestion

Ingestion is not considered to be a major route of entry of a toxic material in the workplace. But it is a possible route of entry and cannot be completely discounted, and at least some of the material that is inhaled may be swallowed and absorbed into the digestive tract. Ingestion is considered as presenting a significant problem when the contaminant is highly toxic or carcinogenic. The usual method of evaluation is not by environmental sampling but by evaluation of work practices. A toxic material that is deposited on food, on tobacco, or in liquids may be ingested. Possible sources of contamination, and the possibility of ingestion caused by poor work habits, must be evaluated. For example, ingestion of lead dust by a worker who held his sanding disc between his teeth when not grinding on automotive bodies presented evidence of lead intoxication, although the environmental levels for lead were essentially negative.

3.3 Skin Absorption

Skin absorption, along with ingestion, is not considered a major route of entry for industrial contaminants. Not all chemicals that cause skin irritation effectively penetrate through the skin barrier and reach the bloodstream. In the 1977 list of TLVs, the ACGIH indicates that about 25 percent of the chemicals included may make a potential contribution to the overall exposure by absorption through the skin, including the mucous membranes and the eyes (28). Included on that list are such highly toxic chemicals as parathion and tetraethyl lead.

It is not possible to make environmental measurements for the effect from skin absorption, as with ingestion, and any evaluation of the contaminant must be by observation of work habits, including personal hygiene, changes of clothes, and use of personal protective equipment. The use of wipe samples of the skin area has been suggested as a method for evaluating skin absorption, but the approach has not been widely accepted. As with ingestion, the compounds that penetrate the skin usually may also be inhaled. When making any assessment of the occupational environment, it is important that the possibility of an additional exposure resulting from skin absorption and/or ingestion be considered in addition to the results of samples collected in the breathing zone of the worker.

4 WORK EXPOSURE EVALUATION

Sampling of the occupational environment, as outlined under Section 2 is done for a number of reasons, all of which, except for sampling to measure the effectiveness of control programs, and animal experimentation, are directly related to worker exposure. In all cases the biologic reaction of the worker or former workers to the chemical insult must be measured.

The environmental evaluation for compliance and surveillance may be related to the worker only in the abstract, that is, to an existing occupational health standard not to a direct effect in a worker. Environmental assessment for clinical cases and epidemiologic purposes are directly related to the evaluation of the worker's health, even if the assessment of the worker's health is by evaluation of past medical records.

Much of the same is true of medical assessments; that is, many of the medical examinations are for medical evaluations of nonoccupational and preventive purposes not related to occupational exposure. Occupationally related medical assessment may also be done only in the abstract—not directly related to actual environmental exposure levels, but for compliance and routine medical surveillance and related directly to existing standards or "normals" for clinical tests and examinations. Medical assessment for epidemiologic studies, and for specific chemical hazard evaluations, are directly related to worker exposure and biologic response. It is in this area of epidemiology and clinical case studies that the closest scrutiny is given to the medical and environmental records that have been maintained and to the medical examinations and environmental assessments that are to be completed.

The probable cause for the poor quality of many of the records available today is that at the time of collection, it was not intended that these data be used for epidemiologic or clinical case evaluation. For this reason environmental and medical data should be collected and maintained as if they were going to be used for clinical and epidemiologic research purposes. It is in relation to epidemiologic and clinical case studies that the direct connection between the worker and exposure is most likely to exist and the data collected must satisfy not only the objectives of environmental or medical monitoring but the interactions between the two sets of data. The basic differences between epidemiologic studies and clinical cases from the standpoint of the assessment of the data involve the number of workers covered in the study, whether it is necessary to draw a sample of the population, and whether inferences can be made to a larger universe.

The techniques and uses of epidemiology for occupational health assessment, and the effect of the routes of entry, have been discussed in detail in chapters 5 and 6, of Volume I and are not repeated here.

Choice of specific measurement technique for worker exposure evaluation many times is limited by the information that is available and/or can be collected, and the technique chosen may not be the preferred method. For example, the lack of any or adequate past medical records or a change in sampling or analytical procedures may limit the value of any retrospective study with the objective of defining a dose–response relationship. The choice of measurement techniques must be based first on the best epidemiologic method to answer the main objective of a study and second, on the information that is available or can be collected and whether it allows for valid conclusions to be drawn. If it does not appear that the data would lead to valid conclusions, the next best approach must be considered, and so on, until data availability and evaluation goals are compatible.

The final selection of a study protocol may be further modified by a reexamination of the evaluation goals to determine whether the end result will satisfy all the questions raised by the limitation of availability of data. For example, if the objective was to study the relative risk in specific job classifications by differences in exposure between "disease" and "no disease" groups and the most detailed job classification information available was classification by industry (chemicals, textile mills, lumber, tobacco, etc.), the study would give no information by specific job classifications. In fact, it might mask serious health risk problems in a specific job classification by comparing it grouped with all other job classifications within its overall industrial classification. Under these circumstances different methods would have to be employed to compare the relative risk of disease by job classification.

The problem with the recognition of potential interference from other factors has been covered in Chapter 9, Volume 3A, and in this chapter (Section 2.3.1). However, it is important to emphasize several of the compounding factors that have resulted in invalid interpretation of data. Unfortunately the erroneous interpretations and conclusions may not be apparent when the results of the specific study are published. These factors of special concern are the reliance on inadequate environmental data and not treating all medical data with equal scrutiny. For example, evaluating both primary and secondary causes of death for the exposed group but not the unexposed group. If the epidemiologic measurement technique is not compatible with the information that will be available and the study objectives, either consideration should be given to another approach or potential sources of error of the study should be acknowledged. It must always be considered that because of lack of data, animal experimental studies may be more appropriate than the utilization of the epidemiologic approach.

In prospective studies it is possible to decide which specific biologic parameters should be measured, the type of environmental samples that should be collected, and the intervals at which the measurements should be made. It is possible to include in such investigations specific biologic parameters that are precursors to disease and can be used to measure subclinical manifestations. This, of course, can occur only after development of the correlation between biologic test levels and environmental exposure levels.

The methods of measurement will change with the mode of entry and biologic action and must be considered in the protocol design. For example, in a protocol for the study of a pulmonary irritant, the medical examinations could be designed to measure the possibility of short-term (acute) and long-term (chronic) effects in the worker, and the environmental evaluation could be designed to detect short-term excursions (ceiling values) and long-term (time-weighted average) exposure levels. In the case of an irritant, use of detailed medical history and/or questionnaires may be necessary in addition to the medical examination procedures, to determine the incidence of cough, sputum, dyspnea, and cigarette smoking, to obtain information that may not be evident on physical examination. In addition, careful scrutiny must be given to the action of any other chemicals that many times are found in the workplace in conjunction with irritants.

The prospective assessment of exposure to a carcinogen is usually a follow-up study, after the carcinogen has been identified by retrospective epidemiologic or animal studies. Long-term tests in experimental animals are valuable in assessing the potential carcinogenesis of a chemical used in industry. These animal studies do not necessarily confirm that a chemical is carcinogenic in humans, but they are valuable corroborating evidence.

Several approaches are possible in a carcinogenic prospective study. Traditionally investigators have examined differences between exposed and nonexposed groups, with follow-up studies to determine the effect of exposure level on the occurrence of disease and/or the latent period from first exposure to onset of disease. Further evaluation of specific tests to measure subclinical signs or symptoms of disease may be necessary, and additional animal experimentation and chemical tests may be made to confirm the results of the epidemiologic study. Specific attention must be paid to other possible exposures, particularly for any compound that may act as a cocarcinogen or a potentiator. The complexity of the chemical structure of many of the known carcinogens requires specific analytical procedures. As a result, new field sampling and analytical methods may have to be developed.

The identification of chemicals having mutagenic potential has traditionally been from retrospective studies. The more recent development of tissue culture techniques and use of animal experimentation has resulted in the use of these techniques for screening for mutagenic potential.

Clinical studies in workers and epidemiologic methods are very difficult to use effectively, since the effect of mutagenic reaction may not be apparent for several generations. However, workers exposed to chemical compounds that cause sterility or semisterility, depression of the bone marrow, teratogenic or carcinogenic effects, inhibition of spermatogenesis or oogenesis, inhibition of mitosis, inhibition of immune response, or chemicals whose structure is closely related to known mutagens, should be closely followed.

Systemic poisons, on the other hand, may be ideal candidates for prospective, as well as retrospective, studies. Chapter 6, Volume I, lists examples of systemic poisons and the organs that are injured by exposure to them. The latent period for extensive damage to an organ system may be long; however, early signs and symptoms, and results of specific tests, indicate some biologic reaction prior to permanent damage. Thus it is logical to subject systemic poisons to prospective assessment. The standardization of medical examinations, which can be achieved, will minimize inconsistencies between examinations and can be directed for in-depth evaluation of the organ system affected. Careful recording of the data collected also allows for the determination of the significance of subjective signs and work complaints between exposed and nonexposed groups. It is also possible to correlate the results of specific and nonspecific tests with the results of environmental exposure. The epidemiologic method of intervention also can be used effectively with chemicals that cause systemic poisoning if removal from exposure causes a reversal of evidence of the disease in the worker. This requires the development of sufficient medical and environmental data for in-depth analysis and correlation between exposure and effect. In the use of these techniques the possibility of some other mode of entry, as well as biologic action, must be considered—for example, ingestion of lead and absorption through the skin of aromatic hydrocarbons, in addition to inhalation.

Unusual physical conditions and physical agents can affect the assessment of the biologic insult caused by a chemical agent. These confounding factors must be considered and adjustments made in the analysis of the data, just as adjustments are made for age, smoking, and so on. Heat, cold, altitude, increased pressure, high demands for physical activity, and work schedules are examples of some of these physical conditions that are addressed later in this chapter. Another example of the interaction between a chemical and a physical agent is the photosensitivity effect on the skin of certain chemicals when the skin is exposed to ultraviolet electromagnetic energy. If at all possible, any combined effect should be eliminated in the study population; otherwise such factor(s) must be given careful attention when conclusions are based on the data.

The establishment of an equilibrium between the amount of a chemical agent in the body tissue with environmental levels is covered in detail in Chapter 6, Volume I and Chapter 3, this Volume. It is important to recognize that in clinical cases and in morbidity studies, this equilibrium of chemicals in blood, urine, tissues, and expired air and its rate of detoxification and elimination is critical to the determination of the proper time interval between exposure and collection of biological specimens. The effect of intermittent exposures must also be considered.

5 RELATING ENVIRONMENTAL EXPOSURE AND CLINICAL DATA

Defining the relationship of environmental exposure data to clinical measurements is the basis for the practice of occupational health by the specialties of industrial hygiene and occupational medicine, as well as the contributions made by many other scientific disciplines, including chemistry, toxicology, epidemiology, and engineering.

The importance of developing the relationship between the environmental measurement and clinical reaction cannot be overemphasized, for without establishment of this relationship, the collection of data is of little value. This relationship is imperative for the evaluation of exposure regardless of whether the data are being used for research, routine plant surveillance, or engineering control. The objectives of any evaluation program must be to assess this basic relationship clearly and concisely. Unfortunately, it is not possible to control all the parameters that can affect this relationship in surveillance of worker exposures. The two basic questions in the final assessment of the data collected must be: Does the environmental exposure and clinical data fit a predictable pattern? And is there a rational basis for any aberrations from the predictable relationship? Consideration must be given in depth to each of these questions before any final conclusions or recommendations for corrective action are made.

5.1 Do Environmental Exposure and Clinical Data Fit a Predictable Pattern?

The absence of any expected sign of symptom (negative data) can be as troublesome as the finding of clinical signs (positive data) that were not expected. In the evaluation of a single or a few clinical cases, the variation in human tolerance to the chemical stress must be considered. This is also true in epidemiology studies but perhaps is a less significant point than in individual clinical cases. The worker or workers are more often than not exposed to more than one contributing cause, both on and off the job, that result in a more complex evaluation of the effect of workroom exposures than may have been anticipated initially. In these cases chemical exposure may only be an additional contributor to an already existing disease or clinical symptom. For example, the consideration of food intake is particularly important when the symptoms are being evaluated in relationship to environmental exposure to inorganic arsenic and to urinary arsenic levels. The urinary arsenic levels may be excessively high when the worker has eaten seafood within 48 hr of the collection of urine samples (29).

NIOSH, in the criteria document on inorganic mercury (30), also pointed out several other pitfalls in attempting to correlate clinical data and environmental levels, including a lack of substantive data to support a relationship between mercury excretion with signs and symptoms, and environmental levels. NIOSH based these conclusions on the belief that it is impossible to establish a level at which no signs and symptoms are observed because the signs and symptoms of mercury exposure, which are also prevalent in the general population, are nonspecific. In addition, it was pointed out that the validity of sampling and analytical methods for environmental monitoring was open to question. The use of biologic sampling for mercury has resulted in a wide variance in correlation factors with environmental and clinical results. As a consequence, the use of a biologic standard for mercury would be meaningless for assessment of the general worker populations. Biologic monitoring of several biologic media (urine, blood, breath, and hair being the most common: see also Chapter 3) has been used with a variety of results for other chemicals, including organic and inorganic lead, benzene, arsenic, fluorides, toluene, carbon monoxide, tetraethyl lead, DDT, organic phosphate insecticides, halogenated hydrocarbons, hydrocarbons, aromatic nitrogen compounds, ketones, aldehydes, and alcohols. For additional details on the interpretation of these biological tests in relation to environmental exposures, see *Biological Monitoring for Industrial Chemical Exposure Control* (31).

The correlation of biologic analysis and environmental data was considered to be so significant for carbon monoxide that NIOSH (32), in its recommendation for a standard, based the proposed standard on the correlation between the carboxyhemoglobin levels and a time-weighted average exposure to carbon monoxide for an 8-hr period (33).

The assessment of the total life experience of the worker in conjunction with the clinical picture and the environmental exposure information is necessary, particularly when assessing a few clinical cases—for example, atypical clinical effects resulting from physical effort on the job may be the expected results if it is known that the employee has a cardiac problem or is working in a hot environment.

Interpretation of epidemiologic-related environmental exposure data may not follow the predicted pattern. This may be caused by only a slight excess in relative risk, use of an inappropriate control group, lack of measurement of other possible chemical, physical, or emotional stresses, and other conditions (smoking habits, nutrition, disease, family history, marital status, economic status, age, sex, race, work schedule, off-the-job stress, etc.) that may affect the response of the workers in the cohort under investigation. The increase in the number of workers included in the sample may help reduce the effect of the variability of some of these parameters, but the increase in the number of workers studied may result in inability to investigate in depth all possible causes of the variation from the expected pattern of correlation. Inconsistent results from different epidemiologic methods are not uncommon, such as negative results between the exposed and nonexposed groups and positive results between first-time exposed and disease development. This may suggest the need for additional analysis and possibly a follow-up study to establish a dose–response relationship.

Nonenvironmental parameters can have a major impact on the results and conclusions drawn from any study. The effect of some of these nonenvironmental conditions have been reported in epidemiologic studies by using wives of the workers for controls (34). Kalačić (35), in an attempt to determine the effect of domestic and nonoccupational factors on cement workers, studied workers in nondusty trades and wives of the cement workers and the nonexposed workers, who were also used as controls. He concluded that there was no significant difference between the two groups of wives and suggested that domestic and nonoccupational factors probably were not significant in the worker exposure. He was not able to draw any conclusions on the effect of smoking habits, probably because of the lack of similarity in smoking habits between the workers and the wives. Higgins (36), in the study of coal miners in England, used wives as a control and by comparing differences between the workers and the wives was able to determine a difference in the incidence of bronchitis between those who worked and their wives.

Changes in environmental measurement techniques and diagnostic procedures, with the passage of time, have made it difficult in some cases and impossible in others to develop any meaningful retrospective studies, and these changes have a direct impact on the establishment of any predictable pattern between chronic occupational diseases and environmental exposure measurements.

The validity of an observed correlation between environmental exposure and clinical data may be open to question until the work has been replicated, if the observed relationship is not the one expected or has not been previously reported in the literature.

5.2 Is There a Rational Basis for the Aberration of the Predictable Relationships?

The search for the cause of deviations from the expected relationships between environmental exposure and clinical change would start with a reevaluation of the environmental sampling strategies discussed in Section 2 and of the clinical examination objectives and procedures. Special consideration should be given to the parameters of the evaluation strategies that were designed to hold constant as many as possible of the parameters that could affect the measurement of the occupational exposure and to determine whether the methods employed did "in fact" hold constant these interfering factors (e.g., were the analytical results correct, and was there a possibility of an additional exposure to some other chemical contaminant?). Any new information that may have become available from other studies or sources should be assimilated.

A number of specific parameters discussed in the following sections should be dealt with in the reevaluation of the results of an occupational health study if the expected correlation of environmental and clinical data is not realized. If careful consideration is not given to these parameters as well as others that are specific for the hazard being evaluated, the relationship between exposure and effect cannot be well defined.

5.2.1 Unknown Past Exposures and Medical Histories

In many cases the medical histories of employees are of questionable value, particularly when medical information has been obtained from several different employers. In some large industrial concerns and in many smaller ones, no preplacement medical examinations or periodic follow-ups have been given, or such exams have been given only over the last few years. In these cases it may not be possible to make a meaningful correlation of exposure to clinical evidence of chronic occupational disease.

In other circumstances involving unknown past exposure records, unknown work histories, and retired or deceased employees, or persons who have moved to other work (i.e., cannot be followed), these employees may be lost to a study, adversely affecting the validity of the correlation of the clinical and environmental results. The presence of workers or former workers with poor work or medical histories that lack such data as age, sex, race, and family medical history, can also adversely affect the correlation of exposure and clinical data. This lack of data may be the result of poor recordkeeping and/or failure to recognize what information may be of value in the future. For example, the history of smoking habits today is a very important and necessary piece of information for many studies in which the route of entry is by inhalation and chronic results are expected. Detailed smoking histories were obtained in the past as part of only a few medical histories, and the information that may be available is usually too superficial to meet present requirements.

The collection of environmental data and the maintenance of environmental records was even less well organized in the past than was the maintenance of medical records. Environmental data usually were kept a few years and then destroyed; only occasionally were the data kept with the employee's medical file. These recordkeeping practices were justified at that time by acknowledging that the constant changes in medical procedures and environmental sampling techniques reduced the value of old records to the questionable level at best.

Two examples of changes in procedures that improved the practice of industrial hygiene and occupational medicine but reduced the value of existing records are found in the change in the collection and analysis for asbestos (a) from counts for dust particles to (b) collection of fibers and counting fibers that have a certain length-to-diameter ratio. Even though there have been several attempts to correlate the two procedures, insufficient data have been collected to be convincing (13, 37). The change in the radiologic classification system for pneumoconioses and the improvement in the techniques for taking and developing X-rays have had a similar effect on the evaluation of chest films.

Records of both medical and environmental data should be retained even after changes in sampling and/or medical procedures, since these records may have at least some value in future years for research purposes and will certainly be of value for compliance and legal use.

5.2.2 Off-the-Job-Stresses

Off-the-job stresses must be investigated as possible sources of deviation from the expected relationship between environmental data and clinical experience and can be considered in three basic classifications: additional exposure off the job to the same chemicals used on the job, exposure to other hazardous chemicals, which may complicate the exposure on the job, and off-the-job stresses that are not related to chemical exposure but may be the result of physical stress or infectious disease. An example of continuation of exposure while off the job is provided by a worker who upon medical examination was found to have been overexposed to uranium (radon daughters), presumably from the work of removing uranium paint from aircraft instruments. The work was done in a completely enclosed glove box. Removal from the job did not cause a change in the worker's condition, and investigation of off-the-job exposure revealed that the employee painted luminous street markers as a private enterprise from home.

In addition, a number of cases of toxicologic effects in the community or in the worker's family suggest that continued exposures may occur to workers while away from the workplace. In one reported case (38), a mother of a beryllium worker died of beryllium-induced disease. The only known exposure had been at home, from the beryllium dust that had been brought in on the clothes of the worker. As early as 1966 a total of 60 neighborhood cases of beryllium disease had been reported to the Beryllium Registry (39). Similar family and community cases have been reported from asbestos exposure (40).

Community exposures must also be considered. For example, high levels of lead are known to be in both community air and water supplies in some metropolitan areas.

The complicating effect of alcohol on workers exposed to trichloroethylene is a classic example of the effect of off-the-job stress in addition to workplace exposures. Exposures to home cleaning solvents, degreasing agents, photographic chemicals, paint pigments and solvents, and isocyanates are all recognized as possible sources of exposure at home or in the workshop.

Infectious and parasitic diseases, including anthrax and brucellosis, have long been associated with agricultural workers. These conditions could also contribute to a worker's combined effect of at-home and on-the-job exposures if the worker lives on a farm. It is also possible that poor nutrition, substandard housing, and poor economic conditions can reduce resistance to workplace exposures and result in the development of other diseases. Air pollutiuon, physical stress, heat, and altitude can have some effect on workers who may have progressive respiratory and circulatory difficulties brought about by occupational exposures. The major off-the-job, and in some cases on-the-job, stress is the smoking habits of the worker.

5.2.3 Work Schedule

Little research has been done in this country relating the effects of the work schedule or shift work to occupational disease; however, the physiological effects caused by shift work can certainly influence the health of the worker. El Batawi and Noweir (41) indicated that the stress of night work adversely affected the blood pressure of garage workers. Such stress could result in peptic ulcer, fatigue, nervousness, irritation, and insomnia. In another study (42) it was reported that night workers may have a higher incidence of peptic ulcer than those working on the day shift. Peptic ulcers were also found in (night) shift workers in Egypt, as well as chronic gastritis. The authors suggested that this may be due to the stress of shift work, as well as nutritional factors (41).

An excellent review of the physiological and psychological aspects of night and shift work has been published (46), including the effects on the circulatory functions, ventilatory capacity, circadian rhythm, and so on. If unexpected physiological responses are noted in shiftworkers that may not be directly attributable to work conditions, considerations should be given to the work cycles, break time, environmental conditions at home during sleeping time, sleep patterns, and working on more than one job. Chapter 5 covers this subject area in detail.

5.2.4 Additive, Synergistic, and Antagonistic Effects

The TLVs for chemical substances (28) contain in Appendix C a formula for estimating the combined or additive effect of two or more chemical hazards. This formula treats the combined exposure as if the maximum effect of the combination could not be larger (no synergistic effect) than the maximum of each chemical. This reduces the exposure proportional to the concentration and the TLV of each compound. Even assuming that the mechanism of action is the same, this approach can underestimate the effect of combined exposure. In addition to the formula for mixtures, ACGIH suggests (28), (Appendix C) the use of the formula for independent biologic effects when the toxicologic response is different. In these cases the formula assumes that there is no interactive effect and that it is possible to be exposed up to the TLV for each compound without exceeding the TLV for the combined exposure.

It is unfortunate that it is necessary to use such an approach in the case of exposures to two or more chemicals when it is not known whether the action in the body is additive (as suggested by the mixture formula), synergistic (which would not give sufficient protection), or antagonistic (affords more protection than is necessary). There is a synergistic effect from the combination of cigarette smoking and asbestos inhalation in the production of bronchial carcinoma (26), and there is increasing evidence of the synergistic effect of cigarette smoking in other worker populations, such as uranium miners. There is some evidence of a synergistic reaction between arsenic and sulfur dioxide (43). Laskin et al. (44) reported a synergistic effect to rats inhaling benzo[a]pyrene and sulfur dioxide in combination in the production of bronchial mucosal changes and tumors of bronchogenic origin. Both these chemicals are present in some industrial operations (e.g., in the production of coke from coal). Antagonistic effects result in a reduction in the biologic response over what would have been normally expected, such as the response that occurs when the combined fumes of nitrogen oxides and iron oxides are found, as in some welding operations (45).

5.2.5 Abnormal Temperature or Pressure

Individuals who must work in elevated temperatures usually become acclimatized or move to other jobs that are less demanding. This natural selection reduces the effects that would be expected from increased temperature; however, the effects of temperature and heavy work may produce heart failure in workers with cardiac problems. Arteriosclerosis and enlarged hearts have been found in excess of that expected in workers in steel and glass industries. Organic heart defects are reported in another hot industry, foundries.

What the combined effect of chemical exposures may have been is not known. In all probability the effect from excessive heat is a more significant contributing factor to the inability of a worker (who already has some medical complications) to resist any additional insult from a chemical or physical stress. Linemen, loggers, surveyors, and construction workers are also exposed to a variety of plant and wood hazards; the most common of these are poison oak and ivy, in addition to exposure to chemical hazards and heat stress.

Workers exposed to cold temperatures may work out of doors and not be exposed to high concentrations of chemical agents. Some workers, such as packing house, construction, and highway workers, and mail carriers and firefighters, are likely to be exposed during a workday to a combination of heat and cold that may adversely affect the ability to resist infectious diseases. Other workers, such as police officers and highway construction and tunnel workers, may be exposed in addition to chemical agents such as carbon monoxide, and exhaust gases from internal combustion engines. Liquefied gas and oil field workers may be exposed to hydrogen sulfide as well as to cold. Little is known about the effect of cold in combination with chemical agents. It is known, however, that cold as well as hot environments result in increased frequency of minor industrial accidents. This suggests that under adverse temperature conditions the employee does not follow normal expected safety rules and in all probability may not be as conscientious in using personal protective devices and procedures as under more normal work conditions.

The primary concern with increased pressures has been associated with tunnelers, caisson workers, and divers. Recently compression and decompression problems have been intensified by the development of offshore drilling for crude oil, which is requiring divers to go to greater depths. There has been increased concern about exposure to compressed air because of the growing number of skin divers (off-the-job stress) and tunnel workers. In the case of compressed air, concern is not only of the consequences of breathing compressed air for long periods but also the possibility of contamination of the breathing air. The possibility of contaminated compressed air requires controls of the air similar to those required for use with air-supplied respirators. The increased adverse effects of both carbon dioxide and oxygen under pressure are well known. These subject areas are covered in detail in Chapters 9 and 11.

5.2.6 Extreme Job Energy Requirements

Energy requirements may have a very significant effect on the expected relationship between environmental data and the clinical response observed. Extreme or excessive energy requirements can result from work periods longer than the normal 8-hr workday, and higher than normal physical activity during the regular workday. The expected clinical response is normally based on a time-weighted average exposure over an 8-hr workday. In the preface to the TLVs for chemical substances (28), the ACGIH defines the conditions that are to be used for the evaluation of workroom exposures by a time-weighted average as being for a normal 8-hr workday over a 40-hr workweek. This defines the conditions by which comparisons are made between environmental data and clinical response.

The concentration of the toxic agent that reaches the site of the biologic response in the body can be increased by extending the time of exposure, or by increasing the breathing rate by increasing the physical activity (energy demands) of the job within the 8-hr period. The ventilation rate of the lungs increases about twentyfold from measurements taken at rest to those taken during heavy exercise. As a consequence, it is expected that heavy work demands will increase the amount of toxic material that is inhaled and deposited in the body with a resultant increased toxic action. Although work conditions for the most part do not normally require heavy energy demands, when such demands are made even for short periods, they could result in a departure from predictable patterns of biologic response if the energy demands have not been taken into consideration in the initial evaluation.

5.2.7 Altitude

The effects of high-altitude flying are minimized by the use of pressurized aircraft and supplementary oxygen. The effect of high altitude on humans, when it is the living environment, is one to which the worker is subjected not only on the job, but at home as well. There are physiological differences between populations native to high altitudes and those that recently have moved to the high areas. Workers who are native to the mountainous area may have working capacity similar to those who work at sea level, whereas those who have been acclimatized may hyperventilate during exertion or exercise. This difference in reaction to increased altitude may result in some inconsistencies between the observed and expected response to a specific toxic chemical, if the results are considered together for both the native and acclimatized workers. The hyperventilation in the acclimatized worker will result in the same type of reaction that would be expected from a heavy energy demand job, with additional effects due to lowering of carbon dioxide tension.

Decreased pressure also requires calibration of field sampling equipment at the altitude at which environmental sampling is done. Also, see Chapter 9 for additional information on this subject.

5.2.8 Indoor Air Pollution

Interest has grown in the number of illnesses and complaints from office workers related to working conditions. In these cases the investigation of the traditional sources of exposures may not result in finding the cause of the complaint or illness. As a result, it may be necessary to investigate further. In the past many of these conditions were considered as complaints without merit and no corrective action was taken. It may be necessary to examine the ventilation system, make-up air intake, and other factors for possible sources, not only chemical contamination, but also for molds and other microorganisms.

Chemicals from smoking, synthetic building materials and furniture, copying machines, and other office equipment may add to the problem.

As an example of another source of contamination, bacteria in a water heater was found to be the source of Legionnaires' disease for heart–lung transplant patients at Stanford Medical Center. In this case it was not believed that other patients or staff were at risk, but only those in the high risk group of the heart–lung transplants. This does indicate the degree and intensity required to find the sources for these types of exposures.

5.2.9 Need for New Data Assessment

The data and information necessary to assess properly the effect of the specific parameters that can cause deviations from the predicted relationship, as well as others such as differences in sex and race, are not readily available, and much more research must be done. Many of these conditions, such as the altitude effect, will affect the evaluation of relatively few working environments; however, others such as synergism apply to a much wider segment of the working population, and all can cause the development of erroneous information.

6 WORKROOM ENVIRONMENTAL EXPOSURE LIMIT

Any procedure or evaluation of worker exposure to noxious chemicals carries the assumption that at some level of exposure there will be some type of biologic effect, and it is necessary at some level of exposure to initiate engineering or administrative control of that exposure.

The development of such exposure guides had been based on their use by a competent industrial hygienist familiar with daily exposure levels, an understanding of the basis of the exposure limit, and a subjective evaluation of all factors to determine the appropriate control action necessary.

A number of exposure limits that do not have the status of law have been used for many years in the United States; these include the Hygienic Guide Series of the AIHA, the standards of the American National Standards Institute (ANSI), Z.37 Committee, the TLVs for Chemical Substances of the ACGIH, the standards of the National Academy of Sciences–National Research Council (NAS–NRC) and NIOSH, as well as the legal standards developed by OSHA, MSHA, and some state authorities where the basis for regulating is not derived from federal law, as in Pennsylvania. These limits include time-weighted averages, ceiling values, maximum peaks above ceiling, short-term limits, and emergency exposure limits. Richard S. Brief, in *Basic Industrial Hygiene—A Training Manual,* developed a table covering most of these standards (Table 7.1).

Both ACGIH TLVs and the ANSI Z.37 guidelines have been adopted, at least in part, as OSHA standards (3). Neither of these exposure guides will fill the requirement for standards as outlined in the Occupational Safety and Health Act, but they were promulgated as OSHA standards either as national concensus standards (ANSI) or as established federal standards (ACGIH TLVs) because they had already been adopted as federal standards under the Walsh-Healy Act. In cases where there were both federal standards and national consensus standards, the Secretary of Labor was to promulgate the standard that assured the greatest protection to the employee. The ANSI standards as consensus standards were adopted in preference to the TLVs of ACGIH when there was a choice to be made.

The largest list of environmental exposure limits has been developed by the ACGIH's Chemical Agent Threshold Limit Committee, then adopted by the membership of that organization. The ACGIH publishes a new list of TLVs each year, including revisions to already existing TLVs, and adds chemical compounds when it has determined that limits are needed because of exposures in the workplace. It has been estimated that approximately 95 percent of the

Table 7.1 Organizations Recommending Occupational Health Limits[a]

Source	Limits	Applicable Daily Time Period	Comment
American Conference of Governmental Industrial Hygienists (ACGIH) P.O. Box 1937 Cincinnati, Ohio 45201	Threshold limit value (TLV) Ceiling value	8-hr/day time-weighted 15-min	For both chemical and physical agents; limits are published annually
American National Standards Institute (ANSI) Z.37 Series[b] 1430 Broadway New York, New York 10018	Time-weighted average Ceiling value Maximum for peaks above ceiling	8-hr Related to 8-hr average Time specified	Published irregularly
Pennsylvania Department of Health P.O. Box 90 Harrisburg, Pa. 17120	Short-term limits	5, 15, and 30 min	Last published April 24, 1970
National Academy of Science—National Research Council (NAS–NRC) 2101 Constitution Avenue NW Washington, DC 20418	Short-term public limits (STPL) Public emergency limits (PEL)	10, 30, 60 min and 4–5 hr/day, 3–4 days/month 10, 30, 60 min	Published irregularly
American Industrial Hygiene Association (AIHA) 475 Wolf Ledges Parkway Akron, Ohio 44313	Emergency exposure limits Hygienic guides Maximal atmospheric concentration Short exposure tolerance Atmospheric concentration immediately hazardous to health	5, 15, 30, and 60 min 8 hr Time specified Time specified	Published irregularly
National Institute for Occupational Safety and Health 5600 Fishers Lane Rockville, Maryland 20852	Time-weighted average Short-term	8-hr/day Time specified	Published as guidance for OSHA

[a] *Source*: R. S. Brief, *Basic Industrial Hygiene—A Training Manual*, Exxon Corp., New York, 1975. Reprinted by permission.
[b] The (ANSI) Z.37 Committee is no longer developing standards.

exposures in industry in the United States is covered by an ACGIH TLV. Any new TLVs or any changes in the limits are publicly announced to the ACGIH membership and are placed on a tentative list of TLVs to allow interested professionals to present additional information and comment on the proposed limit.

The ACGIH TLVs have been adopted in other countries, even though the organization has strongly opposed the use of its TLVs in countries where working conditions differ from those in the United States (28). ACGIH also pointed out that the TLVs should be interpreted only by a person trained in industrial hygiene; and the TLVs should not be used as a relative index of hazard, or toxicity, for the evaluation or control of community air pollutants, for evaluating toxic effect, for continuous uninterrupted exposures or for other extended periods, or to verify or disprove the existence of a physical condition or disease (28).

This wide acceptance of ACGIH TLVs has resulted in these limits being used as the basic working standard for the measurement, evaluation, and control of workplace hazards for many years. The important differences between these suggested guides and the legal standards being developed by OSHA involve the ability of a professional committee of ACGIH to act in a reasonably short time to subjectively evaluate data and to propose TLVs, and for the industrial hygienist, who uses the TLVs, to exercise professional judgment in evaluating a work hazard.

The preface to the TLVs for Chemical Substances in workroom air should be read each year for the minor changes that occur in the preface from year to year. The following quotation from the 1977 TLV list (28) of chemical substances gives a good indication of how the TLV Committee believes that the TLVs should be used, and their intent.

Threshold Limit values refer to airborne concentrations of substances and represent conditions under which it is believed that nearly all workers may be repeatedly exposed day after day without adverse effect. Because of wide variation in individual susceptibility, however, a small percentage of workers may experience discomfort from some substances at concentrations at or below the Threshold Limit; a small percentage may be affected more seriously by aggravation of a preexisting condition or by development of an occupational illness.

A separate ACGIH publication, "The Documentation of Threshold Limit Values for Substances in Workroom Air," gives the rationale for the TLV and quotes the primary references used to support the TLV. This publication and supplements to the documentation are published at irregular intervals; in the interim periods the annual proceedings of ACGIH contain the recommendations and discussions of the TLV by the membership of ACGIH at the time of their adoption at the annual meeting. The basis for the TLVs may be epidemiologic studies in industrial, clinical experience, or animal studies.

Recently there has been a significant effort by corporations to develop and utilize corporate workroom environmental exposure limits after decades of using exposure limits developed by both governmental and professional organizations. The primary objective of this effort has been to fill a void created by the increasing number of new commercial chemicals that are coming on the market, and increasing the use of mixtures of chemicals in any one commercial formulation. Also, this action has been stimulated by the significant increase in the availability of new toxicologic and epidemiologic research data from both governmental and industrial research on chemicals established in commerce for a number of years, as well as those chemicals whose commercial applications have been developed recently. The inflexibility of federal regulators to quickly develop new or change existing occupational exposure limits, and the professional organizations' approach of addressing mainly those chemicals whose use has been well established and have broad applications. In addition, these organizations seldom review research data and make recommendations for exposure limits for mixtures of chemicals.

The requirement for research data information for material safety data sheets to meet governmental regulations prior to approving commercial application for new chemicals or new applications for old chemicals and the necessity of making this information available to their own workers and those of their customers have had a significant impact on the decision of industry to assume part of the responsibility for the development of corporate workroom environmental exposure limits that have traditionally been the providence of governmental and professional organizations.

The position on exposure levels that must be taken by industry is limited by the necessity to always recommend workroom environmental exposure limits that are equal to or lower than governmental permissible exposure limits regardless of what the data may suggest. When the data do suggest a higher exposure limit, the only recourse is to petition the regulating governmental organization. Although these industrial efforts are commendable, the occupational exposure limits developed by industry also may have some limitations that must be considered in the evaluation of their suitability for use. Confirmation of the original studies on which the exposure limit is based may not be completed for several years or may never occur by similar testing or by more sophisticated testing protocols. In addition, the evaluation of the raw data and the recommendation for the exposure limit does not usually receive the same degree of scientific peer review and open public discussion as occurs with governmental standards development. The recommendations are normally developed by a smaller number of scientists than is possible in the governmental review process. This can result in greater dependence on the evaluation by one or several scientists rather than from a more broadly based scientific group. Finally, the acceptance of these industrial exposure limits may be questioned by some professionals.

The limitations of the use of industrially developed workplace exposure limits can be equally applied to the use of exposure limits developed by other organizations, including those developed by professional organizations. As a result of these considerations, each industrial hygienist or occupational health professional must evaluate the basic data and assure what can reasonably be adopted as an exposure limit for the purpose intended.

As a result of the inflexibility of the federal regulators to develop workroom exposure limits in a timely manner and give higher priority for some standards based on a higher potential for exposure in some state jurisdictions than in the nation as a whole, some state Occupational Safety and Health organizations have developed state workplace environmental exposure limits. In many ways these state organizations face some of the same limitations as discussed previously as faced by their professional and industrial colleagues. The California State Division of Occupational Safety and Health is an example of a state that has been very active in the development of workplace exposure limits.

These developing standard activities at the state level are an indication of another source that the industrial hygienist must check to be assured that the appropriate regulatory exposure limit is being followed and to assure themselves of the validity of an environmental workplace exposure level.

REFERENCES

1. Occupational Safety & Health Act, Public Law 91-596 S2193, 91st Congress, December 29, 1970, Superintendent of Documents, Government Printing Office, Washingnton, DC.
2. A. D. Hosey and H. L. Kusnetz, "General Principles in Evaluating the Occupational Environment," in: *The Industrial Environment—Its Evaluation and Control,* 2nd ed., C. H. Powell and A. D. Hosey, Eds., Public Health Service Publication 614, Superintendent of Documents, Government Printing Office, Washington, DC, 1965, p. B-1-1.
3. Department of Labor, Occupational Safety and Health Administration, "Occupational Safety and Health Standards," *Federal Register,* Vol. 39, No. 125, Part II, Subpart G-1910.93, June 27, 1974, p. 23540.
4. (American Conference of Govermental Industrial Hygienists,) "Threshold Limit Values of Airborne Contaminants for 1968, Recommended and Intended Values," ACGIH, Cincinnati.
5. N. A. Leidel and K. A. Bush, "Statistical Methods for the Determination of Noncompliance with Occupational Health Standards," Department of Health, Education, and Welfare, Publication (NIOSH) 75-159, Superintendent of Documents, Government Printing Office, Washington, DC, 1975.
6. A. E. Russell, R. H. Britten, L. R. Thompson, and J. J. Bloomfield, "The Health of the Workers in Dusty trades. II. Exposure to Siliceous Dust (Granite Industry)," Public Health Bulletin 187, Superintendent of Documents, Government Printing Office, Washington, DC, 1929.
7. G. P. Theriault, W. A. Burgess, L. J. DiBerardinis, and J.M. Peters, *Arch. Environ. Health,* **28,** 12 (1974).
8. S. J. Reno, H. B. Ashe, and B. T. H. Levadie, "A Comparison of Count and Respirable Mass Dust Sampling Techniques in the Granite Industry," presented at the Annual American Industrial Health Conference, Pittsburgh, 1966.
9. N. A. Talvitie, *Anal. Chem.,* **23,** 623 (1951).
10. N. A. Talvitie and F. Hyslop, *Am. Ind. Hyg. Assoc. J.,* **19,** 54 (1958).
11. N. A. Talvitie, *Am. Ind. Hyg. Assoc. J.,* **25,** 169 (1964).
12. S. Milham, "Cancer Mortality Patterns Associated with Exposure to Metals," in: *Occupational Carcinogenesis, Ann. NY Acad. Sci.,* **271,** 243 (1976).

13. H. E. Ayer, J. R. Lynch, and J. H. Fanney, *Ann. NY Acad. Sci.*, **132**, 274 (1965).
14. E. V. Olmstead, *AMA Arch. Indust. Health*, **21**, 525 (1960).
15. Department of Labor, Occupational Safety and Health Administration, *Federal Register*, Vol. 42, No. 192, Part VI, [29 CFR Part 1990], October 4, 1977, p. 54149.
16. R. A. Kehoe, "The Metabolism of Lead in Man in Health and Disease," *The Harben Lectures, 1960*, Reprinted from J. Royal Inst. of Public Health and Hyg., (1961), by McCorquodale and Co. Ltd., London.
17. J. J. Bloomfield and W. Blum, *Public Health Rep.*, **43**, 2330 (1928).
18. American Conference of Governmental Industrial Hygienists, "Documentation of the Threshold Limit Values for Substances in the Workroom Air," 3rd ed., ACGIH, Cincinnati, 1971.
19. R. J. Vernon and R. K. Ferguson, *Arch. Environ. Health*, **18**, 894 (1969).
20. A. Ahlmark and S. Forssman, *Arch. Ind. Hyg. Occup. Med.*, **3**, 386 (1951).
21. American Conference of Governmental Industrial Hygienists, *Industrial Ventilation—A Manual of Recommended Practice*, 14th ed., ACGIH, Cincinnati, 1971.
22. "Criteria for a Recommended Standard . . . Occupational Exposure to Hydrogen Sulfide," Department of Health, Education, and Welfare Publication (NIOSH) 77-158, Superintendent of Documents, Government Printing Office, Washington, DC, 1977, p. 3.
23. B. Weiss, and E. A. Boettner, *Arch. Environ. Health*, **14**, 304 (1967).
24. "Criteria for a Recommended Standard . . . Occupational Exposure to Inorganic Lead," Department of Health, Education, and Welfare Publication (HSM) 73-11010, Superintendent of Documents, Government Printing Office, Washington, DC, 1972, p. VII-1.
25. "Criteria for a Recommended Standard . . . Occupational Exposure to Chromic Acid," Department of Health, Education, and Welfare Publication (HSM) 73-11021, Superintendent of Documents, Government Printing Office, Washington, DC, 1973, p. 64.
26. "Criteria for a Recommended Standard . . . Occupational Exposure to Asbestos," Department of Health, Education, and Welfare Publication (HSM) 72-10267, Superintendent of Documents, Government Printing Office, Washington, DC, 1973.
27. "Criteria for a Recommended Standard . . . Occupational Exposure to Crystalline Silica," Department of Health, Education, and Welfare Publication (NIOSH) 75-120, Superintendent of Documents, Government Printing Office, Washington, DC, 1974, p. 101.
28. American Conference of Governmental Industrial Hygienists, "TLV's: Threshold Limit Values for Chemical Substances and Physical Agents in the Workroom Environment with Intended Changes for 1977," ACGIH, Cincinnati.
29. H. H. Schrenk and L. Schreibeis, Jr., *Am. Ind. Hyg. Assoc. J.*, **19**, 225 (1958).
30. "Criteria for a Recommended Standard . . . Occupational Exposure to Inorganic Mercury," Department of Health, Education, and Welfare Publication (HSM) 73-11024, Superintendent of Documents, Government Printing Office, Washington DC, 1973, pp. 75, 79.
31. A. L. Linch, *Biological Monitoring for Industrial Chemical Exposure Control*, CRC Press, Cleveland, 1974.
32. "Criteria for a Recommended Standard . . . Occupational Exposure to Carbon Monoxide," Department of Health, Education, and Welfare Publication (HSM) 73-11000, Superintendent of Documents, Government Printing Office, Washington, DC, 1973, p. V-5.
33. R. F. Coburn, R. E. Forster, and P. B. Kane, *J. Clin. Invest.*, **44**, 1899 (1965).
34. P. E. Enterline, *Public Health Rep.*, **79**, 973 (1964).
35. I. Kalačić, *Arch. Environ. Health*, **26**, 84 (1973).
36. I. T. T. Higgins, "An Approach to the Problem of Bronchitis in Industry: Studies in Agricultural, Mining, and Foundry Communities," in: *Industrial Pulmonary Diseases*, E. King and C. M. Fletcher, Eds., Little, Brown, Boston, 1960.
37. W. C. Cooper and J. L. Balzer, "Evaluation and Control of Asbestos Exposures in the Insulating Trade," *Proceedings of a Working Conference on the Biological Effects of Asbestos*, Dresden, 1968.

38. L. B. Tepper, H. L. Hardy, and R. I. Chamberlin, *Toxicity of Beryllium Compounds,* Elsevier, New York, 1961.
39. H. L. Hardy, E. W. Rabe, and S. Lorch, *J. Occup. Med.,* **9,** 271 (1967).
40. P. Champion, *Am. Rev. Respir. Dis.,* **103,** 821 (1971).
41. M. A. El Batawi and M. H. Noweir, *Indr. Health* (Kawasaki) **4,** 1 (1966).
42. A. Aanonsen, *Ind. Med. Surg.,* **28,** 422 (1959).
43. "Criteria for a Recommended Standard . . . Occupational Exposure to Inorganic Arsenic," Department of Health, Education, and Welfare Publication (NIOSH) 75-149, Superintendent of Documents, Government Printing Office, Washington, DC, 1975.
44. S. Laskin, M. Kuschner, and R. T. Drew, in: *Inhalation Carcinogenesis,* M. G. Hanna, Jr., P. Nettesheim, and J. R. Gilbert, Eds., Atomic Energy Commission, Washington, DC, 1970, p. 321.
45. H. E. Stokinger, in: *Occupational Diseases,* W. M. Gafafer, Ed., Public Health Service Publication 1097, Superintendent of Documents, Government Printing Office, Washington, DC, 1964, p. 15.
46. J. Wojtozak-Haroszowa, *Physiological and Psychological Aspects of Night and Shift Work,* Public Health Service Publication 78-113, Superintendent of Documents, Government Printing Office, Washington, DC, 1978.

CHAPTER EIGHT

Applied Ergonomics

ALEXANDER COHEN, Ph.D., and
FRANCIS N. DUKES-DOBOS, M.D.

The content of this chapter on ergonomics differs from the one found in the Patty's Vol. 1 1978 revision of this same *General Principles* in at least two ways (1). First, and as admitted by its author (E. R. Tichauer), the earlier chapter focused largely on a subarea of the overall topic, namely, biomechanics. In scope, the material presented herein is much broader, befitting the array of human, task, equipment, work station, and environmental factors that need to be addressed in considering ergonomic issues in the workplace. Second, the earlier treatment tended to have a basic orientation with some applied references. Indeed, structures and mechanisms underlying human function were explained, as were fundamental methods for measuring human forces and actions. The approach adopted here is much more applied—concentrating on end points in problem-solving rather than a study of basic integral processes. Taken together, the two chapters should give the reader both an in-depth as well as cross-sectional view of the ergonomics subject area.

1 INTRODUCTION

1.1 Definition of Ergonomics

In general terms, ergonomics represents the study and design of job demands and working conditions from the standpoint of the person(s) involved (2). Its focus is on interactions—between the capabilities and limitations of workers and the work requirements imposed. In theory and practice ergonomic efforts are aimed at ascertaining those forms of human–work interactions that can yield the highest levels of productivity, health, safety, comfort, and job satisfaction. As can be noted, ergonomics is an interdisciplinary science, elaborating on human attributes of critical importance to the design of job tasks, tools, equipment, work stations, and environmental conditions. Objectives and benefits of research in ergonomics include:

1. Definition of the nature of job demands and work regimens that can be effectively met by individuals with minimal stress.
2. Development of design principles for machines, equipment, and tools for safe, error-free operation.
3. Identification of workplace or work station configurations to ensure natural, healthful body postures and efficient, nonfatiguing performance.
4. Acknowledgment of environmental conditions in the workplace, such as heat, noise, vibration, and light that can suit the workers' comfort needs as well as pose no significant hazard.

In ergonomics, as in other applied disciplines, real problems cannot be solved by retaining a science perspective. Rather, the techniques and knowledge generated through research must become translated for use by the practitioner. For this purpose, manuals and handbooks have been developed containing essential data and guidelines for best fitting working conditions to people (3–7). Design information ensuring (a) proper allocations of functions between people and machines in a system or process, (b) adequate workspace layout to accommodate the range of size in the worker population, (c) environmental conditions meeting safety and other criteria, and (d) ease of understanding informational displays and performing related control operations are examples of such contents. Digests of this material are also covered here in amplifying on ergonomics as a technology for effecting a healthier and more efficient work situation.

1.2 Ergonomics in Research and in Practice

Ergonomic research requires an integration of principles and knowledge gained from several disciplines and fields of science (8). The most important fields applied in ergonomics are those of the biological and engineering sciences. The biological sciences are needed for the assessment of human dimensions, limits of motility, performance capacity, and tolerances to environmental factors, whereas the engineering sciences are utilized in the assessment of the mechanical properties of the body as well as in the measurement of physical demands of tasks performed in different jobs.

Derived from anatomy, physiology, and psychology, there are a large number of subspecialties that directly address certain problems of ergonomics. For instance, the techniques of *anthropometry* are utilized for measuring human dimensions and motility; knowledge available in *exercise physiology* is being adapted for assessment and enhancement of physical work performance; assessment of human tolerance to the physical factors of the environment, such as thermal conditions, noise, vibration, electromagnetic radiation, gravity, and other parameters are covered by the discipline of *environmental physiology*; and finally, the data generated in *applied experimental psychology and engineering psychology* are utilized for determining whether the requirements of a task on the worker's sensory, manual, and mental capacities are compatible with the worker's available capabilities.

Similarly, disciplines belonging to engineering science are applied for research in ergonomics, as in the case of *biomechanics*. Biomechanics made available the methods for human motion analysis and strength measurement and for establishing the effect of forces acting on the body, such as in lifting work, where it is important to know how much weight the human skeleton can support without becoming damaged.

Considering ergonomics from the point of view of practical application, it becomes evident that when occupational physicians have to determine whether a patient's health problem is related to job conditions, they must understand some of the engineering aspects of ergonomics, whereas the industrial engineers and hygienists who are involved in designing the job with all its implements must be aware of the biological and medical implications of human factors engineering.

1.3 Purpose of Chapter

Reflecting the aforementioned theme, the intent of this chapter is to offer a digest of information on ergonomic considerations at the workplace in a form suitable for use by occupational safety and health practitioners. The organization and elaboration of the subject matter is, in fact, keyed to items typically found on ergonomics checklists as a means of facilitating this aim. Thus the material is geared to factors guiding the collection and eventual evaluation of observations that may identify problems or design needs for improving the worker–work interface with respect to health, safety, and productivity concerns.

2 MAJOR FACTORS IN APPLIED ERGONOMICS

2.1 Characteristics of an Ergonomic Checklist

Ergonomics (or human factors) checklists have been devised as aids for those surveying work situations for real or potential problems (7, 9–11). Recognizing the broad and diverse nature of factors both affecting and being affected by worker–job–equipment–environmental interactions, the checklists provide reminders as well as offer a means for systematizing the data collection process. Checklists found in different ergonomics texts vary in specificity, level of required quantification, ordering of inquiry (from general to specific, hierarchical) to eliminate nonrelevant items early, and other features. Common to most of them, however, are certain subject categories marking the major concerns of ergonomics. These are acknowledged in the following list with some added description for emphasis. The main factors in ergonomics checklists are:

Workspace design	Issues of human dimensions as applied to work station design and layout, including reaches, clearances, working heights, and distances
System–equipment design	Display modes for ensuring accurate information transmission and control system for effecting required actions with reliability and ease; tool and equipment design, accommodating size factors in user population as well as minimizing biomechanical stress
Job demands	Nature and levels of physical workloads imposed (e.g., heavy lifting, static muscle loading) with respect to risks of overexertion, undue fatigue; repetitive and manipulative operations posing risks of chronic trauma and means for its alleviation; perceptual, mental demands and issues of concentration, information handling, complexity of decision making
Environmental factors	Conditions of illumination, noise, temperature, humidity, vibration, and air quality that will satisfy comfort and work efficiency needs as well as meet limits for safe exposure
Work organization	Aspects of alternative work schedules and shiftwork with respect to health, safety, performance issues; work–rest regimens that are optimal for different kinds of job requirements; stress factors associated with repetitive work under the control of others

2.2 Workspace Design

2.2.1 Human Dimensions

In order to adapt the workspace to human dimensions the first question is how to establish values that truly reflect these dimensions. Rarely is it necessary to perform actual measurements on workers because good anthropometric data are available in the literature (12, 13). As a matter of fact, some of the sources have data on people of different national and ethnic origin (14, 15). When comparing data from different sources, however, it is important to make sure that the reference points used in the measurements were identical. Actual measurements may be necessary for designing special protective clothing or tools used by a small group of workers. This can be performed by relatively simple conventional methods, with the use of a measuring tape and different calipers, such as the anthropometer and the spreading caliper (Figure 8.1). For surveying large groups, more sophisticated methods have been described with the use of photographic techniques (andrometry, stereophotogrammetry, IR imagery, or laser beam techniques). Anthropometric data are usually presented

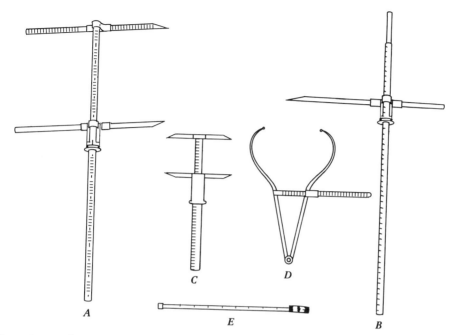

Figure 8.1 Anthropometric instruments: (A) Anthropometer; (B) Beam caliper; (C) Sliding caliper; (D) Spreading caliper; (E) Steel tape. (After J. T. McConville and W. W. Leubach (cited in 14).)

for different percentiles of a population (Figure 8.2). In many instances it suffices to know the 50th percentile values, that is, the median. For instance, most furniture, machines, and tools are available in one size only. In the case of furniture, however, allowance has to be made for the larger person, whereas for machinery, allowance should be made for the smaller person who must be able to reach and handle the controls. In building design the 99th percentile value must be considered in order to ensure that the door openings and passageways will accommodate the tallest and largest person. In some other applications, however, where the clearances are not of critical consequence from the point of view of safety and efficiency, the 95th percentile values can be adapted. Figure 8.3 shows the dimensions of the body most often needed for job design.

2.2.2 Space Allowances

For the purpose of job design, two kinds of body envelope must be considered. The first envelope is determined by means of static anthropometry and is based on the actual physical size of the worker's body. The second envelope is called a *functional envelope* and is determined by dynamic anthropometry. The functional envelope is defined by the limits of the area that the workers can reach with different parts of their bodies while standing or sitting at the work station.

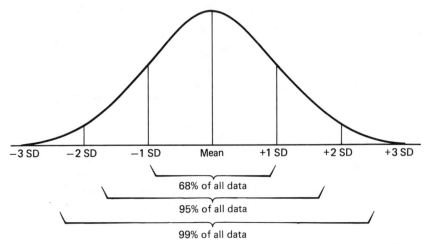

Figure 8.2 Areas under the normal curve. About two-fifths of a set of data fall between $\bar{X} - 0.5$ SD and $\bar{X} + 0.5$ SD; about two-thirds of a set of data fall between $\bar{X} - 1.0$ SD and $\bar{X} + 1.0$ SD; about 87 percent of a set of data fall between $\bar{X} - 1.5$ SD and $\bar{X} + 1.5$ SD; about 95 percent of a set of data fall between $\bar{X} - 2.0$ SD and $\bar{X} + 2.0$ SD; almost all of a set of data fall between $\bar{X} - 3.0$ SD and $\bar{X} + 3.0$ SD.

As an example, Figure 8.4 shows the mean sitting grasp reach envelope in the horizontal plane at shoulder height of the U.S. Air Force personnel. Wherever possible, the job design must allow free space for both standing and sitting envelopes so that the worker is not inhibited in moving freely without colliding with objects or persons. When workers are crammed in by machines or fellow workers in a limited space, they must maintain a certain fixed position for an extended time. This requires static muscle contraction that impedes the blood circulation in the muscles involved, leading to early fatigue and the development of pain.

A mock-up of the work station and an anthropometric dummy are very good tools for establishing space allowances (16). A more advanced technique is a computerized biomechanical model which can simulate the functions of the human skeletal system (17).

2.2.3 Working Heights and Distances

Whereas it is important to consider the workers' static and dynamic dimensions when developing a workplace design, other functional and sensory limitations must be accounted for. If workers have to reach out for a tool, a control knob or any object that is located at the maximal reach distance, their ability both to apply skilled movements and force will be limited by the phenomenon described by Brunnstrom (18) as muscular insufficiency. Visual control of movements is also less efficient at full reach distance. The larger the angle between the torso and upper arm during work, the greater will be the static contraction in the

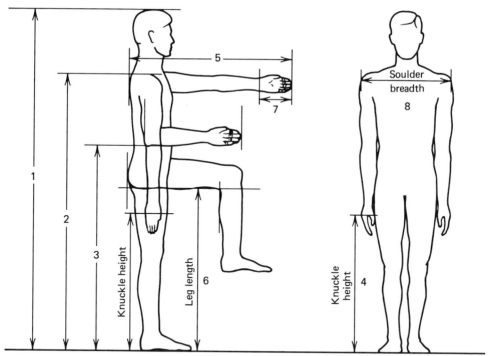

Figure 8.3 Dimensions of the body most often needed for job design: (1) standing height (stature); (2) shoulder height; (3) elbow height; (4) knuckle height; (5) arm reach; (6) leg length; (7) hand length; (8) shoulder breadth.

shoulder musculature and the faster will the worker feel fatigued. Performance of any task above the head constitutes one of the most strenuous exertions and should be avoided if possible.

For deciding what the optimal height of a work bench should be, it must be established what is more important for performing the task: skill or force. The more skill is required the higher the working surface should be to assure good visual control and support for the elbow and arm. However, if force is more important, the working surface should be lowered so that the worker can lean on the object, thus enhancing the muscular force exerted by the weight of the torso and head. More details on working heights and distances can be found in Grandjean's ergonomics handbook (19).

2.2.4 Postural Factors

The most commonly occurring complaint of workers at the end of the workday is back, shoulder, and neck pain, as the muscles located in these regions of the body are involved in maintaining the posture of the torso, arms, and head. This requires static muscular contraction for extended periods and, as men-

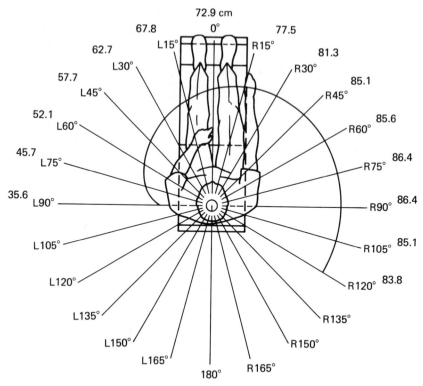

Figure 8.4 Man's right hand: mean sitting grasp reach envelope in horizontal plane at shoulder height (L = left; R = right) (Air Force Personnel).

tioned before, reduces blood circulation in these muscles, resulting in lack of oxygen supply and the accumulation of metabolic breakdown products in the muscle fibers. Therefore, a properly designed workplace must minimize the need for static contraction of the postural muscles.

Whenever possible, the work should be performed in a sitting position on a well-designed chair, and the worker should have the opportunity to stand up and walk occasionally. If the work demands continuous sitting for extended periods, development of fatigue pain in the postural muscles can be prevented by allowing short rest breaks each hour for standing up and stretching. In some industries such breaks are filled in with group exercise, designed by physical education experts or physical therapists. The aim is to provide rhythmic contractions for the muscles that are statically loaded during work. Rhythmic contractions increase blood circulation by a pumping action on the blood vessels. A recent study (20) produced evidence that these exercises prevent the development of fatigue pain and the resulting decrease in performance.

2.3 System–Equipment Design

2.3.1 Display Factors

Dial indicators, light panels, horns, and buzzers typify different means for conveying information to workers about the functional states of equipment they may be operating or processes in which they are engaged. The basic function of such signals is to notify the operator of any required actions. This, in turn, dictates two major objectives in designing informational displays, namely, that the presented signals are easily detected and readily understood. Guidelines for meeting these objectives, starting with the early focus on cockpit consoles at the start of World War II, have become generated on the basis of a variety of principles and research efforts. Dominant themes here have been an assessment of:

1. The detection–discrimination–response time capacities of the human observer for signals presented in different sensory modes.
2. Limitations in the ability of the human observer to sustain constant attention over time.
3. Length and complexity of the information to be comprehended relative to the display mode.
4. The nature of adverse conditions that could hamper particular modes of information transmission.

Ergonomics texts and human factors design guides have condensed much of the literature on these and related subjects to recommendations for display conditions meeting a variety of operational requirements (3, 5–7, 21). Nearly all interest here has been focussed on visual and auditory display modes as these are the primary sensory channels for receiving information input.

The following are illustrations of decision rules regarding when it is best to use visual and auditory displays:

Use Visual Display	Use Auditory Display
Receiver is at one location	Receiver must move about
Message is complex	Message is simple
Message is long	Message is short
Message needs to be referred to later	There is no need to refer to message again
Receiving location is noisy	Receiving location is dark
Message does not require prompt action	Message requires prompt action
Auditory channels for information are overburdened	Visual channels for information are overburdened

Use of visual or auditory displays for different applications can dictate a number of design features to ensure maximal effectiveness. For example, such

Table 8.1 Recommended Visual Display Indicators for Different Types of Information Transmission

Type of Information	Preferred Indicator	Rationale	Workplace Application
Quantitative readings	Digital readout or counters	Minimum reading time least error potential	Units of production
Check readings	Moving pointer (dial) or graphic reading	Trends, deviations easy to detect	Temperature, pressure monitoring on a console
Status reading	Lights (color coded)	Simple, suggestive meanings, easy referral and monitoring	Consoles production lines
Operating instructions	Annunciator lights, electron displays	Highlighting critical information for required follow-up actions	Use on a control panel where many functions are monitored

displays for warning or emergency purposes must include those characteristics that enhance detectability. For visual displays, this would include:

1. Red-colored signals because this color signifies danger to most people.
2. Display located within or at observers line of sight.
3. Flashing light or indicator to gain attention, especially for urgent forms of warnings.
4. Brightness of display for high contrast against background and all other expected light conditions.

For auditory displays, design features for maximum detectability are:

1. Use of high-intensity, low-frequency sounds, to reduce attenuation–interference factors posed by distance or intervening barriers.
2. Use of tonal alarm frequency(ies) for which expected background noise will cause little masking.
3. Modulation or pulsing of the alarm signal to increase detectability.

Auditory displays have better attention-attracting properties than visual ones, and one's reactions to sound are faster than to visual stimuli. On the other hand, visual displays may be more effective in instructing the receiver regarding assorted adjustment–control–protective actions that may be needed. Table 8.1 describes certain visual display indicators that are preferred when transmitting specific kinds of information to operators along with some sample applications found at the workplace.

Conditions necessitating speech rather than tonal signals for information transmission include those where (1) a rapid two-way exchange of information is needed, (2) listeners lack special training in a tonal code or where situations of stress might cause one to forget the code, (3) needs exist for determining sources of information.

Dial design and layout have received much attention in visual display research and application. Figure 8.5 summarizes some important features in ensuring accuracy in reading and information transfer.

2.3.2 Control Factors

Control devices represent the means for operators to effect some change in a machine, work process, or system usually in conformance with information presented by means of the display modes just described. Controls have varied forms, with pushbuttons, knobs, switches, cranks, levers, wheels, pedals, and keyboards common in different workplaces. Prescriptions for appropriate control designs for different operational requirements have emerged from a body of experimental and applied studies. For example, some of the basic considerations regarding the selection of a control are:

1. Does the control fit its intended function? A control for effecting a simple on/off action would be different from one needed for a few discrete settings or one requiring continuous movements. Figure 8.6 depicts the preferred types of controls for different system responses.
2. Does the control facilitate the task requirements? Needs for precision or speed in control action and force would be important factors here. Table 8.2 summarizes information on the choice of different controls given these considerations and others acknowledged in the following paragraphs.
3. Does the control need to be made discriminable from others around it? Design information on knob shape–size coding and minimum separable spacing to reduce confusability and hindrance in control movement has been specified (3, 4).
4. Are there restrictive factors on control operation? Limitations in space, placement, or location of the control relative to the operator's position need to be addressed along with any requirements for its association with a particular display. Whether control actions can be affected by environmental factors (darkness, vibration) also need to be considered.

Apart from these considerations, control design must respect population stereotypes. That is, a characteristic of human behavior is to expect certain control movements to cause certain changes or outcomes. Throwing a toggle switch in an upward direction, for example, is expected to actuate, whereas pressing it down is expected to do the reverse. Moving a lever in a forward

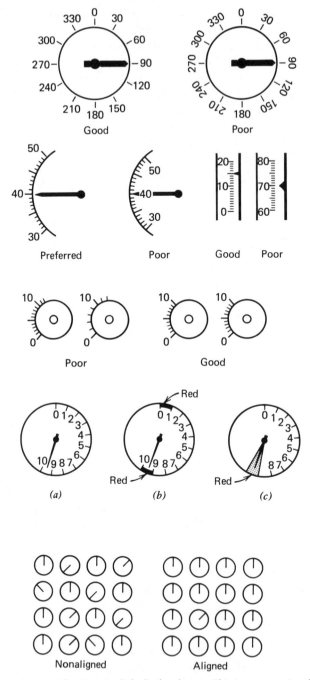

Figure 8.5 Some major considerations in dial–display design. (This is a composite of illustrations taken from References 5 and 6.)

APPLIED ERGONOMICS

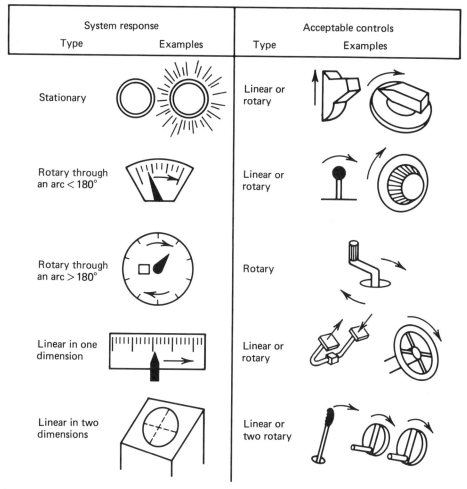

Figure 8.6 Choice of controls for different forms of system response. (These illustrations are taken from References 3 and 7).

direction is expected to increase the thrust, intensity, and speed of a machine or process, as is turning a knob or wheel in a clockwise direction. When control movements conform to stereotypes, they are likely to be learned faster, made more quickly and in a precise manner, and be less prone to error even under stressful conditions. Figure 8.7 describes some direction of movement relationships for different control devices which take account of population stereotypes.

Other principles having application to both control and display designs can be summarized as follows:

1. Wherever possible, arrangements for control and displays should be standardized. The same groupings of control and displays should be used in

Table 8.2 Recommended Control Features for Different Tasks[a]

For *Small* Forces and	Use
2 discrete settings	Hand pushbutton, foot pushbutton, toggle switch, or rocket switch
3 discrete settings	Finger pushbutton, toggle switch, or rotary selector switch
4 to 24 discrete settings	Bank or array of finger pushbuttons, rotary selector switch, or detented thumbwheel
25 or more discrete settings	Bank or array of finger pushbuttons
Small range of continuous settings	Knob, lever, or thumbwheel
Large range of continuous settings	Crank

For *Large* Forces and	Use
2 discrete settings	Detent lever, large hand pushbutton, or foot pushbutton
3 to 24 discrete settings	Detent lever
Small range of continuous settings	Handwheel, rotary pedal, or lever
Large range of continuous settings	Large crank

[a] Table adapted in part from Reference 7.

all similar models of equipment or operations. All controls that look alike should behave alike.

2. Emergency controls require special attention because of their importance. They need to be clearly distinguished from all others by being physically separated from them. At the same time they must be located within easy reach regardless of where the operator's momentary position might be. They must be distinctly coded to avoid confusion and within the operator's immediate visual field. Emergency controls should be recessed, covered with shields or otherwise protected to prevent inadvertent actuation.

3. Controls and displays that are identical in function or are used together in interacting with one part of a system should be grouped together. A variant of this idea is to arrange such displays and controls in the same order as they would be used in performing a machine function or regulating a work process. For operation of displays and controls in a fixed order, it is recommended that they be arranged in sequences either from left to right or from top to bottom. If rows are used, the sequential arrangement should be from top to bottom and from left to right within the rows.

4. When a number of controls appear with paired displays on a panel, each control should be located below its corresponding display. An alternative would be to group all the displays above and all the controls below conforming to the same layout. Controls associated with particular displays should be located so that the operator's hands do not obscure the readings.

APPLIED ERGONOMICS

Figure 8.7 Recommended relationships between control movement and system response. (This is a composite of illustrations taken from References 3 and 4.)

An analysis of pilot error in dial reading and mistakes in aircraft operations during the World War II years created the major thrust for much of the design information on displays and controls that is now available. The need for reinforcement in the application of such material was evident in the inquiry of the near meltdown of a nuclear power plant at Three Mile Island (TMI) in

Pennsylvania. A major finding was the deficiency in control–display design and layout. The following problems at this as well as other nuclear facilities have been observed (22):

1. Controls were located far from instrument displays that show the condition of the system. For instance, the high-pressure injection throttle valves are operated from a front panel but the flow indicator is on a back panel and thus cannot be read from the throttle operating position.
2. Instruments that appeared identical and were placed side by side controlled widely different functions.
3. Contradictory systems of lights, levers, and knob controls (e.g., a red light meant that a valve was open in one plant area and closed in another).
4. Poor panel layout made systems misalignment difficult to detect. It was ascertained that in the TMI accident a paper tag hanging from a handle on the control panel obscured an indicator light that would have shown the operators the position of one of the blocked valves shutting out the needed emergency feedwater system.

Other human factor elements referencing needs for better training of reactor operators, especially in emergency situations or equipment malfunctions, were noted. The massive influx of computerized operations in industry and offices have focused much attention on the video display terminal (VDT), as this is the point of operator interaction with the computer system. A VDT exemplifies a visual display–control unit, and manufacturers of such equipment are being sensitized to design features for enhancing productivity and reducing fatigue when using such equipment. Kodak (5) and the NRC (23) provide a condensation of data indicating critical image projection factors and keyboard charcteristics for optimum performance. Reported operator problems with this equipment, indicating eyestrain and postural discomfort, can transcend display and control features, such as locating the unit near glare sources and limited adjustability of VDT workstation to accommodate the body size of the user. The latter issues are addressed in other sections of this chapter.

2.3.3 Handles of Controls, Tools, and Other Equipment

The handles of controls must be designed with the same ergonomic approach as it is required with handles of tools; that is, the kinesiologic and strength capacities of the worker must be accounted for in order to avoid unnecessary fatigue or acute and chronic musculoskeletal injuries. This is particularly important if the controls or tools are handled over a long period of time, such as for hours, in a repetitive or continuous manner.

The kinesiologic capacities are well described by Tichauer (1) in the 1978 Patty's Volume 1. The emphasis is on the limitations caused by certain "unnatural" positions on the strength, endurance, and mobility of the fingers, hands, arms, neck, and torso, all of which may be involved during handling of tools and controls. Also described are some of the tool designs that make it possible to avoid these unnatural positions as well as the overexertion of certain weaker muscles of the finger–hand–arm aggregate. The examples for biomechanically corrected tool designs presented by Tichauer include pliers, scrapers, and trigger-operated tools. These and other "ergonomically" designed tools are now commercially available from several tool manufacturers.

The weight of certain tools by themselves can be the cause of overexertion or excessive fatigue. Even light tools can become burdensome if held for a long time, in view of the fact that holding an object requires static contraction of the muscles. This problem can be overcome by suspending the tools on tool balancers. This is particularly important in the case of vibrating tools because the tighter the handle of a vibrating tool must be held, the greater will be the coupling between the tool and the hand, and the greater will be the transmission of the vibration energy from the tool to the hand, resulting in greater tissue damage.

Tool handles that have to be held in the palm should be made of, or covered with, nonslippery material, should have no sharp edges, and should be oval or round shaped. The greatest torque can be exerted on a handle if its diameter is about 2 in. whereas for precision work its diameter should be 0.3 to 0.6 in. The distance between the two handles of pliers, cutters, clippers, or similar tools, called the *grasp span*, should not exceed 3.5 in. (24) in order to accommodate the closing of the tool handles with satisfactory power by a high percentage of both male and female workers. The length of tool handles should be 4 in. for most operations, but for greater force application they should be made longer (e.g., 5 in.). When the workers wear gloves, the tool handles should be made longer by about 0.5 in. Formfitting handles with finger recesses should be avoided.

Quantitative design criteria for handles of controls, valves, cranks, pushbuttons, toggle switches, levers, handwheels, and foot pedals can be found in the ergonomic design book published by the Human Factors Section of the Eastman Kodak Company (25). The critical parameters for optimum design are the dimensions and shapes of the handles, the direction and resistance of handling the controls, and the amplitude and frequency of motions required.

2.3.4 Repetitive Manipulations

Manipulative tasks usually belong to the light work category because they do not require great physical exertion. The performance of these tasks is not limited by either maximum oxygen uptake capacity $\dot{V}O_{2max}$, also termed *aerobic power* (AP) or by muscular strength. If certain motions of the hand or arm are performed in a continuously repetitive manner for long periods of time, however, such as for 8 hr daily for many months or years, certain tissues involved in these motions can suffer chronic injury (26). Most vulnerable to such injury are tendons of the muscles that move the fingers and the median nerve, which provides sensory-motor functions to two-thirds of the palm area on the thumb side, to the first three and one-half digits, and over the base of the thumb. When the wrist is flexed or extended and the fingers are moved, the tendons and the median nerve are constantly rubbing against each other and against the bones and ligaments surrounding them. Depending on the frequency and type of motions and the resistance the motion has to overcome, the sheets covering the tendons and the median nerve may become irritated or swollen and the median nerve pinched or entrapped. The condition resulting from this phenomenon is called "carpal tunnel syndrome" because the tendons and the median nerve pass through the "carpal tunnel" from the arm to the hand. People with narrow carpal tunnels may be more susceptible to this syndrome.

The symptoms vary depending on whether the tendon sheets or the nerve is affected. The inflammation of the tendon sheets is called *tenosynovitis* and is connected with great pain when an attempt is made to move the fingers. The nerve injury may result in tingling, numbness, and pain in the fingers, particularly at night and in loss of skill, weakness, and atrophy of the finger muscles innervated by the median nerve. Prevention of this incapacitating condition consists of:

1. Reduction in exposure to highly repetitive motions by rotating the jobs.
2. Early recognition of the syndrome and placement of the worker to another job before the condition becomes incapacitating.
3. Redesign of tools and controls to eliminate the need for wrist flexion.
4. Reduction in the forces demanded from the hand and fingers by the task by using tools, mechanical aids, or automation.

More details on this subject can be found in the Chapter 22 of Volume 1 of the Patty Series (E. R. Tichauer, "*Ergonomics*").

3 JOB DEMANDS

3.1 Nature and Level of Workload

When the task to be performed requires intense physical work, it is important to know how much work can be performed by an individual. If the workload exceeds the worker's performance capacity, it may lead to overexertion and/or excessive fatigue, both of which have harmful immediate or long-range health effects and increase the worker's susceptibility to accidents.

Recent investigations (27) suggest that a person's subjective feeling of fatigue is a fairly reliable indicator of physical fatigue but the feeling of fatigue is not necessarily correlated with decreased physical or mental performance. Nevertheless, simple questionnaires have been used successfully (28) to establish the fatiguing effects of jobs, provided the group of workers tested performed similar tasks and the group was large enough to apply appropriate tests for statistical significance.

Table 8.3 Tolerance Time for Physical Work at Different Percentages of AP and HR[a]

AP (%)	Tolerance Time (min)	Maximum HR (%)
100	4	100
75	20	84
50	100	66
33	480 (including 30 min. rest)	54

[a] From data of Christensen (31), Bink (32), and Bonjer (33).

Physical work may be excessive either because it exceeds the workers' physical performance capacity (PPC), muscular strength, or both. The magnitude of the PPC depends on the amount of oxygen that can be absorbed from the inhaled air into the circulating blood and transported to the muscle fibers, where it is needed for converting stored nutrients to kinetic energy and heat. The kinetic energy is manifest in muscle contractions and the heat is utilized for maintaining constant body temperature. Surplus heat is dissipated to the environment. The heavier the work performed, the greater the amount of oxygen used. Maximum PPC depends on one's AP. On the average, men have about 25 to 30 percent greater AP than women, mainly because they usually are larger. If AP is calculated for a unit of linear body mass (1 kg or 1 lb), men and women have the same AP, however, a woman's body usually contains a higher percentage of fat (29).

The AP changes with age, peaking between the ages of 18 and 20, and generally showing an almost linear decrease with age. At the age of 65 the average AP is only 70 percent of what it was between the ages of 20 and 30 (19). Physical training can increase AP only 10 to 20 percent because the determining factors are inherited. The AP can be expressed in kilocalories (kcal), the same units that are used for identifying the energy values of nutrients. In order to maintain constant body weight, one must consume enough food to replace the nutrients used up in physical work and in rest. The AP of an average, middle-aged, moderately trained man weighing 70 kg is 3 liters of oxygen per minute, corresponding to 15 kcal/min. The length of time for which physical work rate can be maintained at different percentages of AP and the concomitant percent of maximum heart rate (HR) are shown in Table 8.3.

If the physical activity of a mixed worker population is at the 50 percent level of their AP, their HR will be about 120 to 150 beats per minute and they will be able to maintain their activity for 8 hr with only 10 min of rest each hour. However, such a regime will be perceived to be very strenuous, and after about 3 weeks the workers' performance will start to decline. If well-motivated workers are permitted to work at their own speed, their energy consumption will be bout 33 percent of their AP in an 8 hr average and their HR will be about 95 to 125 beats per minute. However, the interindividual variability of

HR values will be great, not only because of differences in work rate but also because of differences in age, sex, physical fitness, and psychological factors that also affect the HR.

In terms of energy requirement, it is immaterial whether a certain total amount of work is performed at a low rate in continuous manner or at a high rate but interrupted with rest pauses. In regard to HR, however, it is much better to take often short rest pauses than to take fewer long rest pauses after longer or more intense work bouts because HR will remain lower in the former case. The energy cost of static work is low but is accompanied with relatively high HR and blood pressure, and, as mentioned, should be avoided or kept to the minimum.

The amount of rest allowance can be calculated by the equation due to Spitzer (30):

$$\text{Rest allowance (\%)} = \frac{(\text{kcal/min} - 1)}{4} \times 100$$

A good estimate of maximum HR can be obtained by the equation due to Arstilla et al. (34):

$$\text{Maximum HR} = 200 - 0.66 \times \text{age}$$

The energy cost of different jobs can be estimated by the use of tables available from the literature. The most up-to-date and pertinent table can be found in the *Bioastronautics Data Book* (35).

3.2 Manual Material Handling

Tasks consisting of lifting, pushing, or pulling objects without the assistance of mechanical devices are called *manual materials handling* (MMH). The weight a worker can lift, push, or pull at a single occasion or repeatedly but infrequently depends mainly on muscular strength and not on the person's AP. On the other hand, tasks consisting of frequent repeated lifting or continuous pushing or pulling of objects that are within the strength capacity of the worker fall in the same category as endurance work, and the limiting factor will be the worker's AP.

Manual materials handling tasks carry a high risk of injury not only because of the interaction between the worker and the object (e.g., sharp edges of an object, dropping an object, slipping, stumbling, or falling while carrying an object) but also because there is a potential for overloading the body's supporting structure, the musculoskeletal system. The low-back pain syndrome is often caused by overexertion during performance of a MMH task and is the most costly among occupational injuries, not only because of its high prevalence but also because often it incapacitates the worker for a long period of time.

Epidemiologic studies (36) confirmed, what could be a conclusion simply based on logic, that the closer the weight of the lifted objects was to the maximum lifting capacity of the workers and the more frequently they lifted heavy loads, the greater was the frequency and severity of musculoskeletal and contact injuries. As Tichauer (37) so clearly pointed out, the "heaviness" of a load to be lifted depends not only on its weight but also on its distance from the body and on the body position during lifting. The more distant an object is held from the body during lifting, the greater will be the load on the spinal column and, in particular, on the lumbro-sacral joint which is the site of injury in many patients suffering from lower back pain.

As to the body position during lifting, Tichauer (37) recommends to follow as closely as possible the golden rule: "Knees bent—back straight—head up." He also emphasizes the need for using common sense for determining when to abandon the golden rule. In fact, few workers apply in practice the "golden rule" method for lifting because it is more fatiguing than other methods. The reason for this is that every time they would squat they would have to lower and then lift again their own bodies, in addition to lifting the object. Furthermore, in squat lifting there is substantial dynamic loading on both the back and the knees as a result of the inertia of the moving body and the object, whereas in stooped lifting position the load on the lifting muscles and on the lumbrosacral joint is less by about 50 percent (38).

Ballistic lifting is a method for converting the static load, bearing down on the lumbosacral disk, to a dynamic load that is distributed along the vertebral column as well as other parts of the body. In addition, in ballistic lifting the static muscular contractions are replaced by dynamic contractions. The principle of ballistic lifting is that the object is moved horizontally prior to lifting, thus giving the object an inertia that can be utilized for lifting by changing the direction of movement gradually from the horizontal to the vertical. Ballistic lifting is often accompanied by a sudden muscle contraction at the instant when the object is moved, which can lead to disk hernia or torn ligaments. Ballistic lifting should be performed with a slow, gradual start and ending as well. Sidestepping and rotating the body at the hip (torsional movement) should also be avoided during lifting (37) because these movements put great stress on the vertebral column. Anderson (39) suggests that it is safer to allow workers to use their own common sense for finding the easiest and safest way of performing their MMH tasks. Experiments performed by Brown (40) and Garg and Saxena (41) on the metabolic energy cost of lifting the same object by different methods support Anderson's suggestion, inasmuch as the energy cost is the lowest when the subjects were permitted to choose freely their lifting method. Nevertheless, if lifting heavy objects from the floor is performed only infrequently (once an hour or less) the fatiguing effect of squatting loses its importance. If the object is not wider than the distance between the knees during straddling of the object with the feet, the method of choice should be the squat lifting because this permits one to bring the object closest to the body.

Table 8.4 F_{max} Table

Period (hr)	Average Vertical Location cm (in.)	
	$V > 75$ (30) Standing	$V \leq 75$ (30) Stooped
1	18	15
8	15	12

According to Ayoub (42), a good rule of thumb is that continuously performed MMH should not exceed 30 percent of the maximum forces the muscles are capable of generating. Occasionally, forces up to 50 percent are acceptable when only maintained for short duration (~5 min). The Ergonomic Guide of the American Industrial Hygiene Association (43) describes the method recommended by an ad hoc committee for measuring the static strength of workers. Dynamic strength measurement methods have been recently tested by several investigators, but no generally accepted method has been developed as yet. It is assumed that there is no great discrepancy between a person's static and dynamic strength capacities. To be meaningful, however, the static test equipment must imitate as much as possible the body position and direction of muscular exertion that is used while performing the MMH task on the job.

Lifting work can also be affected by factors other than muscle strength. For instance, with lifting of an object by straightening out a bent back, an increased pressure in the abdominal cavity can assist the back muscles to straighten out the body. Increased abdominal pressure can be brought about by contracting the abdominal and diaphragmatic muscles, taking a deep breath simultaneously, and retaining the air in the lung by closing the glottis. The resulting increased pressure will push against the front and rear walls of the abdominal cavity, forcing them to straighten. This mechanism is triggered involuntarily during lifting, as a protective reflex, reducing some of the load on the back muscles and the vertebral column. This reflex can be voluntarily inactivated or accentuated. Persons who have a hernia, high blood pressure, an aneurism, or severe myopia should avoid lifting of heavy objects because increasing the abdominal pressure may aggravate the condition. The maximum attainable intra-abdominal pressure has been used as a criterion for establishing maximum allowable weights for lifting work (44).

The *Work Practices Guide for Manual Lifting* published by the National Institute for Occupational Safety and Health (45) recommends action limits (AL) and maximum permissible limits (MPL) for lifting work. These limits are based on the results of epidemiological, biomechanical, physiological, and psychophysical studies.

Action and maximum permissible limits can be calculated by the following equations:

$$\text{AL (kg)} = 40 \left(\frac{15}{H}\right)(1 - 0.004\,|V - 75|)$$
$$\times \left(\frac{0.7 + 7.5}{D}\right)\left(\frac{1 - F}{F_{\max}}\right) \quad \text{(metric units)}$$

or

$$\text{AL (lb)} = 90 \left(\frac{6}{H}\right)(1 - 0.01\,|V - 30|)$$
$$\times \left(\frac{0.7 + 3}{D}\right)\left(\frac{1 - F}{F_{\max}}\right) \quad \text{(U.S. Customary units)}$$

$$\text{MPL} = 3\,(\text{AL})$$

where H = horizontal location (cm or in.) forward of midpoint between ankles at origin of lift
V = vertical location (cm or in.) at origin of lift
D = vertical travel distance (cm or in.) between origin and destination of lift
F = average frequency of lift (lifts/min)
$F_{\max}$ = maximum frequency that can be sustained (see Table 8.4)

1. The value of H is between 15 cm (6 in.) and 80 cm (32 in.). Objects cannot, in general, be closer than 15 cm (6 in.) without interference with the body. Objects further than 80 cm (32 in.) cannot be reached by many people.
2. The value of V is assumed between 0 and 175 cm (70 in.), representing the range of vertical reach for most people.
3. The value of D is assumed between 25 cm (10 in.) and $(200 - V)$ cm $[(80 - V)$ in]. For travel less than 25 cm, set $D = 25$.
4. The value of F is assumed between 0.2 (one lift every 5 min) and $F_{\max}$ (see Table 8.3). For lifting less frequently than once per 5 minutes, set $F = 0$.

According to the NIOSH Guide for MMH (45), when lifting objects that weigh more than the AL but less than the MPL, it is necessary to apply administrative controls for prevention of injuries. Such administrative controls include selection of workers by assessment of their muscular strength and physical performance capacity and by teaching the workers how to avoid unnecessary stress when lifting and to be aware of the health hazards of MMH work. The guide gives a detailed description of how to perform the different tests and what topics to cover in the training course. If the weight of objects to be lifted is more than the MPL, the lifting task should be performed by two or more workers or by application of mechanical aids.

Recommendation of safe limits for pushing and pulling tasks is more complex because these limits are also affected by the person's body weight, which can be used to assist the muscles in performing the task or by leaning against the object to be pushed or against the handle of a cart to be pulled. Other important factors in pushing and pulling are the friction between the shoe sole and the floor, the hardness and smoothness of the floor and its angle of slope, and the type of cart. Thus overexertion and other injuries resulting from pushing and pulling work must be prevented, taking into consideration all these factors. However, the critical factors, the force required to push or pull an object or cart and the pushing or pulling force a worker can exert can be measured. The principles of prevention of overexertion are similar to those in lifting work.

3.3 Perceptual Demands

Increasing automation in industrial operations can reduce the physical labor once required of workers but increase the perceptual and cognitive or mental demands. Perceptual tasks can include watchkeeping or monitoring functions, requiring sustained attention to equipment displays and control system indicators to ensure normal conditions in ongoing work processes and/or timely corrective actions in the event of deviations. Alternatively, the job may involve inspection, checking the quality of products emerging from a production process such as to exclude defective items. One's ability to maintain a high level of alertness or vigilance as warranted by the aforementioned task has been the subject of much research. Laboratory studies of vigilance have been particularly voluminous, examining an assortment of factors influential to detecting signals over an extended time period (46, 47). Grandjean (7) in summarizing this literature notes the following:

1. Sustained alertness (measured by the number of missed signals) decreases with longer periods of vigil. For a simple monitoring task, decrements in vigilance performance may occur after the first 30 min.
2. Within certain limits, the observational performance is relatively improved if:
    a. The signals requiring response occur more frequently. The optimum range appears to be 100 to 300 per hour. Other factors being equal, signal frequencies well below or above this range are associated with vigilance decrements.
    b. The individual signals are strengthened relative to the background in which they appear.
    c. The observer is given feedback on performance.
    d. The signals are made more distinct in shape or other features to enhance their discriminability.
3. Vigilance performance is worse if:
    a. Intervals between signals requiring response vary a great deal.

    b. The observer has previously been under a great deal of stress or aroused from sleep.
    c. The observer is performing the task under adverse environmental conditions of noise, temperature, and humidity.

Losses in vigilance or monitoring behavior are believed to be attributable to either a reduction in the observer's sensitivity to the signal occurrences, that is, a change in sensory threshold for responding, or the adoption of a poorer criterion as to what constitutes a real versus imagined signal. The latter factor is particularly cogent in alarm detection and inspection tasks. Indeed, a liberal observer would be responding to too many false alarms that can create a nonresponsive (or "cry wolf") disposition toward a real alarm when it does occur. Too strict a criterion for detection could also result in a failure to take action when, in fact, an alarm situation exists. A criterion that maximizes the rejection of false alarms while at the same time limiting the rejection of real alarms is a sought-after goal. In terms of inspection work, the desired criterion would be one that enhances the rejection of faulty material but minimizes the rejection of good material (48, 49). Field studies to improve inspection performance have indicated the following practical aids, which could also be generalized to alarm detection situations:

1. Instruct the observer to search specifically for those parts of the object most likely to contain nonconformities; otherwise increase the time allowable for item inspection.
2. By use of magnification, increase the visual size of the nonconformity or enhance contrast through use of overlays.
3. Have limit standards available for decision-making as they change the task from one of absolute judgment to a more accurate comparative one.
4. Provision of rapid feedback from sample reinspection of items can improve the fineness of the discrimination.
5. Recognizing limitations in vigilance efficiency over time, it is recommended that such work is rotated with other jobs not requiring similar concentration.
6. Inspection error can be further reduced through reinspection by others.

3.3.1 Sensory Overloading

The control room in mechanized or automated plants typically contain banks of displays and controls revealing the operating status of the ongoing processes. When all is normal, there is literally nothing to do; consequently, limited personnel to handle the monitoring function would seem justified. On the other hand, one person cannot simultaneously attend to two or more visual displays, especially if immediate and appropriate responses must be made to signal indicators on either display. Given this situation, the reaction time to the presented information will be longer than if only one had to be observed.

Similarly, several loudspeakers can be monitored, provided only one presents information at a time. If both are actuated together, the listener may become distracted, or the messages may mask each other or become fused. People usually deal with simultaneous sensory inputs by attending to one while waiting on the others (21, 48). If the delays and errors that this introduces cannot be tolerated, the tasks have to be allocated to separate observers.

3.3.2 Time-Sharing

Alternating attention from one display to another is referred to as *time-sharing*. Having to shift attention in fairly rapid form as dictated by many inputs arriving near simultaneously imposes a time or load stress on the observer (21, 22, 48). Research to evaluate subject's performance in coping with two tasks presented simultaneously has been utilized to evaluate behavior under such conditions. Subjects may adopt strategies under these conditions, giving priority attention to either of the two tasks because of salience, difficulty, or nature of instructions. Uncertainty about how subjects will allocate attention to simultaneously presented tasks suggests the following cautions:

1. Where possible, reduce the number of potentially competing sources of information.
2. Provide the observer with some notion of priority inputs in order to give immediate or primary attention to some relative to others.
3. Where possible, allow the observer to control the input rate.
4. Where a choice of sensory modalities is feasible in a situation where there are competing sensory inputs, the auditory sense is generally more desirable and less influenced by other inputs.
5. Training and extended practice can improve one's ability to handle dual-task situations.

Where it is possible to have identical information presented on two or more sensory channels such as vision and hearing, the result has a reinforcing effect on the reception. Hence, redundant coding of information may have benefits. One exception may be when one of the sensory modes is already overloaded such that it would be unwise to further burden it for redundant transmissions.

Task conditions requiring responses to either too few or too many sensory inputs can yield suboptimal performance, as has been explained in terms of arousal theory (48, 50). This theory asserts that there is a level of activation, representing a neurological or physiological state of the individual, which is conducive to effective functioning. Task demands and environmental as well as situational factors control this level of activation. Those conditions generating little excitation (e.g., simple tasks, minimal environmental stimulation, little performance incentive) would result in lowered levels of alertness and laxity in response. Those conditions generating excessive activation (complex job demands, extreme environmental conditions, significant performance pay-offs) could also show a decrement, but as a result of a loss of behavior control or disorganization in the response process. Between these two extremes there exists an optimum level of arousal and performance effectiveness. In recognition of these outcomes, it is important to enrich those work conditions that otherwise are characterized as boring or dull. Job enlargement or rotation and added performance feedback with incentives represent means for introducing extra stimulation. When overload or overarousing conditions are evident, task simplification measures and more frequent rest pauses would seem in order.

3.4 Mental and Cognitive Demands

As already noted, an observer can only respond to one input at a time. The possibility of a range of different sensory inputs, each requiring differential or alternative responses, makes matters more difficult. This was demonstrated in a classical study of reaction time of subjects in making a single response to a single visual signal versus as many as ten different responses to ten different visual stimuli (48, 51). The difference in response time between the simplest and most complex case was 0.5 sec, which represented the decision-making time with regard to the alternatives presented. Further work has illustrated how direct correspondence between the geometry of the stimulus and response arrays can reduce response time and error rates in such tasks. With greater incompatibility in the arrangement of the two, more training and practice would be needed to assure a similar degree of efficiency.

The capacity of humans to store and process information is limited. This is most evident in tasks involving short-term memorization and recall. For example, one's memory span—that is, the ability to exactly reproduce a series of digits after a single reading—rarely exceeds 12, with the average around 8 (48, 51). For more meaningful or connected materials such as sentences reproduced verbatim after a single reading, the number of words recalled can be about 15. As the amount of material presented exceeds the memory span, the absolute number of items recalled increases but at a decreasing rate. Unless operated on, this information fades away in no more than a few seconds. To ensure longer retention, there must be opportunities for active rehearsal that, in effect, enables some or perhaps all of the material to enter the long-term memory store. The degree of forgetting in the latter case depends on the amount of interference between the original and subsequent sets of related information being retained and the extent of disuse of such stored information. In recognition of these problematic aspects of both short- and long-term memory processes, it is recommended that the amount of memorization of task or emergency procedures required among workers be minimized. Workers should be provided with task and memory aids to remind them of appropriate actions to undertake when needed. Memory aids can take varied forms— presented instructions as labels on specified equipment, equipment operation checklists, posters calling attention to task or area related hazards, and prescribed safe work practices to be followed. Obviously, drills, training, or practice sessions offer added means for ensuring the retention of the required procedures (52).

4 ENVIRONMENTAL FACTORS

Environmental conditions impose an important set of factors in considering worker–job interactions having consequences for health, safety, productivity, and comfort. Indeed, adverse illumination, acoustic, thermal, and air quality conditions may create problems in themselves, complicate the operational demands of a work process, or limit the functional capabilities of the worker so engaged (48). The reader is referred to comprehensive treatments of different environmental factors with respect to industrial hygiene and broader ergonomic issues as found in the earlier issues of the Patty series as well as other human factor and industrial engineering source books (3–5, 7, 53). The material here will be quite selective and summary in nature, emphasizing the most important ideas to remember in acknowledging the influence of environmental factors common to many work situations.

4.1 Illumination and Color

Illumination of the work environment enables the worker to perform those jobs involving visual tasks as well as provides for the visibility of objects and spatial awareness necessary for safety. The adequacy of illumination in a workplace must take into account the following considerations, which can be framed as a series of questions:

1. Is the illumination appropriate to the task being performed?
2. Is the light uniformly distributed so as to minimize needs for continual eye adaptation that can prompt fatigue?
3. Does the illumination afford suitable visual contrasts such as to facilitate discrimination of objects and important details?
4. Is the potential for glare and undue reflections from light sources and work areas minimized?
5. Are the quality and color of the illuminants and the work areas compatible?

Illumination, also called *illuminance*, refers to the amount of light falling on an object or surface from both ambient and local light sources. Incident light can be read by an illumination meter that expresses an output, either as lumens per square meter (or lux) or lumens per square foot (or footcandles). *Luminance* refers to the amount of light reflected off a surface or object and is associated with the sensation of brightness. Units of luminance are termed *footlamberts* or candelas per square meter and are measured by photometers. *Reflectance* refers to the ratio of the luminance from a surface to the amount of light falling on it.

4.1.1 Task Performance and Illumination

The difficulty of visual tasks at work varies as a function of their target size, the contrast between their detail and the background, the time available for viewing, and their luminance (5, 53–55). Without altering illumination, one could facilitate visual tasks by increasing their size (or decreasing the viewing distance), increasing the contrast of important detail, allowing more time to make the required determinations. Recommendations for illumination levels respect different task features and visual performance requirements as just noted as well as characteristics of the workforce, notably age, which can be a factor in visual performance. Selected ranges of illumination for work areas in which certain kinds of tasks are to be performed are noted in Table 8.5. See Crouch (55) for a more extensive set of recommended interior lighting levels. Illumination levels are shown to increase to accommodate demands for viewing finer details or poorer contrast and sustained types of visual efforts.

Uniformity in Lighting. A more even distribution of luminance within one's visual environment, including that for the task in question, eases the extent of eye adaptation that may be needed, given shifts in viewing between the immediate task and surrounding areas from time to time. For this purpose, maximum luminance ratios between tasks and surrounding surfaces and areas have been suggested, as have reflectance values for room walls and ceilings and desk and bench tops. Table 8.6 offers such information. Indirect or diffuse lighting provides for more uniform dispersion than direct lighting, although the latter is more efficient in terms of the amount of light actually projected. Direct lighting may also produce shadows and reflection in some cases, which can obscure objects warranting attention. On the other hand, some visual tasks can be aided by direct light by utilizing the shadows and bright areas to enhance needed contrast.

For energy conservation, a balance can be struck in regard to workplace lighting by maintaining a uniformly low level of ambient illumination, and as necessary, using local task lighting to meet difficult visual work requirements (5). Caution in the latter applications need to be taken with regard to glare.

4.1.2 Glare Control

Dazzle or glare is produced by light sources of relatively high intensity that are directly in the individual's visual field or by bright surfaces that reflect glare into one's eyes (specular glare). However produced, glare of any kind decreases visual acuity markedly, and evidence of discomfort or irritation is shown by watering of the eyes, excessive blinking, and headaches. According to Grandjean (7), glare in the industrial environment is one of the most important ergonomic considerations in designing a work station. Methods for glare control specific to direct and reflected (specular) glare problems have been cited. The following

Table 8.5 Recommended Ranges of Illuminance for Different Tasks or Types of Activities

Type of Activity or Area	Range of Illuminance[a]	
	Lux	Footcandles
Public areas with dark surroundings	20–50	2–5
Simple orientation for short temporary visits	>50–100	>5–9
Working spaces where visual tasks are only occasionally performed	>100–200	>9–19
Performance of visual tasks of high contrast or large size: reading printed material, typed originals, handwriting in ink, good xerography; rough bench and machine work; ordinary inspection; rough assembly	>200–500	>19–46
Performance of visual tasks of medium contrast or small size: reading pencil handwriting, poorly printed or reproduced material; medium bench and machine work; difficult inspection; medium assembly	>500–1000	>46–93
Performance of visual tasks of low contrast or very small size: reading handwriting in hard pencil on poor-quality paper, very poorly reproduced material; very difficult inspection	>1000–2000	>93–186
Performance of visual tasks of low contrast and very small size over a prolonged period: fine assembly, highly difficult inspection, fine bench and machine work	>2000–5000	>186–464
Performance of very prolonged and exacting visual tasks: the most difficult inspection, extra-fine bench and machine work, extra-fine assembly	>5000–10,000	>464–929
Performance of very special visual tasks of extremely low contrast and small size; some surgical procedures	>10,000–20,000	>929–1858

[a] The choice of a value within a range depends on task variables, the reflectance of the environment and the individual's visual capabilities. [See Bennett (54) for making such estimations.]

Table 8.6 Recommended Maximum-Luminance and Reflectance Ratios

Recommended Maximum Luminance Ratios[a]

	Environmental Classification[b]		
	A	B	C
Between tasks and adjacent darker surroundings	3/1	3/1	5/1
Between tasks and adjacent lighter surroundings	1/3	1/3	1/5
Between tasks and more remote darker surfaces	10/1	20/1	—[a]
Between tasks and more remote lighter surfaces	1/10	1/20	—[a]
Between luminaires (or windows, skylights, etc.) and surfaces adjacent to them	20/1	—[a]	—[a]
Anywhere within normal field of view	40/1	—[a]	—[a]

Recommended Reflectance Values[c] Applying to Environmental Classifications A and B

	Reflectance[c] (%)
Ceiling	80–90
Walls	40–60
Desk and bench tops, machines and equipment	25–45
Floors	≥20

[a] Luminance ratio control not practical.
[b] Classification: A—interior areas where reflectances of entire space can be controlled in line with recommendations for optimum visual conditions; B—areas where reflectances of immediate work area can be controlled but control of remote surroundings is limited; C—areas (indoor and outdoor) where it is completely impractical to control reflectances and difficult to alter environmental conditions.
[c] Reflectance should be maintained as near as practical to recommended values.
Note: These tables are taken from Kaufman as cited in Reference 53.

offers generally accepted ideas for glare reduction [after Grandjean (7)]:

Reducing Direct Glare	Reducing Specular Glare
Position light sources as far from line of sight as feasible	Keep luminous light level as low as feasible
Use more low-intensity light sources instead of few, brighter ones to attain prerequisite illumination level	Use diffuse light, indirect light, baffles
Use light shields, hoods, visors where glare sources cannot be reduced	Use surfaces that diffuse light such as flat paint, matte finishes; avoid bright metallic, glass finishes
Increase luminance in areas around glare source so that the luminance ratio is less	Position light source and/or work area so that reflected light will not be directed toward the worker's eyes

4.1.3 Color

Color rendering, that is, the degree to which the perceived colors of objects illuminated by various light sources match the same object when illuminated by natural light, is another important factor in designing proper workplace lighting. Artificial illuminants that approximate daylight or that emit light throughout the entire visible spectrum are preferred in workplaces to any strong chromatic source. As a general rule, colored filters should not be used over most lamps for ordinary illumination purposes. A colored filter always reduces the total amount of light produced, and this loss in light intensity usually does more to reduce the ability of the eye to see than the color does to increase it (5). Special care is necessary in specifying illumination for work environments where precise color matching for color discriminating tasks are to be performed. Indeed, the use of incandescent, fluorescent, and high-intensity lamps of different types (mercury, metal halides, sodium) can introduce variations in color rendering that can enhance or complicate the required judgments. For example, incandescent (tungsten) lamps tend to emphasize yellows and reds while deemphasizing blues and greens. Warm-white fluorescent lamps enhance yellows and reds. Mercury lamps produce light rich in yellow and green tones, which would render blues and reds more prominently. Obviously the same colored object when illuminated by these different sources would acquire different colorations.

By judicious use of color, an interior can be made attractive and pleasing to workers. Light shades are appropriate to large surface areas (walls, ceilings). Green and blue tints suggest a cooling effect, whereas ivory and cream are associated with warmth. Ceilings and walls closest to windows should have light colors with a high reflectance value such as to have a moderating effect on the bright window area. This could be achieved by painting the walls that have windows a lighter shade of the main color in the room. A large area can be functionally divided by color to give identity to different functions and facilitate supply and servicing needs. Bright colors for highlighting objects and areas having critical functions are also important. In this regard red is the danger–hazard color, and signs and equipment having this connotation are so colored. Yellow, usually in contrast with black, conveys warnings. Green implies safety and is used to identify safety exits, rescue services, and related first-aid equipment. Blue is typically used for directional (sign) purposes.

4.1.4 Eye Fatigue and Eyestrain

Although certain work conditions may include lasers, welding, or the sun which pose a threat to vision, indoor lighting and illumination concerns are directed mainly to issues of discomfort and productivity. Complaints of eyestrain and eye fatigue remain ill defined, having no obvious correspondence to physiological or clinical disorders (23, 55). The prevalent notion is that such reports are related to oculomotor adjustments of the eyes. Faulty lighting conditions (e.g., glare, low-level illumination) cause greater frequency and amplitude of saccadic movements, indicating that the oculomotor adjustment system is working harder under the adverse conditions and is therein subject to more fatigue. Transient changes in convergence, or temporary myopia, have also been noted for workers engaged in close visual work, such as microscopists and VDT operators (23). Whether such changes could lead to a permanent dysfunction given typical spans of service in these jobs is not known. Presumably, incorporation of rest breaks in the course of such daily tasks should retard potential for these types of alterations.

4.2 Thermal Conditions

The effect of severe heat exposure on the worker's health and the methods of monitoring heat stress and preventing heat illnesses in industry has been described in great detail by Mutchler in Chapter 20, in Patty's Volume I, 3rd revised edition. However, recent studies by Ramsey et al. (56) have shown that temperatures that hardly exceed the comfort zone, either on the warm or the cold side, tend to increase the number of unsafe acts. This is in agreement with the studies of Hancock (57), who reviewed the literature on the effect of heat exposure on performance decrement and found evidence that the higher demand the task has on the mental performance of the subject, the lower will be the heat stress exposure needed for increasing errors on lowering task performance. It is thus important to maintain comfort conditions in offices or industrial control rooms where errors made by the managers or controllers of industrial processes may be hazardous for many workers in the plant as well as very costly for the company. For instance, on the shop floor sweaty hands can cause the tools to slip out from the worker's hand, or vision can become blurred by drops of sweat running in the eyes. Cool temperatures, on the other hand, can cause numbness in the fingers, reducing the accuracy of skilled work. The ASHRAE Standard (58) for thermal environmental conditions for human occupancy describes the comfort zones limits and the methods of monitoring the indoor climate conditions.

4.3 Noise

Occupational noise exposures are in all likelihood the most severe and prolonged of any experienced in daily living. Hence one would expect the full array of adverse noise effects to be displayed in worker populations so exposed. Hearing loss is recognized as the major health hazard posed by excessive noise exposure, and current standards for limiting noise exposure in the workplace are geared to minimizing hearing loss risk (59–62). Noise can also interfere with speech communication and warning signals, and current exposure limits satisfying hearing conservation criteria are not sufficiently stringent to mitigate these kinds of problems. Moreover, questions persist as to whether occupational noise conditions typical to many mechanized jobs can produce extra-auditory effects, that is, noise-induced problems that are apart from or extend beyond hearing processes per se and that could be of consequence to one's health, safety, or productivity (63, 64). It must be determined whether, for example, chronic stress reactions may result from such noise exposure with associated physical and mental health problems or whether such noise through distraction, over-arousal, or interference can cause performance errors and accidents. Granted that current occupational noise standards effectively place a lid on noise exposures from the standpoint of safeguarding hearing, it must be emphasized that these limits still permit high levels of noise for intermittent or interrupted exposure conditions, which are the conditions most likely to induce stress and reduce chances for adaptation. Critical points bearing on the aforementioned issues are briefly mentioned here. Their intent is to sharpen the reader's appraisal of noise as a problem in the work environment. Additional information on this subject is presented in Chapter 13 of this volume and Chapter 10, Patty's, Vol. I, Second Edition.

4.3.1 Noise and Hearing Loss

The OSHA limit of 90 dBA (a sound of pressure level determination made on the A network of a standard sound level meter) is set for continuous 8-hr exposures with higher levels being permitted for shorter daily exposures, allowing a 5-dBA increase for each halving of exposure time up to a ceiling value of 115 dBA. No exposure is to be permitted above this level for a continuous or steady-state type of noise exposure, and a 140-dB peak sound pressure level limit is invoked for impulse-type sounds. These limits are intended to protect against hearing impairment to that portion of the hearing range critical to speech perception (and not the full range of audible frequencies), and even here it is recognized that not everyone can be protected. In that regard, it is prescribed that the action point for hearing conservation measures be set at least 5 dBA lower than the limit. Indications are that the ideal action point for the most susceptible persons would be around 70 dBA, which could be impractical considering that many nonoccupational exposures can be of that magnitude. The 5-dBA trading rule between exposure time and intensity of the noise is still a matter of controversy (65). Some advocate a 3-dB adjustment factor, presuming that the ear acts as an energy integrator. The impulse noise limit remains to be substantiated. Indeed, the data base to support an impulse noise limit represents a glaring gap in the field of noise and hearing loss research. Instrumentation and measurement procedures to characterize impulse noise conditions in industry as contrasted with steady-state ones are still considered problematic.

4.3.2 Noise and Speech Interference

The most demonstrable descriptive effect of noise is masking or the interference with the reception of desired sounds, notably speech. Noise exposure conditions

Table 8.7 Nature of Speech Reception Possible Under Noise Conditions Rate in dBA[a]

Noise Level (dBA)	Voice Level and Distance	Nature of Communication	Telephone Use
55	Normal voice at 10 ft	Relaxed communication	Satisfactory
65	Normal voice at 3 ft; raised voice at 6 ft; very loud voice at 12 ft	Continuous communication	Satisfactory
75	Raised voice at 2 ft; very loud voice at 4 ft; shouting at 8 ft	Intermittent communication	Marginal
85	Very loud voice at 1 ft; shouting at 2–3 ft	Minimal communication (restricted, prearranged vocabulary desirable)	Impossible

[a] Table adopted in part from *Bioacoustics Data Book*, NASA Report SP-3006, National Aeronautics and Space Administration, Washington, DC, 1964, page 301.

judged safe for hearing may still interfere with needed sound transmissions. This is seen in Table 8.7, which describes the nature of speech communications possible under different ambient noise conditions. Even moderate noise conditions are shown to require a raised voice or shouting to communicate effectively, especially for distances of 10 ft or more between talker and listener. Telephone use can also be affected.

In industry, lack of adequate speech reception due to noise masking can degrade efficiency in those jobs dependent on such functions. Special measures for rating or predicting the masking effects of noise have been developed that take account of the acoustic energy found within certain frequency bands of the noise in question. One such measure, called the *speech-interference level* (SIL), simply averages the sound levels for three octave bands of the incidence noise, namely, those centered at 500, 1000, and 2000 Hz. Noise limits for offices or other areas having critical sound reception needs as expressed in terms of SIL and equivalent dBA levels are shown in Table 8.8. More definitive acoustical design criteria encompassing speech interference as well as loudness or comfort considerations for functional activity areas are cited in Beranek et al. (66). Noise masking is a major factor in annoyance reactions, as is the fact that communications under noisy conditions may demand raised voices and strained listening, which adds to a general discomforting experience.

4.3.3 Noise and Productivity

The effects of noise on human performance, although an active area of research, continues to yield highly variable results. Critical reviews and analyses

Table 8.8 Range of Background Noise Levels Deemed Acceptable for Select Work Areas Having Specified Listening Requirement as Expressed in SIL and dBA Values[a]

Type of Space	Listening Requirement	SIL	dBA
Concert and recital halls	Listening to faint musical sounds	10–20	21–30
Large auditoriums, theaters	For excellent listening to speech, musical sounds	≤20	≤30
Executive offices, large meeting or conference rooms for 50 people	For excellent listening of speech	≤25	≤34
Private or semiprivate offices, small conference rooms, libraries, classrooms	Good listening for speech at relaxed levels	30–40	38–47
Reception areas, retail shops, cafeterias and restaurants	Moderately good listening, speech privacy	35–45	42–52
Laboratory work areas, drafting, engineering rooms, secretarial areas	Fair listening conditions, no telephone interference	40–50	47–56
Office and computer equipment rooms	Moderately fair listening conditions	45–55	52–61

[a] Adapted from Beranek et al. (66). The SIL values represent average of octave band sound levels centered at 500, 1000, and 2000 Hz as found on preferred noise criterion curves defining the total noise spectrum for the specified listening condition. The dBA levels have been computed from such spectra.

of such work have attempted to sort out acoustic–environmental factors, task factors, and individual subject factors that can influence outcomes (64, 65). The following general observations seem to apply in situations where noise may affect performance:

1. *Acoustic factors.* Adverse effects of noise on performance are typically reported where the noise levels are in excess of 90 dB, are impulsive, intermittent or variable in level and content, and possess significant energy in the high frequency region (above 1000 Hz). More moderate level noise, either continuous or rhythmic in occurrence, could in fact have a positive effect on performance, especially on simple, repetitive tasks through providing added arousal.

2. *Task factors.* Tasks involving complex, mental demands (memory, arithmetic, calculations), forced pace of work, dual or multiple operations will be most vulnerable to noise degradation (as they probably would to any stress insult).

3. *Personal or individual subject factors.* Persons perceiving no control over the noise or believing that it will adversely affect their performance will be most affected. Also those showing maximal anxiety and physiological arousal seem most sensitive.

Combinations of the aforementioned factors will increase the likelihood for performance losses in noise. It bears stating that gross output measures such as total items completed and total number correct or in error can conceal more subtle effects of noise on performance. For example, noise may induce greater variability in performance level, indicating a destabilizing effect not captured in typical productivity measures.

4.3.4 Noise and Health–Safety Issues

Noise can induce physiological arousal characteristic of a generalized stress reaction (61, 64). Although more pronounced for high-level noise exposures, even moderate levels (70 to 75 dBA) can trigger such responses. The effect includes a variety of acute changes—increased blood pressure, elevated skin conductance, rapid and shallow breathing, and excretion of hormones indicative of a heightened autonomic nervous system activity. Taken together, these changes are symptomatic of a spreading state of alarm in which the body becomes mobilized to ward off danger. These reactions dissipate or are subject to adaptation with recurring encounters to the same noise conditions, which suggests no lasting harm. Yet epidemiologic studies in industry demonstrate linkage between prolonged exposures to noise and health problems. The strongest case seems indicated for cardiovascular functions where increased cases of hypertension have been reported for workers chronically exposed to workplace noise over 85 dBA (68). Caution is urged in considering these findings as conclusive as they represent correlational data, in many instances lacking for a control group to rule out other etiologic factors. A similar situation exists in evaluating workplace noise as a cause of accidents. Two field studies have shown that issuance of personal ear protectors to reduce noise conditions harmful to hearing has been accompanied by fewer numbers of job injuries (69, 70). Although the improvement in safety could be attributed to lower noise exposure, correlations between actual ear protector use and injury frequency were not notable. Moreover, how a reduced noise level could counter certain factors otherwise contributory to accident occurrence in these investigations is not explained. Descriptions of accident scenarios depicting noise as a disruptor of critical task function or interfering with those having safety significance would be helpful in providing such insights. Clearly, the matter of relating noise to general health and well-being deserves more definitive study.

4.3.5 Noise Control and Hearing Conservation

A vast literature has emerged offering a range of options for coping with excessive noise conditions at the workplace. Examples of possible control measures categorized as engineering, administrative, and worker self-protective are presented in the following paragraphs. Examples of cases illustrating the efficacy of many of these applications to noise reduction are available in a number of sources (53, 71, 72). Although self-protective measures involving use of personal ear protectors represents a simplistic, economic solution, it is not a preferred approach and, in fact, has a number of drawbacks. The majority of users find personal ear protective devices to be uncomfortable and irritating such that widespread worker acceptance and regular utilization of this equipment can be a problem. Motivational schemes to overcome resistance have been tried with some success. On the other hand, the amount of attenuation offered by such protectors, as typically worn by workers, is only 50 percent of the noise reduction expected based on laboratory certifications of performance (73). Hence pre-employment and follow-up audiograms in a hearing conservation program take on greater importance. Such monitoring must provide for the early detection of any undue shifts in hearing thresholds that would warrant added noise protection measures for affected workers.

The following *engineering control measures* are recommended:

1. Substitution of machines:
   a. Larger, slower machines for smaller, faster ones.
   b. Step dies for single-operation dies.
   c. Presses for hammers.
   d. Rotating shears for square shears.
   e. Hydraulic for mechanical presses.
   f. Belt drives for gears.
2. Maintenance of equipment:
   a. Replacement or adjustment of worn and loose or unbalanced parts of machines.
   b. Lubrication of machine parts and use of cutting oils.
   c. Properly shaped and sharpened cutting tools.
3. Substitution of processes and techniques:
   a. Compression for impact riveting.
   b. Welding for riveting.
   c. Hot for cold working.
   d. Pressing for rolling or forging.
4. Damping vibration in equipment:
   a. Increase mass.
   b. Increase stiffness.
   c. Use rubber or plastic bumpers or cushions (pads).
   d. Change size to change resonance frequency.
5. Reducing sound transmission through solid materials.
   a. Flexible mounts.
   b. Flexible sections in pipe runs.
   c. Flexible shaft couplings.
   d. Fabric sections inducts.
   e. Resilient flooring.
6. Include noise-level specifications when ordering new equipment.
7. Reducing sound produced by fluid flow:
   a. Intake and exhaust mufflers.

b. Fan blades designed to reduce turbulence.
   c. Large, low-speed fans for smaller, high-speed fans.
8. Isolating operator:
   a. Provide a relatively soundproof booth for the operator or attendant of one or more machines.
9. Isolating noise sources:
   a. Completely enclose individual machines.
   b. Use baffles.
   c. Confine high-noise machines to insulated rooms.

The following administrative control measures, involving changes in job schedules, are suggested:

1. For workers reaching permissible noise exposure dose during the workday, interpose work periods in environments well below the levels posing any threat to hearing.
2. For work involving high-level noise exposures, divide the exposure time among as many workers as needed so as to keep individual noise exposure within permissible limits.
3. Perform necessary high-level noise-producing operations during the night or at other times where there is a minimum of workers exposed.
4. If possible, arrange for high-level noise producing operations to be performed during only a portion of each day rather than all day, and for only a few days each week.

The following self-protective measures involving establishment of a hearing conservation program are recommended:

1. Educate workers in needs for conserving hearing in noise.
2. Instruct workers as to proper use of personal ear protectors in noise.
3. Establish audiometry program involving preemployment and follow-up monitoring of hearing levels to ensure no significant threshold losses attributable to workplace noise exposure.
4. Inform workers of status of hearing.

4.3.6 Vibration

Brief mention was made of vibration transmission to hand–arm areas and the whole body in examining issues of the worker–equipment interface in Tichauer's ergonomics chapter in the third edition of Patty's volumes (1). Reference was made to "white finger syndrome" or intermittent blanching and numbness of the fingers resulting from long-term exposure to vibrating hand tools such as chain saws, pneumatic chip hammers, grinders, and jack hammers. Follow-up work in recent years has evolved a grading system for defining the severity of vibration white finger disorders based on the extensiveness of the reported symptoms (number of fingers affected) and the amount of interference with job or hobby activities (74). Although differential diagnosis of vibration white finger disorders remains troublesome, improved tests of peripheral vascular function (plethysmography, skin thermography), neurological function (two-point discrimination, depth sensation), and radiographs allow for more sensitive characterizations of the disorders. Recommendations for reducing the risk of vibration-induced white finger syndrome include:

1. Workers subject to frequent use of vibratory hand tools should undergo pre-employment and follow-up physical examinations to detect any signs of white finger disease. Evidence of such symptoms or a prior history would make further exposures inadvisable.
2. Vibratory tools should be regularly maintained and kept in good working order per manufacturer's recommendations.
3. Workers using vibratory tools are advised to:
   a. Wear gloves that yield limited attenuation to hand but more importantly, prevent cooling of the hands, which can enhance the risk.
   b. Let the tool do the work; thus grip it lightly.
   c. Reduce smoking since nicotine acts as a vasoconstrictor, reducing blood supply to the fingers.
   d. Use the tool only when necessary and where possible in an interrupted fashion to avoid continuous, sustained vibratory exposure.

Morbidity studies of workers experiencing whole-body vibration, such as bus drivers and heavy equipment operators, demonstrate an excess of renal, bowel, general musculoskeletal, and back problems. It has been suggested that whole-body vibration in combination with postural problems due to poor seats, long working hours, coupled to poor dietary habits are responsible for such ailments. Human performance studies have found vibration conditions within specified frequency ranges to upset tracking ability ($\sim$5 Hz), visual acuity (1–25 Hz). Whole-body vibration in the 0.1 to 0.7 Hz range is associated with motion sickness. A data base to define safe and healthful levels for whole-body vibration is currently not available. In the interim, it is recommended that vibration levels and exposures be kept to a minimum by:

1. Maintaining equipment to prevent excessive vibration from developing through undue wear.
2. Limiting time spent on any vibrating surface to the minimum necessary to perform the job safely.
3. Where possible, isolating worker areas from vibrating sources.

Additional information on this subject is contained in Chapter 15.

4.4 Chemical Agents

Toxicity issues aside, exposures to chemicals even at safe levels may elicit reactions that require attention, if not concern. Indeed, chemical odors have become triggers for near contagious outbreaks of symptomatic complaints (dizziness, headaches, nausea), although environmental measures indicate no contaminant at a harmful level (75). Investigations suggest that such episodes may serve as outlets for much pent-up stress and frustration connected with the boring, repetitive jobs performed by the affected workers, excessive workload demands necessitating much unwanted overtime, and poor employee–management relations. The presence of an odorous substance serves as the basis for attributing the symptoms displayed. Chemicals below present permissible limits can be shown to induce alterations in behavioral performance, measures whose implications for production and safety in the job operations require closer scrutiny. Clearly, ergonomics demands that the industrial hygienist and toxicologist take a more expanded view of chemical effects at the workplace.

5 WORK ORGANIZATION

5.1 Machine-Paced Work

Machine-paced work refers to jobs in which workers perform set tasks at rates governed by the machine such as to ensure a continuous flow of products through the system (76, 77). Invariably the tasks are well defined and cyclic in nature in meeting the system requirements for high productivity. Although they have economic advantages, machine-paced jobs also can be a source of significant worker dissatisfaction and stress. Such job routines do not allow for individual differences in performance such that the rate set may be too fast for some and too slow for others, or if satisfactory for an individual at one time may not be all the time. Machine-paced operations, being repetitive and often simplistic in nature, present problems of monotony and boredom to the workers engaged. The literature cites evidence for hormonal changes associated with demanding machine-paced work suggesting significant stress response (76, 78). It is also evident that tasks requiring forceful arm–wrist–hand–finger movements, if repeated, can result in chronic "wear and tear" injuries to the musculoskeletal structure (80). Approaches to reducing these problematic aspects of machine-paced work can take various forms. Obviously, provision of options for self-paced operations could alleviate the temporal stress factors, and performance motivated by adopting output goals to be met daily or incentives to increase productivity have been shown to be successful in this regard. The use of buffers between work stations in a machine-paced process can also serve to reduce the time pressure. The buffers make more than one component of the feeding position available to the operator at the same time. A variant of this is discrete pacing, where index periods are provided between job cycle periods for the workers to do preparatory work for the next cycle or remain idle. Still other ideas include rotating workers through different jobs on the process line to alleviate boredom.

Machine-paced work has raised a more fundamental issue in the job stress literature where there is increasing evidence that loss of job control or absence of decision latitude in one's job constitutes a major stress-producing factor. Karasek et al. (81) have shown that workers having low decision authority in their jobs and subjected to high work-load demands are most prone to psychological strains and prevalence of coronary heart disease. It is believed that productivity enhancement based on older notions of narrowing the skill requirements and the judgmental capacities needed in a job are no longer tenable. The fact is that the level of formal training of the work force has risen wherein greater utilization of skills and responsibilities for job decisions is an important factor in worker satisfaction and performance. Attempts of enriching jobs are consistent with this view.

5.2 Psychosocial Stress Factors

Degree of job control is one of a mix of psychosocial factors in a work situation that loom as significant sources of stress. Others include work load; referring to cognitive as opposed to physical task demands; responsibilities for people (as opposed to things); clarity of job functions and role in organization; relationships with one's boss, co-workers, and subordinates; and job security (79, 81, 82). There is a growing literature linking adverse conditions relatable to these factors to emergent stress and strains, ranging from negative worker attitudes and moods (anxiety, depression), through maladaptive behaviors (e.g., excessive drinking, smoking, absenteeism) to somatic complaints (recurring headaches, stomach indigestion, assorted aches and pains) and diagnosed disease or illness (notably, hypertension, ulcers, neuroses) (76–82). Indicated in Table 8.9 is a summary of surveys and worksite investigations reporting adverse psychosocial factors found at work and correlated outcome measures reflecting stress–strain consequences. Admittedly, such problems fall outside the conventional expertise of the industrial hygienist or health–safety professional when called on to investigate worksite problems and prescribe control measures. Moreover, the methods for data collection and analysis of psychosocial factors are less exact and require careful interpretation. Still, these types of problem factors can interact with the more apparent physical and environmental ones of concern and produce a more serious or complicated situation defying the usual solutions.

Alerting the industrial hygienist and safety–health professional to situations warranting a broadened assessment of workplace concerns is the major motive for including such material here. It is also important to note that psychosocial factors represent more potent worksite problems in connection with office, managerial, and professional occupations than with blue collar groups (83).

6 NEW DEVELOPMENTS AND TRENDS

The workplace as well as the workforce in modern, well-developed industrial nations such as the United States are changing. New advances in automation

Table 8.9 Selected Research Literature Illustrating Adverse Psychosocial Factors at the Worksite and Related Stress–Strain Outcomes

Target Group(s)	Methodology	Results		Reference
		Psychosocial Factors Found Significant	Related Significant Stress–Strain Outcome	
Federal workers	Questionnaire, interviews, physiological monitoring	Overload role conflict	Excessive cigarette smoking—risk factor in CHD;[a] increased heart rates	84
Representative national sample	Questionnaire	Overload	Drinking, absenteeism from work	85
Air traffic controllers	Periodic clinical exams, psychiatric interviews	Overload, responsibility for people	Hypertension; greater illness rates	86
Varied white collar groups	Questionnaire, interview	Role conflict	Lower job satisfaction; higher job tension	87
Select occupational groups in communal setting	Questionnaires, clinical–bioassay	Role conflict	Increased CHD risk factors, especially in white collar groups of workers	88
Workers in sawmills	Questionnaires, hormonal assays	Repetitive machine-paced jobs	Increased catecholamine levels	78
Automobile factory workers on assembly line and in tool room	Interviews	Paced and repetitive work	Repetitive work without pacing increases job content; with pacing, anxiety level is increased	99
Assorted occupational groups—morbidity data	Analysis of national health survey data	Lack of job control and high productivity demands	Increased risk of CHD	89

417

Table 8.9 (*Continued*)

Target Group(s)	Methodology	Results		Reference
		Psychosocial Factors Found Significant	Related Significant Stress–Strain Outcome	
23 occupational groups including blue and white collar	Structured interviews	Underutilization, repetitive work, job insecurity, poor social support	Job dissatisfaction, depressive tendencies	90
Workers in rubber, tire, plastics manufacturing	Questionnaire survey	Social support from supervisors and wives and perceived stress from job conditions	Found to buffer self-reports of neurotic symptoms, digestive disorders	91
Seven plants involved in automobile manufacturing	Interviewing, health diary keeping	Job termination	Changes in emotional state, self-esteem; increase in arthritics, hypertension, and attacks of patchy baldness, especially just before job loss	

a Coronary heart disease.

driven by computer applications to industrial processes and office operations are most evident. Robots on the shopfloor and VDTs in office environments represent the most outward signs of this technology. The composition and nature of the U.S. work force is also in a state of transition. The percentage of white collar workers is increasing significantly whereas those engaged in blue collar jobs is remaining the same or decreasing. As of 1980, the number of white collar workers exceeded those employed in all other types of employment (i.e., blue collar, service, farming). More women have entered the workforce, a segment now assuming jobs in mining, construction, and manufacturing occupations, which traditionally were held only by men. Also, the retirement age is being lifted or set back such that more workers now remain employed into their later years. These different developments can redefine the thrust of worker–work interactions, requiring renewed attention to ergonomic issues.

6.1 Industrial Robots

Elimination of workers from operations or processes posing toxic hazards and other dangers along with reduced labor costs and increased production efficiency are acknowledged as primary motives for introducing robots into work processes. Currently, most robots in the United States and other countries are being deployed in the following operations: welding, materials handling, assembly, and spray painting. The robots essentially relieve workers from exposures to the toxic fumes and UV radiation in welding, the repeated physical (lifting, carrying, manipulative) demands in materials handling and assembly work, and the exposures to solvents in spray painting. The cyclic nature of these tasks, a source of monotony or boredom, is also spared the worker by these roboticized operations. At the same time, however, the worker is not entirely excluded from robot operations. Worker–robot interactions occur, raising the following issues, which have ergonomic significance:

1. Robot production lines will require servicing and maintenance to be performed by workers. Current installations where a robot is closely surrounded by other feeder–support–transport equipment give considerably less room for maintenance and repair work than was the case with more conventional processes.

2. The production parameters for a roboticized process as a rule will operate at higher feeding speed, higher pressure and temperature, and shorter cycle times. These factors, together with needs for fewer machine guards and lesser environmental concerns, carry a high hazard risk, given occasions where workers may be called on to back-up or run the system in a manual mode.

3. Related to the preceding points, robot design and engineering may dictate operational tasks and procedures that are not optimal in terms of human capabilities. Hence, having humans work roboticized operations in cases of robot malfunction can present problems in human adaptability and endurance. Salvendy and Smith (76), in addressing human factor issues in robotics, advocate that a robot production line should be designed to allow humans and robots to be used interchangeably. That is, the system should respect human factor considerations such that switchovers to manual work can be handled easily.

4. As indicated by reports from Japanese experiences with robots, accidents involving robot contact with workers occurred because (a) the worker was trying to remove material with the robot still in operation; (b) a work cycle was unintentionally started by a worker; or (c) a robot started to behave erratically during worker inspection, programming, or service work. As a consequence, prescriptions for additional sensing, lock-out, interlocking, emergency stopping, and reset devices have been made. These measures, in addition to mandatory safety training for workers who may be connected with robotics, are contained in a set of ordinances and guidelines promulgated in Japan.

6.2 Work Force Change

The growth of the white collar work force will cause more attention to be directed to examining health, safety, and productivity issues connected with office-type jobs. In this regard, case reports of symptomatic outbreaks among office worker groups have reached notable levels. One reporting source, a federal agency that responds to requests for evaluating real or alleged worksite hazards, has indicated that 25 out of every 100 incoming requests refer to an office-type work area. Symptoms such as eye and throat irritation, headache, and fatigue are reported with regularity in these cases, and indications are that poor ventilation and inadequate air circulation may be the principal causal factor (83). The expressions "indoor air pollution" and "stuffy office syndrome" connote such thinking. But two other factors have also been recognized as contributors to this kind of outbreak or have raised other concerns. One is the advent of new technology in office production that some feel rival the mechanized assembly lines introduced in factories early in the twentieth century. Video display terminals linked to computer systems are preempting the more traditional office equipment. It is clear that the use of this equipment imposes added visual and postural demands on operators; the literature already has documented complaints of eyestrain, blurred vision, and assorted back, neck, and shoulder complaints of those who work these units for substantial parts of the workday. This, in turn, has dictated added environmental and workstation design requirements of which the industrial hygienist along with others concerned with worker health must be cognizant. Control of glare sources, proper near field–far field illumination, provisions for variable height and tilt of VDT screen and brightness contrast, detachable keyboard and adjustable chairs, and other supports are ergonomic elements to be included in an evaluation of modern office work conditions. A treatment of these considerations can be found in a number of sources (23, 76, 83).

The other factor represents job stress, which has been alluded to earlier. Video display terminal–computer-based operations in office work can mean elimination of jobs, requirements for new skills, loss of job control to a computer, and closer monitoring of one's performance, all of which can be perceived as threatening to affected workers. It has been noted that the stressors here can add to the other problems mentioned previously and therein intensify the overall adverse reactions. Indeed, it has been suggested that there can be a psychological component in cases of stuffy office syndrome where the reactions are driven, in part, by stress factors reflecting underlying discontent in the work situation. Thus our understanding of the psychosocial factors operating in an office situation is needed insofar as they may influence apparent symptomatic effects attributed to environmental agents. One can envision greater interaction among industrial hygiene, ergonomics, and psychology specialists in addressing the problems that may emerge here.

The influx of more women into the work force and into nontraditional blue collar occupations in heavy industry, mining, and construction poses added ergonomic problems. Job tasks through design or through custom have been fitted to a male work force having strength, size, and other features that become burdensome for the average women. Personal protective equipment for use in such work is also sized for a male population. Hence, instances of women having to work with ill-fitting gloves, safety shoes, and hard hats are not infrequent. Fortunately, manufacturers of safety equipment have recognized this problem and are expanding the range of available apparel sizes to better accommodate women. Similarly, new tool or workstation designs or retrofit of existing ones must recognize human factor characteristics of both male and female workers. Dealing with older workers can raise problems given changes in job routines, especially involving added time pressures and short-term memory demands.

7 CONTROL–INTERVENTION APPROACHES

7.1 Systems Approaches

As in problem-solving work in other areas, the early design stage in the development of an industrial process or operation is the ideal time to deal with ergonomic issues. Once the process has reached the implementation or installation phase, opportunities are far more limited.

A systems approach for addressing ergonomic concerns typically would begin with some form of analysis aimed at defining the objectives of the production process or operation and the functions to be performed in attaining them. The analysis may focus on the organizational make-up of the system, the activities of different components, and/or sequences of operations. The functional requirements derived from this analysis would present options for choosing human or physical (machine) components. In some instances, the nature of the process demands, economic factors, or other considerations may indicate obvious superiority of one over the other and therein predetermine the choice. On the other hand, the designer can also be confronted with a range of functions for which either humans or physical components may serve. In these cases, added knowledge about the relative capabilities of humans versus machines could aid the function allocation process. A set of such guidelines patterned after versions found in the human factors engineering literature appears in the following list. Aside from their utility to function allocation decisions in process design, they serve to sustain a consciousness of the general characteristics of people and machines as components in a system. The latter message is important in considering ergonomic issues at whatever stage they are encountered.

Humans are generally better than machines in their ability to:

1. Recognize patterns of signals that may vary from situation to situation.
2. Sense unusual and unexpected events in the environment.

3. Remember principles and ideas from long-term storage of masses of information.
4. Draw upon varied experience in making decisions and adopting decisions to different situation requirements. (In short, humans do not require previous programming for all situations.)
5. Select alternative modes of operation, if certain modes fail.
6. Reason inductively, generalizing from observations.
7. Apply principles to solutions of varied problems or develop new solutions.
8. Concentrate on most important activities when overload conditions require.

Machines are better than humans in their ability to:

1. Sense stimuli that are outside normal range of human sensitivity (X-rays, ultrasound).
2. Monitor for prespecified or programmed events.
3. Store and retrieve coded information in substantial quantities.
4. Process quantitative information following specified instructions.
5. Perform repetitive acts reliably.
6. Exert and sustain considerable force in a highly controlled manner.
7. Maintain performance over extended periods of time.
8. Measure or count physical quantities.
9. Perform several activities simultaneously.
10. Maintain efficient operations under heavy load and/or under distracting or environmentally stressful conditions.

In reducing the preceding guidelines to a short-hand statement, Jenkins states: "Men are flexible but cannot be depended upon to perform in a consistent manner, whereas machines can be depended on to perform consistently but have no flexibility whatsoever."

There are some qualifiers to the use of the guidelines that must be borne in mind. One is that certain comparative advantages may change. Computer advances to problems of pattern recognition may dictate their consideration for such tasks relative to humans. Also, function allocation decisions may have to consider more than performance capabilities. Cost, weight, power, and maintenance may also be weighing factors. In addition, function allocation should consider social and cultural issues. Indeed, functions should not be assigned to humans that are degrading or hold no reasonable intrinsic satisfaction. Jobs need to be allocated humanely based on what the people can do and want to do. The latter brings in the human value aspect of job design and earlier references to job enrichment, job enlargement ideas.

Once functions in a system are assigned to people and machines, the interfaces between the two must be scrutinized to ensure compatibility between the machine design and operation and the capabilities of the human operator. The subjects for consideration here are those already discussed in the major sections of this chapter.

7.2 Behavioral Approaches

Design of production systems or their physical components to ensure optimal coupling with worker attributes with respect to productivity, hazard control, and comfort considerations is a primary ergonomic goal. Also playing an important role in achieving this objective are the workers' actions in the work environment and in performing their jobs. For example, are they engaging in the production processes in ways to provide for maximum work efficiency? Do their work habits fully exploit the capabilities of the hazard control systems that are in place? Room- or booth-type ventilation controls with spray mist arrestors represent one common solution to reducing exposure hazards in spray painting. However, without appropriate employee actions such as making sure the ventilation is engaged, the applications are within the capture range of the ventilation, and the mist arrestors are not saturated, the controls would provide little benefit. Bearing on this point, Conrad (93) found in a study of laminated plastics operations that, although employees actuated the exhaust ventilation in a spray area before they began to spray toxic polyester resins (styrene), they often turned it off as soon as they finished the application. They did not realize that vapors created during the curing process, which remained for nearly half an hour, could disperse into other plant areas.

Is employee behavior while on the job actually increasing the risk of hazard? Williams et al. (94) noted that employee exposures are frequently higher than those found in the general work environment because, in moving from task to task, they tended to spend much of their time near the exposure sources. Employees have many options under their direct control that can influence the level of exposures to hazards. Cohen and Jensen (95) found in analyzing driver behavior of forklift truck operators that they tended not to signal or yield to co-workers at intersections, thus increasing the risk of a mishap.

Similarly, do workers regularly wear personal protective equipment, if these are needed to augment the engineering control system in place or to serve as interim measures until more effective, feasible physical controls can be developed or installed? Use of personal protective clothing and devices are regularly recommended and just as regularly acknowledged as problematic in terms of worker acceptance. Respirator devices are the best known example of personal protection and present particular difficulties in terms of encumbrance and discomfort.

Are housekeeping and maintenance measures being followed such as to prevent any unnecessary exposures to workplace containments or safety hazards? Hopkins (96) observed that failure to replace lids on containers on styrene-coated scraps in a laminated plastics plant added to the concentrations of this toxic material in the worker's environment. Alternatively, use of disposable floor coverings to prevent an accumulation of styrene overspray in the forming areas was believed to be an important factor in reducing the exposure level.

Do workers know what to do in case of emergency situations such as chemical spills, fires, control equipment failures, machine malfunction, and explosions? Employee behavior is critical in these instances; employees must be informed, trained, and rehearsed in procedures enabling them to effectively cope with the apparent danger. This would include actions to prevent harm to oneself or co-workers and to contain the source of the problem. Finally, are workers practicing good personal hygiene, that is, washing, showering, changing, and laundering clothes regularly such as to reduce further potential contact with toxic materials?

The industrial hygienist will recognize many of these considerations as falling into the realm of work practices. Although such prescriptions are made in the context of plant operational procedures or hazard control programs, little systematic thought has been given to the manner of their development or adoption by the work force. Indeed, they are typically recommended on a common-sense basis and through usual employee training and orientation are expected to be acquired.

There has now emerged a behavior management technology that is cogent to efforts aimed at enhancing worker productivity, health, and safety. The behavior management program approach can vary in complexity; however, there are four key components that are common to most of them. It must be emphasized that rather than offering recommendations with loosely defined objectives, behavior management programs emphasize the targeting of specific behaviors for change and then set a course to alter those actions in a very direct manner. The four components are:

1. *Assessment.* In selecting the behavioral targets, a first analysis is undertaken in which observations are made of the problematic work conditions or operations of concern and related worker behaviors. Particular attention is paid to worker behaviors that can have some effect on the desired outcome—that is, increased productivity, reduced exposure risk to a chemical, and increased safety margins.

2. *Objective formulation.* The list of behaviors and job conditions are then prioritized such that those behaviors deemed to have the greatest impact on the outcome sought are selected for consideration as behavioral objectives. Final decisions here need take account of a number of practical matters. How difficult will the behavior change be to the workers? Will management support the work practice? How difficult will it be to monitor the work practice?

3. *Select behavior change methods.* The methods here can utilize any number of ideas. Typically, training supplemented with opportunities for practice with motivational reinforcement and feedback to mark progress in attaining the behavioral objectives are carried out.

4. *Evaluation.* Before–after evaluations to determine extent of behavior change is a basic feature of a behavior management program. Measures of the

targeted work practices are taken repeatedly before, during, and after the employees have been trained, and the results here correlated with independent measures depicting the outcomes of concern, specifically, production volume, exposure levels, and job safety measures. The ultimate objective of the behavior management program is to change critical work practices, which, in turn, impact the latter outcomes in a desired way.

Work by Hopkins (96) in evaluating work practices for reducing worker exposures to styrene in manufacturing laminated plastic products exemplifies these different steps.

Specifically, Hopkins and co-workers first undertook an analysis of the basic processes found in manufacturing the product with styrene, providing an indication of the operations causing the highest styrene levels and job factors that could influence these levels. Work procedures and housekeeping measures were then devised, representing the most practical and effective ideas for limiting the exposures. Examples of the select production procedures and housekeeping measures arrived at in this process are noted in the following list:

Work Procedures	Housekeeping Practices
Stay out of spray booths except when spraying or setting up	Cover floors in work areas with disposable material allowing frequent changing
Spray toward exhaust ventilation	Remove cured parts to areas isolated from primary worker areas
Keep all skin below neck, including hands, covered	Keep waste cans of excess resin close to exhaust ventilation ports

These measures became the behavioral end-goals for a training and motivational effort established at the worksites in question. The techniques followed cardinal rules of learning and reinforcement. Videotape sessions to depict the work practices were held, ranging from 20 to 30 min in length or short enough to maintain attention. There was worker involvement in the training and on-the-job tests to certify worker knowledge and conformance to the work practices and housekeeping measures set forth. Frequent training, feedback, praise, and monitoring rewards for passing the on-the-job tests were used to motivate learning and adherence to the prescribed procedures.

Sampling of the workers' actions before, during, and after the training program was handled by specially trained observers to determine compliance with the prescribed work practices. These observations indicated a 25 percent average improvement in compliance with the prescribed work procedures and a 40 percent average improvement in the housekeeping practices that was evident 4 months following training. In terms of limiting styrene exposures, the aforementioned changes in work practices were coincident with changes in breathing zone concentration of styrene of about 40 percent.

This study provided a major illustration of the value of work practices in attaining hazard control objectives and means for their establishment. It could be argued that the training and motivation methods used by Hopkins and co-workers were so elaborate and time-consuming that they would have little likelihood of being voluntarily adopted by industry. Less formidable variations of these ideas are possible. Indeed, other examples of such types of intervention are much simpler in nature and still yield substantial positive outcomes. Specifically, Cohen and Jensen (95) used a behavior management program to effect changes in the driving behaviors of forklift truck operators. The targeted changes were derived from analyzing accident reports involving such truck operations. Similarly, Zohar (97) has used a behavior sampling approach and feedback plus other incentives to show how worker resistance to wearing ear protectors for hearing conservation in noise can be overcome.

The important fact is that a technology does exist for altering worker behavior such as to enhance a worker's health, productivity, and well-being (98).

NOTATION

The views expressed in this chapter are those of the authors and do not necessarily reflect those of the agency with which they are affiliated.

REFERENCES

1. E. R. Tichauer, Patty's *Industrial Hygiene and Toxicology*, 3rd Rev. Ed., Vol. 1, *General Principles*, Wiley, New York, 1978, Chapter 22.
2. L. Parmeggiani, Technical Ed., *Encyclopedia of Occupational Health and Safety*, International Labor Office, Geneva, Switzerland, 1983, p. 775.
3. C. T. Morgan, J. Cook, A. Chapanis, and M. W. Lund, Eds., *Human Engineering Guide to Equipment Design*, McGraw Hill, New York, 1963.
4. A. Chapanis, *Man–Machine Engineering*, Tavistock Publications, London, 1965.
5. *Ergonomic Design for People at Work*, Vol. 1, *Human Factors Section*, Eastman Kodak Company, Lifetime Learning Publications, Belmont, CA, 1983.
6. W. E. Woodson, *Human Factors Design Handbook*, McGraw-Hill, New York, 1981.
7. H. P. Van Cott and R. G. Kinkade, Eds. *Human Engineering Guide to Equipment Design*, (Revised Edition), U.S. Government Printing Office, Washington, D.C., 1973.
8. F. Dukes-Dobos, "The Place of Ergonomics in Science and Industry," *Am. Indust. Hyg. J.*, **31**(5), 565–571 (1970).
9. Ergonomic System Analysis Checklist, International Ergonomic Association, 2nd International Congress on Ergonomics, Dortmund, Germany, September 26, 1964.
10. R. D. G. Webb, *Industrial Ergonomics*, Industrial Accident Prevention Association, Ontario, Canada, 1982.
11. J. H. Ely, H. M. Bowen, and J. Orlansky, Man–machine dynamics in Joint Services Human Engineering Guide to Equipment Design, USAF Wright Air Development Center, Wright-

Patterson Air Force Base, Ohio, WADC Technical Report 57-582, November 1957, Chapter VII.
12. A. Damon, H. W. Stoudt, and R. A. McFarland, *The Human Body in Equipment Design*, Harvard University Press, Boston, 1971.
13. N. Diffrient, A. R. Tieley, and J. C. Bardagy, *Human Scale 1/2/3*, MIT Press, Cambridge, MA, 1974.
14. Webb Associates, Anthropology Research Project, Eds., *Anthropometrics Source Book. A Handbook of Anthropometric Data*, NASA Reference Publication 1024.
15. S. Konz, *Work Design: Industrial Ergonomics*, 2nd ed., Grid Publishing, Columbus, OH, 1983.
16. J. A. Roebuck, K. H. E. Kroemer, and W. G. Thomson, *Engineering Anthropometry Methods*, Wiley, New York, 1975.
17. J. W. McDaniel, "Computerized Biomechanical Man Model," in: *Proceedings of the 6th Congress of the International Ergonomics Association*, College Park, MD, 1976, pp. 384–389.
18. S. Brunnstrom, *Clinical Kinesiology*, 3rd ed., Lea and Febriger, Philadelphia, 1972.
19. E. Grandjean, *Fitting the Task to the Man*, 3rd ed., Taylor and Francis, London, 1980.
20. R. Kukkonen, T. Luopajarvi, and V. Riihimaki, "Prevention of Health Hazards of Key Punchers with Application of Ergonomy and Occupational Physiotherapy" (Abstract), *Ergonomics*, **23**(8), 685 (1980).
21. E. J. McCormick, *Human Factors in Engineering and Design*, 5th ed., McGraw-Hill, New York, 1982.
22. B. H. Kantrowitz, "Interfacing Human Information Processing and Engineering Psychology," in: *Human Performance and Productivity*, W. Howells and E. A. Fleischman, eds., Lawrence Erlbaum Associates, Hillsdale, NJ, 1981.
23. *Video Displays, Work and Vision*, Natural Research Council Panel on Impact of Video Viewing on Vision of Workers, National Academy Press, Washington, DC, 1983.
24. L. Greenberg and D. B. Chaffin, *Workers and Their Tools*, Pendell Publishing Co., Midland, MI, 1977.
25. Eastman Kodak Company, *Ergonomic Design for People at Work*, Vol. 1, Lifetime Learning Publications, Belmont, CA, 1983.
26. T. J. Armstrong, *An Ergonomics Guide to Carpal Tunnel Syndrome*, American Industrial Hygiene Association, 1983.
27. F. N. Dukes-Dobos, G. Wright, W. S. Carlson and H. H. Cohen, "Cardiopulmonary Correlates of Subjective Fatigue," in: *Proceedings of the 6th Congress of the International Ergonomics Association*, College Park, MD, 1976, pp. 24–27.
28. H. Yoshitake, "Relations Between the Symptoms and the Feeling of Fatigue," in: K. Hashimoto, K. Kogi and E. Grandjean, Eds., *Methodology in Human Fatigue Assessment*, Taylor and Francis, London, 1971.
29. P. O. Astrnad and K. Rodahl, *Textbook of Work Physiology*, McGraw-Hill, New York, 1970.
30. H. Spitzer, "Physiologische Grundlagen fur den Erholungszuschlag bei Schwerarbeit." *REFA-Nachrichten* (Darmstadt) No. 2, (1951).
31. E. H. Christensen, *Man at Work*, ILO, Occupational Safety and Health Series No. 4, ILO, Geneva, 1964.
32. B. Bink, "The Physical Working Capacity in Relation to Working Time and Age," *Ergonomics*, **5**, 25–28 (1962).
33. F. H. Bonjer, "Actual Energy Expenditure in Relation to the Physical Working Capacity," *Ergonomics*, **5**, 29–31 (1962).
34. M. Arstilla, H. Wendelin, I. Vuori, and I. Valimaki, "Comparison of Two Rating Scales in the Estimation of Perceived Exertion in a Pulse-Conducted Exercise Test," *Ergonomics*, **17**, 577–584 (1974).

35. J. F. Parker, Jr. and V. R. West, Eds., *Bioastronautics Data Book*, 2nd ed., NASA SP-3006, Washington, DC, 1973.
36. D. B. Chaffin, G. D. Herrin, W. M. Keyserling, and J. A. Foulke, *Pre-Employment Strength Testing*, DHEW (NIOSH) Publication No. 77-163, National Institute for Occupational Safety and Health, Cincinnati, OH, 1977.
37. E. R. Tichauer, "Ergonomics Aspects of Biomechanics" in: *The Industrial Environment—Its Evaluation and Control*, DHEW (NIOSH) Publication No. National Institute for Occupational Safety and Health, Cincinnati, Ohio, 1973, pp. 431–492.
38. D. B. Chaffin and K. S. Park, "A Longitudinal Study of Low Back Pain as Associated with Occupational Lifting Factors," *Am. Indust. Hyg. Assoc. J.*, **34** 513–525 (1973).
39. T. N. Anderson, "Human Kinetics in Strain Prevention," *J. Occup. Safety*, **8,** 248 (1970).
40. J. R. Brown, *Lifting as an Industrial Hazard*, Labour Safety Council, Ontario, Canada, 1971.
41. A. Garg and U. Saxena, "Effects of Lifting Frequency and Techniques on Physical Fatigue with Special Reference to Psychophysical Methodology and Metabolic Rate," *Am. Indust. Hyg. Assoc. J.*, **40,** 894–903 (1979).
42. M. M. Ayoub, "Work Place Design and Posture," *Human Factors*, **15,** 265–268 (1973).
43. D. B. Chaffin, "Ergonomics Guide for the Assessment of Human Static Strength," *Am. Indust. Hyg. Assoc. J.*, **36,** 505–511 (1975).
44. P. R. Davis and D. A. Stubbs, "Safe Levels of Manual Forces for Young Males," *Appl. Ergonom.*, **8,** 141–150 (1977); ibid., **8,** 219–228 (1977); ibid, **9,** 33–37 (1978).
45. National Institute for Occupational Safety and Health, *Work Practices Guide for Manual Lifting*, DHHS (NIOSH) Publication No. 81-122, Cincinnati, OH, 1981.
46. D. N. Buckuer and J. J. McGrath, Eds., *Vigilance: A Symposium*, McGraw-Hill, New York, 1963.
47. F. R. Mackie, Ed., *Vigilance—Theory, Occupational Performance and Physiological Correlates*, Plenum Press, New York, 1976.
48. E. C. Poulton, *Environment and Human Efficiency*, Charles C. Thomas, Springfield, IL, 1972.
49. C. G. Drury, "Improving Inspection Performance," in: *Handbook of Industrial Engineering*, G. Salvendy, Ed., Wiley, New York, 1982, Chapter 84.
50. D. E. Broadbent, *Decision and Stress*, Academic Press, New York, 1971, pp. 411–440.
51. R. S. Woodworth and H. Schlosberg, *Experimental Psychology*, Henry Holt, New York, 1954, Chapter 23.
52. I. L. Goldstein, in: *Training in the Human Side of Accident Prevention*, B. L. Margolis and W. H. Kroes, Eds., Charles C. Thomas, Springfield, IL, 1975.
53. *The Industrial Environment—Its Evaluation and Control*, National Institute for Occupational Safety and Health, U.S. Government Printing Office, Washington, DC, 1973.
54. C. A. Bennett, "Lighting," in: *Handbook of Industrial Engineering*, G. Salvendy, Ed., Wiley, New York, 1982, Chapter 6.11.
55. C. L. Crouch, "Lighting for Seeing" in: *Patty's Industrial Hygiene and Toxicology*, 3rd rev. ed., G. D. Clayton and F. Clayton, Eds., Wiley, New York, 1973, Chapter 13.
56. J. D. Ramsey, C. L. Burford, M. Y. Beshir, and R. C. Jensen, "Effects of Workplace Thermal Conditions on Safe Work Behavior," *J. Safety Res.*, **14,** 105–114 (1983).
57. P. A. Hancock, "Task Categorization and the Limits of Human Performance in Extreme Heat," *Aviation, Space, Environ. Med.*, **53,** 778–784 (1982).
58. ASHRAE Standard, "Thermal Environmental Conditions for Human Occupancy," No. 55-74 (1974).
59. J. V. Tobias, G. Jansen, and W. D. Ward, Eds., "Noise as a Public Health Problem," *Proceedings of the Third International Congress*, American Speech and Hearing Association (ASHA) Report No. 10, Rockville, MD, 1980.
60. *Public Health and Welfare Criteria for Noise—July 27, 1973*, Technical Report No. 550/9-73-002, U.S. Environmental Protection Agency, Washington, DC, 1973.

61. W. D. Ward and J. E. Fricke, Eds., *Noise as a Public Health Hazard* (Proceedings of the Conference), American Speech and Hearing Association (ASHA) Report No. 4, Washington, DC, 1969.
62. *Proceedings of the International Congress on Noise as a Public Health Problem* (Dubrovnik, Yugoslavia), 1973, Technical Report No. 550/9-73-008, U.S. Environmental Protection Agency, 1973.
63. *The Effects on Human Health from Long-Term Exposure to Noise. Committee on Hearing, Broacoustics and Biomechanics*, Assembly of Behavioral and Social Sciences. National Academy of Science, Washington, DC, 1981.
64. A. Cohen, "Extrauditory Effects of Acoustic Stimulation," in: *Handbook of Physiology, Section a: Reaction to Environmental Agents*, D. H. K. Lee, H. L. Falk, S. D. Murphy, and S. R. Geiger, Eds., American Physiological Society, Bethesda, MD, 1977, Chapter 3.
65. D. H. Eldredge, "The Problems of Criteria for Noise Exposure" in: *Effects of Noise on Hearing*, D. Henderson, R. P. Hamernik, D. S. Dosanjh, and J. H. Mills, Eds., Raven Press, New York, 1976, pp. 3–20.
66. L. L. Beranek, W. E. Blazier, and J. J. Figure, "Preferred Noise Criteria Curves and Their Application to Rooms," *J. Acoust. Soc. Am.*, **50**, 1223–1228 (1971).
67. M. Loeb, "Noise and Performance—Do We Know More Now," in: *Noise as a Public Health Problem* J. V. Tobias, G. Jansen, and W. D. Ward, Eds., American Speech and Hearing Association (ASHA) Report No. 10, Rockville, MD, 1980.
68. B. L. Welch, *Extra-auditory Health Effects of Industrial Noise: Survey of Foreign Literature*, AMRL Technical Report, 79-41, Wright-Patterson Air Force Base, Ohio, 1979.
69. A. Cohen, "The Influence of a Company Hearing Conservation Program on Extra-Auditory Problems in Workers," *J. Safety Res.*, **8**,(4), 146–162 (1976).
70. J. W. Schmidt, L. H. Royster, and R. G. Pearson, "Impact of an Industrial Hearing Conservation Program on Occupational Injuries for Males and Females," *J. Acoust. Soc. Am.*, **67**, (Suppl. 1), 5–59 (1980).
71. V. Salmon, J. S. Mills, and A. C. Petersen, *Industrial Noise Control Manual*, HEW Publication No. (NIOSH) 75-183, National Institute for Occupational Safety and Health, Cincinnati, OH, 1975.
72. C. M. Harris, *Handbook of Noise Control*, McGraw-Hill, New York, 1957.
73. B. Lempert and R. G. Edwards, "Field Investigations on Noise Reduction Afforded by Insert-Type Hearing Protectors," *Am. Indust. Hyg. Assoc. J.*, **44**, (12), 894–902 (1983).
74. D. Wasserman and W. Taylor, "Occupational Vibration," in: *Environmental and Occupational Health*, W. Rom, Ed., Little Brown and Company, Boston, 1982.
75. M. J. Colligan and L. R. Murphy, "A Review of Mass Psychogenic Illness in Workplace Settings," in: *Mass Psychogenic Illness*, M. J. Colligan, J. W. Pennebaker, and L. R. Murphy, Eds., Lawrence Erlbaum Associates, New Jersey, 1982, Chapter 3.
76. G. Salvendy and M. J. Smith, Eds., *Machine Pacing and Occupational Stress*, Taylor and Francis, London, 1981.
77. M. J. Dainoff, J. J. Hurrell, Jr., and A. Happ, "A Taxonomic Framework for the Description and Evaluation of Machine-Paced Work," in: *Machine Pacing and Occupational Stress*, G. Salvendy and M. J. Smith, Eds., Taylor and Francis, London, 185–190 1981.
78. M. Frankenhaeuser and B. Gardell, "Underload and Overload in Working Life: Outline of a Multidisciplinary Approach," *J. Human Stress*, **2**, 35–46 (1976).
79. C. L. Cooper and J. Marshall, "Occupational Sources of Stress: A Review of the Literature Relating to Coronary Heart Disease and Mental Ill Health," *J. Occup. Psychol.*, **49**, 11 (1976).
80. T. Armstrong, *An Ergonomics Guide to Carpel Tunnel Syndrome*, Ergonomics Guides, American Industrial Hygiene Association, Akron, OH, 1983.
81. R. Karasek, D. Baker, F. Marxer, A. Ahlbom, and T. Theorell, "Job Decision Latitude, Job Demands, and Cardiovascular Disease, A Prospective Study of Swedish Men," *Am. J. Public Health*, **71**(7), 694–705 (1981).

82. R. R. Holt, "Occupational Stress," in: *Handbook of Stress: Theoretical and Clinical Aspects*, L. Goldberger and S. Breznitz, Eds., 1982, pp. 419–444. Free Press, New York, 1982, pp. 419–444.
83. B. Cohen, *Human Aspects in Office Automation*, Elsevier, Amsterdam, 1984.
84. J. R. French, Jr. and R. Caplan, "Psychosocial Factors in Coronary Heart Disease," *Indust. Med.*, **39**, 383–397 (1970).
85. B. Margolis, W. H. Kroes, and R. P. Quinn, "Job Stress—An Unlisted Occupational Hazard," *J. Occup. Med.*, **16**, 659–661 (1974).
86. R. M. Rose, C. D. Jenkins, and M. W. Hurst, Air Traffic Controller Health Change Study, Report No. FAA-AM-78-79, Federal Aviation Administration, Washington, DC, 1978.
87. R. Kahn, D. M. Wolfe, R. P. Quinn, J. D. Snoek, and R. A. Rosenthal, *Organizational Stress: Studies in Role Conflict and Ambiguity*, Wiley, New York, 1964.
88. A. Shirom, D. Eden, S. Silberwasser, and J. J. Kellerman, "Job Stress and Risk Factors in Coronary Heart Disease Among Five Organizational Categories in Kibbutzim," *Social Sci. Med.*, **7**, 875–892 (1973).
89. R. Karasek, "Job Decision Latitude, Job Design and Coronary Heart Disease," in: *Machine Pacing and Occupational Stress*, G. Salvendy and M. J. Smith, Eds., Taylor and Francis, London, 1981, p. 45–55.
90. R. Caplan, S. Cobb, J. R. P. French, R. V. Harrison, and S. R. Pinneau, *Job Demands and Worker Health: Main Effects and Occupational Differences*, DHEW (NIOSH) Publication No. 75-160, National Institute for Occupational Safety and Health, Cincinnati, OH, 1975.
91. J. S. House and J. A. Wells, "Occupational Stress, Social Support and Health," in: *Proceedings of Conference in Reducing Occupational Stress*, A. McLean, Ed., DHEW (NIOSH) Publication No. 78–140 National Institute for Occupational Safety and Health, Cincinnati, OH, 1977.
92. S. Cobb and S. Kasl, *Termination: The Consequences of Job Loss*, DHEW (NIOSH) Publication No. 77–224 National Institute for Occupational Safety and Health, Cincinnati, OH, 1977.
93. R. J. Conrad, *Employee Work Practices*, NIOSH Contract Report No. 81-2905, National Institute for Occupational Safety and Health, Cincinnati, OH, 1983.
94. T. Williams, R. Harris, E. Arp, and M. Symons, "Worker Exposures to Chemical Agents in the Manufacture of Rubber Tires and Tire Particulates," *Am. Indust. Hyg. Assoc. J.*, **41,** (1980).
95. H. H. Cohen and R. J. Jensen, "Demonstration of the Effectiveness of an Industrial Lift Truck Operator Safety Training Program Utilizing a Behavior Sampling Procedure," *J. Safety Res.* **15**(3), 125–135 (1984).
96. B. L. Hopkins, *Behavioral Procedures for Reducing Worker Exposure to Carcinogens*, NIOSH Contract Report No. 210-77-0042, National Institute for Occupational Safety and Health, Cincinnati, OH, 1981.
97. D. Zohar, "Promoting the Use of Personal Protective Equipment by Behavior Modification Techniques," *J. Safety Res.*, **12**, 75–85 (1980).
98. A. Cohen, "Perspectives on Self-Protective Behaviors and Workplace Hazards," in: *Proceedings of Workshops—Encouraging Self-Protective Behavior*, N. Weinstein, Ed., Rutgers, State University of New Jersey, New Brunswick, NJ, 1984.
99. D. E. Broadbent and W. Gath, "Some Relationships Between Clinical and Occupational Psychology," in: *Proceedings of 20th International Congress of Applied Psychology*, Edinburgh, Scotland, 1982.

CHAPTER NINE

Abnormal Pressure

JEFFERSON C. DAVIS, M.D.

1 INTRODUCTION

The pressures to be considered are decreased barometric pressure at altitude in the atmosphere and in space flight and barometric pressure greater than sea level in undersea work, exploration, and recreation. An emerging use of pressure greater than sea level is the use of high dose, short-term oxygen inhalation therapy (hyperbaric oxygen) in treatment of a limited group of medical disorders. Once the realm of a limited group of flight surgeons, diving medical officers, divers, and tunnel and bridge building engineers, altered barometric pressure today affects the lives of workers, passengers in aircraft, and recreational scuba divers and aviators. Excursions of sea level dwellers to mountainous regions for business or recreation is another important exposure.

Medical uses of hyperbaric oxygen have introduced a new dimension to the need for full engineering safety aspects related to exposures of patients and medical staff to altered barometric pressure as well as associated gas purity standards and fire and structural safety concerns. Because Chapter 9, Volume I, presented the biomedical factors of undersea work so completely, this chapter concentrates on the altitude environment and medical hyperbaric chambers in more detail. The central theme is represented by the interrelated physiological effects of the continuum of altered barometric pressure—higher or lower than sea level.

1.1 Physics of Abnormal Environmental Pressure

The weight of atmospheric air and water at any point on or above the surface of the earth or beneath the surface of the water may be expressed in various units of pressure. The major difference between pressure changes at altitude and at depth is the curvilinear change of altitude compared to the linear pressure changes at depth. On ascent to altitude, we pass through the blanket of air that extends from sea level to an altitude of about 430 miles, where the number of molecules is so small that no collisions occur. At sea level the atmosphere weighs 14.7 pounds per square inch (psi) or 760 mm Hg (760 torr). Because pressure at any altitude is simply the weight of air above, pressure changes are greater at low altitudes. For example, with ascent from sea level to 3048 m (10,000 ft), a pressure reduction of 237 mm Hg from 760 to 523 mm Hg is seen, whereas the same 3048 m altitude change from 9146 m (30,000 ft) to 12,192 m (40,000 ft) gives a pressure reduction of only 85 mm Hg from 226 to 141 mm Hg. At depth pressure change is linear because each foot of seawater weighs 0.445 psi; thus 14.7 ÷ 0.445 = 33 ft of seawater or about 10 m, is equivalent to a full atmosphere of pressure.

The barometric pressure (P_B) at depth can be referred to in terms of gauge pressure (that in excess of sea level pressure) or more commonly, as absolute pressure (the total of sea level atmospheric pressure plus the water pressure). The most commonly used pressure units are millimeters of mercury (mm Hg), feet of seawater (FSW), meters of seawater, or atmospheres absolute (ATA). An example of the linear pressure changes at depth is that at 20 m (66 FSW), the total pressure is 3 ATA or 2280 mm Hg and at 30 m (99 FSW), the total pressure is 4 ATA or 3040 mm Hg. Barometric pressure at any altitude is available from standard altitude charts and pressure at any depth can be easily derived.

2 ALTITUDE PHYSIOLOGY

Air at altitude maintains a relatively constant percentage of gases in the mixture with oxygen about 21 percent and nitrogen about 79 percent. Traces of carbon dioxide, water vapor, and other inert gases are so small that they are included with the nitrogen percentage for calculation. Dalton's law states that in a mixture of gases, the total pressure is the sum of partial pressures of gases in the mixture. For example, in air at sea level, partial pressure of oxygen (P_{O_2}) equals 0.21 × 760 mm Hg or 160 mm Hg and partial pressure of nitrogen (P_{N_2}) equals 0.79 × 760 mm Hg or 600 mm Hg. The key point is that at altitude, gas percentages remain constant but partial pressures of gases change. For example, at 18,000 ft where barometric pressure is 380 mm Hg or one-half an atmosphere, P_{O_2} = 0.21 × 380 mm Hg or 80 mm Hg and P_{N_2} = 300 mm Hg.

2.1 Changes in Inspired Gas Composition

At normal body temperature the vapor pressure of water is 47 mm Hg; thus by the time inspired dry air reaches the trachea, it has equilibrated at a P_{H_2O} of 47 mm Hg, which then remains constant in the alveolar air. Venous blood returns carbon dioxide from the body's normal metabolism to the pulmonary circulation at P_{CO_2} of 47 mm Hg. Admixture of carbon dioxide diffused across the alveolocapillary membrane into the alveolus results in an aveolar P_{CO_2} of about 40 mm Hg in the normal person. Thus to calculate the approximate alveolar partial pressure of oxygen (P_{AO_2}) at sea level or at any altitude, the simplified alveolar gas equation may be used if we assume the respiratory quotient (RQ) to be 1:

$$RQ = \frac{CO_2 \text{ output (ml/min)}}{O_2 \text{ consumption (ml/min)}}$$

The simplified alveolar gas equation is

$$P_{AO_2} = F_{IO_2}(P_B - P_{AH_2O}) - P_{ACO_2}$$

where P_{AO_2} = alveolar P_{O_2}
F_{IO_2} = fraction of inspired O_2
P_B = total barometric pressure
P_{AH_2O} = alveolar P_{H_2O}
P_{ACO_2} = alveolar P_{CO_2}

For example, at sea level breathing air,

$$P_{AO_2} = 0.21(760 - 47) - 40$$
$$P_{AO_2} = 109 \text{ mm Hg}$$

and breathing air at 18,000 ft

$$P_{AO_2} = 0.21(380 - 47) - 40$$
$$P_{AO_2} = 30 \text{ mm Hg}$$

The same equation can be used to determine the F_{IO_2} needed to remain above hypoxic levels at altitude.

For example, breathing 100 percent O_2 at 18,000 ft

$$P_{AO_2} = 1.0(380 - 47) - 40$$
$$P_{AO_2} = 293 \text{ mm Hg}$$

Nitrogen is inert and maintains equilibrium between alveolar, blood, and tissue P_{N_2}. At sea level, for example,

$$\text{Alveolar } P_{N_2} = 760 - (P_{AO_2} + P_{ACO_2} + P_{AH_2O})$$
$$P_{N_2} = 760 - (109 + 40 + 47)$$
$$P_{N_2} = 564 \text{ mm Hg}$$

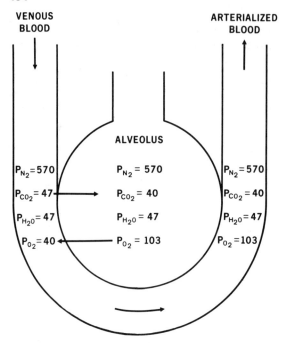

Figure 9.1 Model of alveolus and pulmonary capillary.

2.2 Oxygen Transport—Lung

The transfer of gases at the alveolocapillary membrane is by the process of diffusion and can be represented by Figure 9.1, a schematic model of alveolus and capillary.

From ambient P_{O_2} of 160 mm Hg, the alveolar P_{O_2} of 103 mm Hg remains after dilution with water vapor and CO_2 returned by venous blood. With passage through pulmonary capillaries, arterialized blood leaves the lungs with P_{AO_2} (arterial P_{O_2}) about 100 mm Hg.

Oxygen diffusion occurs as the alveolar P_{O_2} is exposed to the respiratory membrane of some 0.36 to 2.5 μm thickness over its 70 m² of surface area. By means of the almost solid network of capillaries that fill the respiratory membrane, alveolar oxygen is exposed to the 140 ml of blood contained in the pulmonary capillaries at any one time. Red blood cells "squeeze" through the narrow 8-μm-diameter capillaries, providing close proximity to alveolar gases for rapid diffusion from the higher 104 mm Hg alveolar P_{O_2} to the lower O_2 concentration of about 40 mm Hg in returning venous blood.

2.3 Oxygen Transport from Lungs to Tissue

The unique properties of hemoglobin and the way it combines with and releases oxygen constitute a key element in altitude and diving physiology. The presence of hemoglobin allows blood to transport 30 to 100 times as much oxygen as

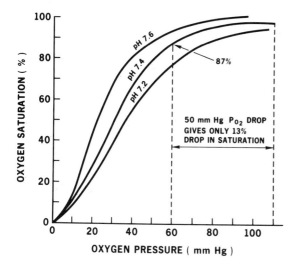

Figure 9.2 Hemoglobin–oxygen dissociation curve.

would be possible by physical solution of dissolved oxygen alone. Normally, 97 percent of oxygen carriage is by hemoglobin and 3 percent is dissolved in plasma and cells. The so-called oxygen buffer function of hemoglobin is based on the shape of the oxygen–hemoglobin dissociation curve (Figure 9.2). Thus with hypoxia at altitude P_{AO_2} can drop to 60 mm Hg and hemoglobin saturation is maintained at about 90 percent and no matter how high the P_{AO_2} goes, with oxygen breathing at depth, only 100 percent hemoglobin saturation is possible and the only added oxygen carriage possible is the inefficient physical solution in plasma.

Hemoglobin concentration is expressed as grams-percent or grams of hemoglobin per 100 ml of blood (1). An average, healthy adult male has about 15 g of hemoglobin per 100 ml of blood or 15 g-percent. Each gram of hemoglobin can combine with 1.34 ml of oxygen. Thus a person with 15 g-percent of hemoglobin can carry 1.34 × 15 = 20.1 ml of oxygen per 100 ml of blood or 20.1 volumes-percent.

Hemoglobin uptake and release of oxygen is described by the hemoglobin–oxygen dissociation curve (Figure 9.2) whose features are of great significance in altitude physiology (2).

The flat portion of the curve between 60 and 110 mm Hg allows a significant drop in alveolar P_{O_2} before there is a physiologically important drop in blood oxygen saturation. It is this characteristic of the curve that allows ascent to 3048 m (10,000 ft) altitude breathing air, where alveolar P_{O_2} is 60 mm Hg, without symptoms of hypoxia. Further ascent above 3048 m (10,000 ft) results in a rapid drop in arterial oxygen saturation as small drops in oxygen pressure result in large decreases in arterial saturation along the steep portion of the curve.

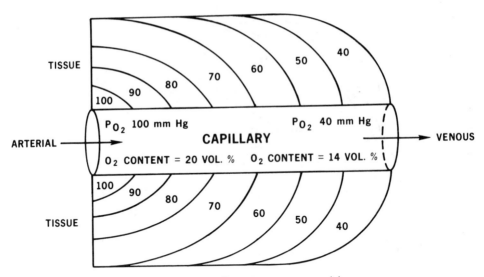

Figure 9.3 Capillary–tissue oxygen model.

The curve also shows the profound effect of acidosis or alkalosis as the shape of the curve is altered by changes in pH. Thus, for example, hyperventilation, a normal compensatory mechanism to hypoxia, results in a lowering of P_{CO_2} and a resultant rise in pH. The shift of the curve to the left illustrates the body's attempt to increase efficiency of oxygen uptake at the lung level and release of oxygen to the tissues at higher oxygen pressures. On the other hand, the elevated P_{CO_2} and acidosis seen in exercise at altitude aggravates hypoxia by shifting the curve to the right.

Figure 9.3 is a simplification of the Krough–Erlang tissue oxygen model (3). The cylinder of tissue supplied by a given capillary is seen as a spectrum of oxygen tensions regulated by the arterial oxygen partial pressure supplied, distance from the capillary, and location along the capillary as oxygen is extracted during flow of blood toward the venous end.

3 ACUTE ADAPTATION TO ALTITUDE EXPOSURE

On rapid exposure to altitude, whether in an unpressurized aircraft or balloon or because of accidental decompression of a pressurized aircraft, decompression in an altitude chamber, or rapid ascent of a mountain, a sequence of physiological mechanisms attempts to compensate.

With decrease in arterial P_{O_2} there is hyperventilation caused by hypoxic stimulus to the chemoreceptors, the carotid and aortic bodies (4). Hyperventilation gives a decrease in alveolar and arterial P_{CO_2}, tissue alkalosis, and the favorable shift to the left of the oxygen–hemoglobin curve. There is also unfavorable cerebral vascular vasoconstriction as arterial P_{CO_2} drops, since carbon dioxide is a potent cerebral vasodilator (5). This effect is overriden by the hypoxia-induced vasodilation that occurs when venous P_{O_2} drops to about 30 mm Hg (5).

Also, with decreased P_{AO_2}, the carotid and aortic bodies mediate bradycardia, vasoconstriction in the extremities, systemic hypertension, and increased pulmonary vascular resistance. The net effect is hyperventilation, pulmonary hypertension, increased heart rate, and beneficial redistribution of blood from extremities to the brain and heart.

4 HYPOXIA

Without supplemental oxygen, the physiological effects of the reduced atmospheric partial pressure of oxygen (P_{O_2}) at altitude are described as hypoxic hypoxia. Other factors that lead to tissue hypoxia can coexist and be additive. Thus a patient whose oxygen transport capacity is impaired by anemia or carbon monoxide bound hemoglobin ("hypemic hypoxia") or whose tissue oxygen utilization is impaired by cyanide or ethanol ("histotoxic hypoxia") will be more susceptible to hypoxia at altitude. A special potentiating factor in aviation is "stagnant hypoxia" in an aviator exposed to high head-to-foot acceleration forces as in recovering from an aircraft dive (6). The increased weight of the column of blood reduces cerebral perfusion with resultant cerebral hypoxia. Although these factors can be important contributors to hypoxia in aviation, this discussion centers on hypoxia causes by reduction in P_{O_2} at the decreased barometric pressure at altitude.

As previously discussed, the shape of the oxygen–hemoglobin dissociation curve is beneficial up to 3048 m (10,000 ft). Except for a decrement in night vision at about 1220 to 1829 m (4000 to 6000 ft) and above, there are no significant effects of hypoxia in healthy people below 3048 m (10,000 ft). In general, pressurized aircraft cabins are maintained well below that equivalent air pressure, and no supplemental oxygen is required. However, patients with reduced cardiac reserve may have difficulty tolerating even this small drop in arterial oxygen saturation and will require supplemental oxygen during flight.

4.1 Manifestations

Table 9.1 summarizes the major manifestations of altitude hypoxia with corresponding P_{AO_2} breathing ambient air.

The time required for the onset of symptoms listed in Table 9.1 varies depending on age, fitness, acclimatization, and individual susceptibility. For example, at the altitude of 3658 m (12,000 ft) the symptoms shown may not appear in a subject at rest for several hours. At higher altitudes the progress of symptoms to the point of inability to perform useful functions [time of useful consciousness (TUC)] is quite rapid and within a narrower range of individual variation. For example, the TUC at 9146 m (30,000 ft) is about 90 sec and at 13,106 m (43,000 ft), about 15 sec or less.

Table 9.1 Responses to Hypoxic Hypoxia

Altitude		Alveolar P_{O_2} (mm Hg)
3,048 m (10,000 ft)	Impaired judgment and ability to perform calculations; increased heart rate and respiratory rate	60
3,658 m (12,000 ft)	Shortness of breath, impaired ability to perform complex tasks, headache, nausea, decreased visual acuity	52
4,573 m (15,000 ft)	Decrease in auditory acuity, constriction of visual fields, impaired judgment, irritability; exercise can lead to unconsciousness	46
5,486 m (18,000 ft)	Threshold for loss of consciousness in resting unacclimatized individuals after several hours exposure	40
6,706 m (22,000 ft)	Almost all individuals unconscious after sufficient exposure time	30
12,802 m (42,000 ft)	Inability to perform useful function in 15 sec or less	—

4.2 Prevention

The use of pressurized cabins to stay within the physiologically comfortable zone below 3048 m (10,000 ft) is the main preventive factor in modern aviation. When aviators must be exposed to higher cabin altitude, the percentage of supplemental oxygen required to maintain an alveolar P_{O_2} no less than that at 3048 m (10,000 ft) breathing air can be derived by varying the F_{IO_2} in the simplified alveolar gas equation. For example, 100 percent oxygen is required at 12,192 m (40,000 ft) where the barometric pressure is 141 mm Hg and P_{ACO_2} is somewhat lower because of hyperventilation:

$$P_{AO_2} = 1.0(141 - 47) - 35$$

$$P_{AO_2} = 59 \text{ mm Hg}$$

This is equivalent to breathing air at 3048 m (10,000 ft) and is the maximum altitude at which 100 percent oxygen breathing is adequate to prevent hypoxia. At higher altitudes positive pressure breathing and, finally, pressure suits are required (7).

5 AIRCRAFT CABIN PRESSURIZATION

The most commonly employed method to avoid the risks of exposure of aircrews and passengers to the low barometric pressure and partial pressures of oxygen at altitude is to pressurize the aircraft cabin. There are two basic factors in cabin pressurization: (1) air compressors powered by the aircraft engines force outside air continuously into the cabin, whose structural integrity (2) is able to safely withstand differential pressures of 8 psi or more, with compensatory relief valves to continuously dump air overboard (8). A commonly used example in aviation is a differential pressure (cabin above ambient pressure) of 8.19 psi, so that at flight level (FL) 12,000 m (40,000 ft; 2.72 psi) the addition of the 8.19 psi pressurization gives a total cabin pressure of 10.91 psia (pounds per square inch absolute) equivalent to 2400 m (8000 ft). From previous discussions, it is clear that 2400 m (8000 ft) is within the physiologically safe zone where no supplemental oxygen is required to maintain sufficient arterial oxygen saturation.

5.1 Limitations of the Pressurized Cabin

As the atmosphere density decreases at great altitude, a limitation is finally reached where it is no longer feasible to compress the ambient gas molecules efficiently to pressurize the aircraft cabin. This practical limitation is reached at altitudes of about 21,000 to 24,000 m (70,000 to 80,000 ft), where only very high aircraft velocity with sufficient ram pressure could achieve adequate aircraft pressurization (8). Even so, adiabatic compression of gas so rare results in very high cabin temperatures so that at FL above 24,000 m (80,000 ft), the sealed cabin with self-contained life support systems becomes necessary. Thus spacecraft carry onboard gas supplies to pressurize the craft to habitable cabin altitudes and gas composition.

5.2 Cabin Decompression

The remote but ever-present risk in aircraft flying at high altitude with cabins pressurized as just described is that of accidental decompression. This could occur in case of loss of aircraft structural integrity as in failure of a door or window. The severity of cabin decompression is dependent on several factors, including volume of the cabin, area of the opening into the pressurized compartment, and cabin pressure–ambient pressure differential pressure. Hazards include physical harm to occupants by being sucked out of the opening or being struck by loose objects flying about the cabin because of the rush of pressurized air out of the cabin, hypoxia on exposure to ambient pressure, and low temperature.

Excellent quality control in construction of modern aircraft has made this event quite rare, but provision of emergency oxygen supplies for crew and passengers, wearing of restraint belts or harnesses during flight, routine securing of loose objects, and training of flight crews in emergency procedures are necessary backup provisions.

Table 9.2 Comparison of Mean Alveolar and Blood Oxygen Partial Pressures (mm Hg)

Parameter	Lima	Morococha
Alveolar P_{O_2}	96	46
Arterial P_{O_2}	87	45
Mixed venous P_{O_2}	42	35

6 ACCLIMATIZATION TO ALTITUDE AND MOUNTAIN SICKNESS

The acute physiological adaptations to altitude exposure presented earlier are ineffective in preventing serious symptoms of hypoxia upon acute exposure to altitude. Therefore, the main protection lies in ensuring adequate oxygen equipment to provide supplemental oxygen.

Yet people born in mountainous regions of the world live and work for a lifetime at altitudes that would produce serious symptoms in the sea level dweller exposed to such altitudes. An understanding of practical aspects of altitude acclimatization as well as risks to workers and travelers who must do business in mountainous cities of the world is pertinent to this text.

6.1 Altitude Acclimatization

The goals of physiological adaptive processes can be summarized (9):

1. To facilitate acquisition of oxygen by venous blood entering the lungs.
2. To increase the capacity of blood to transport oxygen to tissues.
3. To promote oxygen utilization at the cellular level.

Hurtado (10) did extensive studies on natural acclimatization among natives living at 4540 m (14,000 ft) in Peru. He compared 50 subjects living in Lima (150 m or 500 ft) to 40 subjects living in Morococha at 4540 m (14,000 ft).

He found that the Morococha residents had a mean resting ventilation (liters per minute) of 18 percent above the Lima subjects. Table 9.2 summarizes his findings of alveolar and blood oxygen partial pressures. The increased efficiency of gas exchange in the lung among acclimatized mountain residents is indicated by the resting differential of alveolar to arterial P_{O_2}. In the Lima group this difference was 8 to 12 mm Hg, whereas in the Morococha acclimatized subjects, it was only about 1 mm Hg. The mechanism that accounts for this efficient diffusion of alveolar oxygen may be dilatation of pulmonary capillaries.

Table 9.2 indicates that Morocochan natives live on the steep slope of the oxygen–hemoglobin dissociation curve. Thus more oxygen is delivered to tissue with only a small change in oxygen tension. Table 9.3 summarizes hematologic factors in altitude acclimatized natives (10).

Table 9.3 Summary of Hematologic Findings by Hurtado Among Morococha Natives and Lima Natives[a]

Hematocrit (percent)	46 (40–52)	59 (49–79.2)
Hemoglobin (g-percent)	15.6	20
Reticulocytes (1000/mm^3)	17.9	45.5
Total blood volume (liters)	4.77	5.70
Plasma volume (liters)	2.52	2.23

[a] Mean of 41 to 250 subjects in each observation.

Acclimatization of near-sea-level dwellers to altitude never is as complete or efficient as it is for those who reside at altitude. The level of ability to perform work at altitude seen in natives does not occur in newcomers even after prolonged residence. One reason may be the higher number of capillaries in muscle observed in animals and natives of high altitude (10).

For newcomers to a high altitude region, the time required and best schedule for acclimatization is of importance. A summary of the literature on this subject can be stated:

1. There is individual variability in the ability for acclimatization.
2. Acclimatization starts at low altitudes and continues to about 18,000 ft.
3. The limiting altitude for sea-level dwellers to achieve significant acclimatization is about 18,000 ft. Again, there is individual variability.
4. It is possible for mountaineers to achieve partial acclimatization up to 23,000 ft, but deterioration begins at about 20,000 ft.
5. One recommended schedule (9) for acclimatization is:

$$6,000-7,000 \text{ ft}$$
$$9,000-10,000 \text{ ft}$$
$$12,000-13,000 \text{ ft}$$

Spend 10 days at each level with planned exercise and daily excursions to 2000 ft higher, but reside at the prescribed altitude for time indicated.

6. Drug therapy to improve altitude tolerance has been the subject of many studies. Ammonium chloride to increase blood acidity and offset respiratory alkalosis has been used (11), but there is no convincing evidence of its efficacy. Acetazolamide (Diamox®) is a carbonic anhydrase inhibitor that causes acidosis. Cain and Dunn (12) showed that the arterial P_{O_2} of dogs pretreated with Diamox before 21,000-ft exposure was 9 mm Hg higher than untreated controls. The mechanism for these changes is accelerated renal excretion of base and correction of respiratory alkalosis. In a human study (13) they showed that low doses of Diamox did increase tolerance to exposure to 16,000 ft.

7 MOUNTAIN SICKNESS

Acute exposure of near sea level, unacclimatized people to high altitude may produce a spectrum of serious clinical problems.

7.1 Acute Mountain Sickness

Soon after arrival at high altitude, the sea-level dweller may note subjective difficult respiration, and increased rate and depth of breathing can be measured because of hypoxic stimulus to carotid and aortic body chemoreceptors (9). There may be Cheyne–Stokes respiration and persistent cough. The pulse rate is elevated. The subject may have decreased visual discrimination, impaired judgment, emotional lability, and even hallucinations. Headache, insomnia, giddiness, eyeball pain, visual blurring, and decreased night vision are noted. There may be loss of appetite, intolerance to fatty foods, nausea, vomiting, and dehydration. Patients with underlying disorders like hypertension or angina pectoris may have aggravation of their condition.

7.2 High-Altitude Pulmonary Edema (HAPE)

HAPE is the most serious of the mountain sickness syndromes. It can occur in the first few days of residence at altitude if acclimatization has not been followed (9). It can occur in acclimatized persons who undertake very strenuous exercise at altitude, or even in natives who return to their mountain home after a sojourn at lower altitude. During their low-altitude residence, about 4 to 8 weeks is required for full deacclimatization, and individuals experience a drop in hemoglobin and red blood cell count. On return to the mountain home, the hypertrophied heart muscle cannot receive adequate oxygenation because of the drop in hemoglobin.

The mechanism of HAPE is "leakiness" of pulmonary capillaries with resultant pulmonary edema. Several factors may contribute. The increased pulmonary artery pressure seen in acute hypoxia is coupled with hypoxic capillary damage and loss of integrity and depressed amount of surfactant, lowering osmotic pressure in alveoli with fluid drawn into alveoli.

HAPE may be recognized by onset within a few hours after arrival at altitude of insomnia; chest discomfort; cough productive of frothy, blood-stained sputum; shortness of breath, becoming worse on recumbency; nausea and vomiting; irritability; delirium; and finally coma. The patient will by cyanotic, with rapid pulse rate and systemic blood pressure elevated and fine rales heard on auscultation of the chest.

The major preventive measures are gradual altitude acclimatization and regulating exercise at altitude to avoid breathlessness for the first few days.

7.3 Chronic Mountain Sickness ("Monge's Disease")

After apparently complete acclimatization and perhaps after months of residence at high altitude, low-grade symptoms of dimming of vision, loosening of teeth, weight loss, decreased ability to work, loose stools, indigestion, and difficulty with mental concentration may occur (9). These symptoms of chronic mountain sickness clear within 3 to 4 weeks of returning to lower altitude.

8 DECOMPRESSION SICKNESS

Decompression sickness refers to the plethora of clinical manifestations first described in the middle 1800s among caisson workers who were decompressed to sea level after working in compressed air to keep water and mud out of tunnel construction. (Hence the previous term, "caisson disease.") Military, commercial and, more recently, sport diving interests brought emphasis to studies of this complex disorder to determine safe decompression schedules and treatment of casualties. Since the 1930s it has been recognized that decompression from sea level to altitude can produce the same disorder. Since 1960 treatment of altitude decompression sickness persisting at ground level has been compression to greater than sea level pressure in chambers, as has been the treatment of divers and caisson workers for decades.

8.1 Pathophysiology

Chapter 9, Volume I, contains a very complete discussion of all aspects of decompression sickness; hence only a summary is given here.

The compressed gas respired at increased atmospheric pressure (diving) must be a mixture of oxygen and an inert gas diluent to avoid oxygen toxicity. This inert gas is usually nitrogen for shallow depths and helium at greater depths to prevent the narcotic effects of nitrogen at high partial pressures. When compressed gas is breathed underwater to balance the hydrostatic pressure of water on the thoracic wall or from the environment of a compressed air pressurized chamber (hyperbaric chamber or compression chamber), the alveolar partial pressure of inert gas rises according to barometric pressure (e.g., at 6 ATA or 165 FSW, absolute barometric pressure is 4560 mm Hg and breathing air the inspired P_{N_2} is $0.79 \times 4560 = 3602$ mm Hg). Allowing for water vapor, carbon dioxide, and oxygen ($P_{O_2} = 0.21 \times 4560 = 948$ mm Hg). the alveolar P_{N_2} approximates 3525 mm Hg.

According to Henry's law (the amount of gas in solution varies directly with the partial pressure of gas in contact with the solution), there is rapid diffusion of nitrogen across the alveolocapillary membrane to physical solution in plasma. Upon reaching the tissues, the high P_{N_2} dissolved in arterial blood is diffused into tissues at a rate dependent on the perfusion rate of various tissues. The total nitrogen uptake by a given tissue is determined by its composition, with lipid-rich tissues taking on larger amounts of nitrogen. The body's uptake of inert gas can thus be seen as a complex of uptake curves that results in variable loading of different tissues, increasing with time of exposure to the elevated pressures until full equilibration is achieved at approximately 24 hr (saturation).

On decompression from the high-pressure environment to lower pressures and finally to sea level, the reverse set of gradients is established for offloading of inert gas from tissues to venous blood, to the lung. Inert gas will stay in solution in tissues and blood within strictly defined pressure reductions, but if critical limits are exceeded, it comes out of solution into the gas phase as bubbles in blood and tissues. The limits of safe decompression are discussed in Chapter 9, Volume I. Once formed, bubbles in tissues and blood cause a series of pathophysiological events that can culminate in permanent paralysis or death. Local tissue distortion by bubbles may cause pain, and venous gas emboli may produce congestive infarction of the spinal cord or gas pulmonary embolism. Platelet aggregation has been seen at the blood–bubble interface, and with platelet damage there is release of vasoactive substances such as serotonin and epinephrine, complicating the picture with vasoconstriction. Furthermore, alteration of the platelet membrane makes phospholipid (platelet factor 3) available and accelerates clotting. Untreated, bubble-induced tissue ischemia and hypoxia result in loss of capillary integrity with resultant edema and hemoconcentration. Therapy is directed toward this entire spectrum of events.

The sea-level dweller is in equilibrium (saturated) with the nitrogen partial pressure in the atmosphere. Upon rapid exposure to decreased barometric pressure at altitude, a series of events comparable to that just described can occur, with evolution of nitrogen bubbles. Because of the relatively higher proportion of carbon dioxide and water vapor at altitude, these gases diffuse into bubbles and play a larger role than in diving. The precise critical altitude for bubble formation has not been defined, but it is of clinical importance that the lowest documented case of altitude decompression sickness occurred at 5640 m (18,500 ft). The incidence rises sharply with increasing altitude and with time at altitude. Most cases recover upon recompression to sea level, but those that do not require emergency treatment in a compression chamber to resolve bubbles that persist at sea level because of growth while at altitude. Combining the two environments (i.e., ascent to altitude after diving and while tissues still contain excess inert gas) can produce bubbles at much lower altitudes. Cases of decompression sickness have been seen at as little as 1220 m (4000 ft) after safe decompression from diving.

8.2 Clinical Manifestations

The onset of symptoms and signs of decompression sickness tend to be gradual and progressive, beginning minutes to hours after an inciting dive or exposure to altitude. With certain exceptions, to be noted, the manifestations resulting from both environments are the same and can be divided into type I (minor manifestations) and type II (serious manifestations).

Type I (minor manifestations) include bends, involving mild to severe deep, boring pain in single or multiple joints. The pain is usually aggravated by exercise and is relieved by application of local pressure over the joint. There may be local swelling due to bubble blockade of lymphatic drainage. The other type I manifestations include unusual profound fatigue following diving or exposure to altitude and itching or mottling of the skin.

Type II (serious manifestations) include the very dangerous "chokes" with shortness of breath, substernal pain, cough, and cyanosis. The basic mechanism of massive bubbling in the venous blood, with obstruction of the pulmonary circulation, hemoconcentration, and platelet damage, may lead to profound circulatory collapse or shock.

Neurological manifestations may involve the brain or spinal cord, with the former more common in altitude decompression sickness and spinal cord involvement more common in divers. Brain manifestations include visual disturbances, spotty numbness or weakness, paralysis, speech impairment, decreased mental alertness, severe headaches, or seizures. Spinal cord manifestations are usually heralded by severe low back or abdominal pain followed by varying degrees of numbness, weakness, or paralysis below the level of the cord involvement. There may be loss of anal sphincter tone and inability to urinate. Untreated, patients in these cases can remain permanently paralyzed.

A late manifestation is the so-called dysbaric osteonecrosis of divers and caisson workers. The exact etiology of this aseptic necrosis of bone in divers is still unclear, as is its relationship to untreated previous episodes of decompression sickness. This subject is covered in Chapter 9, Volume I.

8.3 Prevention and Treatment

Decompression sickness usually responds rapidly to early and adequate treatment by compression. Descent from altitude produces sufficient recompression to abolish the early manifestations but slowly developing serious forms of decompression sickness can develop later. In such cases widespread bubble formation presumably already has occurred, and descent is entirely analogous to inadequate compression therapy during which symptoms subside only temporarily.

Divers can prevent decompression sickness by knowing and following established limits for depth and time at depth. Adequate denitrogenation by breathing 100 percent oxygen before and during ascent is an effective preventive measure in fliers.

9 HYPERBARIC CHAMBERS USED FOR OTHER MEDICAL INDICATIONS

Hyperbaric oxygenation is defined as a mode of medical treatment in which the patient is entirely enclosed in a pressure chamber breathing oxygen at a pressure greater than 1 atm. Treatment may be carried out either in a monoplace chamber pressurized with pure oxygen or a larger, multiplace chamber pressurized with compressed air, in which case the patient receives pure oxygen by mask, head tent, or endotracheal tube. Breathing 100 percent oxygen at 1-atm pressure or applying oxygen topically to parts of the body without the use of a pressurized chamber that encloses the patient completely is not considered hyperbaric oxygenation (14).

The expanding use of hyperbaric chambers to provide short-term, high-dose oxygen inhalation therapy calls for a description of chamber types and safety provisions for their use. Besides their role in treatment of altitude and diving decompression accidents, hyperbaric chambers are finding use in treatment of gas gangrene, carbon monoxide poisoning, and certain wound healing problems (15). Two general classes of hyperbaric chambers are in use:

1. The multiplace hyperbaric chamber in which the patient and medical attendants are pressurized together with compressed air to greater than sea-level pressure and the patient is given oxygen by mask, head tent or endotracheal tube.

2. The monoplace hyperbaric chamber, where the patient is compressed in oxygen in a small chamber and there is no attendant. The patient breathes therapeutic oxygen from the chamber atmosphere.

9.1 Multiplace Hyperbaric Chambers

All multiplace hyperbaric chambers are, in principle, only a simple hermetically sealed volume (Figure 9.4). However, they are complicated, and their safe operation requires that a number of rules be meticulously observed. Hazards arise from (a) the enormous pressures exerted on the chamber walls (at 3 ATA this is 20 tons/m^2), (b) the increased concentration of oxygen and the risk of fire, (c) the danger of decompression sickness among air breathing attendants (patients breathing oxygen by mask are not at risk), (d) air embolism, (e) oxygen toxicity, (f) contamination of the breathing mixtures, and (g) barotrauma of middle ears and paranasal sinuses.

The design of a chamber depends on its purpose, and there are numerous designs and sizes. Usually the chambers are steel cylinders with hemispheres on the ends, equipped with numerous penetrations to provide entrance and exit of breathing gas, water, air, electrical cables, and viewing ports. Alteration of the chamber walls requires complete retesting and safety certification of the chamber.

Most frequently used are chambers with a cylinder whose axis is on a horizontal plane. A "long" chamber is convenient, since strong hermetic partitions can be used to divide it into several intercommunicating lock compartments.

These locks serve primarily for the entry and exit of personnel from the chamber without reducing pressure in the main chamber. Hatches connect the lock with the main pressure chamber and with the outside environment. To enter the main pressure chamber when it is at high pressure, one enters the lock, the outer hatch closes, and the pressure is raised in the lock compartment. When pressure in the lock becomes equal to that in the pressure chamber, the connecting hatch opens to allow passage from the lock to the main chamber.

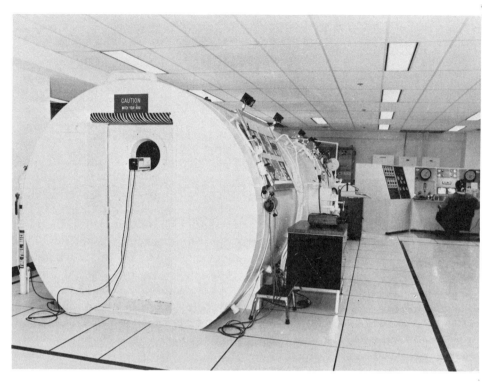

Figure 9.4 A stainless-steel, 25-ft-long, double-lock compressed-air multiplace hyperbaric chamber. Note the external lighting system to avoid any risk of electrical ignition source inside the chamber.

Each compartment in the pressure chambers must be equipped with permanent safety valves set to prevent the possibility of accidental overpressurization of the chamber.

Each compartment in a chamber has air ducts to allow air to enter and to exit. Since the noise level of the air entering or leaving these ducts is high, and since the chamber itself is a good resonator, the ducts should have mufflers. If the noise level is still excessive, chamber occupants must be provided with noise protection devices for the ears. If muff-type ear defenders are used, a ventilation hole must be drilled in them to prevent a relative vacuum between muff and ear when the pressure is increasing.

Air compressors for hyperbaric chambers must provide exceptionally pure air and must be free from any possibilities of oil contaminating the air. As an emergency backup to the compressor, there should be cylinders containing enough compressed air to complete any normal treatment or decompression schedule. An alternative is to have two air compressors on line, either of which has the capacity to maintain operations and an emergency power source. The breathing oxygen enters the chamber independent of the chamber pressuri zation sources and is supplied to the individuals by a well-sealed face mask or hood. Exhaled oxygen must be dumped outside the chamber to avoid buildup in the chamber with increased fire risk.

Strict fire prevention procedures must be developed because of the increased oxygen and the concomitant increased danger of fires. All possible ignition sources must be removed. All but absolutely essential electrical connections must be outside the chamber. Lighting must be either isolated or, preferably, located outside the chamber. Sources of static electricity need to be eliminated, and metal objects that could strike the chamber wall and produce sparks are not used. Combustibles inside the chamber should be kept to a minimum, and fire retardant material is used for clothing and bedding. There are several fire-resistant materials that can be used, such as PBI, Durette Gold, and fiber glass fabrics.

The oxygen percentage in the chamber should be monitored continuously and kept as close to normal as possible; it must never exceed 25 percent. This can be controlled by adequate and frequent air ventilation of the chamber. Ventilation is accomplished by letting air into and out of the chamber simultaneously and in equal amounts. This method is also used to control the carbon dioxide level and to maintain optimum temperature and humidity. In case all precautions fail and a fire does ensue, a rapid, adequate fire suppression system is essential. A pressurized water deluge system forces water into the chamber through numerous nozzles to cover all areas.

Suction for removing secretions is provided by the difference of pressure inside the chamber and the outside world. A simple penetration through the chamber wall provided with tubing, a control valve, and a gauge are used to avoid excess suction.

Close observation of those inside the chamber must be maintained by outside operators. This can be done through viewing ports, over closed circuit television, and by instruments measuring physiological parameters. Besides visual contact with those inside the chamber, voice communication must be maintained.

All large pressure chambers need to have a medical lock for the passing in and out of medications, dressings, instruments, food, drinks, and other small objects.

All greases, fats, and oils should be kept out of the chamber, since these may become explosive in oxygen-enriched environments.

To ensure the safety of all personnel in the chamber, each person should have a breathing mask, and there must be a breathing mixture supplied from outside the chamber for use in case of smoke or other chamber atmosphere contamination.

The relative humidity in the chamber should be left high, at least 60 percent for comfort and to reduce the possibility of static electricity. Since the humidity of the breathing mixture is low, on-line humidification for patients should be provided.

9.2 Monoplace Hyperbaric Oxygen Chambers

Most of the safety principles described in Section 9.1 apply to monoplace chambers. The major difference is that these chambers are totally pressurized and purged with pure oxygen. Thus it is absolutely essential to prevent at all times the admittance of volatiles, greases, ointments, or oil (15).

REFERENCES

1. F. M. G. Holmstrom, "Hypoxia," in: *Aerospace Medicine*, 2nd ed., H. W. Randel, Ed., Williams & Wilkins, Baltimore, 1971, p. 60.
2. F. J. W. Roughton, "Respiratory Functions of Blood," in: *Handbook of Respiratory Physiology*, W. M. Boothby, Ed., USAF School of Aviation Medicine, Randolph Air Force Base, Texas, 1954, pp. 51–102.
3. A. Krough, *J. Physiol.*, **52**, 409 (1919).
4. F. M. G. Holmstrom, "Hypoxia," in: *Aerospace Medicine*, 2nd ed., H. W. Randel, Ed., Williams & Wilkins, Baltimore, 1971, p. 63.
5. A. C. Guyton, *Textbook of Medical Physiology*, 5th ed., Saunders, Philadelphia, 1976, pp. 373–374.
6. F. M. G. Holmstrom, "Hypoxia," in: *Aerospace Medicine*, 2nd ed., H. W. Randel, Ed., Williams & Wilkins, Baltimore, 1971, p. 57.
7. J. Ernsting, "The Principles of Pressure Suit Design," in: *A Textbook of Aviation Physiology*, J. A. Gillies, Ed., Pergamon Press, London, 1965, pp. 374–405.
8. R. W. Bancroft, "Pressure Cabins and Rapid Decompression," in: *Aerospace Medicine*, 2nd ed., H. W. Randel, Ed., Williams & Wilkins, Baltimore, 1971, pp. 337–363.
9. B. Bhattacharjya, Ed., "Problems for Consideration, Acclimatization," in: *Mountain Sickness*, John Wright and Sons, Ltd., Bristol, 1964.
10. A. Hurtado, "Mechanisms of Natural Acclimatization," School of Aviation Medicine Reports 56-1, Randolph Air Force Base, Texas, 1956.
11. A. L. Barach, M. Eckman, E. Ginsbury, A. E. Johnson, and R. D. Brookes, *J. Aviat. Med.*, **17**, 123 (1945).
12. S. M. Cain and J. E. Dunn, II, *J. Appl. Physiol.*, **20**, 882 (1965).
13. S. M. Cain and J. E. Dunn, II, *J. Appl. Physiol.*, **21**, 1195 (1966).
14. *Hyperbaric Oxygen Therapy: A Committee Report*, Undersea Medical Society, Bethesda, MD, 1983.
15. J. C. Davis and T. K. Hunt, Eds., *Hyperbaric Oxygen Therapy*, Undersea Medical Society, Bethesda, MD, 1977.

CHAPTER TEN

Biological Agents

JOHN S. CHAPMAN, M.D.

1 INTRODUCTION

1.1 Definition of Industry

The narrower definition of industry, a process or manufacture that takes place in some type of enclosed space, disregards such activities as oil well drilling and operations, highway construction, seismographic and geologic exploration, and real estate development. Operations of this type expose employees to hazards similar to those encountered in farming, ranching, and dairying. Very considerable numbers of people are employed in all of these processes. If they are to be regarded as industries, industrial hygiene has to account for a variety of hazards not commonly regarded as industrial. The problem is rendered more complex by the fact that activities of these kinds are not confined to the United States but lead to longer or shorter residence in very diverse regions of the world.

1.2 Types of Agent

Agents under consideration in this chapter are of two types: (a) nonliving substances of biological origin, products of fungal or bacterial metabolism plus perhaps degradation products of natural fibers altered by fungal or bacterial activity and (b) living organisms encountered in the course of activity in a specific industrial setting. These agents typically consist of spore formers and fungi, zoönotic organisms, vector-borne groups of microorganisms, and a few free-living organisms of soil or water.

1.3 Modes of Exposure

1.3.1 Cutaneous

Continuous exposure of the skin to some of the agents just cited is likely to occur under usual working conditions. Dusts and particulates of various kinds settle out and stick to sweaty skin, where they may set up various inflammatory and allergic processes. The immediate agent may not be readily identifiable. Viable pathogenic organisms may be inoculated through minor wounds or through preexisting abrasions or other breaks in the integument.

In specific conditions the mode of inoculation is by means of stings and bites, as in the many vector-borne diseases or in envenomization.

1.3.2 Gastrointestinal

The gastrointestinal portal of entry is probably the least likely under industrial conditions. Improper cleaning of the hands or exposure of food or water to some of the agents may permit entry by way of the gastrointestinal tract. In addition, larger particulates caught on the mucous lining of the respiratory tract may be swallowed as a result of ciliary activity. Nonliving products of microbial metabolism, however, would probably be destroyed by digestive enzymes and rendered harmless.

1.3.3 Respiratory

The respiratory tract constitutes the main portal of entry for most of the biological agents, either as viable organisms or as products of their multiplication on various fibers. Larger particles are filtered through the nose or deposited on the mucous membranes of the bronchi. Particles of 5 μm or less mean diameter will reach alveoli, whose surface, approximately 70 m^2, provides a huge area of exposure to allergenic material. Dependent on the size of particulate matter, therefore, the site of reaction in the respiratory tract may be primarily bronchial, with symptoms suggestive of asthma, or pulmonary, with various types of response within the alveoli.

1.4 Sources of Biological Agents

1.4.1 Soil

Soil constitutes the reservoir from which infections with *Histoplasma capsulatum* and *Coccidioides immitis* occur. The source of the first, however, appears to be bird or bat droppings, whereas that of the second is contamination of the ground by decaying bodies of small rodents. Likewise, infections caused by tetanus and the clostridial organisms originate in soil that has already been contaminated by the manure of sheep, horses, and cattle. In the case of the two fungi, distribution of the infectious particles into the atmosphere occurs as a result of human activity in part, but also winds may be a sufficient cause of disturbance. Infections by the spore formers and other fungi most commonly occur in connection with puncture wounds or injuries that devitalize tissue.

1.4.2 Water

Mycobacterial infections such as that caused by *M. marinum* and possibly other mycobacteria arise from contact with contaminated water. In the instance of *M. marinum* it appears that the organism may survive and propagate in interstices of concrete and stone and be inoculated by minor abrasions. The source of the contamination of water is unknown, but it has been observed that organisms are present in the mouths of fish. There is evidence that wind-driven turbulence of water may nebulize organisms of the *M. avium–intracellulare–scrofulaceum* complex.

1.4.3 Air

Human disease of the general type of Farmer's Lung Disease is airborne. Particles containing antigenic substances of fungi constitute the damaging material. Although contaminated soil constitutes the reservoir of *C. immitis* and *H. capsulatum*, their distribution and human infections result from the airborne state. A similar statement applies to most of the infections acquired from diseased animal carcasses in preparation for human consumption, as mentioned in Section 1.4.4, although inoculation through the skin undoubtedly represents an alternate pathway.

1.4.4 Carcasses and Excreta of Animals

The role of bird droppings and bat guano in contamination of soil by *H. capsulatum* has been mentioned. The treatment of fields and gardens by manure of larger quadrupeds produces heavy contamination with spore-forming organisms, such as tetanus and the clostridia.

The carcasses of freshly slaughtered animals have resulted in specific infections in employees of abattoirs. In the case of fowl, the result has been outbreaks of ornithosis, whereas brucellosis has developed among similar employees who deal with swine and Q fever among those dealing with cattle. The processing of fowl may also give rise to variable allergenic manifestations from airborne dissemination of material derived from feathers.

Decaying carcasses of larger animals may contaminate soil with clostridia and tetanus, just as their excreta do. The decay of small rodents of the Lower Sonoran Life Zone is regarded as the most probable source of contamination by *C. immitis*.

1.4.5 Insect Bites

All blood-sucking insects possess the potential of transmitting infectious agents. Ticks, lice, fleas, and mosquitoes are the commonest vectors. The variety of diseases transmitted in this manner is extensive. The more familiar diseases are malaria, Rocky Mountain Spotted Fever, typhus, yellow fever, and dengue. Many other diseases of more limited geographic distribution are encountered elsewhere in the world.

1.5 Contributing Factors

As this chapter reveals, symptoms and injury appear to be dose dependent. Thus the amount of material per unit volume, the amount of increased pulmonary ventilation required by the type of physical activity, and the number of years of exposure enter into any expression of risk. In addition, there is variability in individual susceptibility, for which we possess no useful criteria of predictability.

In all studies the smoker appears to be more prone to develop symptoms than the nonsmoker. Since smoking itself results in bronchitis, the importance of this factor is obvious. But with respect to particulates, the smoker is at further disadvantage, since the efficiency of the mucociliary escalator for removal of larger particles is less than normal (1); hence retention of injurious material is both greater in amount and longer in duration than in the nonsmoker. At the alveolar level, moreover, the impairment of phagocytic pneumocytes that results from smoking leads to less efficient removal of particulates.

The presence of inert particulates may also affect the response of the host, whether the particles are viable microorganisms or allergenic materials (2). Such material as carbon or other nonfibrogenic dusts may impair defense mechanisms at either the bronchial or the alveolar site, and irritant gases such as hydrocarbons may affect the protective mechanisms of the bronchi or alveoli.

Because these secondary factors are so numerous and variable—and often unquantifiable or not measured—the following discussion deals with the principal agent of disease as if it were present alone. It should always be remembered, however, that in a specific individual secondary factors may be of considerable importance.

1.6 Preventive Measures

1.6.1 Introduction

The variety of agents is so great and their loci in nature as well as their own biological character so diverse that general statements with regard to all of them are impossible. Methods of control must be particularized with respect to the agent and its reservoir, mode of dissemination, and effects on humans. Methods suitable for the control of histoplasmosis cannot be effectual to prevent

BIOLOGICAL AGENTS

vector-borne disease or brucellosis in an abattoir. In the following discussion appear the classic modes of control: the specific agent of disease determines which of these may be applicable in a given circumstance.

1.6.2 Medical Programs

Medical programs require adaptation to the particular problem involved. They are further complicated by the fact that under the broader definition of industry small numbers of employees in remote areas of the United States or in little known areas of undeveloped countries may be very much on their own. A means of direct communication should be provided if at all possible, since some of the conditions threaten high mortality and represent acute problems. In all instances, either in large plants in an industrial city or among a small group in some remote place, the aim is to recognize trouble at its first appearance and to limit its extent.

Periodic medical examinations serve useful purposes when employees are available. The frequency depends on the condition under study. For inhalational disease produced by inert material, the examination should certainly include tests of pulmonary functions. Among new employees in industries these tests may be necessary at the beginning and end of each shift, or alternatively on Monday mornings and Friday afternoons. The development of wheezing in the chest and of a drop of 100 ml or more in FEV_1 should warrant removal from the area. Other signals of problems in this area may be inflamed conjunctivae, unexplained loss of weight, or the occurrence of flulike symptoms in the more susceptible.

In some circumstances the use of skin tests for atopy may be a part of the general examination, whereas more limited and special situations may require examination of serum for antibodies against *H. capsulatum* or *C. immitis*. The frequency with which any of these may be required depends on circumstances of activity and location of the area of work. Among individuals living abroad in underdeveloped countries, it is essential to obtain a baseline Mantoux test with purified protein derivative (PPD). If it is negative it should be repeated at yearly intervals as a routine, but much more frequently if experience indicates a high rate of infection.

Immunizations should be carried out as indicated. For all employees in the field it is necessary to maintain current immunization against tetanus. Other immunizations required in specific situations are those set forth in the annual *Health Information for International Travelers* from the Center for Disease Control. Bacillus Calmette-Guérin (BCG) may be considered for employees and families who are to be in areas of known high prevalence of tuberculosis. Alternatively, the tuberculin negative among them should receive tests with PPD as frequently as every 3 months, depending on the prevalence of disease, the presence of servants in the house, or extent of contact with native workers. If the skin test becomes positive, the new reactor should be placed on prophylactic isoniazid. In certain areas atabrin as prophylaxis against malaria may be required.

Antimicrobial agents as prophylaxis against diarrheal disorders have not been notably successful, and the most effective prevention lies in preparation of food and water, as well as avoidance of uncooked fruit or vegetables unless one prepares it personally. The customary use of native servants practically assures a breakdown of hygienic measures.

1.6.3 Sterilization

Heat, the traditional method of dealing with bacteriologically contaminated materials, is not well adapted to large industrial use. Many substances may be damaged by dry heat, especially fibers to be used in fabrication. Steaming is known to reduce the problem in byssinosis without damage to fibers but is not known to be effective in other immunogenic fiber-borne conditions.

Chemical sterilization in relation to the agents of this section has been confined largely to the use of formalin in direct application to soil for the eradication of *H. capsulatum*, and as vapor, after complete sealing of the building, to disinfect an anthrax-contaminated plant.

Irradiation with either γ or UV rays is known to be effective but in an industrial setting is not likely to be efficient or practical. It is known that UV, a much less expensive method, is effective against airborne, well-dispersed living agents, such as *M. tuberculosis*. Otherwise its effects are surficial only. Careful maintenance is essential: even in an airconditioned building tubes require removal of dust at least once weekly.

1.6.4 Engineering Controls

If a process that is known to produce hazards is contemplated, original design of the plant may prove the most effective means of control of industrial disease. The processes known to generate the greatest amount of dust should be isolated from the cleaner areas and provided with separate ventilation. Unbaling, chopping, hackling, and carding of fibers, depending on the fiber and the process, generate the largest efflux of particles. These activities should be isolated and, when possible, robotized. Rapid change of atmosphere is desirable, with various treatments of the exhausted air. Electrostatic precipitation has been found to be effective with respect to cotton fibers.

In outside problems, such as land contaminated by pathogenic fungi, solutions may vary according to expected use. Repeated spraying with oil was shown to reduce the incidence of coccidioidal infections in a very arid, dusty area. One piece of land heavily contaminated with *H. capsulatum* was treated with a formalin solution and removal of top soil. In other situations, addition of top soil with sowing of grass might constitute a sufficient means of enclosure. Areas known to be contaminated by spore formers probably can be best rendered safe by burning over, although this measure will probably not be adequate for organisms deeper in the soil.

The general principles of dust control within factories and plants are the same:

1. Reduce generation of dust by modification of process.
2. Confine unavoidable dust generation to enclosed areas.
3. Clean settled dust regularly and thoroughly, with special protection of the cleaning crew, including individual air supply.
4. Maintain rapid turnover of atmosphere in dusty areas, with some form of removal of particles from the exhaust.
5. Maintain careful and regular inspection and cleaning of air-conditioning intake areas (the Legionella experience).
6. If air conditioners are of the window unit type, they must be protected against service as bird roosts or nests. Use of evaporative types of air cooling requires constant attention to the water bed for control of fungi and bacteria.

1.6.5 Personal Protection

Employees generally prefer comfort to protection, unless a hazard is obvious and imminent. Individual air-filtering devices of all types are poorly tolerated under conditions of work that call for increased respiration and are still less acceptable in the presence of heat. When protection is imperative, as in housekeeping, individual air supply under positive pressure with an open hood constitutes the best protection against inhaled agents.

Protective clothing is of little benefit except in those occupations that involve the field, where heavy boots and thick socks provide about the only protection against bites by snakes, scorpions, and tarantulas. In areas of great heat, clothing should be loose and composed of absorbent material such as cotton or wool rather than synthetics and be closed at the necks and wrists to prevent access by insects.

Against insect-borne disease the clothing may be impregnated with benzoyl benzoate or free application of a repellant such as diethyl toluamide is desirable.

Employee education is desirable under all conditions, with special instructions to report wheezing and shortness of breath, if the worker is engaged in activities dealing with inert agents. Flulike illnesses should also be reported at once, as well as the onset of arthralgias and skin eruptions.

In no group is education as important as among small teams that operate in remote places, either in this country or abroad. They require solid information with respect to the particular infections they are apt to encounter, as well as every possible means of avoiding them. If the team is of some size, one member ought to receive full instruction in first aid, including the management of envenomizations. A radio link with the nearest hospital is desirable, and there should be a prepared system for rapid evacuation and hospitalization in the face of more serious problems.

2 NONLIVING MATERIALS

2.1 Nature of the Materials

The products of bacterial and fungal metabolism are complex and numerous. Chemically they consist of a wide variety of split proteins, lipoproteins, and mucopolysaccharides, the relative amounts of which are affected by temperature, humidity, and possibly by the kind of substrate (in this case, fibers) on which organisms grow. Although proteins probably produce the greater number of immunologic effects, terminal sugars may also be antigenic (3), and bacterial and fungal material and vegetable fibers contain a wide array of exotic mono- and polysaccharides (4).

2.2 Modes of Action of These Substances

2.2.1 Direct Effects

Some of the substances produced by bacteria or molds may have a pharmacological effect. Pernis and associates demonstrated substances in samples of cotton that seemed to resemble bacterial endotoxin and postulated that these compounds probably produced fever and tightness of the chest (5). Cavagna et al., after sensitizing rabbits to cotton dusts, concluded that an endotoxinlike substance in the challenge material released serotonin and resulted in a type of chronic bronchitis with acute bronchospasm (6). This view of the mechanism of production of byssinosis was further amplified by Rylander and associates (7). Nicholls, however, disputed this mechanism on the basis of the very small amount of endotoxin that could be demonstrated, maintaining that a histaminelike material in the stems and pericarps of cotton more adequately accounted for symptoms (8).

Bouhuys and co-workers also recovered substances from hemp and cotton dust that seemed to release histamine in experimental preparations (9). Subsequently they isolated from cotton brachts a compound they identified as methyl piperonylate. This compound released histamine and produced symptoms that could be inhibited by isoproterenol (10). Whatever the precise compound or mechanism of its action, it is evident that a direct pharmacological effect may be produced by chemical substances contained in cotton, and—as the various investigators found—other vegetable fibers as well.

2.2.2 Immunologically Mediated Effects

Pepys (11), in a review of "farmer's lung," emphasized that in individuals who manifest this reaction, it is possible to detect at least two types of immune reaction to an extract of moldy hay or products of *Thermopolyspora faeni*. The first was atopic, as manifested by immediate wheal. The second was an Arthus phenomenon with induration 4 to 6 hr after injection. The Arthus response was associated with demonstrable precipitins in serum against his antigenic materials. In some asthmatics with pulmonary infiltrations, in cows with "fog fever," and in humans with bird fancier's disease, Pepys was also able both to elicit an Arthus reaction and to demonstrate precipitins against antigens of *Aspergillus fumigatus*. In many patients biopsies revealed also the presence of epithelioid tubercles (11). It is evident that cell-mediated immunity is involved in a number of instances. Other features of the infiltrations include the presence of fair numbers of eosinophiles in some cases, of large numbers of lymphocytes and fibrocytes in others. Regardless of the substrate, when pulmonary infiltrations develop as a result of these various immune mechanisms, the reaction should be regarded as a hypersensitivity pneumonitis or, in Pepys's term, "extrinsic allergic alveolitis" (12).

It appears, moreover, that neither textile nor cordage fiber is essential to growth of molds. Fink et al. (13) report the isolation of *Micropolyspora faeni* from ducts of a circulating warm air system to which a humidifier had been added. It is thus evident that the mold rather than compounds derived from the fiber is of paramount significance, although fiber-derived substances may alter or enhance the reaction of the alveoli or may participate in transition of skin reactions from wheal to tuberculin-type induration (13).

Since the range of reaction of tissue is so varied in all these responses to allergenic material, some kind of relationship may exist between these hypersensitivity reactions and the much less common diffuse fibrosing alveolitis (14). Alternatively, a similarity between these responses and so-called thesaurosis is manifest, although in thesaurosis (15, 16) it should be possible to observe particulate foreign material.

In any event it is clear that the range of reaction to nonliving biological substances extends from acute reactions, similar to that to methyl piperonylate, to the other extreme of tubercle formation and fibrosis.

2.3 Textile and Cordage Fibers

2.3.1 Cotton (Byssinosis)

Byssinosis is marked by a sensation of tightness of the chest and cough, usually worse on return to work after the weekend, but also worsening in the course of the workday. Workers in cotton mills are particularly at risk, although other workers involved in the processing of cotton, from gin to finished goods, may occasionally manifest symptoms. In cotton mills the highest incidence has been found among those in areas of preparation (opening, blending, picking, and carding) and in workers with more than 18 years of exposure. Males are affected about twice as often as females, and smokers much more than nonsmokers (17).

Byssinosis has been classified by Schilling, and a national conference has accepted the following grades of severity, which should be employed in every survey of a population at risk.

0. No symptoms of respiratory difficulty.
$\frac{1}{2}$. Occasional tightness of the chest on the first day of the working week.
1. Chest tightness on the first working day of each week.
2. Tightness of the chest on every working day.
3. Grade 2, plus evidence of permanent (irreversible) loss of exercise tolerance, with or without reduced ventilatory functions (18).

It is obvious that except in grade 3, evaluation of impairment from exposure to cotton dust is on the basis of complaints or symptoms. The degree to which workers in a particular mill may be affected will be reflected in absenteeism, complaints, or respiratory insufficiency in those with greatest seniority. Preliminary evaluation may take the form of a questionnaire.

Quantitative evaluation depends on dust concentrations in various parts of the building and on the results of simple pulmonary ventilatory tests, which should be performed under specified conditions. Imbus and Suhs used FEV_1 (forced expiratory volume, 1 sec) as the most suitable test under operating conditions (17). Hamilton et al. also employed this measurement and found that byssinotic workers exhibited a decrease of 10 percent from morning values in the course of the working day, whereas only 3 percent nonbyssinotic workers experienced a decrease of similar proportion (19). Merchant and associates, after comparing several ventilatory tests, also concluded that FEV_1 furnished the most consistent and accurate discrimination between byssinotic and unaffected workers (20). McKerrow and others, however, preferred the indirect MBC (maximum breathing capacity), which they found to decrease linearly through the working day (21). Airways resistance also varied in accord with the complaint of tightness of the chest. In studies under other conditions of exposure Bouhuys et al decided that the forced expiratory flow (FEF_{25-75}) provided the most sensitive evidence of change of function (22).

Probably for most industrial surveys the vital capacity (VC) and the FEV_1 will be found to be effective and useful. Regardless of the test that may be selected, it should be carried out under identical conditions, with the same instrument (which must be maintained in good order and calibrated from time to time), and by the same technicians, who should be well trained to urge maximum effort and to recognize poor performance. Tests ideally ought to be carried out at the beginning of each first working day of the week and repeated at the conclusion of the shift.

All the recommended tests for pulmonary function are based on the presumption that nonliving biological agents produce, as their first effect, varying degrees of bronchospasm. They further imply that continued exposure results in chronic changes in the bronchi, with progressively irreversible obstruction and ultimate fibrosis. The real test of obstruction is the rate of flow of gases during expiration, and FEV_1, (forced expiratory volume, 1 sec), MVV(MBC) (maximum voluntary ventilation, maximum breathing capacity), and FEF_{25-75} (forced expiratory flow at 25 to 75 percent of vital capacity) all deal with this aspect of ventilation; FEF_{25-75} is regarded as representing more specifically obstruction of smaller bronchial branches.

Note that these remarks with respect to tests of pulmonary function apply to all exposures to nonliving biological agents discussed in this section.

In surveys of various stations for amount of cotton dust in the atmosphere, particles smaller than 15 μm are of primary concern. McKerrow and associates (21) found that improvement in symptoms and in measured function was apparent when concentration of these particles was reduced to 2.4 mg/m^3. However a national conference has recommended that a suitable threshold limit value (TLV) for such particles be 1 mg/m^3.

Various measures of control have been advised and attempted. Filtration of atmosphere is both difficult and expensive, and McNall advocates electrostatic precipitation as an economical and feasible measure (23). Prewashing of cotton before other processing was effective but interfered considerably with spinning (24). Steaming, which may be carried out as early as ginning, materially reduces byssinotic effects and does not alter characteristics of the fiber for spinning and weaving (24). McKerrow and colleagues (25) recommend vacuum extraction of dust, which reduces fiber content of the atmosphere by nearly 50 percent. In the recommendations of a national committee, the most easily achieved and satisfactory means of maintaining an acceptable TLV consisted of partial isolation of the procedures established as the most dusty. A necessary feature of partial isolation is effective local exhaust ventilation (17).

In individual instances it is recommended that workers who present grade 2 or 3 symptoms, or who show a decrease of 10 percent in FEV_1 during a working day, should be removed from further exposure to cotton dust. It is also recommended that applicants who present initial abnormal functions, who have abnormal roentgenograms of the chest, or who show a decrease in FEV_1 after a short period of employment be assigned to nondusty procedures.

There is disagreement over the possibility of progression after exposure ceases. It seems reasonable to conclude that in a few individuals changes may be irreversible, and even that further progression of fibrosis may occur (17, 22). If this view is correct, either the pharmacological effect results in permanent alteration of the bronchi, or additional immunologic responses are involved.

2.3.2 Hemp, Jute, and Sisal

In their study of workers in hemp, Bouhuys and co-workers (9) concluded that hemp fiber produced a disease similar to byssinosis. It occurred particularly among workers engaged in bolting and hackling by hand. They found only a fair correlation between FEV_1 and symptoms, and workers exhibited a wide variability in sensitivity to the fiber. Heavily exposed workers produced urines with a high content of histamine metabolites. As they studied hemp workers in 1966 the authors concluded that many changes in the airways might prove to be irreversible, and the follow-up work in 1974 confirmed this earlier impression (22).

Masks afforded no protection to the exposed individuals, who had more cough and greater dyspnea than controls. However removal of gums from the raw product materially reduced respiratory complaints. The authors, however, do not make it clear that degumming represents technologically feasible or acceptable procedure in the industry (9).

Jute seems to be less noxious than hemp, at least in biological tests (7), and sisal, another fiber employed in production of cordage, seems not to produce symptoms of byssinosis or to result in significant alterations of ventilatory indices (25).

Flax chiefly affects workers engaged in preliminary preparation, such as hackling. Reported symptoms resemble those of byssinosis (26). Correction of the situation depends on enclosing the dustier procedures and providing effective local exhaust ventilation.

2.4 Farmer's Lung

The term "farmer's lung" is applied to a syndrome marked by chills and fever following disturbance of moldy hay (or other moldy material.) Associated with the clinical syndrome are evanescent infiltrations of the lung. Usually the first episode subsides without appreciable effect on pulmonary function, but as attacks recur on subsequent exposures, portions of the acute infiltrate fail to resolve and gradually organize. Eventually extensive fibrosis of the lung, marked impairment of function, and respiratory insufficiency develop.

The pathogenesis of the condition consists of the inhalation of airborne substances produced by the growth of molds in contaminated hay. Classically these molds are thermophilic actinomycetes. The process does not involve invasion of tissue but results from immunological reactions against the fungal antigens. In Great Britain the organism responsible seems to be *Micropolyspora faeni*. Pepys has demonstrated that the reaction involves an Arthus phenomenon, with circulating antibodies against extracts of moldy hay (11). Tissues reveal the presence of noncaseating tuberculoid granulomas; the process clearly is not a foreign body response. Although Pepy's concept of pathogenesis has been rather widely accepted, Harris and co-workers have presented findings that suggest that cell-associated rather than circulating antibodies may better explain the response in tissue (27).

Apparently the molds responsible for farmer's lung grow in hay that is improperly cured or subject to repeated moistening; hence the disease is much more common in cold and humid areas such as Great Britain and the region around the Great Lakes. However conditions adequate for the growth of thermophilic organisms may result from storage of improperly cured or wet hay in almost any area. Since the intensity of response to the antigen is proportionate to dose, storage in a tight barn offers the best conditions for development of the syndrome. It should be noted, however, that thermophilic molds do not require the presence of vegetable fiber, since Fink et al. have reported finding these organisms in the duct of a warm air heating system provided with a humidifier (13).

Diagnosis of farmer's lung depends on adequate occupational and environmental history supplemented by appropriate skin tests and the demonstration of specific precipitins, or other antibodies (28). Since suitable antigens are not commercially available, recognition rests largely on history, characteristic roentgenograms of the chest, sometimes biopsy of the lung, and the exclusion of other causes of a fibrosing and granulomatous disease.

Treatment depends on the severity of manifestation. In acute exacerbations, full doses of corticosteroids are necessary, and measures applied in any form of acute respiratory distress may be required. In late fibrotic disease it is probable that no measures will be effective other than such as may be required for intercurrent bronchitis.

Prevention, of course, calls for interdiction of further contact with hay. Complete curing of hay before storage might be effective, but in many parts of the world this measure would not be possible. In nonagricultural exposures such as that described by Fink and colleagues, the removal of the humidifier, together with disinfection by formaldehyde, might be adequate; otherwise an entirely new system of ductwork would be required.

2.5 Bagassosis

The symptoms of bagassosis are entirely similar to those of farmer's lung; the exposure, however, is to certain batches of bagasse, the dried stalks of sugarcane. Since only certain batches appear to be associated with symptoms, it seems probable that the mode of storage, perhaps the degree of extraction of juice, and the opportunity for wetting after storage, may determine the growth or nongrowth of molds (29).

Bagasse usually is shipped in bales, quite like baled hay. The material serves as a binder for many products, such as wallboard and insulating boards. Processing begins with the breaking open of bales and the grinding of the fiber. In both processes a very large amount of dust is generated, and it is at these sites that workers receive the heaviest exposure.

The many similarities to farmer's lung suggest a similar pathogenesis, but Pepys reported that precipitins against his antigens were far less consistently present. It is possible that other types of antibody or different varieties of molds or both may account for his lack of success. The pulmonary pathology is identical.

Progression with recurrent bouts of fever and pulmonary infiltrations resembles the pattern observed in farmer's lung. Restrictive defects in ventilation appear to be most characteristic, for Weill and associates reported a decrease in both vital capacity and residual lung volumes in 17 of 20 repeatedly exposed workers. Diffusing capacity of lung by carbon monoxide method DL_{co} was correspondingly reduced in these workers; FEV_1 was reduced in most patients, and it is interesting that among those who had persistent respiratory difficulty, the predominant manifestation was obstruction (29). In a limited outbreak workers were seen within 2 to 4 weeks after the first occurrence of the condition. There was a wide range of symptoms, from the slightest respiratory complaints to severe cyanosis with acute right ventricular failure. Those most severely ill had residual respiratory functional abnormality (30).

Management of the acute and chronic forms of the disease does not differ from corresponding stages of farmer's lung. Affected workers must be removed from further exposure. Prevention would apparently call for isolation of the original processes of opening, chopping, or grinding from the rest of the plant and the provision of adequate exhaust ventilation.

2.6 "Grain Fever"

Apparently similar in pathogenesis and pathology to byssinosis, grain fever has been observed in handlers of grain in all stages of production from harvest and combining to drying and storage in elevators. DoPico and co-workers consider both the fever and the pulmonary infiltrations to be mediated by antibodies (31). Wheal response was observed by this group and by Darke and associates (32), but the latter found no consistent correlation between the presence of precipitins in serum and the presence or absence of symptoms or of pulmonary infiltrations. In all exposed workers they demonstrated the presence of circulating antibodies, regardless of whether symptoms were present.

DoPico's group found that symptoms occured more frequently among smokers than nonsmokers, with an overall incidence of 37 percent. In the Darke study in England, examination was limited to men engaged in harvesting, in whom the incidence was 25 percent and development of the syndrome was related to duration of exposure. The English study further demonstrated an exposure during harvesting to a very high concentration of molds, as many as 200 spores and fragments of hyphae per cubic meter. Cultures were not reported, and no specific agent can be incriminated (32).

Physiological effects on the lung seem to consist especially of bronchospasm, or possibly of infiltrations of the submucosa of finer bronchioles, since the study of DoPico et al. revealed that the midflow rate best demonstrated early effects of exposure.

Since most of the work takes place in the open, unless there is entry into elevators or storage bins, the most feasible control consists of the removal of sensitive workers from further exposure.

2.7 "Bird Fancier's Disease"

2.7.1 Pathology and Pathogenesis

The name "bird fancier's disease" has been applied to extrinsic allergic alveolitis as observed in some individuals who have been exposed to birds. The condition must be distinguished from invasive aspergillosis of the lung, particularly that reported from France, and it probably differs from the type of granuloma Pepys has encountered in some cases of asthma. Bird fancier's disease is encountered both in an acute form indistinguishable from the acute form of farmer's lung, and also in a chronic phase manifested by extensive fibrosis. Such reactions have occured not only in those individuals directly involved in the care and feeding of birds (in England, chiefly pigeons and budgerigars) but even in members of the household who have had no direct association with birds at all (33).

In considering pathogenesis one has to take account not only of the feed (usually grain), but also antigens of avian origin. Hensley et al. were able to show in sensitized monkeys a specific pneumonitis induced through aerosol challenge with pigeon serum (34). In addition, in experiments involving passive transfer of lymphocytes, they were able to prove the presence of cell-associated antibodies. Circulating antibodies in high titer also are known to be involved in the reaction (33).

Pathological changes in the lung of bird fancier's disease seem to differ from those of other forms of allergic alveolitis. The monkeys in the experiments of Hensley et al. revealed hemorrhagic alveolitis 6 hr after challenge. Undoubtedly the degree of original hypersensitivity of the individual, the duration of exposure, and the concentration (and perhaps the character) of the antigenic material affect the kind of response observed in tissue.

In two patients with milder exposure and insidious onset, Riley and Soldana found the alveolar spaces filled with desquamated septal cells and foamy macrophages, while lymphocytes and plasma cells infiltrated the interstitial tissue. One of the tissues revealed a much more striking interstitial infiltration, with little alveolar material.

In a study of pulmonary tissue from 33 patients Hargreave and associates reported that the acute disease appears as a histiocytic granuloma of the interstitium, with occasional nonnecrotic tubercles containing multinucleated giant cells. Within these giant cells they observed lanceolate spaces not observed in other forms of extrinsic allergic alveolitis (35). In the more chronic forms of the disease there were dense fibrosis and honeycombing.

To Hensley and his associates, however, it seemed that the most striking feature was the presence of remarkable numbers of foamy histiocytes. In a study of the histology of the disease these investigators observed in one instance many foamy histiocytes in the alveoli and intersititial tissue, while in another workman histiocytes were associated with considerable infiltration of lymphocytes. In sections from both sources they observed occasional sarcoidlike granulomas. In their second patient, in whom exposure had been much heavier and more protracted, they found severe obliterative bronchitis and dense intersitital fibrosis (36).

2.7.2 Symptoms

The onset of symptoms may be either dramatic or insidious. Allen and others (37) believe that the age of the patient and the duration of exposure constitute the principal factors affecting the degree of distress and the reversibility of changes. In all forms of the disease the principal complaint is breathlessness or tightness of the chest, with more or less cough. Wheezing, an expression of the degree of bronchitic or bronchiolitic hypertrophy observed histologically, may be a prominent feature.

2.7.3 Roentgenographic Findings

In some roentgenographic studies nodulation is regarded as the earliest evidence of disease. Later in the course the finer nodulation becomes confluent, contraction and fibrosis become very prominent, and honeycombing is evident (35).

2.7.4 Physiological Effects

Physiological effects differ with the type of pathology. Allen et al. (37) observed that four of nine patients, who had exhibited mild obstructive and restrictive ventilatory defects early in the course of exposure, regained normal function after exposure ceased. Three of the remaining five showed progressive obstruction of the smaller airways, and in one greatly reduced elastic recoil of the lung was observed (37). Hargreave and associates (35) encountered reduction of all pulmonary volumes in their patients, and associated with this a reduction of DL_{co}. Riley and Soldana (33) observed both restrictive and obstructive defects in one of their patients, a pure and severe restrictive defect in the other.

In one study it was observed that removal from further exposure resulted in improvement in 37 of 41 patients (35). In severely symptomatic patients, such as those of Riley and Soldana, the administration of adrenal corticosteroids resulted in considerable improvement. The principles of treatment for acute symptoms and withdrawal from further exposure are therefore indicated in bird fancier's disease as in most other causes of extrinsic allergic alveolitis.

2.7.5 Specific Exposures

Turkeys. Reports establish that individuals employed in the raising and processing of turkeys face the same problems as those exposed to pigeons. In 142 of 205 subjects Boyer et al. (38) observed symptoms within one hour after commencing a shift. The principal complaints were tightness of the chest and cough. In 13 workers, 4 to 8 hr after the reexposure began, symptoms progressed to include myalgia, fever, and dyspnea. These individuals, as well as some with less severe symptoms, demonstrated both positive skin tests and circulating antibodies against antigens derived from feathers or serum (38).

Chickens. Very similar symptoms and objective findings occur in workers engaged in processing chickens. They exhibit both immediate and Arthus responses to skin tests; precipitins are directed against antigens derived from feather, serum, and droppings, Warren and Tse found precipitins in bronchial mucus to be especially impressive; these were most manifest in response to the antigens contained in feathers, which may constitute the most important inhalant (39).

2.8 Wood Dust Disease

2.8.1 General Features

Pulmonary disease, in some respects resembling farmer's lung, in others predominantly of an asthmatic type, has been observed in individuals engaged in woodworking industry, but not in logging. Direct chemical effect, particulate structure, and immunological responses all play greater or less roles in the production of symptoms and structural changes in the lung. Sometimes the effect is related to the type of wood or the presence of various fungal contaminants.

A direct effect of wood dusts (type not specified but presumably from several species) appears in a report by Michaels (40) of disease in two workers, one a carpenter, the other employed in wood pulping industry. The tissues of each revealed peribronchiolar fibrosis and histiocytes and foreign body giant cells in alveolar spaces. Basophilic inclusions that had properties resembling those of wood were found in the giant cells in each instance.

2.8.2 Sequoiosis

Cohen et al. (41) encountered similar histological changes in a patient exposed to sawdust from redwood, a pulmonary reaction to which they gave the name sequoiosis. Although they observed in giant cells particles apparently derived from wood, they also demonstrated precipitating antibodies in the patient's serum. Chemically the antigens of redwood were identified as polysaccharides (42). Roentgenograms revealed diffuse pulmonary infiltrations, and the principal physiological impairment consisted of a restrictive defect associated with decreased DL_{co}.

2.8.3 Cork

Similar roentgenographic changes, with nodulation or occasional reticular patterns, appeared in the examination of workers exposed to cork dust. Most of these abnormalities occurred only after 15 to 20 years of exposure. Physiological findings included both restrictive and obstructive defects, with an especially striking increase in residual volumes.

Pathological findings consisted of histiocytes, interalveolar fibrosis, and chronic inflammation. There is no mention of inclusions or giant cells.

2.8.4 Maple Bark Disease

According to Wenzel and Emanuel (44), workers engaged in the production of chips from maple logs develop a pattern of illness in most respects identical to that of farmer's lung. In so-called maple bark disease the source of sensitizing antigen was shown not to be the wood itself but a heavy infection of the bark by *Cryptosoma corticale*, which was dispersed in the atmosphere in clouds. The process is marked by the usual symptoms of cough, tightness of the chest, chills, and fever, by patchy infiltration of the lung, and by precipitins against the fungus. As in most such exposures, a number of the individuals at risk developed precipitins without clinical or roentgenographic abnormalities.

2.8.5 Red Cedar

In workers exposed to the dust produced in processing of red cedar, Gandevia (45) observed asthma in three persons who previously had been nonasthmatic, and rhinitis in four men. In the remaining workmen FEV_1 decreased by at least as much as 100 ml in the course of the working day.

2.9 Carcinoma in Furniture Manufacturing

Finally it should be mentioned that adenocarcinoma of the nasal passages and paranasal sinuses has been observed in a disproportionately high percentage of workers engaged in the manufacture of furniture, but not in carpenters or cabinetmakers. A similar high incidence was observed among leather workers (46).

2.10 Miscellaneous Problems

2.10.1 Tobacco

Among nonsmoking women who worked in a tobacco-processing plant there were complaints of chest tightness and wheezing. Respirable dust ranged from 0.3 to 3.6 mg/m^3. These workers exhibited no chronic changes in pulmonary function, but acute decreases in FEV_1 took place during the shift (47).

2.10.2 Coffee

Van Toorn encountered among coffee workers in a plant a case of extrinsic allergic alveolitis that closely resembled farmer's lung. A 46-year-old male who complained of breathlessness was found to respond to skin tests prepared from coffee beans, and precipitins were also demonstrable. Roentgenograms revealed nodulations that became confluent in the lower lung fields. Pulmonary biopsy presented vasculitis, and alveoli were lined with granular histiocytes and the alveolar walls were infiltrated with round cells (48).

2.10.3 Mushrooms

Stewart reported the occurrence of a syndrome of tightness of the chest among six workers who were exposed to pasteurized compost in the spawn sheds for the culture of mushrooms (49). In some workers onset of symptoms occurred within a few hours of exposure, but in others exposure was much longer. Neither skin tests nor precipitins provided useful information. Roentgenograms revealed pulmonary infiltrations, but no tissue was available for histological examination.

2.10.4 Subtilin

Certain manufacturers added enzymes of *Bacillus subtilis* to various types of washing powders. Tightness of the chest and breathlessness were common complaints among workers exposed to this substance. Roentgenograms of the chest were normal, and neither skin tests nor search for antibodies proved useful (50, 51). In nearly half the employees at one such plant, pulmonary ventilatory functions were decreased, as was DL_{co} (50). In the study of Dijkman et al. bronchial challenge resulted in prompt reduction of VC and FEV_1, and dyspnea persisted for 6 to 8 hr after challenge (51). This condition seems to be more nearly related to papain-induced bronchitis than to either pharmacological or immunological effects.

2.11 Recognition of a Problem

Probably many other substances of biological origin may play a role in industrial processes that lead to illness. Mechanisms may vary from direct injury, through atopy, to pulmonary Arthus response to tubercle formation. Some agents may produce direct pharmacological effects, whereas others may produce response through histamine-releasing properties. Some substances may have all these effects sequentially, dependent in part on the substance and, perhaps more importantly, on the host.

Symptoms of tightness of the chest, wheezing, and breathlessness are indicators of some deleterious substance in the workplace. These complaints may be just as characteristic of chemical substances as of those of biological origin, and the physician must immediately determine which of several possibilities best explains the conditions produced in workers. The development of symptoms may be insidious and the complaints may be rather vague, easily confusable with "asthma" or "flu." This is especially true of those individuals who also have arthralgias and fever. If the fever is recurrent ("Monday morning fever"), if the frequency is considerably greater than in the general unexposed population, and if "relapse" takes place on return to duty after an interval of sickness, there should be strong suspicion that some type of material of biological origin is the agent.

Prompt recognition of the possibility is important, since earlier changes are reversible and no permanent pulmonary damage will result, whereas prolonged exposure of susceptibles is almost certain to result in pulmonary insufficiency.

2.12 Approaches to the Problem

The initial investigation should be a questionnaire directed to all employees. A modification or shortened form derived from the questionnaire of the British Medical Research Council would be suitable. The next procedure should consist of simple auscultation of the chest for wheezing, crackles, and rhonchi. It is important to note on the record of examination which individuals are smokers.

Standard 6-ft roentgenograms of the chest should follow the preliminary investigation. The first sample of the population might consist of worst cases—those who are males, have the longest record of employment and are smokers. Subsequently it may be necessary to obtain X-rays of all employees who are symptomatic or who present persistent abnormalities on physical examination.

If these procedures suggest that a problem exists, the severity of damage that may result must be measured. Ventilatory tests, properly conducted under standard conditions by well-trained technicians, constitute the appropriate next step. These tests should include as a minimum VC, FEV_1, and probably FEF_{25-75} at the beginning and conclusion of the shift. Results obtained should be tabulated separately for smokers and nonsmokers and for males and females, and—if the group is at all large—arranged by decade of age. (Due allowance for smaller ventilatory figures for blacks should be kept in mind.) Other more sophisticated measurements of pulmonary function may be desirable, depending on the types of ventilatory defect discovered.

Skin testing materials, methods for bronchial challenge, and studies of serum for the presence of antibodies are seldom available under conditions of industrial surveys. In some circumstances, however, it may be possible and important to arrange a limited application of some of these techniques. If the problem appears to be quite troublesome, the advice and assistance of pulmonary physiologists or immunologists may become necessary.

3 LIVING AGENTS OF DISEASE

Infectious diseases related to occupational activity are sporadic and uncommonly affect a high proportion of workers. Three types of agent may appear in special situations and affect considerable numbers of the exposed labor force. The first class is composed of those organisms capable of developing vegetative states that permit long survival and transportation over considerable distances: notably, the spore formers and the fungi. The second class consists of diseases transmissible from animals to humans and, except for occasional cases among veterinarians and animal husbandry workers, appears especially in meat processing plants. The third type consists of vector-borne diseases: workers are not much more exposed to these diseases than is the general population, except as their activities may require extensive work in the field or travel and exposure in parts of the world where endemic diseases of this type are prevalent.

In addition, laboratory employees, veterinarians, physicians, and other medical personnel undergo special risk by virtue of their occupations.

3.1 Vegetative Organisms

3.1.1 Coccidioides immitis

Exposure to *C. immitis* takes place most frequently in the southwestern United States and the adjacent states of Mexico. Distribution of the organism in this area is not uniform, but is largely restricted to the Lower Sonoran Life Zone where vegetative forms survive in the soil. As a result of high wind or mechanical disturbance of the soil, the organism becomes airborne and may be inhaled. Operators of graders, bulldozers, and other heavy equipment are particularly at risk.

Inhalation of a viable unit results in a pulmonary infiltration, followed very soon by enlargement of bronchial and tracheobronchial lymph nodes. After 2 or 3 weeks cavitation of the central part of the pulmonic infiltration may occur. Symptoms are similar to those of influenza and include fever, malaise, and weakness, which persist usually for 3 or 4 months. In dark-skinned ethnic groups the primary infection may rapidly lead on to widespread dissemination with lesions of bone, skin, and meninges (52).

In the endemic area infections have occurred in oil field workers, in highway maintenance and construction crews, among drivers of heavy trucks, and occasionally among train crews on frequent runs through this area. Passive transport of the organism to distant areas may occur on trucks or automobiles; thus repairworkers and mechanics may encounter the organism.

The prevalence of coccidioidal infection in a population may be measured by the coccidioidin skin test. Complement fixation studies are generally available through state health departments. Titers ranging in one dilution around 1:16 suggest recent infection, but those of 1:64 or higher almost always denote hematogenous dissemination. It is important that serum for complement fixation be drawn in advance of the skin test, which tends to produce an elevation of the complement fixation. If the rate of positive reactions to coccidioidin 1:100 exceeds 1 to 2 percent among nonresidents in a group of workers in the endemic area, a problem of occupational infection exists. (In the Air Force during World War II, in some sites, infections as measured by the skin test ran as high as 20 to 25 percent.)

The only measure found to be of any value in reducing the rate of infection appears to be oiling of the soil (53). This form of control may be undertaken when an oil field crew is likely to be at work in the same site for several weeks or months, but obviously no method can affect transient or irregular exposure. Since it was shown that the highest rate of incidence takes place in summer and autumn (53), the safest measure for maintenance and construction crews of highways and railroads would be to schedule work in such a way that most of it could be done during the winter and spring.

3.1.2 Histoplasma capsulatum

H. capsulatum survives in soil in much the manner of *C. immitis*, but its geographic distribution is quite dissimilar. Infection occurs endemically along the valleys of the Mississippi River and its main tributaries, but histoplasmosis is also associated with exposure to chicken and bird droppings (54, 55). The guano of bats, in such places as caves and rock shelters, also provides suitable conditions for the survival of the organism (56).

H. capsulatum produces clinical symptoms and signs like those of *C. immitis*. Roentgenographic changes are similar but often are very extensive and may resemble scattered bronchopneumonic infiltrations or even assume a miliary pattern. Histologically the lesions resemble those of tuberculosis, although there is usually less necrosis; with suitable stains the organism can be identified in the lesions. Skin tests and serologic reactions provide additional information—subject to very much the same restrictions and features as those described for similar tests for *C. immitis*.

Conditions of exposure are similar to those for *C. immitis*, except for geographic differences, although raising of large flocks of poultry poses a different and equal risk. Any type of industrial activity that results in disturbance of soil may be followed by an outbreak of histoplasmal infection among workers. Residents of adjacent areas will also be at risk.

The skin test with 1:100 histoplasmin affords a rapid and fairly effective means of screening a population but has the disadvantages of coccidioidin with respect to effect on serologic results. If the skin test is employed, the frequency of response in the test population should be compared with figures for the region as a whole, usually obtainable from health departments. Cultures of soil can reveal sites of very heavy contamination. For such sites, if they are not very large, Powell and associates have recommended three sprayings of 3 percent formalin (55).

3.1.3 Bacillus anthracis

Inhalation anthrax, known for many years as "wool sorter's disease," has become exceedingly rare, but it still occurs occasionally in textile mills where imported goat hair is prepared for incorporation into fabrics of mixed fibers (57). On the basis of airborne infection of monkeys, Brachman and associates conclude that exposure to 1000 spores over a period of 3 to 5 days should correspond with an infection rate of 10 percent in a plant (58). Fortunately conditions for so serious an outbreak occur very rarely.

The result of inhalation of an adequate number of organisms is a highly lethal pneumonia. The occurrence of such a disease in an employee of a plant in which imported goat hair is used calls for immediate shutdown and disinfection of the entire plant.

Cleanup of an infected mill is difficult and expensive. Young et al. (57) removed and burned all wooden material, dismantled all machinery and cleaned it with hydrocarbon solvents, and sanded and repainted all painted metal surfaces, first having exposed the entire building to vaporized formalin for 2 days. In Great Britain all goat hair is treated with formalin before it leaves the pier (57).

3.2 Zoönoses

3.2.1 Brucella

Although traditionally brucellosis has been associated with farm workers and particularly with consumption of unpasteurized milk, this mode of infection has diminished very strikingly in the United States. In Texas a review of cases reported to the Health Department establishes that after 1957 this mode of infection applied to only one-fifth of the cases (59). In Scotland, however, 62 percent of infections occurred among farm workers (60), and in Great Britain as a whole a review in 1975 still showed 70 percent of brucellosis to occur in farm workers (61).

In contrast in 1969 Busch and Parker reported that 68 percent of all cases in the United States had involved packing house employees; infections were derived from both swine and cattle (62). In a study of an outbreak of the disease in an abattoir in Illinois, Schnurrenberger and associates found 13 clinical cases of brucellosis during a 6-month period. In addition there were 71 serological reactions of 1:25 or more among 551 personnel. In this outbreak infected swine were the source of the human infection.

The review by Buchanan et al. (64) underscores the magnitude of the problem: among packing house employees between 1960 and 1971 there were 1644 symptomatic cases. In Iowa and Virginia the attack rate was highest, more than 20 human cases per million slaughtered animals. Study of the activities and the histories of the patients suggested that some cases developed as a result of infection through breaks in the skin, that probably the majority were airborne, and that perhaps a number may have occurred through conjunctival inoculation.

The development of acute illness among employees at a slaughter site calls attention to the possibility of brucellosis. Unfortunately the symptoms of this disease are quite nonspecific and the diagnosis may be overlooked unless a high degree of suspicion exists. Since brucellosis appears somewhat sporadically, and since its symptoms are so variable, the figures above may underrepresent actual rates of infection.

Although serological studies are of limited clinical value, they serve a useful purpose in surveys. The discovery of titers in an intermediate range, 1:32 to 1:128 among a group of employees specifically exposed would indicate that a problem exists. Office and sales personnel might serve as controls for the particular region of the country, since their jobs lead to little or no exposure.

Control is most difficult, since *Brucella suis* infection among pigs is often undetectable. Eradication of brucellosis from cattle herds is possible but is not yet complete. If the principal mode of infection is airborne or conjunctival, there is no conceivable protection under operating conditions. The use of protective goggles might be considered, and exhaust or dilution ventilation of the atmosphere of the slaughter area might afford some protection. Complete eradication of infections among animals represents the only final protection for exposed humans.

3.2.2 Ornithosis

Employees engaged in poultry packing plants are at risk of ornithosis, which may be transmitted by birds of many kinds. At highest risk are workers engaged in plucking and evisceration. Durfee found that in the United States in 1971 71.7 percent of all reported cases occurred among the employees of two turkey processing plants, most of them during a period of only 2 months (65). In their review in 1976 Dickerson and associates found that 13 outbreaks of ornithosis occurred in the United States between 1948 and 1972, all of them in association with the slaughter and processing of turkeys. Of the 494 verified cases that developed, seven were fatal. Infection was traced back to specific flocks, from which birds were transported widely, and small epidemics occurred in Nebraska and Missouri (66). In Sweden Jernehus et al. traced two outbreaks in poultry plants to infected flocks of geese and ducks (67).

These reports suggest a higher attack rate and a much closer grouping of human cases than is characteristic of brucellosis. Evidence of a small epidemic should put one on guard, but the symptoms are such that the real nature of the problem might be masked by such diagnoses as "flu" or "atypical pneumonia."

Presumably most infections with ornithosis take place through inhalation, although inoculation cannot be entirely excluded. Control of the human infection depends on eradication of the disease among birds destined for human consumption. When it is appreciated that a flock poses a problem, quarantine and antibiotic treatment are necessary. Whether this management will later permit preparation of the birds for human consumption is uncertain. Complete eradication of the disease among domestic fowl is probably a practical impossibility: often infection of the birds is inapparent. Reinfection of flocks by wild birds makes control most difficult.

3.2.3 Q Fever

Q fever is a rickettsial infection caused by *Coxiella burnetti*, and it has appeared sporadically in the United States, chiefly as a disease of packing house employees. The organism has been shown to become airborne (69); as a particulate in the atmosphere it retains its viability and infectivity for long periods (70). Epidemiological studies in California also indicated that the organisms may survive on surfaces or in soil for long periods, only to become airborne again (71). Direct contact with infected animals (sheep, goats, and cattle) is not essential (69, 71); many cases evidently arise from organisms in the environment.

Infection among dairy cattle may be demonstrated by serological examinations for antibodies. In the Milwaukee area it was possible to isolate the organisms from individual animals in 42 of 50 serologically identified herds (72). Human infection, however, was not well correlated with the dairy or the cattle-raising industry. An outbreak in Texas, on the other hand, was marked by infection of 55 of 136 exposed workmen and was limited to handlers of cattle in the holding pens and to workers in the slaughter area of the abattoir (74).

Recognition of the disease may be delayed. It should be suspected if cases of severe pneumonia, sometimes associated with hepatosplenomegaly, occur among individuals engaged in the handling or slaughter of cattle, sheep, or goats.

Control measures ultimately depend on eradication of the infection among domestic animals. Since it appears possible that premises may remain infected for a long time after original contamination, immunization of employees at risk may become necessary. There is no experience with any method of environmental decontamination.

3.2.4 Mycobacterium marinum

Mycobacterium marinum, an atypical mycobacterium, produces nodular, slowly necrosing lesions of the skin, usually that of the extremities, where inoculation takes place readily through current or previous abrasions and minor wounds. Most infections have occurred among children bathing in certain swimming pools or in adults who clean aquariums for tropical fish. However, infections of commercial fishermen have occurred, both in the Gulf of Mexico and in Scandinavian waters (75).

Although the organism may produce fatal illness in tropical fish, it apparently results in no detectable disease in larger fish (or the cause may have been overlooked or not considered). Whether *M. marinum* may colonize the skin, fins, or mouths of fishes is uncertain, but a Swedish case suggests that the source of human infection was an inoculation from a fin (75) and another human infection followed the bite of a dolphin (76).

The disease should be suspected if commercial fishermen or workers at the dock develop slowly progressive nodular lesions of the extremities. These ultimately drain a small amount of material and heal spontaneously, but new crops of nodules may develop beyond the original sites. Tissue reveals necrotizing granulomas containing acid-fast bacilli.

Since lesions of the hands and forearms are so common among commercial fishermen, it is possible this infection has been overlooked. A survey of Gulf fishermen demonstrated that 10 percent reacted with positive skin tests to an antigen prepared from *M. marinum* (75). Even so, it appears that infections are probably quite rare, and since infection is limited to the skin and never spreads to deeper structures, it is of limited importance.

4 VECTOR-BORNE DISEASES

4.1 The Vectors

The vectors capable of transmitting disease are so numerous and vary so much geographically that a full account of diseases they transmit can be better studied elsewhere. In North America the principal vectors are mosquitoes, ticks, and fleas. In other areas lice and bedbugs may constitute important transmitters of disease. Biting flies may also carry and inject organisms of varying types. Any arthropod that depends on blood for its sustenance is a potential vector. The prevalence of a disease in a given area—and hence the potential vector—is available from the Center for Disease Control in Atlanta.

4.2 The Infections

The diseases conveyed by these vectors are usually acute and may be associated with severe morbidity and high mortality. Rarely is it possible to provide adequate care in the field, and prompt removal to a hospital is desirable.

4.3 Prevention

A complete plan for a crew operating in an infested area calls for the following preparations and activities:

1. Extensive indoctrination of the crew with respect to vectors and the habits of those vectors.
2. Instructions in selection of camp sites to minimize presence of vectors and proper application of pesticides around the camp site (when return to base is not possible).
3. Use of screens, nets, or other barriers around beds, bunks, cots, or sleeping bags.
4. Proper use and application of repellants on clothing and/or skin.
5. Selection of protective clothing, necessarily variable according to climate, terrain and type of vector.
6. Familiarization of crew with early symptoms and prearranged methods of communication and evacuation of the ill.

5 ENVENOMIZATION

5.1 Sources of Venoms

On land the principal bearers of venoms consist of snakes, scorpions, and spiders. In aquatic environments poisonous snakes are to be found in fresh water, while there are numerous poisonous fish in saltwater. In North America the dangerous snakes are rattlesnakes, copperheads, moccasins, and coral snakes. Habitats and habits vary with the different species; elsewhere in the world many other species of venomous snakes exist and constitute hazard to employees and crews working in the field.

Scorpions are very widely dispersed over the entire globe. Their poisonous secretions vary in potency among the various species. Since these creatures characteristically inhabit rock crevices, they possess a remarkable ability to penetrate under doors, around windows, and through ceilings to reach human prey.

Spiders abound, of course, and their various species are widely distributed. Their bites are usually relatively harmless, although some types are neurotoxic.

5.2 Mode of Action of Venoms

The possibility always exists that in the highly susceptible individual a bite from even a bee or a wasp may produce acute anaphylaxis and may prove fatal. These and other bites and stings may result in locally painful and even severe tissue damage, dependent on the amount injected. Many secreted substances injected into the bite contain proteolytic enzymes and result in local necrosis.

The more dangerous venoms are usually either neurotoxic or hemolytic. If not promptly and adequately treated, these envenomizations produce grave illness and may prove lethal.

5.3 Prevention

Awareness of the possibility of encounter is essential. In areas in which infestation is expectable, the crew must exert unremitting vigilance. General measures similar to those described in Section 4.1.3 are useful with respect to prevention of envenomization. At least one member of a field team should have received complete instruction in the immediate treatment of snake bite and the use of such antivenins as may be available for the specific animal.

REFERENCES

1. A. Wanner, J. A. Hirsch, D. E. Greeneltch, E. W. Swenson, and T. Fore, *Arch. Environ. Health*, **27,** 370 (1973).
2. A. Tacquet, A. Collet, V. Macquet, J.-C. Martin, C. Gernez-Rieux, and A. Policard, *C. R. Acad. Sci. Paris*, **257,** 3103 (1963).
3. J. K. N. Lee, E. A. Pachtman, and A. M. Frumin, *Ann. N. Y. Acad. Sci.*, **234,** 161 (1974).
4. E. A. Kabat, in: *The Chemistry and Biology of the Mucopolysaccharides, A Ciba Symposium*, G. E. Wolstenholme and C. M. O'Connor, Eds., Little, Brown, Boston, 1958, p. 230; also G. F. Springer, in: *ibid.*, p. 230.
5. B. Pernis, E. C. Vigliani, C. Cavagna, and M. Finulli, *Br. J. Ind. Med.*, **18,** 120 (1961).
6. C. Cavagna, V. Foa, and E. C. Vigliani, *Br. J. Ind. Med.*, **26,** 314 (1969).

7. R. Rylander, A. Nordstran, and M. C. Snella, *Arch. Environ. Health*, **30,** 137 (1975).
8. P. J. Nicholls, *Br. J. Ind. Med.*, **19,** 33 (1962).
9. A. Bouhuys, A. Barbero, S.-E. Lindell, S. A. Roach, and R. S. F. Schilling, *Arch. Environ. Health*, **14,** 533 (1967).
10. M. Hitchcock, D. M. Piscitelli, and A. Bouhuys, *Arch. Environ. Health*, **26,** 177 (1973).
11. J. Pepys, *Ann. Intern. Med.*, **64,** 943 (1966).
12. J. Pepys, in: *New Concepts in Allergy and Immunology*, V. Serafini, A. W. Frankland, C. Musala, and J. M. Jamar, Eds., Amsterdam, Excerpta Medicine, Foundation, 1971, p. 136.
13. J. N. Fink, E. F. Banaszak, W. H. Thiede, and J. J. Barboriak, *Ann. Intern. Med.*, **74,** 80 (1971).
14. J. G. Scadding and K. F. W. Hinson, *Thorax*, **22,** 291 (1967).
15. G. W. H. Schepers, *JAMA*, **181,** 635 (1962).
16. J. M. Gowdy and M. J. Wagstaff, *Arch. Environ. Health*, **25,** 101 (1972).
17. H. R. Imbus and W. M. Suh, *Arch. Environ. Health*, **26,** 183 (1973).
18. "The Status of Bissynosis in the United States, A Summary of the National Conference on Cotton Dust and Health," *Arch. Environ. Health*, **23,** 230 (1971).
19. J. D. Hamilton, G. M. Halprin, K. H. Kilburn, J. A. Merchant, and J. R. Ujda, *Arch. Environ. Health*, **26,** 120 (1973).
20. J. A. Merchant, G. M. Halprin, A. R. Hudson, K. H. Kilburn, W. N. McKenzie, Jr., D. J. Hurst, and P. Bermazohn, *Arch. Environ. Health*, **30,** 222 (1975).
21. C. B. McKerrow, M. McDermott, J. C. Gilson, and R. S. F. Schilling, *Br. J. Ind. Med.*, **15,** 75 (1958).
22. A. Bouhuys and E. Zuskin, *Ann. Intern. Med.*, **84,** 398 (1976).
23. P. E. McNall, Jr., *Arch. Environ. Health*, **30,** 552 (1975).
24. J. A. Merchant, J. C. Lumsden, K. H. Kilburn, V. H. Germino, J. D. Hamilton, W. S. Lynn, H. Byrd, and D. Baucom, *Br. J. Ind. Med.*, **30,** 237 (1973).
25. C. B. McKerrow, J. C. Gilson, R. S. F. Schilling, and J. W. Skidmore, *Br. J. Ind. Med.*, **22,** 204 (1965).
26. P. C. Elwood, J. D. Merrett, G. C. R. Carey, and I. R. McAulay, *Br. J. Ind. Med.*, **22,** 27 (1965).
27. J. O. Harris, D. Bice, and J. E. Salvaggio, *Am. Rev. Respir. Dis.*, **114,** 29 (1976).
28. G. A. doPico, W. G. Reddan, F. Chmelik, M. E. Peters, C. E. Reed, and J. Rankin, *Am. Rev. Respir. Dis.*, **113,** 451 (1976).
29. H. Weill, H. A. Buechner, E. Gonzales, S. J. Herbert, E. Aucoin, and M. M. Ziskind, *Ann. Intern. Med.*, **64,** 737 (1966).
30. D. P. Nicholson, *Am. Rev. Respir. Dis.*, **97,** 546 (1968).
31. G. A. doPico, W. Reddan, D. Flaherty, A. Tsiatis, M. E. Peters, P. Rao, and J. Rankin, *Am. Rev. Respir. Dis.*, **115,** 915, (1977).
32. C. S. Darke, J. Knowelden, J. Lacey, and A. M. Ward, *Thorax*, **31,** 294 (1976).
33. D. J. Riley and M. Saldana, *Am. Rev. Respir. Dis.*, **107,** 456 (1973).
34. G. T. Hensley, J. N. Fink, and J. J. Barboriak, *Arch. Pathol.*, **97,** 33 (1974).
35. F. Hargreave, K. F. Hinson, L. Reid, G. Simon, and D. S. McCarthy, *Clin. Radiol.*, **23,** 1 (1972).
36. G. T. Hensley, J. C. Garancis, G. D. Cherayil, and J. M. Fink, *Arch. Pathol.*, **87,** 572 (1969).
37. D. H. Allen, G. V. Williams, and A. J. Woolcock, *Am. Rev. Respir. Dis.*, **114,** 555 (1976).
38. R. S. Boyer, L. E. Klock, C. D. Schmidt, L. Hyland, K. Maxwell, R. M. Gardner, and A. D. Renzetti, Jr., *Am. Rev. Respir. Dis.*, **109,** 630 (1974).
39. C. P. W. Warren and K. S. Tse, *Am. Rev. Respir. Dis.*, **109,** 672 (1974).
40. L. Michaels, *Canad. Med. Assoc. J.*, **96,** 1150 (1967).
41. H. I. Cohen, T. C. Merigan, J. C. Kosek, and F. Eldridge, *Am. J. Med.*, **43,** 785 (1967).

42. H. I. Cohen, T. C. Merigan, and F. Eldridge, *Clin. Res.*, **13,** 346 (1965).
43. L. De Carvalho Cancella, *Ind. Med. Surg.*, **32,** 435 (1963).
44. F. J. Wenzel and D. A. Emanuel, *Arch. Environ. Health*, **14,** 385 (1967).
45. B. Gandevia, *Arch. Environ. Health*, **20,** 59 (1970).
46. E. D. Acheson, R. H. Cowdell, and E. Rang, *Br. J. Ind. Med.*, **29,** 21 (1972).
47. F. Valić, D. Ceritić, and D. Butković, *Am. Rev. Respir. Dis.*, **113,** 751 (1976).
48. D. W. Van Toorn, *Thorax*, **25,** 399 (1970).
49. C. J. Stewart, *Thorax*, **29,** 252 (1974).
50. T. Franz, K. D. McMurrain, S. Brooks, and I. L. Bernstein, *J. Allergy*, **47,** 170 (1971).
51. J. H. Dijkman, J. G. A. Borghans, P. J. Savelberg, and P. M. Arkenbout, *Am. Rev. Respir. Dis.*, **107,** 387 (1973).
52. M. L. Seviers, *Am. Rev. Respir. Dis.*, **109,** 602 (1974).
53. C. E. Smith, R. R. Beard, H. G. Rosenberger, and E. G. Whiting, *J. Am. Med. Assoc.*, **132,** 833 (1946).
54. D. J. D'Alessio, R. H. Heeren, S. L. Hendricks, P. Ogilvie, and M. L. Furcolow, *Am. Rev. Respir. Dis.*, **92,** 725 (1965).
55. K. E. Powell, K. J. Kammerman, B. A. Dahl, and F. E. Tosh, *Am. Rev. Respir. Dis.*, **107,** 374 (1973).
56. H. F. Hasenclever, M. H. Shacklette, R. V. Young, and G. A. Gelderman, *Am. J. Epidemiol.*, **86,** 238 (1967).
57. L. S. Young, J. C. Feeley, and P. S. Brachman, *Arch. Environ. Health*, **20,** 400 (1970).
58. P. S. Brachman, A. F. Kaufmann, and F. G. Dalldorf, *Bacteriol. Rev.*, **30,** 646 (1966).
59. S. J. Lerro, *Tex. Med.*, **67,** 60 (1971).
60. D. Reid, *Scot. Med. J.*, **21,** 125 (1976).
61. Anon., Editorial, *Lancet*, **1,** 436 (1975).
62. L. A. Busch and R. L. Parker, *J. Infect. Dis.*, **125,** 289 (1972).
63. P. R. Schnurrenberger, R. J. Martin, P. R. Weaver, and G. G. Jelly, *Arch. Environ. Health*, **24,** 337 (1972).
64. T. M. Buchanan, S. L. Hendricks, C. M. Patton, and R. A. Feldman, *Medicine*, **53,** 427 (1974).
65. P. T. Durfee, *J. Infect. Dis.*, **132,** 604 (1975).
66. M. S. Dickerson, W. R. Bilderback, and L. W. Pessarra, *Tex. Med.*, **72,** 57 (1976).
67. H. Jernelius, B. Pettersson, J. Schvarcz, and A. Vahlne, *Scand. J. Infect. Dis.*, **7,** 91 (1975).
68. C. Gale, *Proceedings of the 64th Annual Meeting, U.S. Livestock Sanitary Association*, 1960, p. 223. (No publisher or location given.)
69. E. H. Lennette and H. H. Welsh, *Am. J. Hyg.*, **54,** 44 (1951).
70. W. W. Spink, in: *Tropical Medicine*, 5th ed., Hunter, Schwarzwelder, and Clyde, Eds., Saunders, Philadelphia, 1976, pp. 188–192.
71. W. H. Clarke, E. H. Lennette, and M. S. Romer, *Am. J. Hyg.*, **54,** 319 (1951).
72. H. J. Wisniewski and E. R. Krumbiegel, *Arch. Environ. Health*, **21,** 58 (1970).
73. H. J. Wisniewski and E. R. Krumbiegel, *Arch. Environ. Health*, **21,** 66 (1971).
74. N. H. Topping, C. C. Shepard, and J. V. Irons, *J. Am. Med. Assoc.*, **133,** 813 (1947).
75. W. C. Miller and R. Toon, *Arch. Environ. Health*, **27,** 8 (1973).
76. D. J. Flowers, *J. Clin. Pathol.*, **23,** 475 (1970).

CHAPTER ELEVEN

Hot and Cold Environments

STEVEN M. HORVATH, Ph.D.

1 INTRODUCTION

A number of environmental stressors have been found to tax one's ability to perform in the working environment. Aside from environmental controls, in responses to these situations the individual has utilized both physiological and psychological (behavioral) systems to successfully accommodate the imposed stress. Some common stressors faced are heat, cold, pollutant gases and particles, toxic chemicals, and a variety of psychological stressors such as boredom and high achievement demands. Each of these stressors induces a series of adaptive responses that protect the individual from their potentially adverse effects. In this chapter the emphasis is placed on the simpler of the environmental stressors, that is, human responses to thermal environments, since the adaptation to these stressors provides a clear indication of one's potential to respond favorably.

The responses to cold and hot environments have been studied extensively. Humans are considered to be subtropical animals who have developed physiological adaptations and mechanisms that result in maximal increases in their body heat content. This ability to acclimatize to cold environments is limited. Even at an ambient temperature of 25°C an unclothed person cannot maintain body heat stores. However, the individual does have a much greater capacity to acclimatize to hot environments. One's response to thermal changes is dependent on a number of factors.

In any consideration of these responses to either cold or hot environments, the possibility that the individual has either adapted, acclimated, or acclimatized must be taken into account. "Adaptation" implies changes of a genetic nature that have occurred because of natural selection processes. "Acclimation" refers to physiological changes in response to ambient temperature changes. "Acclimatization" refers to any change in physiological mechanisms that better enables an individual to perform and/or survive in a particular environment. In general, most studies on humans are concerned with acclimatization processes, and the primary emphasis of this chapter is on these processes and the manner in which they enable individuals to improve their performance.

2 PHYSIOLOGICAL RESPONSES TO COLD

The successful response to a cold ambient environment is dependent on behavioral and physiological mechanisms. Behavioral factors involve a complex interplay utilizing other elements of the environment to protect against the impacting one. The employment of external energy sources, properly designed insulative clothing, reduction of effective area for radiation, convective and conductive losses, limited exposure times, and so on, illustrate the many behavioral responses utilized. Physiological responses, that is, modification of internal adaptive mechanisms, are also involved. These physiological alterations have been studied during short (acute) and prolonged exposures to lowered ambient temperatures. The experimental approaches utilized to study this problem have been made on three types of exposure: subjects living in polar regions, those exposed to cold (minutes to weeks) in temperature-regulated chambers, and those exposed to normally occurring seasonal environmental cold.

Several groups such as the Australian aborigines and the Kalahari Bushmen appear to demonstrate some genetic selection. Their responses to a cold stress differ from those of Caucasians in that the former apparently are capable of allowing their core temperature to fall while keeping their surface (skin) temperatures relatively high, with no significant increase of metabolic rate (energy output). This is in contrast to Caucasians, who have a high metabolic rate, a maintained or higher core temperature, and a decreased surface temperature under similar cold stress.

The responses to cold environments are integrated in the thermoregulatory centers in the brain (hypothalamus). In general, the brain operates through closely coordinated actions mediated through three effector outputs: the endocrine system, the autonomic nervous system, and the skeletomuscular system. These actions are:

1. Behavioral responses by way of the skeletomuscular system provide numerous options: moving out of the cold, building a fire, wearing heavier and warmer clothing, reducing surface area.

2. The responses of the autonomic nervous system include diminishing peripheral blood flow, thus minimizing heat loss.

3. The endocrine system can trigger increased liberation of catecholamines, thyroxin, and other substances involved in regulating metabolic processes.

4. Further examples of skeletomuscular activity are shivering, pilomotor activity, and active muscular work.

Involuntary muscular activity during cold exposure appears to be of a consistent pattern for a particular individual. The activity observed varies from fibrillation in single muscles or small muscle groups to clonic and tonic spasms that induce marked skeletal movement. Most of this muscular activity is limited to the thorax and upper portion of the lower extremities. The cooler distal portions of the body show only pilomotor activity. It is rather suggestive that the areas suffering most markedly from the local chilling are not involved in the protective reaction of shivering.

Shivering originates either in the upper thorax (pectoral muscles) or in the thigh (gluteal and vastus group). The spread of shivering activity may be localized to the area of origin or extend throughout the area of shivering muscles. Such extension, a so-called march of shivering, is observed only rarely. If massive shivering occurs, it originates almost simultaneously in both the thorax and the thighs. Periodic fluctuations in shivering activity are always observed. The intervals separating shivering bouts are far from consistent. They can vary from seconds to minutes.

Respiratory movements (end of expiration) or yawning appear to be related to the occurrence of shivering bouts. In some individuals this relationship is limited to the earlier, more rapid cooling phase, although some subjects retain this correlation even when further heat loss from the body and changes in magnitude of peripheral vasoconstriction were minimal or nonexistent. Body size or obesity (fat) shows no relation to shivering activity.

The increased oxygen uptake (metabolism), which may be as high as three to five times normal standard resting rates during cold exposure, is related to either (a) shivering, (b) increased voluntary movement caused by cold discomfort, or (c) nonshivering thermogenesis. Shivering probably represents the first line of defense against cold, and nonshivering thermogenesis a defense that goes into action as a result of prolonged, continuous exposure to cold. Shivering is a relatively inefficient process in comparison to active muscular work (1) and results in an impression of fatigue beyond its metabolic cost. A severe cold stress, probably more than one would be willing to tolerate, is required to induce nonshivering thermogenesis (2).

Davis and Johnson (3) have suggested that in cold adaptation the individual seems to rely less on shivering and more on nonshivering thermogenesis and a decrease in the metabolic rate. When young men, dressed only in shorts and sneakers, were continuously exposed to an 8°C environment for 3 to 10 days (4), they initially responded by extensive and intensive shivering, which persisted more or less continuously night and day despite sleeping under a single blanket at night. The increased heat production (twofold to fourfold) remained high even during sleep. Body temperature remained normal. In some instances the subjects ceased vigorous shivering during the night and their body temperatures fell to low levels. These subjects doubled their metabolic rate, had an increased heart rate, and had increased urinary nitrogen losses. It was suggested that these changes were the resultant of hormonal changes brought about by the cold stress. Some of these subjects suffered ischemic cold injury of their feet, although the ambient temperature did not approach the freezing temperature of the tissue. This increased metabolism may be related to the calorigenic effects of the catecholamines, which are known to be released during cold exposure, but only one study has been made on men who have been subjected to acclimatization procedures.

Whereas increased heat production is dependent on metabolic processes, the heat loss from the body is largely dependent on the physical factors of the environment. The laws of heat flux from the skin to the environment have been well described by Hardy (5). Burton and Edholm (6) have emphasized the importance of the applicability of Newton's law of cooling, which states that the rate of body heat loss from the exposed skin is directly proportional to the temperature gradient between the skin and its environment and is independent of the humidity. Engineers refer to the proportionality constant as the heat transfer coefficient C and have described coefficients for any number of substances and forms over a wide range of air velocities. The reciprocal of C is the thermal insulation of the system. The physiological analogue to the cooling constant is the combined radiative and convective heat transfer coefficient h, which describes the rate of dry heat exchange between the skin and the environment that can occur at any constant air velocity. The effect of such factors as ambient water vapor pressure, air velocity, and effective surface area, which also modify the rate of total heat loss, have been more accurately defined. Relative humidity had no significant effect on heat loss and the subjective sensation of cold in naked or lightly clothed men. Damp cold is not colder than dry cold (7, 8), although subjectively damp cold appears to be colder.

Precise measurement of the rate of evaporative heat loss is not of great importance during nude or nearly nude cold exposure, since in nonsweating, resting conditions, loss from the skin has a relatively constant value between 6 and 10 W/m^2 and loss from the respiratory tract can vary between 3 and 6 W/m^2. However during conditions of exercise, even in very cold environments, total evaporative heat loss can be increased as a result of the active secretion of sweat onto the skin surface (9, 10) and the increased ventilation subsequent to the elevated oxygen uptake.

Tolerance to cold exposure can be considered either as an aspect of cold acclimatization or merely as the consequence of the physiological response to a cold stress. Cold tolerance can be assessed by one of several means, including the time at which shivering begins, a change in the magnitude of shivering, the time at which metabolic heat production becomes significantly elevated, the state of discomfort of the subject, and/or some evaluation of peripheral resistance, causing a reduced peripheral blood flow and a reduction in convective heat transfer from the body core to the skin. There is consequently a reduction in heat loss from skin to environment by minimizing the skin-to-ambient thermal gradient. The second line of physiological defense is the increase in heat production. Another most important factor in cold tolerance is related to the quality and consistency of behavioral adaptations, which relate to knowhow, experience, and probably the state of physical fitness.

The importance of the level of physical fitness as a modifier of human resonse to cold was shown by Adams and Heberling (11). The responses of men to a standardized cold stress before and after a 3-week period of intensive physical training were compared. In the posttraining test average levels of heat production were 15 kcal/(hr)/(m^2) higher, mean rectal temperatures were 0.5°C lower, average skin temperatures were 1.0°C higher, and foot and toe temperatures were 3.0 and 4.0°C higher, respectively, with no significant differences in average body temperatures. Although these observations were suggestive in that they imply some improvement in tolerance to cold, they need further confirmation.

Saltin (12) has found that submaximal and maximal work during short-term cold exposure (-5°C) can be performed at the same levels of circulatory capacity as observed in 20°C ambients. Pulmonary ventilation, however, was somewhat higher when working in the cold.

Because the heat production response is an important one, measurement of oxygen uptake for determination of the metabolic heat production above basal level will be related to the intolerance of the cold stress. The test protocols with the use of oxygen uptake have tended to involve several variations. One variation has been the determination of the basal metabolic rate (BMR) in neutral conditions to evaluate whether previously cold-exposed subjects or populations differ from an appropriate control group. Any adaptation affecting the BMR (hence stimulating a form of nonshivering thermogenesis) has yet to be adequately shown.

Kang et al. (13) have examined Korean Ama (women divers), previously noted as having an elevated shivering threshold, greater tissue insulation index, and elevated BMR (14). They found no differences in BMR between Ama and their matched controls during summer and winter. Furthermore, there was almost no calorigenic response to norepinephrine in either group, indicating either that nonshivering thermogenesis is difficult to demonstrate in humans or that the Ama were not cold acclimatized. Although Hammel (15) has described the Alacaluf Indians, a primitive group living on Tierra del Fuego, as demonstrating a metabolic adaptation to cold (i.e., elevated metabolic rate during warm exposure), he did not indicate whether this could be considered to be nonshivering thermogenesis.

Finally, several investigators have shown a significant seasonal variation in BMR. Yoshimura (16) has concluded that the elevated BMR in the winter months seen in Japanese subjects was due to an increase in thyroid activity permitting the Japanese to endure cold more comfortably. However, the role of the endocrine system during cold stress remains to be clarified. Davis and Johnson (3) and Girling (17) have reported that hourly cold exposures induced a greater heat production response to the cold stress in the summer than during the winter, suggesting that some degree of cold acclimation had occurred. Both these studies were performed under conditions of marked seasonal variation in ambient conditions. In Santa Barbara, California, however, where the average ambient temperature variability is small over the year, monthly exposures of 2 hr to 4.5°C revealed no seasonal variation in metabolic response to cold in seven subjects (18). Previously Hammel et al. (19) had demonstrated that no metabolic adaptation occurred in the presumably cold-acclimatized Australian aborigine. Thus the question of the individual's ability to become cold adapted by way of some metabolic process or processes remains to be answered.

Another index of cold intolerance has involved the determination of the oxygen uptake response during prolonged cold exposure. This method has clear advantages since comparisons in metabolic response between an experimental and a control group can readily be made at any period of cold exposure, at any level of internal body temperatures, or at any combination of body temperatures. Raven and Horvath (20) have recently emphasized that an attempt to assess the metabolic response to cold requires continuous measurements over greater than a 1-hr exposure. They found that the metabolic heat production response became relatively constant only during the final 30 min of a 2-hr exposure to 5°C, and it was only during this period that the rate of change in mean body temperature became constant and minimal. Wyndham et al. (21) have also proposed that a 2-hr cold exposure should serve as a standard procedure. Yoshimura and Yoshimura (22) found that 2.5 hr was required to attain consistent values of metabolic rates and internal temperature in a 10°C ambient. O'Hanlon and Horvath (23) also reported that a 2-hr cold exposure was necessary to reach a steady state.

The scarcity of information on other physiological changes during cold exposure induced O'Hanlon and Horvath (23) to evaluate the heart rate response. They found that the heart rate remained constant or was slightly increased during cold exposures. This constancy of heart rate and the marked increase in oxygen uptake during these cold stresses suggested that this high oxygen uptake could be accomplished only by an elevation in cardiac output, greater oxygen extraction, or a combination of both. Raven et al. (24), whose subjects were exposed to 5°C for 2 hr, reported that the cardiac output was elevated 95 percent above levels found in a neutral environment. This increased cardiac output was due primarily to an increased stroke volume.

There have been only a few investigations regarding physiological adaptations to long-term cold exposure, mainly because of the inherent difficulties in conducting such experiments. Convincing evidence for acclimatization to cold has been obtained from studies on small animals (rats, etc.). Unfortunately, such evidence has not been adequately documented for the human. Horvath et al. (25) studied five subjects who had lived continuously for 8 days in an ambient temperature of −29°C. They could find no evidence for acclimatization after that period. Carlson et al. (26) exposed seven subjects to −6°C for 16 to 18 hr daily for 14 days and reported that their subjects had reduced their heat elimination rate after this period. Davis (27) reported that men artificially acclimatized to cold in the summer had decreased heat production and lower skin and rectal temperatures as a consequence of a series of 8-hr daily exposures to an ambient temperature of 13.5°C. Other studies by Davis and Johnson (3) exposed nude subjects to a constant cold stress (14°C) for a period of 1 hr each month over a 1-year period. Metabolic heat production and shivering decreased as a consequence of this habituation. Girling (17) similarly demonstrated a decreased metabolism and shivering with no change in mean skin or rectal temperature. There was a time lag in the metabolic response when compared with the minimal environmental temperature.

Yoshimura (16) has also reported seasonal changes in acclimatization to cold in Japanese subjects. Cold acclimatization was accomplished by lowering mean skin temperature and consequently decreasing the heat loss, which, in turn, minimized the metabolic cost of the thermal stress in winter. A second type of adaptation (i.e., an acceleration of metabolic heat production in cold environment) was evident in the work of Scholander et al. (28). Their subjects were able to maintain a warm skin by raising their heat production.

Local, primarily peripheral acclimatization to cold is much better documented than the previously discussed total body acclimatization. Deep-sea fishermen can work with their hands immersed in cold water. Similar exposure of the hands of normal (sic) individuals would result in incapacitation because of numbness and pain. Fishermen apparently have a greater blood flow than do other people in this situation (29, 30). Functional capacity was attained at the expense of additional heat loss. Similar adaptations have been reported for racial groups accustomed to local cold stress [Eskimos (31), Arctic Indians (32), British fish filleters (33), and polar explorers (34)]. This adaptation is probably an acquired characteristic.

Additional studies on local acclimatization have utilized the "hunting" reaction (35). The "hunting" response represents the periodic fluctuation in blood flow to an extremity exposed to low temperatures (usually cold water). Alterations in the intensity of the hunting response have been utilized also to investigate the differences between ethnic groups (36) and groups exposed to cold (37). Meehan (38) compared the hunting reaction of Caucasians, Negroids, and Alaskan natives. Krog et al. (39) studied this reaction in Lapps, Norwegian fishermen, and Norwegian controls. Adams and Smith (40) demonstrated an acceleration of the hunting reaction of the index finger that had been exposed repeatedly to cold for long periods and emphasized the existence of "local cold conditioning."

Most methodology employed for the study of cold tolerance has emphasized the relationship between a particular response variable and the time of cold exposure. This approach has tended to create ambiguous conclusions regarding cold tolerance. The optimal evaluation of cold tolerance should be attainable from a complete understanding of the means for physiological regulation of body temperature. The body has elements that are capable of sensing cold (i.e., thermosensors near or on the skin surface and in the hypothalamus) and an integrating center that receives these neural messages and directs an appropriate thermoregulatory output. More recently, however, experimentation involving both water bath immersion (41) and exposure to a wide range of ambient air environments (42, 43) has shown that the regulation of the metabolic heat production response is a function of both skin and internal temperatures. Brown and Brengelmann (41) used different driving functions for skin temperature to identify both transient and steady-state relationships between skin and internal temperatures in the determination of metabolic heat production. They concluded that the regulation was resultant primarily from the multiplication of thermal signals from the body core and skin.

Nadel et al. (42, 43) required their subjects to ingest different volumes of ice cream and/or hot pudding to change internal temperature by different amounts. They performed these studies in a wide range of ambient environments, thereby investigating the response to a change in internal temperature at many different levels of skin temperature. They extended the Stolwijk and Hardy (44) mathematical model, which described the primary thermal signal as multiplicative in the determination of the response, as follows:

$$\Delta M = 42(36.5 - T_{in})(32.2 - \overline{T}_s) + 8(32.2 - \overline{T}_s) \quad (W/m^2)$$

where ΔM represents a metabolic increase from resting level (W/m^2).

Sympathicoadrenal hormones have been implicated as playing an integral part in homoisotherms' response to cold exposure (45). Release of catecholamines from the sympathicoadrenal medullary system has been associated with both metabolic and cardiovascular responses to cold. Apparently adrenal coricosteroids in association with other hormones act directly on the thermoregulatory centers of the brain. Wilkerson et al. (46) have recently shown that acute cold stesses caused significant alteration in the plasma concentration of the sympathicoadrenal hormones. An ambient temperature of 15°C or less appeared to be necessary to induce increased plasma levels and changes in urinary excretion rates of the catecholamines and cortisol. These data imply that adrenal medullary and adrenal cortical functions are interrelated with peripheral sympathetic nervous system functioning to provide the necessary substrates to allow augmented heat production in response to cold exposure and to reduce peripheral heat loss in these conditions. Earlier Joy (47) studied the responses of cold-acclimatized men to an infusion of norepinephrine. His results indicated that after long-term cold exposure there was a change in sensitivity to norepinephrine, with a decrease in vasopressor response and the development of a calorigenic response. His data suggest that norepinephrine may be a mediator of a nonshivering thermogenesis occurring with cold acclimatization.

The sensitivity of skin receptors is diminished when tissue temperatures drop. Irving (48) has shown that sensitivity was reduced sixfold when skin temperature dropped from 35 to 20°C. Stuart et al. (49) found that muscle spindles show an increased sensitivity at moderately lower muscle temperatures, but when muscle temperatures were 27°C, the activity in response to a standardized stimulus was 50 percent of normal; it was completely abolished when temperatues reached 15 to 20°C. In general numbing and loss of tactile sensitivity occurs at finger temperatures of approximately 8°C, and manual performance may be decreased because of loss of finger and hand dexterity. This probably explains the loss of fine coordinated movements noted to occur in individuals attempting to work in cold environments, as well as their lack of appreciation of the degree of cold experienced.

Mills (50) reported that the loss of tactile discrimination was roughly inversely proportional to the change in mean finger skin temperature. Rewarming of the hands resulted in a return of tactile sensitivity, but the degree of recovery lagged behind the return of skin temperature. Clark and Cohen (51) demonstrated a marked reduction in finger dexterity (to about 75 percent) when hand skin temperature was quickly cooled from 22.2 to 7.3°C, but dexterity was further reduced (to about 50 percent) if cooling time was extended. This finding reflects the more extensive cooling of deeper tissues consequent to the prolonged period of vasoconstriction.

Dexterity of the fingers and grip strength (hand) was markedly diminished by exposure to low ambient temperatures (52). Grip strength decreased some 28 percent, even though cold exposure was of relatively short duration. Subjective impressions of the severity of cold hands were not related to the degree of impairment that occurred, even when hands felt relatively comfortable. Some variability in loss was observed, with some subjects demonstrating a 70 percent decrement in strength. Davies et al. (53) have shown that hypothermia produced by immersion in cold water results in a reduced capacity to perform work. When body temperatures were approximately 35°C, maximal aerobic capacity was reduced and the oxygen cost of submaximal work increased, suggesting a considerable loss in efficiency.

Older individuals appear to exhibit a slightly different response to a cold environment; they allow their deep body temperature to fall and do not increase their metabolic rate to the same degree, in contrast to younger subjects, who raise their oxygen consumption and maintain normal or elevated rectal temperatures (54, 55). Older subjects also appear to be less aware that they are cold. This may indicate that they have carried the normal physiological process of habituation to cold to the stage where it becomes detrimental to their welfare.

It has been occasionally reported that some individuals have an extraordinary sensitivity to cold exposure manifest by an essentially allergic response. This is exhibited by the symptoms of urticaria and syncope. An early review of selected cases was given by Horton et al. (56) and a more recent one by Kelly and Wise (57). The types of reaction vary from localized urticaria at the point of contact to systemic reactions, including headache, flushing, fall in arterial pressure, and even syncope. The latter symptom has been related to the death of some individuals who have been swimming in cold water. Another important facet of cold exposure is related to the aggravation of anginal symptoms in patients with angina pectoris (58). These investigators suggest that the increase in peripheral resistance consequent to cold exposure would result in augmenting myocardial oxygen requirements and thus more readily would provoke an attack of angina.

Much of the concern regarding the combined effects of cold and alcohol on the human organism are based on theoretical considerations of the effects of alcohol on subjects in thermally neutral environments. Anderson et al. (59) reported that moderate doses of alcohol had no deteriorative effect on heat balance during prolonged mild cold exposures. The ambient temperatures were 20 or 15°C, and the naked subjects were studied while sleeping. The typical vasodilating effects of alcohol were not observed in these subjects, probably because the vasoconstrictor drive consequent to cold counteracted the vasodilating drive of the alcohol. It should be noted that hypothermia, followed by death, has been frequently reported in highly intoxicated individuals exposed concurrently to a cold environment.

The potential development of hypothermia in humans exposed to cold is not discussed, although it can be a problem under certain conditions. An extensive literature (60, 61) on this topic is available in the clinical literature, both in the surgical area (where hypothermia has been extensively employed) and in the internal medicine field (where it has been utilized in treatment of renal disorders, burns, etc., and more particularly in the occasional instances where a combination of alcohol ingestion and a cold ambient has resulted in hypothermic states). This problem is one of degree of cold exposure and usually reflects a marked drop in deep body temperature. Physiological actions to combat cold are not effective when deep body temperatures fall below 31 to 29°C. Hypothermia is often observed in older individuals (62), reflecting their inability to maintain normal thermal homeostasis. A recent review (63) has been published that provides the interested reader with additional insight into human ability to work in cold environments.

3 PHYSIOLOGICAL RESPONSES TO HEAT

The individual lives and functions in a physical environment wherein thermal and other stessors influence the function of the body. This physical environment may produce significantly intense strains, resulting in either adaptation or altered health status. Perception of the environment may affect one's performance, efficiency, and adaptative potential as much as the actual environment itself. Nonetheless, the most important factor(s) in response to environmental conditions are related to the basic physiological mechanisms necessary for human adjustments and adaptations to the ambient stressor. The manner and degree to which these physiological mechanisms are involved in response to heat stress are dependent on the degree, duration, and intensity of the exposure. Individuals who work continuously in the heat each day benefit from the acclimatization process. Other persons who are only sporadically exposed will develop less effective acclimatization, but they may tolerate short heat exposures. Individuals who are only occasionally exposed to heat stress may be able to withstand the influence of increased body heat stress, but a limitation to their performance capacity will be determined by the degree to which their body temperature is elevated. In all cases the final determination is made by the effectiveness of thermoregulatory processes and the interplay with the circulatory, sweating, hormonal, and nervous systems.

According to the principle of thermoregulation, body temperature is maintained constant within narrow limits. However, it is probable that the body is never in complete thermal equilibrium. An individual's core temperature exhibits a diurnal rhythm wherein the rectal temperature usually reaches a peak in the late afternoon or early evening and falls to its lowest level during the early morning hours. Other physiological parameters that may be involved in temperature regulation also exhibit 24-hr cycles (i.e., skin temperature, peripheral blood flow, metabolic rate, and heart rate). The exact and precise correlations of these various rhythmic alterations have been only roughly related to each other. Phase shifts can also be modified by altering the daily routine, by long distance flights, or by entraining the core temperature to other environmental conditions.

Environmental heat leads to definite reactions in humans. The increased body heat content leads to alterations in the circulatory, respiratory, sweating, endocrine, and temperature-regulating systems. One's thermoregulatory system serves to maintain equilibrium of the body core within a narrow limit. If this core temperature is to be so precisely maintained, the amount of heat gained by the body must be equal to the amount of heat lost. Ultimately, the amount of heat exchanged between the body and its environment depends on the differences in temperature and of vapor pressure present between the skin and its environment. Calculation of this exchange can be obtained from the heat balance equation: $(M - W) \pm C \pm C_0 \pm R - E = \pm S$, where M = metabolic heat production; W = work performed; $C, C_0, R,$ and E = convective, conductive, radiant, and evaporative heat exchanges, respectively; and S = amount of heat stored in or lost from the tissues. If the body maintains thermal equilibrium, S is zero.

The foregoing physical relationships are affected by physiological mechanisms regulating primarily the cardiovascular, fluid balance, and sweating systems. The physiological strain will be related to the total heat stress (environmental and work loads). Heart rate has been frequently utilized to determine the magnitude of the cardiovascular stress. However, this simple measure provides an inadequate indicator of the stress. The cardiac output for a given work load is similar in hot or thermally neutral environments. However, stroke volume is markedly reduced in hot situations. Furthermore exercise reduces perfusion of internal organs such as the splanchnic and renal beds. This usually reflects the need to provide more blood to active muscles. In hot environments additional blood flow to the skin is needed to assist in thermoregulation.

The other major factor involved in one's ability to work in hot environments is the sweating mechanism. The ability to wet the skin and provide for evaporative cooling determines much of the individual's capability to perform effectively in hot conditions. Sweating also involves consideration of the water and salt balance of the body, as well as the ultimate capacity of the sweat glands to continue to function. It is common for males to sweat at a rate of 1.5 to 2.0 liters/hr. Observations on acclimatized industrial workers show that they can produce at least a liter of sweat per hour over an 8-hr workday. The highest sweat production, 2 liters/30 min, was recorded during a laboratory experiment. In environments with high vapor pressure, the effectiveness of evaporative cooling is impaired. It should be emphasized that clothing worn by individuals must be wetted to enable evaporative cooling to occur. The degree to which clothing of different types can be wetted will influence the capability of the body to attain and maintain thermal equilibrium.

The signs and symptoms that can be expected from water and salt depletion have been documented by Adolph (64) and Marriott (65). Water deficits of approximately 2 percent of body weight lead primarily to symptoms of thirst. However, a water deficit of 6 percent of body weight leads to thirst, oliguria, irritability, and aggressiveness. Water deficits in excess of 6 percent result in marked impairment of physical and mental performance. Salt depletion produces serious symptoms. A deficit of 0.5 g per kilogram of body weight results in lassitude, giddiness, fainting, and mild muscle cramps. A continuous slow decease in body weight should arouse concern that a chronic salt deficit is developing. The safest procedure to prevent water and salt deficits is to provide adequate replacement of water and additional salt at meal times. Most diets contain sufficient excess of salt to meet most needs if water is adequately replaced. In situations where salt may be in deficit, replacement by drinking 0.3 percent saline is preferable to the indiscriminate use of salt tablets.

Repeated exposure to hot environments induces changes in a number of physiological systems, resulting in the development of improved tolerance to the heat load. Heat acclimatization consists of a series of physiological adjustments that occur in individuals exposed to a hot climate (specifically if they are working) allowing them to maintain equilibrium at a higher environmental temperature than was the case before acclimatization. The main features are a less marked increase in the heart rate while working, lower skin and deep body temperatures, a greater production of sweat, increased efficiency of evaporation, and subjectively a lessened sense of discomfort. A more fully (evenly distributed) wetted skin may be the most important factor. Subtle but important aspects of heat acclimatization may require weeks or even months to develop (rather than the few days apparently needed for the previously mentioned effects). These are:

1. Volume of sweat.
2. Threshold for onset of sweating.
3. Blood volume shifts.
4. Retention of acclimatization from 5 to 21 days.
5. Metabolic adaptations.

6. Reacclimatization required if not continuously exposed. If not exposed for 1 week, an additional 1 to 2 days of acclimation is necessary.

Individuals with no recent history of exposure to hot environments show marked differences in their tolerance to a standardized heat exposure. Some individuals cannot develop an adequate level of acclimatization. The process by which this acclimatization to heat occurs has been considered by many investigators, but the mechanisms involved are still not clear. The symptoms most commonly evaluated are the circulatory, sweating, and body fluid control systems. Classical acclimatization responses observed are decreases in rectal temperature, skin temperature, and working heart rates, whereas sweat rates and work time increase and orthostatic tolerance improves (66, 67). Improved heat tolerance is obtained by increased cardiovascular efficiency early in heat exposures (68, 69). Changes in heart rate, stroke output, and rectal temperature occur before sweat rates significantly increase. Maximal venoconstriction and minimized arteriolar dilatation were observed by the third or fourth day of acclimatization (70). The chloride concentration of sweat decreases with continued heat exposure, and the total salt loss by sweating may be decreased with time. Plasma volume increases concomitantly with these early cardiovascular changes, and Senay et al. (71) believe that the preexposure increases in plasma volume observed are responsible for this increased cardiovascular stability. The 23 percent increase in plasma volume could be sufficient to raise right atrial pressure to account for an increased stroke output. However, Rowell et al. (72) postulate that the drop in body temperature leads directly to the decreased heart rate and consequently to an increased stroke volume. These differences of opinion may be attributable to the type of heat stressor employed by these investigators (i.e., wet humid heat vs. dry heat).

It appears that acclimatization to heat occurs in stages. Initially the high body temperatures result in maximum dilatation of the cutaneous blood vessels, which facilitates an expansion of the plasma volume. This, in turn, results in a restabilization of the cardiovascular system with decreased heart rates, larger stroke volumes, and lowering of body temperature. At the same time sweat production increases, but there may not be a more efficient evaporation of available sweat. With continued heat exposure, evaporation rate increases despite no further increase in sweat production. Sweating occurs earlier in the heat exposure as acclimatization proceeds. These processes occur respectively in a time frame of 1 to 4, 6 to 10, and 10+ days of daily consecutive exposure to the hot environment.

Mere exposure to heat without work confers little acclimatization. Acclimatization to heat is the dramatic improvement in ability to work in a hot environment. It is possible to accelerate this acclimatization process. Acclimatization to 1 hr of work at 48.9/33.9°C (dry-bulb/wet-bulb ambient temperatures) enabled men to do 4 hr of work at 48.9/32.2 or 48.9/31.1°C (73). Thus acclimatization to work at a higher environmental load ensures full acclimatization to a similar work load for 4 hr at any lower heat stress. Robinson et al.

(74) restudied men (44 to 60 years of age) who had been evaluated 21 years earlier for their responses to heat exposure. These older individuals exhibited the same degree of strain to their first exposure to heat and acclimatized as well as they had done when they were younger *but* required a longer period of exposure.

Acclimatization to work in a hot environment is not retained indefinitely. There can be a rapid loss over so short a period as a weekend, with recovery to the prior level of acclimatization by the second day of heat exposure. One week without heat exposure may require 4 days of exposure to reattain acclimatization. After 1 week of no heat stress there was a 50 percent loss in terms of the heart rate and sweat production responses. It appears that acclimatization is almost completely lost after 3 to 4 weeks in a cool environment and that a full reacclimatization program is required to return the individual to full acclimatization.

Various factors may modify the processes of acclimatization. Individuals who are engaged in hard physical work that induces a rise in deep body temperature are apparently partly heat acclimatized. They require a smaller number of repeated heat exposures to develop full acclimatization. Any situation that leads to dehydration of an individual will result in significant losses of acclimatization. Convalescence from debilitating illness, alcoholic hangovers, loss of sleep, and so on, have been shown to result in such losses. Strydom et al. (75, 76) have suggested that ascorbic acid supplementation has a significant and beneficial influence on body temperature response during heat stress, and additionally it also enhanced the rate of heat acclimatization. Simply exposing resting individuals to hot environments results in a rapid heart rate and cutaneous vasodilatation. When individuals are preheated by resting in such an environment prior to engaging in work leading to exhaustion in a hot environment (heart rate of about 200), their walk time decreases linearly with the increased starting heart rate. Similarly, if rectal temperature is elevated prior to engaging in exhausting work, the time to exhaustion is linearly related to the initial temperature.

Sex differences in response to thermal stress are thought to exist, but there are many questions about the precise nature of these differences. The number of studies that have investigated heat strain on females (i.e., their physiological reactions and adaptive capabilities) are few in number, and their results are inconclusive. Kawahata (77) has reported that females have a greater number of sweat glands than males but that the single male gland has greater activity. The onset of sweating to reflex heat stimulation or general heat stress is slower in young females than in young males. Haslag and Hertzman (78) reported that when resting males and females were exposed to progressively increasing ambient temperatuers, 25°C for 1 hr and then an increase of 6.6°C/hr the next 3 hr, there was no significant difference in regional sweat rates or the level of body temperature at the time of onset of sweating.

Hardy and DuBois (79) found that men sweat more than women and begin sweating at a lower environmental temperature; men at 29°C and women at 32°C. These investigators also reported that in a warm environment there was a fall in heat production in women but not in men. This observation was not confirmed by Cleland et al. (80), who also discussed the various reasons for a similar confusion in the results obtained by various investigators in studies on young men. The working metabolism of women has been said to increase relatively more than men in the heat (81), but others (82) have not found this to be the case. Various investigators (80, 84) have examined the response of men and women to a single bout of exercise in the heat (variable degrees of heat strain) and have found lower sweat rates, higher rectal temperatures, and higher heart rates in females. Some of these observations suggest that women have a reduced tolerance to heat stress.

Young adult women apparently follow the same pattern of acclimatization to heat that has been observed for young males. Lowered body core temperatures and lowered heart rates were observed during work in women during acclimatization. However, the magnitude of these decreases varied considerably in the various studies (82, 85–88). A similar discrepancy was noted for total body sweating. Different investigators have reported levels of sweating to be similar to, less than, or equivalent to those observed in male subjects. The small total number of subjects studied and the widely different ambient hot environments (from hot–dry to hot–wet) might explain these discrepancies, but it appears obvious that certain striking physiological differences need to be explained. Wyndham et al. (88) have reported some rather bizarre behavior of women during heat exposures. Whether these responses were related to the stage of the menstrual cycle, where somewhat similar psychological behavior patterns have been reported, remains to be determined.

Age has a very definite effect on the responses of women to heat, as Cleland et al. (80) demonstrated in a study of young (mean age 21.3 years) and elderly (mean age 67.6 years) women doing light work at 35°C T_a and 28°C T_{wb}. Cardiorespiratory differences were similar to those that had been noted previously at normal ambient temperatures: a higher systolic, diastolic, and pulse pressure in the elderly, as well as higher values for ventilatory efficiency and the respiratory exchange ratio. The most striking differences were in the temperature responses. The elderly women had a higher core temperature and a lower mean skin temperature than the young girls. At what point in life these differences become apparent and why they occur is not known.

Kawahata (77) studied the relationship between menstruation and the ability to perspire. Since he had earlier found that testosterone was sudorific and estradiol inhibitory, he thought that it was highly probable that the menstrual cycle affected the ability to perspire in females. He found rather large changes in the latent period for thermal sweating measured at different times during the menstrual cycle. The latent period was much shorter near the time of ovulation. These observations were not confirmed by Sargent and Weinman (89) since they failed to find any significant differences in the activity of the eccrine sweat gland during the menstrual cycle. The rate of sweating and the concentrations of sodium, potassium, and chloride in sweat were studied in natural desert environments (90) utilizing girls 9 to 18 years of age and two young women (in addition to boys and men), and it was shown that the rate of sweating depended on body surface, metabolic rate, and ambient temperature, not on sex or age. The sweating response apparently is directly related to the absolute metabolic rate (91).

The role of aerobic power has been generally ignored when studies on heat tolerance of females were conducted. Most investigators assigned the same work task to both men and women. Since core temperature and heart rate, two commonly used indicators of heat strain, vary in proportion to relative work load, it was not surprising that women generally appeared to be under greater stress. Even when both sexes work at the same relative work load, individuals with a high degree of cardiovascular fitness perform more adequately. The fit individual is better able to cope with acute exposure to work in the heat since the cardiovascular system can meet the competing demands for muscular and peripheral blood flow while maintaining an adequate venous return to the heart.

Highly conditioned female athletes (92) have an earlier onset of sweating during heat exposures. This response results in lower body temperatures, thus delaying the increase in cutaneous venous capacitance and diverting less blood from the central to the peripheral circulation. Senay (93) indicated that females do not hemodilate (i.e., expand their plasma volume) to the same extent as males, ascribing this to inherent differences in skin surface to blood volume ratios and to the inability of females to maintain their vascular volume during heat exposure. Nonetheless, the capacity of females to respond to a heat stress at percentages of their aerobic capacity equivalent to that of males appears to be very similar. However since females generally have a maximum aerobic capacity 20 to 25 percent less than males, when a work load requiring a fixed caloric expenditure must be performed, females may be at a disadvantage. That is, females may have to utilize a greater percentage of their capacity to accomplish this fixed task.

There is a need in industry for some general indication of the comparative stresses of different warm environments. Heat stress indices have been developed by a number of investigators in an attempt to reduce the thermal components of an environment to a single figure reflecting the heat stress on the individual. Combining the physiological characteristics of humans with the physical factors of the environment is not a simple task. None of the proposed heat stress indices meets all the requirements. A comprehensive analysis of the most popular indices has been made by Kerslake (94), and those interested should refer to this source. Any hope of an index simple enough for everyday use awaits future research.

Various types of inadequate response to heat stress have been described in the medical literature. These have been categorized by the World Health Organization (Table 11.1). No description of the various symptoms associated with these disorders is given here; a complete and precise description has been presented by Minard (95). A more recent presentation has appeared by Dinman and Horvath (97). It should be noted that heat deaths are a common occurrence

Table 11.1 Classification of Heat Disorders

ICD[a] Code	Clinical Designation	Etiologic Category
992.0	Heat stroke	Thermoregulatory failure
—	Heat hyperpyrexia	
992.1	Heat syncope	Orthostatic hypotension
992.2	Heat cramps	Salt and water imbalance
992.3	Heat exhaustion, water depletion	
992.4	Heat exhaustion, salt depletion	
992.5	Heat exhaustion, unspecified	
992.6	Heat fatigue, transient	Behavioral disorder
—	Heat fatigue, chronic	
705.1	Heat rash	Skin disorders and sweat gland injury
—	Anhydrotic heat exhaustion	

[a] International Classification of Diseases.

when a sudden natural heat wave develops. The individuals most susceptible to heat morbidity and mortality are the young and the aged. These are the individuals who have the most difficulty in attaining an adequate and rapid acclimatization to heat. Ellis et al. (96) suggested that the inability to sweat sufficiently and to benefit from evaporative cooling may be a factor in the high mortality of the very young and the elderly.

Heat stress remains as a potent stressor to the human body, not only in the natural environment but in the artificial environment presented by certain industrial situations. Successful resistance to this stressor depends on the development and utilization of adequate physiological mechanisms.

REFERENCES

1. S. M. Horvath, G. B. Spurr, B. K. Hutt, and L. H. Hamilton, *J. Appl. Physiol.*, **8**, 595 (1956).
2. A Hemingway, *Physiol. Rev.*, **43**, 397 (1963).
3. T. R. A. Davis and D. R. Johnson, *J. Appl. Physiol.*, **16**, 231 (1961).
4. K. Rodahl, S. M. Horvath, N. C. Birkhead, and B. Issekutz, Jr., *J. Appl. Physiol.*, **17**, 763 (1962).
5. J. D. Hardy, in: *Physiology of Heat Regulation and the Science of Clothing*, L. H. Newburgh, Ed., Saunders, Philadelphia, 1948.
6. A. C. Burton and O. G. Edholm., *Man in a Cold Environment*, Edward Arnold Ltd., London, 1955.
7. A. C. Burton, P. A. Snyder, and W. G. Leach, *J. Appl. Physiol.*, **8**, 269 (1955).
8. P. F. Iampietro and E. R. Buskirk, *J. Appl. Physiol.*, **15**, 212 (1960).
9. E. R. Nadel, J. W. Mitchell, and J. A. J. Stolwijk, *Int. J. Biometeorol.*, **15**, 201 (1971).
10. B. Slatin, A. P. Gagge, and J. A. J. Stolwijk, *J. Appl. Physiol.*, **28**, 318 (1970).
11. T. Adams and E. J. Heberling, *J. Appl. Physiol.*, **13**, 226 (1958).

12. B. Saltin, in: *The Physiology of Work in Cold and Altitude,* C. Helfferich, Ed., Arctic Aeromedical Laboratory, Fort Wainwright, Alaska, 1966.
13. B. S. Kang, D. S. Han, K. S. Paik, Y. S. Park, J. K. Kim, D. W. Rennie, and S. K. Hong, *J. Appl. Physiol.,* **29,** 6 (1970).
14. S. K. Hong, in: *Physiology of Breath-hold Diving and the Ama of Japan,* H. Rahn, Ed., National Academy of Sciences-National Research Council, Washington, DC, 1965.
15. H. T. Hammel, in: *Adaptation to the Environment, Handbook of Physiology,* Section 4, D. B. Dill, Ed., American Physiological Society, Washington, DC, 1964.
16. H. Yoshimura, in: *Essential Problems in Climatic Physiology,* H. Yoshimura, K. Ogata, and S. Itoh, Eds., Nankoda Publishing, Kyoto, 1960.
17. F. Girling, *Can. J. Physiol. Pharmacol.,* **45,** 13 (1967).
18. S. M. Horvath, in: *Advances in Climatic Physiology,* S. Ito, K. Ogata, and H. Yoshimura, Eds., Igaku Shoin Ltd., Tokyo, 1972.
19. H. T. Hammel, R. W. Elsner, D. H. LeMessurier, H. T. Anderson, and F. A. Milan, *J. Appl. Physiol.,* **14,** 605 (1959).
20. P. B. Raven and S. M. Horvath, *Int. J. Biometeorol.,* **14,** 309 (1970).
21. C. H. Wyndham et al., *J. Appl. Physiol.,* **19,** 583 (1964).
22. M. Yoshimura and H. Yoshimura, *Int. J. Biometeorol.,* **13,** 163 (1969).
23. J. F. O'Hanlon, Jr. and S. M. Horvath, *Can. J. Physiol. Pharmacol.,* **48,** 1 (1970).
24. P. B. Raven, I. Niki, T. E. Dahms, and S. M. Horvath, *J. Appl. Physiol.,* **29,** 417 (1970).
25. S. M. Horvath, A. Freedman, and H. Golden, *Am. J. Physiol.,* **150,** 99 (1947).
26. L. D. Carlson, H. L. Burns, T. H. Holmes, and P. P. Webb, *J. Appl. Physiol.,* **5,** 672 (1953).
27. T. R. A. Davis, *J. Appl. Physiol.,* **17,** 751 (1962).
28. P. F. Scholander, H. T. Hammel, J. S. Hart, D. H. LeMessurier, and J. Steen, *J. Appl. Physiol.,* **13,** 211 (1958).
29. J. Leblanc, J. A. Hildes, and O. Heroux, *J. Appl. Physiol.,* **15,** 1031 (1960).
30. J. Leblanc, *J. Appl. Physiol.,* **17,** 950 (1962).
31. G. M. Brown and J. Page, *J. Appl. Physiol.,* **15,** 662 (1960).
32. R. W. Elsner, J. D. Nelms, and L. Irving, *J. Appl. Physiol.,* **15,** 662 (1960).
33. J. D. Nelms and J. G. Saper, *J. Appl. Physiol.,* **17,** 444 (1962).
34. J. F. G. Hampton, *Fed. Proc.,* **28,** 1129 (1969).
35. T. Lewis, *Heart,* **15,** 177 (1930).
36. K. Hirai, S. M. Horvath, and V. Weinstein, *Angiology,* **21,** 502 (1970).
37. H. Yoshimura and T. Iida, *Jpn. J. Physiol.,* **1,** 147 (1950).
38. J. P. Meehan, *Mil. Med.,* **116,** 330 (1955).
39. J. Krog, B. Fokon, R. H. Fox, and K. L. Anderson, *J. Appl. Physiol.,* **15,** 654 (1960).
40. T. Adams and R. E. Smith, *J. Appl. Physiol.,* **17,** 317 (1962).
41. A. C. Brown and G. C. Brengelmann, in: *Physiological and Behavioral Temperature Regulation,* J. D. Hardy, A. P. Gagge, and J. A. J. Stolwijk, Eds., Thomas, Springfield, IL, 1970.
42. E. R. Nadel and S. M. Horvath, *J. Appl. Physiol.,* **27,** 484 (1969).
43. E. R. Nadel, S. M. Horvath, C. A. Dawson, and A. Tucker, *J. Appl Physiol.,* **29,** 603 (1970).
44. J. A. J. Stolwijk and J. D. Hardy, *Pflugers Arch. Ges. Physiol.,* **291,** 129 (1966).
45. E. L. Arnett and D. T. Watts, *J. Appl. Physiol.,* **15,** 499 (1960).
46. J. T. Wilkerson, P. B. Raven, N. W. Bolduan, and S. M. Horvath, *J. Appl. Physiol.,* **36,** 183 (1974).
47. R. J. T. Joy, *J. Appl. Physiol.,* **18,** 1209 (1963).
48. L. Irving, *Sci. Am.,* **214,** 94 (1966).

49. D. G. Stuart, E. Eldred, A. Hemingway, and Y. Kawamura, in: *Temperature: Its Measurement and Control in Science and Industry*, Vol. 3, J. D. Hardy, Ed., Reinhold, New York, 1963, p. 545.
50. A. W. Mills, *J. Appl. Physiol.*, **9**, 447 (1956).
51. R. E. Clark and A. Cohen, *J. Appl. Physiol.*, **15**, 496 (1960).
52. S. M. Horvath and A. Freedman, *J. Aviat. Med.*, **18**, 158 (1947).
53. M. Davies, B. Ekblom, U. Bergh, and I. L. Kanstrup-Jensen, *Acta Physiol. Scand.*, **95**, 201 (1975).
54. S. M. Horvath, C. E. Radcliffe, B. K. Hutt, and G. B. Spurr, *J. Appl. Physiol.*, **8**, 145 (1955).
55. A. J. Watts, *Environ. Res.*, **5**, 119 (1972).
56. B. T. Horton, G. E. Brown, and G. M. Roth, *JAMA*, **107**, 1265 (1936).
57. F. J. Kelly and R. A. Wise, *Am. J. Med.*, **15**, 431 (1953).
58. S. E. Epstein, M. Stampfer, G. D. Beiser, R. E. Goldstein, and E. Braunwald, *New Engl. J. Med.*, **280**, 7 (1969).
59. K. L. Anderson, B. Hillstrom, and F. V. Lorentzen, *J. Appl. Physiol.*, **18**, 975 (1963).
60. J. H. Talbott, *New Engl. J. Med.*, **224**, 281 (1941).
61. *Cold Injury*. Transactions of the First to Sixth Conferences, Josiah Macy Foundation, New York, 1952–1958.
62. S. M. Horvath and R. D. Rochelle, *Environ. Health Perspect.*, **20**, 127 (1977).
63. S. M. Horvath, *Exercise Sport Sci. Rev.*, **9**, 221, 1981.
64. E. F. Adolph, *Physiology of Man in the Desert*, Wiley-Interscience, New York, 1947.
65. H. L. Marriott, *Water and Salt Depletion*, Thomas, Springfield, IL, 1950.
66. S. Robinson, E. S. Turrell, H. S. Belding, and S. M. Horvath, *Am. J. Physiol.*, **140**, 168 (1943).
67. N. Nelson, L. W. Eichna, S. M. Horvath, W. Shelley, and T. F. Hatch, *Am. J. Physiol.*, **151**, 626 (1947).
68. H. L. Taylor, A. F. Henschel, and A. Keys, *Am. J. Physiol.* **139**, 583 (1943).
69. C. H. Wyndham, G. G. Rogers, L. C. Senay, and D. Mitchell, *J. Appl. Physiol.*, **40**, 779 (1976).
70. J. E. Wood and D. E. Bass, *J. Clin. Invest.*, **39**, 825 (1960).
71. L. C. Senay, D. Mitchell, and C. H. Wyndham, *J. Appl. Physiol.*, **40**, 786 (1976).
72. L. B. Rowell, K. K. Kraning, J. W. Kennedy, and T. D. Evans, *J. Appl. Physiol.*, **22**, 509 (1967).
73. S. M. Horvath and W. B. Shelley, *Am. J. Physiol.*, **146**, 336 (1946).
74. S. Robinson, H. S. Belding, F. C. Consolazio, S. M. Horvath, and E. S. Turrell, *J. Appl. Physiol.*, **20**, 583 (1965).
75. N. B. Strydom, H. F. Kotze, W. H. van der Walt, and G. G. Rogers, *J. Appl. Physiol.*, **41**, 202 (1976).
76. H. F. Kotze, W. H. van der Walt, G. G. Rogers, and N. B. Strydom, *J. Appl. Physiol.*, **42**, 711 (1977).
77. A. Kawahata, in: *Essential Problems in Climatic Physiology*, H. Yoshimura, K. Ogata, and S. Itoh, Eds., Nankodo Publishing, Kyoto, 1960.
78. W. M. Haslag and A. R. Hertzman, *J. Appl. Physiol.*, **20**, 1283 (1965).
79. J. D. Hardy and E. F. DuBois, *Proc. Natl. Acad. Sci. (USA)*, **26**, 389 (1940).
80. T. S. Cleland, J. C. Bachman, and S. M. Horvath, *J. Gerontol.*, (in press).
81. T. Morimoto, Z. Slabochova, R. K. Naman, and F. Sargent, II. *J. Appl. Physiol.*, **22**, 526 (1967).
82. T. S. Cleland, S. M. Horvath, and M. Phillips, *Int. Z. Angew. Physiol.*, **27**, 15 (1969).
83. L. Brouha, P. E. Smith, Jr., R. DeLanne, and M. E. Maxfield, *J. Appl. Physiol.*, **16**, 133 (1960).
84. R. H. Fox, B. E. Lofstedt, P. M. Woodward, E. Eriksson, and B. Werkstrom, *J. Appl. Physiol.*, **26**, 444 (1969).
85. O. Bar-Or, H. M. Lundegren, and E. R. Buskirk, *J. Appl. Physiol.*, **26**, 403 (1969).

86. B. A. Hertig, H. S. Belding, K. K. Kraning, D. L. Batterton, C. R. Smith, and F. Sargent, II, *J. Appl. Physiol.*, **18,** 383 (1963).
87. K. P. Weinman, Z. Slabochova, E. M. Bernauer, T. Morimoto, and F. Sargent, II, *J. Appl. Physiol.*, **22,** 533 (1967).
88. C. H. Wyndham, J. F. Morrison, and C. G. Williams, *J. Appl. Physiol.*, **20,** 357 (1965).
89. F. Sargent, II and K. P. Weinman, *J. Appl. Physiol.*, **21,** 1685 (1966).
90. D. B. Dill, S. M. Horvath, W. Van Beaumont, G. Gehlsen, and K. Burrus. *J. Appl. Physiol.*, **23,** 746 (1967).
91. B. L. Drinkwater, I. C. Kupprat, T. S. Talag, and S. M. Horvath, *J. Appl. Physiol.*, **41,** 815 (1976).
92. B. D. Drinkwater, J. E. Denton, I. C. Kupprat, T. S. Talag, and S. M. Horvath, *Ann. NY Acad Sci.*, **301,** 777 (1977).
93. J. C. Senay, Jr., *J. Physiol. (London)*, **2325,** 209 (1972).
94. D. McK. Kerslake, *The Stress of Hot Environments*, University Press, Cambridge, England, 1972, p. 317.
95. D. Minard, "Heat Disorders: A Tabular Presentation," in: *Standards for Occupational Exposures to Hot Environments—Proceedings of a Symposium*, S. M. Horvath, Ed.-in-Chief, and R. C. Jensen, Ed., National Institute for Occupational Safety and Health, Cincinnati, Ohio, 1976, pp. 21–25.
96. F. P. Ellis, A. N. Exton-Smith, K. G. Foster, and J. F. Weiner, *Isr. J. Med. Sci.*, **12,** 815 (1976).
97. B. D. Dinman and S. M. Horvath, *J. Occup. Med.*, **26,** 489 (1984).

CHAPTER TWELVE

Ionizing Radiation

ROBERT G. THOMAS, Ph.D.

1 INTRODUCTION

Radiation has been known to be a toxic agent for a relatively short time (approximately 85 years). It has received considerable attention in recent decades, however, primarily as a result of military applications during World War II. Compared to chemically toxic agents, the properties of ionizing radiation generally enable its detection at much lower levels and with much greater accuracy. This condition has contributed greatly to the broadly held awareness of the toxic potential of radiation, but in many (or perhaps most) cases this is mainly because its presence can be detected at very low levels. It appears that much of the concern over health hazards due to radiation is the direct result of this.

There are two distinct possibilities for human radiation exposure: irradiation from sources external to the body and irradiation from sources deposited internally that have entered the body by various routes, such as inhalation, ingestion, and on or through the skin. This chapter devotes considerably more emphasis to the internal source of potential radiation health problems, primarily because accumulation through the internal routes is often less controllable. Internally acquired radionuclides also represent a more classic toxic poisoning, as we have become to know it. Also, measurement of radiation from external sources is much more reliably related to body exposure than is, for instance, the reconstituted air concentration of a radionuclide to the calculated long-term dosage that may be received by an internal organ. There are many difficulties associated with the latter estimates—determining such factors as the particle size breathed, the amount deposited in various sections of the respiratory tract, solubility in body fluids, elemental position in the periodic table—and naturally, the differential metabolism of foreign materials by exposed individuals. The latter factor may involve such basic parameters as sex differences, age, and body and organ weight. Because the degree of injury is probably related to the concentration of a radioactive material in an organ, it is easy to see how organ weight could become such an important factor.

1.1 Types of Radiation Exposure

1.1.1 External Radiation

External radiation, in its less exotic forms, is generally considered to be either from an electromagnetic (wavelike) or a neutron source. There are other sources of external ionizing radiation, but these rarely lead to industrial exposure of the worker. With electromagnetic radiation (gamma-rays, X-rays), ionization in a medium may be caused by one of three primary processes: the photoelectric effect, Compton scattering, and pair production. These three types of ionization are illustrated in elementary textbooks on nuclear physics.

The photoelectric effect generally prevails when tissues are affected at photon energies of less than 100 keV. In this process an electron is ejected from an atom in the medium, and it subsequently becomes the major source of further ionization within the medium. Ionization by electromagnetic radiation is somewhat "inefficient"; therefore, pathways may be very long compared to the length of ionization path of the secondary electron released. Hence ionization events may not occur in close proximity to each other, and effects on tissues may be diffuse. As the secondary electron traverses the tissue, it promulgates additional ionizing electrons known as *delta rays*. These are discussed in the paragraphs that follow. In the photoelectric ionization process all the incident energy is imparted to the ejected electron, which explains the effectiveness of this process at the lower energies.

Between the low ($\sim$100 keV) and very high ($\sim$10 MeV) energy regions, the ionizing process known as Compton scattering prevails. Under these conditions the incident photon is partially degraded by interacting with an orbital electron in an atom of the medium being traversed. The energy remaining in the photon carries it further through the medium, to continue interaction with other atoms in various molecules. The ejected electron (delta ray) continues with its imparted energy (the fraction obtained from the incident photon) to ionize other atoms near to the original event. This "clustering" of events in the immediate vicinity of the first ionization is due to the relatively heavy ionizing density of the particulate radiation (ejected electron), described in Section 1.1.2. Because the principal range of energies for Compton scattering spans most of the common external radiation sources, this ionizing process is most prevalent in radiobiological effects.

Pair production is confined to high-energy photons ($>$1.2 MeV). This event has little place in radiobiological reactions because of the energies involved and the atomic structure required. In this process the photon energy, in the proximity of a highly charged nucleus, is transformed into mass because of the strong nuclear electromagnetic forces, thus forming one negatively charged electron mass and one that is positively charged. Each of these newly formed particles is capable of ionizing atoms to an extent that is dependent on their energies, as described.

Neutrons, uncharged particles in the simplest sense, damage the traversed medium in a different manner. The neutral nature of neutrons allows them to have large penetrating distances. Because they are neutral, they generally penetrate the outer electron clouds of atoms and, in essence, strike the nucleus. This is due to the relative size of the target. If the neutron is of sufficiently high energy (fast), the nucleus is dislodged and actually is driven some distance through the surrounding tissue, causing damage as it goes. The nucleus ultimately picks up enough electrons to once again become its original "self" as it comes to rest. The elements of greatest abundance in soft body tissues are carbon, hydrogen, oxygen, and nitrogen. The abundance of hydrogen nuclei make this element the most vulnerable target for fast neutron interaction; therefore, hydrogen will have a much greater absorbed proportion of the incident neutron than will the less abundant larger atoms. Slower (low energy) neutrons may be captured on encounter with a nucleus. This can result in an unstable condition in which the nucleus will release energy in many forms, thus becoming a primary source of ionizing energy. This generally limits the radiobiological action to the proximity of the initial event.

Protons and other charged particles are also a source of external radiation and their ionization is similar to alpha and beta particles, as described in Section 1.1.2.

1.1.2 Internal Radiation

Internal radiation sources, as far as industrial exposure is concerned, are generally comprised of alpha or beta particle emitters. Beta particles, electrons by charge and weight, may interact with orbital electrons of atoms within molecules or with the nucleus. When the beta particle approaches an orbital electron, energy is imparted, the orbital electron may be ejected as a delta ray, and the incident beta particle will continue at a lowered energy and with an altered direction. This procession by the incident electron will continue until the electron dies by capture somewhere in the medium (tissue). The delta rays (secondary electrons) will traverse the surrounding area and operate in a manner similar to that of the original primary incident electron. Because of the charge and the mass, compared to a photon (Section 1.1.1), the ionizing events associated with electrons are clustered and occur within a very small volume of tissue.

Alpha particles, comprised of helium nuclei of 2+ charge and 4 mass units, are much more heavily ionizing than are electrons (beta particles). Thus their paths are straighter because there is less deflection upon energy loss, and their ionizing events are confined to much smaller volumes of tissue. A 5-MeV alpha particle in soft tissue has a linear path length of approximately 40 µm. Since beta particles from the nucleus of radioactive atoms have a spectrum of energies, their path length is variable. However, the range of the emitted beta particle is many times that of an alpha particle of the same incident energy. Emitted alpha particles are released with discrete energy(ies).

1.2 Radiation Measurement

1.2.1 Electromagnetic Radiation

Electromagnetic radiation represents the easiest form to measure where radiobiological exposures are concerned. It is usually uniform within the location of possible worker exposure, and in most cases one could probably argue that the energy characteristics from a relatively stable source, whether spectral or quite narrow, would fall within a consistent range. This may be speculated because of the more common nature of the source of such "leakage" radiation. If the electromagnetic radiation is from a source of gamma emission, such as from ^{60}Co, the energy distribution is consistent, making interpretation of any measurements, even of a gross nature, much easier and probably more accurate.

Ionization in air has been accepted as the primary measuring standard on which all other (secondary) assessments are based. The amount of ionization in a specific volume of air under precise conditions of temperature and pressure serves to determine this primary standard. When electronic equilibrium is achieved (secondary electron influx into a specific volume of air is exactly equal to secondary electron efflux) within a given measuring device, the amount of current generated by the ionization is proportional to the amount of incoming energy of the photons. This incident energy is directly proportional to the defined units used in radiation studies. The roentgen, which is a measure of this ionization under primary standard conditions, is defined as the quantity of X or gamma radiation such that the associated corpuscular emission per 0.001293 g of air produces, in air, under standard conditions of temperature and pressure, ions carrying one electrostatic unit of quantity of electricity of either sign. With this primary basis for measuring incident electromagnetic radiation, it is not difficult to conceive that many instruments have been devised to measure ionization in media that may be related to the primary standard.

In recent years crystalline and solid-state detectors have been most successful for measuring external radiation dosage. The former will yield spectral energy data if sophistication is desired but may be used as constantly recording, full-energy monitors without this sophistication. Electronic windows may be applied to the detection device to limit detection to a rather specific range of energies. The solid-state variety works extremely well in sharp spectral resolution, but the need for such equipment in routine monitoring for the worker is often questionable. This is not to imply that solid state detectors are not widely used, particularly in personnel monitoring.

1.2.2 Neutrons

Neutrons are generally separated by energy into thermal, intermediate, and fast, as indicated in Section 1.1.1. The classical scheme of energies associated with these is less than 0.5 eV, 0.5 eV to 10 keV, and 10 keV to 10 MeV, respectively. Thermal neutrons are primarily degraded in matter by capture, and the capture cross section is inversely proportional to the velocity of the neutron. For instance, when hydrogen ($_1H^1$) captures a neutron ($_0n^1$) the result is a $_1H^2$ atom, plus a gamma ray as excess energy. The intermediate neutrons fall into a range in which there are resonant peaks in the capture cross section, leading to a slowing down of the neutron in matter. This process is generally inversely proportional to the energy of the particle. Fast neutrons interact by scattering and may be elastic or inelastic, depending on whether the energies are toward the low or high end of the range, respectively. As mentioned in Section 1.1.1, the most important interaction in tissue is with hydrogen, each particle having essentially the same mass. Relativistic neutrons, another category, are not really of importance to this chapter.

Methods of detection and measurement of neutron fluxes have been based on neutron properties, as noted earlier. Calorimetry, the measurement of temperature rise due to ionization in a medium, is always very accurate, but it is difficult to measure the small changes in temperature achieved. Ionization depends on many factors, and the medium being ionized within the instrument should have very precise and determinately known qualities. The size (dimension) and structure of the ionization chamber cavity are very sensitive parameters and must conform to criteria that are related to the energy of the incoming neutron. For instance, an instrument designed to work reasonably well with thermal neutrons will not perform for fast neutrons. As with electromagnetic radiation, chemical means of detection and measurement are choices that give extreme accuracy. The problem in either solid (e.g., photographic) or liquid media is that the dose–response curve is not linear, and impurities play an important role in the resulting estimation. It is difficult to obtain a liquid chemical detection system that is pure enough to give no spurious neutron interactions. Perhaps one of the most reliable methods of detecting and measuring neutron radiation is through its secondary gamma rays following an event. Depending on the energy of the neutron and hence the reaction involved, secondary gamma rays will undoubtedly be emitted. Accurate determination of time of irradiation and magnitude of the gamma rays produced in the reaction of neutrons with the absorbing medium allows calculation back to the incident flux.

All these general principles are operable in their own right. Since some require equipment that is not suitable for field work, however, bulkiness becomes an important drawback with most. In recent years the solid-state detection systems have become popular and probably will ultimately replace all other means for detection on personnel. These include the thermoluminescent detectors used routinely in many laboratories for personnel monitoring. They have many advantages and are still under development for more precise detection characteristics.

1.2.3 Particles Other than Neutrons

Assessment of the impact from particle radiation is perhaps the least gratifying of all measurements that can be made in the case of accidental exposure to the worker, although recent advances are encouraging. The use of standard filter sampling at constant volume through small pore filters is simple and relatively inexpensive. The filter may be constantly monitored, the buildup of activity recorded, and when a certain predetermined radiation level is reached, the result is announced by an alarm. This represents a steady, consistent method of detection of leakage of radioactive materials from a given operation. This type of sampling is usable for both alpha and beta emitters. The difficulty inherent in this method of monitoring arises when an inhalation accident occurs.

Until recently little time and effort have been devoted to methods of determining solubility and particle size of the sample that is collected and is presumed to be representative of the atmosphere breathed by the worker. In the case of external radiation one has some reasonable chance of mocking up the exposure conditions, as was attempted in the Lockport exposures (1), and to closely estimate the worker dose. One seldom has the opportunity to do this when accidental inhalation exposures to radioactive aerosols are involved: this has been the case with most accidental exposures to plutonium in the atomic energy industry (2, 3).

The routine use of continuous air monitors that will allow determination of particle size must be a compromise between sophistication and economics. Ideally, five- or seven-stage cascade impactors of the Anderson (4) or Mercer (5) variety could be used in a constant monitoring mode so that if an accident occurred, the exposure atmosphere could be carefully determined retrospectively with regard to particle size distribution. However, the constant problem of sophisticated samplers becoming clogged with room dust makes routine sampling sometimes impractical. More important with the use of this type of sampler, coupled with the particle size data, is the ability to do solubility measurements on each size fraction. The combination of these two determinations would enable a reasonably accurate estimation of the quantity deposited in the worker's lower respiratory tract and the extent to which the deposited particles would be soluble in body fluids. The type of therapy to be applied after an accident could be much more judiciously determined if this kind of information were available at an early time postexposure. If the deposited particles were completely insoluble, one would consider lung lavage (6); but if they were soluble and entering the blood, chelation therapy (see Section 4.2.1) would become an option.

Perhaps the most practical air monitors are those that attempt a relatively crude separation of "respirable" and "nonrespirable" sizes (7–11). Solubility studies from such samplers can be made, and the results perhaps are equally satisfying as those obtained by use of the more complicated particle sizing instruments, given the spectrum of errors inherent in the measurements.

For certain operations a more specialized system has been devised that enables the detection of various daughter products, by types of emission or energies involved, as in the radium series (12). Collectors may be continuously monitored for various energies and types of emissions so that ratios of two different daughters can be determined, or a total sample can be captured and total daughter activity at a given time determined. Many types of sampler for the uranium mining industry have been developed along these lines with various methods of analysis (13–15). In all cases any counts above "background" may be instrumented to be indicative of an accidental release or an abnormal working environment. These systems have a clear application in the uranium mining industry.

1.3 Dosimetry

1.3.1 Electromagnetic Radiation

As implied, dosimetry with electromagnetic radiation in the human exposure case may be more straightforward than for exposure to airborne radionuclide particles or vapors, primarily because of the ability to mock up exposure conditions. Historically, the determination of radiation dose has proceeded through a series of changes since the discovery of X-rays by Roentgen in 1895. For quite some time the erythema dose was used to determine proper treatment in radiation therapy. In the 1920s it was recognized that ionization in air was the best approach to estimating dosage for therapeutic purposes. The roentgen was defined in 1928, and this ultimately led to widespread usage of the R unit, followed by the rep, the rem, and finally, in current use today, the rad. The rad is a measure of the energy deposited in any medium, and one rad is equivalent to the deposition of 100 ergs per gram of that medium.

Various kinds of instrumentation have been used to define dosimetrically the energy arising from a gamma- or X-ray source. These include air ionization chambers, semiconductors, thermoluminescent devices, and other instruments that use heat, light, and chemical changes as means to quantitate the ionizing events being produced. The most practical and common personnel dosimeters in use are the film badge (optical) and the thermoluminescent dosimeter. In any event, although no dosimetry appears simple and straightforward, electromagnetic radiation lends itself to giving the most consistent results where human exposure is concerned.

1.3.2 Neutrons

The accurate determination of neutron dosage is fraught with problems that entail many factors. In general, neutron dosimetry has actually represented a measurement of accompanying protons through interaction of the latter with some medium that is effective in giving a proportionate relationship to the incident neutron flux. Neutron dosimetry is obviously very energy dependent because the secondary "reactant" serving as the primary detector (of protons) is highly dependent on the type and efficiency of the reaction from which it was derived. Thus, neutron detectors per se are very energy dependent, and awareness of the source term is mandatory when detection instrumentation is selected.

The International Commission on Radiation Units and Measurements report of 1977 (16) on neutron dosimetry in biology and medicine separates dosimetric methods and instrumentation into several general types: (1) gaseous devices, (2) calorimeters, (3) solid-state devices, (4) activation and fission methods, and (5) ferrous sulfate dosimeters. The details of these devices may be obtained from Reference 16, and this chapter does not duplicate that fine summary of the problems inherent in accurate determination of dosage from incident neutrons. The types of device and instrumentation described are quite broad, however, and deserve mentioning here by specific name, for those who may wish to pursue a given type of detection. The more specific devices mentioned are (1) ionization chambers, (2) proportional counters, (3) Geiger–Müller counters, (4) photographic emulsions, (5) thermoluminescent devices, (6) scintillation devices, (7) semiconductors, and (8) nuclear track recorders. Some of the more general categories listed above (e.g., calorimetry) represent an overall methodology and do not require the individual specificity of a given type of instrumentation.

Neutron dosimetry is not as straightforward as one would desire, and it is certain that research in this field will continue.

1.3.3 Particulate Radiation

The most common dose calculation for internal emitters comprised almost solely of alpha and beta radiation utilizes the "average dose" concept. In the classical sense, if the amount of radioctivity in an organ can be estimated by any means, it is assumed for simplicity that the radioactivity is distributed uniformly thoughout that organ. This assumption is almost always in error with soft beta emitters and alpha particles, because of their short range in tissues (35 to 40 μm for the average alpha from heavy radioelements). However, when one attempts to estimate dose–effect relationships on a microdistribution or "hot spot" premise, errors are equally forthcoming. For purposes of this chapter it is assumed that average organ or tissue dose is sufficient, with full knowledge that this concept may be somewhat in error.

To arrive at an average tissue dose to an organ from the estimated air concentration breathed, many steps are required. These steps are described in the following paragraphs.

Particle Distribution. Airborne particles in real life are generally log normally distributed by mass and can be described by a mass median diameter—the particle size above and below which lies one-half of the total mass of the sample collected. The aerodynamic diameter of a particle size distribution is becoming a more popular descriptor than it was about 30 years ago, and it incorporates the density of the material and a correction for very small particles (17). The size spread of the log normally distributed aerosol particles is defined by the geometric standard deviation, symbolizied by σ_g. Thus a distribution of particles collected on a cascade impactor in an industrial situation would be described by $\bar{\mu} \pm \sigma_g$, where $\bar{\mu}$ is the median diameter defined in mass or aerodynamic terms and is commonly expressed in micrometers.

Respiratory Tract Deposition. It has been long known that site of particle deposition and retention in the respiratory tract during and following inhalation is dependent on particle size. Very simply, the larger particles deposit in the upper areas of the tract (nasopharyngeal region, trachea) and are quickly removed by ciliary action. The smaller particles traverse past the tracheal bifurcation and deposit all the way to the alveolar region. Deposition characteristics were described in detail in 1966 (18), and despite some modifications this review remains the most comprehensive.

Respiratory Tract Retention. Respiratory tract retention parameters are also elaborated very thoroughly in the 1966 reference (18). Ciliary activity obviously accounts for a large, rapid clearance of deposited material from the upper respiratory passages. This material does not present a major problem from the viewpoint of toxicity unless the material is readily absorbed during passage through the gastrointestinal tract or is present in very large quantities (high doses). It is the material that lingers in the deeper (alveolar) spaces of the lung that seems to dictate problems such as carcinogenicity. If the deposited material is very soluble, it will enter the bloodstream through the lung or the gastrointestinal tract and will proceed by way of the circulatory system to deposit in the organ dictated (primarily) by its chemistry. If the material is totally insoluble and nothing really is in biological fluids, three alternatives present themselves: to remain in the alveoli indefinitely (life of individual), gradually to be transferred upward by ciliary action through an initial step of phagocytosis by macrophages, or to be transported to the pulmonary lymph nodes. Numerous mechanisms and kinetics describing these alternatives have been reported in the literature, and individual subjects of interest can be found in the proceedings of some recent symposia (19–22).

Uniform Organ Retention. Once the inhaled material has entered and has been deposited in the organ of choice, its kinetics of loss (retention function) can easily be described if a few facts are known. These include rate of loss from the tissue to blood, and the rate of reentry to the same types of tissue versus rate of excretion in urine or feces. This is rather simply described in terms of a general whole-organ concept, but the next section discusses the more complicated situation of highly localized dose, using bone as an example. The problem is that the rate constants described are seldom known accurately, particularly for humans. If first-order kinetics prevail for loss from the organ, and the organ concentration of the radionuclide is relatively well estimated, average dose calculations are quite straightforward. Many such data are described for animals, but the extrapolation to humans is, at best, satisfactory. An example of a dose calculation is shown in Equation 1, using specific units for purposes of demonstration:

$$D(\text{rads}) = C_o \times k_1 \times \overline{E} \times k_2 \times k_3 \times \int_0^t R_t \tag{1}$$

where C_o = organ concentration of the radionuclide at $t = 0$ (µCi/g of tissue)
k_1 = conversion factor (2.22×10^6 dpm/µCi)
$\overline{E}$ = average energy (MeV/disintegration)
k_2 = conversion factor (1.62×10^{-6} MeV/erg)
k_3 = conversion factor (10^{-2} for the definition, 100 ergs/gram = 1.0 rad)
R_t = retention function in the organ of deposition

This equation assumes that all the associated particulate energy is absorbed in the tissue involved. The retention function is obviously the key to accuracy in this equation, and it is the most difficult to estimate for the exposed industrial worker. Only estimations of the assumed inhaled atmosphere, the inhalation-related parameters, and the metabolism of the material involved, based on animal work or suitable related publications (23–25), can be made.

Reversal of Equation 1, using the total dose allowable or recommended in standards, will permit solution for other parameters of interest such as C_o, and working backward through the processes just described, a recommended maximum air concentration may be determined. This process is not as simple as it seems, but it does allow for computer computation for any radioelement about which some biological parameters are known.

Localized Organ Retention. Skeletal deposition may serve as a good example of localized retention with regard to internal dosimetry problems. The dosimetry of radionuclides deposited in bone is very complicated where alpha- or soft beta-emitting radionuclides are involved. The complications are related to the long-discussed "hot spot" problem. With radionuclides (as with any metallic element) the ionic chemistry of the isotope dictates the site in which the particle will locate. Durbin has written an excellent review of the effect of ionic radius on site of deposition (26). One of the chemical groups of elements associated with bone is the bivalent alkaline earths: calcium, barium, strontium, and radium. One of the most important structures in bone is the hydroxyapatite crystal into which mineral calcium is laid down. Alkaline earths are termed *volume seekers* because they are primarily incorporated into the mineral portion of the bone tissues, where they readily become a part of compact bone. The turnover of these elements is slow once they have become incorporated, even though remodeling of bone occurs throughout life. Thus if an area surrounding a haversian canal is not in the process of remodeling at the time one of the radioactive alkaline earth elements enters the plasma, the chances of that isotope entering that particular haversian system are reduced considerably. The unique quality of bone, however, allows a given haversian system to be in the process of remodeling while at the same time the surrounding systems are in a state of rest. In this case, with a relatively high concentration in the plasma of, say strontium or radium, there is a finite probability that the ionic form or the alkaline earth will be deposited in the remodeling sites. This general description indicates that "hot spots" would readily form in the bone as a result of the remodeling system's seeming inconsistency. Autoradiograms of bone observed under these circumstances indicate a very nonuniform distribution, the pattern of which depends on the microanatomic nature of the bone at the time of deposition.

Whereas the alkaline earths are termed *volume seekers,* the transuranic elements such as plutonium and americium are commonly referred to as *surface seekers.* These descriptive terms are not entirely all-inclusive because the radionuclides are not strictly confined to the described areas but are deposited to some extent throughout all portions of the bone. The surface seekers tend to attach to the membrane of the osteogenic cell or to enter that cell, attaching to proteins, collagen, or other molecular structures, and remaining in the matrix, particularly if that area is not in a remodeling stage. Thus surface seekers tend to affect the most sensitive part of the bone, that is, the epithelial lining or the areas containing osteogenic cells. Osteogenic cells are stem cells, and because they are proliferative they have a high turnover rate. With surface seekers emitting alpha radiation, damage to osteogenic cells is highly probable. Conversely, if an alpha emitter such as radium is buried fairly deep in the mineral portion of bone where the path length of the alpha particle is extremely small (high density), the probability of an ionization occurring in an osteogenic cell to create that particular type of associated radiobiological damage is small.

The hot particle problem in bone, as described, is extremely important. Uneven distribution is what leads to the difference in quality factors between some of the elements like radium and plutonium, but the biological effects are very difficult to interpret on this basis.

Some of the foregoing material may be found summarized in an International Commission on Radiation Protection (ICRP) Task Group report published in 1980 (27).

2 GENERAL BIOLOGICAL EFFECTS

2.1 Electromagnetic Radiation and Neutrons

Acute and long-term biological studies with external sources of irradiation have been rather extensively defined, and this is one area in which there is also a sizeable amount of human data. One of the best sources of information is from the persons exposed to the atom bomb radiation at Hiroshima and Nagasaki. These data are updated periodically by various United States and Japanese research groups. Recent publications review findings through the first 35 years (28, 29), and the BEIR III report uses data from these sources to describe many of the general biological effects of radiation on human beings (30). Additionally, the epidemiologic studies on radiologists have supplied considerable first-hand information (31). A few isolated accidental exposures to gamma- and X-rays or neutrons may also be cited (29). These human data are considered along with supporting or pertinent animal research data in the material to follow.

2.1.1 Acute External Radiation Effects

The acute radiation syndrome has been described by many authors, and Table 12.1 from Upton (32) provides a clear description. He lists doses of 400, 2000, and 20,000 rems and indicates that with the second two doses death has occurred by at least the third and fourth weeks. Certain symptoms (i.e., nausea, vomiting, and diarrhea) are present at all doses immediately following exposure to these higher levels. The gastrointestinal tract is one of the most sensitive organs to electromagnetic irradiation.

2.1.2 Intermediate Dosage Effects

There is a radiation dosage range, quite individually variable and quite large, at which many deleterious symptoms may appear, but at which recovery may occur and the individual may return to relatively normal health. The hematopoietic system is also one of those most sensitive to radiation, and its symptoms often indicate whether death is imminent or recovery will occur. The lymphocytes are perhaps the most sensitive blood cells and are affected within the circulatory system as well as at the precursor level. (Most standard texts dealing with radiation biology discuss in detail the effects of external irradiation on the hematopoietic system.) Other blood cells are affected, but this is primarily the result of the radiation damage to precursors in the bone marrow. The circulating lymphocytes show a decrease immediately following dosage and, within limits, may serve as an indicator of the severity of the radiation exposure. This phenomenon has been considered by many as a possible dosimeter for radiation exposure. Figure 12.1, from Arthur C. Upton, New York University Medical Center (32), shows a typical pattern of hematopoietic response with subsequent recovery at lengthy times postirradiation.

The testes and ovaries are also very radiosensitive and react to very low doses (a few rems). The testes recover to normal eventually, the time for recovery and severity of damage depending on the dose. The ovary is in a different category because its oocytes are present in entirety, never to be replaced after they have been permanently damaged. Most other tissue cells,

Table 12.1 Effects of Radiation on Humans: Major Forms of Acute Radiation Syndrome (32)[a]

Time After Irradiation	Cerebral and Cardiovascular Form (20,000 rems)	Gastrointestinal Form (2000 rems)	Hematopoietic Form (400 rems)
First day	Nausea Vomiting Diarrhea Headache Erythema Disorientation Agitation Ataxia Weakness Somnolence Coma Convulsions Shock Death	Nausea Vomiting Diarrhea	Nausea Vomiting Diarrhea
Second week		Nausea Vomiting Diarrhea Fever Erythema Emaciation Prostation Death	
Third and fourth weeks			Weakness Fatigue Anorexia Nausea Vomiting Diarrhea Fever Hemorrhage Epilation Recovery (?)

[a] Reproduced, with permission, from *The Annual Review of Nuclear and Particle Science*, Volume 18. Copyright © 1984 by Annual Reviews, Inc.

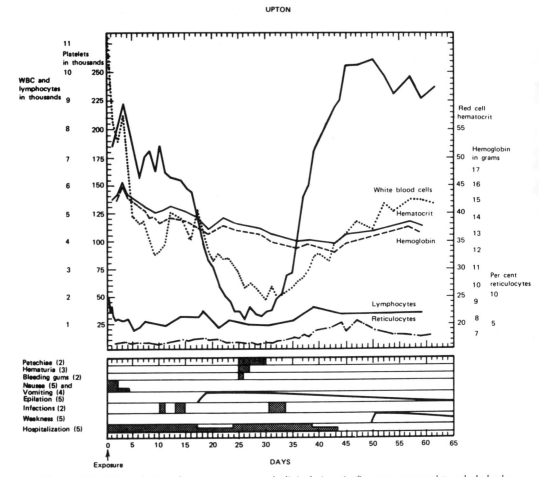

Figure 12.1 Hematologic values, symptoms, and clinical signs in five men exposed to whole-body irradiation in a criticality accident. The blood counts are average values; the figures in parentheses denote the number showing the symptoms and signs indicated (32). Reproduced, with permission, from the *Annual Review of Nuclear and Particle Science*, Volume 18. © 1984 by Annual Reviews, Inc.

such as those of the central nervous system, heart, and lung, are relatively radioresistant.

2.1.3 Late Effects

Low-level, chronic or repeated exposure to ionizing radiation is the primary cause of industrial concern. This type of exposure is often difficult to measure accurately because of its quantitative similarity to background radiation, but more important, it is difficult to control. Some types of occupation, such as those involving radiographical usage or those associated with reactor sites, inherently are accompanied by a certain low-level radiation. Industry has done a remarkable job in controlling such problems, but reduction of levels from a fraction above normal background to natural background cannot be feasibly accomplished in many cases.

Late effects of radiation in mammals have generally been categorized as a generalized accelerated aging process or a specific carcinogenic process. "Life shortening" is a relatively nonspecific term and is considered to mean a premature death resulting from a variety of causes, most of which are specifically unidentified in the given individual. Classical animal experiments, such as early work of Lorenz (33), indicate that, in mice, life span may be shortened as a "function of many factors, some of which are unknown." This type of radiation effect is not well understood and merely represents a mammalian system dying before its time. This type of effect on survival would appear to be minimal compared to the more readily diagnosed causes of death, such as carcinogenesis.

The negative life shortening, or lengthening of normal lifespan, as a result of exposure to subharmful levels of ionizing radiation, has also been shown to be a real radiobiological phenomenon (34). A recent monograph by Luckey (35) gives many (over 1000) references to this stimulating effect of low levels of radiation, defined as "hormesis." Spalding et al. have recently published results of an extensive study showing lengthening of normal life span in mice of various ages exposed to ^{60}Co delivered at many doses and dose rates (36). This stimulation of biological processes by low levels of ionizing radiation merely parallels the effects of many chemical toxicants that are harmful at high doses.

The types of cancer observed with low-level radiation have generally been associated with skin or the hematologic system. Radiation dermatitis was common among workers who had chronic irradiation of the skin, primarily of the hands. After latent periods of many years (e.g., 20) it was not uncommon to observe the formation of skin tumors primarily squamous and basal cell carcinomas (32). Similar lesions have been obtained experimentally with beta particles in rats (37) in which damage to the hair follicles was observed to be implicated in the formation of skin cancer. With modern-day industrial hygiene practices, the observance of radiation-induced skin cancer in workers has essentially disappeared.

Many types of leukemia have been associated with radiation in the human population, mainly in persons who received radiation from nuclear weapons or from radiotherapy techniques (29, 32). Many factors seem to be important in the onset of radiation-induced leukemia, including age at exposure, sex, dose, and perhaps dose rate. One of the greatest sources for study of the leukemias is the Hiroshima–Nagasaki survivors, as indicated, although the patients receiving radiation therapy for ankylosing spondylitis (38–41) and from other radiation sources also add significance to the overall interpretation of induction of leukemia from radiation (42). Summaries of many of the related radiobiological findings in the survivors of the bombings in Japan have been compiled (28, 29) and cover the major categories of dosimetry, biological effects, future research, and health surveillance. Ishimara and Ishimara stated in the

Okada supplement (28) that the intense study of the relationship between radiation dose and incidence of leukemia provides an important link to the effects of external radiation on humans. They quote from other sources (43) that "the apparent excess incidence (leukemia) of A-bomb survivors is about 1.8 cases per million person-year rads for the period 1950–1970." This reference also states that "those who were in either the youngest or oldest age brackets at the time of exposure were more sensitive to the leukemogenic effects of radiation" (0 to 10 and more than 49 years of age). These leukemias were generally of early formation, however, and the risk of chronic granulocytic leukemia among survivors appeared to be greatest at 5 to 10 years postexposure. Doses calculated to be as low as 50 rads were expressed as being associated with the onset of leukemia, with those exposed at Hiroshima having a higher incidence. This is attributed to a higher relative biological effectiveness (RBE) for neutron than gamma rays, and a greater abundance ratio of the former in Hiroshima. These dose estimates were acceptable prior to their report in 1975 (28) but have since been altered, at least in a lowering of the Hiroshima neutron contribution. Beebe's more recent summary (29) considers the data through 1974 and discusses the relevance of the type of modeling used from high to low dose ranges, and compares the cancer incidences found in Hiroshima and Nagasaki with control incidences in the United States. He states that "Even with 550 leukemia cases among the A-bomb survivors in the two cities over the period 1946–1974, it has not been possible to determine what the low-dose risk really is. The question of how to extrapolate from high to low doses in assessing biological effects still remains academically interesting, and uncorroborated with real data."

2.1.4 Dose–Response Relationships

Perhaps one of the most controversial issues in radiobiology has recently centered around the shape of curve (actually, the function describing the curve) to be used in extrapolating from detectable insult (carcinogenesis, life shortening) at intermediate doses to anticipated consequences of very low doses. These low doses are currently thought by most to be of potential hazard to humans, and there is no way in which meaningful biological effects data versus dosage can be collected in this region. The statistical requirements would necessitate a practically infinite population of mammals. It is possible to use cellular and subcellular systems at these dosage levels to study particular radiobiological effects, but extrapolation of these results to the mammalian organism, with the complications and interactions of a simple circulatory system and hormonal interplay, becomes academic in nature. Thus there has been great debate regarding a linear versus nonlinear extrapolation, with neither side being supported directly by experimental evidence. One of the treatises on this subject, and perhaps one of the most lucid, was part of a symposium held in 1976 to elucidate the problems (44). In this, Brown discusses the linear quadratic models that have been used to describe various radiobiological

IONIZING RADIATION

responses as a function of dose. Another simplistic description of these proposed processes has been presented in NCRP Report No. 64 (45); Figure 12.2, describing the mathematical formulations, is from this reference.

The linear and quadratic equations in Figure 12.2 are often interpreted in terms of the single or multihit target phenomenon described years ago by Lea (46). The exponential terms are used to describe the decreased incidence resulting from death at higher doses at a time before the studied end point (cancer) has time to develop. Of course, if hormesis (35) is an accepted phenomenon at very low doses, none of these extrapolative processes to zero dose are meaningful. Upton (47) has associated relative dose ranges and related general effects with these curves and indicated that the lower dose term (αD) would be expected to predominate over the quadratic (βD^2) term for mutational or chromosomal effects in mammalian cells at <50–100 rad at low dose rates. With high linear energy transfer (LET) radiation, the linear term (αD) would predominate and be less dose rate dependent. For carcinogenesis, the relationships differ quantitatively from one type of cancer to another, and no data have been sufficiently extensive to give confidence in any extrapolation to the very low dose region of interest. At this point in time, it is not feasible to reject any of the curvilinear relationships described in Figure 12.2 as being inadequate to fit existing data.

2.1.5 Risk Factors

The acceptance of a given cancer incidence versus dose relationship for Figure 12.2 allows one to proceed one step further in assessing an end point for the exposure conditions. From a linear relationship one can obtain a value for the cancer incidence, say, fractional incidence, for a given quantity of radiation, in rads. For example, one may have a fraction of five cancers per 10,000 individuals in the population exposed to 10 rads of some radiation. Thus the incidence would be $5/10,000 = 5 \times 10^{-4}$; this may be equilibrated to the 10 rads, as $5 \times 10^{-4}/10$ rads $= 5 \times 10^{-5}$ cancers per rad. Generally, this type of expression is reported as 50 cancers per million rads-people or 50 cancers per 10^6 people-rads. This is a risk factor, then, to express the carcinogenic effect for this type of cancer following exposure to the prescribed type of radiation. Risk factors have outmoded other means of expressing radiation damage, particularly for use in radiation protection practices. It was pointed out earlier that life span shortening is perhaps not a good end point for assessing low-level damage because the subjects may well live longer than controls. This should surprise no one. Then, the use of carcinogenesis allows one to overlook the life shortening (whether negative or positive) aspect and deal in terms of a more independent term. This can also be extrapolated to large populations if one believes in the basic premise of the risk per person-rad. From the example just used, of five cancers per 10,000 persons, one could extend this to answer the question of the effect of, for instance, an additional amount equal to background irradiation, or 100 mrad to the individual per year. Using the risk factor derived above,

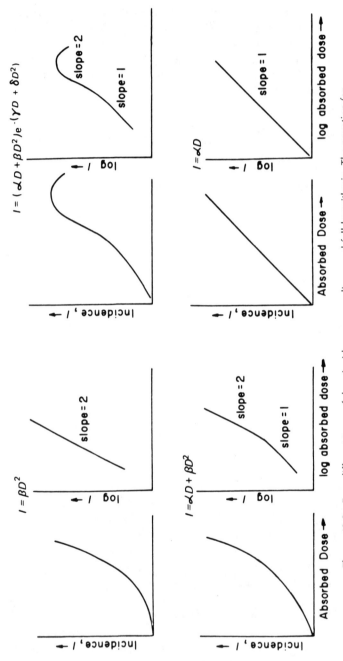

Figure 12.2 Four different types of dose-incidence curve, linear and full logarithmic. The equation for each is also shown. (Integer values for slopes are approximations since a single overall function is described with contributions from terms having different exponents.) Reprinted with permission from National Council on Radiation Protection and Measurements, Influence of Dose and Its Distribution in Time on Dose–Response Relationships for Low-LET Radiations, NCRP Report No. 64, 1980.

one could expect in a population of 200 million people, cancers in 1000 persons per year, if the underlying premises were correct. Thus one in every 200,000 persons exposed to 100 mrad per year above background could expect to develop this hypothetical cancer. One can play interesting games with this concept, and it has the advantage of being applicable to unlimited populations at unlimited smaller and smaller doses. Its practical significance is left to the reader.

2.2 Radionuclide Radiation

As stated in Section 1, the internal exposure to radionuclides is probably the most pertinent to practical toxicity problems in the industrial world. Radionuclides are difficult to control, particularly if their emissions do not allow for detection at reasonable distances from the source. Plutonium is a good example of the latter in that it has alpha particle emission and only a very weak X-ray component, too weak to detect through the ordinary glove box. Section 1.3.3 covered the difficulties in determining biological dosage following exposure. There has been a wealth of information collected over the past 30 years in animal studies, but extrapolation to humans is not straightforward. We do have the advantage of the excellent epidemiologic studies of the radium dial painters (48) and the radium-treated ankylosing spondylitic patients (38, 41). Other studies may also prove useful from the point of view of assessing human exposure to radionuclides [e.g., thorotrast (49) and plutonium (50, 51)]. The recent Snowbird Actinide workshop (52) represents an attempt to bring together analytical data from man and animals and to use these in formulation of usable models. The problem with studies such as those with the plutonium workers is that since no effects have been observed, there is no way to evaluate dose–effect relationships. Laboratory experimental data are presented in the following sections, and where possible, the pertinence to the human situation is discussed.

The problems inherent with radiation dose from internally deposited radionuclides (internal emitters) are much more involved than with whole-body electromagnetic radiation. Compartmentalization of the various radioelements is responsible for the complications of dose estimation, and it is the interplay between these sites of deposition in the body that results in the kinetic modeling carried out with internal emitters. The ultimate goal is to arrive at a time-integrated value, in ergs expended per gram of a tissue, and to relate this through the proper constant(s) to the desired radiation dosage unit. The dose arrived at in this manner is an average over the entire tissue or organ and is generally, today, characterized by the rad, equivalent to 100 ergs per gram of exposed medium. For biological samples of soft tissue, where most beta emitters are involved, this is a reasonably good estimate for comparative purposes. This type of radiation distributes rather uniformly because of the relatively long path length in soft tissue; hence average dose is perhaps a reasonable approach to assessing biological damage or effect. With alpha emitters the pathway is very small (Section 1.3.3), and average dose is rarely as suitable for comparison

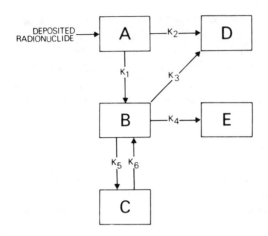

Figure 12.3 Example of a first-order kinetic modeling from the description equations derived: A = lung; B = blood; C = an internal organ; D = gastrointestinal tract excretion; E = urinary excretion.

of effects. Except where intricacies of localized dose are of concern, however, the average is still the most acceptable (often the only) means of relating dose to effect.

In biological modeling to arrive at radiation dose and interplay between organs, one generally uses first-order kinetics to define the parameters. This approach allows one to think in terms of the quantity leaving or entering a given depot as being directly related to the amount in that repository at any one time. Through a series of rate constants representing the various compartments of deposition in the body, half-lives (0.693/rate constant) of retention may be calculated, and investigators in this area of radiation biology or biophysics have learned to think and write in these terms. This is an acceptable concept and will be in use in this field for a long time, although many researchers have preferred the power function method of fitting retention kinetics (53–55).

Figure 12.3 gives an example of gross modeling to arrive at the time course of a radionuclide residing in the body. Here the simple model consists of organs A, B and C, with excretion occurring through two routes to external compartments D and E. The development of equations to fit this scheme proceeds according to the following reasoning. Because inhalation appears to be the most practical route of entry to the body, this is a simplified version of a model in which the lung (A) is one of the organs involved and B is the blood; C may represent any other internal organ (kidney, liver, spleen, etc.), and D and E are gastrointestinal tract and urinary excretion, respectively.

1. The amount of radionuclide entering compartment D, gastrointestinal excretion (feces), from the lung is straightforward and is represented by

$$\frac{dD}{dt} = k_2 A \qquad (2)$$

2. In compartment B, the blood represents a much more complicated

IONIZING RADIATION

scheme. An expression for the rate of accumulation in the blood may be derived through a series of differential equations as follows:

$$\frac{dB}{dt} = k_1 A = \text{to blood from lung} \tag{3}$$

$$\frac{dB}{dt} = k_6 C - k_5 B = \text{to blood from tissue } C \tag{4}$$

$$\frac{dB}{dt} = -k_3 B - k_4 B = \text{to } D \text{ and } E \text{ from blood} \tag{5}$$

An additional rate constant and equation are applicable for absorption from gastrointestinal tract to blood, but this is assumed to be negligible here for purposes of simplicity.

3. The ultimate goal is to determine the time course of radiation energy released into organ C. The process is initiated by combining some of the preceding equations. The amount in C at any time is the integral of the rate of buildup or decline:

$$\frac{dC}{dt} = k_5 B - k_6 C \tag{6}$$

An expression of B may be obtained from the integral of

$$\frac{dB}{dt} = k_1 A + k_6 C - k_5 C - k_3 B - k_4 B \tag{7}$$

An expression for A at any time may be obtained from a form of equations

$$-\frac{dA}{dt} = k_1 A + k_2 A \tag{8}$$

and

$$A = A_0 e^{-(k_1 + k_2)t} \tag{9}$$

Equation 7 now becomes

$$\frac{dB}{dt} = k_1 A_0 e^{-(k_1 + k_2)t} + C(k_6 - k_5) - B(k_3 + k_4) \tag{10}$$

This may be solved using linear differential equation techniques, giving an expression for B in terms of A_0 and C. The form generally recognized for integration is

$$\frac{dB}{dt} + B(k_3 + k_4) = k_1 A_0 e^{-(k_1 + k_2)t} + (k_6 - k_5)C \qquad (11)$$

$$B = [\exp - \int_t^0 (k_3 + k_4)dt] \int_t^0 \exp \int_t^0 (k_3 + k_4)$$

$$[k_1 A_0 e^{-(k_1 - k_2)t} + (k_6 - k_5) \, dt + \text{constant} \qquad (12)$$

4. By proper substitution from Equation 12, an expression for C can be obtained through integration. To quantitate the scheme, one needs to know the initial quantity of radionuclide deposited in the lung A_0 and the various rate constants k_1 through k_6. These may be obtained by use of the time course of measured radionuclide in blood and excreta at any time. An analogue computer is almost a necessity when the number of compartments is large.

Thus through a series of first-order kinetic manipulations, an expression for dose to an organ may be obtained with which to match biological effect. Although this modeling scheme might appear to belong under Section 1.3.3, it seemed logical to place it in a relationship to the ensuing biological effect.

2.2.1 Effects on Lung

Many radiation effects studies on the respiratory system have been carried out following intratracheal injection (IT) and inhalation (INH). There is always some question of the validity of using IT data inasmuch as this route is a nonpractical means of acquiring a radionuclide; thus the tendency is to utilize information from INH whenever possible (IT instillation in the lung is generally nonuniform, and the material is in a solution or suspension; an aerosol, used for the INH route, is more uniform and realistic). Major INH studies using dogs have been carried out in recent years at Battelle Pacific Northwest Laboratories (56) and at the Lovelace Foundation [the Inhalation Toxicology Research Institute (ITRI)] (57). Doses have been expressed as average energy released to the tissue (rads) or in terms of radioactive unit of material deposited per gram of organ (e.g., nCi/g). Either manner of presentation is acceptable.

In both laboratories the inhalation exposures have been carried out in a manner in which only the nasal region is subject to the aerosol. Such exposures are commonly referred to as "nose-only." With whole-body exposure during inhalation, the animal is covered with particles and obtains extremely large amounts of external contamination. Even when the aerosol is essentially insoluble in body fluids, the small amount that may be absorbed through the gastrointestinal tract because of the animal's licking (cleaning) itself can result in a deposit in internal organs. If the half-time of residence in that organ system is extremely long compared to that in lung, the possibility for appreciable radiation dose to that organ exists. Thus nose-only exposures are desirable. At both laboratories the primary radioactive inhalation exposures now involve alpha emitters, but the ITRI work initially dealt chiefly with beta emitters. At the Los Alamos National Laboratory (LANL) similar studies were carried out using alpha emitters in small animals (58).

Typical dose–effect data from PNL and ITRI for a certain end point, such as pulmonary neoplasia, indicate that the dose from beta emitters is up to a factor of 20 times that calculated for a comparable effect (carcinogenesis) from alpha emitters (59–61). These carcinogenic doses are in the hundreds and thousands of rads for alpha emitters and in the thousands and tens of thousands of rads for beta emitters. Similar findings (lung carcinogenesis) in the Syrian hamster following inhalation of plutonium particles requires at least 2000 rads for induction of lung adenocarcinomas (62). Assuming the LET concept (or RBE) to be based on firm premises, this difference between types of radiation is to be expected. Some of the types of lesions found do not appear to vary remarkably with the quality of the radiation. Typical of such lesions are fibrosarcoma, squamous cell carcinoma, bronchiolar carcinoma, adenocarcinoma, hemangiocarcinoma, and bronchioloalveolar carcinoma. Some of these findings described by various authors may be morphologically similar and depend on the individual pathologist's interpretation.

In humans, a limited number of lesions have been attributed to the deposition of radionuclides in the respiratory system. One of the most widely known occupations resulting in such findings is uranium mining. Internal biological lesions in this occupation are primarily the result of breathing radon (^{222}Rn), hence its ensuing decay products (63). The early miners in Schneeburg, Germany and Joachimsthal, Czechoslovakia suffered from a disease that was ultimately determined to be lung cancer (64, 65). In the United States the uranium miners in the early part of this century have been found to have a significant increase in lung cancer (66). The proceedings of a recent symposium dealing with the hazards associated with the uranium mining industry carry much of this early history and brings the state of the art up to date (67). This includes analytical methods and measurements as well as biomedical fundings. The earlier results had shown the lung cancer incidence to correlate with smoking, and a greater incidence of lung cancer has been reported for smokers than for nonsmokers, among the miners (66–70). The lesions are bronchogenic and have been shown to correlate with estimated radiation doses (71). The incidences that have been reported for the increase over that expected from a "normal" nonsmoking population of individuals are variable, and no specific value is significant at this time. It has been estimated that a dose of 360 rads will double the incidence of normal lung cancer, as evidenced in these workers (71). Recently the correlation between lung cancer and smoking in uranium miners has been questioned. In fact, Martell and Sweder's results (72) would imply an association of some of the increased lung cancer in smoking uranium miners to the alpha irradiation from polonium and other isotopes contained in cigarette tobacco smoke. Saccomanno et al. (73), however, have presented convincing evidence that the adenocarcinoma incidence did increase with increasing working-level months (WLM) of exposure and with smoking intensity. They also showed (73) an increase in frequency of adenocarcinomas in miners with younger ages at diagnosis and lower ages at the start of mining as an occupation and state that the type of bronchogenic carcinoma observed with nonsmokers differs from that seen with smoking miners. It would appear that persons who may have added an additional insult, such as poorly controlled mining atmospheres, to their smoking habit, would have created a synergistic background for an increased incidence of pulmonary cancer.

Saccomanno and associates (68, 74) have derived a unique method for detection of cancerous lesions in the lung. His method will indicate, in cells taken from sputum samples, the predisposition to an invasive cancerous lesion in the bronchiolar tree. Squamous cell metaplasia of the bronchi is generally accepted as the forerunner to bronchogenic carcinoma. The average time for development of epidermoid carcinoma has been quoted to be approximately 15 years (75). The time of development from early metaplastic changes to marked atypia may be an average of 4 years. Thus cytological investigation of sputum samples taken during this earlier phase may reflect this atypia and may lead to detection of a carcinomatous condition before it reaches the invasive state. This may allow performance of localized surgery, to eliminate the malignancy that would eventually lead to death.

2.2.2 Effects on Bone

"Bone seekers" are generally classified as two types, volume or surface, as described earlier. The primary cancerous lesion observed following radionuclide deposition in the skeleton is osteogenic sarcoma. One of the most thorough studies of this lesion versus estimated radiation dose has been carried out at the University of Utah over the past 30 years (76). The beagle dog has been the chief experimental subject in that laboratory, allowing for comparison with the inhalation studies described previously (56, 57). In the Utah work the radionuclides (^{90}Sr, ^{241}Am, ^{228}Th, ^{226}Ra, or ^{239}Pu) have been injected intravenously (IV) in a soluble form, primarily the citrate. This route has desirable properties because it gives a greater probability of deposition in the organs of "choice" than perhaps would be predicted to occur with material entering the blood from the lung, and the doses thus delivered are administered with great accuracy. The choice of deposition site is determined by the physicochemical properties of the radionuclide, as is the determination of whether a given material is to be a volume or surface seeker in bone.

Mays et al. have described the finding of bone sarcomas in mice and dogs (77) and have attempted to estimate risk to the skeleton using a linear model. Their paper compares the relative risks between ^{239}Pu and ^{226}Ra in experimental animals; Figures 12.4 through 12.6 plot the results. Also shown are data from the radium dial painters, with the predicted linear extrapolation to what the risk of bone sarcoma may be in humans from ^{239}Pu (Figure 12.7). An estimate

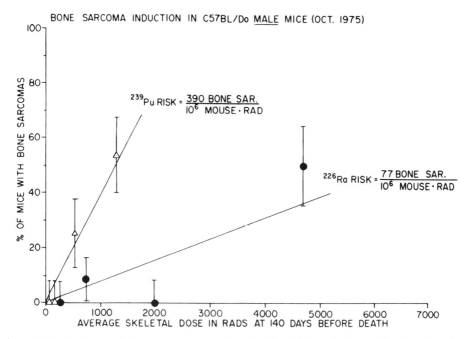

Figure 12.4 Incidence of bone sarcomas in male mice. The risk from ^{239}Pu was five times that from ^{226}Ra. Reprinted with permission froim C. W. Mays, H. Spiess, G. N. Taylor, R. D. Lloyd, W. S. S. Jee, S. S. McFarland, D. H. Taysum, T. W. Brammer, D. Brammer, and T. A. Pollard, "Estimated Risk to Human Bone from ^{239}Pu," in: *The Health Effects of Plutonium and Radium*," W. S. S. Jee, Ed., J. W. Press, University of Utah, Salt Lake City, 1976, p. 343.

of the cumulative risk from ^{239}Pu using a linear model is about 200 bone sarcomas per 10^6 person-rads. This bone sarcoma risk estimate is based on the results of the German ankylosing spondylitic patients that received ^{224}Ra as therapy (Figure 12.8). The paper by Spiess and Mays (38) is a thorough summary of bone sarcoma incidence versus radiation dose from many sources and is a recommended reference, not only for the value of the authors' interpretation but for the complete bibliography associated with this field. A supplement to this review, including discussions of Figures 12.7 and 12.8, is contained in another paper by Mays and Spiess (78).

2.2.3 Effects on Soft Tissues

As stated earlier, radionuclides will localize in an organ or tissue to an extent dictated by their physicochemical properties. Thus by choosing an element, one could feasibly obtain a specific concentration in an organ of choice and confine the resulting biological effect to that area. In fact, if this relationship were as simple as stated, the medical field would be able to localize the proper radionuclide at the site of a tumor and, with the ensuing radiation, destroy the

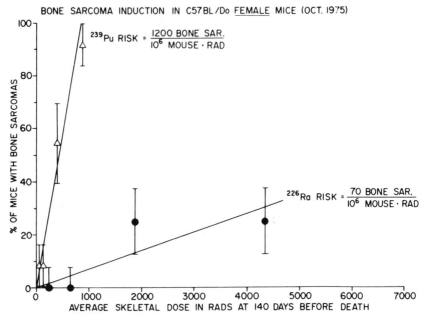

Figure 12.5 Incidence of bone sarcomas in female mice. The risk from ^{239}Pu was 17 times that from ^{226}Ra, assuming negligible control incidence. Reprinted with permission from C. W. Mays, H. Spiess, G. N. Taylor, R. D. Lloyd, W. S. S. Jee, S. S. McFarland, D. H. Taysum, T. W. Brammer, D. Brammer, and T. A. Pollard, "Estimated Risk to Human Bone from ^{239}Pu," in: *The Health Effects of Plutonium and Radium*," W. S. S. Jee, Ed., J. W. Press, University of Utah, Salt Lake City, 1976, p. 343.

abnormal cells. However, localization is rarely relegated to the tissue of concern alone but also to other, nonaffected tissues. Localization in the tumor-bearing organ subjects normal cells to the radiation as well, and the radiation dose that may cause subsidence of the malignancy will also be destructive to the normaly functioning tissue. With this diversion on the behavior of internal emitters as background, it is now feasible to discuss a few examples of specific tissues that are primarily damaged by the entrance of specific radionuclides to the body.

The Reticuloendothelial System (RES). The RES broadly consists of the liver, spleen, bone marrow, lymph nodes, and lung. One normally thinks of the RES components of these tissues as containing cells that are available to the circulation (lymph or blood) and have the ability to react to any material that is recognized as foreign to the body. Colloidal or particulate substances are particularly subject to being acted on by cells of the RES. Phagocytosis is the primary means of detoxifying such foreign bodies, with the destructive action occurring within the phagocytic cell. For instance, when colloidal polonium is injected IV into experimental animals, it follows a body distribution pattern that is essentially analogous to the organs of the RES (79). The larger the particulate entity, the more rapidly it is removed from the circulation. and this can take place during

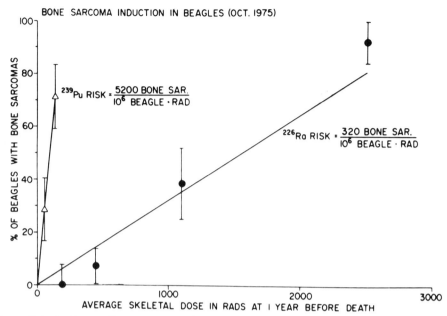

Figure 12.6 Incidence of bone sarcomas in beagles. Linear fits contrained to pass through zero incidence at zero dose were made from 0 to 135 rads for ^{239}Pu and from 0 to 2500 rads for ^{226}Ra. The derived risk from ^{239}Pu was 16 times that from ^{226}Ra. Reprinted with permission from C. W. Mays, H. Spiess, G. N. Taylor, R. D. Lloyd, W. S. S. Jee, S. S. McFarland, D. H. Taysum, T. W. Brammer, D. Brammer, and T. A. Pollard, "Estimated Risk to Human Bone from ^{239}Pu," in: *The Health Effects of Plutonium and Radium,*" W. S. S. Jee, Ed., J. W. Press, University of Utah, Salt Lake City, 1976, p. 343.

the first pass through the circulatory system. The spleen and the bone marrow, for example, are two organs that are highly affected following ^{210}Po administration in this manner (79). The effects on these organs are somewhat comparable, and because they are considered to be at least partly a segment of the hematopoietic system, the resulting damage is included under that discussion.

The lymph nodes are a major sink for the deposition of foreign materials that find their way into the lymphatic channels. Drainage from the lung is one of the more common examples of the defense mechanism exercised by the lymphatics, especially for particles deposited in the alveolar region of the lung. These particles are filtered out in the regional lymph nodes, where they reside for a very long period (19). Although large deposits in lymph nodes have been observed after inhalation of a radioactive aerosol (80), little biological damage of significance to life span, including carcinogenesis, due to the radioactive content of the regional nodes, has been observed. In uranium inhalation studies (81) the lymph nodes of monkeys were seen to concentrate far greater than 50 times the concentration in lung after repeated exposure to UO_2 aerosol. A similar pattern has been described for many other radionuclide particles (19). The effect is generally one of fibrotic change, with some alteration of the

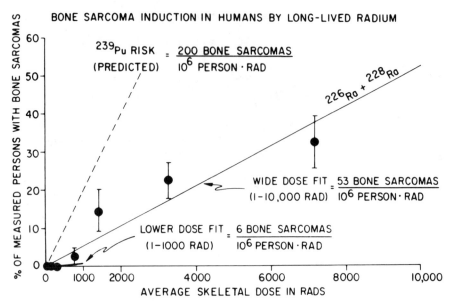

Figure 12.7 Incidence of bone sarcomas in persons (mostly dial painters) exposed to ^{226}Ra and ^{228}Ra. Below 10,000 rads the fitted slope is 53 bone sarcomas/10^6 person-rad, but the slope is based on only one reported sarcoma. If the true risk from low doses of long-lived radium is between these values, the predicted risk from ^{239}Pu in humans is between 4 and 33 times that from long-lived radium. Reprinted with permission from C. W. Mays, H. Spiess, G. N. Taylor, R. D. Lloyd, W. S. S. Jee, S. S. McFarland, D. H. Taysum, T. W. Brammer, D. Brammer, and T. A. Pollard, "Estimated Risk to Human Bone from ^{239}Pu," in: The Health Effects of Plutonium and Radium," W. S. S. Jee, Ed., J. W. Press, University of Utah, Salt Lake City, 1976, p. 343.

cellularization of the germinal centers. However, the cells involved have shown little tendency for tumor formation, and the lymph nodes under this circumstance are not considered by most to be a critical organ for radiation damage.

The Hematopoietic System. It is known that the blood cell forming tissues are subject to radiation when the nuclide is administered in such a form as to deposit in the reticuloendothelial cells associated with this function. There is also a tendency for some elements to localize in the functional cells directly related to blood cell production. Whatever the mode of localization, the general effect is essentially the same. The lymphocytes are the most sensitive of the circulating blood cells, but the precursor cells in the bone marrow and spleen (particularly under conditions of extramedullary hematopoiesis) are also extremely sensitive to radiation. The erythrocyte precursors in the marrow cavity are the key to any subsequent reduction in red cells. Thus what one may readily observe in circulating blood as a result of radionuclide deposition in the hematopoietic system is a rapid decline in the white cell population followed by a much slower decline in circulating erythrocytes (see Figure 12.1). Reduction in the former, naturally, allows the onset of infection, as the ability to cope

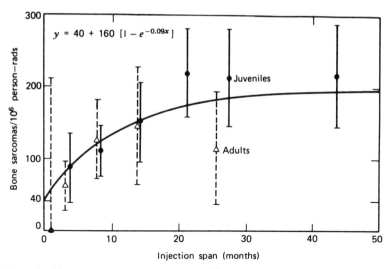

Figure 12.8 Risk of bone sarcomas in humans versus ^{224}Ra protraction. The risk rises to about 200 bone sarcomas/10^6 person-rad at long protraction. The standard deviation of each point is shown. Reprinted with permission H. Spiess and C. W. Mays, "Protraction Effect on Bone-Sarcoma Induction of ^{224}Ra in Children and Adults," in: *Radionuclide Carcinogenesis*, proceedings of the 12th Annual Hanford Biology Symposium, 1973, p. 437.

with outside disturbances is dramatically curtailed. If the animal (or person) survives this ordeal, the lack of red cell production will result in an anemia and, if the dose is sufficiently high, in the death of the subject.

There are many detailed biological effects inflicted on the hematopoietic system by deposited radionuclides, but the general pattern is as described.

The Immune System. The immune system is obviously affected by ionizing radiation, as observed overtly through the effects on the hematologic system. The lymphacitic portion of the circulating defense mechanism is one of the most sensitive to whole-body irradiation. Basic questions arise concerning carcinogenesis and the immune system, as to whether a decreased immune function makes way for carcinogenesis, or whether the formations of cancerous growth depletes the immune system. It is clear, however, that injury to the immune system from radiation, resulting in reduced capability to respond to foreign body infiltration, leads to illness and death, when the situation attains a serious state. More on the immune system is discussed in Section 6.3.

Specific Tissue Localization. The thyroid is an excellent example of a tissue that will localize a particular internal emitter, in this case, iodine. A fine review of this subject with particular regard to the practical aspects of deposition of radioiodine in the thyroid gland has been published by the National Council on Radiation Protection and Measurements (NCRP) (82). The effects of ^{131}I

Table 12.2 Relationship Between Low Dose Exposure to ^{131}I in Children and Subsequent Hypothyroidism[a]

Number of Subjects	Thyroid Absorbed Dose Range (rads)	Estimated Mean Thyroid Absorbed Dose (rads)	Number of Hypothyroid Subjects	Incidence of Hypothyroidism (%/year)
146	10–30	18	0	0
146	31–80	52	3	0.15
151	81–1900	233	5	0.23

[a] From National Council on Radiation Protection and Measurements, *Protection of the Thyroid Gland in the Event of Releases of Radioiodine*, NCRP Report 55, 1977. Preliminary results: Hamilton and Tompkins (83).

irradiation of the thyroid are aptly described in this report and compared to those observed with X-rays. Because radiation effect on this particular organ has been so thoroughly studied, it is valuable to present two tables from the NCRP reference that generally summarize the findings to date (Tables 12.2 and 12.3). These tables list the risk estimates for various thyroid insults and compare the effects between children and adults.

3 STANDARDS AND GUIDELINES

3.1 NCRP Recommendations

The establishment of guidelines for the protection of workers and the public is a complicated endeavor involving many philosophies and some facts. This section includes the most relevant subdivisions in a relatively recent report by the NCRP (85). A brief statement is made regarding each subdivision, as put forth by the committee that generated the general reference (85). Most of the material derives from step-by-step analysis of the information presented in Part II of that handbook. The comments to follow are the authors' interpretation and may not represent that of the full committee.

3.1.1 Sources of Information

Low-dose effects are based on some selected type of extrapolation from biological effects observed at high doses, to dose levels approaching background, where no significant data exist or will be obtained in experimental mammals. For conservative planning, for engineering purposes, it is acceptable to think in terms of an essentially zero dosage threshold, but this approach is not acceptable for predicting radiobiological effects. Genetic effects caused by radiation are known to depend on many factors such as dose rate and

Table 12.3 Absolute Individual Risk of Thyroid Abnormalities After Exposure to Ionizing Radiation[a]

Type of Abnormality and Population Surveyed	Mean Absorbed Dose or Dose Range for Which Data Were Available (rads)	Absolute Individual Risk (10^{-6} rad^{-1} year^{-1})	Statistical Risk Range[b] (10^{-6} rad^{-1} year^{-1})
Internal Irradiation (^{131}I)			
Thyroid nodularity			
Children	9000	0.23	0–0.52
Adults	8755	0.18	0.13–0.23
Thyroid cancer			
Children	9000	0.06	0–0.158
Adults	8755	0.06	0.044–0.075
Hypothyroidism			
Low-dose children	10–1900	4.9[c]	3.9–22.9
High-dose adults[d]	2500–20,000	4.6[c]	2.8–7.8[d]
External Irradiation			
Thyroid nodularity in children	0–1500	12.4	4–47.4[e]
Thyroid cancer in children	0–1500	4.3	1.6–17.3[e]
Hypothyroidism in adults	1640	10.2	0–24.8

[a] From National Council on Radiation Protection and Measurements, *Protection of the Thyroid Gland in the Event of Releases of Radioiodine,* NCRP Report 55, 1977.
[b] Unless otherwise indicated, the range of risk was determined by assuming that the number of cases n out of the population at risk represents the true mean of a Poisson distribution. The range is then estimated by using $\pm 2n$ as the 95 percent confidence level.
[c] Threshold of 20 rads.
[d] See Figure 2 of Reference 84.
[e] In these cases the risk was determined from the slope of the linear regression line. The range was estimated from the extreme data points, which provide the lowest and highest slopes.

fractionation, and relating these into the human population presents an astronomical task. Genetic effects in the population entail geographic factors, types of employment, the quality factor of the radiation, the age of the subject, and a host of other complicated variables. Incorporated into these multifaceted factors (and not specifically stated in the reference) is the factor of acceptance: these factors fall under the concept of what comprises an acceptable risk to the individual and the population.

3.1.2 Dose Level and Priorities

Many terms have been used to define the so-called acceptable levels or the recommendations of such bodies as the NCRP and ICRP. These terms include tolerance, tolerance dose, permissible dose, maximum permissible dose, dose limits, and lowest practical level. The current tendency in radiation is toward the last term, which means the lowest level attainable under a specific set of conditions. The potential source of radiation should become, along with other less easily definable factors, the guiding influence in the use of this criterion.

3.1.3 Administrative Philosophies

A number of factors have to be used when the radiation worker is compared to the population at large. The worker is usually accepted in employment in suitable physical condition; on the other hand, the population represents a spectrum of health conditions. Many attributes of the worker are included in employee selection for obvious reasons, and these restrictions may automatically involve possible genetic effects on the worker as well as with generations to follow, although these may not be applicable to the population at large.

3.1.4 Interpretation by Professionals

There are many reasons for basing guidelines on some sort of time span, such as the protraction of dosage over fractions of a one year potential exposure period. Limits such as 5 rems/year or 3 rems/quarter are arbitrary values, just as 18 years is arbitrary as the starting age for a worker in atomic energy related industries. One of the problems with such values is that the practicing administration (primarily health physics oriented) often places credibility in these numbers as if they were determined from some accurate base. This tends to happen in all fields where such limitations and guidelines are established, and some individuals may at times assume that these numbers, generally derived after excellent and very well-informed thought, are to be used as if they were unequivocal, instead of realizing that they represent only the best numbers that can be derived at the time.

3.1.5 Critical Organ Concept

The critical organ concept has been under discussion for many years. As the reference (85) suggests, it might be more intelligent to consider the "critical" organ as the one most highly radiosensitive , and the "dose-limiting" organ as the one that may receive the greatest radiation dose. Thyroid is used as an example of the latter. The NCRP (85) states:

For the development of protection guides, the concept of identifying the minimum number of organs and tissues that are limiting for dose consideration is the essential

IONIZING RADIATION

simplifying step. For general irradiation of the whole body, such critical organs and tissues are:

1. Gonads (fertility, genetic effects).
2. Blood-forming organs, or specifically red bone marrow (leukemia).
3. Lens of the eye (cataracts).

For external irradiation of restricted parts of the body, an additional critical organ may be:

4. Skin (skin cancer).

For irradiation from internally deposited sources alone or combined with irradiations from external sources, additional limiting organs are determined more by the metabolic pathways of invading nuclides, their concentration in organs, and their effective residence times, than by some inherent sensitivity factors. They include:

1. Gastrointestinal tract.
2. Lung.
3. Bone.
4. Thyroid.
5. Kidney, spleen, pancreas, or prostate.
6. Muscle tissue or fatty tissues.

This is a well-constructed and brief description of the "critical organ concept," and it indicates the problems involved in proper organ selection criteria under the many conditions of possible radiation exposure.

3.1.6 Modifying Dosage Factors

The terms quality factor (QF), relative biological effectiveness (RBE), linear energy transfer (LET), and dose equivalent (DE) represent quantities that tend to compound the potential individual radiation situation and should be studied thoroughly before use.

3.1.7 Dose-Limiting Recommendations and Guidance for Special Cases

The topic of dose-limiting recommendations in routine and extreme situations is covered very well in the reference handbook (85) and can be well understood by referring to the main points indicated in the summary table (Table 12.4).

3.2 ICRP Recommendations

In 1977 the ICRP recommended new objectives for radiation protection (86). These entail a new nomenclature and set of terms on which to base their recommendations. They state that protection of humans must be from detrimental effects against the individual (somatic effects) or against their descendants (hereditary effects). They define "stochastic" effects as those for which the probability of occurring is a function of dose, with no threshold; "nonstochastic" effects are those for which the severity is a function of dose and for

Table 12.4 Summary of Dose-Limiting Guidelines[a]

Maximum permissible dose equivalent for occupational exposure	
Combined whole-body occupational exposure	
Prospective annual limit	5 rems in any 1 year
Retrospective annual limit	10–15 rems in any 1 year
Long-term accumulation to age N year	$(N - 18) \times 5$ rems
Skin	15 rems in any 1 year
Hands	75 rems in any 1 year (25/quarter)
Forearms	30 rems in any 1 year (10/quarter)
Other organs, tissues, and organ systems	15 rems in any 1 year (5/quarter)
Fertile women (with respect to fetus)	0.5 rem in gestation period
Dose limits for the public, or occasionally exposed individuals	
Individual or occasional	0.5 rem in any 1 year
Students	0.1 rem in any 1 year
Population dose limits	
Genetic	0.17 rem average 1 year
Somatic	0.17 rem average 1 year
Emergency dose limits: life saving	
Individual (older than 45 years if possible)	100 rems
Hands and forearms	200 rems, additional (300 rems, total)
Emergency dose limits: less urgent	
Individual	25 rems
Hands and forearms	100 rems, total
Family of radioactive patients	
Individual (under age 45)	0.5 rem in any 1 year
Individual (over age 45)	5 rems in any 1 year

[a] Reprinted with permission from National Council on Radiation Protection and Measurements, *Basic Radiation Protection Criteria*, NCRP Report 39, 1971.

which a threshold may occur. Carcinogenesis is considered the major somatic effect and is considered stochastic. At the very low doses, hereditary changes are also considered stochastic. The aim of the ICRP in radiation protection is to prevent nonstochastic effects by maintaining exposure levels below a threshold dose and to limit stochastic effects by maintaining exposure levels at "as low as reasonably achievable (ALARA)."

3.2.1 Basic Concepts

The Commission (ICRP) has defined a number of terms, each of which requires understanding before they may be applied. It is not the purpose of this chapter to put forth the detailed parts of the recommendations, but rather to describe the overall philosophy. They define the "detriment" concept as a way of

IONIZING RADIATION

identifying and quantifying deleterious effects. It is generally defined as the expected harm that may be incurred in an exposed population, taking into account not only the type of effect but also the severity. The "dose equivalent" is used to predict the degree of the deleterious effect or the probability of attaining it. The dose equivalent involves the QF, the absorbed dose, and other modifying factors defined by the ICRP. The QF values recommended are roughly the same as those used by the NCRP (1 for X-rays and gamma-rays, 20 for alpha particles). The "collective dose equivalent" is really the sum of the dose equivalents in the individuals within a population either in the whole body or the organ system involved. Thus it is the sum of the number of individuals receiving a per capita dose equivalent in the whole body or individual organ system. The "dose-equivalent commitment" is the infinite time exposure at the per capita dose-equivalent rate in a given tissue or organ system for the given population. The "committed dose equivalent" is the dose equivalent that will be accumulated or integrated over the 50 years of exposure following the intake of a radionuclide into the body.

These definitions are perhaps not useful within the chapter context, but they do serve to indicate the presently accepted guideline boundaries as set forth by the ICRP.

3.2.2 Recommended Limits

The ICRP recommends limits of 50 rem in one year to all tissues (except the lens of the eye) as the dose-equivalent that would prevent nonstochastic effects; it recommends 30 rems for the lens (86). These limits apply regardless of mode of exposure (single organ or summed doses to organs). For stochastic effects, they recommend that the dose equivalent be from whole body exposure or the sum of various organ systems. For this they use a weighting system which relates the risk attached to individual systems to the total combined risk. In other words, the weighting factors proportion the stochastic risk to that system to the total risk, as if it were related to the whole body being irradiated uniformly. Their list of weighting factors is given in Table 12.5.

The weighting factor for the "remainder" tissues is really a recommendation of 0.06 for each of the five remaining tissues or organs that are receiving the highest dose equivalents. This could be, for instance, the spleen, kidney, and pancreas. It is recommended that these weighting factors be used unless the more straightforward approach of limiting whole-body exposure to an annual dosage of 5 rems in any year is feasible. This may be clearly expressed as

$$\Sigma_T W_T H_T \leq H_{\text{WB},L}$$

where H_T is the annual dose equivalent in tissue T, W_T is the weighting factor for T, and $H_{\text{WB},L}$ is the recommended annual dose-equivalent limit for uniform whole body radiation, or 5 rems.

Table 12.5 Weighting Factors Used for Dosage Calculations, as Recommended by the ICRP (86)[a]

Tissue	w_T
Gonads	0.25
Breast	0.15
Red bone marrow	0.12
Lung	0.12
Thyroid	0.03
Bone surfaces	0.03
Remainder	0.30

[a] Reprinted with permission from International Commission on Radiological Protection, "Recommendations of the International Commission on Radiological Protection," ICRP Report 26, *Ann. ICRP*, **1**(3) (1977).

3.2.3 Special Cases

The ICRP also has recommendations (86) for special potential exposure cases such as (1) occupational women of reproductive capacity, (2) occupational pregnant women, (3) individual members of the public, (4) populations, and (5) accidents and emergencies.

3.2.4 Applications of ICRP Recommendations

The general concepts outlined previously have been utilized to generate guidelines for limits of intake of radionuclides by the potentially exposed working individual (87). Annual limits on intake of radionuclides and derived air concentrations are calculated for many radioelements and their isotopes with half-lives greater than 10 min (87).

4 THERAPEUTIC MEASURES

It is possible to modify the effects of external irradiation by means of certain sulfur-containing compounds, including cysteine and cysteamine, provided the compounds are administered prior to irradiation (88). It is also considered possible to ameliorate radiation effects by bone marrow transfusion, preferably by using bone marrow provided by the exposed individual prior to exposure. The practicality of measures such as these is so severely limited, because human exposure or contamination is accidental, that they cannot be considered as viable therapeutic measures.

IONIZING RADIATION

The decision to apply therapeutic measures to decrease the effects of exposure to radiation usually must be made promptly and under the supervision of a physician and must always be made carefully. This decision must incorporate information on the potential risk due to the radiation and, often, on the potential risk due to therapy. The very best measurements of exposure conditions can serve only as good estimates of the radiation dose received, in the case of external radiation, or likely to be received, in the case of internal contamination. Therapeutic risks that must be considered include physical injury due to invasive techniques such as injection or surgery, the toxic effects of certain chemicals, the risk associated with general anesthesia or blood transfusion, and the psychological impact of therapy (particularly heroic measures) on exposed individuals and their families.

The basis of therapy is quite different for radiation from external sources as contrasted with internal sources. In the former case therapy is initiated after the total radiation dose has been received. The dose cannot be lessened; therefore, the objective of therapy is modification of effect by treatment of radiation sickness, prevention of secondary infection, or supplementation of dwindling hematapoietic elements. When the radiation dose will be protracted because of the internal presence of the radiation emitter, the goal of therapy is reduction of the quantity of the emitter. This may be accomplished by enhancing excretion of the emitter or by other physical means of removal.

4.1 External Radiation Exposure

The use of therapeutic means to combat exposure to external radiation is an area in which there has been little opportunity to gain human experience. Personnel involved in worker protection in this segment of toxicology should be proud of their record for maintaining such a low number of exposure incidents. A summary of the Lockport, New York, accidental exposures (1) will give sufficient insight into the type of care and therapy that appear to be workable for external whole-body or partial-body irradiation.

The Lockport radiation incident provides knowledge of medical treatment following external radiation exposure that is classical insofar as almost everything, therapeutically, appears to have been done properly. Nine persons were exposed to radiation from an unshielded klystron tube at a U.S. Air Force radar site (1). Three of these personnel received doses of X-rays in the range of 1200 to 1500 R over certain areas of the body. Because exposure occurred during a period of 60 to 120 min in the working area, it was extremely difficult to estimate the dose to a given region of any individual.

Table 12.6 lists the basic symptoms shown by the exposed individuals in descending order of estimated dose received. At the highest exposure level most of the classical signs of acute radiation damage are indicated. Nausea and vomiting were prevalent as well as fatigue and drowsiness. Erythema was a positive indicator of radiation exposure in every case. The first step in therapy was to admit the patients to hospital ward rooms that had been thoroughly

Table 12.6 Signs and Symptoms in Victims of the Lockport Incident[a]

Patient	Tinnitus, Parotid Swelling	Temperomandibular Tenderness	Forehead Swelling	Headache	Abdominal Pain	Nausea	Vomiting	Anorexia	Chills and Fever	Erythema	Conjunctival Reddening	Dry Mucous Membrane	Lassitude and Somnolence
1	+	+	0	+	0	+	0	0	+	+	+	+	+
2	0	+	0	+	0	+	+	0	+	+	+	+	+
3	0	0	0	0	+	+	+	0	0	+	+	+	+
4	0	0	0	+	0	+	+	0	+	+	+	+	+
5	0	0	0	0	0	0	0	0	0	+	0	0	+
6	0	0	+	0	0	0	0	0	0	0	+	+	+
7	0	0	0	0	+	+	0	0	+	+	+	+	+

[a] Reprinted with permission from J. W. Howland, M. Ingram, H. Mermagen, and C. L. Hansen, Jr., "The Lockport Incident: Accidental Partial Body Exposure of Humans to Large Doses of X-Irradiation," in: *Diagnosis and Treatment of Acute Radiation Injury*, proceedings of a scientific meeting jointly sponsored by the International Atomic Energy Agency and the World Health Organization, Geneva, October 17–21, 1960, WHO, Geneva, 1961, p. 11.

IONIZING RADIATION

cleaned, including the culturing of samples from throughout the room (air, furniture, etc.) for pathogens. This procedure was necessary because of the suspected lowered bacterial resistance of the patients.

All personnel entering the ward were required to wear face masks and to observe other contagion precautions. Visitors were limited to the immediate family. Only laboratory studies deemed necessary for making crucial decisions were allowed. Early after admission the decision was made to postpone any bone marrow transplants because the exposures were of the partial body only. Also, no antibiotics or transfusions were administered; it was thought best to hold off these procedures until the advent of a complication of infection or bleeding.

This represents a proper and sensible approach to the medical treatment of individuals following accidental exposure to relatively high levels of external radiation. The methods available to protect the exposed individual if administered prior to exposure, previously mentioned, are hardly applicable to the accidental situation.

4.2 Internal Radiation Exposure

Therapy for exposure from internally deposited radiation emitters consists of reducing the quantity, or body burden, of the deposited radioactive material. Until recently, the use of chelating agents, introduced primarily by IV injection, has been the only productive method of such reduction. Because the effectiveness of chelation therapy is limited to use with internal emitters in a form that is soluble in body fluids, this approach has not been successful for the removal of insoluble materials such as inhaled insoluble particles that are deposited in the lung. In 1972 an accidentally exposed individual underwent bronchopulmonary lavage for removal of inhaled deposited ^{239}Pu (89). The positive results of that treatment, together with results of experiments with animals, indicate that bronchopulmonary lavage is a promising procedure for removing inhaled insoluble radioactive substances. With a choice of effective therapeutic measures, a sometimes difficult problem is that of determining the relative solubility of the internal contaminant to ensure that the most beneficial treatment is chosen.

4.2.1 Chelation Therapy

Chelation is natural to biological systems. Many metabolically formed compounds, including citric acid and gluconic acid, form chelates, and several amino acids are active chelators. Citric acid chelates calcium and is used by blood banks; the citrate–calcium chelate helps to prevent blood coagulation. Chelation therapy for heavy-metal contamination has been in use for more than 30 years, and a variety of chelating agents have been tried (88, 90). Recent reviews of the subject of chelation therapy express the problems with decorporation and evaluate the more current compounds in use or under investigation (91–94). Whereas citric acid is a very efficient binder of calcium, the well-known chelator ethylenediamine tetraacetic acid (EDTA) has a greater affinity for polyvalent metals such as lead, zinc, tin, yttrium, and plutonium. The effectiveness, and thus the proper choice, of a chelator depends on its biological stability and the stability of the materials to be removed. Diethylenetriamine pentacetate (DTPA) forms a more stable chelate with the rare earths than does EDTA and has a residual effect that is desirable in terms of lowering the frequency of treatments.

Chelators are generally poorly absorbed from the gastrointestinal tract; thus they are not of much use when administered orally except in cases of ingestion of the contaminant. Although chelator absorption may be quite good following intramuscular (IM) injection, use of the IM route of administration is rare, partly because of the greater likelihood of local irritation associated with this route. Chelator administration by IV injection is most often reported because of its greater efficiency; however, much greater potential for toxicity exists when the IV route is used. The decisions regarding risk benefit of introducing chelation therapy is not clear-cut; the subject is covered well for plutonium, as an example, in NCRP-65 (92). The greater stability of DTPA and its greater efficiency for removal of heavy metals from the body than EDTA render DTPA more useful at lower dosages and hence less potentially toxic than EDTA. For treatment of internal contamination of humans with polyvalent radionuclides, DTPA has become the chelator of choice. The cation of choice for this chelator may be zinc (Zn–DTPA) as opposed to calcium, as pointed out by Mays et al. (94). The zinc complex appears to be less toxic to the patient by virtue of not depleting necessary cations from the body. Information on the behavior and effects of chelating agents in humans is available because numerous experiments have been carried out by use of tracer quantities of radioactive rare earths in humans (95, 96).

Chelation therapy has been useful in the removal of plutonium from contaminated individuals, as described in the following brief case studies. In one reported case (97) a worker was sprayed with an acid solution of plutonium chloride and plutonium nitrate. Inhalation and ingestion, as well as skin contamination, occurred. The skin, except for burned areas, was decontaminated with dilute sodium hypochlorite solution. Eleven 1-g, IV, DTPA treatments were administered, beginning 1 hr after the accident and at intervals through 17 days. Burn scabs were removed 2 weeks following the accident and they were found to contain most of the plutonium. The combination of treatment methods used in this case was considered highly effective. Similar effectiveness of prompt DTPA treatment was reported for another plutonium-contamination incident involving an acid burn (98). Twenty-seven daily 1-g, IV, DTPA treatments, beginning 1 hr after the accident, resulted in elimination of more than 96 percent of the estimated systemic burden. As in the previous case, much of the contamination was removed with the burn scabs.

The effectiveness of DTPA treatment with the use of a regime similar to those just reported, was considered inconclusive in the case of a wound to the thumb from a plutonium-contaminated metal sliver (97). Initially the sliver was removed without tissue excision. Subsequently tissue excisions were performed at the points of entrance and exit of the sliver; excised tissue from the point of sliver entrance contained about 98 percent of the plutonium removed by excision. Wound counting performed over several months indicated movement of the remaining embedded plutonium toward the skin surface. Nodules that formed concurrently with increased wound counts were excised, and these were found to contain essentially all the plutonium estimated to remain in the thumb. Although DTPA was effective in removing plutonium from the blood, the possible influence of DTPA in mobilizing the plutonium from the wound into the blood was noted.

In another puncture wound episode (99), DTPA was administered IV 4 days per week for about 11 weeks followed by a 30-week period of no treatment, then by a 90-week period in which DTPA was administered either IV or by aerosol 2 or 4 days per month. During the early treatment period, oral (tablet) EDTA was substituted for DTPA as a matter of convenience to the employee; the EDTA was found to be very ineffective in enhancing urinary plutonium excretion. Additional cases of puncture wound injury have been reported (98, 100). In each instance of puncture wound, the greatest efficiency of treatment has resulted from prompt DTPA therapy and one or more tissue excisions. In burn cases surface decontamination except in immediate burn areas, and prompt DTPA therapy, are recommended, followed by careful removal of the burn scabs. In all cases wound counting and/or whole-body counting have been used to determine the plutonium burden.

Several instances of human plutonium contamination by inhalation have been reported. In an early study (101) exposed individuals were treated with calcium EDTA, administered IV in 1-g doses twice a day, for the initial treatment, and then administered orally. The IV treatment resulted in a tenfold increase in urinary plutonium excretion; however, the oral doses caused only minor increases and were not considered to be of value. Chelation therapy was deemed successful, although the total urinary excretion amounted to only 10 percent of the body plutonium content. The key to effective chelation therapy is promptness.

4.2.2 Pulmonary Lavage

Pulmonary lavage is a recent technique that is applied primarily to patients with alveolar proteinosis. Kylstra worked for years in this field and became one of the world's experts (102, 103). In recent years the application of pulmonary lavage to animals has been successful in removing radionuclides after inhalation exposure (6, 104). The combined use of chelating agent and lung lavage has proved very successful as shown in animal studies (105). Some of these studies with dogs utilized inhalation of aerosols from practical uses in the nuclear industry. In addition, one human case of radionuclide exposure has been somewhat successfully treated (89).

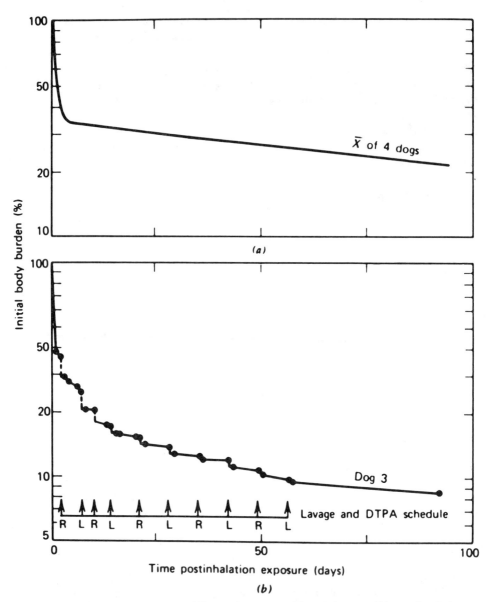

Figure 12.9 Whole-body retention of ^{144}Ce in dogs exposed by inhalation to ^{144}Ce in fused clay: (a) dogs received no postexposure treatment—dog 3 was lavaged 10 times during the first 56 days after exposure; (b) treated dogs—DTPA was given intravenously after each lavage: solid points represent whole-body count. The vertical steps represent the ^{144}Ce removed in the lavage fluid with each treatment. Data are uncorrected for physical decay of ^{144}Ce. From Boecker et al. (107). Reproduced from *Health Physics*, Vol. 26, page 509 (June 1974) by permission of the Health Physics Society.

The manner of treatment is to place a Carlens catheter (106) into the patient's respiratory tract such that one lung is supplied with oxygen while the other has access to washing fluid. In general, physiological saline is sufficient for the washing process. Approximately a tidal volume of saline is introduced into the open lung and then removed. This procedure may be repeated up to 5 to 10 times on the same lung during one session. The catheter is withdrawn, and after 24 to 48 hr the procedure is repeated on the other lung. With this procedure the surfactant that lines the alveolar regions is partially removed, and with it, any cells or other materials that reside there. Thus an insoluble material (radionuclide) in the deep lung, whether free or within macrophages, is removed to some extent. The following experimental data illustrate the effectiveness of the procedure.

At the Inhalation Toxicology Research Institute in Albuquerque, New Mexico, many studies have been performed on the effectiveness of lavage in removing inhaled particles from experimental animals. One aerosol used for this purpose in beagle dogs has been ^{144}Ce in fused aluminosilicate aerosol particles. A study by Boecker et al. (107) showed how effective multiple lavage can be in removing the particles deposited in the respiratory tract. Figure 12.9 gives an example. The reduction in radiation effect on the lung by systematic removal of the radiation source (particles of ^{144}Ce in this case) is obvious. The lavage fluid was physiological saline. DTPA treatment was also used in conjunction with the lavage procedure with some of the dogs, but the overall effectiveness of the added chelator was minimal. A similar inhalation experiment was performed with ^{144}Ce in fused clay aerosol particles in beagle dogs, and those data are expressed in terms of reduction in radiation dose to the lung in Figure 12.10. According to Silbaugh et al. (108), no impairment of health was seen at long times postexposure (14 months) in the lavaged dogs, whereas untreated dogs receiving those initial doses would have been sick or died. An extension of this report to 2600 days (7 years) indicates continued positive results and stresses the lengthening of life span that resulted from use of the combination lavage and chelation therapy (109).

5 BIOASSAY TECHNIQUES

It is difficult to treat irradiated individuals for potential damage if the extent of the projected biological insult cannot be roughly estimated. Bioassay has been one of the descriptive terms applied to this field of making such dosimetric approximations. As will be seen, this is a very difficult task and needs much more experimental advancement before bioassay becomes the exact science that is desirable. It seems to be more effective with internal emitters, as discussed in Section 5.2.

5.1 External Radiation Exposures

There are few techniques for detection of human external radiation exposures, although many have been explored and attempted during the past few decades.

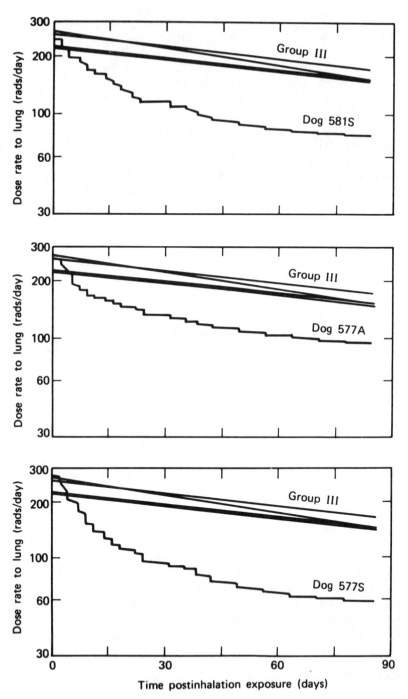

Figure 12.10 Reduction, with time, in the beta dose rate to the lungs of unlavaged group III dogs and lavaged group II dogs 581S, 577A, and 577S. From Silbaugh et al. (108). Reproduced from *Health Physics*, Vol. 29, page 86 (July 1975) by permission of the Health Physics Society.

Biological indicators of exposure must be sensitive to low-level radiation if the practical "in-plant" situation is to be monitored. This criterion eliminates chemical reactions that are quickly restored or require large doses to bring about measurable reaction products. It also essentially eliminates the old practice in therapy of using an erythema reaction to estimate the amount of radiation delivered.

An extremely troublesome factor in biological systems is the tremendous number of different chemical species present in any one finite sphere of interaction (ionization). If one had an ideal chemical oxidation–reduction reaction that would respond to low radiation doses, and if the primary constituents were present in necessary abundance in the volume of tissue of interest (e.g., a cell), the chemical entities would still be overwhelmed by other chemical species. If these "foreign" substances were reactive either to the radiation or to the products being formed and measured in the primary event, the accuracy would immediately be challenged. Such is the case with chemically induced bioassay schemes for detecting external exposure to ionizing radiation *in vivo*.

Perhaps of some potential are schemes that measure actual effects on the biological macrosystems. The genetic materials, chromosomes, offer one possibility as a dosimeter for practical radiation exposure. Although measurement of chromosomal damage through breakage is tedious, this damage nevertheless may prevail for periods of time, lending it practical significance. One drawback is the recombination of chromosomes that have been broken by the radiation to a normal condition. This can readily occur under the influence of ionizing events or of other toxic reactions that affect chromosomes. However background quantitation for the probability of recombination is possible (and feasible) and can become part of the overall scheme for dosimetry purposes. Samples of peripheral blood are readily obtainable from workers for use in this technique of quantitating chromosome damage due to ionizing radiation.

Lymphocytes are very sensitive to radiation and prove to be good indicators of damage. In Figure 12.1 it is obvious that many hematological parameters (cell numbers) show a dramatic change as a result of ionizing radiation. Therefore, one may be able to estimate the radiation dose in a rough manner, as was attempted with the Lockport cases (1). One of the more sensitive indicators using blood samples is the increased abundance of bilobed lymphocytes, a technique first explored and implemented by Ingram (110, 111). This rather specific indicator of radiation damage was first applied to workers associated with a cyclotron at the University of Rochester. Bilobed lymphocytes tend to be slightly larger than normal, and they contain two distinct nuclear masses. Each is a typical nucleus of normal lymphocytes, but they occur two to a cell. The main drawback to this type of personnel monitoring is the tedium of microscopic examination of so many cells in a preparation. The low incidence of these altered cells requires that statistical analyses be performed for the detection of low doses. However, the accuracy is such that this indicator provides a suitable bioassay procedure for low doses of either electromagnetic radiation or neutrons (110–112).

5.2 Internal Radiation Exposure

Radionuclides deposited in the body also point up the gap between ease of detection and interpretation of results. The radioactivity of a material present in a biological system provides one of the most sensitive means that will ever be known for localizing an element involved in the body's metabolic scheme. Measurement of a radionuclide's effect and its metabolic parameters, as related to the overall *in vivo* system, is a difficult problem, however, and almost insurmountable under some circumstances. As an example, detection (quantitation) of an alpha particle emitter in a section of tissue can be carried out with great precision. Equally credible measurements can be made on the contaminated excreta resulting from deposition of this material. However, the intervening biological scheme at play between the deposit of the radionuclide and its elimination is generally an unknown entity. It is this factor that leads to frustration where practical bioassay programs for determination of industrial exposures are concerned.

5.2.1 Inhalation Exposures

Many problems are inherent in bioassay of materials deposited in the lung following inhalation. The solubility of the inhaled radionuclide has an immediate, important role in the effectiveness of use of bioassay techniques. A "completely soluble" (in body fluids) radionuclide may enter the blood and either deposit in the tissue of choice or be excreted, primarily in the urine. If the release from the designated site of localization is with constant and known kinetics, appearance of the material in the urine may be an excellent source of information for extrapolation to an estimated body burden. An example of such a material is tritium. When inhaled, this isotope of hydrogen enters the normal body pool and in a short period becomes part of the body's metabolic scheme. Measurement of $_1H^3$ in the urine and application of isotope dilution principles as related to body characteristics (e.g., weight, height), serves as a suitable bioassay technique (113–115). Tritium assay presents some difficulties, which are not unusual with the bioassay of internally deposited radionuclides, in that tritium leaves the body with multiple exponents of excretion that may depend on environmental factors such as season of the year (115). However, tritium represents the ideal case (soluble), and this bioassay method is generally not applicable in inhalation exposure cases.

When the inhaled radionuclide is not in a soluble form, the situation with regard to bioassay becomes nightmarish. If the radionuclide is "completely insoluble" (a misnomer, since it holds true for no material in body fluids), there is no bioassay technique that is quantitative through normal metabolic processes. A radionuclide in this form, which has some sort of detectable radiation (X-ray or gamma-ray) outside the body, may be quantitated by chest counting. This is a reasonably accurate method except as the energy of emission approaches 100 keV and lower. At these low energies the rib cage becomes a troublesome factor because it absorbs the photon. The degree of absorption depends on the size of the rib cage and associated tissue, in both thickness and variable density, as well as the photon energy. Accurate calibration for a specific individual is essentially impossible, since mockups rarely match the subject. Thus estimates of the biological parameters are made through height, weight, lean mass:body weight ratios, and so on. Ultrasonic measurements are made at many locations about the chest cavity, but determination of the degree of self-absorption for an individual is still close, at best. Whole-body counting does remain the only practical scheme, however, for estimating lung burden of an insoluble radionuclide. One problem encountered with this and all chest counting is the inability to distinguish lymph node accumulation of radionuclide from that in lung tissue. Some experimentation has been carried out to evaluate the use of minute detectors that can be lowered into the respiratory tract by use of a device similar to a bronchioscope, but to date this technique has not found widespread use (116). Pinpoint detection will remain a problem until schemes for precision of measurements have been more thoroughly tested.

The case of exposure to somewhat soluble or slowly solubilized radionuclides is perhaps the most difficult from the viewpoint of bioassay (see discussion of the kinetics of loss from the lung to internal organs in Section 1.3.3). If one knows all the parameters of loss from the lung to blood to urine, a backcalculation readily produces the lung burden. Unfortunately, however, this is rarely the case, particularly in that an aerosol breathed by the same or different persons, at the same location, under seemingly identical conditions, will lead to variable results. Thus any bioassay procedure attempted under conditions of partial solubility will meet with frustration and, probably, erroneous results and erroneous interpretation.

Many references exist on whole-body counting and excretory techniques for determining the radioactivity in humans, and these represent a starting place for those interested in pursuing this field (117, 118). Also, directories have been published that contain information on the laboratories that operate whole-body monitors, with information on the technical design and performance of the individual apparatus (119).

5.2.2 Puncture Wounds

One of the most common means of receiving accidental body burdens of radionuclides is through skin absorption or puncture. The situation is similar to inhalation exposure and can be considered in the same light. The wound burden can be assumed to be comparable to the lung burden, and rate of loss from the site is given the same consideration. If there is a detectable radioactive emission, a wound counter can be utilized in the same manner as a chest counter (97). However, the advantage with wounds is that excision or amputation of the contaminated body part can be put to use in relieving the problem; this is less feasible in the case of inhalation exposure. Estimation of the wound radioactivity even can be made through biopsy, a technique that is not as satisfactory where the lung is concerned. Urine analysis may be satisfactory under certain conditions of wound contamination, but external counting, where feasible, still remains the most practical and successful method of assaying for body burdens.

6 RADIATION AND ENERGY SOURCES

Present predictions indicate that fossil forms of energy will allow the United States to satisfy its energy needs for several decades. Also, there appears to be considerable optimism that solar, geothermal, or other sources may increase this period indefinitely, and many research teams are working in these areas. In addition to these energy sources, however, are those from which radiation is an obvious by-product. The sections to follow briefly indicate the possible problems associated with them.

6.1 Radiation Sources

One of the more promising future sources of energy is the practical application of nuclear fusion. If this source is exploited to an economically feasible state, it can be the answer to the future's energy problems. In the fusion process atomic nuclei (light atomic weight) are brought together in such a manner that energy above the amount of the binding energy of the particles is released and can be captured and utilized as a practical energy source. Since nuclear fusion would not produce as much atmospheric contamination with radioactive materials as would nuclear fission, the process itself might be acceptable to the factions in the public that are alienated by the term "radiation," regardless of the context in which it is used. However, the fusion process, if and when it is feasibly developed as a practical energy source, may not be without radiation problems. There will be waste disposal problems as with nuclear reactors, and the neutrons that are emitted as part of the energy release could present an industrial radiation problem. The practicability of fusion as a source of energy is becoming a political and socioeconomic topic for discussion, and some relatively recent articles face many of these issues (120–123).

Nuclear fission is the most realistic source of energy at present, and it is currently being used in many countries of the world. An article that appeared in 1976 indicates that France planned an increase from 8 to 70 percent electricity production from nuclear power by 1985, as a result of the embargo by major petroleum producers and increased oil prices (124). Various types of power reactor may be used as sources of energy, some more appealing than others from the standpoint of potential release of airborne radioactivity. Modern design of most proposed nuclear power reactors is such that even an internal malfunction (i.e., within the reactor) would not result in atmospheric contamination. Reutilization of the repaired initial plant, or substitution of a new one, could take place to restore the same site as a source of power. Engineering accomplishments to maintain any radiation problems within the confinement of the reactor complex itself indicate that any exposure to the surrounding area would be of little significance, probably zero. Because this energy source represents a horrifying (Hiroshima-like) scene to some of the American public, it will take considerable time and public education before the nuclear reactor is accepted as a safe major source of energy in this country. Our most realistic view on all of this has come from the incidents surrounding the well-known Three Mile Island (TMI) nuclear reactor accident. Much has been learned from this mishap. An aspect of concern to proponents of nuclear sources of energy is that 10 to 20 years may be required from the time that a new power station is begun until it can assume routine operation.

6.2 Mixed Potential Hazards

Potential hazards may be involved in the event that a radiation source combines with an environmental chemical source in the public atmosphere. One must consider the possibility, however slight, that a nuclear power source could give rise to some atmospheric contamination as the result of an accidental release. This release could be in the form of particulates and could be acted on by such factors as local meteorology, temperature, terrain, and altitude.

Synergism in inhalation toxicity, particularly concerning a source of radiation and a potential chemical cocarcinogen, has not been studied in depth. A study with experimental animals at the Los Alamos National Laboratory recently indicated a sizable increase in lung cancer incidence when Fe_2O_3 was added to PuO_2 particles deposited in the respiratory tract (125). The latter insult alone (PuO_2) was only slightly more tumorigenic than controls under similar conditions of administration to the animal species used (Syrian golden hamsters) (126, 127). The public is informed almost daily in newspapers and journals of the possible atmospheric contamination arising from common sources. Materials such as asbestos from automobile brakes, oxides of sulfur, oxides of nitrogen, carbon monoxide, dust, and high humidity are popular subjects for news media and other concerned factions. These materials are generally measured and indexed by local (state and city) environmental regulatory organizations, and when certain limits are surpassed, an attempt is made to reduce these to some established "allowable" level. Radiation is treated in the same manner, but the measured values are required to be essentially zero (natural background for the area), or the lowest practicable.

6.3 Mining as a Factor

Hazards associated with the mining of uranium have been recognized for centuries. Within the past 50 years the origin of lung cancer among these miners has been quite firmly attributed to the radiation associated with the radioactive daughters of radon, the noble radioactive gas from uranium. There is also an external source of radiation associated with the uranium decay products, and this can range from 0.02 to 4 mrem/hr. As Holaday (128) points out, these rates of gamma radiation have been reported for mines in France, Japan, Mexico, Spain, Australia, and the United States. It was determined a few years ago that the respiratory tumorigenicity associated with the miners was closely related to cigarette smoking in conjunction with the deposited radioactive particles of the radon daughters (see Section 2.2.1).

One of the most interesting studies in recent years concerning the mining industry dealt with the immunologic assessment of neoplasia and metaplasia in the lungs of miners (129). The main purposes of the study were to answer the following questions:

1. Can alterations in immunologic function be detected in individuals with preneoplastic pulmonary lesions?
2. If so, will these changes help to identify those individuals who may develop invasive lung cancer?
3. What immunologic test is the most sensitive indicator of abnormalities?
4. Can the relative contributions of carcinogens be determined, such as the relative contributions of radon daughter products and cigarette smoke to lung cancer in uranium miners?

The results of this study, performed on blood samples from the mining population of the Western Slope of the Colorado Rockies, were positive in showing defects in the immune system associated with the early development of lung cancer. Rosette inhibition testing was the most sensitive indicator found that would relate to the presence of malignancy in the miners with known neoplasia. Thus early testing of this nature may be done to detect early changes that are perhaps undetectable by other than simultaneous sputum cytological testing, with the possibility of performing early immunotherapy. There was no evidence in the populations studied (miners and controls) that smoking impaired immune competence. This study, in combination with results of sputum cytological testing (68, 74), shows great promise in associating immune system abnormalities with potentially carcinogenic industrial toxicants. It is included in this chapter because of the obvious relationship to the carcinogenic risks associated with irradiation of the pulmonary system by radon daughters.

Mining problems have assumed new importance since the energy crisis has directed industry into evaluating the extensive procurement, refinement, and use of potential sources of underground fossil fuels. Oil shale mining is an excellent example, and certainly there are potential chemical carcinogens associated with the material mined, processed, refined, and utilized for energy. Radon and radon daughters are prevalent in coal mines, and there is every reason to believe that any underground mining, in the Rocky Mountain region particularly, will be a source of radon and its daughters. With newer, stricter regulations and guidelines, the cost of eliminating any such measurable materials can become enormous. In addition to the potentially small radiation problem, the dusty atmospheres associated with mining may contain potential chemical carcinogens that will be inhaled along with the radioactive sources. In fact, it is likely that if workers or the population receive any exposures to potentially toxic levels, the toxicants will be of chemical origin, not the low levels of radiation that are present in the mines.

REFERENCES

1. J. W. Howland, M. Ingram, H. Mermagen, and C. L. Hansen, Jr., "The Lockport Incident: Accidental Partial Body Exposure of Humans to Large Doses of X-Irradiation," in: *Diagnosis and Treatment of Acute Radiation Injury*, Proceedings of a Scientific Meeting Jointly Sponsored by the International Atomic Energy Agency and the World Health Organization, Geneva, October 17–21, 1960, WHO, Geneva, 1961, p. 11.
2. D. M. Ross, "A Statistical Summary of United States Atomic Energy Commission Contractors' Internal Exposure Experience 1957–1966," in: *Diagnosis and Treatment of Deposited Radionuclides*, Proceedings of a Symposium, Richland, WA, May 15–17, 1967, p. 427.
3. H. C. Hodge, J. N. Stannard, and J. B. Hursh, Eds., *Uranium–Plutonium–Transplutonic Elements*, Springer-Verlag, New York, 1973, p. 643.
4. E. C. Anderson, *J. Bacteriol.*, **76,** 471 (1958).
5. T. T. Mercer, M. I. Tillery, and H. Y. Chow, *Am. Ind. Hyg. Assoc. J.*, **29,** 66 (1968).
6. B. A. Muggenburg, S. A. Felicetti, and S. A. Silbaugh, *Health Phys.*, **33,** 213 (1977).
7. K. J. Caplan, L. J. Doemeny, and S. D. Sorenson, *Am. Ind. Hyg. Assoc. J.*, **38,** 162 (1977).
8. K. J. Caplan, L. J. Doemeny, and S. D. Sorenson, *Am. Ind. Hyg. Assoc. J.*, **38,** 83 (1977).
9. M. Lippmann, "Aerosol Sampling For Inhalation Hazard Evaluation," in: *Assessment of Airborne Particles*, T. T. Mercer, P. E. Morrow and W. Stöber, Eds., Thomas, Springfield, IL, 1972, p. 449.
10. Anonymous, *Am. Ind. Hyg. Assoc. J.*, **31,** 133 (1970).
11. M. Lippmann, *Am. Ind. Hyg. Assoc. J.*, **31,** 138 (1970).
12. D. A. Holaday, "Uranium Mining Hazards," in: *Uranium–Plutonium–Transplutonic Elements*, H. C. Hodge, J. N. Stannard, and J. B. Hursh, Eds., Springer-Verlag, New York, 1973, p. 301.
13. R. F. Droullard, "Instrumentation for Measuring Uranium Miner Exposure to Radon Daughters," in: *Radiation Hazards in Mining: Control, Measurement, and Medical Aspects*, M. Gomez, Ed., Kingsport Press, Inc., Kingsport, TN, 1981, p. 332.
14. H. D. Freeman, *An Improved Radon Flux Measurement System for Uranium Tailings Pile Measurement*, ibid, p. 339.
15. T. B. Borak, E. Franco, K. J. Schiager, J. A. Johnson, and R. F. Holub, *Evaluation of Recent Developments in Radon Progeny Measurements*, IBID, p. 419.
16. International Commission on Radiation Units and Measurements, *Neutron Dosimetry for Biology and Medicine*, ICRU Report 26, 1977.
17. O. G. Raabe, "Generation and Characterization of Aerosols," in: *Inhalation Carcinogenesis*, AEC Symposium Series 18, 1970, p. 123.
18. Task Group on Lung Dynamics, *Health Phys.*, **12,** 173 (1966).
19. R. G. Thomas, "Uptake Kinetics of Relatively Insoluble Particles by Tracheobronchial Lymph Nodes," in: *Radiation and the Lymphatic System*, proceedings of symposium, Richland, WA, ERDA Conference CONF-740930, 1974, p. 67.

20. P. Nettesheim, M. G. Hanna, Jr., and J. W. Deatherage, Jr., Eds., *Morphology of Experimental Respiratory Carcinogenesis*, AEC Symposium Series 21, 1970, p. 417.
21. R. G. Thomas, "Tracheobronchial Lymph Node Involvement Following Inhalation of Alpha Emitters," in: *Radiobiology of Plutonium*, B. J. Stover and W. S. S. Jee, Eds., J. W. Press, University of Utah, Salt Lake City, 1972, p. 231.
22. W. S. S. Jee, Ed., *The Health Effects of Plutonium and Radium*, J. W. Press, University of Utah, Salt Lake City, 1976, p. 169.
23. U.S. Department of Commerce, National Bureau of Standards, *Maximum Permissible Body Burdens and Maximum Permissible Concentrations of Radionuclides in Air and in Water for Occupational Exposure*, NBS Handbook 69, Government Printing Office, Washington, DC, 1959.
24. International Commission on Radiological Protection, Committee II, *Health Phys.*, **3** (June 1960).
25. R. G. Cuddihy, "Modeling of the Metabolism of Actinide Elements," in: *Actinides in Man and Animals*, M. E. Wrenn, Ed., RD Press, Salt Lake City, Utah, 1981, p. 617.
26. P. W. Durbin, *Health Phys.*, **8,** 665 (1962).
27. International Commission on Radiological Protection, *Biological Effects of Inhaled Radionuclides*, ICRP Publication 31; *Ann. IRCP* **4**(1) (1980).
28. S. Okada, Ed., *J. Radiat. Res., Jpn. Radiat. Res. Soc.*, **16** (Suppl.) (1975).
29. G. W. Beebe, *Am. J. Epidemiol.*, **114,** 761 (1981).
30. "The Effects on Populations of Exposure to Low Levels of Ionizing Radiation," Committee on the Biological Effects of Ionizing Radiations, National Research Council, National Academy Press, National Academy of Sciences, Washington, DC, 1980.
31. G. M. Matanoski, R. Seltzer, P. E. Sartwell, E. L. Diamond, and E. A. Elliott, *Am. J. Epidemiol.*, **101,** 188 (1975).
32. A. C. Upton, "Effects of Radiation on Man," in: *Annual Review of Nuclear and Particle Science*, Vol. 18, Annual Reviews, Palo Alto, CA, 1968, p. 495.
33. E. Lorenz, *Am. J. Roetgenol.*, **63,** 176 (1950).
34. H. F. Henry, *JAMA*, **176,** 671 (1961).
35. T. D. Luckey, Ed., *Hormesis with Ionizing Radiation*, CRC Press, Inc., Boca Raton, FL, 1981.
36. J. F. Spalding, R. G. Thomas, and G. L. Tietjen, *Life Span of C57 Mice as Influenced by Radiation Dose, Dose Rate, and Age at Exposure*, Los Alamos National Laboratory Report, LA-9528, October, 1982.
37. R. E. Albert, F. J. Burns, and P. Bennett, *J. Natl. Cancer Inst.*, **49,** 1131 (1972).
38. H. Spiess and C. W. Mays, "Protraction Effect on Bone-Sarcoma Induction of ^{224}Ra in Children and Adults," in: *Radionuclide Carcinogenesis*, Proceedings of the 12th Annual Hanford Biology Symposium, 1973, p. 437.
39. W. M. Court Brown and R. Doll, *Leukemia and Aplastic Anemia in Patients Irradiated for Ankylosing Spondylitis*, Medical Research Council Special Report Series 295, Her Majesty's Stationery Office, London, 1957, p. 21.
40. W. M. Court Brown and R. Doll, *Br. Med. J.*, **II,** 1327 (1965).
41. N. H. Müller and H. G. Ebert, Eds., *Biological Effects of ^{224}Ra. Benefit and Risk of Therapeutic Application*, Martinus Nijhoff Medical Division, The Hague, 1978.
42. R. H. Mole, *Br. J. Radiol.*, **48,** 157 (1975).
43. S. Jablon and H. Kato, *Radiat. Res.*, **50,** 649 (1972).
44. M. Brown, *Radiat. Res.*, **71,** 34 (1977).
45. National Council on Radiation Protection and Measurements, *Influence of Dose and Its Distribution in Time on Dose–Response Relationships for Low-LET Radiations*, NCRP Report No. 64, 1980.
46. D. E. Lea, *Actions of Radiations on Living Cells*, Macmillan, New York, 1947.

47. A. C. Upton, *Radiat. Res.*, **71**, 51 (1977).
48. R. E. Rowland, A. T. Keane, and H. F. Lucas, Jr., "A Preliminary Comparison of the Carcinogenicity of ^{226}Ra and ^{228}Ra in Man," in: *Radionuclide Carcinogenesis*, AEC Symposium Series 29, CONF-720505, 1973, p. 406.
49. J. D. Abbatt, "Human Leukemic Risk Data Derived from Portuguese Thorotrast Experience," in: *Radionuclide Carcinogenesis*, AEC Symposium Series 29, CONF-720505, 1973, p. 451.
50. R. E. Rowland and P. W. Durbin, "Survival, Causes of Death, and Estimated Tissue Doses in a Group of Human Beings Injected with Plutonium," in: *The Health Effects of Plutonium and Radium*, W. S. S. Jee, Ed., J. W. Press, University of Utah, Salt Lake City, 1976, p. 329.
51. G. L. Voelz, *Health Phys.*, **29**, 551 (1975).
52. "Actinides in Man and Animals," in: *Proceedings of the Snowbird Actinide Workshop, October 15–17, 1979*, M. E. Wrenn, Sci. Ed., RD Press, Salt Lake City, 1981.
53. J. H. Marshall, J. Rundo, and G. E. Harrison, *Radiat. Res.*, **39**, 445 (1969).
54. W. P. Norris, T. W. Speckman, and P. F. Gustafson, *Am. J. Roentgenol*, **73**, 785 (1955).
55. C. E. Miller and A. J. Finkel, *Am. J. Roentgenol.*, **103**, 871 (1968).
56. Pacific Northwest Laboratory Annual Report for 1981 to The DOE Office of Energy Research, Part I, *Biomedical Science*, PNL-4100 PT 1, February 1982.
57. *Inhalation Toxicology Research Institute*, Annual Report, 1980–1981, to the United States Department of Energy, LMF-91, December 1981.
58. R. G. Thomas and D. M. Smith, *Int. J. Cancer*, **24**, 594 (1979).
59. R. O. McClellan, S. A. Benjamin, B. B. Boecker, F. F. Hahn, C. H. Hobbs, R. K. Jones, and D. L. Lundgren, "Influence of Variations in Dose and Dose Rates on Biological Effects of Inhaled Beta-Emitting Radionuclides," in: *Biological and Environmental Effects of Low-Level Radiation*, Proceedings of the International Atomic Energy Agency Symposium, Chicago, November 3–7, 1975, Vol. 2, 1976, p. 3.
60. C. L. Sanders, G. E. Dagle, W. C. Cannon, D. K. Craig, G. J. Powers, and D. M. Meier, *Radiat. Res.*, **68**, 349 (1976).
61. F. F. Hahn, S. A. Benjamin, B. B. Boecker, T. L. Chiffelle, C. H. Hobbs, R. K. Jones, R. O. McClellan, and H. C. Redman, "Induction of Pulmonary Neoplasia in Beagle Dogs by Inhaled ^{144}Ce Fused-Clay Particles," in: *Biological and Environmental Effects of Low-Level Radiation*, Proceedings of the International Atomic Energy Agency Symposium, Chicago, November 3–7, 1975, Vol. 2, 1976, p. 201.
62. R. G. Thomas, G. A. Drake, J. E. London, E. C. Anderson, J. R. Prine, and D. M. Smith, *Int. J. Radiat. Biol.*, **46**, 605 (1981).
63. D. A. Holaday, "Uranium Mining Hazards," in: *Uranium–Plutonium–Transplutonic Elements*, H. C. Hodge, J. N. Stannard, and J. B. Hursh, Eds., Springer-Verlag, New York, 1973, p. 296.
64. C. D. Stewart and S. D. Simpson, "The Hazards of Inhaling Radon-222 and Its Short-lived Daughters: Consideration of Proposed Maximum Permissible Concentrations in Air," in: *Radiological Health and Safety in Mining and Milling of Nuclear Materials*, IAEA Symposium Series, Vol. 1, 1964, p. 333.
65. D. A. Holaday, *Health Phys.*, **16**, 547 (1969).
66. V. E. Archer, J. K. Wagoner, and F. E. Lundin, *Health Phys.*, **25**, 351 (1973).
67. *Radiation Hazards in Mining: Control Measurement, and Medical Aspects*, Proceedings, Conference held at the Colorado School of Mining, Golden, Colorado, October 4–9, 1981. Kingsport Press, Inc., Kingsport, TN, 1981.
68. G. Saccomanno, V. E. Archer, R. P. Saunders, O. Auerbach, and M. G. Klein, "Early Indices of Cancer Risk Among Uranium Miners with Reference to Modifying Factors," in: *Occupational Carcinogenesis*, Vol. 271, U. Saffiotti, and J. K. Wagoner, Eds., New York Academy of Sciences, New York, 1976, p. 377.

69. V. E. Archer, J. K. Wagoner, and F. E. Lundin, *J. Occup. Med.*, **15,** 204 (1973).
70. F. E. Lundin, J. W. Lloyd, E. M. Smith, V. E. Archer, and D. A. Holaday, *Health Phys.*, **16,** 571 (1969).
71. V. E. Archer and F. E. Lundin, *Environ. Res.*, **1,** 370 (1967).
72. E. A. Martell and K. S. Sweder, "The Roles of Polonium Isotopes in the Etiology of Lung Cancer in Cigarette Smokers and Uranium Miners," in: *Proceedings of International Conference on Radiation Hazards in Mining: Control, Measurement and Medical Aspects*, M. Gomez, Ed., Kingsport Press, Inc., Kingsport, TN, 1981, p. 383.
73. G. Saccomanno, V. E. Archer, O. Auerbach, M. Kuschner, M. Egger, S. Wood, and R. Mick, *Age Factor in Histological Type of Lung Cancer Among Uranium Miners, A Preliminary Report*, IBID, p. 675.
74. G. Saccomanno, *Lab. Med.*, **10**(9), 523 (1979).
75. R. A. Lemen, W. M. Johnson, J. K. Wagoner, V. E. Archer, and G. Saccomanno, "Cytologic Observations and Cancer Incidence Following Exposure to BCME," in: *Occupational Carcinogenesis*, Vol. 271, U. Saffiotti and J. K. Wagoner, Eds., New York Academy of Sciences, New York, 1976, p. 71.
76. B. J. Stover and C. N. Stover, Jr., "The Laboratory for Radiobiology at the University of Utah," in: *Radiobiology of Plutonium*, J. W. Press, University of Utah, Salt Lake City, 1972, p. 29.
77. C. W. Mays, H. Spiess, G. N. Taylor, R. D. Lloyd, W. S. S. Jee, S. S. McFarland, D. H. Taysum, T. W. Brammer, D. Brammer, and T. A. Pollard, "Estimated Risk to Human Bone from ^{239}Pu," in: *The Health Effects of Plutonium and Radium*, W. S. S. Jee, Ed., J. W. Press, University of Utah, Salt Lake City, 1978, p. 343.
78. C. W. Mays and H. Spiess, "Bone Sarcoma Risks to Man from ^{224}Ra, ^{226}Ra, and ^{239}Pu," in: *Proceedings of Symposium on Biological Effects of ^{224}Ra*, W. A. Müller and H. G. Ebert, Eds., Martinus Nijhoff Medical Division, The Hague, Boston, 1978, p. 168.
79. R. G. Thomas and J. N. Stannard, *Radiat. Res. Suppl.*, **5,** 16 (1964).
80. W. J. Bair, J. E. Ballou, J. F. Park, and C. L. Sanders, "Plutonium in Soft Tissues with Emphasis on the Respiratory Tract," in: *Uranium–Plutonium–Transplutonic Elements*, H. C. Hodge, J. N. Stannard, and J. B. Hursh, Eds., Springer-Verlag, New York, 1973, p. 503.
81. L. J. Leach, E. A. Maynard, H. C. Hodge, J. K. Scott, C. L. Yuile, G. E. Sylvester, and H. B. Wilson, *Health Phys.*, **18,** 599 (1970).
82. National Council on Radiation Protection and Measurements, *Protection of the Thyroid Gland in the Event of Releases of Radioiodine*, NCRP Report 55, 1977.
83. P. Hamilton and E. A. Tompkins, 1975, personal communication (Bureau of Radiological Health, Department of Health, Education and Welfare, Food and Drug Administration, Washington, and Oak Ridge Associated Universities, Oak Ridge, TN).
84. U.S. Nuclear Regulatory Commission, *Reactor Safety Study: An Assessment of Accident Risks in U.S. Commercial Nuclear Power Plants, Appendix VI, Calculations of Reactor Accident Consequences*, Report WASH-1400, NUREG-75/014 (U.S. NRC, Washington, DC), 1975. (Available from National Technical Information Service, Department of Commerce, Springfield, VA.)
85. National Council on Radiation Protection and Measurements, *Basic Radiation Protection Criteria*, NCRP Report 39, 1971.
86. International Commission on Radiological Protection, "Recommendations of the International Commission on Radiological Protection," ICRP Report 26, *Ann. ICRP*, **1**(3) (1977).
87. International Commission on Radiological Protection, *Limits for Intakes of Radionuclides by Workers*, ICRP Report 30, *Ann. ICRP*, **2**(3), 4 (1979).
88. A. Catsch, *Radioactive Metal Mobilization in Medicine*, Thomas, Springfield, IL, 1964.
89. R. O. McClellan, H. A. Boyd, S. A. Benjamin, R. G. Cuddihy, F. F. Hahn, R. K. Jones, J. L. Mauderly, J. A. Mewhinney, B. A. Muggenburg, and R. C. Pfleger, *Health Phys.*, **23,** 426 (1972).

90. A. Catsch and A. E. Harmuth-Hoene, *Biochem. Pharmacol.*, **24,** 1557 (1975).
91. V. Volf, *Treatment of Incorporated Transuranium Elements,* International Atomic Energy Agency Report 184, Vienna, Austria (1978).
92. National Council on Radiation Protection, *Management of Persons Accidentally Contaminated with Radionuclides,* NCRP Report 65, Washington, DC, 1980.
93. G. N. Stradling and R. A. Bulman, "Recent Research on Decorporation Therapy at National Radiological Protection Board" in: *Actinides in Man and Animals,* M. E. Wrenn, Ed., RD Press, Salt Lake City, 1981, p. 369.
94. C. W. Mays, G. N. Taylor, R. D. Lloyd, and M. E. Wrenn, "Status of Chelation Research: A Review," ibid., p. 351.
95. A. Soffer, *Chelation Therapy,* Thomas, Springfield, IL, 1964.
96. H. Spencer and B. Rosoff, *Health Phys.,* **11,** 1181 (1965).
97. C. R. Lagerquist, S. E. Hammond, E. A. Putzier, and C. W. Piltingsrud, *Health Phys.,* **11,** 1177 (1965).
98. C. R. Lagerquist, E. A. Putzier, and C. W. Piltingsrud, *Health Phys.,* **13,** 965 (1967).
99. L. Jolly, H. A. McClearen, G. A. Poda, and W. P. Walke, *Health Phys.,* **23,** 333 (1972).
100. F. Swanberg and R. C. Henle, *J. Occup. Med.,* **6,** 174 (1964).
101. W. D. Norwood, P. A. Fuqua, R. H. Wilson, and J. W. Healy, "Treatment of Plutonium Inhalation: Case Studies," in: *Experience in Radiological Protection,* Proceedings of the Second United Nations International Conference on the Peaceful Uses of Atomic Energy, Geneva, September 1-13, 1958, Vol. 23, 1958, p. 434.
102. J. A. Kylstra, D. C. Rausch, K. D. Hall, and A. Spock, *Am. Rev. Respir. Dis.,* **103,** 651 (1971).
103. J. A. Kylstra, W. H. Schoenfisch, J. M. Herron, and G. D. Blenkarn, *J. Appl. Physiol.,* **35,** 136 (1973).
104. K. E. McDonald, J. F. Park, G. E. Dagle, C. L. Sanders, and R. J. Olson, *Health Phys.,* **29,** 804 (1975).
105. B. A. Muggenburg, J. A. Mewhinney, and R. A. Guilmette, "Removal of Inhaled Plutonium and Americium from Dogs Using Lung Lavage and DTPA," in: *Actinides in Man and Animals,* M. E. Wrenn, Ed., RD Press, Salt Lake City, 1981, p. 387.
106. E. Carlens, *J. Thorac. Surg.,* **18,** 742 (1949).
107. B. B. Boecker, B. A. Muggenburg, R. O. McClellan, S. P. Clarkson, F. J. Mares, and S. A. Benjamin, *Health Phys.,* **26,** 505 (1974).
108. S. A. Silbaugh, S. A. Felicetti, B. A. Muggenburg, and B. B. Boecker, *Health Phys.,* **29,** 81 (1975).
109. B. A. Muggenburg, R. O. McClellan, B. B. Boecker, J. L. Mauderly, and F. F. Hahn, "Long-Term Biologic Effects in Dogs Treated with Lung Lavage after Inhalation of ^{144}Ce in Fused Aluminosilicate Particles," in: *Actinides in Man and Animals,* M. E. Wrenn, Ed., RD Press, Salt Lake City, 1981, p. 395.
110. M. Ingram, "Lymphocytes with Bilobed Nuclei as Indicators of Radiation Exposures in the Tolerance Range," in: *Legal, Administrative, Health and Safety Aspects of Large-Scale Use of Nuclear Energy,* Proceedings of the International Conference on the Peaceful Uses of Atomic Energy, Geneva, August 8-20, 1955, Vol. 13, 1956, p. 210.
111. M. Ingram, "The Occurrence and Significance of Binucleate Lymphocytes in Peripheral Blood After Small Radiation Exposures," in: *Immediate and Low Level Effects of Ionizing Radiations,* A. A. Buzzati-Traverso, Ed., Proceedings of Symposium June 22-26, 1956, UNESCO, IAEA, and CNRN, Venice, 1960, p. 233.
112. R. Lowry Dobson, "Binucleated Lymphocytes and Low-level Radiation Exposure," ibid., p. 247.
113. H. G. Jones and B. E. Lambert, "The Radiation Hazard to Workers Using Tritiated Luminous Compounds," in: *Assessment of Radioactivity in Man,* Proceedings of the Symposium by Inter-

national Atomic Energy Agency, the International Labour Organization, and the World Health Organization, Heidelberg, May 11–16, 1964, Vol. 2, 1964, p. 419.
114. F. E. Butler, "Assessment of Tritium in Production Workers," ibid., p. 431.
115. A. A. Moghissi, M. W. Carter, and E. W. Bretthauer, *Health Phys.*, **23,** 805 (1972).
116. K. L. Smith, J. F. Park, and P. J. Moldofsky, *Health Phys.*, **22,** 899 (1972).
117. International Atomic Energy Agency, *Assessment of Radioactive Contamination in Man*, Proceedings of a Symposium by the IAEA and the World Health Organization, Stockholm, November 22–26, 1971, 1972.
118. G. R. Meneely and S. M. Linde, Eds., *Radioactivity in Man*, Proceedings of the Symposium on Whole-Body Counting, International Atomic Energy Agency, Vienna, June 12–16, 1961, 1962.
119. International Atomic Energy Agency, *Directory of Whole-Body Radioactivity Monitors*, IAEA, Vienna, 1970.
120. P. H. Abelson, *Science*, **193** (4250), 279 (1976).
121. W. D. Metz, *Science*, **192** (4246), 1320 (1976).
122. W. D. Metz, *Science*, **193** (4247), 38 (1976).
123. W. D. Metz, *Science*, **193** (4250), 307 (1976).
124. J. Walsh, *Science*, **193** (4250), 305 (1976).
125. D. M. Smith, "Respiratory-Tract Carcinogenesis Induced by Radionuclides in the Syrian Hamster," in: *Proceedings of Symposium on Pulmonary Toxicology of Respirable Particles*, C. L. Sanders, F. T. Cross, G. E. Dagle, and J. A. Mahaffey, Eds., CONF-791002, Technical Information Center, U.S. Department of Energy, 1980, p. 575.
126. E. C. Anderson, L. M. Holland, J. R. Prine, and C. R. Richmond, "Lung Irradiation with Static Plutonium Microspheres," in: *Experimental Lung Cancer: Carcinogenesis and Bioassays*, E. Karbe and J. F. Park, Eds., Springer-Verlag, New York, 1974, p. 432.
127. L. M. Holland, J. R. Prine, D. M. Smith, and E. C. Anderson, "Irradiation of the Lung with Static Plutonium Microemboli," in: *The Health Effects of Plutonium and Radium*, W. S. S. Jee, Ed., J. W. Press, University of Utah, Salt Lake City, 1976, p. 127.
128. D. A. Holaday, "Uranium Mining Hazards," in: *Uranium–Plutonium–TransplutonicElements*, H. C. Hodge, J. N. Stannard, and J. B. Hursh, Eds., Springer-Verlag, New York, 1973, p. 300.
129. R. L. Gross, D. M. Smith, R. G. Thomas, G. Saccomanno, and R. Saunders, "Immunological Assessment of Patients with Pulmonary Metaplasia and Neoplasia," in: *Proceedings of Conference on Safe Handling of Chemical Carcinogens, Mutagens, Teratogens and Highly Toxic Substances*, Vol. 1, D. B. Walters, Ed., Ann Arbor Science Publishers, Inc., Ann Arbor, MI, 1980, p. 259.

CHAPTER THIRTEEN

Noise

ARAM GLORIG, M.D.

1 INTRODUCTION

The importance of sound to human and animal life is evident throughout the history of the world. Noise and its consequences are by no means peculiar to our generation. Earliest humans, and certainly animals, depended on auditory signals for survival. Hearing and its relation to preservation of life is basic. Even a plan that includes a question of the survival of the human race must include a place for warning signals received through the auditory system. If we examine biblical history, we find a good example of the use of noise by one of the early biblical characters when the walls of Jericho were caused to fall by the blasts of many hundreds of trumpets and shouts from many persons. In the history of the development of civilization, we find that the tide of battle was often turned because of the shouting of the participants. In the history of our own nation the record shows that attacks made by American Indians were always accompanied by loud yelling designed to terrorize the foe. As recently as the time of World War II noise was used to add to the stress and fear produced by bombing raids.

Although humankind has been associated throughout its history with sound of one kind or another, it was not until relatively recently that examples of the adverse organic effects of sound in the form of noise have been recorded. This fact is not unreasonable since, although isolated examples of noises existed, the widespread noise exposure that accompanied the onset and development of modern industry was not present. As mass production increased and heavy mechanization progressed, the problem of exposure to noise increased. In spite of this, no recognition of the effects of noise on humans was recorded until about 1830, when Fosbrooke of England recorded hearing loss in blacksmiths. Some 30 years later Weber made the first report on hearing loss in boilermakers and railway workers. Four years after Weber's report someone who signed himself "volunteer" asked the question, "Can any one of your readers suggest a remedy for preventing hearing loss caused by rifle shooting?" This question was inserted in *Lancet,* and the answer appeared in a later issue, "Use cotton in the ears." Barr in 1886 and in 1890 published what were then considered complete records of the injurious effects of loud sounds on hearing. Several reports which appeared in the latter part of the nineteenth century recorded "that railroad workers suffer hearing loss due to noise." Bauer in 1926 first called attention to airplane noise as a factor in hearing loss.

A universally acceptable definition of noise has been difficult to attain. A popular early definition was "sound without agreeable musical quality." In this sense noise has been both bane and boon to humanity. Loud shouts, horn blasts, wailing sirens, shrieking whistles, and so on, are effective and far-reaching warning signals.

In a somewhat less clamorous way our daily life is pervaded by noise, sometimes unwanted and at other times intentionally introduced to accomplish a given end. On the one hand, noise may interfere seriously with speech; applied in the proper amounts, however, noise may make a room seem quiet. In the hush of a library reading room the sudden sound of a page being turned, the thump of heavy footsteps, the scrape of chair legs on the hard floor, the rustle of clothing, and whispers are annoying and distracting. We have learned that these rooms will seem quieter if they are made just a bit noisier—that is, if enough continuous noise is introduced to mask out the distracting intermittent sounds.

This early definition of noise "as sound without agreeable musical quality" has proved to be quite unacceptable, partly because of the difficulty of defining agreeable musical quality in consistent terms, but mostly because it does not take into account the fact that even agreeable musical quality may be undesirable at times. What we call musical quality in the concert hall may be just plain "noise" when it intrudes on our sleep by way of the neighbor's "hi-fi." Attempts to define noise in terms of the physical character of sound itself have failed. The definitive characteristic of noise is its *undesirability*. Consequently, the currently accepted definition of noise is "any unwanted sound." For centuries people have been aware of noise and of the unpleasant behavioral effects that noise can produce.

To better understand the effects of noise on hearing, a brief coverage is given on the anatomy and physiology of the ear, definition of hearing loss, types of hearing loss, measurement of hearing, and causes of hearing loss.

2 ANATOMY AND PHYSIOLOGY

For the purposes of simple discussion, the ear can be divided into three distinct parts—the external, middle and inner ear.

2.1 External Ear

The external ear is demarcated externally by the auricle (or, as it is sometimes known, the *pinna* or *ear flap*); internally it is demarcated by the outside layer of the drumhead (the tympanic membrane). The external canal is about 25 mm long and has an S-shaped curve directed inward, forward, and a little upward. In order to see the entire canal and drumhead, it is necessary, in most cases, to pull the auricle upward and backward.

The function of the external ear is to conduct sound to the drumhead.

2.2 Middle Ear

The middle ear is bounded externally by the inside layer of the drumhead and internally by the membrane covering the bony wall of the inner ear. The middle ear contains three tiny bones (ossicles)—the hammer, anvil, and stirrup, or more properly, the malleus, incus, and stapes. There are two small muscles connected to the ossicles—the stapedius, which is attached to the stapes; and the tensor tympani, which is attached to the larger of the three ear bones, the malleus. The middle ear presents four openings, the oval and round windows, the opening of the eustachian tube, and an opening into the mastoid cavity. The mastoid cavity is situated just behind the ear in the mastoid process of the large temporal bone. The oval window (vestibular window) is fitted with the flattened, bony footplate of the stapes and is held in place by a fibrous membrane that allows the footplate to move as a hingelike structure, similar to the movement of a trapdoor. The round window (cochlear window) is a tiny opening just below and in front of the oval window. It is covered by a thin membrane resembling the drumhead. The middle ear has three main functions:

1. The efficient transmission of sound vibrations in an air medium to a fluid medium in the inner ear. The electronic counterpart is called an *impedance matching transformer*.

2. Protection for the inner ear. This protective mechanism is not too well understood, but recently it has been determined that the stapedius muscle tenses reflexly and stiffens the ossicular chain. The stiffening of the ossicular chain reduces the shock to the inner ear caused by exposure to sudden, large amplitude, low-frequency sounds.

3. To maintain equal pressure on each side of the drumhead. The eustachian tube (auditory tube), which connects the inner ear with the posterior nasal cavity, allows air bubbles to enter the middle ear, thereby maintaining equal pressures on each side of the drumhead. The operation of this function is most evident with changes in altitude.

2.3 Inner Ear

The inner ear consists of two main parts: (a) the semicircular canals, which contain part of the balance mechanism and (b) the cochlea, which contains the end-organ of hearing. The cochlea, so-called because of its shape, is a snail-like tube with two and a half turns. It houses a system of membranous canals within which lies the end-organ of the auditory nerve (acoustic nerve). One canal travels upward in a spiral from the oval window and without a break in continuity, reverses direction and returns to the round window. The second canal, essentially a closed tube, is located in the space produced by the doubling of the first canal. When the cochlea is cut cross-sectionally, the canals appear somewhat like a pie through the middle with one cut in the top half dividing it into two parts in a ratio of one-third to two-thirds. The organ of Corti, or the special sense organ of the auditory nerve, is located in the closed tube. These canals contain a viscous fluid, perilymph in the doubled tube, and endolymph in the closed tube.

Simply, the mechanism of hearing operates as follows: when an air-conducted sound impinges upon the drumhead, it vibrates much as a microphone diaphragm does. The vibrations are transmitted by way of the ossicular chain (ossicles) to the oval window. The motions at the oval window produce waves in the fluid of the doubled canal that are transmitted to the closed tube. These vibrations move the organ of Corti and produce a bending or shearing, or both, of the hairlike projections from the hair cells. When the hairlike projections change their relative position, the hair cells are activated and initiate an electrochemical impulse that travels along the nerve to the brain, where the sound is interpreted into what we know as hearing. Obviously this is an oversimplified explanation but it is correct in principle.

In general, the human ear can hear frequencies from about 20 to 20,000 Hz. Frequencies below 20 are called *subsonic* or *vibration*, and those above 20,000 are called *ultrasonic*. Perhaps the most important frequencies are those that relate to communication by speech. For practical purposes, the speech frequencies range from 500 to 2000 Hz. There are some fricative consonants that have frequencies as high as approximately 3500 Hz, but when it is necessary to communicate only by everyday speech under ordinary circumstances, hearing in the frequencies above 2000 Hz is not as important.

Auditory sensitivity is theoretically defined as the point on a sound pressure level scale at which 50 percent of the stimuli presented to the ear are heard. The auditory threshold varies as a function of frequency; ear is most sensitive between 1000 and 4000 Hz and less sensitive above and below these frequencies. The threshold also varies with the stimulus used. It is different for pure tones, for speech in the form of words or sentences, and for complex sounds. Whenever the auditory threshold is recorded, the type of stimulus used must also be indicated since each class of stimulus has its own auditory reference level.

3 WHAT IS A HEARING LOSS?

Before we discuss hearing loss, let us first decide what constitutes normal hearing! From that point we can define "hearing loss" and then discuss its various aspects.

At the present time, there is an accepted "normal" hearing reference zero. The American Medical Association (AMA) and the American National Standard Institute (ANSI) approved this reference zero as a point of departure for determining hearing loss. "Normal hearing" is defined as the average hearing of a group of persons considered essentially representative of the normal hearing population between 18 and 24 years of age.

In order to achieve a universal zero reference level that would apply all over the world, an International Standards Organization (ISO) zero is now accepted with slight changes and is called ANSI (American National Standards Institute) 1969. The ANSI Standard has been accepted by the American Academy of Ophthalmology and Otolaryngology and the American Speech and Hearing Association.

Normal hearing has been designated only for the following frequencies: 64, 125, 250, 500, 1000, 2000, 4000, 8000 and 12,000. The three frequencies 64, 125, and 12,000 are not now used because they have been found to be unreliable as well as unnecessary. For any one of the chosen frequencies, hearing is measured in units called "decibels of hearing loss."

The decibel (dB), is a measure of sound pressure level. It expresses a relation between the sound pressure of a given signal and that of a reference or standard. The relationship is not a simple ratio. Because of the extremely wide range of sound pressures to which the ear responds, it is not convenient to describe them as being three times, four times, and so on, greater than the reference pressure; too often they are millions of times greater. Although the simple ratio does not provide a suitable way of describing or comparing sound pressures, the logarithm of the simple ratio does. The decibel scale for sound pressure level is then a logarithmic scale.

The normal hearing reference point is an average value obtained by testing a large group of individuals. An average figure implies that many of the individuals tested are either better or worse than average. Generally, hearing tests are taken for one or more of three reasons: (a) as an aid in diagnosis, (b) for determination of amount of handicap, and (c) for determination of the need for rehabilitation.

Diagnostic testing requires precise methods that will record the status of the external, the middle and the inner ear at specific points of the audible spectrum. Accurate testing and interpretation are extremely delicate operations and should not be performed by other than medically trained specialists or medically supervised audiologists. Only expertly produced data can afford the information necessary to diagnose and treat hearing loss.

How severe must a hearing loss be before it constitutes handicap? What is the degree of handicap? And finally, what is handicap itself? Before a scale of handicap can be formulated, it must relate to an aspect of hearing loss that can be used as a common denominator. So we come inevitably to the realization that communication by speech is the most important function of the ear in our present civilization.

A measure of handicap through hearing loss, then, must be based on the effect of hearing loss on communication by speech. This effect can be quite adequately estimated by averaging the losses at 500, 1000-, 2000-, and 3000-Hz (cycles per second) or at the so-called "speech frequencies." A scale of handicap due to hearing loss has certain points that are important. These points may be designated by averaging the values in dB at 500, 1000, 2000, and 3000 Hz. This is the average hearing level (AHL).

The first important point on our hearing loss scale is 25 dB. In other words, a person who needs greater volume than this is to hear may be said to have impaired hearing, if the loss is between 25 and 40 dB, most hearing-impaired patients realize that they are missing some of the key items in ordinary conversations. Hearing loss at this level becomes evident in the daily activities of the afflicted person.

As the hearing loss progresses above 40 dB, the individual will reach a third stage at 55 dB. At this level a person has considerable difficulty with communication under most circumstances. Unbelieveably, most patients wait until their hearing loss reaches this stage before they seek help. The majority should seek help long before this.

The next level of consequence is at 70 or 75 dB. With losses of this magnitude, the patient cannot communicate without serious difficulties. The final point on our scale is 90 to 95 dB. A hearing loss of that magnitude totally precludes any practical speech communication. Naturally, the ear will respond to sounds that are louder than this, but the human voice is not capable of continuous speech at this level for more than very short periods.

4 TYPES OF HEARING LOSS

A hearing loss may be one of three types, conductive (obstructive), inner ear (sensorineural), or mixed. Conductive or obstructive hearing loss is due to pathology in the external and/or the middle ear. Inner ear hearing loss results from disease or injury to the inner ear or to the end organ of the auditory nerve. Mixed hearing loss is a combination of both conductive and inner ear hearing loss.

Knowledge of the type of hearing loss is important because many conductive losses are amenable to treatment. In most cases, either an improvement or an arrest of progression can be obtained. On the other hand, inner ear (sensorineural) losses are very rarely benefited by therapy. Exceptions are Ménière's disease and some minor vascular problems. Because of this, it is necessary to make an accurate diagnosis of the type of hearing loss as early as possible.

5 HEARING MEASUREMENT

Measurement of the auditory function presents problems, but none is insurmountable. Tests for measuring auditory acuity have been available for many years. Some of the early tests included the spoken and whispered voice, coin clicks, and other methods. Later on, the tuning fork was used and proved to be a valuable aid in the determination of type of hearing loss.

Not until the vacuum tube was invented, however, did we obtain an accurate means of measuring hearing. Accuracy is possible with the pure tone audiometer, provided it is used by adequately trained personnel. Although the audiometer seems to be a simple instrument to operate, it actually requires practical and theoretical instruction and considerable study, and since it is a delicate electronic device, consistent care and calibration.

Use of the instrument by other than a well-trained individual can result in misleading measurements and may even be dangerous. Not only must an audiometer be given the utmost care and be calibrated at frequent intervals (at least once a year) but, also, the environment for its use must be suitable. Testing hearing with an audiometer in a room filled with extraneous noise will not produce valid test results.

6 CAUSES OF HEARING LOSS

The causes of impaired hearing may be divided into two main groups: congenital and acquired. Under congenital causes, the following are important:

1. *Heridtary.* These include otosclerosis, predisposition to early degeneration of the auditory nerve, and anatomical malformations. Otosclerosis is a familial biological defect that causes a bony sclerosis between the footplate of the stapes and the oval window. The eventual result is a fixation of the stapes and a conductive-type hearing loss. As progression continues, the auditory end organ is affected and mixed-type hearing loss results. The process undoubtedly begins *in utero* but usually symptoms (hearingloss and sometimes tinnitus) are delayed until the third decade. It is more common in women than men. The diagnosis of otosclerosis can be made on the basis of the audiogram, absence of physical findings, absence of history or a slowly progressive hearing loss, and the presence of progressive loss in the family. The ear that always shows a predominantly conductive type of hearing impairment when there is no obvious cause must be considered to be otosclerotic. Early diagnosis and treatment are imperative. Predisposition to early degeneration of the auditory end organ is seen in children of various ages from shortly after birth to 5 or 6 years of age. For some unknown reason and without apparent cause, these children show varying degrees of sensorineural hearing loss that may progress to severe impairment. Congenital malformations may vary from a slight defect to the external ear to complete absence of the entire otic mechanism. For example, many individuals have a loss at a single frequency in the area of 4000 Hz.

2. *Toxic.* Virus diseases, notably Rubella in the first trimester of pregnancy; to a lesser extent, mumps and influenza may cause severe sensorineural

Table 13.1 The Causes of Acquired Deafness

Condition	Type of Loss: Sensorineural and/or Conductive
Brain conditions	
Meningitis	Sensorineural
Encephalitis	Sensorineural
Tumors, circulatory diseases	Sensorineural, central
Concussion, central auditory area damage	Sensorineural, central
Fracture of the temporal bone	Sensorineural
General infectious disease	
Scarlet fever	Both
Measles	Sensorineural
Mumps	Sensorineural
Pertussis	Sensorineural
Vericella	Sensorineural
Influenza	Sensorineural
Pneumonia, virus, and pneumococcic	Sensorineural
Thyphoid fever	Sensorineural
Diphtheria	Sensorineural
Syphilis	Sensorineural
Common cold	Both
Any disease causing high fever	Sensorineural
Infections of the ear	
External otitis	Conductive
Otitis media, acute and chronic	
Nonsuppurative	Both
Suppurative	Both
Serous	Conductive
Mastoiditis, acute and chronic	Conductive
Physical agents	
Impacted cerumen	Conductive
Foreign body impaction	Conductive
Trauma, accidental	Both
Noise exposure	Sensorineural
Barotrauma	Conductive
Excessive growth of lymphoid tissue in nasopharynx	Both
Surgical interference	Both
Toxic agents	
Quinine	Sensorineural
Nicotine (tobacco)	Sensorineural
Aspirin (salicylates)	Sensorineural
Streptomycin	Sensorineural
Dihydroxystreptomycin	Sensorineural
Hydroxystreptomycin	Sensorineural
Neomycin	Sensorineural
Kanamycin	Sensorineural

Table 13.1 *(Continued)*

Condition	Type of Loss: Sensorineural and/or Conductive
Miscellaneous	
Functional	
Psychogenic	
Hysteria	
Malingering	
Advancing age (presbycusis)	Sensorineural

impairment. Any severe acute illness, particularly if accompanied by a high fever, is likely to injure the cochlear nerve endings prenatally or in the neonate. Smoking, drinking, or drug use during pregnancy may produce hearing impairment in the child.

The causes of acquired deafness are many; however, the most significant ones are listed in Table 13.1. They may be divided into six main groups.

Infections of the external ear are often quite resistant to treatment. A general physician who fails to resolve the condition in 2 or 3 weeks, should seriously consider consultation with a specialist. Chronic external otitis may cause extreme discomfort and lead to a permanent hearing loss. Foreign bodies in the external auditory canal should be removed with as little trauma to the canal wall as possible. Early detection and removal of foreign bodies or cerumen will prevent serious complications. Many of the consequences of diseases of the external ear may be eliminated if the ear is examined routinely at the time of any general physical examination. There seem to be very few parts of the human body that are examined as infrequently as the ear. The present author has read hundreds of reports of examinations where information concerning the physical appearance of the external ear and drumhead was vital, only to find that no one had even bothered to glance at the ear. This does not refer to testing the hearing; usually no doctor but an otolaryngologist ever does this, despite its importance to the patient's daily life.

Diseases of the middle ear are usually responsive to therapy. The results, however, depend on early diagnosis and specific treatment. Infections of the middle ear respond much better to antibiotics and chemotherapy in the early, presuppurative stages. Once suppuration has begun and bulging of the drumhead occurs, the effectiveness of these agents is reduced considerably. If the infection is allowed to progress until the drumhead ruptures spontaneously and suppuration continues for more than 4 to 6 weeks, surgical interference should be considered and consultation sought. It is wiser to incise the drumhead before spontaneous rupture occurs. Drainage is better and the chances of improvement are enhanced. Remember, though, that the drumhead and middle ear are extremely delicate, important parts of a very sensitive and important organ, and although myringotomies are necessary, they should be performed only with great care and proper justification.

Chronic, suppurative middle ear disease is always associated with mastoiditis of some degree. Continuous suppuration may result in irreversible damage to the middle ear structures, mastoiditis, osteomyelitis, and, frequently, brain abscess. Hearing acuity in the chronic running ear is always reduced. The reduction may be mild, severe, or any degree in between. It is extremely important to seek specialized consultation for any case with chronic middle ear disease. Early application of proper therapy may mean the difference between a dry middle ear with good function and one that shows a severe hearing loss. The treatment may be medical, surgical, or a combination of both. Usually, the guide to treatment is the status of the hearing and the possible effects of the treatment on hearing. The decision in favor of a given treatment should be made only after careful study and specialized consultation.

The treatment of otosclerosis is strictly surgical. There are two procedures open to the surgeon, the fenestration or "window" operation, and stapedectomy. Fenestration is no longer used except in special cases. There are several other procedures for various middle ear diseases, but most of them lead to reconstruction of the diseased middle ear and reestablishing of the connections between the three little bones of the middle ear. All procedures are very difficult to perform and should be tried by only an adept, specialized surgeon. If surgery fails, rehabilitation with a hearing aid must be tried.

Impaired hearing due to pathology of the inner ear is a very serious matter because it is generally irreversible. There are some diseases of the inner ear, however, that may be reversible. Ménière's disease, which is actually a dysfunction of the hydrodynamics of the inner ear, can be treated with some success. Frequently, if started soon enough, therapy directed toward reestablishing the fluid balance in the inner ear will result in a reduction of the auditory threshold. Early treatment may also reduce sensorineural hearing loss due to vascular spasm. Antispasmodic drugs frequently will improve hearing provided they are administered early enough. Again, we are confronted with the necessity for early detection and diagnosis.

Perhaps one of the most common inner ear hearing losses are sensorineural hearing losses and so called presbycusis, which really means a loss due to aging. The problem with determining the amount of loss that is due strictly to aging is always confounded by the fact that as one ages physiologically, one also lives longer in the environment, which influences auditory threshold. It is, therefore, very difficult to come by hearing values for presbycusis that are not tainted by auditory hazards.

There have been many studies made to determine exactly what may be expected as a function of age, and these have resulted in some suggestions by NIOSH and the ISO, and even the present author has published some material on presbycusis curves.

One problem that is raised, as far as industry is concerned, regarding presbycusis is what to do about presbycusis correction when disability is being calculated according to whatever formula is used. Most formulas for converting pure tone thresholds to disability have what is called a *low fence*, and these are approximately around 25 dB at the frequencies involved, which are usually 500, 1000, 2000, and 3000 Hz. If one looks at most of the presbycusis curves, one finds that the hearing loss as a function of frequency related to about 65 or 70 years of age, has not exceeded the 25 dB fence by a sufficient amount to influence the average of the four frequencies. On this basis, the present author believes that no correction for presbycusis should be made when calculating disability with the use of a low fence of 25 dB or thereabouts. If one is going to correct for presbycusis, it should be done starting at 0-dB hearing level since presbycusis starts from this point. The simplest way is not to make any correction for presbycusis, and the present author believes one is being fair to both sides when this is done.

7 TYPES OF NOISE EXPOSURE

7.1 Steady Noise

Steady noise is noise that remains more or less at the same level when root mean square (RMS) values are determined.

7.2 Impulsive Noise

Impulsive noise and impact noise are not interchangeable. Impulsive noise includes impact noise but not vice versa. Impulsive noise may be defined as a noise having peaks that can be distinguished as *separate* peaks by the ear. When peaks recur too rapidly for the ear to define definite peaks, the noise is steady noise as far as the ear is concerned.

7.3 Intermittent or Continuous Noise

As stated previously, noise may be steady or impulsive in character. Each of these can be said to be intermittent or continuous. This is a very important distinction.

7.4 Noise-Induced Hearing Loss or Acoustic Trauma

Properly defined, noise-induced hearing loss is the loss that results from continued exposure over a number of years. It is a gradual shift in threshold first in the high frequencies and then in the lower frequencies, whereas "acoustic trauma" should be used only to designate the loss produced by a single well-defined incident, such as an explosion or a blow to the ear or head. It is important to recognize this difference since there are several differences found in the audiograms and the pathology produced.

7.5 Discussion

One of the most difficult problems associated with a study of noise exposure is that of devising a method of describing and categorizing in a meaningful way the multitudinous types of noise exposure encountered. To describe noise exposure adequately for correlation purposes, some numerical method of classification must be evolved. This method of classification must provide a way of combining the measurement of quantity of sound energy with measurements of its time distribution. The result might be called a "noise exposure index." One approach to the total problem would be to index noise exposure first with respect to its physical dimensions and then as a further step to index the effects of noise exposure on humans. The problem of assessing noise exposure is very complex, and large amounts of data must be gathered and analyzed before we can say what a practical and useful assessment of noise exposure is.

Recent measurements indicate that sufficiently detailed information about the energy content of noise exposure is not provided by RMS measurements. One report shows that although the RMS values of overall levels of certain so-called steady-state noises were in the region of 100 dB, the instantaneous levels were as high as 120 dB for 10 or 15 percent of the time (1). Evidence such as this must not be overlooked. We may find that peak factors in noise exposure correlate better with resulting hearing loss than do some of the factors we have already considered, such as the frequency distribution of energy in the noise exposure. In spite of the variability of the industrial processes and the accompanying noise exposures, large numbers of industrial audiograms show the same type of hearing loss; the loss begins in the region of 4000 Hz and spreads in both directions as exposure time increases. It seems, then, that the energy content of noise, its peaks, its time distribution, and its total duration, regardless of spectral distribution, are factors that bear close scrutiny.

This report does not discuss the techniques of noise measurement—only that there are, in general, two reasons for noise measurement. One is to determine whether a hearing conservation program is needed, and the other is to determine the parameters of noise control. As a group, we should not be concerned with noise control only with respect to its need and not to its details of accomplishment.

For hearing conservation purposes, the general measurement scale is sound level A (dBA). Some of the literature states that dBA is not adequate for this purpose, that it does not give us the information needed to conduct an overall hearing conservation program in industry. The purist will argue that dBA is inadequate because it does not give the status of low-frequency content in industrial noise. Most industrial noises have a sloping spectrum from the low to the high frequencies of 3 to 6 dB per octave. This would argue that the low-frequency content of noise is important and, therefore, should be measured.

However, it is well known that low frequencies do not produce hearing loss unless the levels are very high, that is, over 125 or 130 dB. Very few noises in industry reach this level at any frequency, let alone the low frequencies. A study by Karplus and Bonvallet (2) showed that only about 5 percent of the noises in industry exceed 110 dBA.

In considering the exposure, there are some important parameters that one must take into account while assessing the noise exposure and its relationship to its effects on hearing. These can be divided into two classes: steady and impulsive noise and continuous and intermittent noise. Steady noise that is continuous has different effects on the ear than when it is intermittent. Both steady noise and impulsive noise can be classified as continuous and intermittent exposures. Impulsive noise would seem to be an intermittent exposure because of its characteristics, but it can be just as intermittent as a steady noise. Several impulses can occur over a specific period of time. The noise can then be turned off, and a period with no impulses will follow. When this is true, the impulsive noise becomes an intermittent exposure.

When one considers these parameters as a function of exposure rather than merely noise level, it is not difficult to see that these classifications are important when assessing noise exposure. Any relationship to the effect of noise exposure on hearing must be considered as a function of the character of the daily and the long-term exposures.

From this discussion it is evident that time measurements are just as important as level measurements or, as a matter of fact, more important. One problem in assessing the effects of noise exposure is that it is usually measured in terms of noise level made at any time during a day. This is almost never the level of that noise over time. The variability of industrial noise is well known, and before valid assessment of noise exposure can be made, measurements should be made at different times during a day and over a period of several months. Unless the range of levels is too large, the average of the noise levels can be useful. One or two measurements taken randomly give little or no information. With this in mind it is not too difficult to explain why so many reported studies have shown poor correlation between noise levels and hearing loss. Usually the noise level measurements were made only once or twice during the period of study, and no real assessment of the effective noise exposure was obtained.

Dosimeters are available whose purpose is to assess level and time. With restrictions, they do reasonably well in assessing noise exposure during a daily work period or for longer periods if necessary. However, there are some factors about dosimeters that limit their use for research purposes. If we are to have more accurate information on the effects of noise exposure on hearing, we must know more about the bursts in intermittent noise. Available dosimeters do not furnish this information. Furthermore, the dosimeter does not quantify the time periods of the high levels reached in many industrial noises during any daily period. If these higher levels are distributed over an 8-hr period where there are only a few of them and they are well separated by low-level periods, the reading of the dosimeter would be the same as if these high levels had not been intermittent. The effects of the high-level noise exposure would be different than would be indicated by the dosimeter. However, for compliance purposes as they are being used by the Occupational Safety and Health Administration (OSHA), they give us a reasonable approximation of what the daily exposure is. But when such numbers are used to determine the effects of noise on hearing, the correlations are not as good as they should be.

To sum up, then, noise evaluation should be changed to *noise exposure* evaluation. We are proceeding in the right direction with the use of dosimeters, particularly when dosimeters are designed to assess the noise exposure and not necessarily to determine whether a particular exposure is in compliance with the OSHA regulations.

8 NOISE EXPOSURE MEASUREMENT

8.1 Level

Noise level can mean different things. The meaning of the numbers obtained depends on the sound level meter circuit specifications. All quality meters provide three circuits. These are known as the A, B, and C scales. The A scale provides a filter system that is primarily a high-pass filter. The frequencies from about 500 Hz are allowed to pass. The B scale allows more low-frequency energy to pass, and the C scale allows most of the energy at all frequencies to pass. The A scale is used for hearing conservation purposes since we are most interested in the frequencies above 500 Hz. However, because low-frequency energy can be important if that energy is high, it is wise to measure both A and C scale to establish the difference. If C minus A is greater than 15 dB, an octave band analysis should be performed to examine for the energy in the low end of the spectrum. It would probably be safer if the C scale level were used to determine the permissible level in this case since the low frequencies are undoubtedly contributing to the hazard.

8.2 Character of the Noise

In our definitions we stated that noise can be steady or impulsive and that each of these may be continuous or intermittent. As a matter of fact, most industrial noises are a mixture of steady and impulsive energy. It is important to recognize this fact because the effects on the ear are not a direct function of total energy. Hamernik et al. (3) have shown that the biological effects of combined steady noise and impulsive noise are not the same as when they are not combined. Although clinical audiometry tends to support the position that equal energy can be used to predict the effects of each or the effects of combining the two, the present author does not believe this to be true. If one looks at the mechanical action of steady versus impulsive energy on any membrane system such as the basilar membrane, there is no doubt that membrane motion from steady noise

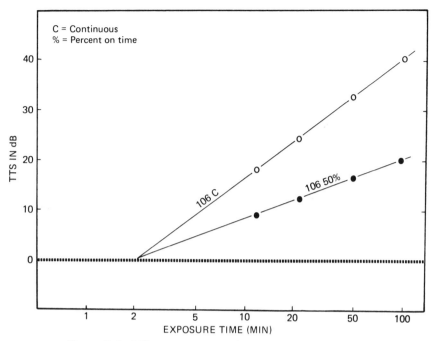

Figure 13.1 Difference in TTS as a function of time distribution.

is different from that from impulsive noise. Hamernik's data support this concept. Nevertheless, from a practical standpoint it is probably reasonable to treat them on a total energy basis for hearing conservation purposes since this is more conservative.

8.3 Duration

It turns out that time or exposure, if you will, has a great deal to do with the biological effects of noise exposure—so much so that a measurement of level without some method of quantifying time is a useless procedure. There are no industrial noises that maintain a constant level as a function of time. The usual industrial exposure fluctuates infinitely at the exposed ear over time, either because the procedure changes and consequently changes the output or the individual changes position with respect to the source, accomplishing the same thing.

Duration may be discussed under two headings, continuous and intermittent and total time. Many studies (4, 5) have shown that the temporary effects of noise exposure do not correlate linearly with equal energy. It is quite obvious that if the ear is given a chance to rest and recover from time to time, the effects on hearing are not quantitatively the same as when energy is continuous (Figure 13.1). This is a very important distinction to recognize. It is one of the principle reasons we have not correlated hearing loss and noise exposure very well. Few noises are continuous over any work period, whether over 1 day or 20 years. All industrial noise exposure is not only intermittent, but it is intermittently intermittent. Consequently, it is next to impossible to quantify noise exposure over almost any time period. Any attempts to quantify the effects of long-term noise exposure by assessing short periods such as one day or one week usually meet with failure.

Since it is next to impossible to quantify noise exposure with any degree of accuracy, it might be more practical to develop some sort of indexing categories that relate to job classifications. This allows us to predict how much hearing loss to expect, on the average, for certain job classifications.

In order to simplify this problem, the original Air Force Regulation 160-3 specified that time and level could be traded on the basis of equal energy, that is, the so-called 3-dB rule. However, temporary threshold shift studies have shown that when noise exposure is interspersed with silence or innocuous noise periods, hearing loss does not increase in direct proportion to the energy. Hence the hedging in the OSHA standards, that is, the 5-dB rule.

Whether temporary threshold shift (TTS) can be used to carry over to permanent threshold shift (PTS) is a moot question at present. There is much discussion in the literature on this point, and we have not settled the matter by any means. There are many studies attempting to show the relations of noise exposure to noise-induced permanent threshold shift (NIPTS), but none can be said to meet all the criteria necessary to provide valid data from which to derive satisfactory conclusions. There are many problems with field studies of this relationship. Perhaps the best attempt was made by Paschier-Vermeer (6, 7). She combined large amounts of data from several studies demonstrating a reasonable relationship of growth of PTS at 4 kHz to sound level (dBA) over a 10-year period. But when attempts are made to compare TTS with PTS, the correlations are poor indeed. However, we can say that the following relationships of PTS and TTS are quite reasonable assumptions:

1. If there is no TTS_2 at the end of a workday, there will be no PTS regardless of exposure duration.

2. If TTS_2 at the end of 8 hr recovers before the next exposure period (approximately 16 hr), there will be no PTS.

Any further attempts to correlate PTS and TTS cannot be properly supported with the extant data. We will probably never know just what this relationship is. It has been suggested by Ward (8) that we might learn more from long-term controlled animal studies. It is not surprising that we cannot determine a direct relationship between PTS and TTS on the basis of present data. We are attempting to compare apples with oranges. The two phenomena must be completely different from a pathophysiological standpoint, if in no other way. Consequently, we must be content with what we know at present and formulate principles that prove both protective and practical. Some sort of relationship must exist, but up to the present, it is unknown.

9 EQUIPMENT

Equipment for measuring noise level has been available for many years. Only recently have we had equipment to assess noise exposure. It is not the purpose here to discuss the equipment necessary for noise reduction. This is a highly technical matter and the equipment for measurement of noise levels for this purpose should be left to the expert acoustical engineer. Our interest should be in assessing noise exposure for hearing conservation purposes.

9.1 Sound Level Meter and Octave Band Analyzer

For our purposes, noise is usually measured to determine sound level and frequency or spectrum. This requires the use of a sound level meter (SLM) and usually an octave band analyzer (OBA). Early studies of noise always included sound pressure level (SPL) data and spectrum analysis usually in octave bands from about 75 to 10,000 Hz. It soon became evident that these measurements meant little without exposure time, and methods of combining time and level were needed. The first attempts to combine level and frequency resulted in the adoption of sound level A. It was reasoned that since the frequencies above 500 Hz were most noxious and that sound level A is weighted to emphasize these frequencies, the information gained by measuring sound level A (dBA) would result in a simple single number reading that could readily be combined with time to establish what became known as level equivalent (Leq) or more recently, time-weighted average (TWA). Hence sound level A (dBA) was adopted to determine the need for hearing conservation measures. Assessment of exposure, however, did not lend itself to such a readily available simple solution. Eventually the accumulated dose concept was applied and we had the birth of the dosimeter.

9.2 Dosimeter

A dosimeter is nothing more than a sound level meter plus a computer which allows the energy to be stored and read out as an accumulated total dose over a predetermined time period. There are several dosimeters available, but none is completely satisfactory for the assessment of all types of noise exposure. The principal problem is related to the electronic specifications, particularly dynamic range and time constants. Probably the real problem is that we do not know enough about the effects of noise on hearing to instruct the designers properly. Kamperman says, "The personal dosimeter should be the preferred instrument for measuring a worker's full shift noise exposure" (9). Kamperman's statement is acceptable provided conclusions regarding total dose over a lifetime is not predicted from the data acquired over 1 or 2 days. As has been stated previously, noise exposure is intermittently intermittent and expected total dose should be based on numerous daily assessments made at various times. In other words, adequate sampling should be done to obtain an expected average for a 10-year exposure period.

Dosimeters were first designed to integrate noise and time from 90 dBA up. This threshold was chosen to comply with the initial OSHA permissible TWA. However, when the most recent version of the OSHA regulation proposed that 85 dBA be included as an action level requiring hearing tests and ear protection when appropriate, many of the dosimeters became outdated since the dynamic range necessary changed to 80 to 130 dBA. Then OSHA argued that noise from 80 dBA must be integrated into the TWA because of the rapidly changing dynamic range of many industrial noises. An 80-dBA threshold for integration is certainly on the conservative side. It is logically argued that if the integrator does not include energy below 85 dBA, there are frequent chances for missing energy in the immediate area of 85 dBA due to unforeseen time constants in the noise, especially impulsive noise.

Until such time as dosimeters have been in use for a number of years and audiometric changes have been correlated with dosimeter readings, we will not really know just what is needed to properly assess the energy in noise. Until then, we should be flexible in interpreting the results of dosimeter evaluations. It stands to reason that at the present state of the "art" SLM readings should be coupled with dosimeter readings before hard and fast numbers are called for. Certain judgment values must be recognized and used.

10 BIOLOGICAL EFFECTS OF NOISE EXPOSURE

The biological effects of noise are usually given as (1) the auditory effects and (2) the nonauditory effects.

10.1 Auditory Effects of Noise Exposure

These effects are logically discussed as temporary effects and permanent effects. They are usually discussed in terms of hearing threshold change. The effects may result in noise-induced temporary threshold shift (NITTS) or noise-induced permanent threshold shift (NIPTS). The former is important for its relationship to (a) NIPTS, which was discussed earlier; (b) the efficacy of ear protection; and (c) industrial audiometry.

If properly dealt with, TTS can be very helpful or detrimental to a hearing conservation program. For example, studies of threshold shift found in industrial audiometry can certainly indicate whether ear protection is effective. Annual audiograms will provide data from which threshold shift can be extracted. When the shifts are found in spite of an ear protection program, one can conclude that the ear protectors are not as effective as they should be.

The average noise exposures encountered in industry can be sufficiently attenuated by most ear protectors if they are worn properly.

TTS should certainly be avoided in baseline audiograms. This brings up the question of whether annual audiograms should be done on rested ears. One could argue that annual hearing tests should be made after the man has been exposed for four to six hours with ear protection in use. If there is no TS, it can be assumed that the ear protector has been effective at least on that particular day. If a TS is found and a repeat test verifies the shift, there has been a permanent shift in that ear. If not, it can be assumed that there was a TTS in spite of ear protection, and corrective action should be taken.

The course and growth of NIPTS is well known. Although the actual correlations of the amount of permanent threshold shift and the amount of noise exposure are poor, we do have some idea of what to expect from long-term exposure to hazardous noise.

The evaluation of noise exposure must include an assessment of hearing loss as a function of group averages. In this way one can predict within one or two standard deviations what the individual will show. It is important to point out that the effects of noise on the individual cannot be predicted except as a function of group averages. Certain points are evident when studies of long-term noise exposure are analyzed:

1. Most of the hearing loss occurs at 3, 4, and 6 kHz. There is some ambiguity about whether 4 or 6 kHz begins to show change first. In the opinion of this author, the first changes generally occur at 4 kHz. A very close second is 6 kHz, followed by 3 kHz.

2. Most of the loss at 3, 4, and 6 kHz occurs within the first 10 to 15 years. By the fifteenth year these frequencies reach an asymptote provided conditions remain the same (10) (Figure 13.2).

3. The hearing loss produced by several industrial exposures is about equal in both ears. If there is a difference between ears of more than 30 dB average at 0.5, 1, and 2 kHz or 40 to 50 dB average at 3, 4, and 6 kHz, causes of hearing loss in the worse ear other than noise exposure should be suspected.

4. If hearing loss continues to progress after the noise exposure has ceased, other causes must be considered. Perhaps the most common of these is presbycusis, that is, hearing loss due to aging.

5. These same principles apply to 0.5, 1, and 2 kHz, except the time period is much longer.

6. Few, if any, noise exposures found in industry will produce total or even profound hearing loss, even when exposures are continued for as much as 40 to 50 years.

10.1.1 Pathology of Noise-Induced Hearing Loss

There appears to be no question that the damage due to noise exposure occurs in the organ of Corti, particularly in the hair cells. It is also well documented

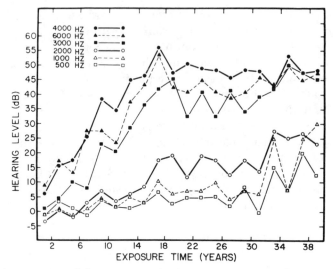

Figure 13.2 High frequencies reach asymptote in approximately 15 years.

that the outer rows of hair cells are the first to show damage. The exact mechanism of the damage has not yet been conclusively demonstrated. We know it can be purely physical (mechanical) or physiochemical. Microscopic studies show loss of hair cells with eventual loss of nerve fibers, particularly in the basal turn of the cochlea. The damage is permanent once it occurs. We know of no treatment, medical or surgical, that will reverse the process.

10.2 Nonauditory Health Effects

The literature is full of reports about this problem. If one believes the literature, noise causes about every illness known to humans. A review of this literature reveals very few, if any, properly designed studies. Borg (11) has mentioned a few of the inadequacies of most of the past studies. Some of these are:

1. The acoustic stimuli have been poorly defined.
2. Experimental subjects and controls (if any) differ not only with respect to normal exposure, but also with respect to other factors, such as intentional exposure to light or vibration or unintentional exposure to handling or housing in the case of animal studies.
3. The physiological state of the subjects differs—normal diurnal or annual variations in physiological parameters have not been considered.
4. Only a few biological parameters have been investigated in each experiment and the question of whether observed reactions are to be regarded as normal physiological response or part of a process leading to damage or disease has not been tested or adequately considered.

Few, if any, of the studies on human subject groups have considered the epidemiologic effects.

Field studies, at best, leave much to be desired. The present state of the situation cannot and does not lead to any adequately supportable conclusions. The general trends of the better studies indicate that if auditory damage is controlled, nonauditory effects of noise will be negligible.

Kryter (12) puts it very well when he says:

In spite of the very large gaps in our knowledge and the existence of some apparently conflicting research results, the following conclusions are put forth with, of course, the usual admonition that more research is needed before they can be accepted with great confidence:

1. There is no likely damage to a person from the possible unconditioned stress responses to noise that are mediated by the autonomic system.
2. Noise may often be concomitant with danger and adverse social environmental factors that are more important than the noise itself as a cause of apparent greater incidences of various physical and psychological diseases and accidents in industry.
3. Autonomic system stress responses could conceivably be a contributing factor to ill health in some persons as the result of noise in their living environment, directly interfering with auditory communications and sleep, and, thereby, creating the feelings of annoyance and anger that serve as the direct cause of stress responses.
4. It would appear that controlling meaningless noise to levels that permit auditory communication and sleep behavior for a given work or living environment would obviate the occurrence of any extraauditory responses in the body of a stressful nature.

11 HEARING CONSERVATION

The need for hearing conservation is found anywhere where human ears are exposed to loud sounds. This involves many walks of life and is not confined to noisy industries by any means.

However, it is true that the majority of people exposed to hazardous noise are found in industry, and logically most of the propaganda and programs are directed toward industry. There is no question that industrial noise produces more hearing loss than all other causes combined and thus deserves the most attention.

11.1 Hearing Conservation Standards

Both federal and state standards are directed specifically at industry. Perhaps the most important of these is the recent OSHA regulation finally published in the Federal Register of March 8, 1983. It took 15 years to bring this about, but it appears that the hard work that went into its preparation paid off. The intent of the regulation is clear. It is direct and reasonably simple and should be effective if implemented properly.

Briefly, the requirements are as follows:

1. Noise-exposure criteria: one at 85 dBA TWA, known as an action level; another at 90 dBA TWA, known as a mandatory level.
2. Noise monitoring at appropriate times and places.
3. Audiometry to establish a valid baseline audiogram and annual hearing tests thereafter.
4. A referral criteria called a standard threshold shift (STS) of an average difference from baseline of 10 dB at 2000, 3000, and 4000 Hz.
5. Noise control where feasible.
6. Ear protection made available and worn where and when appropriate.
7. Employee training to acquaint employees with the hazards of noise exposure and the benefits of a hearing conservation program.
8. Adequate record keeping.

In summary, the need for a more valid method of noise evaluation is great and certainly deserves more research. In the meantime, we know enough to establish reasonably effective hearing conservation programs.

REFERENCES

1. A. Glorig, *J. Laryngol. Otol.*, **75,** 447–448 (May 1961).
2. H. B. Karplus and G. L. Bonvallet, *Am. Ind. Hyg. Assoc. Quart.*, **14,** 235–263 (1953).
3. R. P. Hamernik, D. Henderson, J. J. Crossley, and R. Salvi, *J. Acoust. Soc. Am.*, **55,** 117–121 (Jan. 1974).
4. W. D. Ward, A. Glorig, and D. L. Sklar, *J. Acoust. Soc. Am.*, **31,** 791–794 (1959).
5. W. D. Ward, A. Glorig, and D. L. Sklar, *J. Acoust. Soc. Am.*, **30,** 944–954 (1958).
6. W. Paschier-Vermeer, Delft, Netherlands, 1968 (IG-TNO Report 35).
7. W. Paschier-Vermeer, "Noise-Induced Hearing Loss from Exposure to Intermittent and Varying Noise," in *Proceedings of the International Congress on Noise as a Public Health Problem*, Ward, Washington, DC, 1974 (EPA Report 550/9-73-008).
8. W. D. Ward, "Effects of Noise Exposure on Auditory Sensitivity, in: *Reactions to Environmental Agents* (*Handbook of Physiology*, Vol. 9), D. H. K. Lee, Ed., American Physiological Society, Bethesda, 1977.
9. G. W. Kamperman, *Sound Vibration*, **15,** No. 9, 5 (September 1981).
10. R. Gallo and A. Glorig, *Am. Ind. Hyg. Assoc. J.*, **25,** 237–245 (1964).
11. E. Borg, *Acta Otolaryngologica*, Suppl. 381 (1981).
12. K. Kryter, "Extra Auditory Effects of Noise," in: *Effects of Noise on Hearing*, D. Henderson, R. P. Hamernik, D. S. Dosanjh, and J. Mills, Eds., New York, 1976.

CHAPTER FOURTEEN

Nonionizing Electromagnetic Energies

SOL M. MICHAELSON, D.V.M.

1 INTRODUCTION

During the last half century there has been a marked development and increased utilization of equipment and devices for military, industrial, telecommunications, consumer use, and medical applications that emit a large variety of nonionizing radiant energies. These include ultraviolet (UV), infrared (IR), visible light, and radiofrequency (microwaves), which are components of the spectrum of electromagnetic waves. With passage in the United States of the Radiation Control for Health and Safety Act of 1968 (PL-90-602) and the Occupational Safety and Health Act of 1970 (PL-91-596) there has been a resurgence of interest in the biologic effects and potential hazards of exposure to these electromagnetic energies.

The Radiation Control for Health and Safety Act provides for the protection of the public from unnecessary exposure to radiation from electronic products. The Secretary of Health, Education, and Welfare [now Department of Health and Human Services (HHS)] is authorized by the act to promulgate performance standards for electronic products and regulations pertaining to record keeping, reporting, certification, and notification. Electronic products capable of emitting ionizing or nonionizing radiation, sonic, infrasonic, and ultrasonic energies are subject to radiation control under the act. Television receivers, microwave ovens, X-ray machines, lasers, UV lights, diathermy units, IR heaters, ultrasonic cleaners, and particle accelerators are examples of electronic products included in the control program. Manufacturers, dealers, and distributors of electronic products each have responsibilities identified in the act and the regulations. The Bureau of Radiological Health [now the National Center for Devices and Radiological Health (NCDRH)] is responsible for administering the act.

The Occupational Safety and Health Act of 1970, designed to assure safe and healthful working conditions for the nation's working men and women, provides broad authority to the Department of Labor and the Department of Health, and Human Services. The major provisions are:

1. To promulgate, modify, and improve mandatory occupational safety and health standards.
2. To prescribe regulations requiring employers to maintain accurate records and reports concerning work-related injury, illness, death, and employee exposure to potentially toxic substances, including physical agents.
3. To develop criteria for dealing with toxic materials and harmful physical agents, indicating safe exposure levels for workers for various periods of time.

It is apparent that a growing concern has developed in the military and civilian government agencies, industry, and professional societies regarding health hazards associated with the development, manufacture, and operation of devices that emit nonionizing radiant energies.

2 BIOPHYSICS

The energies of the electromagnetic spectrum (Table 14.1, Figure 14.1) are propagated in the form of waves that act as small bundles of energy with many of the properties ordinarily ascribed to particles. These are called *photons* or *quanta*. The energies residing in these photons (E) are directly proportional to the frequency v of oscillation of the specific electromagnetic radiation associated with them by the formula $E = hv$, where h is Planck's constant. The photon energy and the frequency of electromagnetic waves are inversely proportional to the specific wavelength. The longer the wavelength, the lower is the photon energy; the shorter the wavelength, the higher the photon energy. The energy is measured in electron volts (eV).

To provide some perspective on the order of magnitude of the electron volt, it may be noted that 1 MeV (1,000,000 eV) is equivalent to the energy required to lift a 1-mg weight a height of 10 nm. The thermal energy or motion of molecules at room temperature is about 1/30 eV; the energy of bonds holding atoms together in molecules of chemical compounds ranges from fractions of 1 to approximately 4 eV, whereas nuclear binding energy holding protons and neutrons together is millions of electron volts.

As the frequency decreases, the energy of the emitted photons is insufficient, under normal circumstances, to dislodge orbital electrons and produce ion pairs. The minimum photon energies capable of producing ionization in water and atomic oxygen, hydrogen, nitrogen, and carbon are between 12 and 15

Table 14.1 Nonionizing Electromagnetic Radiation

Radiation	Wavelength Range	Frequency Range	Energy per Photon (eV)	Effects of Absorption
Far and near UV	180–400 nm	10^6 GHz	7–3.1	Excitation of subshell and valence electrons
Visible	400–780 nm	5×10^5 GHz	3.1–1.5	Excitation of valence electrons
Near IR	780–2.5×10^4 nm	3×10^5–10^4 GHz	1.5–0.04	Increased kinetic energy of rotation and vibration (increased temperature)
Far IR	2.5×10^4–1.25×10^5 nm	3×10^4–2×10^3 GHz	0.04–0.008	Increased kinetic energy of rotation and vibration (increased temperature)
Microwaves	3 mm–100 cm	10^5–300 MHz	4×10^{-4}–1.2×10^{-6}	Increased kinetic energy of rotation (increased temperature)
Radio and television (broadcast)	1 m–1000 m	300 kHz–300 MHz	1.2×10^{-6}–1.2×10^{-9}	Unknown

eV. Inasmuch as these atoms constitute the basic elements of living tissue, 12 eV may be considered the lower limit for ionization in biological systems. Although weak hydrogen bonds in macromolecules may involve ionization levels less than 12 eV, energies below this value can, on a biological basis, be considered as nonionizing (1).

Radiant energy can produce an effect only when it is absorbed by the body directly or as a surface stimulation that elicits a response. Interaction with matter follows the physical laws of nature. The primary modes of action of these radiations are either photochemical or thermal. The basic law of Grotthus and Draper states that no photochemical reaction can occur unless radiant energy is absorbed. Such absorption requires that the energy of the photons be transferred to the absorbing molecules (1).

The nonionizing radiant energies absorbed into the molecule either affect the electronic energy levels of its atoms or change the rotational, vibrational,

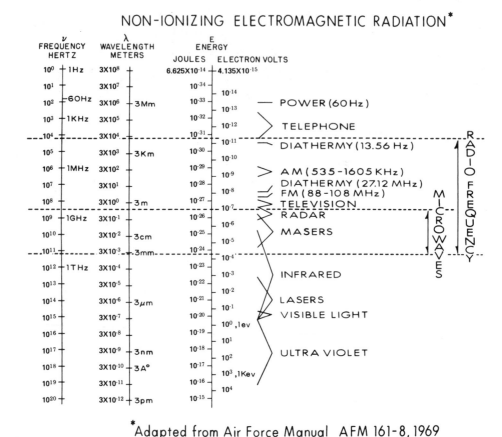

*Adapted from Air Force Manual AFM 161-8, 1969

Figure 14.1 Nonionizing electromagnetic radiation. Adapted from Air Force Manual AFM 161-8, 1969.

and transitional energies of the molecules. In biological systems, energy transfer produces electron excitation, which can result in dissociation of the molecule if the bonding electrons are involved, dissipation of the excitation energy in the form of fluorescence or phosphorescence, the formation of free radicals and degradation into heat. In the latter situation, absorption changes the vibrational or rotational energy and increases the kinetic energy of the molecule (1). The effects of interaction between radiant energy and biological systems are dependent on the energy of the radiated photons, the degree to which these photons are capable of penetrating into the system, and the ability of specific molecules to undergo chemical changes when these energies are absorbed.

It is also relevant to consider the energy flux or the time rate at which the energy is received (i.e., the power delivered). The designation "volts" applies to the quanta or parcels of energy delivered. Even though high-power levels can be reached at longer wavelengths, ionization does not occur, but heating is possible.

The relative effectiveness of different wavelengths in eliciting a specific photochemical response constitutes the "action spectrum" for that response. Photochemical reactions occur primarily upon absorption of radiant energy in the UV and visible portions of the spectrum. Thermal effects are produced upon absorption of energy in the IR and radiofrequency portions of the electromagnetic spectrum.

Many of the electromagnetic energies at certain frequencies, power levels, and exposure durations can produce biological effects or injury depending on multiple physical and biological variables. Although systems or devices utilizing or emitting electromagnetic waves provide immeasurable benefits to mankind, the energies could constitute potential hazards to persons by virtue of exposure to excessive levels of these energies.

3 PROTECTION GUIDES AND STANDARDS

There is a need to set limits on the amount of exposure to nonionizing radiant energies individuals can accept with safety. Setting protection standards, however, is a complicated process. The objectives of protection are to prevent acute effects and to limit the risks of late effects. Prevention of acute effects can be achieved quite readily when necessary. The second objective, to limit the risk of possible late effects, becomes more difficult.

It is important to consider some of the philosophic as well as practical aspects of standard development and promulgation. Protection standards should be based on scientific evidence but quite often are the result of empirical approaches to various problems reflecting current qualitative and quantitative knowledge. A numerical value for a standard implies a knowledge of the effect produced at a given level of "stress" and that both "effect" and "stress" are measurable. One problem is the definition of what an "effect" is and whether it can ultimately be shown to modify one's life-style or offspring's life-style (2).

If there were a clear-cut relationship between exposure level and pathophysiologic effect, the problem of setting standards would be greatly simplified. Numerous variables must be considered, and it is often difficult or impossible to obtain the necessary data to draw valid conclusions concerning effects of exposure to various radiant energies.

As in all biological processes, there is a certain range of levels between those that produce no observable effects and those that produce detectable effects. Since a detectable effect is not necessarily one that is irreparable or even a sign that the threshold for damage has been reached, the setting of permissible or allowable levels of exposure requires considerable caution and circumspection.

To ensure uniform and effective control of potential health hazards from nonionizing radiant energy exposure, it is necessary to establish uniform effect or threshold values. Ideally, effect or threshold values should be predicated on firm human data. If such data are not available, however, extrapolation from well-designed, adequately performed, and properly analyzed animal investigations is required. It is necessary to establish personnel exposure standards for each part of the electromagnetic energy spectrum that takes into consideration general and specific frequency effects as well as modulation (if such are established), exposure time, power density, field effects, wave polarization effects, and threshold phenomena. In establishment of a standard it is necessary to keep in mind the essential differences between a "personnel exposure" standard and a "performance" standard for a piece of equipment.

It must also be appreciated that in considering standards for different population groupings, one has to use a certain amount of inference, calculation, and judgment. The importance of the application to the individual or groups of individuals with consideration of special characteristics; uses such as defense, industrial and consumer applications, medicine, agriculture; and degree of risk of incurring exposures substantially above the planned limits and their consequences (3) are required.

Taylor (2) has noted that it is reasonable to suppose that an allowable exposure of population groups to radiation originating from a controllable source would be at some level less than that causing an observed effect. For exposures above the minimum, E_0, to cause an observed effect, there may be various dose–effect relationships that will vary with the effect and the manner in which the dose is delivered. The allowable dose might be arbitrarily set as low as zero, where there is clearly no direct or indirect benefit to anyone. A level above zero would be set in consideration of possible needs, benefits, risks, and costs. The principle of acceptable risk, not always explicitly stated, is not new. Many activities in which we indulge involve some risk, however small, which is usually accepted or rejected without a formal balancing of gain and loss.

Before a precise evaluation of the risks involved in exposure to radiant energies can be made, the dose–effect relationship must be considered for exposure if such has been determined. It is important to know the rate and extent of injury and recovery. It is also essential to know whether one can extrapolate from *in vitro* to *in vivo* studies, or from lower animals to humans. Exposure risks must be assessed for appropriate standards to be set so that potential risk does not exceed a level judged to be acceptable. With regard to the nonionizing electromagnetic energies, most of the evidence indicates a threshold for effect that should facilitate risk analysis.

The enactment of the Radiation Control for Health and Safety Act of 1968 and the Occupational Safety and Health Act of 1970 have given impetus to setting product emission and personnel exposure standards to protect workers, consumers, and the public at large. It should be noted, however, that radiation protection matters have been subject to political considerations, and where it is carried out in a nonobjective manner, it can be troublesome. It is important to make sure that unreasonable pressures are not developed in the promulgation of protection standards. For the last quarter century there has been developing increased interest of many groups in problems of radiation protection, in general for all forms of electromagnetic energies. An important trend in the

United States is toward governmental control of all uses of radiation. Applied rationally, this promises to be useful; nevertheless, there is dubious need for frenetic attacks on the problem (4).

It is apparent that much work remains to be done with regard to standards. An estimation of the risk involved must be made, and an estimation of the benefit must also be made, a problem that seems to have been ignored to a large extent (5). This could also be said about the setting of standards for all sorts of risk.

There is no question that standards should be promulgated when need is demonstrated. They should be developed, however, by those with knowledge and concepts of radiant energy interaction with the body as a whole or specific part (critical organs)* of the body. In addition, a clear differentiation must be made between biologic effects per se that do not result in short-term or latent functional impairment against which the body cannot maintain homeostasis and effectiveness and injury that may impair normal body activity by somatic change or induce genetic damage.

4 ULTRAVIOLET

Many materials at color temperatures above about 2500 K, or otherwise excited to corresponding energies, may emit UV radiation as their excited atomic orbital electrons lose energy and return to ground state. The energy range of UV radiation is 3.26 eV (380-nm wavelength) to 123 eV (10-nm wavelength). From a biological standpoint, the transition between excitation and ionization of molecules occurs at about 100 nm. The UV spectrum can be separated into three additional regions: vacuum (100 to 190 nm), which is absorbed by both air and water; far (190 to 300 nm), which is strongly absorbed by major biological molecules and is mutagenic; and near (300 to 380 nm), which is absorbed by some biological molecules but does not appear to be mutagenic.

The nonionizing portions of the UV spectrum are also classified as blacklight region (400 to 300 nm), erythemal region (320 to 250 nm), germicidal region (280 to 220 nm), and ozone production region (230 to 170 nm). The direct biological effects of solar UV radiation occur at wavelengths from about 300 nm and longer. Below 300 nm the UV radiation is effectively removed by filtration by the ozone in the atmosphere. This absorbed radiation, however, can contribute to many indirect effects through photochemical reactions with environmental pollutants (6).

A large number of artifical sources of UV energy are available in a variety of forms for use in the home, laboratory, and medical environments as well as commercial and industrial applications. The principal synthetic UV energy

---

* A critical organ can be defined as the organ that on exposure to radiant energies results in the greatest compromise to the optimal functioning and homeokinesis of the organism. This definition invokes the following considerations: (a) potential for exposure, (b) relative sensitivity to the radiant energy, and (c) the essentialness or indispensability of the organ to the well-being of the individual.

sources include germicidal lamps, various arc processes, and coherent energy sources such as the laser. Ultraviolet lasers that have extreme power capabilities with bandwidth of only a few nanometers are available. The most common industrial source of exposure to UV radiation is from electric arc welding operations and the use of germicidal lamps.

Some common uses include illumination from flourescent lamps; phosphorescence for illuminating instrument panel dials; chemical synthesis and analysis; product or process inspection; photoengraving; crime detection; sterilization of food, water, and air; vitamin production; medical diagnosis and therapy; photocopying processes; photoelectric scanning; electrostatic processes; glass blowing; hot metal operations; and entertainment. Population exposures could result from the wide range of apparatus used in homes, industries, places of entertainment, health clubs, and scientific and medical environments.

The degenerative and bactericidal effects of UV are used in medicine for some therapeutic and diagnostic purposes (7). In addition to the medical applications of UV, there is a growing list of industrial and public health uses (8–11). Most of the applications deal with the germicidal effects of UV against a wide variety of microorganisms such as molds, bacteria, and viruses. This use has been suggested for airborne infections in hospitals, schools, and other places where large groups of people may congregate. The principal human health effects from exposures in this region manifest themselves because of the overlapping erythemal effects.

Several quantities and units have been suggested for characterizing UV. The choice of quantity depends on the effects of interest (7):

Minimal erythema dose (MED)	The shortest exposure at a cerain distance that will produce a perceptible reddening of the skin after 8 hr that disappears within 24 hr (8)
Minimal perceptible erythema (MPE)	A just perceptible erythema that disappears in 24 hr; this is essentially the same as the MED (9)
Erythemal unit (EU)	Twenty $\mu W/cm^2$ of homogeneous energy wavelength 296.7 nm (6, 10)
Germicidal unit (GU)	One hundred $\mu W/cm^2$ of energy of wavelength 253.7 nm (12)
Subvesicular dose (SVD)	A dose large enough to produce skin reddening that, however, does not produce blistering (13)
Minimal color dose (MCD)	The time of exposure of a 1×1 cm^2 test area to a high-pressure mercury quartz lamp that gives a reaction after 48 hr (14)

The indirect effects of UV energy include the photochemical alteration of protein as evidenced by germicidal effects and the ability to interact with chemicals present in air such as the production of ozone and oxides of nitrogen.

Generally, these effects are dependent on wavelength. Health hazards are associated with various welding processes that include photons with energies in excess of 5 eV (250 nm), capable of dissociating molecular oxygen with the subsequent formation of ozone; and photons with energies in excess of 6.5 and 9.5 eV, capable of breaking N—O and N—N bonds, respectively. Under these conditions, oxides of nitrogen can be formed. Investigations into this area have shown that the formation of these gases may occur in the area surrounding welding operations (15).

The blacklight region is generally not considered to be biologically active except in the production of rapid skin pigmentation and photoreactions. Most of the solar UV energy reaching the earth's surface falls within this region. It is in this region that UV produces fluorescence.

4.1 Pathophysiology

For a comprehensive review of the literature on the biomedical aspects of exposure to UV radiation up to 1979, the reader is referred to the publications of World Health Organization (11), the U.S. National Academy of Sciences (12), the *Proceedings of the First International Conference* edited by Urbach (10), and NIOSH (17).

Specific absorbed wavelengths of UV elicit a specific biologic response that constitutes the "action spectra" for that response. These action spectra define the relative effectiveness of different wavelengths in eliciting a specific response when absorbed (Table 14.2).

The injurious effects of UV energy on living systems appear to be related to their ability to be absorbed by either the nucleic acid or unconjugated proteins of the cell and the photochemical reaction that occurs with some receptors in these fractions. In the intact animal, incident UV radiation does not penetrate through the skin. The inability of UV photons to penetrate an appreciable distance into the body limits the areas of concern to the skin and to the eyes (1).

Below 290 nm absorption in humans is entirely in the epidermis. Between 290 and 320 nm, less than 10 percent reaches the dermis; above 400 nm, over 50 percent reaches the dermis. Whole-body exposure to UV radiation is possible; however, common articles of clothing are effectively opaque to UV.

The maximum erythemal effect is produced at 260 nm, with the secondary peak at 290 nm, and with appreciable (60 to 70 percent) effect at 280 nm (18). The erythemal response to wavelengths above 320 nm is relatively poor. This is undoubtedly due to the inability of the low energy photons to produce the specific photochemical reactions required to elicit this response (1).

The sunburn spectrum of sunlight is in the UV zone, 290 to 310 nm, with a peak of about 300 nm. Repeated exposures of fair-skinned individuals to the sunburn spectrum results in actinic skin manifested by a dry, brown, inelastic, and wrinkled skin. "Sailor's," "farmer's," and "fisherman's skin" are names given to actinic skin, which indicates their occupational origin (19). Oil field,

Table 14.2 Summary of Some Biological Effects of UV Radiation[a]

Effect	Radiation
Germicidal	260 nm maximum—effect falls rapidly at shorter or longer wavelengths; effective range associated with absorption band of nucleoproteins
Carcinogenic	200–400 nm—maximum effect 290–320 nm
Ozone production	In germicidal range
Photosensitization	Wavelength at which this occurs varies with absorption characteristic of chemical compounds involved
Pigmentation	280–320 nm stimulates formation of melanin—little tanning; 300–650 nm (maximum 360–500 nm) oxidizes preformed melanin—tanning
Thickening of stratum corneum	In solar range 300–400 nm
Degeneration of collagen	Parallels cumulated exposure in solar range 300–400 nm
Keratoconjuctivitis	Greater effect at shorter wavelengths—0.15×10^{-1} J at 288 nm will produce effect
Antirachitic	Ergosterol to vitamin D; 1 IU of vitamin D formed from ergosterol when 9×10^{-5} J, 249–313 nm absorbed
Erythema	296.7 nm, 25,000 μW-sec/cm^2 minimal amount of power to produce erythema at this wavelength, which is wavelength of maximum sensitivity; erythema can be produced by shorter or longer wavelength (within a limited range), but more power is necessary

[a] Modified from Ferris (30).

pipeline, and construction workers also develop this condition. Actinic skin is not harmful in itself but is a warning to susceptible individuals who tan poorly that certain conditions may develop such as senile keratoses, squamous cell epitheliomas, and basal cell epitheliomas.

The carcinogenic effects of UV radiation have been extensively reviewed by Blum (20, 21) and Epstein et al. (22). The carcinogenic action spectrum for humans is believed to be in the region between 280 and 320 nm. Although there is adequate evidence that wavelengths below 320 nm can produce skin cancers directly or indirectly, it is noteworthy that, with the millions of workerhours of exposure to these radiant energies from welding operations, plasma torches, UV lamps, and so on, no cases of industrially induced skin cancer or keratosis have been reported. This may be because the acute manifestations, namely, skinburn and conjunctivitis, are such painful experiences that steps are taken to prevent exposure of the skin to doses that could be tumorigenic (1).

4.2 Potential Occupational Exposures*

Occupations potentially associated with UV exposure include the following:

Aircraft workers	Iron workers
Barbers	Lifeguards
Bath attendants	Lithographers
Brick masons	Metal casting inspectors
Burners, metal	Miners, open pit
Cattlemen	Nurses
Construction workers	Oil field workers
Cutters, metal	Pipeline workers
Drug makers	Plasma torch operators
Electricians	Railroad track workers
Farmers	Ranchers
Fishermen	Road workers
Food irradiators	Seamen
Foundry workers	Skimmers, glass
Furnace workers	Steel mill workers
Gardeners	Stockmen
Gas mantle makers	Stokers
Glass blowers	Tobacco irradiators
Glass furnace workers	Vitamin D preparation makers
Hairdressers	Welders
Herders	

4.3 Critical Organs

In regard to UV exposure the critical organs are the skin and eyes, resulting in erythema of the skin and skin cancer, rapid skin aging, photosensitization, and keratoconjunctivitis. Excessive exposure to UV energy in the erythemal region can result in severe injury to the skin and eyes. Erythema may range from slight redening of the skin to severe blistering. Photosensitizing agents have action spectra that are frequently in the UV range. Many plants such as figs, limes, parsnips, and pinkrot celery contain photosensitizing chemicals. The combined effect of skin contact and exposure to UV energy results in exaggerated sunburn and frequent blisters. An important industrial photosensitizer is coal tar, with an action spectrum in the visible light region.

Although the effects of UV radiation on eyes and skin have been known for a long time, the mechanism of its interference with biological systems is only generally understood. Considerable information concerning UV effects on the eye has been furnished by Pitts et al. (23, 24).

Harmful effects on the eyes primarily involve the cornea by producing a very painful photophthalmia. Objective findings include excessive lacrimation,

---

* From Key et al. (16).

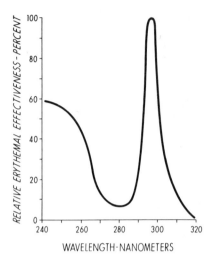

Figure 14.2 Standad curve of erythemal effectiveness.

irritation, and congestion of the conjunctiva; punctate staining of the cornea due to epithelial defects; and ciliary spasm. These symptoms usually become apparent after a latency period of 6 to 12 hr following exposure to UV radiation of sufficient magnitude. This latency period is variable in both directions and it is related to the degree of exposure. The most common cause is probably negligence, such as failure to wear protective goggles when handling industrial welding equipment and using artificial sunlamps in the home. Excessive exposure to sun reflection at the seashore, in the desert, or from snow fields will produce the same reactions. Photochemical denaturation and coagulation of protein structures are the basic mechanisms of cell damage. Nucleoproteins are particularly susceptible, with maximal sensitivity at a wavelength of 265 nm, whereas the proteins of cytoplasm are maximally sensitive to 280 nm. The human cornea is highly sensitive at 288 nm (25).

Production of erythema by UV radiation has been the subject of many reports. On the basis of early work, a standard curve of erythemal effectiveness has been proposed. This curve shows two maxima of erythemal effectiveness, one around 250 nm and another around 300 nm (Figure 14.2). Until such time as more definitive data become available, the American Industrial Hygiene Association (AIHA) Committee on the Nonionizing Radiations believes it prudent to continue acceptance of the Standard Erythemal Action Spectrum Curve in the evaluation of skin hazard potential from UV radiation. The erythemal curves are based on very artificial measures of "redness" indicative of a lack of maturity in experimental approach in this field.

The cornea and conjunctiva of the eye are primary absorbers of UV energy. Mild exposures produce no immediate effect; however, several hours later discomfort is noted as a result of inflammation of the cornea (keratitis). There is some evidence that deleterious effects result to the lens or retina. Photon energies at about 3.5 eV (360 nm) can excite the lens of the eye to fluorescence and produce a diffusion haziness inside the eye that can interfere with visual acuity and produce some eye fatigue. However, when sufficient external luminous flux is also provided, this low-level internal "illumination" is overridden (1).

Snow blindness, which exhibits some of the symptoms of visual blindness, is characterized principally by a burning and scratching of the external surfaces of the eye resulting from exposure to the UV portions of the energy spectrum, which are absorbed in these tissues.

4.4 Threshold

The degree of injury to the skin is a direct function of the initial absorbed dose, weighted in relation to the erythemal effectiveness of the incident wavelength. That necessary to elicit a just barely visually perceived reddening of the skin, based on the standard erythemal curve, is in the order of 3×10^4 μW-sec/cm^2 (3×10^{-2} J/cm^2), ranging from 2×10^4 μW-sec/cm^2 for previously untanned fair skin to 5×10^4 μW-sec/cm^2 for darker Caucasian skin. This value of 2×10^4 μW/cm^2 is variously referred to as the *threshold dose for erythema* (TDE), *the dose for minimum perceptible erythema* (MPE), or the *minimum erythema dose* (MED). If values are determined on the basis of the more recently suggested curves, the MED for the "average" skin may be closer to 8×10^3 μW sec/cm^2. These doses relate to normal skin and skin that has not received exposures to UV in the immediate past. The defensive measures taken by the skin after the symptoms of the first exposure have subsided will necessitate larger than the initial threshold dose to produce similar effects (1).

The action spectrum for UV keratitis, as determined in rabbit eyes, is similar to that of skin erythema, except that the peak effect is shifted somewhat to the lower wavelengths. Maximum keratitic effect is at 288 nm, falling off to about zero at 310 nm at the long-wavelength side of the curve and reaching minimum effect at about 280 nm (26, 27). Threshold keratitic reaction to filtered radiant energy at 288 nm occurs with a dose of 1.5×10^4 μW-sec/cm^2. The thinness of the conjunctival epithelium and the almost complete absence of the stratum corneum and melanin granules deprive the conjunctiva of the protective action these structures afford the skin. There is, thus, no increasing tolerance developed to subsequent irradiation (1).

The investigations by Pitts and associates (23, 24) have contributed fundamental data for the production of photokeratitis. Animal and human exposures were made to establish the photokeratitis threshold and abiotic action spectrum. The abiotic action spectrum ranged from 210 to 320 nm. The peak of the UV threshold curve was at 270 nm for the rabbit and the primate.

Most of the UV rays up to 320 nm are absorbed by the cornea, whereas the lens absorbs those between 320 and 390 nm (25). Since the cornea absorbs almost all of the UV rays up to 320 nm, this form of radiation will produce severe corneal damage before any lenticular changes. Transmittance and absorbance studies of UV radiation including the longer rays (>320 nm)

Table 14.3 Threshold Doses (μW-sec/cm^2) for Keratitis and Erythema

		Wavelength of Maximum Effectiveness 288 nm		Wavelength of Germicidal Lamp 253.7 nm
Keratitis	Rabbit	1.5×10^4	Human	1.2×10^4
	Human	0.3×10^4 (calculated) from 253.7 nm dose and relative effectiveness from rabbit data		
Erythema	Human	296.7 nm, 2.0×10^4 (measured on upper arm, some parts of back)	Human	3.0×10^4

	Duration of Exposure/Day (hr)	Total Dose/Day (253.7 nm)
AMA Council on Physical Medicine	7	1.2×10^3
	24	0.9×10^3

indicate that the intensity of UV energy required for a minimum or threshold response in the lens would be two or three times that necessary for a similar reaction in the cornea (28). Severe corneal burns would have to ensue prior to development of lenticular damage (28). It would seem thus that under ordinary circumstances the deleterious effects of UV radiation with respect to its cataractogenic action is of little significance (25, 28). To protect the eye against overexposure to UV radiation, conscientious use of protective gear recommended for this purpose should be adequate (25).

4.5 Protection Guides

In 1948 the Council on Physical Medicine of the AMA issued criteria for safe exposure to radiant energy from germicidal lamps (29). This group recommended that for the primarily used wavelength, 253.7 nm, exposures should not exceed 0.5 μW/cm^2 for periods of 7 hr or less, nor 0.1 μW/cm^2 in the case of continuous exposure (Table 14.3).

The threshold dose for erythema and its relationship to wavelength vary considerably with the part of the body studied. On the more sensitive skin of the abdomen, the shortest wavelength studied, 250 nm, was the most effective, and the threshold dose for erythema at that wavelength was about 0.6×10^4 μW-sec/cm^2. The AMA recommends an intensity limit of 0.5 μW/cm^2 for exposures of 7 hr or less per day.

A tenfold margin of safety has been allowed in the recommended exposure limits, relative to the threshold dose for production of keratitis by 253.7-nm radiation. It must be recognized that this standard applies only to 253.7 nm radiation, and elevation of the hazard from sources of ultraviolet other than germicidal lamps must take into account the relative effectiveness of the wavelengths produced. Ultraviolet-absorbing goggles are recommended where any possible hazard to the eye exists, and the hazard of erythema should then also be evaluated. This situation may be complicated by photosensitization due to substances on the skin, disease, or drug administration.

With the recognition that many parameters affect threshold values, Matelsky (1) suggested the following threshold doses based on the standard erythemal curve (weighted on the basis of their action spectra):

1. Minimum erythemal dose for previously nonexposed skin—2×10^4 to 2.5×10^4 μW-sec/cm^2 of erythemally weighted UV.
2. Minimal erythemal dose for previously exposed skin—2.5×10^4 to 3.5×10^4 μW-sec/cm^2 of erythemally weighted UV.
3. Minimum keratitic dose—1.5×10^3 μW-sec/cm^2 of keratitically weighted UV.

The American Conference of Governmental Industrial Hygienists (ACGIH)* has proposed threshold-limit values (TLV) for UV radiation:

These threshold limit values refer to levels of physical agents and represent conditions under which it is believed that nearly all workers may be repeatedly exposed day after day without adverse effect. Because of wide variations in individual susceptibility, exposure of an occasional individual, at, or even below, the threshold limit may not prevent annoyance, aggravation of a pre-existing condition, or physiological damage.

Threshold limit values refer to levels of exposure for a 8-hour work day for a 40-hour work week. Exceptions are those limits which are given a ceiling value (C). They should be used as guides in the control of health hazards and should not be used as fine lines between safe and dangerous levels of exposure.

These threshold limits are based on the best available information from industrial experience, from experimental human and animal studies, and when possible, from a combination of the three.

These limits are intended for use in the practice of industrial hygiene and should be interpreted and applied only by a person trained in this discipline. They are not intended for use, or for modification for use, (1) in the evaluation or control of the levels of physical agents in the community, (2) as proof or disproof of an existing physical disability, or (3) for adoption by countries whose working conditions differ from those in the United States of America.

These values are reviewed annually by the Committee on Threshold Limits for Physical Agents for revisions or additions, as further information becomes available.

Ceiling Value—There are some physical agents which produce physiological response from short intense exposure and whose threshold limit is more appropriately based on

---

* American Conference of Governmental Industrial Hygienists (ACGIH), Cincinnati, OH.

this particular response. Physical agents with this type of response are best controlled by a ceiling "C" limit which is a maximum level of exposure which should not be exceeded.

Notice of Intent—At the beginning of each year, proposed actions of the Committee for the forthcoming year are issued in the form of a "Notice of Intent." This notice provides not only an opportunity for comment, but solicits suggestions of physical agents to be added to the list. The suggestion should be accompanied by substantiating evidence.

The TLVs for occupational exposure to UV radiation incident on skin or eye where irradiance values are known and exposure time is controlled are as follows:

1. For the near-UV spectral region (320 to 400 nm), total irradiance incident on the unprotected skin or eye should not exceed 0.1 W/cm^2.
2. For the actinic UV spectral region (200 to 315 nm), radiant exposure incident on the unprotected skin or eye should not exceed the level of 100 mJ/cm^2 for 200 nm or 1000 mJ/cm^2 for 315-nm wavelengths.

These TLVs refer to UV radiation in the spectral region between 200 and 400 nm and represent conditions under which it is believed that nearly all workers may be repeatedly exposed without adverse effect. These values for exposure of the eye or the skin apply to UV radiation from arcs, gas, and vapor discharges and incandescent sources, but not to UV lasers or solar radiation. These levels should not be used for determining exposure of photosensitive individuals to ultraviolet radiation. These values should be used as guides in the control of exposure to continuous sources where the exposure relation shall not be less than 0.1 sec.

Personal protection against the effects from the radiant energies below $\lambda = 320$ nm can be accomplished by the use of eyeglasses or goggles, plastic face shields, protective clothing, or sunscreen creams or lotions. The use of tinted glasses or goggles is seldom indicated for the UV intensities normally encountered in industry but offer added protection and comfort against the very high visible brightness that usually accompanies UV emissions.

5 VISIBLE LIGHT

5.1 Introduction

Visible light is that band of the electromagnetic spectrum from 380 to 750 nm capable of stimulating the photoreceptors of the eye producing a vision. The physical phenomenon of light may be viewed in two differend ways: as energy quanta (photons) or as waves passing through a medium. The interaction of visible light with biological systems can be interpreted as a manifestation of both viewpoints.

Visible light is transmitted through the eye media without appreciable absorption before reaching the retina. Although these photons have relatively low energy values, 3.1 to 1.65 eV, they are unique in that they initiate a photochemical chain reaction in a light-sensitive absorber present in the cells of the retina, the end result of which is the sensation of vision (1). Vision involves the capacity of the eye to adapt to enormous variations in ambient illumination. This adaptation is mediated by the two interspersed photoreceptive systems of the retina. Cone cells are involved with perception of color and vision at higher levels of ambient illumination; rod cells are functional under reduced illumination (31).

The photolabile substances that undergo chemical changes are pigments made up of retinene combined with specific proteins (1). The mechanism of the photochemical reaction that occurs when these "visible" radiant energies are absorbed is not completely known. Several hypotheses have been proposed to explain the mechanism whereby neural stimulation results from absorption of one light quantum by one molecule of visual pigment (32–36). It appears to be an initiating event that, once started, continues as an oxidation–reduction reaction in a series of equilibrium steps involving intermediate degradation products of the pigment and vitamin A (37).

5.2 Sources

Intense sources of radiation in the visible wavelengths, exclusive of lasers, are continuously available from the sun, artificial light sources, highly incandescent bodies, and arc processes such as welding. These are rarely monochromatic, are seldom collimated, and are noncoherent. Light can be produced by a variety of processes:

1. *Thermoradiation.* At temperatures greater than 773 K, many solids and liquids emit radiation with a wavelength of 780 nm. This emission is caused by translation of kinetic energy of the molecules of the substance to electron vibrational or rotational motion.
2. *Electrical discharges.* Light is emitted by spark discharges in gases under reduced pressure or by solids, liquids, or gases bombarded by ionizing radiation.
3. *Chemiluminescence.* Radiation is produced by exothermic chemical reactions such as oxidation of various organic compounds.
4. *Photoluminescence.* A dye or pigment such as chlorophyll absorbs UV light and emits radiation of lower energies and longer wavelengths.
5. *Radiation emitted by high local electrical fields.* Various relatively uncommon types of emission include radiation due to formation or fracture of crystals, cavitations due to sound waves, and recombination of ion pairs.

There are ample data showing that photons of energy in the visible portion of the spectrum are involved in biochemical changes in lower forms of life, that they can photometrically modify or mediate other photochemical processes (i.e., photoreactivation), and that these energies are intimately associated with photodynamic processes (1).

Certain high-intensity light sources (producing absorbed energy levels greater than 50 cal/cm^2-min) may cause eye injury. Items that would probably fall in this category include high-intensity reading lamps, movie and slide projector bulbs, spotlights, floodlights, and other forms of illumination. These items probably number in the tens of thousands or even millions.

5.3 Nomenclature

To better understand the consequences of light absorption, the nomenclature that is used in this field is indicated. The term "luminance" B describes the luminous intensity of an object per unit area of surface. There are many different measures of luminance; the basic unit is the lambert (L), which is equal to one lumen per square centimeter or $1/\pi$ candle per cm^2. Since the L is an inconveniently large unit, luminance is usually measured in millilamberts (1 mL = L $\times$ 10^{-3}), microlamberts (1 µL = L $\times$ 10^{-6}), micromillilamberts (1 µmL = L $\times$ 10^{-9}), or micromicrolamberts (1 µµL = L $\times$ 10^{-12}). In the English system, the unit analogous to the lambert (lumens/cm^2) is the footlambert (lumens/ft^2). The conversion is 1 footlambert (ft-L) = 1.076 mL. Illuminance E is the density of the luminous flux deposited on a surface. A common unit is the foot candle (ft-c). Other units are the mile-candle, centimeter-candle (phot), and meter-candle (lux). In each case the term means that the flux is measured in candles per square foot, per square mile, per square centimeter, or per square meter. The candle is the unit of luminous intensity I. The candle is equal to one lumen per unit solid angle [steradian (sr)]. One lumen is 0.00147 W at 555 nm.

5.4 Exposure

The sun is the major source of environmental exposure. Additional sources to which the general population is exposed include fluorescent and incandescent lamps and televisions. Occupational exposures include use of devices such as welding arcs, motion picture projectors, cathode ray tubes, photocopy machines, microscope illuminators, and high-intensity lamps used for detailed work such as watch repair or product inspection.

5.5 Pathophysiology

Because visible light does not penetrate deeply into most human tissues, the critical organs for this type of radiation are the skin and eyes. Extended exposure to certain wavelengths can lead to thermocoagulation of skin similar to that produced by electrical or thermal burns. Some individuals are sensitive to visible light and develop erythema and edema when exposed to a fluorescent light source. Various lesions are produced in individuals previously exposed to photosensitivity agents. Light rays reflected from an object pass through the cornea at the front of the eye, through the liquid (aqueous humor) in the anterior chamber directly behind the cornea, and then through the lens and vitreous humor onto the retina. The rods and cones of the retina transduce the light into neuroelectrical phenomena. The neural elements of the retina are gathered into the optic nerve at the blind spot and pass by discrete pathways to the highest visual center in the brain, the occipital cortex.

Capacities of the eye are ordinarily measured and defined in terms of thresholds. "Threshold" is defined as a point probability of a target being seen. Any probability of detection may be considered the threshold, but 50 percent is a good value in the absence of a reason to select another. The particular value of the probability selected depends on the use that is to be made of the data.

A detailed review of physiological optics has been prepared by Graham (38). Comprehensive information is available on vision as related to dark adaptation and night vision (31, 39).

The penetrating ability of visible light is slight except for transparent materials such as the lens and humors of the eye. Light entering the eye from a bright source is focused on the retina; therefore, the thermal irradiance is independent of the inverse square law for image sizes greater than the diffraction limit (40).

Because of its narrow depth of penetration, visible light, in general, does not manifest itself as a potential hazard. There are situations, however, in which it can become hazardous. For example, pulsating light at certain frequencies has been reported as a potential source for producing psychological effects.

In some experiments, epileptiform responses have been produced in animals exposed to pulsating light near the alpha rhythm frequency of the EEG. Epileptic persons may experience seizures as a result of exposure to visible light. Such seizures are generally associated with exposure to flickering light (8 to 20 flashes per second), a brightly lit pattern, or sudden appearance of a bright light. Activities triggering photic seizures include watching television (especially from distances of less than 8 ft), reading, driving or riding in a car, walking past a picket fence, swimming, waving the hands between the eyes and a light source, blinking, and observing objects such as rotating helicopter blades, home movies, red and white checked tablecloths, and patterns of sunlight and shadows. The effect is more common in young children, especially females. Some children use the television to intentionally induce seizures because of pleasurable sensations associated with these episodes. Method of treatment include various drugs, tinted eyeglasses to reduce light intensity at certain wavelengths, and covering of one eye.

Probably, the one greatest danger of visible light is that of high-intensity light, which may cause transient loss of visual function (flash blindness) or irreversible thermal injury of the retina. The cornea and lens of the eye are nearly transparent to visible light. It is estimated that approximately 5 percent of the incoming visible radiation is used for vision, with the remainder absorbed by pigment granules in the choroid and pigmented epithelium and converted to heat. Degree of injury depends on pupillary diameter, length of exposure, and intensity and wavelength of light, with blue light producing more damage than other colors. Light intensity in excess of that required for normal vision causes fatigue and discomfort. Exposure to very intense light sources may lead to solar retinopathy (eclipse blindness, retinal burn). The cornea and lens focus incoming light on a relatively small area of the retina. In the case of viewing of a solar eclipse, the pupil is large because of darkness (most of the sun is shielded) and distance of the source from the eye. Intensity of light per unit area at the retina may be 10^5 times that at the cornea, resulting in substantial heat production and thermocoagulation of tissues.

A rare reaction to fluorescent lighting has been observed in individuals who are sensitive to visible light. This sensitivity is manifested by urticaria (urticaria solaris) or by erythema and edema (erythema solare persistans) of exposed areas.

A level of illumination in excess of the amount needed for good vision may produce a feeling of discomfort and eye fatigue. An intensely brilliant light source such as the sun, carbon arc, or welder's arc may produce temporary or permanent blind areas in the retina. This occurs when the retina is subjected to intense light without proper protection and is known as *eclipse blindness* (16). Injuries to the eyes from observing eclipses of the sun have been known since earliest history. Reports of such injuries date back as far as Hippocrates (460 to 370 B.C.), and sporadic references to this have been made throughout the centuries (25). This condition is no doubt due to the retinal heating effect of intense visible light and of infrared rays. Glare may produce a feeling of visual discomfort often due to the work of squinting in an effort to screen reflected nonparallel light rays. If the glare is substantial or frequently induced, it may result in tiredness, irritability, possibly headache, and a decrease in work efficiency (25).

Flash blindness is a relatively new problem occurring only since the development of sources of light brighter than the sun. It is due to glare, which is defined as any degree of light falling on the retina in excess of what enables one to see clearly—that is, any excess of light that hinders instead of helps vision. In modern times, retinal burns have resulted from exposure to lasers and from viewing of nuclear fireballs from distances of up to 42 miles. Damage is usually permanent. It is interesting to note that detached retinas were formerly reattached by photocoagulation with the use of a high-intensity xenon lamp (laser photocoagulation is now the preferred treatment).

Glare can be differentiated into the following types:

1. *Veiling glare.* This is glare created by light uniformly superimposed on the retinal image that reduces contrast and, therefore, visibility.
2. *Dazzling glare.* This is adventitious light scattered in the ocular media so as not to form part of the retinal image.

3. *Scotomatic or blinding glare (flash blindness).* This is glare produced by light of sufficient intensity to alter the sensitivity of the retina.

Although all three types of glare are present in the case of high-intensity light, the effects of the first two are primarily evident only when the source is present. The third type, scotomatic or blinding glare, is especially significant in flash blindness where it produces symptoms (afterimages) that persist long after the light itself has vanished. The afterimage is a prolongation of the physiological processes that produced the original sensation response after cessation of stimulation.

Regardless of whether the glare source is direct or indirect (reflected or specular), it can cause discomfort, can affect the visual performance, or both. The visual discomfort or annoyance from glare is a common well-understood experience and has been confirmed by many experiments. In connection with certain experimental studies, it has been found that people sometimes become more physically tense and restless under glare conditions.

The effect of glare on visual performance can be of serious consequence by itself; however, the visual discomfort brought about by glare can also be a matter of some concern. Although the cause is physical, the discomfort brought about by it is often of a subjective nature. The evaluation of discomfort, then, must make use of subjective responses as criteria. Involved in such procedures is the concept of "borderline between comfort and discomfort" (41).

Flash blindness results from the bleaching of the visual pigments of the eye. Flash blinding can be prevented with proper filtering of light when the occurrence is expected.

In flash blindness the afterimage is essentially a temporary blind area or scotoma in the field of vision. The time duration of this blind area is proportional to the intensity and duration of the light exposure. The greater the intensity and/or the longer the duration of exposure, the more intense and, to a certain extent, the more persistent the afterimage. Ordinarily, the sequence of events following stimulation of the retina by a flash of light is the primary sensation of light followed by a series of positive and negative afterimages. With moderate light intensities, afterimages are not noticed because of the complex action or successive stimulation and continuous movement of the eye. If the original stimulation is of sufficient duration and intensity, however, the sensation will persist with an intensity adequate to reduce or entirely obliterate foveal perception until the effect is dissipated. This is the primary factor in flash blindness.

Irreversible thermal injury of the retina has been known for years. Lookouts for airplanes have on occasion suffered retinal lesions that appear to be caused by heat generated at the sun's image on the retina. Visible light (noncoherent sources) does not seem to be cataractogenic (25, 28).

Long-term exposure to low levels of visible light may result in photochemical damage to the retina. The outer end of the photoreceptor becomes vacuolated and deteriorates; myelin membranes separate and form vesicular structures.

Eventually the photoreceptor is phagocytosed, leaving the underlying layers intact. Damage is more severe when body temperature is elevated during exposure. Such injuries may substantially increase dark-adaptation time. Nearly complete recovery may occur over a period of several months.

5.6 Critical Organs

The visible light range presents little biological hazard except possibly to the eye.

5.7 Threshold

Visual acuity is an important limiting factor in all human detection, target recognition, or other visual tasks. Acuity, like many other visual capacities, is measured and defined in terms of thresholds. One type of visual threshold is a value determined statistically at which there is a 50 percent probability of the target being seen. In most practical situations, a higher probability of seeing, such as 95 or 100 percent, is required.

Because of the optical properties of the eye, the heat energy per unit area on a small part of the retina may be greater by a factor of 10^5 than on the cornea. For visible light a power density of 1 W/cm^2 will exceed the threshold for pain within 1 sec; with a thermal time constant of 0.1 sec, the threshold energy density per pulse will be 0.1 J/cm^2. These factors become exceedingly important in relation to coherent light sources (laser). The sensation of heat, however, serves as an effective warning system under those conditions where there is time to react.

There appear to be three predominant factors controlling potential hazard to the eye: (1) intensity, (2) pupil dilation (i.e., the area of exposure), and (3) length of exposure. If these factors are controlled to keep the absorbed energy below the threshold of burning (40 to 50 cal/cm^2-min), no eye injury should be expected.

Illumination at the eye of 240,000 lumens/ft^2 (0.5 to 1.5 cal/cm^2) represents the proable level required for retinal burns (42). This energy must be delivered at a rate of at least 0.7 cal/cm^2-sec, however, or the rate of heat dissipation in the tissue will be sufficient to prevent elevation of the temperature to a degree where a burn will result. The threshold appears to depend on the time of irradiation and the size of the irradiated area (43, 44, 60–62).

5.8 Protection Guides

The human visual system is a very versatile one, with ample capabilities of adaptation to a variety of environmental changes. This versatility dictates that many variables in the physical and biological environment must be assessed in evaluating human standards for visual performance (45).

Normally, intense and bright sunlight causes maximal constriction of the pupil, thus reducing the energy density on the retina. Bright sunlight, furthermore, causes painful photophobia that will not permit prolonged direct and fixed observation of the sun. The lid reflex (~150 msec) is another mechanism to protect the eye. The continuous action of these measures would be adequate under normal conditions to avoid burn injuries to the retina (25).

Proper placement of light sources in relation to the workplace can provide adequate light while reducing glare. Despite the potential dangers of exposure to high-intensity sources, most standards refer to the minimum light required in the workplace (ranging from 2 footcandles in corridors to 300 footcandles for extremely fine work). In a healthy individual, avoidance mechanisms such as pupil constriction and blinking generally prevent serious eye damage, although these mechanisms may be ineffective when certain mind-altering drugs are used. The ACGIH publishes visible light TLVs. Skin injury may be prevented by wearing of protective clothing. Goggles, facial shields, or tinted eyeglasses may be used to reduce intensity of light entering the eye.

6 INFRARED

Infrared radiation extends from beyond the red end of the visible portion of the electromagnetic spectrum (750 nm) to about 1×10^6 nm. The IR spectrum is frequently arbitrarily divided into three bands: the near IR [wavelength (λ) 750 to 3000 nm], the middle IR (λ 3×10^3–3×10^4 nm), and the far IR (λ 3×10^4 to approximately 1×10^6 nm). The IR spectrum has been subdivided by the International Commission on Illumination (CIE) into three regions: IR-A (0.78 to 1.4 μm), IR-B (1.4 to 3 μm), and IR-C (3 to 1000 μm) (46). Infrared radiation is produced by the rotational and vibrational movements within molecules. All matter above absolute zero (0 K) generates kinetic energy and emits in the IR.

All hot bodies radiate in the IR and radiate IR to other objects with lower surface temperatures. The direct damaging effect of the IR results from an increase in temperature of the absorbing tissue that depends on which wavelengths are absorbed, the parameters involved in heat conduction and dissipation, the total amount of energy delivered to the tissue, and the period over which the energy is supplied.

There is little evidence that photons in the IR (i.e., <1.5 eV) are capable of entering into photochemical reactions in biological systems, probably because they are too low in energy to affect the electron energy levels of these atoms. The interaction that does occur on absorption involves an increase in the kinetic energy of the system, producing a degradation of the radiant energy to heat. Thus, the primary response to exposure to the IR wavelengths is a thermal one.

Infrared radiation is regarded mainly as an occupational rather than environmental hazard. Infrared sources are divided into thermal, luminescent, and electromagnetic sources. Thermal sources are the result of heat radiation emitted from hot objects, and the radiation quantity emitted is a function of the heat input. Luminescent sources, such as the mercury quartz electrical discharge, are the result of absorption of other radiations.

There are both natural and artificial sources of IR radiation. Some of the artificial sources are furnaces, welding torches, flames, heated metal and glass, and incandescent, fluorescent, and high-intensity discharge lamps. Occupations associated with artificial IR exposure include furnace, iron, and steel mill workers, glass blowers, cooks, and welders. The sun is the main natural source of IR radiation and although its energy peaks in the visible region, it extends into the far IR. While the entire population is subject to exposure to the sun, some occupations associated with this natural source are farmers and roofers.

Exposure to 750 to 5000 nm IR involving temperatures from 1000 to 8000 K can occur in many industries from direct IR sources as well as from other heat sources, specifically hot metals, glassmaking, paint and enamel drying, and welding. Infrared is also used in the fields of photography, chemistry, astronomy, criminology, and physiotherapy and in homes and eating establishments for both heating and cooking purposes.

Occupations potentially associated with infrared radiation exposures include the following:

Bakers	Glass furnace workers
Blacksmiths	Heat treaters
Braziers	Laser operators
Chemists	Iron workers
Cloth inspectors	Kiln operators
Cooks	Motion picture machine operators
Dryers, lacquer	Plasma torch operators
Electricians	Skimmers, glass
Firefighters, stationary	Solderers
Foundry workers	Still mill workers
Furnace workers	Stokers
Gas mantle hardeners	Welders
Glassblowers	

In the military forces, passive and active IR systems are employed for tactical communications, beacons, reconnaissance, surveillance, recognition, navigation, airborne proximity warning, direction finding, tracking, homing, fire control, bombing, missile guidance, and similar. Passive IR systems are systems that function to detect IR radiation emitted by objects (targets). Active IR systems function in a manner similar to radar, in that an IR source generates and radiates energy that is reflected from objects and is then detected. The active IR system represents the greatest hazard to personnel since the IR source generally produces a searchlight type of beam, which is filtered to remove any radiation in the form of visible light.

6.1 Pathophysiology

Most biological materials are considered opaque to energies above $\lambda 1500$ nm because of the almost complete absorption of these radiations by water. Radiant energies in the short wavelength region of the near IR can be transmitted into the deeper tissues of the eye.

Skin absorptance is a function of the wavelengths incident upon it. Energies above λ1500 nm are completely absorbed in the surface layers of the skin, where the heat produced is quickly dissipated. The only region of high transmission is between 750 and 1300 nm, with a maximum at 1100 nm. At this wavelength, 20 percent of the energy incident on the surface of the corneum of the epidermis will reach a depth of 5 mm into the dermis (47). This value may be increased in highly pigmented skin (1).

The most prominent direct effects of low wavelength IR radiations on the skin include acute skin burn, increased vasodilation of the capillary beds, and an increased pigmentation that can persist for long periods of time. Under conditions of continuous exposure to high intensities of IR, the erythematous appearance due to vasodilation may become permanent. Many factors mediate the ability to produce actual skin burn, and it is evident that for this immediate effect, the rate at which the temperature of the skin is permitted to increase is of prime importance (1).

The pathological effects of intense IR radiation on the skin of the eyelids are those of an ordinary burn; erythema and blistering are the major signs. Necrosis of the lid has not been reported, primarily because thermal sensation or pain is encountered long before injurious exposure could occur. This automatically gives the individual an indication and thus a chance for self-protection. Repetitive exposure to sunburn doses can, however, give rise to blepharitis, a chronic inflammation of the lids (1).

Lehmann et al. (48) showed that when IR was applied to the area of the ulnar nerve at the elbow, an analgesic effect was found distally in the area supplied by this nerve. This finding is in agreement with previous experimental evidence, obtained in isolated nerves or in small animals, which demonstrated that nerve conduction could be temporarily blocked by heat application.

The heat load imposed on the skin can be removed by the circulating blood or evaporation of moisture. The fact that the eye, and especially the lens, is unable to dissipate the heat because of poor blood supply and the focusing nature of its lens makes it the organ of major concern. Gersten et al. (49) have reviewed the literature on the effectiveness of IR radiation as heating modality.

6.2 Critical Organs

The eye and the skin are the primary organs that are affected by IR radiation. Skin tissue is comprised mainly of water (60 to 70 percent) and absorbs radiation similar to water (46). The skin is a protective barrier for internal organs. It regulates body temperature and fluid equilibrium and possesses a network of nerve endings. These nerve endings protect the individual from considerable exposure. The threshold of pain is dependent on skin temperature. When the pain is perceived, human reaction leads to an avoidance of the thermal sensation. The effects from continuous and excessive IR exposure are burns, increased vasodilation, and a gradual increase in pigmentation, which may become permanent (46).

Radiant energies in the near IR (<1500 nm) is predominantly transmitted into the dermis. The heat is quickly dissipated by the heat regulatory system of the skin. There is a noted, individual variability to the pain threshold. The first detection of pain is when the skin temperature reaches $44.5 \pm 1.3°C$. Cutaneous burns have been produced at this temperature. The tolerance limit of the human body has been determined. An intensity of 0.04 cal/cm^2-sec of short-wave IR can be tolerated by a dermal area of 144 cm^2. Assuming a 75 percent transmittance, 0.03 cal/cm^2-sec is the tolerated intensity (46).

The cornea of the eye is highly transparent to energies between 750 and 1300 nm and becomes opaque to radiant energy above 2000 nm. Thermal damage to the cornea is dependent on the absorbed dose and probably occurs in the thin epithelium rather than in the deeper stroma. A dose of 7.6 W-sec/cm^2 of wavelengths between 880 and 1100 nm was found to elicit minimum regressive corneal damage; whereas only 2.8 W-sec/cm^2 of energy between 1200 and 1700 nm produced this response. These values are consistent with absorption characteristics. With excessive exposure to these critical wavelengths, there may be complete destruction of the protective epithelium, with opacification of the stroma due to coagulation of the protein. Obviously, such denaturization in an area over the pupil would seriously interfere with vision. The probability of incurring such an insult is low except where highly collimated sources can irradiate the eye without producing the sensation of pain in the surrounding skin tissue (1).

The iris is a highly pigmented organ and will absorb practically all the IR radiation reaching it. It is very susceptible to radiations below 1300 nm, as these are appreciably transmitted through the cornea and aqueous humor. The rate at which the heat buildup in the cornea is dissipated is strongly influenced by its contact with the air and the lacrimal fluid. The iris can dissipate its heat only to the surrounding media; therefore, the corneal dose required to elicit damage in the iris would theoretically be more than that for corneal damage by a factor greater than the increment of absorption in the cornea and the aqueous humor. An iris dose of about 4.2 W/sec-cm^2 has been suggested as a threshold for producing damage from a source emitting principally between 800 and 1100 nm, which would require a dose to the corneal surface of 10.8 W-sec/cm^2. Damage to both the cornea and the iris is an acute occurrence, and there are no data available that would indicate that chronic effects in these structures can result from subliminal exposures to IR radiations. The occurrence of aqueous flare as a result of anterior ocular irradiation has not been thoroughly studied in light of its appearance as a criterion for damage of these tissues (1).

It is possible that military personnel might interrupt an IR beam (from an active system) without being aware of the fact. This danger does not exist where only passive systems are in operation. Usually, the danger is not too great because personnel will sense the heating effects of IR and thus be alerted before damage occurs, assuming they have knowledge of the presence of a nearby active IR system. It is possible, however, for the eyes to be damaged before the heating effects provide sufficient warning. It is not advisable to stare into a source of intense infrared radiation, even though the source is equipped with a filter and all visible light is removed.

In the lens, transmission of the shorter wavelengths varies with age and nuclear sclerosis. In the young child the lens transmits about 8 percent of radiation shorter than 300 nm. By 22 years of age, radiation shorter than 300 nm is not transmitted, whereas by 63 years of age the minimum wavelength transmitted is 400 nm. At all ages wavelengths longer than 2700 nm are absorbed. There are selective IR absorption bands from 1400 to 1600 nm and 1800 to 2000 nm. The effective zone of partial absorption is from 400 to 2500 nm (50).

6.3 Cataract

For the sake of convenience, a cataract can be defined as an opacity of the crystalline lens or of its capsule, which may be developmental or degenerative, obstructing passage of light. The degenerative cataract is a manifestation of aging, systemic disease, trauma, or certain forms of radiant energies among others.

Precise definition and meticulous care are required in choosing proper criteria for designation of a cataract. Such criteria may include:

1. The most minute and subtle changes in the crystalline lens recognized by the examiner, regardless of whether they interfere with vision. The term "cataract" for such minute changes is not a very good one and may lead to misinterpretation if not more closely defined.

2. Lens changes that are obvious to any qualified ophthalmologist but do not severely interfere with visual acuity.

3. Lens changes that reduce visual acuity. However, these are by no means in the category of mature "cataracts" (although one depends here on the clinical judgment of the examiner in interpreting changes observed in animal eyes and correlating them with possible deterioration of vision in humans.

4. Mature cataracts, where the lens is milky white and vision is reduced to light projection only (25).

Damage to the lens of the eye from IR has been the subject of considerable investigation for many years. The term "glass-worker's cataract" has become generic for lenticular opacities found in individuals exposed to processes hot enough to be luminous (1).

In 1907 Robinson (51) published the results of his investigations in England on the incidence of opacities on the posterior surface of the lens in the eyes of glass workers that were different than senile cataracts in appearance. On his recommendation, the disease "radiation cataract" became scheduled as occupational in origin in England and by 1921 was copied into the U.S. Workman's Compensation Act.

In 1950 Dunn (52) of Corning Glass Company was unable to find evidence of any ocular disturbance among glass workers who had been exposed to IR for at least 20 years to 2 cal/cm^2-min at 2000 K. But he did recommend more detailed studies be conducted at other industries.

Keatinge et al. (53) were unable to find any posterior cortical changes in the eyes of iron rolling mill workers exposed to 0.02 to 0.1 cal/cm^2-sec for an average of 17 years; however, there was a higher incidence of posterior capsular opacities. The opacity originated in the capsular plane and extended to the cortex. This differs from what is classified as a heat cataract.

Although some serious dissent has arisen as to the validity of the data obtained by earlier investigators (52), the weight of evidence favors the concept that the IR energies emitted from hot sources in industry is the etiological agent responsible for IR cataractogenesis (53–55).

The lens of the eye is avascular, and its position, in suspension behind the iris, removes it from any blood vessels that could assist in dissipation of the built-up heat load. In the normal state, the lens is quite transparent to wavelengths in the IR below 1300 nm that reach it. This question of transparency has raised questions as to whether cataract formation is a direct result of energy absorption by the lens or the indirect effect of temperature increase due to the absorption by the iris. A corneal dose of 7.6 W-sec/cm^2 in the pupillary area of rabbit eyes from IR energy resulted in the production of faint, transient opacities in the anterior subcapsular region of the lens. These were obtained with intensities of 63 W-sec/cm^2 and point out only that even the small amount of thermally effective energy absorbed by the lens is capable of producing an effect if the dose is sufficient (56).

The original mechanism postulated for the formation of radiation cataracts was the focusing of the rays on the posterior surface of the lens causing the maximum damage at that point. Vogt (57) considered changes arising from increase in lenticular temperature and direct effects on the lens fibers. Changes in the lens are secondary to heating of the whole anterior segment of the eye. After the initial damage to the anterior portion, the cataract becomes well defined at the posterior surface of the cortex with successive degeneration of the posterior lens fibers.

Evidence has been presented indicating that heat per se in the absence of IR can produce cataracts. Goldmann (55, 58) noted that elevation of temperature at the anterior portion of the lens is the prime etiologic factor in glassblower's cataract.

Absorption by the retina is strongest at the very shortest IR wavelengths, making it almost impossible to separate the effects from those of the visible radiations. The mechanism of retinal damage involves the absorption of energy by the highly pigmented epithelial layer that separates the retina from the choroid and the rate of conduction of heat from this layer into the adjacent tissues. The production of retinal burn, as with other intraocular tissue, is dose-rate dependent. The reduction in retinal irradiance, due to selective spectral absorption through all the other ocular media, is more than offset by the focusing effect of the lens and the cornea. Therefore, the size or area of the image on the retinal choroid apparatus and the absorbed irradiance are the predominant factors in the production of burn. One millimeter burns have been experimentally produced with 0.1-sec exposures to retinal irradiances ranging from 20 to 40 W/cm². Assuming an iris diameter of 5 mm, the probable corneal dose would have been about 1 W-sec/cm² (1).

There are wide variations in data obtained by different investigators, but some agreement in the relationship between dose, dose rate, and burn diameter exists, especially in the sizes below 1 mm and exposure of 0.1 sec or less (43, 59). *Retinal disability* relates to the area in which the burn occurs, the degree to which the underlying choroid and sclera are involved, and the size of the burn. Obviously, even a threshold burn in the foveal area is more serious than one on the periphery of the retina. The probability of eliciting a retinal burn from industrial processes is quite remote. However, where highly intense and compact sources of radiant energy are being used, an insulting dose can occur in fractions of a second before pain becomes evident (1).

6.4 Perception

Sensations of warmth and cold are evoked in the skin by radiation exchanges between the skin and the environment (60). There seems to be no specific morphological substrate of thermoreceptors. The thermosensitive structures are a network of free nerve endings (61–63). Thermosensitive nerve endings can be defined in two different ways: (a) by the specific temperature sensation aroused by stimulation of a receptor and (b) by the response of a receptor—as indicated by action potentials—to temperature changes (64). These sensations, which are so important to the body economy as thermal detectors for regulating body temperature, are evoked when even the slightest change in skin temperature occurs (65).

At certain constant temperatures, the thermoreceptors show a steady discharge, the frequency of which is dependent on the absolute temperature. The steady discharge of a single fiber has a maximum frequency of 2 to 15 impulses/sec. The temperature of the maximum discharge is between 38 and 43°C for so-called warm fibers (64).

Weber in 1846 proposed that the rate of change of skin temperature dT_s/dt was the effective stimulus for temperature sensation (66). Many investigators subsequently confirmed the fact that, under certain circumstances, a direct relationship could be established between dT_s/dt and temperature sensation, and Hardy and Oppel (67) and Hendler and Hardy (68) made quantitative measures of this relationship. Pulses of IR radiation producing surface dT_s/dt values over 400 times those of longer pulses and 30 times those of longer pulses at a subcutaneous depth of 150 to 200 μm all result in the same level of sensation, namely, threshold (68).

The threshold for warmth perception is reached at a warming of the skin at a rate of about .001 to .002°C/sec at a skin temperature range of 32 to 37°C.

Threshold and intensity of temperature sensation depend to a large extent on the size of the skin area changing temperature. Similarly, the minimal time of warming the skin before a temperature sensation is elicited depends on the size of the area affected and on the density of the specific temperature receptors in that area. The main experimental evidence indicates that temperature sensation is little influenced by the absolute temperature of the skin and is governed by the rate of change of the skin temperature (69). Results of Cook (70), however, indicate that skin temperature is the vital factor in determining pain, although only insofar as this is a measure of the temperature of the thermal pain receptors below the skin surface.

Although spatial summation of warmth sense occurs, some investigators suggest summation is absent in the case of pain sensation; this appears to be based on experiments with irradiated areas of 10 cm^2 and less, in which the pain-producing intensity was independent of exposed area (65). Cook (70), on the other hand, showed that the intensities provoking pain, like those producing sensation of warmth, decrease with increasing exposed area. A physical explanation of both these phenomena, not involving spatial summation, is that the rate of heat transfer from the superficial tissues decreases with increasing exposed area. Thus radiation intensities to produce either warmth or pain decrease with inceasing area is it is assumed that sensations of warmth and pain depend only on the temperature of end-organs (70).

Using IR, Lehmann et al. (48) found that pain occurred at a measured skin temperature of 49.3 to 52.0°C, if the pain threshold was tested over the pad of the little finger. These values are appreciably higher than those determined by Hardy et al. (71). In other areas Wertheimer and Ward (72) found that the temperature in the locus of the pain receptors was between 44.1 and 44.9°C. This discrepancy is probably based on the fact that Hardy et al. (71) and Wertheimer and Ward (72) indirectly determined the temperature that actually caused the irritation or the stimulation of the pain receptors in the skin tissues, whereas Lehmann et al. (48) made a direct measurement of the surface temperature of the skin in an area where no pain receptors are found.

Davson (73) has shown that the cornea and iris exhibit thermal sensitivity to IR. In humans a warm sensation (ca. 10°C elevation in temperature) is experienced in the eye when the cornea is exposed to infrared at a transfer rate of 2 cal/cm^2-sec (74). Radiant stimulation of the cornea produces a greater temperature change in the moderately constricted iris than in the cornea itself.

6.5 Threshold

Ebaugh and Thauer (75) conducted a series of experiments to determine the effect of skin temperature upon the threshold of warmth sensation. The temperature of the surrounding air varied between 15 and 41°C, and thresholds were measured in terms of mcal/cm^2-sec. There was no significant change in threshold of warmth sensation in spite of the change in room temperature and concurrent change in skin temperature.

Among the sensations evoked by irradiation, pain is an important response. To study this pain, Hardy and associates (76) devised an apparatus that consists of exposure control units. The exposure unit consists of a projection lamp, condensing lens, and an aperture behind which the blackened skin of the subject is placed. Immediately in front of the aperture is an opaque shutter that permits exposure of the skin for a fixed period of time to the radiation. The control unit consists essentially of variable resistors and transformers to control the intensity of the lamp. The control unit also contains a meter that reads the intensity of radiation passing directly through the aperture to the skin.

The experimental procedure for measuring pain threshold for thermal radiation is as follows: the subject places the forehead or other skin area against the aperture. The radiation is turned on for a period of 3 sec and the subject is asked when a sensation of pain is perceived. If pain is not perceived, the intensity of light is increased and after a short interval the skin is irradiated again. This procedure is repeated until the subject perceives a sensation of sharp pricking pain just at the end of the 3-sec exposure.

With this method of stimulation it has been found that pain threshold among subjects in the United States was uniform, independent of sex or age, when measured under controlled conditions on the exposed skin of the forehead or back of the hand (65).

Using thermal radiation to determine pain threshold, Hardy et al. (71) observed that the initial skin temperature was important in values obtained. The pain threshold varies linearly with skin temperature and at a skin temperature of 45°C no further stimulus is required. This relationship indicates that pain threshold for any level of skin temperature represents that amount of radiation required to raise the skin temperature to 45°C. Physiologically, this means that the pain threshold is dependent on skin temperature alone and not on the rate of heating of the skin, nor on the rate of change of internal thermal gradients. It has been found that this temperature has a mean value of 44.5 ± 1.3°C (65).

The stimulus giving rise to pain is the same as that giving rise to flexor reflex responses in the spinal animal. When the skin temperature reaches 45°C, reflex responses are obtained. The type of stimulus giving rise as it does to both pain and reflex activity is termed *noxious stimulation* (77). As the intensity of radiation above threshold is increased, increasing intensities of pain are perceived. There is a marked increase in pain between 45 and 52°C after which the increase in painfulness becomes less. Maximal pain is perceived at skin temperature of approximately 65°C (78).

It has been shown that a skin temperature of 45°C is critical for evoking pain and reflex responses. Moritz and Henriques (79) have shown that 45°C is also critical in producing cutaneous burns. Skin temperatures lower than 44 to 45°C rarely produce burns. It may be speculated that noxious stimulation results from chemical reactions in the skin probably involving inactivation of cellular protein (80). The pain threshold is thereby determined as the lowest rate of inactivation of tissue proteins that, if sufficiently prolonged, will cause tissue damage. Pain is related to skin temperature only, whereas tissue damage is dependent on both the skin temperature and the duration of the hyperthermic episode (65).

The IR technique described by Hardy et al. (76) has been widely used in investigations of pain and of the effect of drugs on the pain threshold. Gregg (81) has provided a physical basis for this method of stimulating pain and has also published skin temperatures at which burning pain is felt. The fact that smaller areas can tolerate higher intensities of radiant heat is apparent from the results of workers by use of the Hardy-Wolff-Goodell technique (76). A blackened area (3.5 cm^2) of the forehead can tolerate approximately 0.07 to 0.09 cal/cm^2 sec in prolonged exposures.

It should be pointed out that the intensity of pain sensation does not depend solely on the peripheral excitation pattern sent from the pain receptors but also on many other factors influencing at that time the central nervous system, such as suggestions, attitude, and other psychological factors (82). In addition, subthreshold changes in skin temperature do occur without evoking temperature sensations. Whether these subthreshold changes play a role in thermoregulation is unknown. On the other hand, marked alteration in rate of change and magnitude of change in skin temperature can occur without evoking the temperature sensations usually reported. Thus, precooled skin can be rapidly heated without evoking sensations of warmth.

The tolerance limits of the human body for IR radiations have been determined by several workers. Lloyd-Smith and Mendelssohn (83) found that an incident intensity of 0.04 cal/cm^2-sec of short-wave IR could just be tolerated by epigastric and interscapular skin areas of 144 cm^2. Approximately 25 percent of this energy flux would be reflected, so their result corresponds to a tolerated transmitted intensity of 0.03 cal/cm^2-sec (84). On skin purposely blackened to obtain maximum absorption, maximum temperature increase *in situ*, which produced burns, ranged from 56°C after a 0.5-sec exposure to 5.6 W-sec/cm^2 of filtered radiation to 15°C with a 100-sec exposure to 13 W-sec/cm^2 (85).

Transmission and absorption factors of the ocular media for the IR spectrum and threshold doses to elicit minimum damage have been determined (1, 56):

1. Dose for corneal damage: 7.6 W-sec/cm^2 (880 to 1100 nm); 2.8 W-sec/cm^2 (1200 to 1700 nm).
2. Corneal dose to produce damage in the iris: 10.8 W-sec/cm^2 (800 to 1100 nm).
3. Corneal dose for production of retinal burns: 1 W-sec/cm^2. This value is determined with a 0.1-sec exposure to 20 to 40 W/cm^2 causing a 1-mm burn.

Some early investigators described an "intraocular burn factor" to infrared and designated the most effective region to be between 900 and 1000 nm (86). At these wavelengths, a corneal threshold dose of 0.07 W-sec/cm^2 was required to produce flare, whereas at 1400 nm the threshold dose was 0.18 W-sec/cm^2.

Energies above 1400 nm will not reach the iris, and if it is assumed that the energy per unit area per each 100-nm band between 800 and 1400 nm incident on the cornea is 0.18 W-sec/cm^2, the integrated dose on the cornea would be 1.1 W-sec/cm^2 or approximately one-tenth the dose at which threshold visual burns of the iris are seen (1).

The eye has mechanisms (the blink and pupil reflexes) to protect it from environmental IR radiation, which is usually accompanied by high-intensity visible light. Some industrial IR sources, however, are not found with intense light, so that these protective mechanisms are not always used.

Because of the focusing ability of the lens and its ability to dissipate heat, the eye is of major concern in IR exposure. The cornea is transparent to wavelengths between 0.7 and 1.3 μm and becomes opaque above 2.0 μm. The iris, which is highly pigmented, absorbs almost all IR energies impinging on it and dissipates heat by contact with the air and the surrounding lacrimal fluid. The retina is susceptible to near-IR wavelengths, dissipating its heat by conduction to nearby structures, such as the choroid. If the heat is not dissipated rapidly enough and the temperature of the tissue rises above 45°C, protein denaturation and tissue destruction may occur.

6.6 Protection Guides

Protection guides for IR exposure are designed primarily for protection against ocular effects. The main difficulty, however, in devising protection standards against IR-induced cataract is to correlate the information on the radiation emitted during industrial processes with cataract formation. The dosages of radiation that cause cataract are unknown. Only a small amount of experimentation on animals has been done, but it has provided some knowledge of the way cataract is formed; the numerical data obtained cannot be used in devising standards, because of the relatively massive and frequent doses used in experiments and the possible physiological and anatomic differences between rabbit and human eyes (50).

The nonlinearity of the dose–damage relationship cannot be neglected in accepting the threshold values. Also, the limits were obtained from experiments on animals subjected to relatively massive and frequent doses to IR sources with intensities of 10 to 100 W/cm^2. If workers are subjected to such high intensities (a typical value in glass and steel industries is 1 W/cm^2), it might be appropriate to apply the threshold values to the line workers and accept 0.1 of these values for maximum permissible dose to prevent damage. Otherwise, the experiments must be conducted to simulate conditions in the environment or industries. The extent and localization of damage within the eye depends on the visual size and intensity of the illuminating source (1).

Acute ocular damage from the incandescent hot bodies found in industry can occur with energy densities between 4 to 8 W-sec/cm^2 (1 to 2 cal/cm^2) incident upon the cornea; the ocular tissues involved would depend on the wavelengths that are absorbed. As these relate to threshold phenomena, it would appear that a maximum permissible dose of 0.4 to 0.8 W-sec/cm² (0.1 to 0.2 cal/cm²) could limit the occurrence of these acute effects (1).

7 LASER

The acronym "laser" is commonly applied to devices that use molecular amplification by stimulated emission of radiation operating with an output wavelength of 200 nm to 2×10^4 nm. It includes masers, optical masers, and lasers and refers to coherent light sources emitting visible, IR, or UV light as shown in Figure 14.1.

Various materials have been found that through proper choice allow one to achieve stimulated emission at distinct wavelengths throughout the visible, near-UV, and far-IR spectrum. Output ranges from single pulses as short as 10^{-12} sec to continuous wave (CW). Generally speaking, three types of exposure are encountered from laser beams depending on the mode of operation: (1) Q-switching (exposure time <100 nsec), (2) normal multiple spike mode (exposure time ranging from 100 μsec to 1 or 2 msec), and (3) the CW mode, where the time of exposure may vary from a few milliseconds to seconds or minutes, depending on conditions (87).

The characteristics of lasers that influence their effect on biological systems include the duration of the pulse, the time interval between pulses, the specific wavelength emitted, and the energy density of the beam. The degree of damage produced depends on the absorbing tissue, its absorption characteristics, the size of the absorbing area, and its vascularity (1).

The extremely collimated character and high degree of monochromaticity of the laser beam make this energy of great industrial, military, and communications potential and of physiologic interest. Lasers are used in communications, precision measurements, radar systems (lidar), guidance systems, range finding, metal working, photography, holography, nondestructive testing, and medicine. In medicine, these include eye surgery; in dentistry, cutting hard tooth material; in industry, welding metals and cutting stone; in communications, long-distance transmission; and in geodesy, accurate surveying. Lasers do not constitute an environmental hazard to the general public in the sense that air pollution, noise, radioactive fallout, and other contaminants do, except under rather special circumstances, such as range finding at a military reservation or at a commercial airport, satellite tracking, air turbulence and pollution studies, laser illumination at art exhibits, and holographic public displays. Lasers may be used to beam information over communication channels, ground to air, ship to ship, airport traffic control, highway surveying alignment, satellite tracking, missile guidance, and interception around cities. Lasers may also be installed in public places to count traffic, control various contrivances, and so on. Lasers are multiplying rapidly in the industrial and research fields (87).

It has become common practice to describe the output of pulsed lasers in terms of energy (joules), and that from CW lasers in terms of power (watts).

The joules/cm² unit is used to express absorbed energy density and the watt/cm², to describe power density (1).

7.1 Pathophysiology

Several excellent reviews are available on the subject of biologic effects and hazard criteria for laser exposure (88–91).

Biologic effects can occur through three mechanisms of interaction: (a) a thermal effect, (b) acoustic transients, or (c) other phenomena. The latter two effects are seen only with high-power-density laser pulses. When laser light impinges on tissue, the absorbed energy produces heat. The resultant rapid rise in temperature can easily denature tissue protein. Since tissue is not homogenous, light absorption is not homogenous and the thermal stress it creates around those portions of tissue are the most efficient absorbers. Rapid and localized absorption produces high temperatures. Steam production, evident only at high exposure levels, can be quite dangerous if it occurs in an enclosed and completely filled volume such as the cranial cavity or the eye. A second interaction mechanism is an elastic or acoustic transient or pressure wave. As the light pulse impinges on tissue, a portion of the energy is transduced to a mechanical compression wave (acoustic energy), and a sonic transient wave is built up. This sonic wave can rip and tear tissue and if near the surface, can send out a plume of debris from the impact.

It is generally considered that the biologic effects resulting from exposures to laser energy are primarily a manifestation of a thermal response. Of special interest have been the data available on acute exposures to the eye and skin. The extent of the damage depends primarily on the frequency of the energy, the power density of the beam, the exposure time, and the type of tissue exposed to the beam. The laser is usually a hazard to only those tissues through which the energy can penetrate and that will absorb the wavelength involved. With potential hazard evaluation and safety in mind, the concern is primarily with two organs—the eye and the skin.

7.2 Critical Organs

The primary hazard from laser radiation is exposure of the eye. Radiation levels, if kept below those damaging to the eye, will not harm other tissues and organs of the body. Eye damage can range from mild retinal burns, with little or no loss of visual acuity, to severe lesions with loss of central vision, and total loss of the eye from gross overexposure. Long-term exposure of the retina to wavelengths in the visible spectrum at levels not far below the burn threshold may cause irreversible effects. Long-term exposure of the eye to the near-IR wavelengths can result in opacification of the lens. Overexposure of the cornea to UV wavelengths produces a painful inflammation of the corneal epithelium. Overexposure to a Q-switched laser is especially dangerous and can result in loss of an eye. All structures of the eye can be damaged but the retina is the most sensitive structure. If laser radiation levels in the spectral range 400 to 1400 nm are limited to those considered safe for the eye, biological effects to other parts of the body can be ignored (87).

Light is focused by the cornea and lens onto the fovea of the retina. In this process, the energy density of the light is concentrated by a factor of 10^4 to 10^6 over that falling on the pupil. For this reason, laser energy may pose a serious hazard to the eye. The human eye is relatively transparent to light in a wavelength range of about 400 to 1400 nm. This includes not only the visible range of 400 to 700 nm but also a portion of the IR that is not perceived.

The portion of the eye affected by the laser is dependent on the wavelength of the energy. The ruby laser emits at 694.3 nm. Greater than 90 percent of the energy is transmitted through the ocular media to the retina. Of the energy reaching the retina, about 60 percent is absorbed in the neuroectodermal coat. Almost all of the rest of the energy, 40 percent, is absorbed in the pigment epithelium. Since the pigment epithelium is only 10 nm thick, the greatest absorption per unit volume of energy occurs here, and this layer is the most susceptible to damage.

Neodymium laser of 1060 nm is absorbed to a greater extent in the ocular media with less of its energy reaching the retina than in the case of visible light. Thus there is a greater chance of damage by means of steam production than from other laser types. The aqueous and vitreous bodies are colloidal suspensions in water, and the absorption characteristics of the media are similar to those of water.

Carbon dioxide lasers operate at 10,600 nm. The eye is not very transparent to this frequency range, and danger at low power densities comes from lesions produced in the cornea.

Ultraviolet lasers currently present only a limited hazard to the eye. Up to 300 nm, essentially all the incident radiation is absorbed by the cornea: between 300 and 400 nm the cornea, aqueous humor and lens absorb nearly all the incident radiation. This is particularly true in the adult eye. Thus the energy amplification present with visible laser radiation is not a factor for exposure to UV lasers. Ultraviolet lasers could produce thermal injury to the cornea, conjunctiva, and possibly iris and lens, similar to that sometimes seen among arc welders (92).

The degree of injury to the retina is a function of the physical parameters of power density, exposure time, wavelength, and image diameter on the retina. It also depends on the individual exposed—whether that person is lightly or heavily pigmented in the retina and choroid or has a large or small pupillary diameter, whether retinal exposure is limited by the blink reflex (usually taken as about 150 msec in humans), and many other physical and biological parameters.

The type of damage inflicted on the human eye by laser beams ranges, from a small and inconsequential retinal burn in the periphery of the fundus, to severe damage of the macular area, with consequent loss of visual acuity, up to massive hemorrhaging and extrusion of tissue into the vitreous, with possible loss of the entire eye. The Q-switched laser, because of its high power density and short exposure time, represents potentially the greatest hazard.

Although functional changes in vision have been reported following prolonged exposure of Rhesus monkeys to levels of laser lights as low as 3.1×10^{-5} W/cm^2 on the retina for 3 hr/day over a 7-day period, chorioretinal burns have been found to be the chief eye hazard from lasers operating in the visible region. For purposes of setting laser safety standards, the criterion for damage is a retinal lesion visible through an ophthalmoscope. The magnitude of the damage and the latent period before a visible lesion is seen are functions of the irradiated area, exposure rate, and total exposure. The basic factor in retinal damage is the rate at which heat energy can be moved from the irridiated tissue and the consequent temperature change. A temperature increase a few degrees higher than that experienced during fevers is believed to be capable of producing permanent retinal damage (93).

The effects of laser on the skin cannot be ignored. The large amount of skin surface makes the skin readily accessible to acute and repeated exposures to laser.

In the skin, laser produces damage in varying degrees, depending on the type of laser, the duration of exposure, the area of tissue involved, and the incident energy density (1). Skin pigmentation is a major factor in the severity of injury, although, if sufficient energy is absorbed, even nonpigmented skin will suffer damage. The effects of absorption of high-energy densities by the skin appear to be primarily that of "burn."

The overexposed skin undergoes nonspecific coagulation necrosis whose extent depends on the degree of overexposure. The incident radiant energy is converted to heat that is not rapidly dissipated because of the poor thermal conductivity of the tissue. The resulting local temperature rise leads to denaturation of the tissue proteins. If enough energy is absorbed, the water in the tissue may be vaporized, and the tissue itself may be heated to incandensence and carbonized. The response of the skin to laser light increases as the degree of pigmentation increases. In addition to the inherent optical properties of the skin (reflection, transmission, and absorption), injury to the skin depends on the wavelength of the laser light and the exposure time. Scars may develop when severe lesions from acute overexposure heal. Chronic low-level exposure to laser light generally seldom leads to injury.

7.3 Threshold

Thresholds of retinal damage for different wavelengths are available. Because of the lack of information on the human eye, threshold values are based on data obtained from experimental animals, mostly rabbits and monkeys. As in most other cases, extrapolation to the human eye must be made with caution.

The power density required to reach the threshold of retinal injury has been shown to increase with a reduction in pulse duration. The precise mechanism for producing damage under Q-switched conditions is not yet fully understood.

Table 14.4 Retinal Damage Thresholds[a]

Laser Type	Wavelength	Pulse	Level
Continuous wave	White light	—	6 W/cm^2
Normal pulse	694 nm	200 μsec	0.85 J/cm^2
Q-switched pulse	694 nm	30 nsec	0.07 J/cm^2

[a] *Source:* from D. H. Sliney and W. A. Palmisano, "The Evaluation of Laser Hazards," *AIHA J.*, **20**, 425 (1968).

A thermal model describes reasonably well the observed histological damage, at least for laser pulses down to microsecond duration. Most investigators agree that the retinal injury threshold, defined as that energy or power density required to produce a barely observable lesion after a period of 5 min or more postexposure, would lie in the range of several tenths of a joules/cm^2 for pulsed white light (xenon) and normal-mode pulsed ruby systems and an order of magnitude lower for Q-switched ruby (94, 95). There is some evidence that these thresholds, based on rabbit retina, are somewhat lower than those for humans and other primates (96).

Leibowitz and Luzzio (97) have reported that cataracts can be induced in rabbits exposed to high levels (29.5 J/cm^2) delivered to the cornea from a ruby laser with a pulse length of 1.3 msec. It should be noted that this level is several orders of magnitude greater than the threshold for retinal injury.

Threshold values for retinal damage from visible laser are listed in Table 14.4. Although the values listed are the specific wavelengths, it is reasonable to assume little or no difference over the entire visible spectrum.

The minimal reactive dose (MRD) ranges to the flexor surface forearm of a Caucasian adult from laser light under several different conditions of exposure are shown in Table 14.5.

7.4 Protection Guides and Standards

Some major factors governing laser safety standards are (118):

1. Power density (watts per square centimeter) at the cornea or energy in joules (J) entering the eye.

Table 14.5 Skin Damage Thresholds[a]

Laser	λ (nm)	Exposure Time	Area, cm^2	MRD (J/cm^2)
Ruby, normal pulse	694	0.2 msec	2.4–3.4 × 10^{-3}	14–20
Argon	500	6 sec	95 × 10^{-3}	13–17
CO$_2$	1060	4–6 sec	1	4–6
Ruby, Q-switched	694	10–12 nsec	0.33–1.0	0.5–1.5

[a] *Source:* from L. Goldman, "The Skin," *Arch. Environ. Health*, **18**, 435 (1969).

2. Exposure time.
3. Wavelength or spectral distribution at cornea.
4. Transmission through ocular media (OM) as function of wavelength.
5. Diameter of pupil.
6. Diameter of retinal image.
7. Absorptance or transmission by retinal pigment epithelium as a function of wavelength.
8. Absorptance or transmission by choroid as a function of wavelength.

In setting safety standards the focusing action of the ocular lens must be considered. The total amount of light energy entering the eye is determined by the area of the pupillary opening. The transmitted light energy that reaches the retina is absorbed by the pigmented epithelium, where most of it is converted into heat. Because of the focusing action of the lens, the image of the limiting aperture (the pupillary opening) formed on the retina is very much smaller than the pupillary opening. For light of wavelength λ cm, and an eye whose pupillary diameter is d_p and whose lens has a focal length f cm, the diameter of the image on the retina d_r is

$$d_r = \frac{2.44 \lambda f}{d_p}$$

Since radiant exposure H or irradiance E is related to the illuminated area by

$$H \text{ (or } E) = \frac{\text{energy}}{\text{area}}$$

the ratio of H or E of the cornea to that of the retina varies inversely with the square of the ratio of the pupillary diameter to the diameter of the image on the retina:

$$\frac{H \text{ or } E(\text{retina})}{H \text{ or } E(\text{cornea})} = \left(\frac{d_p}{d_r}\right)^2$$

The maximum permissible exposure of the cornea must allow for the concentration of light energy due to the focusing action of the lens. On the basis of retinal damage thresholds and concentration of light by the crystalline lens, maximum permissible exposure limits have been recommended by the ACGIH (98) and ANSI (99). These two sets of recommended limits are essentially equivalent. For practical application, the complete tables in the latest edition of ANSI or ACGIH recommendations should be consulted.

Approaches to laser safety vary greatly among organizations that have an interest in the problem. Usually, an early consideration in such programs is the institution of periodic eye examinations. Some organizations have written policies and practices outlining the responsibility of management, technical supervision, environmental health, safety, and medical personnel, but such policies are usually broadly defined with a specific provision for attending to problems on an individual case basis. All such policies and procedures should emphasize the primary reliance placed on supervisory personnel for the safe conduct of laser operations; engineering controls rather than personal protective equipment (goggles) should be stressed. Engineering measures should take into account such factors as interlocks, proper layout of room areas, shielding materials, and warning signs. Commercially available protective eyewear is designed for protection against specific wavelengths; no single device protects against all laser wavelengths (94, 95).

The ACGIH has prepared a guide for laser installations (98). In this document it is noted that hazard controls for laser radiation vary according to the type of laser being used and the manner of its use. Only properly indocrinated persons should be placed in charge of laser installations and operations, control of laser hazards should be under the supervision of personnel knowledgeable in laser hazards, and a closed installation should be used when feasible. (The equivalent of a closed installation can be achieved by installing a laser in a lightproof enclosure.)

General precautions (common to every laser installation) are:

1. Personnel should not look into the primary beam or at specular reflections of the beam when power of energy densities exceed the permissible exposure levels.
2. Avoid aiming the laser with the eye to prevent looking along the axis of the beam, which increases the hazard from reflections.
3. Work with lasers should be done in areas of high general illumination to keep pupils constricted and thus limit the energy that might inadvertently enter the eyes.
4. Shatter-resistant safety eyewear designed to filter out the specific frequencies characteristic of the system affords partial protection. Safety glasses should be evaluated periodically to ensure maintenance of adequate optical density at the desired laser wavelength. There should be assurance that laser goggles designed for protection from specific lasers are not mistakenly used with different wavelengths of laser radiation. Distinctively colored frames are recommended, and the optical density should be shown on the filter. Laser safety glasses exposed to very intense energy or power density levels may lose effectiveness and should be discarded.
5. The laser beam should be terminated by a target material that is nonreflective and is fire resistant, and an area should be cleared of personnel for a reasonable distance on all sides of the anticipated path of the beam.
6. Avoid electrical shock from the potentially dangerous electrical sources of high and low voltage.
7. Special precautions should be taken if high voltage tube rectifiers (>15 kV) are used because there is a possibility that X-rays might be generated.

In the United States, laser safety regulations have been promulgated by the Department of Health and Human Services through the Bureau of Radiological Health (BRH). The BRH regulates manufacturers only, not users, through requirements for performance specifications. All lasers are classified into one of four different classes, depending on the level of risk from the laser. Class I is the least hazardous and Class IV is the most hazardous:

Class	Hazardous capabilities
I	Cannot produce hazardous radiation
II	Continuous intrabeam exposure damages the eye; momentary intrabeam exposure (<0.25 sec) is not damaging to the eye
III	Can damage the eye during momentary intrabeam viewing
IV	Can damage the skin as well as the eye during momentary intrabeam exposure or exposure to diffuse reflection

According to the class of the laser, certain engineering and labeling requirements are specified in the regulations. The engineering requirements include:

1. A protective housing, which prevents exposure to laser radiation not necessary for the performance of the intended function of the laser (leakage radiation).
2. Safety interlocks, designed to prevent human access to laser radiation upon removal or displacement of the protective housing.
3. A remote-control connector to allow additional interlocks and remote on–off controls.
4. Key control to prevent unauthorized use of the laser. The key must be removable and the laser must be inoperable unless the key control is turned on by the key.

The labeling requirements include information to be prominently displayed on the appropriate signs, warning of a laser hazard, and information about the laser and its output radiation that must be prominently affixed to the laser.

8 MICROWAVE–RADIOFREQUENCY ENERGIES

Radiofrequency radiation is defined arbitrarily as electromagnetic radiation in the frequency range of 0.3 to 300 MHz; microwaves include electromagnetic radiation whose frequencies range from 300 MHz to 300 GHz. The following microwave frequency bands have been assigned to radar systems:

Band	Frequency (MHz)	λ (cm)
L	1000–1400	27.3–21.4
S	2600–3950	11.5–7.6
C	3950–5850	7.6–5.13
X	8200–12,400	3.66–2.42
K_u	12,400–18,000	2.42–1.67
K	18,000–26,000	1.67–1.16
K_a	26,000–40,000	1.16–0.75

Industrial, scientific, and medical applications of microwave–radiofrequency energies include:

Frequency (MHz)	λ
13.56	66.37 m
27.12	33.19 m
40.68	22.12 m
915	32.8 cm
2450	12.2 cm
5800	5.2 cm
22,125	1.4 cm

Microwave energy, when propagated, is categorized into two discrete modes known as CW and pulsed. CW is associated with communication transmitting devices and consumer products, whereas pulsed microwaves are associated with radar and industrial and medical equipment.

The absorption of microwave energy is directly dependent on the electrical properties of the absorbing medium—specifically, its dielectric constant and electrical conductivity. These properties change as the frequency of the applied electrical field changes. Values of dielectric constant and electrical conductivity and depth of penetration have been determined for many tissues (100, 101). Because of its tissue-penetrating capability, heat generated as a result of the transformation of microwave energy is directly proportional at any depth to the intensity of the energy present, neglecting heat transfer by other methods. The so-called volumetric heating that results therefrom is quite unlike that due to conductive heating.

When microwaves are absorbed by any material, the energy is transformed into increased kinetic energy of the absorbing molecules, which, by increased collision with adjacent molecules, produces a general heating of the entire medium. The energy value of 1 quantum of microwaves (4×10^{-4}–1.2×10^{-6} eV) is much too low to produce the type of excitation necessary for ionization, no matter how many quanta are absorbed.

Knowledge of the electrical properties of a tissue permits direct calculation of the absorption coefficient. Absorption coefficients of 2.5 and 0.6 for 10000 MHz and 30000 MHz, respectively, have been obtained in skin. The reciprocals of these coefficients define the depths at which the incident intensity is reduced to 1/e, or about 37 percent of its initial power. For 10000-MHz microwaves this is about 4 mm, whereas for 3000-MHz microwaves it is approximately 16 mm.

When considering the biological effects of microwave radiation, the wavelength of frequency of the energy and its relationship to the physical dimensions of objects exposed to radiation become important factors. Absorption of energy radiating from a source into space also depends on the relative absorption cross section of the irradiated object. Thus the size of the object with relation to the wavelength of the incident field plays a role. If a small part of the body were subjected to far-field exposure to radiant energy of relatively long wavelength, it is possible that this part could absorb more energy than that falling on its shadow cross section (102). The term "far field" refers to that region away from a source of microwave energy where the power density decreases with increased distance, according to an inverse-square law.

One of the chief difficulties in working with free-field microwave irradiation is to determine how much of the incident energy is absorbed by the tissues. Anything introduced into the microwave field tends to distort the field pattern in a generally unpredictable manner. Therefore, the field strength as measured by even a fine probe at some particular location may be quite changed when the test object itself replaces the probe.

The elucidation of the biological effects of exposure to microwave or radiofrequency (MW/RF) energies requires a careful review and critical analysis of available publications. This entails differentiating established effects and mechanisms from speculative and unsubstantiated reports. Although most of the experimental data support the concept that the effects of microwave exposure are primarily, if not only, a response to heating or altered thermal gradients in the body, there are large areas of confusion, uncertainty, and misinformation.

The organs and organ systems affected by exposure to microwave (300 MHz to 300 GHz) or radiofrequency (300 kHz to 300 MHz) energies are susceptible in terms of functional disturbance, structural alterations, or both. Some reactions to MW/RF exposure may lead to measurable biological effects that remain within the range of normal (physiological) compensation and are not necessarily hazardous or improve the efficiency of certain physiological processes and can thus be used for therapeutic purposes. Some reactions, on the other hand, may lead to effects that may be potential or actual health hazards to health.

The nonuniform, largely unpredictable distribution of energy absorption may give rise to increases in temperature and rates of heating that could result in unique biological effects. Nevertheless, it is important to recognize that the mammalian body normally is not a uniform incubator at 37°C but does contain significant temperature gradients in deep body organs that may act as functional stimuli to alter normal function both in the heated organ and in other organs or organ systems. Thus indirect effects can be elicited in organs far removed from the site of the primary interaction.

Most of the MW/RF responses are explained by thermal energy conversion, almost exclusively as enthalpic energy phenomena.

Specific organ tissue systems may "function" at a significantly different rate if local thermal gradients are altered. Relatively large changes in circulation are provoked by quite small deviations from neutral temperature (103). Body content of heat is equilibrated by approach to equality of two overall processes, gain and loss.

Absorption of MW/RF energy leads to increased temperature when the rate of energy absorption exceeds the rate of energy dissipation. Whether the resultant increased temperature is diffuse or confined to specific anatomical sites depends on (a) the electromagnetic field characteristics and distributions within the body and (b) the passive and active thermoregulatory mechanisms available to the organism, such as heat radiation, conduction, convection, and evaporative cooling. The efficacy of heat convection between a body and its immediate environment is influenced by the environmental conditions.

8.1 Experimental Observations Regarding Chromosomes and Cellular and Genetic Effects

Some investigators have reported chromosome changes in various plant and animal cells and tissue cultures (104, 105). These studies have been criticized because the systems were subjected to a thermal stress; the chosen parameters of the applied field caused biologically significant field-induced force effects in *in vitro* experiments, and many of these experiments have not yet been independently replicated. There is no conclusive evidence for microwave-induced genetic effects (106).

8.1.1 Growth and Development

A few reports suggest that particular combinations of MW/RF wavelength, duration of exposure, and power density produce effects on embryological development and postnatal growth. In almost all instances, however, the reported effects may be ascribed to the excessively increased temperature caused by the exposure (107).

8.1.2 The Gonads

The effect of microwaves on the testes has been studied fairly extensively. Although reports indicate that high power density exposure can affect the testes (108), this can be related to the heating of the organs. The sensitivity of the testes to heat is well known.

8.1.3 Neuroendocrine Effects

Some investigators believe that endocrine changes result from hypothalamic–hypophyseal stimulation due to thermal interactions at the hypothalamic or adjacent levels of organization, the hypophysis itself, or the particular endocrine gland or end organ under study (109–115). According to other investigators, the observed changes are interpreted as resulting from direct microwave interactions with the central nervous system (CNS) (116–119). In either case one should not consider neuroendocrine perturbations as necessarily harmful because the function of the neuroendocrine system is to maintain homeostasis, and hormone levels will fluctuate to maintain organ stability (113).

Before 1970 the investigation of the effects of microwaves on endocrine balance consisted primarily of retrospective studies of exposed microwave workers. In these studies various clinical parameters were analyzed to assess thyroid integrity [thyroid gland size, radioactive iodide uptake, metabolic rate, and plasma protein-bound iodide (PBI)]. The results, however, were not consistent, and diagnostic techniques are questionable.

8.1.4 Immunology

Microwaves have been reported to induce an increase in the frequency of complement receptor-bearing lymphoid spleen cells in mice (120). MW/RF-induced hyperthermia in mice has been associated with transient lymphopenia with a relative increase in splenic T and B lymphocytes (121) and decreased *in vivo* local delayed hypersensitivity. The influence of increased temperature on immune responses is well known (122).

8.1.5 Effects on the Nervous System

Transient functional changes referrable to the CNS have been reported after "low-level" (<10 mW/cm^2) microwave exposure of small laboratory animals. Although some reports describe the thermal nature of MW/RF energy absorption, others implicate nonthermal or "specific" effects at the molecular and cellular level. It should be noted, however, that specific (i.e., nonthermal) MW/RF effects have not been experimentally verified. The first reports of the effect of mirowave energy on conditional response activity in experimental animals was made by investigators in the USSR (123). In subsequent years the study of "nonthermal" effects of microwaves gradually occupied the central role in electrophysiological studies in the Soviet Union.

8.1.6 The Blood–Brain Barrier

In the last few years, considerable interest has been engendered by a report of a transient alteration in the permeability of small inert polar molecules across the blood–brain barrier of rats exposed to microwaves (124). Attempts to duplicate these findings have yielded equivocal results unless the brain is subjected to a large increase in temperature (125, 126).

8.1.7 Behavioral Effects

Studies have been conducted on the effects of MW/RF on the performance of trained tasks by rats and rhesus and squirrel monkeys. Many of these studies indicate that exposure to MW/RF energies can be related to suppressed performance of a trained task and that a power density/dose rate and duration threshold for achieving the suppression exists.

In this context it is important to realize that behavior among animals reflects adaptive brain-behavior patterns and behavioral thermoregulation expresses an attempt to maintain a nearly constant internal thermal environment. Changes in body temperature bring about not only autonomic drives but also behavioral drives (127); thus microwaves can influence behavioral thermoregulation (128, 129).

8.1.8 Hematopoietic Effects

Although several investigators state that the blood and blood-forming system are not affected by acute or chronic microwave exposure (130–132), effects on hematopoiesis have been reported (133–137). The degree of hematopoietic change is dependent on the field intensity, duration of exposure, and induced hyperthermia. In evaluating reports of hematologic changes one must be aware of the relative distribution of blood cells in the population and the susceptibility to thermal influences. Early and sustained leukocytosis in animals exposed to thermogenic levels of microwaves may be related to stimulation of the hematopoietic system, leukocytic mobilization, or recirculation of sequestered cells.

8.1.9 Ocular Effects

During the past 25 years, numerous investigations in animals and several surveys among human populations have been devoted to assessing the relationship between exposure to microwaves and the subsequent development of cataracts. It is significant that of the many experiments on rabbits by several investigators using various techniques, power density above 100 mW/cm^2 for 1 hr or longer appears to be the time–power threshold for cataractogenesis in the tested frequency range of 200 to 10,000 MHz (138, 139). In other species of animals as dogs and nonhuman primates, the threshold for experimental microwave-induced cataractogenesis appears to be higher.

8.1.10 Cutaneous Perception

Perception of microwave energy is a function of cutaneous thermal sensation (Table 14.6) or pain (Table 14.7). Several studies suggest that a threshold sensation is obtained when the temperature of the warmth receptors in the skin is increased by a certain amount (ΔT) (140–142).

8.2 Dosimetry

Detailed discussions that serve as bases for scaling have been presented by several authors (143–146). Maximum absorption during whole-body irradiation of small animals apparently occurs at frequencies between about 0.5 and 3 GHz and for humans at around 60 to 100 MHz with a peak at about 80 MHz. At frequencies below 30 MHz, absorption rate drops off rapidly and is also much less at frequencies above 500 MHz.

An effort is being made to standardize dosimetric measures of MW/RF exposure by using a quantity called the *specific absorption rate* (SAR), the time

Table 14.6 Stimulus Intensity and Temperature Increase to Produce a Threshold Warmth Sensation[a]

Exposure Time (sec)	3000 MHz	10,000 MHz		Far IR	
	Power Density (mW/cm^2)	Power Density (mW/cm^2)	Increase in Skin Temperature (°C)	Power Density (mW/cm^2)	Increase in Skin Temperature (°C)
1	58.6	21.0	0.025	4.2–8.4	0.035
2	46.0	16.7	0.040	4.2	0.025
4	33.5	12.6	0.060	4.2	—

[a] Forehead surface area 37 cm^2. Data from Hendler (140, 141).

rate at which radiofrequency electromagnetic energy is imparted to an element of mass of a biological body (147). The SAR depends on a finite period of exposure to yield the amount of energy absorbed by a given mass of material, which is termed *specific absorption* (SA)—that is, joules per kilogram (J/kg). The SAR is the time rate at which radiofrequency electromagnetic energy is imparted to a component or mass of a biological body. Thus the SAR is applicable to any tissue or organ of interest or is expressed as a whole-body average and specified in SI units of watts per kilogram (W/kg).

Durney et al. (143) have calculated the SAR as a function of frequency for different sizes of laboratory animals and humans. This concept is useful in allowing limited extrapolation from one species to another, but it should be used cautiously in view of its limitations as already noted. Because of the different properties of various tissues, energy deposition may be much more uneven in animals than the models predict.

Whole-body absorption rates approach maximal values when the long axis of a body is parallel to the E-field vector and is four tenths the wavelength of the incident field. At 2450 MHz, (λ = 12.5 cm), for example, a standard man

Table 14.7 Threshold for Pain Sensation as a Function of Exposure Duration (3000 MHz, 9.5 cm^2 area)[a]

Power Density (W/cm^2)	Exposure Time(s)
3.1	20
2.5	30
1.8	60
1.0	120
0.83	>180

[a] From Cook (142).

Table 14.8 Ratio of Specific Absorption Rate to Basal Metabolic Rate for an Average Person Exposed to Far-Field Incident Power Densities of 1 mW/cm^2 and 5 mW/cm^2[a]

Frequency (MHz)	Average SAR/BMR (%)	
	1 mW/cm^2	5 mW/cm^2
10	0.13	0.65
20	0.60	3.00
50	5.80	29.00
60	10.00	50.00
80	16.00	80.00
100	12.00	60.00
200	5.20	26.00
500	3.70	18.50
1,000	2.90	14.50
2,000	2.50	12.50
5,000	2.50	12.50
10,000	2.50	12.50
20,000	2.50	12.50

[a] Based on Stuchly (148).

(long axis 175 cm) will absorb about half of the incident energy. If the human whole-body SAR is divided by the basal metabolic rate for humans, a ratio is obtained that provides a measure of the thermal load incurred as the result of a known incident power density (148). Table 14.8 illustrates the variation of this ratio with frequency at two incident power densities. In the region of human whole-body resonance (60 to 80 MHz), this ratio reaches a maximum value (about 0.16 for an incident far-field power density of 1 mW/cm^2). The ratio drops off rapidly on either side of this peak, and at 10 MHz and below the ratio would be less than 0.001 (149).

At frequencies that result in maximal absorption, which defines whole-body resonance, the electrical cross section of an exposed body increases in area. This increase occurs at a frequency near 70 MHz for a standard man and results, as shown in Table 14.9, in an approximate eightfold increase in absorption relative to that in a 2450-MHz field.

8.3 Survey of Human Exposures

8.3.1 Epidemiologic Studies

There have been several epidemiological studies of MW/RF exposure. Individuals exposed while assigned to the military services or occupationally exposed in industrial settings have been the principal groups studied. A few other

Table 14.9 Specific Absorption Rate for Animals and Humans (W/kg for 1 mW/cm^2 Incident PD)[a]

Species	Maximum Absorption (MHz)	Frequency (MHz)						
		20–30	70	300	1000	2450	3000	10,000
Mouse	2000	8×10^{-4}	0.008	0.06	0.4	1.00	0.965	0.322
		(0.05)	(0.04)	(1.50)	(13)	(36)	(36.60)	(12.40)
Rat	600	1.8×10^{-3}	0.0125	0.3	0.6	0.23	0.26	0.25
		(0.12)	(0.06)	(7.50)	(20)	(8)	(9.60)	(9.60)
Rabbit	320	0.015	0.050	0.80	0.250	0.15	0.08	0.07
		(0.33)	(0.22)	(20)	(8.30)	(5.40)	(2.96)	(2.69)
Rhesus monkey	300	1.7×10^{-3}	0.0125	0.195	0.10	0.07	0.065	0.060
		(0.01)	(0.06)	(5.00)	(3.33)	(2.50)	(2.41)	(2.30)
Dog	200	1.5×10^{-3}	0.010	0.100	0.050	0.040	0.037	0.030
		(0.10)	(0.04)	(2.50)	(1.67)	(1.40)	(1.40)	(1.15)
Human (1 year)	150	0.004	0.040	0.15	0.065	0.055	0.050	0.042
Man (average)	70	0.015	0.225	0.04	0.03	0.028	0.027	0.026

[a] Parentheses ( ) indicate SAR relative to average man.

populations living or working near generating sources or exposed to medical diathermy have been or are being investigated (150–152). Information about health status has come from medical records, questionnaires, physical and laboratory examination, and vital statistics. Sources of exposure data include personnel records, questionnaires, environmental measurements, equipment emission measurements, and (assumed adherence to) established exposure limits. Although there have been advances in measurement, accurate estimates of dose present formidable problems in most epidemiologic studies (150, 151).

An early study on U.S. Navy personnel during World War II did not reveal any conditions that could be ascribed to radar exposure (153). Ten years later a 4-year surveillance of a relatively large group of radar workers in the United States did not reveal any significant clinical or pathophysiologic differences between the exposed and control groups (154, 155). On the other hand, surveys of eastern European workers revealed functional changes in the nervous and cardiovascular systems (156–158).

In an extensive 12-year survey in Poland, workers exposed to microwaves for various periods were examined for the incidence of functional disturbances and disorders considered as contraindications for occupational exposure to microwaves according to the criteria used in Poland (159–161). The population worked under identical conditions except for the exposure levels, of which there were two subgroups. Workers in the first subgroup (507 individuals) were exposed to varying power densities between 0.2 and 6 mW/cm^2; in the second subgroup power densities were below 0.2 mW/cm^2. No dependence of incidence of disorders, such as organic lesions of the nervous system, changes in translucency of the ocular lens, primary disorders of the blood system, neoplastic diseases, or endocrine disorders on exposure level, duration, or work history could be shown. The incidence of functional disturbances ("neurasthenic syndrome," gastrointestinal tract disturbances, cardiovascular disturbances with abnormal ECG) as reported in the Soviet publications (156–158) was found not to be related to the level or duration of occupational exposure. There were no instances of irreversible damage or disturbances caused by exposure to microwave energy (159).

The Medical Follow-up Agency of the U.S. National Academy of Sciences studied mortality and morbidity among 40,000 personnel of the U.S. Navy (162) potentially exposed to radar. There was no indication of any adverse effect due to exposure to microwaves. This study was preceded by a survey to investigate physiological and physical effects among U.S. Navy crewmen who could be exposed to 0.1 to 1 mW/cm^2 aboard an aircraft carrier (163). No significant differences were found with respect to task performance, physiological tests, or biological effects. Hematological findings were within the normal range.

A study of 4388 employees and 8283 dependents of the American Embassy in Moscow, some of whom had possibly been exposed to 5 to 15 $\mu W/cm^2$ of microwaves for variable periods of time (9 to 18 hr/day) up to 8 months, showed no differences in health status as indicated by their mortality experience and various morbidity measures (164, 165). Exhaustive comparative analyses were made of all symptoms, conditions, diseases, and causes of death among employees and dependent groups of adults and children. No differences in health status by any measure could be attributed to microwave exposure.

This study was preceded by a cytogenetic evaluation for possible mutagenesis performed on 250 samples from 71 U.S. State Department employees and family members before, during, and after exposure to microwaves at the American Embassy in Moscow (165). No genetic or other adverse biologic effects among employees and dependents attributable to microwave exposure could be established.

A study was also conducted to determine the blood lymphocyte counts of adult employees and dependents at the American Embassy in Moscow (165). About 350 adults who were embassy employees during the study period were examined; roughly 1000 foreign service personnel in the United States served as a comparison group. There was no correlation between the higher average lymphocyte count found in the Moscow population and microwave irradiation in the embassy. Altered lymphocyte count was believed to be of microbial origin.

Considerable publicity has been given to an alleged death from microwave exposure.* This case was a workmen's compensation award and not subjected to litigation. The individual apparently was suffering from Alzheimer's disease, which is not uncommon in the general population. Alzheimer's disease has been recognized since 1903 and is not related to MW/RF exposure, as it was first recognized before MW/RF emitting devices were available. By necropsy the diagnosis was bilateral lower-lobe bronchial pneumonia; the brain was not examined. The subject was never exposed to significant levels of microwaves. His activities required him to work at the back of the transmitting system. He was never required to work in front of the "antenna dish," although even in front of the antenna dish, the power density would not have been greater than 1 mW/cm^2.

8.4 Ocular Effects

Numerous surveys of ocular effects of MW/RF energies in humans have been made, especially in the United States. Most investigations have involved military personnel and civilian workers at military bases and in industrial settings. The principal factors of interest have been the significance of minor lens changes in the cataractogenic process and cataracts (opacities impairing vision). Several cases of cataract attributed to microwave exposure have been reported, but substantiation has not been established. There is no clinical or experimental evidence that ocular lens damage allegedly due to microwave exposure is morphologically different from lens abnormalities from other causes, including aging. All the reported effects of microwave exposure on the lens can be

---

* In *The New York Times*, March 3, 1981; April 21, 1981 and the *Philadelphia Inquirer*, March 4, 1981.

explained on the basis of thermal injury. It is well established that cataract can be produced in rabbits by exposure to microwaves. Extrapolation of results from animal studies to humans is difficult because the conditions, durations, and intensities of exposure are usually quite different.

Several cases of alleged cataract formation in people exposed to microwaves have been reported, but the precise details of exposure are generally impossible to determine. It is also difficult to relate cause and effect because lens imperfections do occur in otherwise healthy individuals, especially with increasing age. Numerous drugs, industrial chemicals, and certain metabolic diseases are associated with cataracts.

Lenticular defects too minor to affect visual acuity have been studied as possible early markers of microwave exposure or precursors of cataracts. The studies have been mainly prevalence surveys, although the time periods are often variable or not specified; reexamination data rarely permit estimates of incidence. Some generalizations, however, can be made about observations of lens changes in microwave workers and comparison groups (150, 151).

1. Lens imperfections occur normally and increase considerably with age among employed males studied. There is evidence that lens changes increase with age even during childhood (166). By about age 50, lens defects have been reported in most comparison subjects, based on data from various studies.

2. Although a few suggestive differences have been reported (166–168), there is no clear indication that minor lens defects are a marker for microwave exposure in terms of type or frequency of changes, exposure factors, or occupation. The reported earlier appearance of lens defects in microwave workers than in comparison groups is not convincing because there is considerable variation in the type, number, and size of defects recorded; in the scoring methods used by different observers; and in the numbers examined.

3. Clinically significant lens changes, which would permit selection of individuals to be followed up, have not been identified (169).

4. There is no evidence from ophthalmological surveys to date that minor lens opacities are precursors of clinical cataracts; a case-control study of World War II and Koran War veterans was negative for cataract (170).

Neither definitions nor methods of detection of cataract are standardized (150, 151). The common meaning of cataract, a lens opacity that interferes with visual acuity, is open to many interpretations as to degree and nature of the opacity and loss of visual acuity. Specific disorders, physical agents, and injuries are known to cause cataracts, but many cataracts are loosely called "senile" when they occur after middle age. Alleged "microwave cataracts" are not distinguishable from other cataracts in the opinion of most ophthalmologists (171–174).

The most prominent characteristic of cataracts is their age distribution. Although estimates of frequency are not comparable because of differences in the population groups surveyed, as well as nonuniform methods of detection and definition, all point to low frequencies until about the fifth decade of life, when sharp increases occur. Although not comparable with general population figures, recorded mean annual incidence rates are of the order of 2 per 100,000 (150, 151). In a preliminary national estimate by age of the total prevalence of cataracts in the civilian noninstitutionalized population aged 1 to 74 years in the United States, one or more cataracts was found in 9 percent of the population (175). For the various age groups under 45, the frequency of the condition increased gradually from 0.4 percent in those aged 1 to 5 years to 4 percent in those aged 35 to 44. The pronounced increase that occurs after age 45 reaches a maximum in the oldest group examined: of those aged 65 to 74 over half had cataracts. Cataract data for personnel on active duty in the armed services (who are mainly healthy, relatively young men) are available as incidence rates that show similar age dependence up to about age 55 (176). Parenthetically, in the United States as of August, 1981 no alleged microwave-induced cataracts in humans have been ruled in favor of the plaintiff in legal proceedings.

8.5 Nervous System and Cardiovascular Effects

Clinical and laboratory studies of workers in the Soviet Union and other eastern European countries employed in the operation, testing, maintenance, and manufacture of microwave-generating equipment are reported to have shown CNS and cardiovascular reactions to MW/RF exposure (106, 157, 158). Functional disturbances of the CNS have been described as "radiowave sickness"— the neurasthenic or asthenic syndrome. The symptoms and signs include headache, fatigability, irritability, loss of appetite, sleepiness, sweating, thyroid gland enlargement, difficulties in concentration of memory, depression, and emotional instability. The clinical syndrome is generally reversible if exposure is discontinued (150, 151).

Another frequently described manifestation is a set of labile functional cardiovascular changes including bradycardia (or occasional tachycardia), arterial hypertension (or hypotension), and changes in cardiac conduction. This form of neurocirculatory asthenia is also attributed to nervous system influence. Effects indicated by hypotonus, bradycardia, delayed auricular and ventricular conduction, decreased blood pressure, and ECG alterations in workers in RF or microwave fields have been reported (156–158, 177). The identification and assessment of these poorly defined, nonspecific complaints and symptom complexes (syndromes) is extremely difficult (178, 179). These changes, however, do not diminish the capacity to work and are reversible (184). No serious cardiovascular disturbances have been noted in humans or animals as a result of microwave exposure (181).

8.6 Growth and Development

A case-control study of Down's syndrome in relation to exposure to ionizing radiation yielded an unexpected finding regarding paternal exposure to radar

(182). Apparently, fathers of children with Down's syndrome gave more frequent histories of occupational exposure to radar during military service than did fathers of unaffected children, a difference that was of borderline statistical significance. Exposure during military service occurred before the birth of the affected child. After publication of the first report in 1965, expansion of the study group, follow-up of all fathers to obtain more detailed information about radar exposure, and a search of available military personnel records were all undertaken. The suggestive excess of radar exposure to fathers of babies with Down's syndrome was not confirmed on further study (183).

A report of congenital anomalies at a U.S. Army base (184) suggested that during the 3-year period 1968 through 1971 the adjoining communities surrounding the base had a reported number of cases of clubfoot among white babies that greatly exceeded the expected number (based on birth certificate notifications for the state). A more detailed investigation showed that in the six-county area surrounding the base, there was, during the same time period, a considerably higher rate of anomalies (diagnosed within 24 hr after birth) among births to military personnel than in the state as a whole. This base was a training facility for fixed-wing and helicopter aircraft, situated within 35 miles (56 km) of dozens of radar stations. Analysis showed that apparently there were errors in the malformation data on the birth certificates and a probable overreporting from the army base. Thus convincing evidence was lacking that radar exposure was related to congenital malformations (185). The higher malformation rate across a group of counties of the state was presumably environmentally induced, but no specific agent was suggested (150, 151).

A few human data are available from studies in which radiofrequency heating of the pelvic region was used to treat gonorrhea, pelvic inflammatory disease, endometriosis, carcinoma of the uterus, or pelvic peritonitis. In one report (186) the pelvic temperature in women was raised to 46°C. Although the author was concerned about possible harmful effects, he did not allude to specific complications. In another report four women were treated with microwave diathermy (2450-MHz, 100-W output) for chronic pelvic inflammatory disease before or during pregnancy (187). Three women delivered normal infants; the fourth, who received eight treatments during the first 59 days of pregnancy, aborted on day 67 but delivered a normal baby after a subsequent pregnancy during which she again received microwave treatment. The authors concluded that microwaves did not interfere with ovulation, conception, and pregnancy.

Microwave heating has been used to relieve the pain of uterine contractions during labor (188, 189). The analgesic effect was found helpful in 2000 selected patients without obstetric pathology, and the babies were born healthy with good circulation. No evidence of injury was manifest in a 1-year follow-up of the children; there was no evidence of mental retardation. Four cases of chromosome anomalies in controls and two cases in the irradiated group were noted. It is important to note that the human fetus at partuition is almost fully developed; thus gross structural defects at this late stage of development would not be expected.

There are reported case studies of increases in congenital abnormalities in woman working in RF fields in eastern Europe, (190) but there are no unequivocal reports of microwave-induced human teratologies.

8.7 Cancer

Microwave-induced cancer has not been reported experimentally or suspected in medical surveillance examinations of microwave workers or military personnel (150, 151). Two cohort epidemiologic studies (162, 164) that investigated the question systematically did not show an excess of any form of cancer to date that could be interpreted as microwave related (150, 151).

8.8 Eastern European Reports

Nervous system perturbations and behavioral reactions in humans after exposure to microwave energy have been reported mostly in eastern European publications that describe subjective complaints consisting of fatigability, headache, sleepiness, irritability, loss of appetite, and memory difficulties (190–193). Psychic changes that include unstable mood, hypochondriasis, and anxiety have also been reported. Most of the subjective symptoms are reversible, and pathological damage to neural structures is insignificant (194). Several reviewers (138, 139, 150, 151, 195–199) have noted the difficulties in establishing the presence of and quantifying the frequency and severity of "subjective" complaints. Individuals suffering from a variety of chronic diseases may exhibit the same dysfunctions of the central nervous and cardiovascular systems as those reported to be a result of exposure to microwaves (199); thus it is extremely difficult, if not impossible, to rule out other factors in attempting to relate microwave exposure to clinical conditions.

8.9 Critique of Epidemiologic Studies

An important concept of disease causation is that, in general, disease is not caused by a single factor or agent, but rather is influenced by multiple, interactive components, including subjects and their environment. Health effects or manifestations of disease have a spectrum of intensity ranging from the barely discernible and rapidly reversible symptomatic disorders, through an increasing gradient of severity to the point of irreversibility, and, finally, to disease states of such gravity as ultimately to cause death. For electromagnetic fields, in common with most other agents, biological or physical, the trivial end of this severity scale includes detectable physiological effects, which are well within the range of physiological adaptation and do not constitute disease in any meaningful sense.

The validity of application of the epidemiological method to the study of the health impact of an agent is determined largely by the ascertainability and definition of an effect. Perhaps the main limitation of epidemiologic studies of

MW/RF exposure is the lack of recognized pathophysiologic manifestations at realistic levels of exposure as indicators for measuring the effects of the fields on humans.

Eastern European reports describe such symptoms as listlessness, excitability, headache, drowsiness, fatigue, and cardiovascular deficits in persons occupationally exposed to electromagnetic fields. These symptoms are also caused by many other occupational factors, so it is not possible to define a cause–effect relationship. Many other factors in the industrial setting or home environment as well as psychosocial interactions can cause similar symptoms. In addition to smoking and obesity, genetic factors, and emotional stress, psychological personality factors that determine an individual's reactivity to environmental conditions are proven risk factors in the development of cardiac ischemia (199). Thus these factors are important in assessing the effects of environmental insults.

Reports of effects in humans must be put in perspective. Epidemiologic and incidence studies may suffer from inadequate design and examination as well as lack of data on actual power levels and duration of microwave exposure. It is essential to evaluate the multiple environmental factors that may interact among themselves and with personal characteristics of the individual. There is always the danger that real factors may be overlooked, leading to false association with factors of initial interest.

Analysis of occupational exposure to MW/RF energies is fraught with many difficulties. Of utmost importance is the assessment of the relationship bvetween exposure levels and the health status of the examined groups of workers. The problem of adequate control groups is controversial and hinges mostly on what one considers "adequate."

Quantitation of occupational exposure is extremely difficult. This is particularly true when personnel move around in the course of their duties or are exposed to nonstationary fields, such as moving beams or antennae, as well as to near- and far fields at random. The possible role of other environmental factors and of socioeconomic conditions must be taken into account. As often happens in such studies, it is difficult to show a causal relationship between a disease and the influence of environmental factors, at least in individual cases (160).

8.10 Protection Guides and Standards

8.10.1 Exposure Standards

The first standards for controlling exposure to MW/RF were introduced in the 1950s in both the United States and the Soviet Union. The maximum permissible exposure levels proposed then have remained substantially unchanged; for continuous exposure, these are respectively 10 mW/cm^2 and 10 µW/cm^2. Most countries that developed national standards based them on either the U.S.

NONIONIZING ELECTROMAGNETIC ENERGIES

Table 14.10 ANSI Recommendations

(1) Frequency[a] (MHz)	(2) Power Density (mW/cm^2)	(3) E^2 (V^2/m^2)	(4) H^2 (A^2/m^2)
0.3–3	100	400,000	2.5
3–30	900/f^2	4,000 (900/f^2)	0.25 (900/f^2)
30–300	1.0	4,000	0.025
300–1500	f/300	4,000 (f/300)	0.025 (f/300)
1500–100,000	5	20,000	0.125

[a] Frequency, f is in megahertz (MHz).

(200) or the Soviet (201) values. Subsequently, however, some countries have proposed standards intermediate between these extremes.

In the United States there are nongovernment organizations that develop recommended standards and safety criteria, for example, the American National Standards Institute (ANSI), a voluntary body with members from government, industry, various associations, and academic community, which develops consensus standards (guides) in various areas. In 1966 ANSI issued a nonionizing radiation safety standard with maximum permissible exposures of 10 mW/cm^2, as averaged over any 6-min period, for frequencies from 10 MHz to 100 GHz. This standard was reviewed and reissued with minor modifications in 1975. New recommendations (200) are based on frequency dependence and SARs. For human exposure to electromagnetic energy of radiofrequencies from 300 kHz to 100 GHz, the radiofrequency protection guides, in terms of equivalent plane wave free space power density, and in terms of the mean squared electric (E^2) and magnetic (H^2) field strengths as a function of frequency, are those shown in Table 14.10.

These standards do not apply to practitioners of the healing arts. For near field exposure, the only applicable radiofrequency protection guides are the mean-squared electric and magnetic field strengths given in columns 3 and 4. For convenience, these guides may be expressed in equivalent plane wave power density (column 2).

For both pulsed and nonpulsed fields, the power density, the mean squares of the field strengths, and the values of SAR or input power as applicable are averaged over any 0.1-hr period and should not exceed the values given in Table 14.10. Where whole-body exposure is concerned, the radiofrequency protection guide is believed to result in energy deposition averaged over the entire body for any 0.1-hr period of about 144 J/kg or less. This is equivalent to an SAR of about 0.40 W/kg or less, spatially and temporally averaged over the entire body mass.

The ANSI standard applies to the total population. In other words, the standard applies to nonocccupational as well as occupational exposures but is not intended to apply to the purposeful exposure of patients by or under the

Table 14.11 Canadian Standards

Canada	Frequency	Limits			Comments[a]
		Power Density (mW/cm^2)	RMS E Strength (V/m)	RMS H Strength (A/m)	Safety Code 6 (1976)
Occupational	10 MHz–1 GHz	1	60	0.16	Averaged over 1 hr
		25	300	0.8	Averaged over 1 min
	1–300 GHz	5	140	0.36	Averaged over 1 hr
		25	300	0.8	Averaged over 1 min
General public	10 MHz–300 GHz	1	60	0.16	Averaged over 1 min

[a] Additional provisions for exposures shorter than 1 hr and exposure of extremities.

direction of practitioners of the healing arts. The U.S. National Institute for Occupational Safety and Health (NIOSH) is developing a criteria document with recommended standards for occupational MW/RF exposures, which is, except for certain modifications, comparable to that recommended by ANSI. The United Kingdom generally follows the Standards in the United States.

In Canada the maximum permissible occupational levels (MPL) are 1 mW-h/cm^2 average energy flux for whole-body exposure as averaged over 1 hr and a maximum exposure during any 1 min of 25 mW/cm^2 in occupational settings. The MPLs would apply for the frequency range of 10 MHz to 300 GHz. No distinction is made between CW and pulsed waveforms (202–204) (Table 14.11).

The State Committee on Standards of the Council of Ministers of the Soviet Union has promulgated "Occupational Safety Standards for Electromagnetic Fields of Radiofrequency (GOST 12.1.006-76)," effective from January 1, 1977. It specifies the maximum permissible magnitudes of voltage and current density of an electromagnetic field in the workplace. It does not apply to Ministry of Defense personnel. Maximum permissible RF fields in the workplace must not, during the course of the workday, exceed those listed in Table 14.12.

In 1972 and 1977 the Polish Council of Ministers and the Minister of Health and Social Welfare promulgated a change in the Polish Standard (205–207). For the general population the values of 10 and 100 µW/cm^2 were adopted for continuous and intermittent exposures, respectively. These values were taken as the upper limits of a safe zone, in which occupation is unrestricted. Three other zones are also defined, based on power density. For stationary (continuous) fields, these are defined as follows:

1. *Safe zone.* In this zone the mean power density cannot exceed 10 µW/cm^2, and human exposure is unrestricted.

Table 14.12 Soviet Standard[a]

Frequency Range	P (mW/cm^2)	E (V/m)	H (A/m)
60 kHz–1.5 MHz			5
1.5 MHz–3.0 MHz		50	
3.0 MHz–30 MHz		20	
30 MHz–50 MHz		10	0.3
50 MHz–300 MHz		5	
300 MHz–300 GHz	0.01	(entire workday)	
	0.10	(2-hr period during workday)	
	1.00	(20-min period during workday)	

[a] Note 1: Also applies in environments with ambient temperatures above 28°C, in the presence of X-ray radiation, except under these conditions the maximum during a 20-min period is restricted to 0.1 mW/cm^2. Note 2: It appeared that the workday level would be raised from 10 μW/cm^2 to 25 μW/cm^2 in January 1982 (USSR Committee on Science and Technology).

2. *Intermediate zone.* In this zone, a minimal value of 10 μW/cm^2 with upper limit 200 μW/cm^2 occupational exposure is allowed during a whole work day (normally 8 hr, but in principle it can be extended to 10 hours).

3. *Hazardous zone.* In this zone, a minimal value of 200 μW/cm^2 with upper limit 10 mW/cm^2 occupational exposure time per 24 hr is determined by the formula $t = 32/p^2$, where t is exposure time in hours and p represents mean power density in W/m^2.

4. *Dangerous zone.* In this zone, a mean power density in excess of 100 W/m^2 (10 mW/cm^2) human exposure is forbidden.

For exposures to nonstationary fields—that is, intermittent exposure—the following values were adopted:

1. *Safe zone.* In this zone, the mean power density does not exceed 1 W/m^2 (100 μW/cm^2).

2. *Intermediate zone.* In this zone, the minimal value is 1 W/m^2 (100 μW/cm^2) and the upper limit is 10 W/m^2 (1 mW/cm^2).

3. *Hazardous zone.* In this zone, the exposure time is determined by the formula $t = 800/p^2$, where the symbols t and p designate the same values as previously.

4. *Dangerous zone.* In this zone with mean power density in excess of 100 W/m^2 (10 mW/cm^2), human exposure is forbidden.

The Swedish National Board for Industrial Safety promulgated a nonionizing RF standard (Worker Protection Authority Instruction No. 111) effective January 1, 1977 (110). This regulation applies to all work that may necessitate

Table 14.13 Swedish Standard

Frequency Range	Power Density (mW/cm^2)
10–300 MHz	5
300 MHz–300 GHz	1

exposure to radiofrequencies between 10 MHz and 300 GHz. The instruction specifically excludes applications affecting the treatment of patients. Maximum permissible exposures (as averaged over a 6-min period) are shown in Table 14.13.

According to the World Health Organization and the International Radiation Protection Association (IRPA), the exposure range of 0.1 to 1 mW/cm^2 has a high enough safety factor to permit continuous exposure over the whole frequency range (209). Environmental exposure should be based on the ALARA principle—as low as reasonably achievable—with social and economic benefits considered.

The International Radiation Protection Association has approved limits for occupational and public exposures to RF/MW radiation (201). Like ANSI, IRPA's most stringent occupational exposure limit is 1 mW/cm^2. But IRPA mandates this level over a wider range of frequencies. Whereas ANSI specifies a maximum exposure of 1 mW/cm^2 in the 30 to 300-MHz band, rising to 5 mW/cm^2 at 1.5 GHz, the IRPA band is 10 to 400 MHz and the 5-mW/cm^2 limit takes effect at 2 GHz.

In the 100-kHz- to 1-MHz band, the IRPA exposure limit is 10 mW/cm^2, as compared to ANSI's 100 mW/cm^2 for 300 KHz to 3 MHz. For the general population, the IRPA limits are five times more stringent than its occupational limits; thus for 10 to 400 MHz, the limit is 200 μW/cm^2. Whereas ANSI recommends the same limits for workers and the general public, above 2 GHz, the public would be exposed to a maximum of 1 mW/cm^2 under the IRPA guidelines. The IRPA and ANSI limits for frequencies above 10 MHz are based on the same conclusion: specifically, that exposure should not exceed a whole-body SAR of 0.4 W/kg when averaged over 6 min.

8.10.2 Product Emission Standard

The Radiation Control for Health and Safety Act of 1968 (PL 90-602), administered by the U.S. Department of Health, Education and Welfare and the Food and Drug Administration (FDA) (Bureau of Radiological Health), provides authority for controlling radiation from electronic devices. One standard, the BRH microwave oven standard, effective October 1971, declares: "Ovens may not emit (leak) more than 1 mW/cm^2 at time of manufacture and 5 mW/cm^2 subsequently, for the life of the product-measured at a distance of 5 cm and under conditions specified in the standard" (211).

The Canadian standard (204, 212) restricts the maximum leakage to 1 mW/cm^2 at 5 cm from the oven (consumer, commercial, and industrial). The U.S. standard (211) specifies a maximum emission level at 5 cm of 1 mW/cm^2 before purchase and 5 mW/cm^2 thereafter, which is consistent with standards for the general population in the Soviet Union and Poland (209). The standard applies to domestic and commercial ovens but not to industrial equipment. A similar standard has been adopted in Japan and most of Western Europe (213). Computation of human exposure during typical operation of a microwave oven leaking up to 1 mW/cm^2 at a distance of 5 cm indicates that actual exposure of the body is 5 to 20 µW/cm^2. In the Soviet Union an emission level of 10 µW/cm^2 at a distance of 50 cm from the surface of the oven has been adopted.

8.10.3 Problems and Recommendations

Elucidation of the biological effects of microwave exposure requires a careful review and critical analysis of the available publications. Such a review requires the differentiation of the established effects and mechanisms from speculative and unsubstantiated reports. Most of the experimental data support the concept that the effects of microwave exposure are primarily, if not only, a response to hyperthermia or altered thermal gradients in the body. There are, nevertheless, large areas of confusion, uncertainty, and actual misinformation.

There is a philosophical question about the definition of hazard. One objective definition of injury is an irreversible change in biological function as observed at the organ or system level. Thus it is possible to define a hazard as a probability of injury on a statistical basis. It is important to differentiate between the hazard levels at which injury may be sustained and effect per se or perception. All effects are not necessarily hazards. In fact, some effects may have beneficial applications under appropriately controlled conditions, such as diathermy and cancer therapy. Microwave-induced changes must be understood sufficiently so that their clinical significance can be determined, their hazard potential assessed, and the appropriate benefit risk analysis applied. It is important to determine whether an observed effect is irreparable, transient, or reversible, disappearing when the electromagnetic field is removed or after some interval of time. Of course, even reversible effects are unacceptable if they transiently impair the ability of the individual to function properly or to perform a required task.

A critical review of studies of the biological effects of microwaves indicates that many of the investigations suffer from inadequacies of either technical facilities and energy measurement skills or insufficient control of the biological specimens and the criteria for biological change. More sophisticated conceptual approaches and more rigorous experimental design must be developed. There is a great need for systematic and quantitative comparative investigation of the biological effects, with the use of well-controlled experiments. This should be done by using sound biomedical and biophysical approaches at the various levels of biological organization from the subcellular to the whole animal on an integrated basis, with full recognition of the multiple associated and interdependent variables.

Proper investigation of the biological effects of MW/RF requires an understanding and appreciation of biophysical principles and "comparative biomedicine." Such studies require interspecies "scaling," the selection of biomedical parameters that consider basic physiological functions and work capacity, identification of specific and nonspecific reactions, and differentiation of adaptational or compensatory changes from pathological manifestations.

It is important that research be conducted in such a way that all aspects of the study are quantified, the type and magnitude of the effect, whether, the effect is harmful, harmless, or merely an artifact and how it relates to the results obtained by other investigators. For microwave bioeffects, body size of the experimental animal must be taken into account. Since body-absorption cross sections and internal heating patterns can differ widely, an investigator may think that a low-level or a "nonthermal" effect is being manifest in an animal because the incident power is low, whereas in fact the animal may be exposed to as much absorbed power in a specific region of the body as another larger animal exposed to a much higher incident power density. The contrary can hold at low frequencies. In the performance of experimental studies on animals, interspecies scaling factors must, therefore, be used for extrapolation to humans.

Well-designed and appropriately controlled epidemiological and clinical investigations of groups of workers and others exposed to microwaves should be fostered. Studies of workers and individuals exposed to MW/RF energies along with appropriate control groups should include a thorough analysis of the exposure environment, including cofactors as well as electromagnetic fields. There is always the danger that real factors may be overlooked, leading to false association with factors included in the study. Such interacting factors could be heat, cold, toxic agents, hypoxia, noise, other radiant energies such as X-rays, chronic disease, and medication.

Because of the difficulties in extrapolating from animal experiments to humans, epidemiological studies, including appropriate clinical and laboratory examinations, are essential to improve our understanding of possible health hazards from exposure to MW/RF energies. As noted by Silverman (150, 151), it is difficult to identify exposed populations, select suitable controls, and obtain exposure data. Some study groups already characterized can be improved by the acquisition of additional exposure data, some groups should be followed for longer periods, and some should be investigated for additional end points.

The reports from eastern Europe of a wide variety of functional changes and possible nervous system effects have yet to be confirmed. In appropriate epidemiologic studies, medical reports should be augmented to include an assessment of emotional and psychological status.

ACKNOWLEDGMENT

This chapter is based on work performed partially under Contract No. DE-ACO2-76EVO3490 with the U.S. Department of Energy at the University of

Rochester Department of Radiation Biology and Biophysics and has been assigned Report No. UR-3490-2055 and a grant from the National Institute of Environmental Health Sciences, ESO3239.

REFERENCES

1. I. Matelsky, "Non-ionizing Radiations," in: *Industrial Hygiene Highlights*, Vol. 1, L. V. Cralley, L. J. Cralley, and G. D. Clayton, Eds., Industrial Hygiene Foundation of America, Inc., Pittsburgh, 1968, pp. 140–179.
2. L. S. Taylor, "The Development of Exposure Guidelines," in: *Proceedings of Conference on Estimation of Low-Level Radiation Effects in Human Populations*, Argonne National Laboratory, December 1970, Report ANL-7811, May 1971, pp. 27–28.
3. National Council on Radiation, *Protection and Measurements, Basic Radiation Protection Criteria*, NCRP Report 39, Washington, DC, 1971.
4. L. S. Taylor, "Radiation Protection Trends in the United States," *Health Phys.*, **20**, 499–504 (1971).
5. *Radiosensitivity and Spatial Distribution of Dose*, ICRP Publication 14 (reports prepared by two Task Groups of Committee 1 of the International Commission on Radiological Protection), Pergamon Press, London, 1969, p. 115.
6. A. E. S. Green, Ed., *The Middle Ultraviolet: Its Science and Technology*. Wiley, New York, 1966.
7. J. H. Epstein, "Ultraviolet Light in Office Practice," *Postgrad. Med.*, **37**, 170–174 (1965).
8. M. A. Everett, R. M. Sayre, and R. L. Olson, "Physiologic Response of Human Skin to Ultraviolet Light," in: *The Biological Effects of Ultraviolet Radiation (With Emphasis on the Skin)* (proceedings of 1st International Conference, 1966), F. Urbach, Ed., Pergamon Press, New York, 1969, pp. 181–186.
9. L. R. Koller, *Ultraviolet Radiation*, 2nd ed., Wiley, New York, 1965.
10. F. Urbach, Eds., *The Biological Effects of Ultraviolet Radiation (With Emphasis on the Skin)* (proceedings 1st International Conference, 1966), Pergamon Press, New York, 1969.
11. M. J. Suess, Ed., *Nonionizing Radiation Protection*, WHO Regional Publications European Series No. 10, p. 267. WHO Regional Office for Europe, Copenhagen, 1982.
12. Council on Physical Therapy, "Acceptance of Ultraviolet Lamps for Disinfecting Purposes," *JAMA*, **122**, 503–504 (1943).
13. K. J. K. Buettner, "The Effects of Natural Sunlight on Human Skin," in: *Biological Effects of Radiation (With Emphasis on the Skin)*, F. Urbach, Ed., Pergamon Press, New York, 1969, pp. 237–249.
14. H. Brodthagen, "Seasonal Variations in Ultraviolet Sensitivity of Normal Skin," in: *Biological Effects of Radiation (With Emphasis on the Skin)*, F. Urbach, Eds., Pergamon Press, New York, 1969, pp. 459–467.
15. M. Kleinfeld, C. Giel, and I. R. Tabershaw, "Health Hazards Associated with Inert-Gas-Shielded Metal Arc Welding," *Am. Med. Assoc. Arch. Ind. Health*, **15**, 27–31 (1957).
16. M. M. Key, T. H. Milby, D. A. Holaday, and A. Cohen, "Physical Hazards," in: *Occupational Diseases. A Guide to Their Recognition*, W. M. Gafafer, Ed., USDHEW, PHS Publication 1097, 1964, pp. 259–297.
17. G. M. Wilkening, "Non-ionizing Radiation," in: *The Industrial Environment—Its Evaluation and Control*, USDHEW, PHS, CDC, NIOSH, 1973, Chapter 28, pp. 357–376.
18. M. A. Everett, R. L. Olsen, and R. M. Sayer, "Ultraviolet Erythema," *Am. Med. Assoc. Arch. Dermatol.*, **92**, 713–719 (1965).
19. J. M. Knox, R. G. Freeman, and R. Ogura, "The Destructive Energy of Sunlight," *Dermatol. Invest.*, **4**, 205–212 (1965).

20. H. F. Blum, *Carcinogenesis by Ultraviolet Light,* Princeton University Press, 1959.
21. H. F. Blum, "Quantitative Aspects of Cancer Induction by Ultraviolet Light: Including a Revised Model," in: *Biological Effects of Ultraviolet Radiation(With Emphasis on the Skin),* F. Urbach, Ed., Pergamon Press, New York, 1969, pp. 543–549.
22. W. L. Epstein, K. Fukuyama, and J. H. Epstein, "Ultraviolet Light, DNA Repair and Skin Carcinogenesis in Man," *Fed. Proc.,* **30,** 1766–1771 (1971).
23. D. G. Pitts, J. E. Prince, W. I. Butcher, K. R. Kay, R. W. Bowman, H. W. Casey, D. G. Richey, L. H. Mori, J. E. Strong, and T. J. Tredici, *The Effects of Ultraviolet Radiation on the Eye,* USAF School of Aerospace Medicine, Aerospace Medicine Division (AFSC), Brooks Air Force Base (AFB), Technical Report SAM-TR-69-10, February 1969.
24. D. G. Pitts, W. R. Bruce, and T. J. Tredici, *A Comparative Study of the Effects of Ultraviolet Radiation on the Eye,* USAF School of Aerospace Medicine, Aerospace Medicine Division (AFSC), Brooks AFB, Technical Report SAM-TR-70-28, July 1970.
25. W. J. Geeraets, "Radiation Effects on the Eye," *The Sight-Saving Rev.,* **39,** 181–196 (winter 1969–1970), *Ind. Med.,* **39,** 441–450 (1970).
26. A. Bachem, "Ophthalmic Ultraviolet Action Spectra," *Am. J. Ophthalmol.,* **41,** 969–975 (1956).
27. D. G. Cogan and V. E. Kinsey, "Action Spectrum of Keratitis Produced by Ultraviolet Radiation," *Arch. Ophthalmol.,* **35,** 670–677 (1946).
28. S. Lerman, "Radiation Cataractogenesis," *NY State J. Med.,* **62,** 3075–3085 (1962).
29. Report of the Council on Physical Mecicine, *JAMA,* **137,** 1600–1603 (1948).
30. B. G. Ferris, "Environmental Hazards. Electromagnetic Radiation," *New Engl. J. Med.,* **275,** 1100–1105 (1966).
31. K. D. Fisher, C. J. Carr, J. E. Huff, and T. E. Huber, "Dark Adaptation and Night Vision," *Fed. Proc.,* **29,** 1605–1638 (1970).
32. S. L. Bonting and A. D. Bangham, "On the Biochemical Mechanism of the Visual Process," *Exp. Eye Res.,* **6,** 400–413 (1967).
33. D. G. McConnell and D. G. Scarpelli, "Rhodopsin: An Enzyme," *Science,* **139,** 848 (1963).
34. B. Rosenberg, "A Physical Approach to the Visual Receptor Process," *Adv. Radiat. Biol.,* **2,** 193–241 (1966).
35. G. Wald, "The Biochemistry of Visual Excitation," in: *Enzymes: Units of Biological Structure and Function,* O. H. Gaebler, Ed., Academic Press, New York, 1956, p. 355.
37. F. H. Adler, *Physiology of the Eye,* 4th ed., Mosby, St. Louis, 1965.
38. C. H. Graham, Ed., *Vision and Visual Perception.* Wiley, New York, 1965.
39. *A Study of Vision as Related to Dark Adaptation and Night Vision in the Soldier.* Bethesda, MD, Federation of American Society for Experimental Biology, Life Sciences (1969).
40. K. Beuttner and H. W. Rose, "Eye Hazards from an Atomic Bomb," *Sight-Saving Rev.,* **23,** 194–197 (1953); *Excerpts Med. Ophthalmol.,* **8,** 221 (1954); *Abstr. Mil. Aviat. Ophthalmol.,* **5,** 182–188 (1960).
41. M. Luckiesh and S. K. Guth, "Brightness in Visual Field at Borderline Between Comfort and Discomfort," *Illum. Eng.,* **44,** 650–670 (1949).
42. D. W. DeMott and T. P. Davis, "An Experimental Study of Retinal Burns: Part I. The Irradiance Thresholds for Chorio-retinal Lesions. Part II. Entopic Scatter as a Function of Wavelength," *Arch. Ophthalmol.,* **62,** 653–656 (1959).
43. W. T. Ham, Jr., H. Wiesinger, F. J. Schmidt, R. C. Williams, R. S. Ruffin, M. C. Schaffer, and D. Guerry, III, "Flash Burns in the Rabbit Retina as a Means of Evaluating Retinal Hazard from Nuclear Weapons," *Am. J. Ophthalmol.,* **46,** 700–723 (1958).
44. W. T. Ham, Jr., R. C. Williams, H. A. Mueller, R. S. Ruffin, F. H. Schmidt, A. M. Clarke, J. J. Vos, and W. J. Geeraets, "Ocular Effects of Laser Radiation," *Acta Ophthalmol.,* **43,** 390–409 (1965).

45. H. Davson, Ed., *The Eye (Vegetative Physiology and Biochemistry)*, Vol. 1; *The Visual Process*, vol. 2; *Muscular Mechanisms*, vol. 3; *Visual Optics and the Optical Space Sense*, vol. 4, Academic Press, London, 1962.
46. C. E. Moss, R. J. Ellis, W. E. Murray, and W. H. Parr, "Infrared Radiation," in: *Nonionizing Radiation Protection*, M. J. Suess, Ed., World Health Organization Regional Office for Europe, Copenhagen, 1982, pp. 69–95.
47. W. E. Forsythe and F. Christison, "The Absorption of Radiation from Different Sources by Water and Body Tissue," *J. Opt. Soc. Am.*, **20**, 693 (1930).
48. J. F. Lehmann, G. D. Brunner, and R. W. Stow, "Pain Threshold Measurements after Therapeutic Application of Ultrasound, Microwaves and Infrared," *Arch. Phys. Med. Rehab.*, **39**, 560–565 (1958).
49. J. Gersten, K. G. Wakim, R. W. Stow, and F. H. Krusen, "A Comparative Study of the Heating of Tissues by Near and Far Infrared Radiation," *Arch. Phys. Med.*, **30**, 691–699 (1949).
50. C. M. Edbrooke and C. Edwards, "Industrial Radiation Cataracts: The Hazards and the Protective Measures," *Arch. Occup. Hyg.*, **10**, 293–304 (1967).
51. W. Robinson, "On Bottle-Maker's Cataract," *Br. Med. J.*, **2**, 381–384 (1907); *Ophthalmoscope*, **18**, 538 (1915).
52. K. L. Dunn, "Cataract from Infrared Rays. 'Glass Workers' Cataract.' A Preliminary Study on Exposures," *Arch. Ind. Hyg. Occup, Med.*, **1**, 166–180 (1950); "A Preliminary Study on 'Glass Workers' Cataract' Exposures," *Trans. Am. Acad. Ophthalmol. Otolaryngol.*, **54**, 597–605 (1950).
53. G. F. Keatinge, J. Pearson, J. P. Simons, and E. E. White, "Radiation Cataract in industry," *Arch. Ind. Health*, **11**, 305–315 (1955).
54. D. G. Cogan, D. D. Donaldson, and A. B. Reese, "Clinical–Pathological Characteristics of Radiation Cataract," *Arch. Ophthalmol.*, **47**, 55–70 (1952).
55. H. Goldmann, "The Genesis of the Cataract of the Glass Blower," *Ann. Ocul.*, **172**, 13–41 (1935); *Am. J. Ophthalmol.*, **18**, 590–591 (1935).
56. J. H. Jacobson, B. Cooper, N. W. Najac, and A. Kohtiao, "The Effects of Thermal Energy on Anterior Ocular Tissues," 6570th Aerospace Medicine Research Laboratory, Wright-Patterson AFB, Ohio, Technical Report AMRL-TDR-63-53, 1963.
57. A. Vogt, "Fundamental Investigation of the Biology of Infrared," *Klin. Monatsbl. Augenheilk.*, **89**, 256–260 (1932).
58. H. Goldmann, H. Koenig, and F. Maeder, "The Permeability of the Eye Lens to Infrared," *Ophthalmologica*, **120**, 198–205 (1950).
59. J. H. Jacobson, B. Cooper, and H. W. Najac, "Effects of Thermal Energy on Retinal Function," 6570th Aerospace Medicine Research Laboratory Wright-Patterson AFB, Ohio, Technical Report AMRL-TDR-62-96, 1962.
60. J. D. Hardy and T. W. Oppel, "Studies in Temperature Sensation. IV. The Stimulation of Cold Sensation by Radiation," *J. Clin. Invest.*, **17**, 771–777 (1938).
61. T. H. Bullock and F. P. J. Diecke, "Properties of an Infrared Receptor," *J. Physiol.*, **134**, 47–87 (1956).
62. D. C. Sinclair, "Cutaneous Sensation and the Doctrine of Specific Energy," *Brain*, **78**, 584–614 (1955).
63. G. Weddell, "Studies Related to the Mechanism of Common Sensibility," in: *Advances in Biology of Skin, Cutaneous Innervation*, Vol. 1, Pergamon Press, New York, 1960, pp. 112–160.
64. H. Hensel, "Electrophysiology of Thermosensitive Nerve Endings," in: *Biology and Medicine (Part 3. Temperature: Its Measurement and Control in Science and Industry)*, J. D. Hardy, Ed., Reinhold, New York, 1963, p. 191.
65. J. D. Hardy, "Thermal Radiation, Pain and Injury," in: *Therapeutic Heat*, Vol. 2, S. Licht, Ed., Elizabeth Licht, New Haven, CT, 1958, pp. 157–178.

66. E. H. Weber, 1846, cited in E. Hendler, "Cutaneous Receptor Response to Microwave Irradiation," in: *Thermal Problems in Aerospace Medicine*, J. D. Hardy, Ed., Unwin, Surrey, England, 1968, p. 159.
67. J. D. Hardy and T. W. Oppel, "Studies in Temperature Sensation. III. The Sensitivity of the Body to Heat and the Spatial Summation of the End Organ Responses," *J. Clin. Invest.*, **16**, 533–540 (1937).
68. E. Hendler and J. D. Hardy, "Temporal Aspects of Temperature Sensation," ASME Paper 58-A-220 (1958): "Infrared and Microwave Effects on Skin Heating and Temperature Sensation," *IRE Trans. Med. Electron*, **ME-7**, 143–152 (July 1960).
69. E. Fischer and S. Solomon, "Physiological Responses to Heat and Cold," in: *Therapeutic Heat*, S. H. Licht, Ed., Elizabeth Licht, New Haven, CT, 1958, pp. 116–156.
70. H. F. Cook, "The Pain Threshold for Microwave and Infrared Radiations," *J. Physiol.*, **118**, 1–11 (1952).
71. J. D. Hardy, H. Goodell, and H. G. Wolff, "Influence of Skin Temperature upon Pain Threshold as Evoked by Thermal Radiation," *Science*, **114**, 149–150 (1951).
72. M. Wertheimer and W. D. Ward, "Influence of Skin Temperature upon Pain Threshold as Evoked by Thermal Radiation—a Confirmation," *Science*, **115**, 499–500 (1952).
73. H. Davson, *The Physiology of the Eye*, 2nd ed., Little, Brown, Boston, 1963.
74. P. P. Lele and G. Weddell, "The Relationship Between Neurohistology and Corneal Sensibility," *Brain*, **79**, 119–154 (1956).
75. F. G. Ebaugh, Jr. and R. Thauer, "Influence of Various Environmental Temperatures on the Cold and Warmth Thresholds." *J. Appl. Physiol.*, **3**, 173–182 (1950).
76. J. D. Hardy, H. G. Wolff, and H. Goodell, "Studies on Pain, a New Method for Measuring Pain Threshold: Observations on Spatial Summation of Pain," *J. Clin. Invest.*, **19**, 649–657 (1940).
77. J. D. Hardy, "Thresholds of Pain and Reflex Contraction as Related to Noxious Stimulation," *J. Appl. Physiol.*, **5**, 725–739 (1953).
78. J. D. Hardy, H. Wolff, and H. Goodell, "Studies on Pain: Discrimination of Differences in Intensity of a Pain Stimulus as a Basis of a Scale of Pain Intensity," *J. Clin. Invest.*, **26**, 1152–1158 (1947).
79. A. R. Moritz and F. C. Henriques, Jr., "Studies of Thermal Injury. II. Relative Importance of Time and Surface Temperature in Causation of Cutaneous Burns," *Am. J. Pathol.*, **23**, 695–720 (1947); F. C. Henriques, Jr. and A. R. Moritz, "Studies of Thermal Injury. 1. The Conduction of Heat to and Through Skin and the Temperatures Attained Therein. A Theoretical and Experimental Investigation," *Am. J. Pathol.*, **23**, 531–549 (1947).
80. F. C. Henriques, Jr., "Studies of Thermal Injury: V. Predictability and Significance of Thermally Induced Rate Processes Leading to Irreversible Epidermal Injury," *Arch. Pathol.*, **43**, 489–502 (1947).
81. E. C. Gregg, Jr., "Physical Basis of Pain Threshold Measurements in Man," *J. Appl. Physiol.*, **4**, 351–363 (1951).
82. H. G. Wolff and J. D. Hardy, "On the Nature of Pain," *Physiol. Rev.*, **27**, 167–199 (1947).
83. D. L. Lloyd-Smith and K. Mendelssohn, "Tolerance Limits to Radiant Heat," *Br. Med. J.*, No. 4559, 975–978 (May 22, 1948).
84. H. M. Whyte, "The Effect of Aspirin and Morphine on Heat Pain," *Clin. Sci.*, **10**, 333–345 (1951).
85. W. L. Derksen, T. I. Monohan, and G. P. Delhery, "The Temperatures Associated with Radiant Energy Skin Burns," in: *Temperature, Its Measurement and Control in Sciences and Industry*, 3, Part 3, J. D. Hardy, Ed., Reinhold, New York, 1963, p. 171.
86. S. Duke-Elder, *Textbook of Ophthalmology*, Vol. VI, *Injuries*, Mosby, St. Louis, 1954, p. 6476.

87. W. T. Ham, Jr., A. M. Clarke, W. J. Geeraets, S. F. Cleary, H. A. Mueller, and R. C. Williams, "The Eye Problem in Laser Safety," *Arch. Environ. Health*, **20,** 156–160 (1970).
88. A. M. Clarke, "Ocular Hazards from Lasers and Other Optical Sources," *CRC Crit. Rev. Environ. Control*, 307–339 (November 1970).
89. L. Goldman, *Biomedical Aspects of the Laser*, Springer, New York, 1967.
90. D. Sliney and M. Wolbarsht, *Safety with Lasers and Other Optical Sources*, Plenum Press, New York, 1980, p. 1035.
91. L. Goldman, S. M. Michaelson, R. J. Rockwell, D. H. Sliney, B. M. Tengroth, and M. L. Wolbarsht, "Optical Radiation with Particular Reference to Lasers," in: *Nonionizing Radiation Protection*, M. S. Suess, Ed., World Health Organization Regional Office for Europe, Copenhagen, 1982, pp. 39–68.
92. M. B. Landers, "The Laser Eye Hazard," *Surv. Ophthalmol.*, **14,** 338–341 (1970).
93. H. Cember, "Non-ionizing Radiation," in: *Introduction to Health Physics*, Pergamon Press, New York, 1983, Chapter 14, pp. 412–462.
94. G. M. Wilkening, "A Commentary on Laser-Induced Biological Effects and Protective Measures," *NY Acad. Sci.*, **68,** (Part 3), 621–626 (1970).
95. G. M. Wilkening, T. Behrendt, J. A. Carpenter, W. T. Ham, P. W. Lappin, M. Mautner, R. W. Neidlinger, A. E. Sherr, C. H. Swope, A. Vassiliadis, and H. C. Zweng, "Eye," *Arch. Environ. Health*, **20,** 197–199 (1970).
96. H. C. Zweng, "Thresholds of Laser Eye Hazards," paper presented at 33rd Annual Meeting of the Industrial Hygiene Foundation, 1968; cited in J. A. Carpenter, D. J. Lechmiller, and T. J. Tredici, "U.S. Air Force Permissible Exposure Levels for Laser Irradiation," *Arch, Environ. Health*, **20,** 171–176 (1970).
97. H. M. Leibowitz and A. J. Luzzio, "Laser-Induced Cataract," *Arch. Ophthalmol.*, **83,** 608–612 (1970).
98. American Conference of Governmental Industrial Hygienists, *A Guide for Control of Laser Hazards*, ACGIH, 1976.
99. American National Standards Institute, *The Safe Use of Lasers*, ANSI Z-136-1-1976, American National Standards Institute, New York, 1976.
100. H. P. Schwan and K. Li, "Capacity and Conductivity of Body Tissues at Ultrahigh Frequencies," *1953 IRE National Convention Record*, Part 9, pp. 121–128; *Proc. IRE*, **41,** 1735–1740 (December 1953).
101. H. P. Schwan and K. Li, "Hazards Due to Total Body Irradiation by Radar," *Proc. IRE*, **44,** 1572–1581 (November 1956).
102. A. Anne, M. Saito, O. M. Salati, and H. P. Schwan, "Relative Microwave Absorption Cross Sections of Biological Significance," in: *Proceedings of 4th Annual Tri-Service Conference on Biologic Effects of Microwave Radiating Equipments; Biological Effects of Microwave Radiations*, M. F. Peyton, Ed., Plenum Press, New York, 1961, Technical Report RADC-TR-60-180, pp. 153–176.
103. R. Thauer, "Circulatory Adjustments to Climatic Requirements," in: *Handbook of Physiology*, W. F. Hamilton, Ed., American Physiological Society, Washington, DC, Section 2, Circulation III, 1965.
104. J. H. Heller, "Cellular Effects of Microwave Radiation," in: *Biological Effects and Health Implications of Microwave Radiation, Symposium Proceedings*, S. F. Cleary, Ed., U.S. Department of Health, Education and Welfare, Public Health Service, Washington, DC, pp. 116–121 (BRH/DBE 70-2), 1970.
105. D. E. Janes, W. M. Leach, W. A. Mills, R. T. Moore, and M. L. Shore, "Effects of 2450 MHz Microwaves on Protein Synthesis and on Chromosomes in Chinese Hamsters," *Nonionizing Radiat.*, **1,** 125–130 (1969).
106. S. Baranski and P. Czerski, *Biological Effects of Microwaves*, Dowden, Hutchinson and Ross, 234P Stroudsburg, PA, 1976.

107. M. E. O'Connor, "Mammalian Teratogenesis and Radio-frequency Fields," *Proc. Inst. Electr. Electron. Eng.*, **68,** 56–60 (1980).
108. T. S. Ely, D. Goldman, J. Z. Hearon, R. B. Williams, and H. M. Carpenter, "Heating Characteristics of Laboratory Animals Exposed to Ten Centimeter Microwaves," *IEEE Trans. Biomed. Eng.*, **11,** 123–137; U.S. Naval Medical Research Institute, Bethesda, MD; (Res. Rep. Proj. NM 001-056.13.092) (1964).
109. W. G. Lotz and S. M. Michaelson, "Temperature and Corticosterone Relationships in Microwave Exposed Rats," *J. Appl. Physiol. Respirat. Environ. Exercise Physiol.*, **44,** 438–445 (1978).
110. W. G. Lotz and S. M. Michaelson, "Effects of Hypophysectomy and Dexamethasone on Rat Adrenal Response to Microwaves," *J. Appl. Physiol. Resp. Environ. Exercise Physiol.*, **47,** 1284–1288 (1979).
111. S. T. Lu, N. Lebda, S. Pettit, D. Rivera, and S. M. Michaelson, "Thermal and Endocrinological Effects of Protracted Irradiation of Rats by 2450 MHz Microwaves," *Radio Sci.* **12**(6S), 147–155 (1977).
112. S. T. Lu, N. Lebda, S. Pettit, and S. M. Michaelson, "Delineating Acute Neuroendocrine Responses in Microwave Exposed Rats," *J. Appl. Physiol. Resp. Environ. Exercise Physiol.*, **48,** 927–932 (1980).
113. S. T. Lu, W. G. Lotz, and S. M. Michaelson, "Advances in Microwave-Induced Neuroendocrine Effects: The Concept of Stress," *Proc. Inst. Electr. Electron. Eng.*, **68,** 73–77 (1980).
114. R. L. Magin, S. T. Lu, and S. M. Michaelson, "Microwave Heating Effect on the Dog Thyroid Gland," *IEEE Trans. Biomed. Eng.*, **24,** 522–529 (1977).
115. R. L. Magin, S. T. Lu, and S. M. Michaelson, "Stimulation of Dog Thyroid by Local Application of High Intensity Microwaves," *Am. J. Physiol.*, **233,** E363–E368 (1977).
116. H. Mikolajczyk, "Microwave Irradiation and Endocrine Functions," in: *Biologic Effects and Health Hazards of Microwave Radiation*, P. Czerski, K. Ostrowski, and C. Silverman, et al., Eds., Warsaw: Polish Medical Publishers, Warsaw, 1974, pp. 46–51.
117. H. Mikolajczyk, "Microwave-Induced Shifts of Gonadotropic Activity in Anterior Pituitary Glands of Rats," in: *Biologic Effects of Electromagnetic Waves*, C. C. Johnson and M. L. Shore, Eds., DHEW (FDA) 77-8010, Rockville, MD, 1977, pp. 377–383.
118. A. A. Novitskii, B. F. Murashov, P. E. Krasnobaev, and N. F. Markozova, "The Functional Condition of the System Hypothalamus–Hypophysis–Adrenal Cortex as a Criterium in Establishing the Permissible Levels of Superhigh Frequency Electromagnetic Emissions," *Voen. Med. Zh.*, **8,** 53–56 (1977).
119. I. R. Petrov and V. A. Syngayevskaya, "Endocrine Glands," in: *Influence of Microwave Radiation on the Organism of Man and Animals*, I. R. Petrov, Ed., Meditsina Press, Leningrad (NASA TTF-708), 1970, pp. 31–41.
120. W. Wiktor-Jedrzejczak, A. Ahmed, K. W. Sell, P. Czerski, and W. M. Leach, "Microwaves Induce an Increase in the Frequency of Complement Receptor-Bearing Lymphoid Spleen Cells in Mice," *J. Immunol.*, **118,** 1499–1502 (1977).
121. R. P. Liburdy, "Radiofrequency Radiation Alters the Immune System: Modulation of T- and B-Lymphocyte Levels and Cell-Mediated Immunocompetence by Hyperthermic Radiation," *Radiat. Res.*, **77,** 34–36 (1979).
122. N. J. Roberts Jr. and R. T. Steigbigel, "Hyperthermia and Human Leukocyte Functions: Effects on Response of Lymphocytes to Mitogen, Antigen and Bactericidal Capacity of Monocytes and Neutrophils," *Infect. Immun.*, **18,** 673–679 (1977).
123. Z. V. Gordon, Y. A. Lobanova, and M. S. Tolgskaya, "Some Data on the Effect of Centimeter Waves (Experimental Studies)," *Gig. Sanit.*, **12,** 16–18 (1955).
124. K. J. Oscar and T. D. Hawkins, "Microwave Alterations of the Blood–Brain Barrier System of Rats," *Brain Res.*, **126,** 281–283 (1977).

125. J. H. Merritt, A. F. Chamness, and S. J. Allen, "Studies on Blood–Brain Barrier Permeability After Microwave Radiation," *Rad. Environ. Biophys.*, **15,** 367–377 (1978).
126. E. Preston, E. J. Vavasour, and H. M. Assenheim, "Permeability of the Blood Brain Barrier to Mannitol in the Rat Following 2450 MHz Microwave Irradiation," *Brain Res.*, **174,** 109–117 (1979).
127. J. A. J. Stolwijk, "Responses to the Thermal Environment," *Fed. Proc.*, **36,** 1655–1658 (1977).
128. S. Stern, L. Margolin, B. Weiss, S.-T. Lu, and S. M. Michaelson, "Microwaves: Effect on Thermoregulatory Behavior in Rats," *Science*, **206,** 1198–1201 (1979).
129. E. R. Adair and B. W. Adams, "Microwaves Modify Thermoregulatory Behavior in Squirrel Monkeys," *Bioelectromagnetics*, **1,** 1–20 (1980).
130. N. V. Tyagin, "Change in the Blood of Animals Subjected to a SHF UHF Field," *Voyenno-Medit. Akad. Kirov.*, **73,** 116–126 (1957).
131. A. S. Hyde and J. J. Friedman, "Some Effects of Acute and Chronic Microwave Irradiation of Mice," in: *Thermal Problems in Aerospace Medicine*, J. D. Hardy, Ed., Unwin, Ltd., Old Woking, 1968, pp. 163–175.
132. J. F. Spalding, R. W. Freyman, and L. M. Holland, "Effects of 800 MHz Electromagnetic Radiation on Body Weight, Activity, Hematopoiesis and Life Span in Mice," *Health Physics*, **20,** 421–424 (1971).
133. S. M. Michaelson, R. A. E. Thomson, M. Y. E. Tamami, H. S. Seth, and J. W. Howland, "Hematologic Effects of Microwave Exposure," *Aerospace Med.*, **35,** 824–829 (1964).
134. S. M. Michaelson, R. A. E. Thomson, and J. W. Howland, *Biologic Effects of Microwave Exposure*, RADC: ASTIA Document No. AD 824–242, Griffis AFB; also United Stated Senate, Ninetieth Congress, second session on S 2067, S 3211 and HR 10790, 1968, *Radiation Control for Health and Safety Act of 1967*, pp. 1443–1570 (1967).
135. S. Baranski, "Effect of Chronic Microwave Irradiation on the Blood Forming System of Guinea Pigs and Rabbits," *Aerospace Med.*, **42,** 1196–1199 (1971).
136. P. Czerski, "Microwave Effects on the Blood-Forming System with Particular References to the Lymphocyte," in: *Biological Effects of Nonionizing Radiation*, P. E. Tyler Ed.; *Ann. NY Acad. Sci.*, **247,** 232–242 (1975).
137. P. Czerski, E. Paprocka-Slonka, M. Siekierzynski, and A. Stolarska, "Influence of Microwave Radiation on the Hemopoietic System," in: *Biological Effects and Health Hazards of Microwave Radiation*, P. Czerski, K. Ostrowski, C. Silverman, et al., Eds., Polish Medical Publishers, Warsaw, 1974, pp. 67–74.
138. S. M. Michaelson, "Human Exposure to Non-ionizing Radiant Energy—Potential Hazards and Safety Standards," *Proc. Inst. Elect. Electron. Eng.*, **60,** 389–421 (1972).
139. S. M. Michaelson, "Effects of Exposure to Microwaves: Problems and Perspectives," *Environ. Health Perspect.*, **8,** 133–156 (1974).
140. E. Hendler, "Cutaneous Receptor Response to Microwave Radiation," in: *Thermal Problems in Aerospace Medicine*, J. D. Hardy, Ed., Unwin, Ltd., Old Woking, Surrey, 1968, pp. 149–161.
141. E. Hendler, J. D. Hardy, and D. Murgatroyd, "Skin Heating and Temperature Sensation Produced by Infra-red and Microwave Irradition," in: *Temperature Measurement and Control in Science and Industry*, J. D. Hardy, Ed., Part 3, *Biology and Medicine*, New York, Reinhold, 1963, pp. 221–230.
142. H. F. Cook, "The Pain Threshold for Microwave and Infra-red Radiations," *J. Physiol.*, **118,** 1–11 (1952).
143. C. H. Durney, C. C. Johnson, P. W. Barber, et al., *Radiofrequency Radiation Dosimetry Handbook*, 2nd ed., Brooks Air Force Base, TX 1978; 141 (USAF Report SAM-TR-78-22).
144. A. W. Guy, J. C. Lin, and C. K. Chou, "Electrophysiological Effects of Electromagnetic Fields on Animals," in: S. M. Michaelson, M. W. Miller, R. Magin, and E. L. Carstensen, Eds., *Fundamental and Applied Aspects of Non-ionizing Radiation*, Plenum Press, New York, 1975, pp. 167–207.

145. O. P. Gandhi, "State of Knowledge for Electromagnetic Absorbed Dose in Man and Animals," *Proc. Inst. Electr. Electron. Eng.*, **68**, 24–32 (1980).

146. H. Massoudi, *Long Wavelength Analysis of Electromagnetic Power Absorption by Prolate Spheroidal and Ellipsoidal Models of Man*, Ph.D. Thesis, University of Utah, Salt Lake City, 1976, p. 217.

147. National Council on Radiation Protection and Measurements: *Radiofrequency Electromagnetic Fields*, NRCPH, Washington, DC, (NCRP Report No. 67) 1981, p. 134.

148. M. A. Stuchly, *Health Aspects of Radiofrequency and Microwave Radiation Exposure*, Part 2, Department of National Health and Welfare, Ottawa, 1978, p. 107.

149. C. C. Johnson, C. H. Durney, P. W. Barber, H. Massoudi, S. J. Allen, and J. C. Mitchell, *Radiofrequency Radiation Dosimetry Handbook*, SAM-TR-76-35, Brooks AFB, AFSC, AMD, SAM, 1976, p. 125.

150. C. Silverman, "Epidemiologic Approach to the Study of Microwave Effects," *Bull NY Acad. Med.*, **55**, 1166–1181 (1979).

151. C. Silverman, "Epidemiologic Studies of Microwave Effects," *Proc. Inst. Electr. Electron. Eng.*, **68**, 78–84 (1980).

152. P. S. Ruggera, *Measurements of Emission Levels During Microwave and Shortwave Diathermy Treatments*, Bureau of Radiological Health, HHS Publication (FDA) 80-8119), Rockville, MD, 1980.

153. L. Daily, "A Clinical Study of the Results of Exposure of Laboratory Personnel to Radar and High Frequency Radio," *US Naval Med. Bull.*, **41**, 1052–1056 (1943).

154. C. I. Barron and A. A. Baraff, "Medical Considerations of Exposure to Microwaves (Radar)," *JAMA*, **168**, 1194–1199 (1958).

155. C. I. Barron, A. A. Love, and A. A. Baraff, "Physical Evaluation of Personnel Exposed to Microwave Emanations," *J. Aviation Med.* **26**, 442–452 (1955).

156. Z. V. Gordon, *Biological Effect of Microwaves in Occupational Hygiene*, Izvestiya Meditisina, Leningrad (TT 70-50087, NASA TT F-633, 1970), 1966, p. 164.

157. Z. V. Gordon, "Occupational Health Aspects of Radiofrequency Electromagnetic Radiation," in: *Ergonomics and Physical Environmental Factors*, International Labour Office, Geneva, Occupational Safety and Health Series, No. 21, 1970, pp. 159–172.

158. M. N. Sadchikova, "Clinical Manifestations of Reactions to Microwave Irradiation in Various Occupational Groups," in: *Biological Effects and Health Hazards of Microwave Radiation*, P. Czerski, K. Ostrowski, et al. Eds., Polish Medical Publishers, Warsaw, 1974, pp. 261–267.

159. P. Czerski and M. Piotrowski, "Proposals for Specifications of Allowable Levels of Microwave Radiation," *Medycyna Lotnicza* (Polish), **39**, 127–139 (in Polish) (1972).

160. P. Czerski and M. Siekierzynski, "Analysis of Occupational Exposure to Microwave Radiation," in: *Fundamental and Applied Aspects of Non-Ionizing Radiations*, S. M. Michaelson, M. W. Miller, R. Magin, and E. L. Carstensen, Eds., Plenum Press, New York, 1975, pp. 367–375.

161. M. Siekierzynski, P. Czerski, H. Milczarek, et al., "Health Surveillance of Personnel Occupationally Exposed to Microwaves. II. Functional Disturbances," *Aerospace Med.*, **45**, 143–1145 (1974).

162. C. D. Robinette, C. Silverman, and S. Jablon, "Effects upon Health of Occupational Exposure to Microwave Radiation (Radar) 1950–1974," *Am. J. Epidemiol.*, **112**, 39–53 (1980).

163. U.S. Senate, Radiation Health and Safety, *Hearings Before the Committee on Commerce, Science, and Transportation*, 95th Congress, First Session on Oversight of Radiation Health and Safety, June 16, 17, 27, 28 and 29, 1977, Serial No. 95-49, 1977, pp. 284, 1195, 1196.

164. A. M. Lilienfeld, J. Tonascia, S. Tonascia, et al., "Foreign Service Health Status Study: Evaluation of Health Status of Foreign Service and Other Employees from Selected Eastern European Posts, Final Report," July 31, 1978 Contract No. 6025-619073 Department of Epidemiology Johns Hopkins University, Baltimore, NTIS PB-288, 1963, 1978.

165. U.S. Senate, Committee on Commerce, Science, and Transportation, Committee Print, *Microwave Irradiation of the U.S. Embassy in Moscow, April 1979* (43-949), U.S. Government Printing Office, Washington, DC, 1979.
166. S. Zydecki, "Assessment of Lens Translucency in Juveniles, Microwave Workers and Age-Matched Groups," in: *Biological Effects and Health Hazards of Microwave Radiation*, P. Czerski, K. Ostrowski, C. Silverman, et al. Eds., Polish Medical Publishers, Warsaw, 1974, pp. 306–308.
167. S. F. Cleary and B. S. Pasternack, "Lenticular Changes in Microwave Workers, A Statistical Study," *Arch. Environ. Health*, **12**, 23–29 (1966).
168. K. Majewska, "Study of Effects of Microwaves on Visual Organs," *Klin. Oczna.*, **38**, 323–328 (1968).
169. M. M. Zaret, S. F. Cleary, B. Pasternack, M. Eisenbud, and H. Schmidt, A *Study of Lenticular Imperfections in the Eyes of a Sample of Microwave Workers and a Control Population*, final contract report for Rome Air Development Center, RADC-TDR-6310125, March 15, 1963.
170. S. F. Cleary, B. S. Pasternack, and G. W. Beebe, "Cataract Incidence in Radar Workers," *Arch. Environ. Health*, **11**, 179–182 (1965).
171. D. E. Shaklett, T. J. Tredici, and D. L. Epstein, "Evaluation of Possible Microwave Induced Lens Changes in the United States Air Force," *Aviat. Space Environ. Med.*, **46**, 1403–1406 (1975).
172. J. A. Hathaway, H. Stern, E. M. Sales, and E. Leighton, "Evaluation of Ocular Medical Surveillance on Microwave and Laser Workers," *J. Occup. Med.*, **19**, 683–688 (1977).
173. J. A. Hathaway, "The Needs for Medical Surveillance of Laser and Microwave Workers," in: *Current Concepts in Ergophthalmology*, B. Tengroth and D. Epstein, Eds., Societas Ergophthalmologica Internationalis, Stockholm, 1978, pp. 139–160.
174. B. Appleton, *Results of Clinical Surveys for Microwave Ocular Effects*, U.S. Government Printing Office [DHEW (FDA) Publ. No. 73:803], Washington, DC, 1973.
175. Health Examination Statistics U.S. Department of Health, Education and Welfare, National Center for Health Statistics, Medical Statistics Branch, March 2, 1979.
176. L. T. Odland, "Observations on Microwave Hazards to USAF Personnel," *J. Occup. Med.*, **14**, 544–547 (1972).
177. M. N. Sadchikova and A. A. Orlova, "Clinical Picture of the Chronic Effects of Electromagnetic Microwaves," *Gigiena Truda i Professional'nye Zabolvaniya* (Moscow), **2**, 16–22 (1958).
178. C. Silverman, *The Epidemiology of Depression*, Johns Hopkins University Press, Baltimore, 1968.
179. C. Silverman, "Nervous and Behavioral Effects of Microwave Radiation in Humans," *J. Epidemiol.*, **97**, 219–224 (1973).
180. Yu. A. Osipov, *Occupational Hygiene and the Effect of Radio-frequency Electromagnetic Fields on Workers*, Izvestiya Meditsina Press, Leningrad, 1965, pp. 78–103.
181. Z. Edelweijn, R. L. Elder, E. Klimkova-Deutschova, and B. Tengroth, "Occupational Exposure and Public Health Aspects of Microwave Radiation," in: *Biological Effects and Health Hazards of Microwave Radiation*, P. Czerski, K. Ostrowski, C. Silverman, et al. Eds., Polish Medical Publishers, Warsaw, 1974, pp. 330–331.
182. A. T. Sigler, A. M. Lilienfeld, B. H. Cohen, et al., "Radiation Exposure in Parents of Children with Mongolism (Down's Syndrome)," *Bull. Johns Hopkins Hosp.*, **117**, 374–399 (1965).
183. B. H. Cohen, A. M. Lilienfeld, S. Kramer, et al., "Parental Factors in Down's Syndrome—Results of the Second Baltimore Case-Control Study," in: *Population Cytogenetics, Studies in Humans*, E. B. Hook and I. H. Porter, Eds., Academic Press, New York, 1977, pp. 301–352.
184. P. B. Peacock, J. W. Simpson, C. A. Alford, et al., "Congenital Anomalies in Alabama," *J. Med. Assoc. State Ala.*, **41**, 42–50 (1971).
185. J. A. Burdeshaw and S. Schaffer, *Factors Associated with the Incidence of Congenital Anomalies: A Localized Investigation*, Environmental Protection Agency, Final Report, Contract No. 68-02-0791, March 31, 1976.

186. S. Gellhorn, "Diathermy in Gynecology," *JAMA*, **90**, 1005–1008 (1928).
187. A. Rubin and W. J. Erdman, "Microwave Exposure of the Human Female Pelvis During Early Pregnancy and Prior to Conception," *Am. J. Phys. Med.*, **38**, 219–220 (1959).
188. J. Daels, "Microwave Heating of the Uterine Wall During Parturition," *Obstet. Gynecol.*, **42**, 76–79 (1973).
189. J. Daels, "Microwave Heating of the Uterine Wall During Parturition," *J. Microwave Power*, **11**, 166–168 (1976).
190. K. Marha, J. Musil, and H. Tuha, *Electromagnetic Fields and the Life Environment*, State Health Publishing House, 1968 San Francisco Press, Prague, 1971, p. 138.
191. I. R. Petrov, ed., *Influence of Microwave Radiation on the Organism of Man and Animals*, Meditsina Press (NAS TT F-708), Leningrad, 1970.
192. A. S. Presman, *Electromagnetic Fields and Life*, Izd-vo Nauka, Moscow, 1968, p. 332 (translated by Plenum Press, New York, 1970).
193. Z. V. Gordon, "The Problem of the Biological Action of UHF," *Gigyena Truda Akademiya Meditsina Nauk USSR*, No. 1, 5–7 (1960).
194. T. N. Orlova, "Clinical Aspects of Mental Disorders Following Protracted Human Exposure to Super-High Frequency Electromagnetic Waves," in: *Cerebral Mechanisms of Mental Illness; Kazanskiy Meditsinskiy Zhurnal*, 16–18 (1971).
195. S. M. Michaelson, "Radiofrequency and Microwave Energies, Magnetic and Electric Fields," in: *The Foundations of Space Biology and Medicine*, Chapter 1, Vol. II, Book 2, M. Calvin and O. G. Gazenko, Eds., NASA, Washington, DC, 1975, pp. 409–452.
196. S. M. Michaelson and C. H. Dodge, "Soviet Views on the Biologic Effects of Microwaves: An Analysis," *Health Physics*, **21**, 108–111 (1971).
197. C. H. Dodge and Z. R. Glaser, "Trends in Non-ionizing Radiation Bioeffects Research and Related Occupational Health Aspects," *J. Microwave Power*, **12**, 319–334 (1977).
198. R. M. Albrecht and E. Landau, "Microwave Radiation, An Epidemiologic Assessment," *Rev. Environ. Health*, **3**, 44–58 (1979).
199. A. K. Guskova and Y. M. Kochanova, "Some Aspects of Etiological Diagnostics and Occupational Diseases as Related to the Effects of Microwave Radiation," *Gigyena Truda i Professional'nye Zabolvaniya* (Moscow), **3**, 14–17 (1975).
200. American National Standards Institute, *Safety Level of Electromagnetic Radiation with Respect to Personnel*, ANSI, New York, C95.1-1966, C95.1-1974; *Safety Levels with Respect to Human Exposure to Radiofrequency Electromagnetic Fields* (300 kHz to 100 GHz), C95.1-1982.
201. Ministry of Health Protection of the USSR, *Temporary Sanitary Rules for Working with Centimeter Waves*, 1958.
202. Health and Welfare, Canada, *Health Aspects of Radiofrequency and Microwave Radiation Exposure*, Part II, DHEW, Ottawa (Document 78-EHD-22), 1978, p. 107.
203. Health and Welfare, Canada, *Recommended Safety Procedures for the Installation and Use of Radiofrequency and Microwave Radiation Devices in the Frequency Range 10 MHz–300 GHz*, DHEW, Ottawa (Safety Code No. 6, Publication 79-EHD-30), 1979, p. 39.
204. M. H. Repacholi, "Proposed Exposure Limits for Microwave and Radiofrequency Radiations in Canada," *J. Microwave Power*, **13**, 199–277 (1978).
205. Order of Council of Ministers of May 25, 1972, "Concerning Safety and Hygiene when using Equipment Generating Electromagnetic Fields in the Microwave Region," *Official Journal of the Ministry of Health and Social Welfare (Poland)*, PRL No. 21 II, p. 153 (1972).
206. Order of the Ministry of Health and Social Welfare, September 8, 1972, "Concerning the Determination of Electromagnetic Fields in the Microwave Range and Duration of Permissible Work Within the Hazardous Zone," *Official Journal of the Ministry of Health and Social Welfare (Poland)*, No. 17 p. 78 (1972).

207. Order of Ministry of Health, Labor and Social Welfare, "Concerning Health and Safety of Work with Equipment Producing Electromagnetic Fields in the Range of 0.1 MHz to 300 MHz," *Official Journal of the Ministry of Health and Social Welfare (Poland)*, PRL No. 8, March 17, p. 8 (1977).
208. Royal Swedish Academy of Engineering Sciences, *Biological Effects of Electromagnetic Fields*, RSAES, Stockholm, 1976, p. 160. (ISBN91.7082.123.2).
209. World Health Organization/International Non-Ionizing Radiation Committee of the International Radiation Protection Association, *Environmental Health Criteria for Radiofrequency and Microwaves*, WHO/INIRC IRPA, Geneva, 1981, p. 152.
210. USDHEW, *Regulations for Administration and Enforcement of the Radiation Control of Health and Safety Act of 1968*, paragraph 1030.10, Microwave Ovens DHEW Publication No. (FDA) 75-8003 July 1974, pp. 36–37.
211. DHEW, *Radiation Emitting Devices Regulations*, SOR/74-601 23, October 1974; Part III, *Microwave Ovens, Canada Gazette*, Part V, 108, 1974, pp. 2822–2825.
212. Internal Electrotechnical Commission, *Particular Requirements for Microwave Cooking Appliances*, EIC 335-25, Part 2, 1976.

CHAPTER FIFTEEN

Vibration

JOHN C. GUIGNARD, M.B., Ch.B.

1 INTRODUCTION

Mechanical vibration is ubiquitous (1). It occurs in factories; in vehicles on land, at sea, in the air, and in space; and in the vicinity of working machinery of every kind. Like acoustical noise, with which it is commonly associated, mechanical (i.e., structure-borne) vibration at intensities ranging from the barely perceptible to the intolerably severe is encountered on construction sites; in mines and quarries; around manufacturing plant; in equipment used in shipyards; on the farm and in the forest; and indeed, wherever humans harness power to travel or to get work done.

In the context of occupational medicine, mechanical vibration may be defined as any continuous or intermittent oscillating mechanical force or motion that affects humans at work through the mediation of anatomic structures and sensory receptors other than the organ of hearing. (Such a definition excludes present consideration of the effects of noise, i.e., airborne vibration at audible frequencies, which are dealt with in Chapter 13.)

In engineering physics, vibration, which is a universal phenomenon, has received a succinct definition by Crede (1) as a series of reversals of velocity. This definition reminds us that in any vibratory process, both displacement of matter and acceleration (change of velocity, requiring mechanical force) necessarily take place. The reversal of velocity in a vibration may take place at regular intervals, that is, with a specific frequency, in which case the vibration is termed *periodic*, or it may occur irregularly and unpredictably, in which case the vibration is called *nonperiodic*, or *random*.

Occupational exposure to vibration arises in a variety of ways and can reach the worker at intensities variously detrimental to comfort, efficiency, safety, or health through several routes of mechanical transmission (2–4). It may affect the worker primarily by transmission through a supporting or contacting surface that is vibrating, such as the deck of a ship; the seat or floor of a vehicle shaken by ground roughness or engine vibrations; or the surroundings of machinery in a plant. In some occupations, the worker riding on the machine is exposed to vibration intentionally generated to grade, move, or compact such materials as coal, ore, or cement. The foregoing are all examples of exposure to whole-body vibration.

In many industries the chief route of entry of mechanical vibration into the human body is through the hands and arms of the worker while using a hand-held powered tool such as a road breaker, chipping hammer, grinder, power drill, rivet driver, or chain saw. The quality and severity of vibration from such tools depends on several factors, including the nature of the source of power (e.g., an electric, pneumatic, or gasoline motor). As is emphasized in the major divisions of this chapter, it is practical and convenient to treat whole-body and hand-transmitted vibrations separately as distinct provinces of biomedical study and engineering practice. This is recognized both in the domain of research and in the fields of regulatory and engineering approaches to protecting the worker from the adverse effects of occupational vibration exposure. In accordance with international expert consensus, separate international standard guidelines have been issued (and ratified as national standards in many countries) or drafted for the evaluation of human exposure to whole-body (4; 5) and hand-transmitted (6) vibration.

In some working situations, for instance, during operation of heavy equipment or driving tractors and the like, vibration enters the body through several routes, including the operator's seat, the floor or controls on which the feet are placed, and the steering wheel or other hand controls. Sometimes, as in many kinds of aircraft and military vehicles, vibration may be transmitted directly to the head from headrests and sighting devices. In some circumstances, mechanical vibration can be disturbing, fatiguing, and prejudicial to occupational efficiency and safety indirectly, without actually entering the human body, such as when vibration of the dials and pointers of gauges or other instruments in a vehicle or plant makes the displays difficult and sometimes impossible to read.

1.1 Description and Measurement of Vibration

The physical description and measurement of vibration affecting humans have much in common with the description and measurement of noise and, in many applications, make use of similar instrumentation and analytical techniques. Several descriptive parameters are necessary to characterize completely a mechanical vibration affecting humans. The most important ones are frequency, complexity (spectrum), direction of vibration (with respect to anatomical axes), intensity (strength of the vibration), duration, and time course (including intermittency) of the vibration exposure.

VIBRATION

1.1.1 Frequency

The frequency of a periodic (i.e., wavelike) vibration is the number of complete cycles of oscillation occurring in unit time. The international standard unit of frequency is, as in acoustics, the hertz (Hz), which is one complete cycle of oscillation per second. An older unit, the cycle per second (c/sec), is still sometimes used and is found in many of the references cited in this chapter. However, the hertz (1 Hz = 1 c/sec) is now the internationally preferred unit. In some applications, particularly marine engineering, where predominant vibrations are related to the speed of revolution of an engine or drive shaft, frequency may be expressed in cycles per minute (cpm): 1Hz = 60 cpm.

Traditionally, the frequency of vibration (or of any component of a complex vibration) is determined by methods of graphical analysis, that is, inspection and measurement of a plot or recording of the vibration waveform against a time base. In most present-day applications, however, vibration is measured by use of mechanoelectrical transducers and electronic instrumentation to yield a power spectrum or some cognate mathematical representation of the power, energy, or intensity of the vibration as a function of frequency or inverse time.

Although accelerometers or other special transducers that convert mechanical force or motion into proportional electrical signals are needed to detect vibration in the low-frequency range (< 20 Hz) that is mainly of interest in the case of whole-body exposure, the techniques of magnetic tape recording and electronic signal processing permit the kind of equipment now generally available commercially for noise analysis to be used for vibration analysis in a corresponding manner.

When the vibration frequencies of interest lie in the subaudible range, that is, below the range normally handled by analytical equipment designed for acoustical work, the vibration frequencies to be analyzed can be shifted into a higher part of the spectrum for analytical purposes by replaying the tape-recorded data at an appropriate faster speed. Rapidly increasing use is now being made of digital computers to analyze vibration data to generate such informative functions as power spectra and vibration transmission factors. Such computational procedures require the analog-to-digital conversion of the "raw," continuous vibration signals generated by the measuring transducer to a digital form that can be handled by the computer.

1.1.2 Complexity

Vibration affecting workers is almost always complex; that is, it is composed of more than one frequency and may be multidirectional. Just as the pure tone of a tuning fork or an electronic oscillator is rarely heard among the sounds and noises of everyday life, so single-frequency (sinusoidal) vibration is rarely, if ever, encountered outside the laboratory. (Sinusoidal vibration, however, is a common and useful laboratory tool for gaining insight into the human biodynamic and physiological response to vibration, just as the pure tone of the audiometer—which is precisely definable, measurable, and repeatable—is an ideal stimulus for defining the response of the organ of hearing).

Vibration that is approximately sinusoidal can be felt on board ship at frequencies determined by the product of the revolutions per minute (rpm) of the propeller shaft and the number of blades on the screw. Such vibration is particularly strong in single-screw vessels and when the engine revolutions pass through a critical speed, that is, a speed at which the driving frequency coincides with one of the major natural frequencies of structural vibration of the ship's hull or superstructure, which is then set into sympathetic vibration (resonance). Heavy vibration at the rotating wing blade-passage frequency (again, a product of the engine rpm and the number of blades in the lifting aerofoil) can also be a considerable problem in helicopters. In fixed plants, troublesome single-frequency vibration can be transmitted to workplaces when faulty design, installation, or maintenance of a machine permits resonant vibration of the machine or its foundation or vibration of the building frame or structures in the vicinity.

Vibration in vehicles and from industrial machines is often complex, irregular, or essentially random (e.g., the lurching of a tractor or heavy equipment operated over rough ground) and accordingly is lacking in obvious periodicity. Nevertheless, with the use of the techniques of frequency (spectral) analysis, it is still possible and in many applications appropriate to describe and evaluate the motion in terms of frequency or spectrum. A vibration spectrum, by analogy with the spectra of electromagnetic radiations, is a graph or table of vibrational energy, power, or intensity against frequency.

1.1.3 Intensity

Another important descriptive parameter of vibration is its amplitude or intensity, or in other words, the extent or severity of the motion at any given frequency. When the vibration is of the simplest kind, namely, sinusoidal or simple harmonic oscillation resembling the swing of a simple pendulum, the amplitude is defined as the maximum (sometimes called the "peak") displacement of the vibrating system from its midposition of rest or equilibrium. A velocity of vibrational motion and an acceleration due to the restoring or impressed force of the vibration are always proportionally associated with the instantaneous displacement in the vibratory process. For sinusoidal vibration or any single-frequency component of a complex vibration, when the magnitude of the displacement is fixed, the vibrational velocity associated with it rises proportionally with frequency, and the vibrational acceleration increases with the square of frequency. For this reason very large displacement amplitudes of motion (perhaps several meters, in the case of the heave motion of a large ship or an oil-drilling platform induced by ocean waves at frequencies well below 1 Hz) are required to generate appreciable levels of acceleration at low frequencies. Higher up the frequency spectrum, however—such as at vibration frequencies, associated with the peak output of vibrational energy by hand-held power tools (typically ranging from around 30 Hz to several hundred hertz)—enormous accelerations can be generated by vibrations of very small displacement amplitude (fractions of a millimeter).

With the use of the International System of Units (SI Units), vibrational displacement is properly measured in meters, although fractional metric units may be used for convenience to express the tiny displacements characteristic of high-frequency vibrations. In some countries, including the United States, however, the inch or microinch is still in widespread use as the unit of displacement. It is usually convenient, particularly when modern electronic vibration-measuring instrumentation is used, to measure the intensity of vibration in terms of acceleration, which is the second derivative of displacement with respect to time and is directly related to the force of vibration.

Acceleration measurements are becoming standard practice where vibrations affecting humans are concerned (3–6). A variety of small transducers (accelerometers) generate an electrical output directly proportional to the mechanical vibrational acceleration of a vibrating machine or, in field and laboratory research applications, the human body. These instruments permit direct measurement of vibrational acceleration at the point of instrumentation.

In some applications it is convenient to use other kinds of transducer or vibration "pickup" that respond to displacement, pressure, or velocity rather than to vibrational acceleration. Signal processing (e.g., differentiation of a velocity signal with respect to time) is then required to obtain a measurement of vibrational acceleration. It is important to ensure that the measuring system can handle such operations on the signal adequately over the bandwidth of concern in any given application.

Vibrational acceleration is measured in meters per second per second (m/sec^2). For many practical purposes, however, and commonly in biomedical work, it is convenient and acceptable (4) to express the acceleration associated with vibrating forces nondimensionally in terms of g, that is, as multiples or submultiples of the standard acceleration of gravity at the earth's surface, where 1 g is approximately equivalent to 9.81 m/sec^2.

An instantaneous or peak value of acceleration has little meaning in the description of a continuous vibration that is complex or random. It is accordingly common practice to measure or compute an average or root mean square (rms) value of the accelerations recorded instantaneously over a finite data sample. Many electronic instruments used to measure vibration or noise (e.g., the sound level meters widely used in noise surveying and monitoring) are designed to yield an output that is approximately proportional to the rms value of the measured quantity. The rms value of a varying quantity such as vibrational acceleration or sound pressure is mathematically equivalent to the standard deviation (a measure of variance) of the instantaneous values of the quantity. In the case of a sinusoidally varying quantity such as a pure harmonic vibration, the rms value is 0.707 times the maximum (peak) value.

Although rms measurements are an adequate approximation for many applications such as the comparative evaluation of high-frequency vibrations or noises, caution should be exercised in application of such an intensity-averaging technique to the low-frequency domain. This is because the biological effectiveness of mechanical vibrations or whole-body motion in the subaudible range depends on additional factors, such as the relative intensities and the phase relationships of the component frequencies of a complex vibration, which are ignored by rms measurements. Human experimentation on vibration tolerance in the range 1 to 30 Hz (7) and on motion sickness incidence induced by complex harmonic motion in the range below 1 Hz (8) has shown that it can be fallacious to assume that the severity of combinations of sinusoidal motions can be equated solely in terms of the rms acceleration.

1.1.4 Direction

The mechanical response of humans to whole-body vibration, and the corresponding sensory and psychophysiological reactions to the motion, are heavily dependent on the direction in which vibrating forces are applied to the body or its parts. For instance, the subjective perception and tolerance of vibration in the range 1 to 10 Hz and the degree of interference with working efficiency that such vibration can provoke are quite different depending on whether one's seat vibrates up and down (z-axis) or from side to side (y-axis vibration). This is mainly because the human body as a complex vibrating system exhibits resonance at different frequencies, corresponding with different modes of vibration, when vibrated in the transverse as opposed to the longitudinal axis in that range of frequency. The International Organization for Standardization (ISO) has accordingly developed separate guidelines for different directions of human whole-body vibration exposure in the range 1 to 80 Hz (4). The same authority (9) is also drafting a standard on anatomical coordinate systems for use in biodynamics: properly defined coordinate systems are essential frames of reference for the precise specification, measurement, and comparative evaluation of force and motion inputs to humans.

1.1.5 Duration and Time Course

The human response to vibration depends in a complex manner on the duration of a continuous or steady-state vibration and on the time course of a fluctuating (nonstationary), transient, or intermittent vibration. The time factor in morbidity associated with prolonged exposure to whole-body or hand-transmitted vibration repeated occupationally on a daily basis is mentioned in more detail later in this chapter. In setting standards governing human exposure to whole-body vibration on a short-term basis (i.e., discrete exposures or exposures repeated daily but lasting less than a day), the ISO (4) has presumed, for want of definitive data, that a general biological principle applies, namely, that human tolerance of steady-state vibration declines monotonically with increasing duration of exposure. The degree to which such a tendency is mitigated by specific adaptation or general habituation to vibration stress remains an open question, however, for very little definitive research has yet been devoted to it.

2 WHOLE-BODY VIBRATION

Whole-body vibration is experienced universally in land, sea, air, and space vehicles and in buildings and workplaces where the floor, ground, or structures in the vicinity are shaken by the action of machinery or other mechanically disturbing activity. The entire body may be vibrated by powerful hand-held or hand-guided machines (e.g., road breakers, tamping machines, and rock drills) that generate substantial amounts of vibrational energy at low frequencies (<30 Hz). The vibration of handheld machines, however, is of concern mainly as a threat to the hands and upper limb joints of the operator, which is considered later in this chapter. Whole-body vibration is usually considered to enter the body through a supporting surface such as a floor or a vehicle seat or couch, but in some occupations it can be received by other routes, such as through the back and shoulders of workers by use of certain types of motorized appliance (e.g., crop sprayers and dusters, hedge trimmers) supported or carried on the person as a backpack. Not all whole-body vibration is artificially generated: ground or floor vibration, occasionally at alarming or threatening levels (10), can be encountered from natural sources such as the action of wind, water, or seismic forces upon the land or man-made structures.

Vibration at work is generally regarded as, at best, an inescapable irritant or nuisance (akin to noise, with which it is commonly associated) that causes concern or alarm when it reaches such an intensity, constancy, or regularity of action as to prejudice the individual's or a crew's working efficiency, safety, rest, or health. It may be noted in passing that not all mechanical vibration felt by people is unwanted or necessarily harmful. Naturally generated or self-induced oscillations of the body or its parts are normally present for some or all of the time, albeit usually at very low intensities, throughout life. Such oscillations are generated by the pulsatile action of the heartbeat (and are of sufficient strength to excite a characteristic pattern of mechanical response in resonant anatomical structures, recordable as the ballistocardiogram) and, in the region of 10 Hz, by physiological neuromuscular tremor. Strong periodic oscillations of body structures, not infrequently exceeding the internal motions provoked by the vibration of mechanical transportation, are self-induced by natural human locomotion (walking, running, jumping, dancing, and other varieties of cyclical volitional activity). Moreover, artificially induced rhythmic motion or vibration, sometimes at quite intense levels, is used by many people for amusement, relaxation, erotic gratification, or quasitherapeutic purposes, as is shown by the widespread popularity of a variety of motion or vibration generating devices, ranging from fairground machines through coin-operated bed vibrators in hotels to personal vibrators and massagers. Most of the personal applicators apply low-amplitude vibration at the local electrical AC supply frequency (50 to 60 Hz), and whereas the beneficial effects of their use may be debatable, they probably do no harm when properly used.

Nevertheless, whole-body vibration encountered occupationally is a matter of concern in many industries where it occurs at intensities sufficient to undermine working efficiency (hence safety and productivity) and health. When vibration is part of the composite stress of a working environment, distinct patterns of morbidity may be causally associated with chronic, daily exposure to the stress. Whole-body vibration in the range 1 to 20 Hz is the most disturbing and hazardous to workers in many such industries (11, 12). Many types of commercial and working vehicle, as well as mobile heavy equipment driven or ridden over rough surfaces by the operator, produce strong whole-body vibration (predominantly vertical) and repeated jolting with most of the spectral energy of the motion contained within that frequency band (11, 13). In that range, current standards or guidelines for limiting human whole-body vibration exposure may frequently be exceeded substantially (11). It has been estimated (13) that in the United States alone, nearly seven million workers are exposed occupationally to significant levels of whole-body vibration.

Mechanical vibrations applied globally to the whole body at frequencies above 20 or 30 Hz probably do not contribute substantially to morbidity in most occupational circumstances. This is because vehicles and other generators of whole-body vibration do not transmit substantial amounts of vibrational energy to humans at high frequencies (unlike handheld power tools, discussed later in this chapter). Moreover, because high-frequency vibrations are strongly attenuated by the human body surface and by seating and flooring materials, much of the energy of such vibrations does not reach or enter the body. In several respects, mechanical whole-body vibration at frequencies above 20 Hz, although it may contribute to discomfort, fatigue, and distraction from work, is more akin to irradiation by noise and, at higher frequencies, may be regarded as part of the acoustic environment. Because of such considerations, current international (4) and American national (5) standards on the evaluation of human exposure to whole-body vibration are restricted in scope to frequencies below 80 Hz.

Whole-body mechanical oscillation of humans at frequencies below 1 Hz does not induce appreciable internal vibratory movements of the body, which moves in response to such very low-frequency motion essentially as a single mass. Nevertheless, such motion, when sufficiently severe, can disrupt human activity and exert important physiological effects, the most commonly experienced of which is motion sickness (kinetosis) (8, 14–19). Large-amplitude, low-frequency oscillations, particularly in roll, pitch, and heave (vertical translational motion), are a perennial nuisance and often a hazard to the crews of ships, aircraft, and floating structures such as offshore oil-drilling platforms. In heavy seas, the wave-induced motion of ships or floating rigs subjects the crew to whole-body oscillation in the band 0.05 to 2 Hz. This motion renders hazardous and not infrequently impossible the performance of seaborne tasks and even simple locomotion aboard the craft. When a storm at sea is at its height, and often for many hours after it has abated, while the seas remain high, the risk of falls and other motion- and fatigue-induced accidents is greatest. Similar difficulties are experienced, usually for relatively brief periods in most airborne operations, aboard aircraft flying in turbulence.

If sufficiently intense or prolonged, whole-body motion in the band 0.1 to 1 Hz will induce motion sickness in susceptible people. [It is probable that almost all the human population, and most other vertebrate species, provided the vestibular function is normal (14), can ultimately succumb to motion sickness, although there are wide individual and circumstantial variations in susceptibility.] Vertical ("heave") motion in the relatively narrow frequency band 0.1 to 0.7 Hz appears to be the most provocative in humans, with the maximum susceptibility occurring at around 0.17 Hz (8, 15). Such a frequency corresponds to a heave period of some 6 sec, not uncommonly characteristic of medium-sized ships at sea. Motion sickness is considered further in the following sections.

2.1 Effects of Whole-Body Vibration on Humans

Several detailed accounts of the human response to whole-body vibration have been published previously (3, 21–23), and there are also reviews with particular reference to occupational exposure (24, 25). This section opens with a consideration of the human response to vibration as a mechanical system, an understanding of which is crucial to understanding the psychophysiological action of oscillatory motion and to developing methods of protecting people from adverse effects, including vibration injury.

2.1.1 Biodynamics

The physiological, psychological, and pathological effects of mechanical whole-body vibration in humans are caused, directly or indirectly, by oscillatory displacement or deformation of the structures, organs, and tissues of the body so as to disturb their normal functioning and stimulate the sensory organs and the distributed mechanoreceptors that mediate the composite vibration sense. In addition to physiologically mediated responses, vibration, particularly in the range 1 to 30 Hz, can also degrade or disrupt human volitional activity and the performance of tasks by mechanically forcing differential motion to take place between humans and their perceptual or physical points of contact with their surroundings or tasks (Reference 3, Chapter 6). The living body in terms of engineering dynamics is a complex vibrating system capable of resonance at characteristic frequencies in the band 1 to 30 Hz; thus many biological effects of whole-body vibration are strongly frequency dependent.

Human body resonance is the condition in which a forcing vibration is applied to the body at such a frequency that some anatomic structure, part, or organ is set into measurable or sensible oscillation greater than that of related structures (Reference 3, Chapter 4). In terms of physical biodynamics, the body can be visualized (and modeled analogically, with the use of computers or engineering devices) as a complex inertial system comprising several masses linked by elastic and damping elements. To that extent, the body resembles such engineering structures as airframes, ship hulls, or steel-framed buildings. All such structures exhibit resonant modes of vibration when shaken at their

Table 15.1 Biodynamic Factors Influencing the Absorption and Transmission of Mechanical Vibration Entering the Human Body[a]

External Factors

Intensity (force) of vibration
Direction, site, and area of application of vibration
Nature (including contour) of support and restraint of the person
Resilience and damping of seat, cushion, or other structure through which vibration is transmitted
Nature, distribution, and weight (mass) of any external load on body (e.g., work clothes; man-mounted equipment)

Internal Factors

Individual build (height, weight, fatness, etc.)
Nonlinearity of tissue stiffness and damping
Posture (relative position of body segments and limbs)
Activity of the subject
Degree of muscular tone or tension

[a] Adapted from Guignard (2, Chapter 29).

characteristic frequencies. Those frequencies, and the magnification factors (i.e., the extent to which impressed vibration is mechanically amplified by the structure at its resonance frequencies) are determined by the ratios between mass, elasticity, and damping in the responding systems (Reference 3, Chapter 1). A large part of the practice of vibration engineering, and many operational procedures (e.g., the regulation of ship engine speeds), are devoted to preventing or suppressing potentially destructive conditions of resonance; and similar principles can be applied to the protection of humans from the adverse effects of low-frequency whole-body vibration (Table 15.1).

The human body exhibits several modes of resonant vibration in response to whole-body motion in the range 1 to 50 Hz (Reference 3, Chapter 4; Reference 26). The absorption of vibrational energy at low frequencies by the body and the excitation of particular modes of oscillation in humans by vibrating or impact forces depend on several intrinsic and external biodynamic factors. Table 15.1 summarizes the most important of these.

The principal, or most prominent, resonance of the seated, standing, or recumbent human body, when a person is vibrated in the z-axis (cephalocaudally), occurs in the region of 5 Hz, with substantial mechanical amplification of impressed vibrations observable in the band 4 to 8 Hz. Resonance phenomena in the human body can be measured by comparing the motion recorded with the use of lightweight accelerometers attached to defined bony landmarks on the person with the input motion recorded at the vibrating seat; or by measuring the mechanical impedance (the complex ratio of impressed vibratory force to resultant velocity of motion) of the subject at the point of vibration input. Experimental measurements of this kind, and the instrumentation used, on human volunteers are closely analogous to corresponding tests carried out on engineering structures. The principal z-axis resonance in humans is reflected in the frequency dependence of a variety of human physiological and psychological reactions to vibration (3), and it corresponds with a minimum in human subjective tolerance of whole-body vibrational acceleration in the z-axis, particularly at severe levels (27, 28). Accordingly, current standard human whole-body vibration exposure guidelines (4, 5), expressed in terms of rms acceleration as a function of frequency between 1 and 80 Hz, are set at their lowest (i.e., most conservative) in the band 4 to 8 Hz.

When a person is vibrated in the x- (anteroposterior) or y- (lateral) axis, for example, when seated in a lurching or swaying vehicle, the principal mode of resonant vibration, seen typically as oscillatory flexion of the trunk at the hips or lower back, occurs in a different frequency range, namely, 1 to 2 Hz. Such motion (sometimes enhanced by softness in the vehicle seating) is often experienced by passengers in trains, which are apt to sway when passing over uneven track. In contrast to z-axis vibration, which as mentioned is least well tolerated by humans at frequencies in the 4- to 8-Hz band, transverse whole-body vibration is least well tolerated in the lower band below 2 Hz. This difference is again reflected in the current standards (4, 5) for human exposure to whole-body vibration (see Section 2.3.1 and Table 15.2).

During whole-body vibration at frequencies much above 50 Hz, the mechanical response of the human body can be visualized as that of a continuous viscoelastic medium of mechanical energy propagation, rather than as a system of resiliently linked discrete masses (although resonance of finer structures can still be elicited at high frequencies). As the frequency of excitation rises above 50 Hz and up into the kilohertz range, the propagation of mechanical vibration within the body and its tissues progressively becomes essentially acoustic in nature; that is, at high frequencies, most of the vibrational energy entering the body through whatever surface is propagated through the tissues as compressional waves (29).

2.1.2 Physiology of Whole-Body Vibration

The physiological effects of moderately intense whole-body vibration fall into two broad categories (Reference 2; Reference 3, Chapter 5): (1) vibration elicits frequency-dependent responses directly related to the differential oscillatory motion (especially at or near resonance) of body organs, receptors, or structures; and (2) vibration (especially when the exposure is prolonged or repeated) elicits nonspecific generalized reactions of the kind seen as responses to stress in general, that is, not specifically related to the physical nature of the vibration stimulus. Similar reactions appear in response to loud noise and other environmental stressors. Such reactions to vibration are not markedly frequency-dependent but seem to be related rather to the overall cumulative severity of the vibration exposure (a function of time and intensity). Industrial vibration associated with noise in factories and other workplaces is regarded by Soviet

Table 15.2 International Standard (ISO) Values of "Fatigue-Decreased Proficiency Boundary" for Human Whole-Body Exposure to Vibrational Acceleration in the Range 1 to 80 Hz, as Functions of Frequency and Exposure Time[a]

Frequency, or Center Frequency of Third-Octave Band (Hz)	Boundary Values for z-Axis (seat- or foot-to-head) Vibration							
	Root-Mean-Square Acceleration (m/sec^2)							
	Notional Exposure Times on an Occasional or Daily Basis							
	24 hr	16 hr	8 hr	4 hr	2.5 hr	1 hr	25 min	16 min	1 min
1.0	0.224	0.315	0.63	1.06	1.40	2.36	3.55	4.25	5.60
1.25	0.200	0.280	0.56	0.95	1.26	2.12	3.15	3.75	5.00
1.6	0.180	0.250	0.50	0.85	1.12	1.90	2.80	3.35	4.50
2.0	0.160	0.224	0.45	0.75	1.00	1.70	2.50	3.00	4.00
2.5	0.140	0.200	0.40	0.67	0.90	1.50	2.24	2.65	3.55
3.15	0.125	0.180	0.355	0.60	0.80	1.32	2.00	2.35	3.15
4.0	0.112	0.160	0.315	0.53	0.71	1.18	1.80	2.12	2.80
5.0	0.112	0.160	0.315	0.53	0.71	1.18	1.80	2.12	2.80
6.3	0.112	0.100	0.315	0.53	0.71	1.18	1.80	2.12	2.80
8.0	0.112	0.160	0.315	0.53	0.71	1.18	1.80	2.12	2.80
10.0	0.140	0.200	0.40	0.67	0.90	1.50	2.24	2.65	3.55
12.5	0.180	0.250	0.50	0.85	1.12	1.90	2.80	3.35	4.50
16.0	0.224	0.315	0.63	1.06	1.40	2.36	3.55	4.25	5.60
20.0	0.280	0.400	0.80	1.32	1.80	3.00	4.50	5.30	7.10
25.0	0.355	0.500	1.0	1.70	2.24	3.75	5.60	6.70	9.00
31.5	0.450	0.630	1.25	2.12	2.80	4.75	7.10	8.50	11.2
40.0	0.560	0.800	1.60	2.65	3.55	6.00	9.00	10.6	14.0
50.0	0.710	1.000	2.0	3.35	4.50	7.50	11.2	13.2	18.0
63.0	0.900	1.250	2.5	4.25	5.60	9.50	14.0	17.0	22.4
80.0	1.120	1.600	3.15	5.30	7.10	11.8	18.0	21.2	28.0

occupational hygienists as an important centrally acting composite stress in the working environment, which may adversely affect the health of the worker through the mediation of the autonomic nervous system and neuroendocrinologic mechanisms (30–32). This view, however, is not apparently widely accepted elsewhere.

Low-frequency whole-body vibration of moderate intensity (in the region of 2 to 20 Hz at intensities of 0.1 to 0.5 g_{rms}) elicits a general cardiopulmonary response resembling the vegetative manifestations of moderate exercise or alarm, with variable increases in heart beat and respiration rates, cardiac output, pulmonary ventilation, and oxygen uptake (Reference 2; Reference, Chapter 5; Reference 33). Blood pressure may also show slight to moderate elevation, but the response is apt to be unpredictable. Such changes may be attributed to raised metabolic activity associated with increased activity in the skeletal musculature provoked by the vibration. According to Liedtke and Schmidt (34),

Table 15.2 (*Continued*)

Frequency, or Center Frequency of Third-Octave Band (Hz)	Boundary Values for x- or y-Axis (transverse) Vibration							
	Root-Mean-Square Acceleration [m/sec^2]							
	Notional Exposure Times on an Occasional or Daily Basis							
	24 hr	16 hr	8 hr	4 hr	2.5 hr	1 hr	25 min	16 min	1 min
1.0	0.100	0.150	0.224	0.355	0.50	0.85	1.25	1.50	2.0
1.25	0.100	0.150	0.224	0.355	0.50	0.85	1.25	1.50	2.0
1.6	0.100	0.150	0.224	0.355	0.50	0.85	1.25	1.50	2.0
2.0	0.100	0.150	0.224	0.355	0.50	0.85	1.25	1.50	2.0
2.5	0.125	0.190	0.280	0.450	0.63	1.06	1.6	1.9	2.5
3.15	0.160	0.236	0.355	0.560	0.8	1.32	2.0	2.36	3.15
4.0	0.200	0.300	0.450	0.710	1.0	1.70	2.5	3.0	4.0
5.0	0.250	0.375	0.560	0.900	1.25	2.12	3.15	3.75	5.0
6.3	0.315	0.475	0.710	1.12	1.6	2.65	4.0	4.75	6.3
8.0	0.40	0.60	0.900	1.40	2.0	3.35	5.0	6.0	8.0
10.0	0.50	0.75	1.12	1.80	2.5	4.25	6.3	7.5	10
12.5	0.63	0.95	1.40	2.24	3.15	5.30	8.0	9.5	12.5
16.0	0.80	1.18	1.80	2.80	4.0	6.70	10	11.8	16
20.0	1.00	1.50	2.24	3.55	5.0	8.5	12.5	15	20
25.0	1.25	1.90	2.80	4.50	6.3	10.6	16	19	25
31.5	1.60	2.36	3.55	5.60	8.0	13.2	20	23.6	31.5
40.0	2.00	3.00	4.50	7.10	10.0	17.0	25	30	40
50.0	2.50	3.75	5.60	9.00	12.5	21.2	31.5	37.5	50
63.0	3.15	4.75	7.10	11.2	16.0	26.5	40	45.7	63
80.0	4.00	6.00	9.00	14.0	20	33.5	50	60	80

[a] To obtain the ISO "exposure limit," these values of acceleration are multiplied by a factor of 2. To obtain the "reduced-comfort boundary" they are correspondingly divided by a factor of 3.15 (4).

low-frequency whole-body vibration of the dog can induce a vasodilator response in the forelimb, apparently mediated by the vibratory excitation of muscle and tendon stretch receptors. (But note the peripheral vasoconstrictive action of chronic exposure to intense hand-transmitted vibration in humans, discussed in Section 3).

Ernsting and Guignard (35) showed that in certain conditions strong z-axis (vertical) whole-body vibration of a seated person in the band 2 to 10 Hz can induce hyperventilation, sometimes accompanied by symptoms and signs of hypocapnia. The hyperventilation does not appear to be explainable by oscillatory forced ventilation of the lungs as might be supposed but is probably due to the widespread vibratory stimulation of somatic mechanoreceptors, including those in the lungs and respiratory structures (35–37).

Various changes in the cellular and biochemical constituents of urine and blood have been reported in animals and humans in response to moderate or severe whole-body vibration. Generally speaking, these changes appear to reflect a nonspecific response to mechanical vibration as a stressor. Morphological changes in endocrine glands observed experimentally in animals subjected to prolonged vibration (38) likewise appear to reflect a generalized response to stress and are of debatable significance in relation to human morbidity associated with occupational vibration exposure.

Acute vibratory trauma, including contusion and abrasion of internal organs and tissues, with physiological and pathological consequences, can occur. Severe experimental vibration of animals has been shown to cause hemorrhagic damage, sometimes leading to degenerative changes in chronic preparations, in various glandular and other internal organs and systems (39–41). Brief, intense (≤ 3 g peak acceleration) whole-body z-axis vibration of animals, including the dog and the pig (approximately simulating humans as vibration receivers), in the frequency band (3 to 10 Hz) associated with major body resonance phenomena can be shown to produce gross mechanical interference with the hemodynamics of central (aortic) and regional arterial blood flow (42–44). Edwards et al. (42) have postulated that by such mechanisms severe whole-body vibration in rough-riding vehicles could disturb cerebral blood flow in drivers, leading to decrements in performance: such a consequence has yet to be demonstrated in humans, however.

2.1.3 Sensory and Neuromuscular Effects of Vibration

Mechanical vibration is perceived by humans over a much broader range of frequency than that (approximately 16 to 20,000 Hz) spanned by the sensation of hearing (3, 10). Sensations of oscillatory motion and of vibration, unlike the sensation of airborne sound, are not mediated solely by one specialized kind of receptor organ but by several sensory modalities subserved by a variety of receptors distributed throughout the body. The chief vibration-detecting organs in humans are the small but numerous mechanoreceptors arrayed in the skin and in deeper tissues, particularly the muscles, tendons, and periarticular structures and in the viscera and their attachments. Collectively, these receptors subserve the vibrotactile sense, at frequencies ranging from a few hertz to several kilohertz.

Sensations of oscillatory motion at very low frequencies below about 10 Hz are augmented by stimulation of the vestibular (and specifically, the otolithic) receptors. Vestibular stimulation, which may be enhanced by visual cues, becomes paramount at frequencies of oscillatory motion below approximately 2 Hz. The several kinds of diffusely distributed somatic mechanoreceptor, by contrast, respond mainly to vibration at higher frequencies (above about 40 Hz, in the case of Pacinian corpuscles), in various overlapping frequency bands. These receptors and their afferents differ in both the effective bandwidth of their unit responses and the degree of temporal integration of the oscillatory force and motion information that they transmit to the central nervous system (CNS).

The threshold of sensation of whole-body vibration varies systematically with frequency, and the strength of the vibratory sensation as a function of physical intensity of vibration reaching the body at any given frequency appears to vary in a manner somewhat akin to the power law governing the strength of sensation (loudness) of sound (23, 45–50).

Transient disequilibrium and enhanced postural sway and manual tremor have been reported following exposure to whole-body vibration of moderate intensity, both in everyday experience (e.g., lengthy drives over bumpy roads; flights through rough air; time spent at sea) and in laboratory studies (Reference 3, Chapter 5; Reference 51). Workers exposed daily to vibration in factories or in operating mobile equipment such as large cranes have also been reported to experience such symptoms or signs (52, 53). Johnston (54) has observed that whole-body vibration above 2 Hz can adversely affect the speed of orientation. It can be reasoned that such effects in humans could threaten safety at work.

The physiological basis of disturbances of equilibrium and the regulation of posture caused by whole-body vibration is as yet ill-defined. Possibly the effects are due to vibratory overstimulation of the receptors, particularly in muscle, and to competition in the neural pathways that subserve both the regulation of posture and the low-frequency somatic and vestibular vibration senses. However, similar responses (particularly, enhanced manual tremor) are observed during or following conditions other than vibration exposure and thus cannot be regarded as specific to motion or vibration exposure. Enhanced postural sway and tremor are seen in states of high arousal and in fatigue associated with sustained demanding work load and environmental stress not accompanied by motion stimuli (55).

Mechanical vibration of the whole body or of individual postural muscles or their tendons increases tonicity, whereas phasic spinal reflexes (e.g., tendon jerks) sometimes appear to be depressed or inhibited. These phenomena, observable over a wide range of frequency from below 10 to over 200 Hz, have been recorded in humans as well as in animals (including decerebrate preparations). Tendon vibration in humans can interfere with the position sense in the limb, resulting in feelings of weakness, displacement, incoordination, or other distortions of that sense in the arm or leg: this has implications concerning safety at work in vibrating environments (56, 57).

Vibration of limb muscle tendons at relatively high frequencies (100 to 200 Hz) elicits or enhances a tonic stretch reflex (a spinal reflex, as has been demonstrated in the decerebrate cat) (58, 59). The tonic vibration reflex is mediated by receptors in muscle itself, chiefly (but not necessarily solely) the primary spindle endings (58, 60). In humans, the tonic vibration reflex can be shown to gain in strength as the initial length of the muscle is increased (57, 61). (Vibration can be used therapeutically to facilitate volitional contraction in cases of spasticity.) The tonic vibration reflex apparently can be modified by a polysynaptic pathway involving higher centers including the cerebellum; accordingly, it can be influenced by various factors operating supraspinally. Moreover, a degree of voluntary inhibition can be achieved (62). Low-frequency vibration of postural muscles in humans does not appear to alter the reflex excitability of the muscle, nor the character or strength of the maximal volitional response (63).

Varied findings have been reported concerning the effects of vibration on the CNS. Investigators in the Soviet Union in particular maintain that chronic exposure to whole-body vibration causes generalized debilitating effects mediated by the CNS. Melkumova and Russkikh (32), in experimental studies in the dog, cat, and rabbit have reported a syndrome that includes debilitation, lack of coordination, and weight loss. Postmortem examination of the animals following several months of intermittent exposure to 50-Hz vibration was said to have shown cerebral edema and dystrophic changes in nerve fibers. Luk'yanova and Kazanskaya (64) have reported that vibration stress in experimental animals may be associated with fluctuations in the oxygen uptake of cerebral tissue. The lethality of intense (10 g at 25 Hz) whole-body vibration in mice is enhanced by centrally acting stimulants (dexoamphetamine) and reduced by central depressants (chlordiazepoxide, reserpine, barbiturates) (65).

Qualitative observations suggest that vibration can alter the level of arousal in more than one way (as can noise), depending on the physical characteristics of the stimulus and the nature of the subject's activity at the time of exposure. Low-frequency (1 to 2 Hz), whole-body oscillations at moderate intensities (as in swings and rocking chairs) can be relaxing and soporific in humans, and there is a well-established soporific effect of low-frequency ship motion (sopite syndrome). However, higher frequencies, higher intensities, and inconstancy of the stimulus have an arousing effect. A considerable degree of adaptation or habituation to steady-state vibration (e.g., the drone of piston-engined aircraft or the rumble of shipboard noise and vibration) can be achieved, provided the stimulus is regular (7) and is not changed or interrupted. Habituation can be so complete that one is startled and alarmed if the vibration unexpectedly ceases, as when a ship stops at sea. Habituation to low-frequency vibration is probably a central phenomenon, although some adaptation may occur at the receptor level. The habituated person can rest and apparently sleep adequately, if not well, during quite strong vibration and noise in aircraft, ships, and vehicles, once accustomed to the particular motion.

The electroencephalogram (EEG) can be recorded during whole-body vibration at moderate acceleration levels, provided that care is taken in the placement of electrodes and other measures to minimize recording artifacts and electrical noise. However, no specific EEG changes are known to occur in humans during vibration exposure. It can be difficult to distinguish synchronous activity of neuronal origin from mechanoelectrical motion artifacts when the EEG (or other electrophysiological phenomena, particularly when using surface electrodes) is recorded from vibrated subjects (66).

2.1.4 Effects of Vibration on Performance

Intense vibration and low-frequency whole-body oscillatory motion of humans can degrade the performance of tasks or render tasks more difficult to perfect or perform satisfactorily, by both central and peripheral mechanisms (Reference 2; Reference 3, Chapter 6; References 22, 23). Disruption of task performance by motion and vibration is stressful, fatiguing, and sometimes hazardous. Peripherally, mechanical vibration disrupts or interferes with one's application to a task by degrading both the sharpness of vision and the precision of manipulation of tools or controls. Such mechanical effects are immediate and are generally strongly frequency dependent, related to resonance phenomena in the vibrated body (Reference 3, Chapter 6). In tasks of some kinds, requiring precise eye–hand coordination, skill (accuracy of performance) can be maintained during moderate vibration at the expense of speed of performance (67).

It is known from flight experience (3), as well as from various laboratory studies, that heavy and prolonged vibration may also degrade performance centrally, acting in a distracting, thought-disrupting, and fatiguing agent (as does loud noise), but this mechanism is not as easy to demonstrate experimentally (68). The effect is related more to the cumulative severity of the vibration exposure, a function of intensity and duration, and to the demands of the task, rather than simply to the frequency of vibration. Military experience (69) has been extended by Caiger (70) and others to civil air transport flying to show that heavy air turbulence encounters and associated airframe vibrations cause distraction, disruption of instrument-reading and control movements, impairment of quick thinking, and sometimes spatial disorientation. Pilots and navigators find that, in the main, turbulence-induced oscillation below about 2 Hz produces discomfort and progressive fatigue; increased effort by the pilot to avoid or correct inadvertent control movements; difficulty in rapid use of navigational instruments so that fewer "fixes" can be obtained in a given time during rough-air flying; and difficulty in the prompt mental interpretation of information presented by the flight instruments. (To some extent, these problems are mitigated by the increasing use of automation and computer assistance in the modern airline and military flyer's task.) Aboard ships and floating offshore structures such as oil-drilling rigs, wave-induced motion below 2 Hz can be a constant and serious problem, causing disruption of task performance and shipboard activity (Figure 15.1a) by mechanical disturbance, fatigue, and seasickness. These effects are cumulative, reducing the rapidity and efficiency of seaborne operations, increasing the likelihood of accidents, and often causing further delay or temporary cessation of ship function. It is, of course, difficult to gather other than anecdotal data about such effects in flight, at sea, or in other field situations, but the problem can be studied by using motion simulators (71, 72) and laboratory vibration machines (3, 21–23).

Higher frequency (2 to 10 Hz) structural vibrations, which can be particularly troublesome in helicopters, in surface vehicles, aboard ships, in mobile heavy

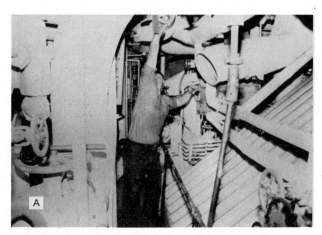

Figure 15.1 (a) A combination of ship motion and deficient human factors engineering can make a simple task very difficult, time-consuming, and hazardous. This man is forced to stand and stretch up (using whatever handhold is available to brace himself), a tricky task in heavy seas, to turn an important badly placed valve. Photograph courtesy of United States Navy. (b) In driving a tractor, whole-body jolting and vibration (mainly vertical) is transmitted to the driver principally through the seat but also through the footrests, steering wheel, and controls. Drivers spend much time twisting around in their seats to watch the rig being drawn. Note that this driver is also smoking at work. Author's photograph.

equipment, and around fixed plants in factories, are associated with difficulty in reading instruments and printed matter; interference with finer manipulative skills, such as setting cursors on navigational aids, making adjustments to electronic equipment, and writing; and also general discomfort, irritability, and fatigue that progressively worsen during prolonged missions or spells of work in vibration. Vibration in the same range can also affect communication (a potential threat to safety in some circumstances) by altering the pattern and degrading the clarity of speech when the speaker is being vibrated (73, 74).

Air pressure oscillations at frequencies too low to stimulate the human organ of hearing in the normal manner to produce the sensation of sound (i.e., below about 16 Hz) are called *infrasound*. Such waves, impinging on the human body surface at sufficient intensity, can set the body into vibration and provoke generalized reactions akin to those caused by whole-body vibration of equivalent intensity and frequency (Reference 3, Chapters 11, 14; References 75, 76). High levels (> 110 dB SPL) of infrasound in the range 1 to 20 Hz can be generated in diesel-engined ships, large trucks and other powerful vehicles, high-performance military aircraft (as a result of turbulent airflow around the cockpit canopy), and fixed industrial plants where heavy machinery is running (77). Infrasound can be a nuisance in any large space in which the air can be set into low-frequency oscillation by cavity or panel vibrations (e.g., in the structures of ships or factory buildings) excited by the action of machinery or turbulent airflow. The latter can be a problem in poorly designed large-duct air-conditioning systems.

Figure 15.1 (*Continued*)

In most circumstances it remains a moot point (for lack of any major research effort in this area) whether infrasound can pose a serious threat to health or safety. However, it can undoubtedly produce unpleasant subjective effects, including body vibrations; "fluttering", gagging, or choking sensations in the chest, abdomen, or throat; occasional feelings of postural unsteadiness and feelings of physical displacement; irritation; and fatigue. Recent animal studies, using genetically deaf subjects (78), indicate that high levels of infrasound can produce effects in the CNS mediated by nonauditory receptors and pathways.

Tentative limits for short-term ($\leq$ 8 min) human exposure to intense infrasound in the band 1 to 20 Hz have been published by von Gierke et al.

(21), based on admittedly meager experimental data. The proposed limiting sound pressure levels (SPL) are as follow: 150 dB from 1 to 7 Hz; 145 dB from 8 to 11 Hz; and 140 dB from 12 to 20 Hz. A comprehensive review of the effects of infrasound on human beings, including other exposure criteria, has recently been published by Backteman, Köhler and Sjöberg (205) in a two-part tutorial paper received by the author while this chapter was in proof.

2.2 The Pathology of Whole-Body Vibration

Intense whole-body vibration can cause pain and injury. Acute traumatic effects, highly dependent on the frequency, intensity, and direction of application of the vibration to the human body, are most likely to occur when severe vibration is applied to the unprotected person at frequencies related to the organ and system resonances of humans. Impact forces (e.g., in vehicle collisions, aircraft crashes, or violent encounters with atmospheric turbulence, and air-, sea-, or ground-borne shocks due to explosions) are extreme cases of transient vibratory forces injurious to humans. Also, prolonged and repeated exposure to less severe and not immediately or obviously damaging levels of vibration in the course of one's work or occupation may nevertheless lead to chronic ailments or injury.

2.2.1 Severe Acute Whole-Body Vibration Exposure

Animal studies of the lethality of intense whole-body vibration (peak acceleration amplitudes up to 20 g at frequencies below 50 Hz) have shown that severe shaking can cause hemorrhagic injury to the viscera and other parts of the living body. Roman (79) demonstrated in the mouse that the lethality of severe whole-body vibration is strongly frequency dependent, being greatest at frequencies of major internal organ resonance. Pathological changes usually seen in animals killed by vibration are hemorrhagic damage to the lung and to the myocardium and bleeding into the gastrointestinal tract; less commonly, there is superficial (subcapsular) bruising of kidney and brain.

As distinct from impact exposure, whole-body vibration of humans at levels sufficient to cause acute internal injury is probably negligibly rare in industrial and even in military experience. However, the present author has heard anecdotal reports of severe spinal injuries resulting from intense vibrations or repetitive impacts sustained recreationally in riders of power boats speeding through choppy seas and snowmobiles raced over rough, snow-covered terrain (80). Impact injury from being thrown or toppled against sharp or hard structures aboard ships in heavy seas or aircraft encountering severe turbulence are a much more familiar experience and an ever-present hazard to those who work or travel at sea or in the air. Here the vibration or oscillatory motion is in a sense the indirect cause of injury, which is consequent on the person being mechanically displaced or losing control of posture or locomotion in the moving environment. It is of interest in this connection that case histories and general observations have been presented by Pichard (81) and Snook (82) to show that poor riding characteristics in ambulances driven at speed over bumpy surfaces can seriously aggravate the injuries or medical disorders already suffered by patients who are on their way to the emergency room and can also make it difficult and hazardous for the attendants to minister to the patient during the ride.

2.2.2 Chronic Occupational Exposure to Whole-Body Vibration

Certain afflictions of the back and internal systems have long been ascribed to chronic occupational exposure to the rough motion of working vehicles such as farm tractors (24, 83), haulage trucks (84), and mobile heavy equipment (13, 85, 86). Rough-riding land vehicles are not the only problem, however. In recent years the heavy vibration experienced in helicopter flying has been recognized as potentially harmful to the spine in professional military and civilian helicopter pilots (87, 88). The etiology of the disorders reported, and the causal link between the mechanical vibration exposure and the supposedly vibration-related disease or disability, remain obscure and have proved difficult to establish either epidemiologically or experimentally.

Certain factors apart from the vibration exposure (e.g., climatically adverse working conditions; poor ergonomic factors in the design, construction, or method of operating a vehicle; fatigue and nonspecific work load stress) may figure prominently, if not predominantly, in the etiology. The extent to which chronic exposure to moderate levels of whole-body vibration, repeated daily over many years, can injure the otherwise healthy body (as opposed to merely aggravating preexisting weaknesses) in the absence of compounding factors remains an open question.

The constant vibration and jolting experienced on agricultural tractors has for many years been blamed by farm workers and physicians for an undue prevalence of gastrointestinal, anorectal, and spinal ailments, especially peptic ulceration and low-back pain. From a systematic study of 371 farm tractor drivers, Rosegger and Rosegger (83) concluded that in many cases the cumulative microtraumatic effects of the vibration and jolting were exacerbated by coincidental factors. These include faulty posture, especially twisting around in the seat to watch the tow (Figure 15.1b); faulty seating, with regard to both seat design and seat maintenance; working overly long and irregular hours during busy seasons and inclement weather, with inadequate rest breaks and mealtimes; and certain secondary factors such as alcohol and nicotine consumption.

Similar uncertainty regarding the etiology of vibration-related occupational disorders exists in kindred industries, such as long-distance bus and truck driving and the operation of mobile heavy equipment. In an analysis of the periodic physical examination records of 1448 interstate bus drivers, Gruber and Ziperman (89) found that a number of chronic disorders occurred more frequently among drivers with more than 15 years of experience than in control populations. Some of the disorders (particularly those affecting the gastrointestinal, respiratory, venous, and muscular systems and the lumbar spine) might have been caused by the drivers' chronic exposure to whole-body vibration and jolting on the job; however, it was not possible epidemiologically to distinguish vibration as a principal etiological factor from several others also present in the working situation. Such coincidental factors included poor seating and driving posture in a cramped cab; questionable dietary habits; long and sometimes irregular hours of work; the general job stress of dealing with the public and being responsible for a busload of passengers in traffic; and possibly synergistic combinations of other environmental stressors such as noise and fumes.

As in the instance of the bus drivers just cited, some ailments appear to be more prevalent in long-distance truck drivers than in control populations subject to comparable job stress (84). Nevertheless, although some of the chronic afflictions reported (low-back pain, spinal deformities, gastrointestinal troubles, hemorrhoids) could be ascribed to the effects of prolonged and repeated exposure to whole-body vibration, other etiologic factors cannot be ruled out as possible contributory or even primary causes. These again include poor driving posture and deficient seat design or maintenance; driving for excessive hours in cramped and uncomfortable conditions; heavy cargo handling (faulty lifting and load-handling techniques can cause or aggravate back ailments, regardless of vibration exposure); rushed meals at odd hours; and the composite job stress of high anxiety and heavy physical demands in managing a valuable vehicle and its payload in traffic and against the clock.

Seris, Auffret, and their colleagues in France, examining the vibration factor in aviation, consider low-back pain to be a relatively common and serious malady of professional helicopter pilots flying prolonged and repeated sorties (87, 88). The risk is attributable to a combination of adverse factors capable of amelioration: these include not only the frequent vertical gust loads (due to low-altitude air turbulence) and heavy low-frequency vibration typical of helicopter flying (Reference 3, Chapter 2) but also such familiar ergonomic factors as poor seating, awkward postures required by some in-flight activities, and the general stress of the occupation, which is in many respects akin to that of the land vehicle operators mentioned previously. A recent paper by Shanahan and Reading (118) has questioned the role of vibration in causing back problems in aviators flying army helicopters and suggested that, for lack of significant evidence that vibration in the machines in question played a causal role, the circumstantial factors in themselves were to blame for the pilots' symptoms.

Epidemiological studies undertaken to identify unusual patterns or incidence of morbidity in the operators of rough-riding vehicles are likely to prove inconclusive for several reasons (84–86, 89). Spear et al. (86), who studied insurance claims for medical services in various categories of disability afflicting heavy equipment operators, have drawn attention to a tendency for the statistics to show an initial increase in the risk of certain afflictions attributed with varying degrees of confidence to rough ride, followed by a decrease with the operators' experience (equated with number of years in the occupation). Spear and his co-workers pointed out that this tendency suggests a selection process by which workers in the study groups exposed to rough riding tend to quit the occupation as they become afflicted with the ailments in question.

Spear et al. (85) made a follow-up study to earlier work on morbidity patterns among heavy equipment operators to test whether self-selection out of jobs involving severe whole-body vibration exposure influenced the incidence of medical service insurance claims for general occupationally related diseases (notably ischemic heart disease and musculoskeletal ailments). They observed that onset of some ailments or conditions may be hastened by vibration but that the overall incidence of these ailments does not differ significantly between vibration-exposed and nonexposed workers in the construction industry. A very experienced and observant heavy equipment operator of the author's acquaintance, who before retirement had become an expert much sought after to repair and test mobile heavy equipment and thus (unlike many operators who restrict themselves to one kind of equipment) had occasion to drive a wide variety of machines, maintained that, while the general operating conditions remained the same, the motion and vibration of different machines had very different effects on his chronic low-back pain; his back would "tell" him at the end of the day which were the "bad" machines. He conceded, however, that the machines he tested differed not only in the amount of vibration and jolting but also in the way in which he climbed aboard and drove them.

In the Soviet Union and other countries in eastern Europe, many investigators have taken the view that continuous whole-body vibration in factories, in various kinds of mobile working equipment such as cranes, concrete compactors, and other machines used in the construction industry, and in ships' engine rooms (90), can produce debilitating stress and malaise mediated centrally by neuroendocrinological mechanisms. Particular attention has been focused on more or less steady-state vibration of the type generated by machinery, mostly at somewhat higher frequencies (above about 10 Hz) than the low-frequency jolting and shaking experienced in rough-riding vehicles, and acting in concert with industrial noise and general job stress (31). Such occupational exposure is apparently recognized in the Soviet Union as the cause of various "vegetative–vascular," polyneuritic, asthenic, and vestibular syndromes having a detrimental effect on working capacity and, in many instances, resulting in a change of job or compensation for reduced earning capacity (91).

Fatigue, irritability, and the impairment of dexterity and postural function have been observed not only in vehicle drivers but in workers in other occupations in which, in addition to whole-body vibration, there is a high job stress due to the degree of manual control skill and responsibility that must be exercised for prolonged periods, such as in the operation of large tower cranes used in construction work (52). Stabilometric measurements at the end of the work shift in dump truck drivers exposed to severe whole-body vibration and jolting have been reported to show increased postural instability (ascribed to "otolithic excitabilty") in comparison with measurements in control groups of miners not exposed to the motion (53). This tendency has been found by

Nurbaev (53) to worsen with increasing length of service (up to 8 years) in dump truck driving. Chernyuk (92), however, has stated that 15-min rest breaks interposed between half-hour stints of rough-riding vehicle driving substantially reduce the vestibular and CNS excitability engendered by the vibration. Eklund (93) has described mechanisms that may be involved in the causation of neuromuscularly mediated vibration effects on the sense of balance and postural stability.

Although vibration has been reported by Soviet investigators (94) to affect renal function in human sufferers from vibration disease and "kidney trouble" (frequently a layperson's synonym for low-back pain) apocryphally attributed by drivers to long hours of truck driving in the United States, the question of whether chronic whole-body vibration exposure in vehicle driving or industrial activity poses a significant threat to the kidney remains open.

Experimental studies have yielded negative or equivocal results. In protracted experimental exposures of rhesus monkeys to an intensity of low-frequency (12-Hz) vibration sufficient to cause hemorrhagic damage to the gastrointestinal mucosa, Sturges et al. (41) failed to find evidence of renal damage. In human volunteers exposed experimentally to sinusoidal whole-body (z-axis) vibration at moderate levels in the range 2 to 32 Hz for up to 8 hr, Guignard et al. (95) found no significant changes in urinary constituents. The intensity of the vibration in those experiments was, however, limited in accordance with the ISO "fatigue-decreased proficiency boundary" (4), and a variety of other measurements of the subjects' physiological state and task performance made concurrently also proved negative.

Specific attention has been drawn by Soviet investigators to the question of the particular susceptibility of women to whole-body vibration in industry. Parlyuk (96) has claimed that the motions and vibrations to which female workers were exposed in their daily work operating bridge cranes led to specific neurocirculatory and vestibular disorders. The symptomatology included vertigo, unsteadiness of gait, and a heightened susceptibility to motion sickness. Lysina and Parlyuk (97) have reported hemodynamic and cardiographic changes that they attributed to the irregular vibration and jolting experienced on the job in a series of 33 female bridge crane operators, all of whom had spent 10 years or more in the occupation.

However, a causal link between the physical agents in the industrial environment, as opposed to general job stress, and the symptoms reported in such cases has yet to be established. When the symptomatology is of a kind not peculiar to one sex, it cannot be presumed that disorders affecting a group of workers who happen to be women are related to sex differences in susceptibility to vibration stress unless it can be shown that comparable populations of male workers exposed to the same conditions are significantly less affected. Comparative data relating to this question are lacking and, accordingly, current standards (4, 5) on human exposure to whole-body vibration are deemed to apply equally to both sexes.

Gratianskaya and her colleagues (98) have postulated a possible risk of increased incidence of disorders of menstruation and parturition in female workers employed in the vibrocompaction of concrete. In that job the women were exposed to whole-body vibration in the range 40 to 55 Hz while placing concrete mix for settlement by vibrocompacting machines in construction work. Apart from this report, claims that vibration affects menstruation or pregnancy remain mostly anecdotal or apocryphal. Although mechanical vibration was attempted many years ago in Germany apparently with some success as a means of inducing labor (99, cited by Guignard in Reference 2). It is probably prudent for women in advanced stages of pregnancy, particularly when the pregnancy has been threatened by obstetric complications, to avoid unnecessary exposure to whole-body vibration at work and in travel.

2.2.3 Vibration Combined with Other Stressors, Especially Noise

It was mentioned earlier that in the Soviet Union whole-body vibration, particularly machinery vibration at moderately high frequencies of the kind experienced in factories, is generally regarded as part of a composite industrial environment that can adversely affect the worker's physiological state, metabolism, and health through the mediation of central mechanisms: these involve the central and autonomic nervous systems and the endocrines (31, 100). Noise, especially, is an important stressor that is almost always present in factories and other work situations where a worker is simultaneously exposed to continuous mechanical vibration. In addition to its possible enhancement of the auditory hazard associated with industrial noise (101, 102), whole-body vibration may also act additively with noise to cause stress and industrial fatigue and to degrade vigilance and the performance of skilled tasks (3, 103).

Workers exposed daily to industrial noise and vibration can be shown to have high stress indices such as enhanced urinary excretion of vanillyl mandelic acid and 17-ketosteroids, when compared with control groups working in quieter environments (104). Work situations remote from the factory may also be contaminated by undesirable levels of combined vibration and noise. Noise and vibration are constantly present in ships, for example, and can be a serious nuisance, with high intensities occurring at relatively low frequencies, on board very large vessels such as oil tankers with poor structural damping qualities (105). A peculiar environmental problem at sea is that ship noise and vibration are present constantly throughout the voyage, so that the ship's crew are exposed to the stress not only when on duty but off duty also, finding little respite from the pervasive rumble of the ship's machinery, or from ship motion, when they retire to their quarters for recreation or sleep. Gibbons et al. (106) have reported enhanced 17-ketosteroid levels in ships' officers at sea, compared with the levels found following shore leave. Of course, such a sign is not specific to vibration and noise stress but may be indicative of the combined stress of the shipboard environment and the demands of responsible duty at sea. It is of interest regarding the special occupation of space exploration that a growing problem of vibroacoustic habitability in large space vehicles may be faced in the near future as more extended missions (sometimes involving astronauts dwelling and working in space for periods of several months) are undertaken. Large space stations with poor structural and acoustical damping, and long sojourns in the environment, are somewhat akin to ships in this regard.

Russian work has indicated that vibration stress may in some circumstances enhance specific damage to the organism from other physical or chemical agents, including damage from ionizing radiation (107, 108) and heavy-metal toxicity, particularly mercury (109). Such observations, however, have been based largely on animal experiments, and firm data establishing a causal association for humans between vibration and other noxious agents hazardous to occupational health are not yet available.

2.3 Protection of Humans from Effects of Whole-Body Vibration

Four main steps must be taken to protect humans from adverse effects of vibration:

1. Predict or measure the vibration exposure at the work station or area of concern. Ground rules for the proper measurement and evaluation of human exposure to whole-body vibration in the range 1 to 80 Hz have been issued as national and international standards (4, 5).

2. Select and apply an appropriate criterion of vibration control and corresponding guidelines regarding human exposure. The current standards already cited distinguish three main criteria: health and safety (i.e., an occupational criterion); maintenance of performance (implicitly an accident-prevention standard in many working contexts); and comfort (mainly applicable to passenger vehicles and nonindustrial buildings).

3. Determine the type and amount of vibration control (usually reduction) that is required: this is a computational step, based on the information acquired in steps 1 and 2.

4. Select and apply the most efficacious and economical means of control available for protection of the worker.

In step 4, adopting the classical approach of vibration or acoustical engineering, three main points can be distinguished at which to attack vibration disturbing to humans (Reference 3, Chapter 9): (1) at the source of the vibration, (2) in the route of transmission of vibration from source to humans, and (3) at the receiver (i.e., in the person in question).

Reduction or minimization of mechanical vibration at source, particularly when the source is machinery running in a factory or propelling a ship, vehicle, or aircraft, is a matter of engineering design and practice (including proper installation, good preventive maintenance, and correct operation). Even the large-amplitude, low-frequency (< 2 Hz) oscillatory motion of ships, aircraft, and vehicles interacting with the environment or surface on which they move can be optimized by basic engineering design or architecture design (e.g., of ship's hulls, airframes, road and rail vehicle structures, running gear, and riding surfaces). Sometimes it is necessary to reduce motion or vibration at source (to protect either the machine or vessel itself or the human occupants) by operational measures, such as changing course or altitude to avoid air turbulence; altering course or speed to reduce a ship's response to wave encounters; or temporarily restricting operations aboard an offshore oil rig. Such solutions, however, often represent a compromise to the detriment of the craft's mission-effectiveness or profitability.

2.3.1 Standards

International Standard ISO 2631-1978 was issued by the International Organization for Standardization (ISO) in 1978 as an attempt to achieve international consensus regarding the measurement and evaluation of human exposure to whole-body vibration. It superseded a confusing multiplicity of guidelines, many based on meager, biased, or unreliable data, that had been written in many different countries for particular industries or applications (Reference 2; Reference 3, Chapter 8). The ISO document prescribes standard methods of evaluating whole-body vibration disturbing to people at work or in transportation and provides guidelines for limiting human exposure to such vibration in the range 1 to 80 Hz, a band selected by practical considerations and the availability of human data suitable for incorporation into a standard. Several countries, including the United States (5), have adopted national standards incorporating essentially the same guidelines. Supplementary ISO standards have been drafted for particular situations, such as fixed buildings and offshore structures.

In Table 15.2, the "fatigue-decreased proficiency (FDP) boundary" values in the current ISO standard (4) are reproduced (by courtesy of the Secretariat), showing guidelines for limiting human exposure to whole-body vibration in either the anatomic z-axis (head-to-foot direction) or in the x- or y- (transverse) axis. As mentioned earlier, the mechanical properties of the human body, and hence the psychophysiological response to vibration in the range of major body resonance phenomena, differ markedly with direction of vibration. The reader will observe that the standard is most conservative (i.e., the acceleration boundary for any given combination of frequency, exposure-time, and direction is lowest) in the frequency bands of the principal body resonance phenomena. The criterion of application here is prevention of decrements in task performance or disruption of human activity by whole-body vibration. The standard provides a guide to acceptable levels of rms acceleration at the point of input to humans (typically the seat or the floor or deck on which they stand) as a function of vibration frequency and exposure time on a daily basis. The text of the standard, which must be consulted in order to interpret or apply these figures, contains instructions for weighting the values when other criteria apply; when the spectrum or level of vibration varies for different times during the daily exposure; or when a measured spectrum contains nonsinusoidal (i.e., multifrequency or distributed) vibration. It has been demonstrated that this standard is protective (perhaps overly so) of human performance of various tasks during z-axis sinusoidal vibration in the range 2 to 32 Hz for exposures of up to 8 hours (95, 110). Current work of the drafting ISO subcommittee (ISO/TC 108/ SC4, *Human Exposure to Mechanical Vibration and Shock*), which is constituted by delegations from some 20 interested countries, including the United States, is directed toward revision and amplification of the standard and extending it to frequencies below 1 Hz.

2.3.2 Controlling Vibration Between the Source and the Individual

The application of guidelines such as ISO 2631-1978 allows for calculation of the type and amount of vibration control for humans. Several engineering solutions can then be sought to minimize the human recipient's vibration exposure (Reference 3, Chapter 9). Approaches include:

1. Prevention or attenuation of the vibration of structures.
2. Isolation of sources of vibration.
3. Location of critical work stations (e.g., on a ship's bridge) away from positions, such as ship structural antinodes, of high vibration (117).
4. Location and orientation of the recipient with respect to pathways of transmission.
5. Isolation of the recipient from vibration.
6. Prevention of flanking transmission.

A detailed discussion of these engineering approaches is outside the scope of this chapter. It should be noted, however, that a knowledge of the frequency dependence and other aspects of the human response to vibration, especially in the biodynamically important range 0.1 to 30 Hz, is most important to the design of motion and vibration alleviating systems and devices to protect humans from seat-, deck-, floor-, or structure-borne oscillatory motion.

Many types of vehicle and some fixed-base machines generate whole-body vibration in the range above 1 Hz. Current standards for limiting human whole-body vibration in that range are frequently exceeded substantially (11, 13). Suspension or isolation systems, based on the interposition of springing or some resilient and damping elements between the source of vibration and the receiver, can materially reduce exposure. In addition to the familiar suspension systems of land and rail vehicles, special spring-mounted seats can be fitted in rough-riding vehicles and aircraft. The main principle of design of such devices is that a resilient suspension attenuates vibration of the suspended mass of the rider or vehicle body at frequencies above the resonant frequency of the system. Most vehicle and seat suspensions in everyday use are of the passive kind; that is, they make use of the inherent mechanical properties (elasticity and damping) of springs and resilient materials (115, 116). For a few special applications (e.g.,

VIBRATION

in rough-riding working vehicles and high-performance military aircraft), where riders are exposed to severe large-amplitude vibrations in the range 1 to 10 Hz, it is feasible to use "active" systems in which a sensor quickly reacts to the impressed forces reaching the vehicle or seat and generates a signal that is used to drive an actuator to oppose the motion (112–114). Systems operating on related principles are used to reduce the nauseogenic and mechanically disturbing low-frequency rolling of ships at sea.

2.3.3 Minimization of Adverse Effects of Vibration Reaching Humans

When an irreducible amount of vibration or disturbing oscillatory motion still gets through to an individual in the vehicle seat or workplace, a number of possibilities remain for mitigating the effects in the human recipient. These include, when practicable (Reference 3, Chapter 9):

1. Minimizing the necessary duration of exposure to the vibration on an absolute or daily basis.
2. Allowing adequate rest and recovery periods between periods of work or duty in the motion environment.
3. Ergonomic design and location of displays, controls, and workspaces for best use during irreducible vibration or motion.
4. Training and experience of workers necessarily exposed to severe motion or vibration. In the case of aircraft and ship motion, dynamic simulators can be used for this purpose.
5. Habituation (especially to nauseogenic motion in the band 0.1 to 1 Hz) and adaptation.
6. Physical fitness and a normal physiological state in the worker (88).
7. Avoidance or reduction of other physical agents that might potentiate susceptibility to motion or vibration effects.
8. Selection procedures and exclusion of persons with unusual susceptibility to or who are medically unfit for motion or vibration exposure.
9. Engineering design or treatment to minimize excessive vibration of working clothing, headgear, or worker-mounted equipment.
10. Biomechanical devices to restrain or suppress body oscillation or excessive head movement.
11. Medication (for motion sickness only: none is known to alleviate other effects of vibration).

Several of these items again entail human factors or general engineering solutions, elaboration of which is outside the scope of this review, but some of them fall within the purview of the industrial or ship medical officer, or the flight surgeon, who may well be in a position to recommend eps to reduce human vibration exposure or alleviate its effects. Regarding item 11 in the preceding list, except for specific remedies for motion sickness, no drugs or medications are known to increase human tolerance of whole-body vibration, and accordingly none can be recommended.

3 HARMFUL EFFECTS OF HAND-TRANSMITTED VIBRATION: VIBRATION SYNDROME

It is widely recognized that the intense vibratory energy generated by handheld electric, pneumatic, and gasoline-engined power tools can, when transmitted to the fingers, hands, and arms of the operator by the tool or workpiece, produce vasospastic, neuromuscular, and arthritic disorders of the hand and upper limb. These effects, particularly the vasospastic manifestations, are collectively referred to as "vibration syndrome" and are associated with the use of handheld or manually supported or guided machines that produce intense vibration mainly (although not exclusively) in the frequency range 10 to 1000

Figure 15.2 Rock-drilling postures. (a) Downward drilling, in which heavy vibration is transmitted through the hands to the upper limb girdle and upper torso. This is also a typical posture in road breaking and ground breaking for construction work. (The cables leading from the handle of this machine carry signals from instrumentation used to measure vibration levels on the job.) (b) Rock drilling underground in subway construction. Note left arm supporting the weight of the tool while it is guided by the right hand. Photographs courtesy of Dr. I.-M. Lidström, Stockholm.

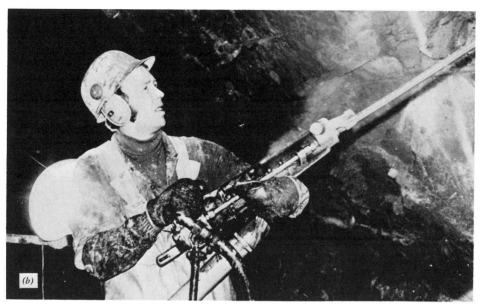

Figure 15.2 (*Continued*)

Hz. (The machines also create a noise hazard for the operator and fellow workers in the vicinity.)

The effects of such disorders on the worker may range from transient, mild, and occasional discomfort or inconvenience to severe social and occupational disability. The problem is of particular importance in certain industries in which large numbers of workers are at risk, using power tools on a daily basis, often in adverse environmental conditions. Industries notable in this respect include forestry and lumberjacking, shipbuilding, construction work, mining and quarrying, foundrywork, vehicle building, and many others.

Vibrating power tools are used so universally in industry and in such a variety of occupations (Figures 15.2 to 15.5) that it is very difficult to estimate the numbers of workers at risk—believed to be more than 1 million in North America alone (122)—and the problem is now regarded as serious by clinical and other authorities familiar with it in many industrialized countries around the world. It has received special research and administrative attention in countries (notably Britain, Canada, Czechoslovakia, Japan, Sweden, and the United States) where substantial numbers of workers use power tools outdoors (e.g., in shipbuilding, construction, quarrying, and forestry) in inclement weather, or in other workplaces, such as some mines, where cold and damp conditions prevail. The urgency of the problem has been highlighted by the recent publication of several detailed treatises (30, 119–125, 200).

However, notwithstanding the conclusions of numerous scientific authors, medical authorities, and administrators in industry and government in several

Figure 15.3 Pneumatically driven tools such as this casting scaler (also pneumatic chippers, riveters, etc.) are powerful sources of hand-transmitted vibration of relatively low frequency. Several factors (e.g., method of handling, proximity of hand or fingers to working point, inclusion of padding) determine the severity of vibration exposure, which frequently differs substantially between the two hands. Photograph courtesy of Dr. W. Taylor, Dundee.

countries regarding the seriousness of the problem, considerable reluctance to recognize vibration syndrome as a compensatable industrial disease persists within industry and among occupational safety and health legislators. This is partly because there is still a paucity of firm epidemiological and experimental evidence to prove a direct and unequivocal causal relationship between cumulative habitual exposure to hand-transmitted vibration at work and the vibration syndrome. Moreover, the chronic disorders complained of by workers with hand-held vibrating power tools are generally difficult to distinguish diagnostically from similar disorders of idiopathic or other etiological origin that are not associated with occupational exposure to hand-transmitted vibration.

Figure 15.4 Occupational use of the motor-driven chain saw is strongly associated with the risk of vibration syndrome (commonly, Raynaud's phenomenon). Unbalance forces from the power unit are the chief source of vibration in these machines, affecting both hands through the carrying and guiding handles. (The machine illustrated is fitted with a specially instrumented handle for measuring vibration exposure at work.) The risk of vibration disease in chain sawyers can be reduced by the introduction of modern saws incorporating antivibration devices. Photograph courtesy of Dr. I.-M. Lidström, Stockholm.

In the United Kingdom, for instance, since the early 1950s, efforts have been made repeatedly by medical advocates to persuade the Department of Health and Social Security to recognize Raynaud's phenomenon of occupational origin—also known as "vibration white finger" (VWF) or traumatic vasospastic disease—that is, in workers with a history of regularly using handheld power tools, as a compensatable occupational disease. However, prescription as such has not yet been enacted, mainly on the grounds that in most cases the symptoms have been deemed to be relatively trivial (although not necessarily from the patient's viewpoint!) and not necessarily an impediment to the worker's continued employment (124).

In certain industries, however, the condition is far from being trivial, with substantial majorities of workers in some jobs (over 90 percent in some populations of foresters using chain saws, before remedial measures were undertaken) succumbing to the affliction within a few years or months of entering the occupation. The occupations of chain sawing and pedestal grinding

Figure 15.5 Working posture, particularly when it is cramped or awkward, may play an important etiologic role in the incidence of vibration-related disorders. The postures adopted by workers on the job are often necessarily far from ideal. (a) Using a grinding wheel to finish the interior of a welded pipe. Photograph courtesy of Dr. I.-M. Lidstrom, Stockholm. (b) Stooping to work on a casting. Photograph courtesy of the Steel Castings Research and Trade Association (United Kingdom), Sheffield.

have provided particularly notorious examples in recent years (120, 125); and this observation has been repeated in several countries.

3.1 Characteristics of Hand-Transmitted Vibration

Most hand-held and hand-guided power tools generate strong mechanical vibration distributed over a broad frequency spectrum. In many types of tool the vibration transmitted to the hands of the operator through the handles of the appliance or through the workpiece is complexly periodic, with the spectrum showing a number of peaks at frequencies related to the cycle of rotation or reciprocation of the moving parts of the tool. In some tools (e.g., concrete breakers, chippers and scalers, riveting tools) the vibration at source is a rapid series of impacts delivered to the workpiece, but the impacts set up a complex

Figure 15.5 (*Continued*)

series of harmonic vibrations with a basic frequency determined by the operating speed of the machine because of ringing at the multiple resonance frequencies of the machine body and moving parts (and sometimes also of the workpiece when it is metallic). In certain kinds of continuously operating tool, notably the chain saw (126), driven by a gasoline engine, the vibration arises mainly from unbalance forces generated by the engine and transmission and can accordingly be ameliorated by engineering treatments. [Contrary to popular belief, most chain saw vibration, at least in properly used and maintained equipment, arises from the power unit and not from the cutting action of the sawteeth (127): the worst vibration can arise when the saw is allowed to run at high speed in free air.]

In most kinds of hand-held or hand-supported tool, the principal components of the vibration spectrum lie in the band 10 to 1000 Hz, within which range frequencies between about 30 and 300 Hz in particular have been pronounced by several investigators to be those most clearly associated with the development of occupational Raynaud's phenomenon and allied vasospastic disorders in susceptible workers (119, 127–131). The reason for that observation appears to be complex: vibrations in the band 30 to 300 Hz not only contain the maximum vibrational energy output of many vibrating tools but also appear to be preferentially absorbed by the tissues of the hand (132). Moreover, it has been postulated (131) that mechanoreceptors (especially Pacinian corpuscles) in the hand and upper limb, which are particularly responsive to vibration in the same frequency band, are important mediators of the repeated vibratory activation of the sympathetic nervous system, which eventually leads to a permanent alteration in regional vasomotor tone that is the precursor of vibration-related traumatic vasospastic disease.

The ISO (6) is accordingly considering a draft international standard that prescribes methods and guidelines for evaluating hand-transmitted vibration in the band 6.4 to 1000 Hz. Outside that band, very low-frequency vibrations below about 8 Hz (of the type generated by the repetitive action of heavy road breakers, compacting machines, etc.) are transmitted through the hands and up the arms to the shoulder girdle and the whole body (Figure 15.2a), with little of the low-frequency energy absorbed in the hands themselves. Such low-frequency components probably are not significant in the causation of vibration syndrome, but may be associated with arthritic afflictions of the arms and shoulders. At the other extreme of frequency, most vibrating tools used in industry probably do not generate significant amounts of structure-borne vibratory energy at frequencies much above 1000 Hz, and such mechanical energy is heavily attenuated at the interface between the handle of the vibrating tool or the workpiece and the skin of the hand.

It should be borne in mind, however, that the vibration spectra of powered handtools extend well up into the audiofrequency range. The moving parts and casings of such machines, therefore, as well as workpieces (especially castings, plates, and other metal workpieces being scaled, drilled, riveted, etc.) act as radiators of intense acoustical noise. Loud noise also emanates from secondary sources, such as the exhausts of pneumatic and gasoline-engined appliances. Apart from hand-transmitted vibration affecting the handler of the tool, therefore, the user and others in the vicinity are subjected to the additional hazards of noise-induced hearing loss (133) and the prejudicial effects on safety at work of the masking of speech and other significant sounds by noise at the workplace.

3.2 Varieties of Vibration Syndrome

Raynaud's phenomenon secondary to an altered vasomotor tone in the hands is the disorder most commonly associated with the occupational use of vibrating power tools and is accordingly considered in some detail in Section 3.3. However, other troublesome and sometimes incapacitating afflictions may be the presenting complaint in power tool workers. These neurovascular, sensory, and musculoskeletal manifestations (including Raynaud's phenomenon of occupational origin), collectively referred to as the *vibration syndrome*, may share a common underlying physical etiology but depend for their variety of clinical manifestations on differing predisposing factors in the nature of the occupation, in the individual, and in the circumstances of the worker's exposure to hand-transmitted vibration (Table 15.3).

The vibration syndrome may include vasomotor disorders not associated with manifest Raynaud's phenomenon, such as erythrocyanotic changes accompanied by pain in the hands (134); neurosensory and neuromuscular disorders, including the loss of tactile sensitivity and manual dexterity (135); pain and stiffness in joints of the fingers and hands; and bone and joint changes, observable radiographically, affecting the wrist and upper limb (136). Some who work with vibrating tools experience lacerating pains in the hands (seldom a feature of classical Raynaud's), often awakening the victims at night and robbing them of sleep (119).

The pattern of disease in the individual can be strongly dependent on the spectrum, intensity, and spatial and temporal distribution of vibrational energy entering the hand–arm system. The vibratory input is, in turn, strongly influenced by physical, mechanical, anatomical, and physiological factors of the types listed in Table 15.3 and considered in more detail in Section 3.3.

3.3 Raynaud's Phenomenon of Occupational Origin

3.3.1 Symptoms and Differential Diagnosis

It is necessary to distinguish Raynaud's phenomenon of occupational origin, that is, the symptom classically associated with habitual exposure to mechanical vibration or repetitive impact applied to the hands, from other causes of secondary Raynaud's phenomenon—which must itself be distinguished from primary or idiopathic Raynaud's disease. When Maurice Raynaud presented his M.D. thesis entitled "Local Asphyxia and Symmetrical Gangrene of the Extremities" in 1862, he described the clinical manifestations of paroxysmal blanching and numbness of one or more fingers of either hand. Similar manifestations are diagnostic of occupational Raynaud's phenomenon or VWF (Figures 15.6 to 15.8).

As Ashe and his co-workers (137, 138) reported from their experimental observations a century later, the finger blanching is characteristically provoked by whole-body rather than local chilling. Up to 8 percent of a general population could be afflicted with the idiopathic disease, according to Raynaud (although a number of factors may have changed that incidence in modern times); and more than 90 percent of sufferers from primary Raynaud's disease (i.e., patients without a history of mechanical injury to the hands or other identifiable etiologic factor) are likely to be female and of middle age or older. It should be remarked in this connection that no definitive study has to date been published that

Table 15.3 Factors Influencing Risk of Vibration Syndrome

External and Occupational Factors

Physical factors
    Intensity of vibration in the range 8 to 1000 Hz
    Directionality (one or more axes) of vibration
    Spectrum/principal frequencies of vibration
    Incorporation or absence of antivibration devices

Operational and ergonomic factors
    Work cycle/temporal pattern of vibration exposure (i.e., frequency and relative duration of periods of work at full power, idling, rest breaks, etc.)
    Method of handling and using tool
    Weight (mass) of tool supported by the user
    Configuration of handle(s)
    Pressure required to guide or advance tool
    Necessity to support, hold, or guide workpiece
    Cramped or awkward working posture
    Use of gloves or padding; clothing worn at work
    State of maintenance or repair of tool (e.g., state of cutting or drilling edges, bits, etc.; tuning of internal combustion engines; condition of exhaust systems of motor-driven and pneumatic tools; balancing of rotating or reciprocating parts)

Environmental factors
    Prevailing climate at worker's home or workplace
    Temperature ranges encountered commuting and at work
    Humidity/dampness
    Noise

Internal and Personal Factors

Experience, training, and exposure history of individual
    Age of entry into work with vibrating tools
    Time since entry into occupation (months or years on the job)
    Level of training, skill, and familiarity with the tool (affecting individual's propensity to tightly grip or easily guide the tool)
    Vibration exposure outside work (e.g., from domestic and recreational tools and appliances)

Vasoconstrictive agents affecting the individual
    Smoking
    Predisposing disease or injury (including previous trauma, vascular, cardiopulmonary, neuropathic, endocrinological, rheumatic or other disorder affecting peripheral nerve or blood flow: cf. Table 15-4)
    Medication
    Dietary factors and meal habits

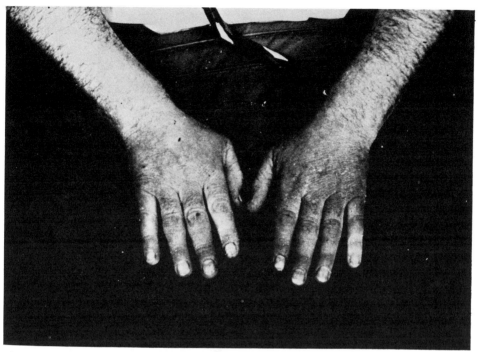

Figure 15.6 A mild case of Raynaud's phenomenon (vasopastic response to chilling). Note particularly the fifth digit of this patient's left hand. Photograph courtesy of Dr. W. Taylor, Dundee.

indicates whether female workers using vibrating tools are more likely than their male counterparts in the same industry to contract VWF.

Raynaud noted incidentally that an emotional or psychosomatic factor may not uncommonly underlie the idiopathic syndrome: some patients with Raynaud's disease report increased frequency or severity of white-finger attacks during periods of emotional or general stress. Toes as well as fingers can suffer similarly in some cases. (In occupational Raynaud's phenomenon, however, the affliction involves digits only of the hands, in a way characteristically related to the manner in which the affected worker has customarily been handling and receiving vibration from power tools and workpieces.) Although the symptoms may remain trivial in many cases, the condition can become severe in a small proportion (up to 3 percent) of sufferers, progressing over several years to skin atrophy, sometimes with peripheral ulceration and, rarely, to digital gangrene requiring surgery (Figures 15.9 and 15.10). In the severe cases of both the primary disease and secondary (including occupational) Raynaud's phenomenon, concomitant microscopic changes become established in the intimal lining and vasoconstrictive muscular layers of the digital arterial wall (Figure 15.11). It is unlikely that such changes in the arterial wall and in the skin of the

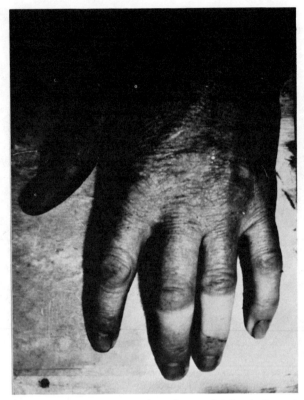

Figure 15.7 Raynaud's phenomenon affecting three digits in left hand of a patient with a history of occupational exposure to hand-transmitted vibration. Note the sharp demarcation of the areas of blanching. Photograph courtesy of Dr. W. Taylor, Dundee.

affected digits, once established, can be completely (if at all) reversed by medical treatment.

3.3.2 Symptomatology and Progression of VWF

The vascular symptoms and signs of secondary Raynaud's phenomenon of occupational origin resemble those of the primary disease. The classical attacks of sudden blanching and numbness of digits distal to a sharp line of demarcation (Figures 15.6 to 15.8) are often precipitated by general body chilling. Except in severe cases, these digital vasoconstrictive paroxysms usually last for some 5 to 15 min and may be relieved by warmth and massage of the afflicted extremity (119).

The occupational affliction is progressive as long as the patient continues exposure to hand-transmitted vibration on a regular or frequent basis. (It

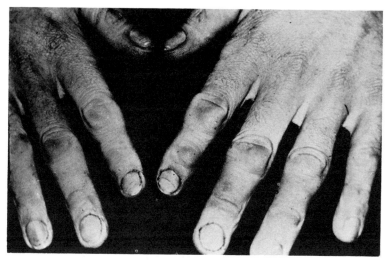

Figure 15.8 Hands of a patient affected by Raynaud's phenomenon involving distal phalanges of fingers, associated with a history of working with percussive and vibrating tools over several years. Note also callus formation over several of the joints of the fingers of the same hands. Photograph courtesy of Dr. W. Taylor, Dundee.

should be borne in mind, however, that not all hand-transmitted vibration is occupational and that the effects of occupational exposure to power tool vibration may be confounded with those of exposure to recreational and other sources such as domestic power tools used in handicrafts and do-it-yourself enterprises and motor-bicycle riding, just as hearing loss ascribed to occupational exposure may be compounded with that acquired in high-noise environments outside the workplace. Such possibilities should be considered in evaluating claims for compensation for occupationally acquired disability.) In the progression of occupational VWF, an initially symptom-free period of hand vibration exposure follows first entry into the occupation, before the first experience of a white-finger attack. This period is called the *latent period* or *interval*. In addition, because there is no known threshold below which exposure to hand-transmitted vibration may be deemed completely harmless, nor any reliable method developed yet to distinguish the unusually susceptible worker, the latent period must be considered to begin on the first day at work with a vibrating tool. According to Taylor and his coauthors (120–122), the length of the latent period is inversely related to the severity of the vibration exposure. Moreover, it is reported that the shorter the latent period (which may range from several years to a matter of weeks in extremely severe instances of occupational exposure), the more severe and rapidly progressive will be the ensuing VWF in the individual patient and the higher the incidence of VWF in the population at risk. During the latent period some patients may first

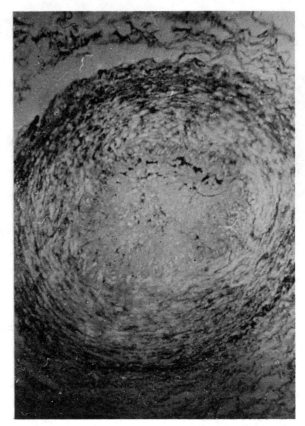

Figure 15.9 Photomicrograph of a section of a digital artery from a case of occupationally related Raynaud's phenomenon, showing degeneration of the intima and hypertrophy with infiltration of the muscular layer of the arterial wall. Photograph courtesy of Dr. W. Taylor, Dundee.

notice early warning symptoms other than blanching, such as numbness or unusual tingling in the fingers following work with a vibrating tool; but others experience no such premonitory symptoms before their first attack of VWF.

Although a variety of diagnostic and screening tests have been proposed for the vibration syndrome, the diagnosis of VWF rests essentially on the patient's history; as Okada et al. (139) have remarked, Raynaud's phenomenon is difficult to diagnose between attacks. A miscellany of clinical conditions enter into the differential diagnosis of secondary Raynaud's (Table 15.4), most of which are disorders that on a permanent or intermittent basis restrict or modulate the regional or digital arterial blood supply to the hand or digits.

Taylor and other authorities have published detailed descriptions of the progression of classical VWF in the individual case (121, 122). After the latent period, the frequency of VWF attacks increases over the years or months of

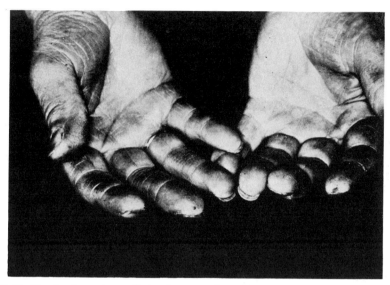

Figure 15.10 Hands of a patient suffering from an advanced case of traumatic vasospastic disease, showing areas of ischemic necrosis of the skin of the fingertips. From a color photograph by Dr. W. Taylor, Dundee.

cumulative exposure to hand-transmitted vibration. At first, attacks occur mainly in the winter in temperate and cold climates, and especially during the early morning, either at home or on the way to work when the worker is exposed to cold conditions (e.g., shaving in a chilly bathroom, bicycling to work on a cold, damp morning, starting the car in an unheated garage, or scraping frost or snow off a windshield). Employees who work outdoors in all types of weather (e.g., foresters) are most prone to early morning attacks. It is of etiologic interest as well as practical value in prevention of suffering and disability that circumstantial changes in the daily habits of workers with established VWF can reduce the incidence of attacks. Futatsuka (140, 141) has reported a diminished incidence in chain sawyers who changed from bicycling to work to using a newly introduced bus service: presumably the warmer environment and shorter journey time in the bus lessened the probability of attacks in the individual workers and also meant that, as a group, they were warmer at the start of the day's work.

Workers suffering from VWF quite commonly report progressive interference with activities outside work (e.g., gardening, home or car maintenance, car washing, hunting or fishing, attending meetings in poorly heated buildings such as church halls) before succumbing to attacks at work. All such activities tend to have as a factor in common a reduced environmental temperature, affecting the person as a whole as well as the hands. Taylor and his coauthors (121, 122) have introduced a scale of stages for gauging the progression or

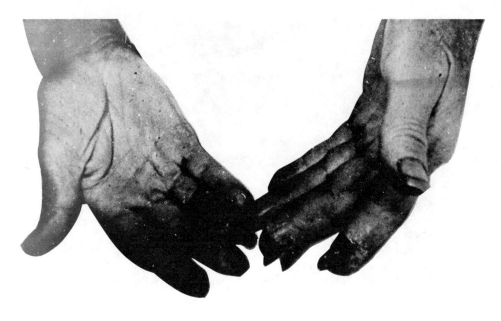

Figure 15.11 Hands of a patient exhibiting severe case of traumatic vasospastic disease, showing ischemic gangrenous necrosis of distal parts of affected fingers. From a color photograph by Dr. W. Taylor, Dundee.

severity (ultimately an index of both social and occupational disability) of cases of VWF (Table 15.5).

In stage 2 (stages 0 and 1 are premonitory stages) there intrudes interference with hobbies, social and recreational activities, whereas in stage 3 some activities are abandoned, which may be interpreted as a curtailment of the patient's enjoyment of leisure time and of life. In stage 3 there may also be interference with work, or with rest breaks at work, particularly in outside jobs such as forestry or shipyard work. In stage 4 the severity of the affliction has become such that serious interference with work, social activities, or hobbies induces the patient to change occupations. By the time that stage has been reached, it is likely that established changes have taken place irreversibly in arteries of the affected digits, of the type illustrated in Figure 15.11.

Jepson (142), in a selective survey of cases of Raynaud's phenomenon associated with various occupations, concluded that the severity of the condition was likely to reach a maximum within a few years of its first appearance, after which it would not worsen with continued vibration exposure, at least in workers who remained in the same occupation. (An apparent limitation of the incidence

Table 15.4 Differential Diagnosis of Secondary (Nonidiopathic) Raynaud's Phenomenon: Conditions Predisposing or Leading to Paroxysmal Digital Vasopastic Attacks on Blanching of Fingers

Disorders resulting from physical agents
   Traumatic vasospastic disease
     Vibration syndrome
     Vasospasm following mechanical injury to hand or digits
   Cold injury/frostbite
Mechanical compression of upper limb vessels
   Thoracic outlet syndromes
   Costoclavicular syndromes
Occlusive vascular disorders
   Arteriosclerosis
   Thromboangiitis obliterans
   Arterial embolism
Primary pulmonary hypertension
Neurological lesions
   Neurological diseases of the central nervous system
     Poliomyelitis
     Syringomyelia
     Hemiplegia (cerebrovascular)
   Peripheral neuropathies: diabetes
Connective tissue disorders
   Scleroderma/progressive systemic sclerosis
   Polyarteritis nodosa
   Rheumatoid arthritis
   Systemic lupus erythematosus
   Dermatomyositis
Endocrinological and metabolic disorders
   Dysproteinemias
   Myxedema
Vasoconstrictive medications, drugs, and chemical intoxications
   Ergot/methysergide
   Polyvinyl chloride
   Nicotine

of VWF with time may be seen on a group basis, however, because severely afflicted workers tend to leave the occupation.)

The traditional view, expressed by Jepson (142), has been that once Raynaud's phenomenon has developed, no abatement of symptoms occurs, even with removal of the patient from work with vibrating power tools. More recent opinion, however (121, 122), has tended to moderate that pessimistic view: some patients in the early stages of disease apparently can respond favorably to removal from vibration exposure and possibly to remedial treatment.

Table 15.5 Taylor's Grading System for the Severity of Vibration-Related Secondary Raynaud's Phenomenon ("VWF")[a]

Stage of Disorder	Symptoms and Signs	
	Response of Digits	Work and Social Interference
0	No blanching	No complaints
0_T	Intermittent tingling without blanching	No interference
0_N	Intermittent numbness without blanching	
1	Blanching of a fingertip with or without tingling and/or numbness	No interference
2	Blanching of one or more complete fingers, usually during winter	Interference at home or in social activities but not at work
3	Extensive blanching of all fingers of one or both hands; attacks during all seasons	Definite interference at work, at home and in social activities; restriction of hobbies
4	Extensive blanching of all fingers, both winter and summer	Occupation changed because of severity of signs and symptoms of VWF

[a] Adapted from Wasserman et al (122).

3.4 Other Varieties of Vibration Syndrome

3.4.1 Vasomotor Symptoms Other than Raynaud's Phenomenon

Although VWF is the most frequently presenting consequence of occupational exposure to hand-transmitted vibration, other vasomotor symptoms can be the presenting and predominant ones in certain industries. Agate and Druett (128, 129) suggested that power tools generating their peak vibrational energy output in the range 40 to 125 Hz were associated particularly with the causation of classical VWF (as is the chain saw—not in widespread occupational use at the time of Agate and Druett's observations—which produces its most intense vibration in the same frequency range). However, lighter, higher-speed electric drills, of a type widely used in the aircraft industry during World War II and whose principal output of vibrational energy lay in the range 166 to 833 Hz, were reported by Dart (134) to be associated with a different kind of vasomotor disorder. The commonest presenting symptom in Dart's cases was pain, associated with swelling and erythrocyanotic discolorations in the hands. The patients did not report blanching of their hands in response to chilling.

3.4.2 Neurological Effects

Apart from vasomotor reactions, other types of disorder or injury are associated with the handling of vibrating or impactng power tools, including both peripheral and CNS symptoms, as well as disorders of the musculoskeletal system. In addition to attacks of VWF, patients in Taylor's stage 3 of the vibration syndrome may experience difficulty in tactile recognition of small objects when picking them up. They may also complain of difficulty in doing up and undoing buttons or carrying out other everyday domestic or hobby activities calling for fine tactile and manipulative skill, because of loss of tactile discrimination, clumsiness of the fingers, and increasing stiffness of the finger joints. Banister and Smith (135) have reported an association of VWF with loss of manipulative dexterity (measured by use of the Purdue pegboard test) as a consequence of using vibrating tools. They reported that in the chain saw users examined the dominant hand showed the most marked loss of skill.

Agate and Druett (128, 129) had shown incidentally that in hammerers and grinders, the left hand, which most commonly was used to guide the workpiece and thus received the most intense vibration, was often the hand more severely afflicted by VWF. It has proved difficult to establish, and accordingly remains an open question, whether the nerve, skeletal, and muscular disorders seen in some varieties of the vibration syndrome develop independently or are secondary to the vasospastic response (143).

Sevčik et al. (144), reporting an incidence of 65.5 percent of vibration disease in 183 chain sawyers in Czechoslovakia, found 83 cases (45.5 percent) of traumatic vasoneurosis in those workers, who had an average age of 36 years and had been using chain saws in forestry work for an average of 6.6 years. Some two-thirds of the cases examined showed neuropathological manifestations, and 24 percent showed radiological changes at the wrist. Drobny (145) also reported neurological changes in foresters using chain saws in Czechoslovakia. Lukas and Kuzel (146) recorded pathological changes in nerve conduction velocity in either or both of the ulnar and median nerves in 47.4 percent of 137 miners who had been using pneumatic tools for an average working life time of 14 years. Bjerker et al. (147) have demonstrated changes in the vibrotactile threshold in the hand following exposures to powerful vibration. These authors have proposed the adoption of tests of vibrotactile temporary threshold shift in the hands of workers with vibrating tools as a diagnostic and screening examination for the vibration syndrome.

3.4.3 Musculoskeletal Changes

Somewhat less commonly than vasomotor or neurological disorders, musculoskeletal injuries are reported in workers who have habitually used vibrating power tools. Cystic or vacuolar areas or radiographically manifest decalcification of hand and arm bones have been described (128, 129, 148), particularly in the region of the wrist joint. Horváth and Kákosy (148) found 45 cases of atrophic decalcifications (particularly involving the lunate bone and the distal epiphysis of the radius) and 75 cases showing "pseudocysts" a few millimeters in diameter in a series of 274 foresters who had used chain saws for up to 15 years. The etiology of such changes is unknown.

Heavy low-frequency vibration is transmitted through the hands to more proximal structures, including the elbow and shoulder joints, where the repeated mechanical assault can induce occupationally related arthritic changes and periarticular disorders. Iwata (136) reported substantial incidences of upper limb joint (46.9 percent) and muscular (36.7 percent) pain in a survey of 529 metal ore miners who used rock drills. The elbow joint was most commonly afflicted, followed by the wrist joint. In many cases articular abnormalities were seen radiographically, although a clear relationship between symptomatology and radiological findings was not established. Pyykkö (119) mentions that working with vibrating tools is likely to provoke pain in joints that have previously sustained an injury (not necessarily work related). Previous injury to a digit may also predispose that digit to attacks of VWF.

Some workers with vibrating tools complain mainly of muscular weakness or stiffness in the hands, particularly on waking in the morning or at the start of a work shift. The present author knows of a foundry worker who at age 32 had used swaging tools on castings for several years. He frequently found on awakening after a night's sleep that his hands were so stiffened in flexion that the fingertips approximated the palm and could not at first be extended; it was necessary for the patient's wife to help him open his hands by gentle extension of the fingers and massage of the hands before he could get ready to go to work. Dupuytren's contracture involving the palmar fascia is sometimes seen in workers with vibrating tools but is not necessarily related to the vibration exposure (119).

3.5 Etiology of Vibration Syndrome

Various hypotheses have been put forward to explain the manifestation of Raynaud's phenomenon of occupational origin and the associated disorders that constitute the vibration syndrome, but none has been universally accepted. Several etiological factors apart from the vibration exposure may be associated with the condition (see Table 15.4). However, some of those factors may be only coincidentally related to the occupation, whereas others may not be related to it at all. That is why it has so far proved difficult to establish a firm causal link between exposure to hand-transmitted vibration at work and compensatable disability, particularly on an individual basis.

Pyykkö (119) has stated succinctly that in traumatic vasospastic disease (Raynaud's phenomenon of occupational origin) vibration is the etiologic factor and cold, the provocative one. After the latent period the cumulative trauma causes an enhanced reactivity of the finer arterial subdivisions of the vascular system of the hand distal to the brachial artery (138), accompanied in the established case by anatomic changes that may be largely irreversible. Ashe and his co-workers (138) found that this begins with muscular hypertrophy in the medial layer of the arterial wall. Later, the intima becomes affected also.

According to Koradecka (149), vibration syndrome results from a disorder of the thermoregulatory function of the circulation in the skin, although that hypothesis is open to question. Thermoregulatory disorders may appear at any stage in the vibration syndrome (150). Climatic conditions undoubtedly influence the rapidity of onset and the pattern and severity of the vasospastic vibration syndrome. In a comparative study of groups of riveters doing similar work in climatically distinct regions of the Soviet Union, Mirzaeva (151) found that, in the hot climate of Uzbekistan, the latent period of the syndrome was about twice as long as in a comparable group of riveters living and working in Moscow. Moreover, the warm climate group experienced a milder syndrome, with neurologiocal rather than vasospastic symptoms predominating, and without classical Raynaud's attacks (finger pallor) occurring during cold weather. Iwata (136) has also noted the absence of classical Raynaud's phenomenon in workers using rock drills in a warm climate. He has drawn attention to the significance of whole-body chilling in a comparative study of the effects of rock drill vibrations on miners in pits having varying degrees of coldness combined with high humidity.

The vasospastic or vascular form of the vibration syndrome (Raynaud's phenomenon or VWF), has been characterized by Langauer-Lewowicka (152) as a "cryopathy of occupational origin"; its sufferers can be shown to exhibit certain immunologic manifestations. On a group basis, patients with occupational Raynaud's showed increases in β-globulins, γ-globulins, and antiglobulin antibodies compared with vibration-exposed subjects not suffering from Raynaud's phenomenon. Serum concentrations of immunoglobulin M (IgM) were also significantly raised in the group with early Raynaud's, as were IgG immunoglobulins in cases with established disease. These findings appear to support those of Okada et al. (139). Both Okada and his colleagues and Langauer–Lewowicka have accordingly postulated that vibration-induced Raynaud's phenomenon may be the result of serum protein disturbances caused by a change in "reactivity" in the body, presumably mediated by the sympathetic nervous system (131).

Nerem (153, 154) has drawn attention to a possible mechanism underlying arterial disease associated with exposure to mechanical vibration. Using ^{131}I-albumen for *in vivo* and *in vitro* studies of blood–arterial wall macromolecular transport in the dog, he has demonstrated that heavy vibration at 10 Hz induces an arterial wall shear stress that enhances a shear-dependent transport process, in turn increasing albumen uptake by the arterial wall. Such a process, if occurring in humans, may underlie endothelial damage and changes in the medial layer of vibration-afflicted hands.

Neuromuscular reflex mechanisms may also play a contributory role. Vibration of the upper limb in humans excites the vibration reflex (155), with an increase in tonicity of vibrated muscle. Jansen (156) has claimed that coincident noise may be a predisposing factor in vibration syndrome, increasing the incidence of occupational Raynaud's. According to Pyykkö (119), vibration, noise, and cold may together act synergistically to maintain peripheral vasoconstriction.

Pyykkö (119) has postulated the following local mechanism for Raynaud's phenomenon: cold causes vasoconstriction, and because of the hypertrophic changes developing in the vessel wall, arterial blood flow in the affected digit falls very low. When the intraluminal pressure falls below the critical closing pressure for that vessel, the vessel collapses and a white-finger attack is precipitated. Stewart (157, 158) has advanced an alternative mechanical hypothesis, not generally supported, namely, that Raynaud's phenomenon of occupational origin results from constriction of the vessels due to increased digital skin thickness and callus formation in workers exposed to hand-transmitted vibration.

3.5.1 Prevalence of Vibration Syndrome

In the long term, changing trends in more than one direction may be seen in the incidence and pattern of vibration syndrome in industry. For instance, Yershova (159) has reported that in the decade 1958 to 1968 the incidence of vibration syndrome in the Ukraine rose for complicated epidemiological reasons, including the increasing use and power of vibrating tools, and also increased awareness and diagnosis of vibration-related disorders. Since 1968, however, the trend has reversed, with the incidence of vibration syndrome reduced by the progressive introduction of preventive programs, improved power tool design and technology; and the introduction of remote control of some industrial processes—which effectively removes the worker from harmful vibration exposure. Some examples of new technology leading to a reduction of the prevalence or severity of vibration syndrome are the introduction of rock drilling rigs and trestles, the displacement of riveted by welded structures in shipbuilding, and the widespread introduction of the antivibration chain saw in forestry.

3.6 Biodynamic Aspects of Vibration Syndrome

Depending on the nature and magnitude of the vibrating forces transmitted by a hand-held tool or workpiece and the frequency content (spectrum) of the impressed motion, vibration and shock loads are transmitted through the hands of the worker and up the arms to the upper limb joints. The transmission and distribution of vibrational energy in the hand–arm system is highly frequency-dependent: as a general rule, high-frequency mechanical energy is strongly absorbed and dissipated in the skin and deeper layers of the fingers and hand grasping the source of vibration, whereas low-frequency vibrations are propagated, relatively unimpeded, up the arm (7, 132, 136, 160–163). Using data obtained from mechanical impedance and cognate measurements of hand-transmitted vibration, it is possible to construct biodynamic analogical models

Table 15.6 Apparent Relationship Between Incidence of Vibration Disorder and Rate of Energy Absorption from Hand-Transmitted Vibration in Three Occupations[a]

Occupation	Frequecy of Vibration Disorder (%)	Rate of Energy Absorption (nm/s)
Rock drilling	72	21.0
Chiseling	53	2.7
Grinding	21	0.07

[a] *Source:* From I.-M. Lidström, "Vibration Injury in Rock Drillers, Chiselers, and Grinders," in Reference 122, pp. 77–83.

of the human upper limb as a vibration receiver for research and test purposes (132, 164, 165).

Vibrational energy absorption can be measured in the human hand operating various types of powered tool, in a standard manner applicable to both laboratory and field studies, using small vibration transducers fixed in specially prepared and fitted handles (131, 166, 167). The principal location of mechanical energy absorption depends on many factors in addition to frequency, for example, the strength of the grip applied to the tool (168, 169). Several such factors are listed in Table 15.4. Such factors probably account in large measure for the high variability of latent period, incidence, severity, and pattern of vibration disorder seen in different industries.

In a carefully conducted comparative study in three occupations (rock drilling, chiseling, and grinding) characterized by different patterns of hand-transmitted vibration exposure, Lidström (166, 167) examined the correlation between vibration disorder (determined, according to specific criteria, from the history and a laboratory test of individual susceptibility to vibration-induced vibrotactile sensory threshold shift) and the vibrational energy absorption per unit time (determined from force and velocity measurements on the tools in question, using a standard instrumented handle). She found that there was a tendency for the incidence of the vibration disorder to rise with the rate of vibrational energy absorption by the hand, depending on the type of tool and the method of use (Table 15.6).

3.7 Diagnostic and Screening Tests for Vibration Syndrome

A variety of clinical tests and measurements have been advocated for use in research or clinical practice for the diagnosis of vibration syndrome and the assessment of disability (119, 121, 149, 170, 171). None has yet been universally accepted for use in industrial medical practice or in epidemiological studies, because all are open to criticism on the grounds that the tests lack reliability (i.e., they yield to many false positive or negative results), specificity, or repeatability (especially on an individual basis); or because the test is taxing or difficult to apply in general practice or field conditions (119). Nevertheless, certain approaches have been established, and some methods have won a substantial measure of acceptance although requiring refinement and further validation.

3.7.1 Tests of Peripheral Vascular Function

Koradecka (149) and other investigators have advocated tests of peripheral circulation, and particularly the measurement of digital blood flow, for the diagnostic screening and periodical monitoring of workers who complain of VWF or who are at risk because of their occupation. Several techniques have been tried, with varying degrees of confidence, including digital (occlusive) plethysmography, skin thermometry, capillary microscopy, skin and muscle rheography, and arteriography.

Mentioning the problem of lack of specificity of tests for peripheral arterial disease, Zweifler (172) has presented case histories showing that finger ischemia (demonstrable by plethysmographic and arteriographic techniques) may result from occupational traumata of varied origin. Cold provocation tests, in which the autonomically mediated peripheral vascular response to sudden general chilling is examined and compared with supposedly normative data, have been found to be disappointingly unreliable (120, 143, 173, 174).

The technique of IR thermography, established in diagnostic use in cases of primary Raynaud's disease, has been adapted by (169) Tichauer for the examination of patients suffering from various forms of "work stress" affecting the back or the upper or lower limbs in jobs involving repeated volitional or imposed motion or vibration (e.g., operating treadles, handling vibrating appliances). According to Tichauer, the technique shows promise for diagnostic and prognostic purposes in relation to vibration syndrome, although Tiililä (174) had previously expressed disappointment with it.

Buzalo (175), in advocating thorough ophthalmologic examination in the course of full clinical evaluation of patients with vibration syndrome, has stated that disorders of eye vessels, retinal and iridal structures, and the visual fields (perimetry) can be observed in such cases.

3.7.2 Neurological Tests

Recognition that neurological symptoms and signs occur (often prodromally) in a substantial proportion of cases of vibration syndrome has led to the use of peripheral nerve function tests in evaluating VWF. Tests advocated include measurement of conduction velocity by the method due to Hopf (176) and of peripheral sensory function. Tests of the latter include tactile discrimination, vibrotactile sensitivity (pallesthesiometry), and vibrotactile sensory threshold shift following exposure of the hands to vibration (166, 167).

Radziukewich and Mikulinskii (178) have pointed out, however, that considerable variance is to be expected in pallesthesiograms even in normal subjects not exposed to occupational vibration and that, even in nonvibrated control subjects, there is a systematic rise in vibration threshold with advancing age between 18 and 60 years, for which a correction factor would have to be introduced if the test were used for the assessment of vibration-related disability. (The problem is somewhat akin to that of presbycusis affecting the audiometric evaluation of noise-induced hearing loss.) Seppäläinen (176) has reported disappointing results with neurological tests that failed to distinguish between vibration-exposed shipyard workers and a comparison group not occupationally exposed to hand-transmitted vibration.

3.7.3 Radiological Examination of Bones and Joints

As mentioned earlier in this chapter, some patients with vibration syndrome reveal radiographical changes in the bones of the upper limb, particularly in the wrist joint (128, 129, 133). Bone densitometry has been used in attempts to evaluate such cases, again, with often disappointingly equivocal results (179). Again, a problem arising is the lack of specificity and quantifiability of the bone density changes observed.

3.7.4 Physical Measurements and Electromyography

Electromyography (EMG) has been explored, chiefly as a research tool, as a method of studying human reactions to hand-transmitted vibration by several investigators (160, 162, 180, 181). The level of EMG activity probably can be correlated quantifiably with the spectrum and intensity of segmental vibration inputs to humans and with mechanical vibration propagation in the hand–arm system (162, 170). It may be indicative of the skill and effort required to handle and operate different kinds of vibrating tool in various work situations and postures and thus be primarily of ergonomic rather than clinical value.

3.8 Prevention and Treatment of Vibration Syndrome

3.8.1 Standards

Following efforts in several countries to set regulatory or consensus standards for the protection of workers exposed to hand-transmitted vibration (182–185), ISO Subcommittee ISO/TC 108/SC4 (*Human Exposure to Mechanical Vibration and Shock*) Working Group 3, undertook to prepare guidelines (6) for the measurement and evaluation of human exposure to hand-transmitted vibration. In addition to presenting an international consensus regarding the proper manner in which to measure hand-transmitted vibration for occupational safety and health purposes, earlier versions of the document (at the time of writing this chapter, a draft proposal for an international standard) included proposed recommendations for limiting the rms acceleration and vibrational velocity of vibration reaching the hands from vibrating tools and workpieces in the range 6.4 to 1000 Hz, deemed to be the hazardous range.*

The boundaries proposed in Reference 6 remain tentative because of the paucity of reliable quantitative data relating risk of vibration syndrome to measured human exposure to hand-transmitted vibration in industry. Some draft standards have been heavily, if not entirely, based on subjective data [such as the work of Miwa and his colleagues (48, 186) on the magnitude estimation of the intensity of hand-transmitted vibration]. Accordingly, their adoption as guidelines for the prevention of occupational vibration syndrome in the absence of supporting biomedical data is open to cogent criticism (132). A comparative review and critique of standardization efforts in the field of hand–arm vibration has been published more recently by Griffin (202).

3.8.2 Engineering Solutions: Vibration Reduction in Power Tools

Assuming that a consensus standard can be established as a guideline to limiting human exposure to hand-transmitted vibration, vibrating tools and appliances can be designed or modified to meet it in several ways. Recent years have seen substantial improvement in the design of the chain saw to reduce the vibrational energy transmitted to the hands of foresters by these motor-driven machines (187–192). The most important innovation has been the introduction of the antivibration saw, in which part of the mass of the machine, including the fuel tank, can be resiliently suspended so as to act as a dynamic vibration absorber. The increasing adoption of the antivibration saw has been reflected in a marked decline in the incidence of VWF among forest workers in countries where state forestry commissions have made the use of such saws the rule in their large work forces, notably Great Britain (177, 189) and Sweden (187).

Regrettably, the same administrative and engineering effort that has so markedly improved the lot of the chain sawyer has not yet been successfully devoted to other kinds of vibrating tool in industrial use, such as pneumatic hammers, grinders, scaling machines, and the like: many such tools remain in use after many years—indeed, decades. In some, the design has changed little, if at all, since the early years of this century (Figures 15.2, 15.3, and 15.5).

3.8.3 Human Factors Engineering and Operational Solutions

Perusal of Table 15.3 will have suggested to the reader several ways in which the problem of vibration syndrome may be attacked. The beneficial influence

---

* In the previous (1979) edition of this book, Volume III, pages 513–514, the author quoted (Table 13.7) the then current draft exposure limits from ISO Draft Proposal 5349 of May 1978, in anticipation of its issuance as an international standard. In doing so, he may have acted prematurely, because that version of the document was not in the event ratified by all the member countries of ISO/TC 108/SC4, who subsequently voted on it. That draft proposal was taken back into committee and has undergone substantial revision. At the time of writing this chapter, it has not yet been ratified as an international standard. Accordingly, the reader is advised to disregard the previous (1979) citation and to await formal ratification of the guidelines as ISO 5349. Pending publication of that standard, the current draft guidelines, although retaining the essential data of earlier drafts, are not reproduced in the present edition of this book.

of adequate breaks from continuous work with tools such as the chain saw has long been recognized and advocated (130, 193). It is also desirable that the unsupported weight of vibrating tools not exceed 10 kg, nor the effort required to guide or advance the tool a force of 20 kg (130, 194, 195).

In some occupations, notably power sawing, in some countries the use of gloves is recommended (127), both to keep the hands warm and to provide some mechanical damping or attenuation of the vibration reaching the hands. In other industries, such as foundrywork where castings are chipped and swaged, padding (even with old rags or cloths) is traditionally interposed between the worker's hand and the vibrating tool or workpiece. The value of such devices is debatable (and probably highly variable) for as a rule, unless carefully designed and fabricated, padding, gloves, or other soft material placed between the source and the hand provides relatively little attenuation of vibration below about 200 Hz (196).

Proper care and maintenance of power tools is always to be strongly recommended throughout industry, both for reasons of safety and good practice and also because the levels of hand-transmitted vibration generated by power tools are apt to rise when the equipment is not regularly overhauled and properly maintained (197).

In cold climates, keeping fit and warm at home and on the job (and avoiding whole-body chilling on the way to work) are generally held to be beneficial. Work practices may usefully include prewarming the hands and person before beginning a stint of work in cold, damp conditions such as in forestry (127, 130, 193). The guidance given to chain saw operators by the United Kingdom Forestry Commission embraces many points of good practice in that industry, including those mentioned previously, and deserves requoting from Taylor et al. (193):

There are a number of ways in which you can reduce the amount of vibration passing into your hands, when you are using any saw, and these points apply whether or not the saw has antivibration handles:

1. Good techniques for felling, cross-cutting, and snedding with lightweight saws include resting the saw as much as possible on the tree (or occasionally on your thigh); this means that some of the vibration is absorbed by the tree or the large muscles of your thigh. Holding the saw as lightly as possible when it is at full throttle, without, of course, reducing effective control of the saw, will also reduce vibration absorbed into your hands.

2. Wearing chain saw gloves spreads the grip over a larger area of your hands. (Often the first sign of white finger is on one or two finger joints which have taken most of the vibration due to too tight a grip.)

3. Good blood circulation to the arms and hands gives maximum protection to the flesh, nerves, and bones of the hands, and this is achieved by warming up *before* starting the saw and wearing suitable clothing and gloves. Thus, it is better to be too warm than cold.

4. Sprockets, guide bars, and chains should be well maintained, and chains should be correctly sharpened with the recommended clearance for the depth gauge. Poor

maintenance increases vibration by as much as one-third of the normal level for the saw.

5. The "safe" limits are based on continuous use of the saw, and every time the saw is idling or stopped gives your hands and arms a chance to recover from the effects of vibration. The more evenly breaks in saw usage can be spread throughout the day, the less the risk of any discomfort in your hands; try to organize the stops for fuel, sharpening, meals, piling of timber, or other work so that the saw is switched off for at least 10 minutes as often as possible during the day rather than a few longer stoppages.

3.8.4 Experience, Training, Skill, and General Fitness

Clutching a power tool or workpiece in an overly tight grip is characteristic of the novice or the anxious person. A tight grip restricts the circulation in the hand and may also enhance the transmission of vibration into the hand (E. Tichauer, comment in Reference 122, p. 208; Reference 188). It is desirable that workers with vibrating tools learn to handle the equipment in a reasonably relaxed manner, avoiding a tight grip or an awkward, rigid working posture. Workers at risk or already suffering from vibration syndrome should also exercise caution regarding exposure to intense hand-transmitted vibration off the job (e.g., in motorcycle riding; riding or guiding power lawn mowers; using power saws, drills, sanders, and the like in hobbies or in car or home maintenance), particularly if the activity is carried on in cold or damp climatic conditions. The industrial medical officer or practitioner responsible for patients exhibiting prodromal or established symptoms and signs of VWF may also wisely advise the patient to consider the benefits of adequate exercise, regular meals and suitable diet, avoidance or reduction of smoking, and general health care.

3.8.5 Medical Treatment for Vibration Syndrome

If it is not in an advanced stage, VWF usually is abated and does not progress after cessation or substantial reduction in the patient's hand-transmitted vibration exposure. Provided the disorder has not progressed to the stage of arterial damage and trophic changes in the digits, there is nothing to treat when an attack is not in progress. Severe or progressive Raynaud's of occupational origin may possibly be alleviated by drugs such as are used to treat primary Raynaud's disease (e.g., reserpine, methyldopa, mild sedatives), but the results are likely to be disappointing, and any success short lived, in cases of secondary Raynaud's phenomenon where the underlying cause is not treated.

It has been claimed that medication to reduce the coagulability of blood can reduce the frequency and severity of symptoms in sufferers from severe vibration-related Raynaud's phenomenon. This is based on the observation that such patients not infrequently show an increased tendency to hemocoagulation. Demin et al. (198), noting that at least one-third of patients suffering from vibration disease showed such a tendency, prescribed heparin therapy. They reported that 43 out of 50 cases showed substantial improvement and the remainder some improvement (at least in relation to the coagulability of the blood, if not symptomatically). In Germany, Ehrly (199) has reported some success with the use of subcutaneous Ancrod (a purified fraction obtained from snake venom) in six patients with severe secondary Raynaud's. The rationale of the treatment was to reduce blood viscosity, thus enhancing peripheral blood flow. Beneficial effects of treatment, including improved skin warming and healing of ulceration, are said to have persisted for some months after short courses of the treatment.

3.9 Supplementary Note: Recent Data on Vibration Syndrome

Since this chapter was drafted, the author has received a number of recent publications that he commends to the attention of the reader interested in vibration syndrome and its prevention. Brammer and Taylor have published (200) an edited collection of papers presented to the Third International Conference on Hand–Arm Vibration, held in Ottawa in 1981. This major compendium covers ground similar to that of Reference 122, to which it may be regarded as a sequel bringing the reader up to date on new methodology, more recent clinical and experimental data, efforts in standardization, and current views concerning the prevention of vibration syndrome. Reference 201 is a concise review of current information on the vibration syndrome. It contains a number of recommendations concerning preventive methods and work practices and a commentary on standards. Reference 202 reviews standards development in this area.

A particularly noteworthy achievement in recent years has been the development of reliable instrumentation enabling precise measurements to be made of the amount and rate of vibratory energy entering the hand from power tools. This is crucial for the determination of dose–response relationships for hand-transmitted vibration and resulting injury (203, 204).

4 VERY LOW-FREQUENCY VIBRATION AND MOTION EFFECTS*

As was mentioned briefly in Section 2, whole-body vibration or passive oscillatory motion of man at very low frequencies (below 2 Hz), although not necessarily thought of as "vibration" in the everyday sense, can impair comfort, working efficiency, and safety in several ways. Like those of higher-frequency vibrations, the effects of such oscillations depend upon many factors, including the frequency content, acceleration, and the direction of action of the motion with respect to the human body; the duration of the disturbance; and the nature of the recipient's activity during the motion.

Author's note: This section has been added in proof to draw specific attention to aspects of human occupational exposure to very low-frequency whole-body oscillation currently subject to a revival of attention in research and development as well as efforts in international standardization.

Oscillatory or repetitive motion at frequencies even below 0.1 Hz (connoting a period of 10 seconds for one complete cycle of motion), exemplified by the heaving of a ship in an ocean swell or the gentle sway of a tall building in high winds, is perceptible by man as oscillatory in nature. Motion below 1 Hz, while not known to be directly causative of any occupational morbidity, can nevertheless be troublesome and sometimes extremely hazardous to safety in the working environment if sufficiently intense and prolonged.

The domain of very low-frequency vibration or motion may be defined for practical purposes as that in which vehicles respond mechanically to the medium in which or the surface upon which they travel essentially as rigid bodies. Accordingly, the motion experienced by the occupant as a result of the interaction depends on the fundamental characteristics of flotation (in the case of fluid-borne vehicles) or suspension (in the case of land vehicles) of the principal mass of the vehicle. Motion of this kind is exemplified by the heave, roll, pitch, and yaw of ships, surface effect craft, and other marine vehicles (72, 206). Ship motion is a complex phenomenon determined by the ship's course and speed in relation to the wind and sea state and many other factors; and it occurs mostly in the band 0.1 to 2 Hz.

Other examples are the analogous "rigid-aircraft" response of an airplane or helicopter to air turbulence [in a similar range of frequency (3)]; and by the motion of suspended land vehicle bodies in response to encounters with surface irregularities in the road, track or terrain passing beneath their wheels or running gear. In such vehicles as haulage trucks, mobile heavy equipment, and trains, the resulting low-frequency motion is a complex function of the vehicle's suspension characteristics and speed, and the spatial frequency of the surface irregularities to which it responds. Of course, potentially troublesome vibration at higher frequencies, having the causes and effects described in Section 2, may be, and often is, generated at the same time, adding to the problems of the induced occupational environment.

Cantilever oscillations of tall buildings (sway) due to the wind (and occasionally to ground motion in seismically active areas), as well as major structural oscillations of other large fixed structures such as suspension bridges and offshore rigs, can enhance the natural hazards of construction and external maintenance. Wind- and traffic-induced oscillation at low frequencies and of relatively large amplitude can sometimes be disconcerting to building occupants and bridge passengers. In high winds, the upper floors of some very tall steel-framed buildings may sway (i.e., oscillate horizontally) several feet at frequencies of the order of 0.1 Hz (periods of 9 to 12 seconds). The motion can manifest itself in bizarre ways (such as the measured sloshing of water in restroom sinks or unpleasant sensations of displacement that distract or nauseate susceptible people) and enhance fear of being in a high-rise building. Occasionally, office workers aware of building sway are nervous about leaving their desks or descending stairs (a potential hazard in emergency) until the motion abates. Part of the solution to this problem is to inform workers in the building that sway during storms and high winds is a normal phenomenon, proper to the construction of the building.

4.1 Human Response to Very Low-Frequency Vibration and Motion

Below about 3 Hz for z-axis vibration and about 1 Hz for x- or y-axis oscillation of seated or standing man, the induced human body motion in the absence of postural activity is essentially that of a rigid mass, without internal resonance or oscillatory deformation of anatomical structures (see Section 2.1.1). Nevertheless, sensory and physiological disturbances, which can in turn affect working efficiency and safety, result from the oscillatory stimulation of the vestibular organs and mechanoreceptors subserving the regulation of posture (Section 2.1.3), as well as from direct mechanical interference. These disturbances can be enhanced by the visual (sometimes conflicting and sometimes illusory) perception of motion.

4.1.1 Motion Sickness

Whole-body oscillation of man in the frequency band 0.1 to 1 Hz causes motion sickness, a physiologically reversible affliction the incidence and development of which depend upon complex environmental, circumstantial, and idiosyncratic factors (17–20, 208, 209). As mentioned in Section 2, the end-result is nausea (the symptom whose very name is derived from the ancient Greek word for a ship), leading ultimately to vomiting, accompanied by other manifestations of physiological distress. These include general malaise, apathy, and the disruption or abandonment of volitional activity (8, 14–20, 72). The malady is in most cases self-limiting, at least with regard to nausea and vomiting, because of habituation and eventual abatement of the provocative motion. However, the tendency toward excessive subjective fatigue or lassitude, and drowsiness (the sopite syndrome) that is often experienced at sea, which is sometimes the principal or sole manifestation of motion sickness (208), is apt to persist indefinitely during continuous ship motion. Except at sea, where seasickness is a perennial problem affecting the manning of ships and where probably very few of the global work force are ultimately immune in severe conditions, motion sickness is much more commonly a problem besetting passengers than those involved in the operation of the vehicle; for preoccupation with a task appears to be to some extent protective against development of the malady. Nevertheless, motion sickness can and does afflict people with critical tasks to perform in relatively moderate motion. Moreover, it can vary from being merely unpleasant or embarrassing to being a danger to safety and, if the susceptibility to sickness is not reduced by training and experience, to being career-threatening in such diverse occupational groups as train crews, ambulance attendants, and student aviators.

The malady is described generically as motion sickness (or kinetosis) but popularly takes many descriptive names from the situations in which it is suffered (e.g., seasickness; airsickness; car sickness and so on). Present physiological consensus is that, regardless of the circumstances, the malady as experienced during exposure to low-frequency mechanical oscillatory motion of man riding or working within the normal gravitational field of the earth has a common causation, normally involving stimulation of the intact labyrinth (14, 17, 209). However, similar maladies, sharing at least the final common path leading to elicitation of the vomiting reflex, occur in certain special working situations in which, in the absence of passive low-frequency motion, unusual vestibular and/or visual stimulation occurs. These include space (motion) sickness, or space adaptation syndrome (the physiological basis of which probably differs from that of terrestrial motion sickness) in astronauts (210–213); and perceptual conflict disorders exemplified by simulator sickness (214). The latter affects aviators undergoing training or research activities in which the subject is required to respond to the visual appearances of flight maneuvers created in a ground-based simulator in the parital or total absence of the real motion that would be felt in an aircraft. Another kind of postulated perceptual conflict disorder affects some members of the rapidly growing work force using video display terminals (VDTs) in work with computers. Some workers with these devices complain of disturbing visual sensations, fatigue, and occasionally nausea associated with the scrolling and flickering of the display. Whether such effects can be enhances by low-frequency motion or vibration, or increase susceptibility to motion sickness (e.g. among operators of VDTs aboard ship) remains a matter for conjecture in the present state of knowledge (215, 216).

4.1.2 Other Motion Effects and Task Disruption

Another physiological effect of oscillatory motion and vibration that develops progressively with the duration and intensity of the motion exposure, and is apt to persist after the causative motion has ceased, is impairment of postural function. Aspects of this ill-understood phenomenon, which has a bearing on both operational efficiency and occupational safety, were reviewed in Section 2.1.3. Disturbances of static and kinetic postural steadiness can result from motion and vibration exposure over a substantially broader band of frequency than motion sickness, which appears to be a response limited to provocation by motion components below 1 Hz (15). Unsteadiness of stance (51) and gait, which can be severe enough to jeopardize safety as well as task-performance for considerable periods following prolonged, intense low-frequency motion exposures, can be observed in motion-exposed subjects regardless of whether or not they experienced motion sickness.

In contrast to the delayed, time-dependent, and persistent (post-motion) effects of low-frequency motion that are physiologically mediated, direct and immediate mechanical interference with human activity and task performance also occurs during severe low-frequency motion. This is due to displacement of the body and disruption of the point or points of contact between man and his task, be it operating a control device, handling a load, or simply walking across a deck.

Aboard ship, for example, wave-induced ship motion mechanically disorganizes human locomotion, making manning of the ship more difficult, time consuming, and fatiguing. Above some threshold (about ±4 degrees of roll, according to a study by Olson (217, 218)), there is a progressive deterioration of human locomotor and manual activity with increasing physical severity of the motion. The risk of falls, stumbles, and load-carrying accidents is increased. Eventually, some ship's tasks become impossible or too dangerous to perform (because of mechanical or electrical hazards), even with additional manning, so that progressive delay is introduced into the work of ship's departments and seakeeping is compromised (217–220).

4.2 Preventive Measures

Fortunately, however, various methods are available or are being evaluated as to their efficacy in preventing or mitigating the adverse effects of very large-amplitude, low-frequency motion of the kind encountered in rough conditions at sea or in the air. A number of approaches are summarized in Table 15.7. Several of the most efficacious methods of protecting man from motion sickness and the mechanical hazards of severe low-frequency motion do not involve biomedical intervention but are mechanical, systems, and human factors engineering techniques. Ideally, these should be based on sound principles of human biodynamics, supported by firm data concerning the human response to low-frequency motion, and applied at the design stage of new ships, aircraft, and other costly systems. In ships, for example, the motion environment [and, concomitantly, motion sickness incidence (221)] may differ markedly from one work station to another in the same vessel. Moreover, differences in hull design and ship arrangements between vessels can substantially influence the comparative seakeeping qualities of different vessels steaming in the same conditions (206, 218–222). These findings suggest several possible approaches in design to minimize motion effects.

General principles of protecting man from adverse effects of vibration at large were reviewed in Section 2.3 earlier in this chapter. Those principles generally apply to very low-frequency vibration also. However, several engineering and biomedical approaches and methods can be identified (Table 15.7) that are particularly appropriate to the very low-frequency domain with its special problems of motion sickness prevention and large-displacment interference with the human locomotion and the performance of manual tasks.

Some of the methods listed in Table 15.7 (by way of illustration only: it is not possible to include a comprehensive treatise on motion effects prevention the present volume) are already proven and widely practiced in aerospace, naval, automotive or civil engineering, while others remain yet experimental or theoretical. Several are the subject of contemporary research and development. Current efforts in this direction are particularly active in naval, military,

Table 15.7 Approaches to Preventing or Mitigating the Effects of Very Low-frequency Vibration and Motion on Man[a]

General and systems engineering
- Design of major structures (e.g., airframes; buildings; off-shore structures; ships' hulls) to avoid or minimize oscillatory response at frequencies particularly disturbing to man
- Location and orientation of critical duty stations and workplaces within structures (e.g., in relation to roll and pitch centers in ships and aircraft) to minimize prevailing motion and vibration levels
- Design and development of motion-attenuating devices (e.g., fin stabilizers in ships, gust-alleviators in aircraft and corresponding devices in surface effect craft) that passively or actively modify the rigid-body response of the vessel or craft

Human factors engineering (ergonomics)
- Selection, design, and placement of work stations, tasks, and task paraphernalia (e.g., displays; controls; seats; working surfaces and furniture; materials, stores, equipment and their stowages, containers and handling equipment) for ease and efficiency of use (including resistance to mechanical disturbances or disruption of the task) in the motion and vibration environment
- Design and furnishing of working and moving spaces (e.g., passageways; corridors; ladders; stairways) for maximum safety and minimum impediment of locomotion in severe motion environment aboard ships, off-shore rigs, aircraft, trains
- Seletion or design of coincident environmental features (e.g., lighting levels and quality; noise and vibration levels; the climatic environment; air quality, i.e., freedom from toxic or nauseogenic contaminants) for optimal working efficiency, safety and general habitability; and for minimal potentiation of adverse motion and vibration effects
- Design of any special clothing (including headgear), man-mounted or man-carried equipment used in the motion environment for maximal ease, safety and efficiency of use; and for minimal encumbrance in the motion environment

Strengthening inherent human resistance to adverse motion effects on a population, group, or individual basis
- Selection of crew for sea or flight duty who are relatively immune to adverse effects of oscillatory motion (particularly motion sickness)
- Habituation of personnel to the actual or simulated motion environment and to activities or tasks to be performed in it
- Specific training of personnel or crews to perform jobs, tasks or missions in the motion environment
- Optimization of work/rest or duty/shore leave cycles for maximal effectiveness of each phase; and for maintenance of mission-related skills and habituation to the motion environment

Modification of the physiological response to motion, or of the physiological state, to enhance resistance to adverse motion effects
- Optimization of physical fitness and morale
- Desensitization of susceptible people using deconditioning techniques
- Optimization of the physiological state at the time of motion exposure (e.g., by adequate sleep; avoidance of alcohol)
- Medication (for motion sickness)

Table 15.7 (*Continued*)

Operational solutions (applying particularly to ships, off-shore operations, or aircraft)
  Strategic, tactical, schedule, and route planning to minimize:
    Routing through areas with a prevailing tendency to rough motion or where adverse conditions can be predicted
    Distance traversed or time necessarily spent in rough conditions
    Number of units in a fleet or task-force necessarily exposed to severe environment at any given time
    Necessity, frequency and duration of resupply operations in heavy seas
    Tactical maneuvering compromises (e.g., altering course and or speed to alleviate vessel response to rough sea, or aircraft response to turbulence)

[a] Adapted from material in References 2, 3, and 223.

and other defense-supported laboratories in the United States, the United Kingdom, and other countries (especially those with a strong maritime interest), because of growing awareness that human factors, and specifically the human response to oscillatory motion, can be critical to mission-effectiveness, and to human safety, in modern high-technology seaborne and aerospace operations.

4.2.1 Standards

Various attempts have been made or are currently in progress to write or strengthen standards for the evaluation of human exposure to very low-frequenccy vibration, where the criterion is the prevention either of motion sickness (224) or of human disturbance in buildings. An existing ISO standard (207) specifically addresses wind-induced horizontal motion below 1 Hz in buildings and off-shore structures. The responsible ISO Subcommittee (ISO/TC 108/SC4, *Human Response to Mechanical Vibration and Shock*) is, at the time of writing this chapter, engaged in a comprehensive revision of Reference 4 with the intention, *inter alia*, of extending the scope of that standard to frequencies in the decade below 1 Hz, associated with human motion discomfort and motion sickness.

ACKNOWLEDGMENTS

The author is indebted to Dr. I.-M. Lidström, Stockholm, and Dr. W. Taylor, Watten, Scotland, for helpful comments and illustrative material and to Mr. D. E. Wasserman, Cincinnati, for illuminating comments and information. Other sources are acknowledged in the credits to the photographs or as personal communications. Views expressed in this chapter are the author's and are not necessarily endorsed by his current employer, the United States Navy.

REFERENCES

1. C. E. Crede, "Principles of Vibration Control," in: *Handbook of Noise Control*, C. M. Harris, Ed., McGraw-Hill, New York, 1957, Chapter 12.
2. J. C. Guignard, "Vibration," in: *A Textbook of Aviation Physiology*, J. A. Gillies, Ed., Pergamon Press, Oxford, 1965, Chapter 29.
3. J. C. Guignard, "Vibration," in: *Aeromedical Aspects of Vibration and Noise*, J. C. Guignard and P. F. King, AGARDograph AG-151, NATO/AGARD, Neuilly-sur-Seine, France, 1972, Part 1, pp. 2–113.
4. International Organization for Standardization, *Guide for the Evaluation of Human Exposure to Whole-Body Vibration*, International Standard ISO 2631-1978, ISO, Geneva, 1978.
5. American National Standards Institute, *Guide for the Evaluation of Human Exposure to Whole-Body Vibration*, American National Standard ANSI S3.18-1979 (ASA 38-1979, published by the American Institute of Physics for the Acoustical Society of America, New York, 1979.
6. International Organization for Standardization, *Guide for the Measurement and the Evaluation of Human Exposure to Vibration Transmitted to the Hand*, Revised Draft Proposal 5349, ISO Technical Committee 108, Subcommittee SC4, committee document, 1982.
7. H. Dupuis, E. Hartung, and L. Louda, *Ergonomics*, **15**, 237–265 (1972).
8. J. C. Guignard and M. E. McCauley, *Aviat. Space Environ. Med.*, **53**, 554–563 (1982).
9. International Organization for Standardization, *Draft Proposal on Standard Biodynamic Co-Ordinate Systems*, ISO Technical Committee 108, Subcommittee SC4, committee document in preparation, Secretariat of ISO/TC 108/SC4, Berlin, 1983.
10. J. C. Guignard, *J. Sound Vib.*, **15**, 11–16 (1971).
11. L. F. Stikeleather, G. O. Hall, and A. O. Radke, "A Study of Vehicle Vibration Spectra as Related to Seating Dynamics," SAE Paper 720001, Society of Automotive Engineers, New York, January 1972.
12. B. Hellstrøm, "Measurement of Occupational Vibration and its Biological Effects," in Reference 123, pp. 89–105.
13. D. E. Wasserman, T. E. Doyle, and W. C. Asburry, *Whole-Body Vibration Exposure of Workers during Heavy Equipment Operation*, U.S. Department of Health, Education, and Welfare (NIOSH) Publication 78-153, National Institute for Occupational Safety and Health, Cincinnati, Ohio, April 1978.
14. R. S. Kennedy, A. Graybiel, R. C. McDonough, and F. D. Beckwith, *Acta Otolaryngol.*, **66**, 533–540 (1968).
15. J. F. O'Hanlon and M. E. McCauley, *Aerosp. Med.*, **34**, 366–369 (1974).
16. D. Goto and H. Kanda, *Motion Sickness Incidence in the Actual Environment*, document ISO/TC 108/SC4/WG2 63, Secretariat of ISO/TC 108/SC4, International Organization for Standardization, Berlin, 1977.
17. K. E. Money, *Physiol. Rev.*, **50**, 1–39 (1970).
18. T. G. Dobie, *Airsickness in Aircrew*, AGARDograph AGARD-AG-177, NATO/AGARD, Neuilly-sur-Seine, France, February 1974.
19. A. N. Razumeyev, "Kinetoses," in: *Pathological Physiology of Extreme Conditions*, P. D. Gorizontov and N. N. Sirotinin, Eds., Meditsina Press, Moscow, 1973, pp. 332–348 (National Aeronautics and Space Administration Technical Translation NASA TT F-15,324, Washington, DC, 1974).
20. J. T. Reason and J. J. Brand, *Motion Sickness*, Academic Press, London, 1975.
21. H. E. von Gierke, C. W. Nixon, and J. C. Guignard, "Noise and Vibration," in: *Foundations of Space Biology and Medicine*, Vol. II, Book 1, M. Calvin and O. G. Gazenko, Eds., National Aeronautics and Space Administration, Joint U.S.A./U.S.S.R. Publication, Washington, DC, 1975, Chapter 9, pp. 355–405.

22. R. J. Hornick, "Vibration," in: *Bioastronautics Data Book,* 2nd ed., J. F. Parker and V. R. West, Eds., NASA SP-3006. National Aeronautics and Space Administration, Washington, DC, 1973, pp. 297–348.
23. R. W. Shoenberger, *Perceptual Motor Skills, Monograph Supplement 1-V34* (1972).
24. J. Matthews, *J. Agric. Eng. Res.* **9,** 3–31 (1964).
25. J. Hasan, *Work-Environ.-Health,* **6,** 19–45 (1970).
26. H. E. von Gierke, *Appl. Mech. Rev.,* **17,** 951–958 (1964).
27. E. B. Magid, R. R. Coermann, and G. H. Ziegenruecker, *Aerospace Med.,* **31,** 915–924 (1960).
28. M. J. Mandel and R. D. Lowry, *One-Minute Tolerance in Man to Vertical Sinusoidal Vibration in the Sitting Position,* 6570th. Aerospace Medical Research Laboratory Technical Documentary Report AMRL-TDR-62-121, Wright-Patterson Air Force Base, Ohio, 1962.
29. H. E. von Gierke, H. L. Oestreicher, E. K. Franke, H. O. Parrack, and W. W. von Wittern, *J. Appl. Physiol.,* **4,** 886–900 (1952).
30. E. Ts. Andreeva-Galanina, E. A. Drogichina, and V. G. Artamaonova, *Vibration Disease (Vibratsionnaya Bolezn'),* U.S.S.R. State Publishing House, Leningrad, 1961.
31. E. Ts. Andreeva-Galanina, *Gig. Tr. Prof. Zabol.,* **8**(8), 3–7 (1964).
32. A. S. Melkumova and V. V. Russkikh, *Byul. Eksp. Biol. Med.,* **1973**(9), 28–31 (1973).
33. W. B. Hood, R. H. Murray, C. W. Urschel, J. A. Bowers, and J. G. Clark, *J. Appl. Physiol.,* **21,** 1725–1731 (1966).
34. A. J. Liedtke and P. G. Schmidt, *J. Appl. Physiol.,* **26,** 95–100 (1969).
35. J. Ernsting and J. C. Guignard, *Respiratory Effects of Whole-Body Vibration,* Royal Air Force Institute of Aviation Medicine, Report RAF-IAM-179, Farnborough, England, 1961. [Also published by J. Ernsting as Air Ministry (London) Flying Personnel Research Committee Report FPRC-1164, 1961.]
36. L. R. Duffner, L. H. Hamilton, and M. A. Schmitz, *J. Appl. Physiol.,* **17,** 913–916 (1962).
37. T. W. Lamb, K. H. Falchuk, J. C. Mithoefer, and S. M. Tenney, *J. Appl. Physiol.,* **21,** 399–403 (1966); T. W. Lamb and S. M. Tenney, ibid., **21,** 404–410 (1966).
38. A. Ya. Rakhimov and V. Sh. Belkin, *Arkh. Anat. Gistol. Entomol. (Leningrad),* **1970**(11), 43–49 (1970).
39. E. Ts. Andreeva-Galanina, *Gig. Tr. Prof. Zabol.,* **15**(12), 22–25 (1971).
40. T. G. Yakubovich and N. M. Zhukova, *Gig. Sanit.,* **1970**(12), 98–100 (1970).
41. D. V. Sturges, D. W. Badger, R. N. Slarve, and D. E. Wasserman, "Laboratory Studies on Chronic Effects of Vibration Exposure," in: *Vibration and Combined Stresses in Advanced Systems,* H. E. von Gieke, Ed., AGARD Conference Proceedings AGARD-CP-145, Paper B10, NATO/AGARD, Neuilly-sur-Seine, France, March 1975. (See also D. W. Gadger et al., *loc. cit.,* Paper B11.)
42. R. G. Edwards, E. P. McCutcheon, and C. F. Knapp, *J. Appl. Physiol.,* **32,** 384–390 (1972).
43. J. Demange, R. Auffret, and B. Vettes, "Effects of Low-Frequency Vibrations on the Human Cardiovascular System," in: *AGARD Conference Proceedings,* AGARD-CP-145 (1975).
44. E. P. McCutcheon, "Effects of Vibration Stress on the Cardiovascular System of Animals," in: *Vibration and Combined Stresses in Advanced Systems,* H. E. von Gierke, Ed., AGARD Conference Proceedings AGARD-CP-145, Paper B9, NATO/AGARD, Neuilly-sur-Seine, France, 1975.
45. A. K. McIntyre, *Proc. Aust. Assoc. Neurol.,* **3,** 71–75 (1965).
46. T. J. Moore, *IEEE Trans. Man-Mach. Syst.,* **MMS-11,** 79–84 (1970).
47. R. W. Shoenberger and C. S. Harris, *Human Factors,* **13,** 41–50 (1971).
48. T. Miwa, *Ind. Health (Jpn.),* **6,** 1–10, 11–17, 18–27 (1968).
49. T. Miwa, *Ind. Health (Jpn.),* **7,** 89–115, 116–126 (1969).

50. T. Miwa, *Ind. Health (Jpn.)*, **8,** 116–126 (1970).
51. J. R. McKay, *A Study of the Effects of Whole-Body + az Vibration on Postural Sway*, 6570th Aerospace Medical Research Laboratory Technical Report AMRL-TR-71-121, Wright-Patterson Air Force Base, Ohio, 1972.
52. A. A. Menshov, D. V. Vinogradov, and A. M. Baron, *Gig. Sanit.*, **35,** 32–36 (1970).
53. S. K. Nurbaev. *Gig. Tr. Prof. Zabol.*, **19**(5), 32–35 (1975).
54. W. L. Johnston, "An Investigation of the Effects of Low-Frequency Vibration on Whole Body Orientation," Doctoral Dissertation (Order No. 70-1474), Texas Technical College, 1969.
55. A. N. Nicholson, L. E. Hill, R. G. Borland, and H. M. Ferres, *Aerospace Med.*, **41,** 436–446 (1970).
56. B. Craske, *Science*, **196,** 71–73 (1977).
57. K.-E. Hagbarth and G. Eklund, *Scand. J. Rehabil. Med.*, **1,** 26–34 (1969).
58. P. B. C. Matthews, "Vibration and the Stretch Reflex," in: *Myotatic, Kinesthetic and Vestibular Mechanisms*, A. V. S. de Rueck, Ed., Section I: "Myotatic and Kinesthetic Mechanisms," Little, Brown, Boston, 1967.
59. D. Burke, C. J. Andrews, and J. W. Lance, *J. Neurol. Neurosurg. Psychiatr.*, **35,** 477–486 (1972).
60. K.-E. Hagbarth and G. Eklund, "Motor Effects of Vibratory Muscle Stimuli in Man," in: *Muscular Afferents and Motor Control*, R. Granit, Ed., Almqvist and Wiksell, Stockholm; Wiley, New York, 1966, pp. 177–186.
61. G. Eklund, *Acta Soc. Med. Upsalien*, **74,** 113–117 (1969).
62. C. D. Marsden, J. C. Meadows, and H. J. F. Hodgson, *Brain*, **92,** 829–846 (1969).
63. J. C. Guignard and P. R. Travers, *Effect of Vibration of the Head and of the Whole Body on the Electromyographic Activity of Postural Muscles in Man*, Air Ministry Flying Personnel Research Committee Memorandum FPRC/Memo 120, London, 1959.
64. L. D. Luk'yanova and Ye. P. Kazanskaya, *Fiziol. Zh.*, **53,** 563–570 (1967).
65. R. Aston and V. L. Roberts, *Arch. Int. Pharmacodyn.*, **155**(2), 289–299 (1965).
66. A. N. Nicholson and J. C. Guignard, *EEG Clin. Neurophysiol.*, **20,** 494–505 (1966).
67. J. C. Guignard, and A. Irving, *Engineering (London)*, **190,** 364–367 (1960).
68. J. C. Guignard, A. C. Bittner, Jr., and R. C. Carter, *J. Low Freq. Noise Vib.*, **1,** 12–18 (1982).
69. G. J. Hurt, *Rough-Air Effect on Crew Performance During a Simulated Low-Altitude High-Speed Surveillance Mission*, National Aeronautics and Space Administration Technical Note D-1924, Washington, DC, April 1963.
70. B. Caiger, *Some Problems in Control Arising from Operational Experiences with Jet Transports*, National Aeronautics Establishment (Canada) Report NAE MISC 41, Ottawa, August 1966.
71. P. Lecomte, Chairman, and Contributors, *Simulation*, AGARD Conference Proceedings AGARD-CP-79-70, NATO/AGARD, Neuilly-sur-Seine, France, 1970.
72. D. J. Thomas, J. C. Guignard, and G. C. Willems, "Problem of Defining Criteria for the Protection of Crewmen from Low-Frequency Ship Motion Effects," in: *Proceedings of the 24th. DRG Seminar on the Human as a Limiting Element in Military Systems*, Toronto, Canada, May 1983, NATO Defence Research Group, DS/A/DR(83) 170, 1983, pp. 189–242.
73. C. W. Nixon and H. C. Sommer, *Influence of Selected Vibrations upon Speech (Range of 2 cps-20 cps and Random)*, 6570th, Aerospace Medical Research Laboratory Technical Documentary Report AMRL-TDR-63-49, Wright-Patterson Air Force Base, Ohio, 1963.
74. C. W. Nixon and H. C. Sommer, *Aerospace Med.*, **34,** 1012–1017 (1963).
75. R. N. Slarve and D. L. Johnson, *Aviat. Space Environ. Med.*, **46,** 428–431 (1975).
76. R. W. B. Stephens, *Rev. Acust.*, **2,** 48–55 (1971).
77. R. A. Hood and J. Leventhall, *Acustica*, **21,** 10–13 (1971).

78. R.-G. Busnel and A.-G. Lehmann, *J. Acoust. Soc. Am.*, **63**, 974–977 (1978).
79. J. A. Roman, *Effects of Severe Whole-Body Vibration on Mice and Methods of Protection from Vibration Injury,* Technical Report WADC-TR 58-107, U.S. Wright Air Development Center, Ohio, 1958.
80. J. S. P. Rawlins; also H. E. von Gierke, personal communications.
81. E. Pichard, *Rev. Corps. Sant. Armees,* **1970**(10), 611–635 (1970).
82. R. Snook, *Br. Med. J.,* **1972**(iii), 574–578 (1972).
83. R. Rosegger and S. Rosegger, *J. Agric. Eng. Res.,* **5**, 241–275 (1960).
84. G. J. Gruber, *Relationships between Whole Body Vibration and Morbidity Patterns Among Interstate Truck Drivers,* U.S. Department of Health, Education and Welfare (NIOSH) Publication 77-167, National Institute of Occupational Safety and Health, Cincinnati, Ohio, 1976.
85. R. C. Spear, C. A. Keller, V. Behrens, M. Hudes, and D. Tarter, *Morbidity Patterns Among Heavy Equipment Operators Exposed to Whole-Body Vibration—1975: Followup to a 1974 Study,* U.S. Department of Health, Education and Welfare (NIOSH) Publication 177-20, National Institute of Occupational Safety and Health, Cincinnati, Ohio, 1976.
86. R. C. Spear, C. A. Keller, and T. H. Milby, *Arch. Environ. Health,* **22**, 141–145 (1976).
87. H. Seris and R. Auffret, "Measurement of Low Frequency Vibrations in Big Helicopters and Their Transmission to the Pilot," paper presented at the 22nd Annual Meeting of the AGARD Aerospace Medical Panel, Munich, September 1965 (translated from the French by John F. Holman & Co., NASA Technical Translation NASA TT F-471, 1967.)
88. R. P. Delahaye, H. Seris, R. Auffret, R. Jolly, G. Gueffier, and P. J. Metges, *Rev. Med. Aeron. Spat.,* **38**, 99–102 (1971).
89. G. J. Gruber and H. H. Ziperman, *Relationship between Whole-Body Vibration and Morbidity Patterns Among Motor Coach Operators,* U.S. Department of Health, Education and Welfare (NIOSH) Publication 75-104, National Institute of Occupational Safety and Health, Cincinnati, Ohio, 1974.
90. G. I. Rumyantsev and D. A. Mikhelson, *Gig. Sanit.,* **1971**(9), 25–27 (1971).
91. V. M. Gornik, *Gig. Tr. Prof. Zabol.,* **20**(4), 24–28 (1976).
92. V. I. Chernyuk, *Gig. Tr. Prof. Zabol.,* **19**(6), 19–23 (1975).
93. G. Eklund, *Upsala J. Med. Sci.,* **77**, 112–124 (1972).
94. V. I. Dynik, *Gig. Tr. Prof. Zabol.,* **19**(9), 32–36 (1975).
95. J. C. Guignard, G. J. Landrum, and R. E. Reardon, *Experimental Evaluation of International Standard (ISO 2631-1974) for Whole-Body Vibration Exposures. Final Report to the National Institute of Occupational Safety and Health,* University of Dayton Research Institute Technical Report UDRI-TR-76-79, Dayton, Ohio, 1976.
96. A. F. Parlyuk, *Vrach. Delo.,* **1970**(7), 122–126 (1970).
97. G. G. Lysina and A. F. Parlyuk, *Vrach. Delo.,* **1**, 124–128 (1973).
98. L. N. Gratianskaya, *Gig. Tr. Prof. Zabol.,* **18**(8), 7–10 (1974).
99. C. J. Gauss, *Zbl. Gynäkol.,* **50**, 50ff (1926) (cited in Reference 2).
100. E. Ts. Andreeva-Galanina, S. V. Alekseyev, A. V. Kadyskin, and G. A. Suvorov, *Noise and Noise Sickness,* Meditsina Press, Leningrad, 1972 (National Aeronautics and Space Administration Technical Translation NASA TT F-748, Washington, DC, July 1973.)
101. A. Begor, *Transp. Med. Vest.,* **14**(2), 22–29 (1969).
102. H. C. Sommer, *The Combined Effects of Vibration, Noise, and Exposure Duration on Auditory Temporary Threshold Shift,* 6570th Aerospace Medical Research Laboratory Technical Report AMRL-TR-73-34, Wright-Patterson Air Force Base, Ohio, 1973.
103. J. C. Guignard, *Combined Effects of Noise and Vibration on Man,* University of Dayton Research Institute Technical Report UDRI-TR-73-51, Dayton, Ohio, 1973.
104. C. Anitesco and C. Contulesco, *Arch. Mal. Prof.,* **33**, 365–371 (1972).

105. A. B. Lewis, *Noise Control Eng.*, **7**, 132–139 (1976).
106. S. L. Gibbons, A. B. Lewis, and P. Lord, *J. Sound Vib.*, **43**, 253–261 (1975).
107. N. N. Livshits, Ed., *Effects of Ionizing Radiation and of Dynamic Factors on the Functions of the Central Nervous System—Problems of Space Physiology*, National Aeronautics and Space Administration technical Translation NASA TT F-354 from the Russian (Nauka, Moscow, 1964), Washington, DC, 1975.
108. T. S. L'vova, "The Influence of Vibration on the Course and Outcome of Radiation Injury in Animals," in: *Problems in Aerospace Medicine*, V. V. Parin, Ed., Moscow, 1966, pp. 347–348, (National Aeronautics and Space Administration translation from the Russian, JPRS-38272, Washington, DC).
109. L. Ya. Tartakovskaya, *Gig. Tr. Prof. Zabol.*, **20**(4), 32–36 (1976).
110. N. N. Malinskaya, *Gig. Tr. Prof. Zabol.*, **19**(7), 16–20 (1975).
111. S. Jaworski, *Pr. Cent. Inst. Ochr. Pr.*, **25**, 325–341 (1975).
112. F. P. Dimasi, R. E. Allen, and P. C. Calcaterra, *Effect of Vertical Active Vibration Isolation on Tracking Performance and Ride Qualities*, NASA Contractor Report CR 2146, National Aeronautics and Space Administration, Washington, DC, November 1972.
113. L. F. Stikeleather and C. W. Suggs, *Trans. Am. Soc. Agric. Eng.*, **13**, 99–106 (1970).
114. R. E. Young and C. W. Suggs, *An Active Seat Suspension System for Isolation of Roll and Pitch in Off-Road Vehicles*, American Society of Agricultural Engineers Paper 73-156, ASME, St Joseph, Michigan, 1973.
115. L. Sjøflot and C. W. Suggs, *Ergonomics*, **16**, 455–468 (1973).
116. D. Dieckmann, *Ergonomics*, **1**, 347–355 (1958).
117. K. L. Merino and R. B. Cooper, *Draft U.S. Merchant Ship Design Standards: Report I*, Eclechtech Associates, Inc., North Stonington, Connecticut, Catalog No. PB83 255653, October 1982.
118. D. F. Shanahan and T. E. Reading, *Aviat. Space Environ. Med.*, **55**, 117–121 (1984).
119. I. Pyykkö, "Vibration Syndrome: A Review," in Reference [122], pp. 1–24.
120. W. Taylor, Ed., *The Vibration Syndrome*, British Acoustical Society Special Volume 2, Academic Press, London, 1975.
121. W. Taylor and P. L. Pelmear, *Vibration White Finger in Industry*, Academic Press, London, 1975.
122. D. E. Wasserman, W. Taylor, and M. G. Curry, Eds., *Proceedings of the International Hand-Arm Vibration Conference*, Cincinnati, Ohio, October 1975, U.S. Department of Health, Education, and Welfare (NIOSH) Publication 77-170, National Institute of Occupational Safety and Health, Cincinnati, Ohio, April 1977.
123. O. Korhonen, Ed., *Vibration and Work, Proceedings of the Finnish–Soviet–Scandinavian Vibration Symposium*, Helsinki, March 10–13, 1975, Helsinki, Institute for Occupational Health, 1976.
124. Department of Health and Social Security (United Kingdom), *Vibration Syndrome*, Report to Parliament by the Industrial Injuries Advisory Council, Cmnd4430, Her Majesty's Stationery Office, London, July 1970.
125. W. Taylor, "Opening Remarks," in Reference 122, pp. 3–4; see also W. Taylor and A. J. Brammer, "Vibration Effects on the Hand and Arm in Industry: An Introduction and Review," in Reference 200, pp. 1–12.
126. D. D. Reynolds and W. Soedel, *J. Sound Vib.*, **44**, 513–523 (1976).
127. M. Allingham and R. D. Firth, *NZ Med. J.*, **76**, 317–321 (1972).
128. N. Agate and H. A. Druett, *Br. J. Ind. Med.*, **3**, 159–166 (1946).
129. N. Agate and H. A. Druett, *Br. J. Ind. Med.*, **4**, 141–163 (1947).
130. S. A. Åxelsson, "Analysis of Vibration in Power Saws," *Stud. For. Suec.*, **59**, 1–47 (1968).
131. J. Hyvärinen, I. Pyykkö, and S. Sundberg, *Lancet*, **i**, 791–794 (1973).

132. D. D. Reynolds, "Hand–Arm Vibration: A Review of 3 Years' Research," in Reference 122, pp. 99–128.
133. H. Rafalski, K. Bernacki, and T. Switoniak, "The Diagnostics and Epidemiology of Vibration Disease and Hearing Impairment in Motor Sawyers," in Reference 122, pp. 84–88.
134. E. E. Dart, *Occup. Med.*, **1**, 515–550 (1946).
135. A. Banister and F. V. Smith, *Br. J. Ind. Med.*, **29**, 264–267 (1972).
136. H. Iwata, *Ind. Health (Jpn.)*, **6**, 28–36, 47–58 (1968).
137. W. F. Ashe, *Physiological and Pathological Effects of Mechanical Vibration on Animals and Man*, Ohio State University Research Foundation Report 862-4, Columbus, Ohio, 1961.
138. W. F. Ashe, W. T. Cook, and J. W. Old, *Arch. Environ. Health*, **5**, 333–343 (1962).
139. A. Okada, T. Yamashita, C. Nagano, T. Ikeda, A. Yachi, and S. Shibata, *Br. J. Ind. Med.*, **28**, 353–357 (1971).
140. M. Futatsuka, *Jpn. J. Ind. Health*, **15**, 371–377 (1973).
141. M. Futatsuka, *Jpn. J. Ind. Health*, **30**, 266 (1975).
142. R. P. Jepson, *Br. J. Ind. Med.*, **11**, 180–185 (1954).
143. A. Okada, T. Yamashita, and T. Ikeda, *Am. Ind. Hyg. Assoc. J.*, **33**, 476–482 (1972).
144. M. Sevčik, *Pracov. Lek.*, **25**, 46–50 (1973).
145. M. Drobny, *Pracov. Lek.*, **25**, 185–189 (1973).
146. E. Lukas and V. Kuzel, *Int. Arch. Arbeitsmed.*, **28**, 239–249 (1971).
147. N. Bjerker, B. Kylin, and I.-M. Lidström, *Ergonomics*, **15**, 399–406 (1972).
148. F. Horváth and T. Kákosy, *Zeitschr. Orthop. Grenzgeb.*, **107**, 482–494 (1970).
149. D. Koradecka, "Peripheral Blood Circulation Under the Influence of Occupational Exposure to Hand-Transmitted Vibration," in Reference 122, pp. 21–36.
150. T. Banaszkiewicz, J. Gwozdziewicz, and J. Waskiewicz. *Biul. Inst. Med. Morsk. Gdansk*, **21**, 147–162 (1970).
151. A. G. Mirzaeva, *Med. Zh. Uzb.*, **1971**(5), 26–29 (1971).
152. H. Langauer-Lewowicka, *Int. Arch. Occup. Environ Health*, **36**, 206–216 (1976).
153. R. M. Nerem, *Arch. Environ. Health*, **26**, 105–110 (1973).
154. R. M. Nerem, "Vibration Enhancement of Blood-Arterial Wall Macromolecular Transport," in Reference 122, pp. 37–43.
155. G. Eklund and K.-E. Hagbarth, *Exp. Neurol.*, **16**, 80–92 (1966).
156. G. Jansen, *Arch. Gewerbepath. Gewerbehyg.*, **17**, 238–261 (1959).
157. A. M. Stewart, *Br. J. Occup. Safety*, **7**, 454–455 (1968).
158. A. M. Stewart and D. F. Goda, *Br. J. Ind. Med.*, **27**, 19–27 (1970).
159. M. A. Yershova, *Vrach. Delo.*, **1970**(7), 116–119 (1970).
160. S. Carlsöö and J. Mayr, *Work-Environ.-Health*, **11**, 32–38 (1974).
161. H. Dupuis, E. Hartung, and W. Hammer, *Int. Archiv. Umweltmed.*, **37**, 9–34 (1976).
162. H. Iwata, H. Dupuis, and E. Hartung, *Int. Arch. Arbeitsmed.*, **30**, 313–328 (1972).
163. D. D. Reynolds and W. Soedel, *J. Sound Vib.*, **21**, 339–353 (1972).
164. C. W. Suggs and J. W. Mishoe, "Hand–Arm Vibration: Implications Drawn from Lumped Parameter Models," in Reference 122, pp. 136–141.
165. L. A. Wood and C. W. Suggs, "A Distributed Parameter Dynamic Model of the Human Forearm," in Reference 122, pp. 142–145.
166. I.-M. Lidström, "Vibration Injury Among Rock Drillers, Chiselers, and Grinders," in Reference 123, pp. 81–88.
167. I.-M. Lidström, "Vibration Injury in Rock Drillers, Chiselers, and Grinders," in Reference 122, pp. 77–83.

168. E. Lukas and L. Louda, *Cesk. Neurol. Neurochir*, **37,** 258–262 (1974).
169. E. R. Tichauer, "Thermography in the Diagnosis of Work Stress due to Vibrating Implements," in Reference 122, pp. 160–168; also a comment on p. 208.
170. S. Samueloff, R. Miday, D. Wasserman, V. Behrens, R. Hornung, W. Asburry, T. Doyle, F. Dukes-Dobos, and D. Badger, *J. Occup. Med.*, **23,** 643–646 (1981).
171. D. Koradecka, *Pr. Cent. Inst. Ochr. Pr.*, **20,** 147–160 (1970).
172. A. J. Zweifler, "Detection of Occlusive Arterial Disease in the Hand and its Relevance to Occupational Hand Disease," in Reference 122, pp. 12–20.
173. B. Hellstrøm, *Int. Z. Angew. Physiol. Einschl. Arbeitsphysiol.*, **29,** 18–28 (1970).
174. M. Tiililä, *Work-Environ.-Health*, **7,** 85–87 (1970).
175. A. F. Buzalo, *Oftal'mol. Zh.*, **25,** 434–437 (1970).
176. A. M. Seppäläinen, "Neurophysiological Detection of Vibration Syndrome in the Shipbuilding industry," in Reference 123, pp. 67–71.
177. W. Taylor, J. C. G. Pearson, and G. D. Keighley, "A Longitudinal Study of Raynaud's Phenomenon in Chain Saw Operators," in Reference 122, pp. 69–76.
178. T. M. Radziukewich and A. M. Mikulinskii, *Gig. Tr. Prof. Zabol.*, **16**(7), 16–20 (1972).
179. T. Kumlin, *Work-Environ.-Health*, **7,** 57–58 (1970).
180. D. B. Chaffin, *Occup. Med.*, **11,** 109–115 (1969).
181. E. R. Tichauer, *Occupational Biomechanics*, New York University Rehabilitation Monograph 51, New York, 1975.
182. Anonymous, "A Czechoslovakian Proposal for a Guide to the Evaluation of Human Exposure of the Hand-Arm System to Vibration," *Work-Environ.-Health*, **7,** 88–90 (1970).
183. Anonymous, *Portable Power Tools—Permitted Levels of Vibration*, USSR Standard (GOST) 17770-72, July 1, 1972.
184. British Standards Institution, *Guide to the Evaluation of the Human Hand-Arm System to Vibration*, Draft for Development DD43, BSI, London, 1975.
185. J. Starck, "A Finnish Recommendation for Maximum Vibration Levels," in Reference 123, pp. 117–120.
186. T. Miwa, *Ind. Health (Jpn.)*, **5,** 182–205, 206–212, and (Hand-Transmitted Vibration) 213–220 (1967).
187. S. A. Åxelsson, "Progress in Solving the Problem of Hand–Arm Vibration for Chain Saw Operators in Sweden, 1967 to Date" (1975), in Reference 122, pp. 218–224.
188. J. R. Bailey, "Chain Saws—Problems Associated with Vibration Measurements," in Reference 122, pp. 187–208.
189. G. D. Keighley, "FAO/ECE/ILO Resolution on Hand-Arm Vibration for Modern Antivibration Chain Saw: Recommendation for Medical Monitoring," in Reference 122, pp. 233–235.
190. U. Naeslund, "The Theory Behind a New Measurement for Hand–Arm Vibration," in Reference 122, pp. 225–229.
191. A. P. Politschuk and V. Oblivin, "Methods of Reducing the Effects of Noise and Vibration on Power Saw Operators," in Reference 122, pp. 230–232.
192. S. Yamawaki, "Reduction of Vibration in Power Saws in Japan," in Reference 122, pp. 209–217.
193. W. Taylor, J. Pearson, R. L. Kell, and G. D. Keighley, *Br. J. Ind. Med.*, **28,** 83–89 (1971).
194. Z. M Butkovskaya, *Gig. Tr. Prof. Zabol.*, **16**(3), 19–24 (1972).
195. P. Lyarskiy, *Sanitary Standards and Regulations to Restrict the Vibration of Work-Places*, Standard 627-66 drawn up by F. F. Erisman Research Institute of Hygiene, Moscow, USSR, 1966 (translated from the Russian by W. Linnard, Forestry Commission, United Kingdom).
196. L. Louda and E. Lukas, "Hygienic Aspects of Occupational Hand-Arm Vibration," in Reference 122, pp. 60–66.

197. A. P. Guk, *Bezop. Tr. Prom-St.*, **1970**(2), 16–17 (1970).
198. A. A. Demin, A. Ia. Khrupina, and G. P. Vasilenko, *Gig. Tr. Prof. Zabol.*, **15**(10), 16–20 (1971).
199. A. M. Ehrly, "Treatment of Severe Secondary Raynaud's Disease," in Reference 122, pp. 44–46.
200. A. J. Brammer and W. Taylor, Eds., *Vibration Effects on the Hand and Arm in Industry* (edited presentations at the Third International Conference on Hand–Arm Vibration, Ottawa, May 1981), Wiley, New York, 1982.
201. Anonymous, *Vibration Syndrome,* Current Intelligence Bulletin 38, U.S. Department of Health and Social Services, National Institute for Occupational Safety and Health, NIOSH Publication 83-110, Cincinnati, Ohio, March 29, 1983.
202. M. J. Griffin, *Vibration Injuries of the Hand and Arm: Their Occurrence and the Evolution of Standards and Limits*, U.K. Safety and Health Executive Research Report No. 9, London, Her Majesty's Stationery Office, 1980.
203. D. D. Reynolds, D. E. Wasserman, R. Basel, and W. Taylor, "Energy Entering the Hands of Operators of Pneumatic Tools Used in Chipping and Grinding Operations," in Reference 200, pp. 133–146.
204. V. Behrens, W. Taylor, and D. E. Wasserman, "Vibration Syndrome in Workers Using Pneumatic Chipping and Grinding Tools," in Reference 200, pp. 147–155.
205. O. Backteman, J. Köhler, and L. Sjöberg, *J. Low Freq. Noise Vib.* **2,** 1–31, 176–210 (1983).
206. M. A. Fisher, *The Impact of Motion and Motion Sickness on Human Performance Aboard Monohull Vessels and Surface Effect Ships: A Comparative Study*, Master's Thesis to the Naval Postgraduate School, Monterrey, California, October 1982 (AD A124614).
207. International Organization for Standardization, *Guidelines for the Evaluation of the Response of Occupants of Fixed Structures, Especially Buildings and Off-Shore Structures, to Low-Frequency Horizontal Motion (0,063 to 1 Hz)*, International Standard ISO 6897-1984, ISO, Geneva, 1984.
208. A. Graybiel and J. Knepton, *Aviat. Space Environ. Med.*, **47,** 873–882 (1976).
209. K. E. Money and B. S. Cheung, *Aviat. Space Environ. Med.*, **54,** 208–211 (1983).
210. A. J. Benson, *Possible Mechanisms in Motion and Space Sickness*, Proceedings of the European Symposium on Life Sciences Research in Space, Cologne, 24–26 May 1977 (ESA SP-130), pp. 101–108, 1977.
211. E. I. Matsnev, I. Y. Yakovleva, I. K. Tarasov, V. N. Alekseev, L. N. Kornilova, A. D. Mateev, and G. I. Gorgiladze, *Aviat. Space Environ. Med.*, **54,** 312–317 (1983).
212. J. R. Lackner and A. Graybiel, *Aviat. Space Environ. Med.*, **54,** 675–681 (1983).
213. J. L. Homick, M. F. Reschke, and J. M. Vanderploeg, "Space Adaptation Syndrome: Incidence and Operational Implications for the Space Transportation System Program," in: *Motion Sickness: Mechanisms, Prediction, Prevention and Treatment,* AGARD Conference Proceedings AGARD-CP-372, Paper 36, NATO/AGARD, Neuilly-sur-Seine, France, 1984.
214. R. S. Kennedy and L. H. Frank, *A Review of Motion Sickness with Special Reference to Simulator Sickness,* Paper to NAS/NRC Committee on Human Factors, Workshop on Simulator Sickness, Naval Postgraduate School, Monterrey, California, 26–28 September 1983 (Revised 9 March 1984).
215. E. J. Rinalducci (Chair), Panel on Impact of Video Viewing on Vision of Workers, Committee on Vision, Commission on Behavioral and Social Sciences and Education, National Research Council, *Video Displays, Work, and Vision,* Washington D.C., National Academy Press, 1983.
216. M. J. Smith, B. G. F. Cohen, L. W. Stammerjohn, and A. Happ, *Human Factors,* **23,** 387–400 (1981).
217. S. R. Olson, *Human Effectiveness in the Ship Motion Environment,* Center for Naval Analyses, Arlington, Virginia, Memorandum (CNA) 77-0665, 13 May 1977.
218. S. R. Olson, *A Seakeeping Evaluation of Four Naval Monohulls and a 3400 Ton SWATH,* Center for Naval Analyses, Arlington, Virginia, Memorandum (CNA) 77-0640, 20 March 1977.

219. J. W. Kehoe, K. S. Brower and E. N. Comstock, *Seakeeping Proceedings, U.S. Naval Institute*, **109**, 63–67 (1983).
220. A. E. Baitis, T. R. Applebee, and T. M. McNamara, *Nav. Engrs. J.*, **96**, 191–199.
221. A. C. Bittner and J. C. Guignard, *Human Factors Engineering Principles for Minimizing Adverse Ship Motion Effects: Theory and Practice, Nav. Engrs. J.* (in press).
222. R. N. Andrew and A. R. J. M. Lloyd, *Trans. Roy. Inst. Nav. Architects*, **123**, 1–31 (1981).
223. J. C. Guignard, *Lecture Notes on Vibration and Man: (2) Criteria for Human Acceptability of Vibration—and Applications*, presented at 1983 John C. snowdon Vibration Control Seminar, Pennsylvania State University, University Park, Pennsylvania, 3 November 1983.
224. M. E. McCauley and R. S. Kennedy, *Recommended Human Exposure Limits for Very-Low-Frequency Vibration*, Department of the Navy, Pacific Missile Test Center, Point Mugu, California, Technical Publication TP-76-36, September 1976.

Index

Abdominal pressure, in lifting work, 396
Abnormal pressure, 431–449
  altitude physiology, 432–436, 440–441
  and decompression sickness, 443–445
  and hypoxia, 437–438
  and mountain sickness, 442
  physics of, 431–432
  and worker exposure, 365–366
Absorption:
  integrated index, 78
  membrane defenses, 145–148
Acceleration, human measurements, 657
Accidents:
  and circadian rhythms, 286
  hospital, 288–289
  with industrial robots, 419
  on night shift, 289
  noise as cause of, 411
  shiftwork and, 176, 288
  temporal patterns in shiftwork, 287
  worker susceptibility to, 392
Acclimatization:
  animal studies, 486
  to cold environments, 484
  to hot environments, 492–493
  human studies, 487
  "hunting" response in, 487
  to thermal environments, 481–482
Acetaminophen, hepatotoxicity of, 163–164
Acetazolamide, and altitude tolerance, 441
Acetylation, amine detoxication by, 164–165
Acetylators, "slow," 85
Acidosis, and hemoglobin-oxygen dissociation curve, 436
Acoustic factors, in work performance, 410
Acoustic trauma, 567
Acquired deafness, causes of, 564–565

Acrophase charts, of circadian rhythms, 217–220, 273
Actinic skin disorders, ultraviolet-induced, 587
Adaptation:
  to altitude exposure, 436–437
  to cold environments, 483–484
  human studies, 487
  to thermal environments, 481–482
Adrenal gland, and circadian rhythms, 208–210, 224
Adrenocorticotrophic hormone (ACTH), in circadian rhythm, 264
Adriamycin, circadian chronotoxicity of, 255–257
Aerobic power:
  age factor, 322, 323, 393
  and heat tolerance, 496
  indirect measurement of, 316–317
  and repetitive manipulations, 391
  test procedures, 318–324
  and work duration, 324, 393
Aerospace research, time-qualified reference values in, 292
Affective illness, in retired shiftworkers, 283
Age:
  and metabolism, 91–92
  and shiftwork intolerance, 274
Air, biological agents in, 453
Airborne sampling, of industrial contaminants, 11–12
Air compressors, for hyperbaric chambers, 447
Aircraft cabin pressurization, 438–439
Airflow:
  calculation methods, 349
  measurements, 350

Airflow (*Continued*)
  particle radiation monitors, 506–507
Air pollution:
  indoor, 367
  oxidant defense mechanisms, 166
Alarm detection, practical aids, 399
Alcohol:
  and aldehyde dehydrogenases, 160
  biological monitoring of, 360
  in cold environments, 490
  glucuronic acid detoxication, 161
  sulfate conjugation of, 162
  toxicity of, 160
Alcoholism, chemical monitoring problems, 95
Aldehydes, biological monitoring of, 360
Alkalosis, and hemoglobin-oxygen dissociation curve, 436
Allergens, 9
Allergic alveolitis, *see* Bird fancier's disease
Allergy:
  biological rhythms and, 225
  susceptibility-resistance rhythms, 261–263
Alpha particles, in internal radiation exposure, 503
Altitude:
  acclimatization, 440–441
  and aerobic power, 324, 325
  and environmental exposure, 367, 436
  hypoxia, 437–438
  physiology and abnormal pressure, 432–436
Aluminum workers:
  energy expenditures of, 319
  sampling techniques, 342
  urine fluoride levels, 115
Alveolar gas equation, 433
Alveolar macrophages, in particle removal, 152
Alveolar proteinosis, pulmonary lavage for, 541
Alveolitis, extrinsic allergic, 459
Alveolus, and pulmonary capillary, 434
Alzheimer's disease, 629
American Academy of Ophthalmology and Otolaryngology, 561
American Board of Industrial Hygiene, 13
American Conference of Governmental Industrial Hygienists (ACGIH), 13
  on chemical mixtures, 364
  exposure limits, 368
  on ultraviolet radiation, 593–594
  unusual workshifts, 188–190

American Industrial Hygiene Association (AIHA), 13, 14
  Ergonomic Guide, 396
  membership growth, 24
  on nonionizing radiation, 590
  on quality control, 336
  on recordkeeping, 337
  "unusual work schedule," 183–184
American Medical Association (AMA):
  on normal hearing, 561
  on ultraviolet radiation, 592
American National Standards Institute (ANSI):
  exposure limits, 368
  microwave-radiofrequency exposure recommendations, 635
  on normal hearing, 561
American Public Health Association, 13
American Speech and Hearing Association, 561
Amino acid conjugation, 162, 165
Ammonium chloride, and altitude tolerance, 441
Amphetamines, circadian-susceptibility rhythms, 248
Analgesia, chronesthesy of, 239–241
Analysis of variance (ANOVA), in biological rhythm detection, 271–272
Anatomical defenses, to particle deposition, 149–150
Anemia, and hypoxia, 437
Angina:
  circadian rhythm in, 231
  and cold exposure, 489
Animals:
  exposure chambers, 346
  microwave-radiofrequency absorption rates, 627
  reliability of studies, 39–42
  standardization of, 259
Annual Review of Nuclear and Particle Science, 513
Anthrax:
  plant disinfection, 456
  in textile mills, 472
Anthropometry:
  in ergonomics, 376
  instruments, 379
  musculoskeletal factors, 380–382
  percentile data, 378, 380
Anticancer agents:
  circadian chronotoxicity of, 255–257
  toxicity in animals, 40

INDEX

Antidiabetic drugs, toxicity of, 155
Antidotes, risk therapy, 79
Antiepileptic drugs, saliva analysis, 136
Antigens, chronograms of circadian susceptibility, 226
Antihistamines, chronokinetics of, 237
Antineoplastic agents, toxicity in animals, 40
Antispasmodic drugs, and inner ear diseases, 566
Antitubercular agents, toxicity differences, 85
Antivibration saw, 706
Area monitoring, validation of, 80
Argonne diet, for jet lag, 284–286
Arousal theory, of sensory inputs, 400–401
Arsenic exposure, food intake and, 360
Arterial disease, vibration-related, 701
Arteriosclerosis, abnormal temperature and, 365
Arthritis drugs:
 and chemical binding, 88
 chronokinetics of, 235–236
Arthus phenomenon, in farmer's lung disease, 458
Asbestos:
 exposure studies, 344, 363
 inhalation of, 353
 lung toxicity from, 152
Ascorbic acid, and heat stress, 494
Aspergillosis, invasive, 465
Aspirin, chronokinetics of, 237
Asthma:
 autorhythmometry and airway function, 293
 biological rhythms and, 210, 227, 231
 nocturnal, 233
 pulmonary infiltration in, 459
 susceptibility-resistance rhythms, 261–263
 temporal variation in airway patency, 233
Astrand-Rhyming nomograms, of exercise heart rate, 322
Astronauts, blood chemistry of, 292
Atabrin, prophylaxis for malaria, 455
Audiograms:
 annual, 574–575
 in noise control, 412
Audiometer, 563
Auditory displays, 383–385
Auditory sensitivity, defined, 560
Australian aborigines, physiological responses to cold, 482
Automation:
 perceptual demands, 398
 and sensory overloading, 399

Autorhythmometry, 269–270
 biological monitoring and, 292–294
 shiftwork studies, 271
Aviation:
 air turbulence and crew performance, 669
 vibration-related morbidity, 673, 674

Background noise, acceptable levels for work areas, 410
Back pain, vibration-induced, 673
Bacteria:
 oxidant toxicity in, 168
 products of metabolism, 458
Bagassosis, 463–464
Ballistic lifting, 395
Barometric pressure, defined, 432
Basal metabolic rate (BMR), in cold environments, 485–486
Basic Industrial Hygiene—A Training Manual, 368
Battelle Pacific Northwest Laboratories, 522
Beam caliper, 379
Bed rest, work performance after, 329–330
Behavioral approaches, to work design, 423–425
Behavioral effects, of microwave-radiofrequency radiation, 623, 633
Bends, in decompression sickness, 444–445
Benzene:
 administrative controls, 351
 enzyme stimulation of, 87
 urinary phenol levels, 102
Benzoic acid, dose effect on metabolism, 89
Beryllium exposure, of nonworkers, 363
Beryllium Registry, 363
Beta particles, internal radiation from, 503
Bicycle ergometers, 318, 321
Bile, toxicant excretion into, 156
Bioassay techniques, for ionizing radiation, 543–548
Bioastronautics Data Book, 394
Biochemical defenses, to toxicant exposure, 156–171
Biochemical problems, in biological monitoring, 81–107
Biological agents, 451–477
 engineering controls, 456–457
 exposure modes, 452
 in living materials, 470–475
 medical programs, 455
 nonindustrial, 451
 in nonliving materials, 458–470
 personal protection, 457

Biological agents (*Continued*)
  preventive measures, 454–457
  recognition of problem, 469
  sources of, 452–454
  in textile and cordage fibers, 459–462
  types of, 451
Biological analysis, *see* Biological monitoring
Biological assays, 95–104
Biological effects:
  of ionizing radiation, 511–530
  of laser exposure, 613
  of microwave exposure, 639–640
  of noise exposure, 574–577
  of radionuclide radiation, 519–530
  of ultraviolet radiation, 588
Biological monitoring:
  analytical methodology, 105–106
  area validation, 80
  and autorhythmometry, 292–294
  biochemical problems, 81–107
  of chemical dosage, 75–136
  chronobiologic aspects of, 289–296
  controversy over, 292
  defined, 76
  direct *vs*. indirect, 80–81
  exposure route of chemical, 76–78, 106–107
  genotype differences in, 85
  history of, 78
  of metabolism, 81–94
  "round-the-clock," 179
  types of, 80–81
  usefulness of, 79–80
Biological Monitoring for Industrial Chemical Exposure Control, 360
Biological rhythms, 13
  acrophase shift in, 267–268
  animal studies, 224
  biochemical oscillators, 225
  categories of, 207–208
  detection and description methods, 271–273
  free-running, 221–223
  generation of, 225
  genetic basis of, 224
  and human illness, 227–234
  human models, 267–269
  illustrative spectrum, 207
  isolation studies, 276
  mechanisms of, 224–225
  medical implications, 225–263
  naturally occurring, 207–208
  night shift adjustments, 276–279

  "nuisance" of, 261
  and occupational health, 175–300
  phase shifting of, 265–269
  significance of, 297
  synchronization of, 217–221
  and toxic responses, 245–261
Biomechanics, in ergonomics, 377
Biomedicine:
  circadian rhythms in, 219
  and microwave radiation, 640
  of nonionizing electromagnetic energies, 580–583
Biorhythm theory:
  and accident rates, 206
  vs. chronobiology, 205–207
  as popular fad, 205
  scientific investigation of, 206–207
Bird fancier's disease, 464–466
  pathology and pathogenesis, 464–465
  specific exposures, 466–467
  symptoms, 465–466
Blacklight region, of ultraviolet radiation, 585
Bladder cells, proliferation of, 64–65
Blinding glare, 599. *See also* Flash blindness
Blood:
  banks, 539
  bound *vs*. free chemical, 87–89, 94
  index chemical concentration in, 118
  levels and exercise, 320
Blood-brain barrier:
  microwave-radiofrequency effects, 623
  to toxicant exposure, 148
Blood cells:
  microwave-radiofrequency effects, 624
  radiation sensitivity of, 512
  radionuclide effects, 528–529
Blood pressure:
  circadian rhythms and, 210–212
  diastolic changes, 211–212
  exercise effect, 315–316
  and whole-body vibration, 664
Blood-testis barrier, to toxicant exposure, 148
Body:
  defense mechanisms of, 143–172
  dimensions in work design, 379–381
  membrane barriers to distribution, 148
  position in manual handling, 395
Bone:
  cancers, 524–525
  densitometry and vibration syndrome, 705
  "hot spot" problem in, 510–511
  radionuclide dosimetry in, 510–511, 524–525

INDEX

"seekers," 524
Boredom, see Job boredom
Botanical hazards, 365
Brain enzymes:
  acrophase maps of, 253
  circadian rhythms in, 252–255
Breast milk, in measurement of industrial exposure, 133–135
Breath analyses, of index chemical, 104
British Medical Research Council, 470
Bronchitis:
  among coal workers, 361
  from smoking, 454
Bronchomotor activity, alterations in, 153
Brucellosis, 453
  in meat packing, 473
Building industry, work capacity studies, 325
Bureau of Radiological Health, 619
Bus drivers, vibration-related disorders, 673–674
Byssinosis, 459–461
  industrial surveys, 460–461
  severity grades, 459–460
  symptoms of, 459

Cabin decompression, accidental, 439
Cadmium:
  hair content of, 132
  protein-binding detoxication, 155
Caffeine, and chronobiologic adjustment, 284
Caisson disease, 443. See also Decompression sickness
California State Division of Occupational Safety and Health, 372
Calorimetry, in neutron measurement, 505
Canada:
  microwave-radiofrequency exposure standards, 636
  unusual workweeks in, 184
Cancer:
  chemical-induced, 43
  in furniture workers, 468
  immunological defense mechanisms, 171
  medications, 247–251
  microwave-induced, 633
  radiation-induced, 43, 517–518
Capillary-tissue oxygen model, 436
Carbon tetrachloride, chronotoxicity of, 254
Carboxylic acids, glucuronic acid detoxication, 161
Carcinogenesis, 57–69
  animal studies, 40–41
  chemical, 43–44
  cytotoxic mechanisms, 59–61
  experimental approaches, 61–69
  genetic and nongenetic mechanisms, 44
  "irritation theory," 59
  mechanisms of, 57–58
  threshold concept, 43
Carcinogens:
  activity spectrum of, 61
  chronobiology of, 247
  classification of, 58
  genetic vs. epigenetic, 58
  ultraviolet radiation, 588
  work exposure evaluation, 357–358
Carcinoma, see Cancer
Cardiac problems, biological rhythms and, 227
Cardiopulmonary effects, of whole-body vibration, 664–665
Cardiovascular activity:
  alterations in, 153, 320
  in hot environments, 491
  microwave effects, 631
Cardiovascular drugs, chronokinetics of, 236, 237
Carpal tunnel syndrome, from repetitive manipulations, 392
Cascade impactors, in radiation monitoring, 506–507
Catalase, as defense mechanism, 169–170
Cataracts:
  age distribution, 630–631
  heat-induced, 606
  infrared-induced, 605–607, 611
  laser-induced, 616
  microwave-induced, 624, 629–630
  susceptibility to, 47–48
Catecholamines, calorigenic effects of, 484
Cattle, brucellosis from, 473
Caucasians, physiological responses to cold, 482
Caves, chronobiological studies in, 222
Cell membranes, and biological rhythms, 224–225
Cellular effects:
  of microwave-radiofrequency radiation, 622–624
  of whole-body vibration, 666
Cement workers, nonenvironment study, 361
Center for Disease Control, 455, 476
Cerebral infarction, circadian rhythm in, 229
Chain saw, vibration syndrome from, 685–686

Chain saw operators, antivibration guidelines, 707–708
Chair, well-designed, 382
Chelation therapy, for internal radiation exposure, 539–541
Chelators, types of, 539–540
Chemical agents:
  acute toxicity of, 34–36
  additive, synergistic and antagonistic effects, 364–365
  adequacy and performance of control program, 348–352
  administrative controls, 351–352
  animal studies, 345–346
  biologically active, 40
  characterization of, 341–348
  clinical cases, 346–348
  data acquisition objectives, 334–335
  enzyme-stimulating, 87
  epidemiology, 343–345
  ergonomic factors, 414–415
  excretion half-life of, 105
  exposure dose effect on metabolism, 89–90
  exposure route of, 106–107
  government regulations, 335–337
  industrial vs. medicinal, 88
  interpreting exposure levels, 333–372
  metabolism of, 86
  modes of entry, 352–355
  "potential hazards," 69
  research methods, 341–342
  "safe" levels, 27
  sampling procedures, 334–335, 341
  sex differences, 93
  surveillance data acquisition, 337–340
  synergistic effects, 364–365
  teratogenic potential of, 247
  testing procedures, 349–350
  toxicodynamics of, 41
  unknown, 342
  unusual sensitivity to, 79
Chemical industry, chronobiology and, 247
Chemical mixtures:
  effects of, 364–365
  evaluation of, 342
Chemiluminescence, 595
Chemotherapy, circadian variation in, 257
Chest counting, in radiation bioassays, 547
Chickens, bird fancier's disease from, 466–467
Children:
  blood lead levels in, 125
  hypothyroidism in, 530

Chloroform, toxicity, 31
Chokes, in decompression sickness, 445
Cholesterol metabolism, and circadian rhythms, 214
Chromic acid exposure, clinical cases, 347–348
Chromium-plating worker, autorhythmometry of, 293–294
Chromosomes:
  microwave-radiofrequency effects, 622–624
  in radiation bioassays, 545
Chronergy, 235
  of medications, 241–245
Chronesthesy, 235
  of chemical substances, 296
  of medications, 238–241
  of unusual shift schedules, 296
Chronic injury, in repetitive manipulations, 391
Chronic toxicity, 37–39
Chronobiologists, importance of, 223–224
Chronobiology, 13, 177, 204–205
  and biological monitoring, 289–296
  criteria for field studies, 270–273
  defined, 177
  detection and description methods, 271–273
  field studies, 269
  light boxes used in, 261, 262
  methodological difficulties in shiftwork research, 270–271
  of shiftwork, 204–225, 263–273
  special medications, 225
  see also Biological rhythms
Chronobiotics, 225
Chronokinetics, 235
  defined, 235
  of medications, 235–238
  of unusual shift schedules, 296
Chronological log, in surveillance data acquisition, 338
Chronopathology, 227, 234
Chronopharmacology, 235–245
Chronotoxicity:
  medical significance, 255–258
  and occupational health, 255–258
  and rapid rotation shift schedules, 294
  and unusual shift schedules, 295–296
Chronotoxicology:
  animal studies, 259–261
  early studies, 245–246
  evaluation methods, 259–261
Chronotypes, 276

INDEX

Cigarette smoke, enzyme stimulation of, 87
Circadian rhythms, 176, 207–208
  ACTH in, 264
  adjustments in shiftwork, 264–265
  adrenal gland and, 208–210, 224
  age factor, 212–214
  in angina, 231
  animal sampling models, 260
  and blood pressure, 210–212
  of cell mitosis, 210, 216
  characteristics of body temperature, 281
  examples of, 208–214
  free-running, 221–223
  of healthy young men, 213–214
  high-amplitude, 210
  and human performance, 286–289
  importance of, 232–233, 297
  individual categories, 212
  individual *vs.* group, 214
  internal desynchronization of, 280–283
  inversion of, 219
  jet lag and, 219
  in liver and brain enzymes, 252–255
  of melatonin, 210
  phase shifting of, 265–269
  of prolactin, 210
  rapid rotation and, 189
  of retired workers, 283
  of serum corticosteroids, 216
  and shiftwork, 257, 264, 286–289
  and shiftwork intolerance, 280–283
  special terminology, 208–209
  in twins, 224
  variations in, 208
  see also Biological rhythms
Cirrhosis:
  and chemical binding, 89
  and metabolism, 92
Clinical medicine:
  biological rhythms in, 177
  time-qualified reference values in, 290
Clostridial organisms, soil sources, 452
Clue (or cue), *see* Synchronizers
Coal miners, nonenvironment studies, 361
Coaptation, of health and environmental stresses, 8–9
Coccidioidal infections:
  oil spraying for, 456
  skin test, 471
Coccidioidomycosis, soil sources, 452
Cochlea, anatomy and physiology, 560
Coffee, and chronobiologic adjustment, 284
Coffee workers, allergic alveolitis in, 468

Cognitive demands, ergonomic factors, 401
Cold environments, 481–497
  adaptation to, 481–482
  age factor, 489
  behavioral factors, 482
  brain reactions to, 482–483
  ethnic considerations, 487
  exposure studies, 482
  extraordinary sensitivity to, 489
  homoisotherm response, 488
  hormonal changes in, 484
  involuntary muscular activity, 483
  long-term exposure studies, 486–487
  physiological responses, 482–490
  prolonged exposure to, 486
  skin sensitivity in, 488–489
  tolerance to, 484
Color:
  ergonomic factors, 406
  and high-intensity lamps, 406
  psychology of, 406
  in work environment, 402–407
Colored filters, disadvantages of, 406
Colorimetry:
  of silica dust, 341
  in urine analysis, 113–114
Coma, hypoglycemic, 155
Compressed workweeks:
  common types, 184
  history of, 187
  by industry and occupation, 185
  main factors affecting usage, 190
  in petrochemical industry, 184, 187
  see also Unusual workshifts
Compton scattering, in external radiation, 502
Congenital malformations, radar-induced, 632
Conjugates, 84
Continuous noise, 567, 569
Control devices:
  population stereotypes, 385–387
  recommended features for different tasks, 388
  selection of, 385
  for system responses, 387
  types of, 385
Control handles, system-equipment design, 390–391
Control programs, adequacy and performance of, 348–352
Copper, content of, in hair, 132
Cordage fibers, biological agents in, 459–462

Cork dust, biological agents in, 467
Cornea, infrared effects, 604
Cortisol:
　biological rhythm studies, 264
　time-qualified reference value, 290–291
Cosinor analysis:
　in chronobiological research, 272–273
　of shiftwork tolerance, 279
Cosmetics, hair analysis and, 128
Cotton, bacterial endotoxins in, 458
Cotton dust:
　biological agents in, 459–461
　control measures, 461
　pulmonary tests, 460–461
Cotton mills, byssinosis in, 459
Coughing, 153
Cows, fog fever in, 459
Creatinine, in biological monitoring, 92
Critical organ concept, 532, 585
Cutaneous absorption, see Skin absorption
Cyanide:
　detoxication of, 166
　and hypoxia, 437
　protein binding in antidotal therapy, 155
Cyclophosphamide, circadian chronotoxicity of, 255–257
Cysteine, in radiation therapy, 536
Cytotoxic mechanism, of carcinogenesis, 59–61

Dairy cattle, Q fever in, 475
Dalton's law, in altitude physiology, 432
Data acquisition:
　adequacy of, 348–352
　in shiftwork research, 270
Dayworkers, retired, 283
Dazzling glare, 598
Deafness, acquired, 564–565. See also Hearing loss
Decompression sickness, 443–445
　neurological manifestations, 445
　pathophysiology, 443–444
　prevention and treatment, 445
Deep-sea fisherman, cold tolerance of, 487
Delta rays, in external radiation exposure, 502
Deoxyribonucleic acid (DNA):
　and carcinogenesis, 44
　chemical interactions with, 61
　excision repair, 170–171
　hepatic alkylation, 68
　postreplication repair, 170–171
　thermodynamic degradation, 59–60
Depression, in retired shiftworkers, 283
Dermatitis, radiation-induced, 515
Detoxication, phase reactions, 158–162
Dial-display design, 386
　principles of, 387–388
Diarrheal disorders, prevention of, 456
Diathermy, radiation hazards of, 628, 632
Diet:
　for chronobiologic adjustments, 284–286
　enzymatic activity and, 86
　hair analysis and, 128
　and metabolism, 86
Diethylene glycol, dose-response, 31
Diethylenetriamine pentacetate (DTPA), 540–541
Diffusion, in respiratory tract, 149
Dilution ventilation, 349
2,4-Dinitrophenol (DNP), linear pharmacokinetics, 47–48
Diseases:
　from animals, 473–475
　biological rhythm symptoms, 225–226
　and chemical binding, 88–89
　chronesthesy of, 296
　circadian rhythms in symptoms, 232–233
　from insect bites, 454
　living agents of, 470–475
　and metabolism, 92
　occupationally induced, 233–234
　from vegetative organisms, 471–472
　and work performance, 329–330
Disequilibrium, vibration-induced, 667
Display factors:
　dominant themes, 383
　in system-equipment design, 383–385
Distribution:
　membrane defenses, 145–148
　nonmembrane defenses, 148–155
　protein binding as barrier to, 154–155
Dose response:
　animal studies, 30–31
　in chemical safety evaluation, 28–32
　curves, 29, 30, 42
　equations, 516–518
Dosimeters, in noise exposure measurement, 569–570, 573–574
Dosimetry, with ionizing radiation, 507–511
Down's syndrome, microwave-induced, 631–632
Drug addicts, hair analysis of, 126
Drugs:
　animal studies, 39–40

INDEX 733

biological rhythms and, 226, 235
  toxicity of, 39–40
Drug therapy:
  chemical monitoring problems, 95
  to improve altitude tolerance, 441
Dust:
  control principles, 457
  inhalation, 353–354
  susceptibility-resistance rhythms, 263
Dynamic exercise, relative work loads, 314
Dynamic strength, ergonomic guidelines, 396
Dysbaric osteonecrosis, in divers and caisson workers, 445
Dyspnea, and circadian rhythms, 227–228

Ear:
  anatomy and physiology, 558–560
  protection, 574
  see also Hearing loss
Eastman Kodak Company, work capacity studies, 326
Eating habits, of shiftworkers, 200–201
Eclipse blindness, in welders, 598
Ecological balance, and public health, 2
Edema, visible light-induced, 596–597
Electrical discharges, radiation from, 595
Electrocardiogram (ECG), in aerobic power measurement, 316
Electromagnetic radiation:
  acute external effects, 512
  biological effects, 511–518
  dose-response relationships, 516–517
  dosimetry, 507, 512
  gastrointestinal effects of, 512
  linear and quadratic equations, 516–517, 518
  measurement of, 504
  risk factors in, 517–519
Electromyography, and vibration syndrome, 705
Electronic products:
  emission standards, 638–639
  radiation exposure from, 579
Electrostatic precipitator, 12
Eli Lilly Company, 12-hour workshift, 187–188
Embryo, microwave effects on, 622
Emergency controls, system-equipment design, 388
Emergency situations, behavior approaches to, 424
Emission standards, for electronic products, 638–639

Employee exposure:
  biological monitoring, 79
  documentation of, 79
  indication of unsuspected, 79
Endocrine system, rhythmic behavior of, 264
Endocrinology, biological rhythms and, 224
Energy conservation, and lighting uniformity, 403
Energy costs, measurement of, 317–327
Energy sources, radiation and, 548–551
Engineering controls:
  of biological agents, 456–457
  continuous monitoring of, 349
  and data collection, 349
  specifications for, 348
  in surveillance data acquisition, 339–340
Engineering psychology, in ergonomics, 377
Enterohepatic circulation, in glutathione conjugation, 163
Entraining agent, see Synchronizers
Entry modes, of chemical agents, 352–355
Envenomization, 476–477
Environmental exposure:
  aberrations in clinical data, 361–367
  biological monitoring of, 360
  and clinical data, 359–367
  compliance with standards, 21
  early field studies, 22
  ergonomic factors, 402–415
  industrial hygiene program, 22–23
  long-term studies, 20–21
  need for new data assessment, 367
  nonworker factors, 361
  predictable patterns, 359–361
  qualification, 20–21
  spot sampling, 21
  total lifestyle and, 360–361
  unknown past, 362–363
  workroom limits, 368–372
Environmental physiology, in ergonomics, 377
Environments:
  and health, 2–4
  hot and cold, 481–497
Enzymes:
  atypical kinetics, 79
  chemical competition for, 90–91
  circadian rhythms in, 252–255
  individual variations, 84–86
  microcosmal, 86
  in toxicant metabolism, 157
Epidemiology:
  of chemical agents, 343–345

Epidemiology (*Continued*)
  methodology, 343–344
  prospective study, 343
  retrospective study, 343
Epigenetic carcinogens, 58
Epilepsy, biological rhythms and, 228, 232
Epileptics:
  circadian pattern of, 232
  menstrual cycle and seizures, 234
  visible light responses, 597
Epoxidation, by microcosmal enzymes, 87
Epoxide hydrolase, in xenobiotic metabolism, 160
Ergometers, *see* Bicycle ergometers
Ergonomics, 375–426
  allied fields, 376
  behavioral approaches, 423–425
  checklist characteristics, 377–378
  chemical agents, 414–415
  control-intervention approaches, 421–426
  defined, 375–376
  environmental factors, 402–415
  illumination and color, 402–407
  industrial robots, 419
  job demands, 392–401
  major factors in, 377–392
  mental and cognitive demands, 401
  new developments and trends, 416–421
  of noise, 407–414
  objectives, 376
  perceptual demands, 398–401
  research and practice, 376–377
  subspecialties in, 376–377
  system approaches, 421–422
  system-equipment design, 383–392
  thermal conditions, 407
  work force change, 420–421
  workspace design, 378–382
Erythema:
  threshold dose for, 591
  ultraviolet-induced, 590
  visible light-induced, 596–597
Eskimos, cold tolerance of, 487
Ethanol:
  chronopharmacology of, 241–245
  detoxication, 160
  excreted *vs.* exposure levels, 94–95
  and hypoxia, 437
Ethnic groups, cold responses of, 487
Ethylenediamine tetraacetic acid (EDTA), 540–541
Europe:
  shiftwork research in, 271

  unusual workweeks in, 184
Evacuated containers, 12
Excretory defense mechanisms, to toxicant exposure, 155–156
Exercise, 6
  dynamic *vs.* static, 314
Exercise physiology, in ergonomics, 376
Exhaled air analysis:
  advantages of, 116–117
  background, 116–117
  decay equations for xenobiotics, 122
  disadvantages of, 120
  experimental studies, 121–124
  of index chemicals, 118–120
  of industrial exposure, 116–124
  sex differences, 119
  of solvents, 121–123
Experimental psychology, in ergonomics, 377
Exposure chambers, animal, 346
Exposure levels:
  administrative controls, 351–352
  analytical methods, 335, 356
  animal studies, 345–346
  blood analysis of, 124–125
  categories of, 347
  of chemical agents, 333–372
  clinical cases, 346–348
  data acquisition objectives, 334–335
  evaluation, 355–359
  exhaled air analysis, 116–124
  government regulations, 335–337
  hair analysis of, 125–133
  "index" of, 342
  index chemical concentrations in urine, 107–108
  industry position on, 371
  interference factors, 357
  milk analysis, 133–135
  sampling equipment, 335
  study protocol selection, 356
  surveillance data acquisition, 337–240
  testing procedures, 349–350
  urinary dose-response relationship, 107–116
  work patterns, 342, 352
External ear:
  anatomy and physiology, 559
  infections and treatment, 565
External radiation, 502–503
  accidental exposure, 537–539
  bioassay techniques, 543–545
  Lockport incident, 537–539
  sources of, 502

therapeutic measures, 537–539
Extrinsic allergic alveolitis, 459
Exxon, 12-hour workshift, 188
Eye:
  infrared radiation effects, 603–604
  laser damage to, 613–615
  radiation hazards, 600
  ultraviolet radiation effects, 589–590
  visible light radiation, 596–598
Eye examinations, and laser safety, 617–618
Eye fatigue, 598
  lighting and, 406–407

Factory workers, vibration-induced ailments of, 675–676
Family life:
  shiftwork and, 176, 191
  unusual work schedules and, 179
Far-field exposure, to radiant energy, 621
Farmer's lung, 462–463
  action mode of substances, 458–459
  defined, 462
  diagnosis and treatment, 463
  immune reactions in, 458–459
  pathogenesis, 462
Farm workers, vibration-induced ailments, 673
Fat, index chemical solubility in, 118
Fatigue:
  shiftwork, 193–197
  vibration-induced, 669
  workload level and, 392
Fat/muscle ratio, and metabolism, 91
Female workers:
  chemical toxicity and, 93
  hair lead levels of, 128, 132
  heat tolerance in, 494–496
Fibers, inhalation of, 353–354
Fibrosing alveolitis, 459
Filter sampling, of particles other than neutrons, 506
Finger dexterity, in cold environments, 489
Fish, mercury contamination of, 126
Fishermen, mycobacterial infections in, 475
Fixed bench test, of physical fitness, 322
Flash blindness, 598–599
Flavin-containing monooxygenase, in xenobiotic metabolism, 160
Flexible work hours, 183
Flight experience, vibration effects on, 669
Fluid balance, in hot environments, 491
Fluorescent lighting, sensitivity to, 598
Fluoride, urine levels for, 115

Fluroxene, toxicity in animals, 39
Fog fever, in cows, 459
Food intake, and environmental exposure, 360
Foodstuffs, benzoic acid in, 100
Forced expiratory volume (FEV), test for byssinosis, 460–461
Forklift operators, behavior of, 423
Formaldehyde:
  detoxication, 160–161
  mucus barrier penetration by, 152–153
  respiratory reflex studies, 153–154
Formalin, in histoplasmosis sterilization, 456
Free-running rhythms, 221–223
Fumes, inhalation of, 353–354
Functional envelope, in workspace design, 379–380
Function allocation decisions, in process design, 421
Fungal metabolism, products of, 458

Gases:
  partial pressure of, 432
  sampling of, 354
Gastric ulcers, among shiftworkers, 202, 203
Gastrointestinal disturbances, in shiftworkers, 201–202
Gastrointestinal exposure, to biological agents, 452
Gastrointestinal tract, as defense mechanism, 147–148
Genetic effects:
  of biological rhythms, 224
  in chemical binding, 88
  of ionizing radiation, 530–531
  of microwave exposure, 629
  of microwave-radiofrequency radiation, 622–624
Genotoxic carcinogens, 44, 59–60
Germicidal lamps, protection guides, 592–593
Germicidal region, of ultraviolet radiation, 585
Glare control:
  direct vs. specular, 405
  ergonomic importance, 403
  in task lighting, 403–405
Glass worker's cataract, 605
Gliding time, in work hours, 183
Glucuronic acid conjugation, of toxicants, 161–162
Glutathione conjugation, of toxicants, 162–164
Glycogen, circadian variation in, 258

Gonads, microwave-radiofrequency effects, 622
Government regulations, for exposure levels to chemical agents, 335–337
Grain fever, 464
Granite industry, research on chemical agents, 341
Grasp span, in tool design, 391
Gravitational settling, in respiratory tract, 149
Grip strength, in cold environments, 489
Group exercise, 382
Growth, microwave-radiofrequency effects, 622, 631

Habituation, to whole-body vibration, 668
Hair analysis, of industrial exposure, 125–133
Hair dyes, and index chemicals, 128
Hair metals, in control populations, 130–131
Handles, of controls and tools, 390–391
Hand-transmitted vibration:
 characteristics of, 686–688
 electromyography, 705
 exposure standards, 705–706
 harmful effects of, 682–709
 incidence of disorders by occupation, 703
Hardy-Wolff-Goodell technique, of infrared radiation, 610
Harvard Fatigue Laboratory, 319, 322
Hay mold, immune reactions to, 458–459
Hazardous materials, worker isolation from, 340
Health, *see* Occupational health
Health Information for International Travelers, 455
Hearing:
 in chronic running ear, 566
 frequencies, 560
 industrial standards, 577–578
 measurement of, 562–563
 noise control and, 411–413
Hearing loss, 560–567
 age factor, 566
 causes of, 563–567
 defined, 560–562
 diagnostic testing, 561
 hereditary, 563
 noise exposure and, 408, 567
 sensorineural, 566–567
 severity of, 561–562
 toxic causes, 563–567
 types of, 562
Heart rate:
 calculations, 394
 cold exposure response, 486
 and oxygen uptake, 317–322
 and work tolerance, 393–394
Heat:
 acclimatization, 492–494
 disorders, 497
 exchange calculations, 491
 exposure, 407
 flux calculations, 484
 infrared-induced, 607–608
 sensation threshold, 625
 stress, 496–497
 see also Hot environments
Heat loss, measurement of, 484
Heat waves, deaths during, 496–497
Heavy equipment operators, vibration-induced ailments of, 674–675
Heavy metals:
 chelation therapy, 539
 recovery time for exposures, 351
Helicopters, vibration in, 669, 673, 674
Hematopoietic system:
 radiation sensitivity of, 512, 624
 radionuclide effects, 528–529
Hemoglobin:
 in altitude physiology, 434–435
 oxygen dissociation curve, 435–436
Hemorrhagic damage, from whole-body vibration, 666
Hemorrhoids, vibration-induced, 674
Hemp:
 biological agents in, 461–462
 histamine-releasing substances in, 458
Henry's law, in decompression sickness, 443
Heparin therapy, for Raynaud's phenomenon, 708–709
Hepatotoxicity:
 abnormal DNA bases in, 60–61
 animal studies, 30–31
Herbicides:
 linear and nonlinear pharmacokinetics, 48–57
 oxidant defense mechanism, 166
Hexobarbital, circadian sleep rhythms, 251–252
High-altitude pulmonary edema (HAPE), 442
High-intensity lamps, radiation hazards of, 596–598
Hippuric acid, 97, 101
Hiroshima survivors, leukemia studies of, 515–516
Histamine aerosols, 241
Histoplasmosis:
 diagnosis and treatment, 472
 in poultry workers, 472

soil sources, 452
Histotoxic hypoxia, 437
Hobby activities, health hazards of, 6–7
Homeostatic theory:
  of biological systems, 177
  chronotoxicity and, 295
  in medical schools, 297
Hormesis, 515
Hormones:
  and biological rhythms, 224
  in cold exposure responses, 488
  time-qualified reference valve, 290–291
Hospitals, accidents in, 288–289
Hot environments, 481–497
  adaptation to, 481–482
  age factors, 495
  individual differences, 493
  nonwork acclimatization, 493–494
  physiological responses, 490–497
  repeated exposure to, 492
  sex differences in responses, 494
Hours of changing resistance concept, 245–246, 297–298
Housekeeping, 423–425
  and plant attitude, 339
Human development:
  biological rhythms of, 215–217
  microwave effects on, 631–632
  and workspace design, 378–379
Human engineering:
  systems approaches, 421
  vibration syndrome, 706–708
  see also Ergonomics
Hunting response, in cold acclimatization, 487
Hydrogen sulfide, protein binding in antidotal therapy, 155
Hydrolysis reactions, as detoxication mechanism, 161
Hydroxyurea teratogenesis, circadian-susceptibility rhythms, 249
Hypemic hypoxia, 437
Hyperbaric chambers:
  medical uses, 445–449
  types of, 446
Hypersensitivity pneumonitis, 459
Hypertension:
  noise-induced, 411
  screening tests, 293
Hyperthermia, microwave-induced, 639
Hyperventilation, 367
  from altitude exposure, 436
  vibration-induced, 665
Hypnotic drugs, in chronobiologic adjustments, 286

Hypothermia, in cold environments, 490
Hypoxia, 437–438
  manifestations, 437
Hypoxic hypoxia, responses to, 438

Illness:
  biological rhythms and, 227–234
  shiftwork and, 197–202
  see also Diseases
Illumination:
  defined, 402
  glare control, 403–405
  guidelines, 402
  recommended levels, 403, 404
  task performance and, 403
  of work environment, 402–407
  see also Lighting
Immune system:
  defenses against toxicants, 171
  microwave effects, 623
  radionuclide effects, 529
Immunization, for biological agents, 455
Impact injury, vibration-induced, 672
Impingement method, of measuring contaminants, 11, 12
Impinger pump, 12
Impulsive noise, 569
  defined, 567
Index chemical:
  biological assays, 105
  conjugated *vs.* free, 105–106, 112
  excreted *vs.* exposure levels, 94–95, 108
  exhaled air levels of, 118–120
  in inhaled air, 118
  nonworkplace progenitors, 100–104, 119
  normal range, 81, 104
  solubility in tissues, 118
  urinary measurement, 95, 107
Indomethacin, chronokinetics of, 235–236
Indoor air pollution, 367, 420
Industrial audiometry, 574–575
Industrial chemicals:
  age and metabolism, 91–92
  blood binding of, 88
  in breast milk, 124
  excreted metabolites, 83–84
  and metabolism, 82–83, 86–87
  organic *vs.* inorganic, 82
  in perspiration, 136
Industrial hygiene:
  acceptable program, 22–23
  background, 1–2
  biological monitoring and, 78–80
  changes in procedures, 363

Industrial hygiene (*Continued*)
  defined, 14–15
  early research, 11–12
  educational involvement, 23–24
  emergence as science, 11–16
  environmental control, 22–23
  objectives of, 76
  practice rationale, 15–16
  professional associations, 13, 24
  program elements, 16
  rationale for, 1–25
  in small companies, 333–334
  technological advances, 9–10
  urine analyses, 76, 107–116
  see also Occupational health
Industrial hygienist:
  certification of, 13
  educational role, 24
  responsibility of, 16–17
  unique role of, 333
Industrialization, health risks of, 10
Industrial Ventilation Manual, 348, 349
Inertial impaction, in respiratory tract, 149
Infectious diseases:
  and occupational activity, 470
  off-the-job, 364
  soil-based, 452–453
  and work performance, 329–330
Influenza, and hearing loss, 563–565
Infradian rhythms, 207
Infrared radiation, 601–611
  critical organs, 603–605
  defined, 601–602
  exposures by occupations, 602
  pathophysiology, 602–603
  perception of, 607–608
  protection guides, 611
  skin effects, 603
  sources of, 602
  threshold dosimetry, 608–611
  tolerance limits of workers, 610–611
Infrasound, 670–672
Ingestion, of toxic substances, 34, 354
Inhalation:
  animal studies, 346
  anthrax in textile mills, 472
  of chemical agents, 32–35, 77, 353
  hyperbaric oxygen therapy, 431, 446
  index chemical concentration, 118
  industrial importance, 77
  mucus barrier to toxicants, 152
  pulmonary tests for disease, 455
  radiation bioassays, 546–547
  respiratory tract reflexes, 153–154
Inhalation Toxicology Research Institute, 522, 543
Inner ear:
  anatomy and physiology, 559–560
  disease and treatment, 566
Inorganic chemicals:
  biological monitoring of, 82
  urine analysis for, 114–115
Insect-borne diseases, 454, 476
Insecticides, 93, 161
Inspired gas, changes in composition, 432–433
Institute of Environmental Stress, 321
Intermittent noise, 567, 569
Internal radiation, 503–504
  bioassay techniques, 546–548
  chelation therapy, 539–541
  inhalation exposures, 546–547
  pulmonary lavage, 541–543
  puncture wounds, 547–548
  sources of, 503
  therapeutic measures, 539–543
International Commission on Illumination, 601
International Commission on Radiation Units and Measurements, 508
International Radiation Protection Association, 638
International Standards Organization (ISO):
  hand-transmitted vibration standards, 688
  on normal hearing, 561
  on vibration exposure, 658, 679, 705
Intracerebral hemorrhage, temporal distribution, 230
Intraocular burn factor, in infrared radiation, 610
Ionizing radiation, 501–551
  administrative philosophies, 532
  application of recommendations, 536
  basic concepts, 534–535
  bioassay techniques, 543–548
  biological effects, 511–530
  critical organ concept, 532–533
  dosimetry, 507–511, 532–536
  exposure types, 502–503
  ICRP recommendations, 533–536
  industrial concerns, 514–515
  information sources, 530–531
  late effects of, 514–516
  measurement, 504–507
  NCRP recommendations, 530–533
  professional interpretation, 532

INDEX 739

recommended limits, 535
special cases, 536
standards and guidelines, 530–536
therapeutic measures, 536–543
thyroid abnormalities, 531
Irritation theory, of carcinogenesis, 59
Isolation chambers, chronobiological studies in, 222–223
Isometric exercise, relative work loads, 315–316
Isoniazid:
 prophylaxis for tuberculosis, 455
 toxicity differences, 85

Japan:
 chemical toxicity in females, 93
 slow acetylators in, 85
Jet lag:
 Argonne diet, 284–286
 circadian rhythms and, 219
 and shiftwork, 178
Job boredom, 415
Job design, see Work design
Job enrichment, 422
Jute, biological agents in, 461–462

Kalahari Bushmen, physiological responses to cold, 482
Keratitis:
 threshold doses, 592
 ultraviolet-induced, 590, 591
Kettering Laboratory, 347
Kidney:
 sulfate conjugation, 162
 toxicant elimination by, 156
 vibration effects on, 676
Kinetosis, see Motion sickness
Kofrangi-Michaelis respirometer, in physical energy measurement, 325
Krough-Erlang tissue oxygen model, 436

Lark-owl chronotypes, 276
Laser radiation, 612–619
 critical organs, 613–615
 pathophysiology, 613
 protection guides and standards, 616–619
 retinal damage thresholds, 616
 skin damage thresholds, 616
Lasers:
 applications, 612
 characteristics of, 612
 engineering requirements, 619
 installation guidelines, 618

risk levels, of, 619
safety standards, 616–617
thermal effects of, 613
types of, 614
Lead:
 blood analysis for, 124–125
 exposure studies, 347, 351
 lung retention models, 353–354
 urine levels for, 115
Legionella, 367
 air conditioning and, 457
Leukemia:
 circadian variation in cure rate, 255–257
 radiation-induced, 515–516
Leukocytosis, microwave-induced, 624
Lidocaine, chronesthesy of, 239–241
Life span, radiobiological experiments, 515
Life-styles, health stresses of, 3–6
Lifting, 395–397
 experiments, 395
 factors affecting, 396
 "golden rule" method, 395
 medical contraindications, 396
 squat vs. ballistic, 395
Light-dark cycle, 215–217, 265–266
Lighting:
 glare control, 403–405
 types of, 403
 uniformity in, 403
 see also Illumination
Lipid peroxidation:
 antioxidant protection, 170
 and tissue damage, 166–167
Liver:
 animal studies, 254
 circadian rhythms, 252, 258
 enzymes, 253–255
 glucuronic acid conjugation, 161
 toxicant excretion, 156
Local exhaust ventilation, evaluation of, 350–351
Localized organ retention, in particulate radiation, 510–511
Lockport incident, 537–539
Longshoremen, performance study of, 313
Los Alamos National Laboratory, 523, 549
Lovelace Foundation, 522
Low-back pain syndrome, in manual material handling, 394–395
Luminance, 402–405
 defined, 402
Lung:
 alveolar macrophages in, 152

Lung (*Continued*)
  defense mechanisms of, 148–150
  mucus protective barrier, 150–152
  "nose-only" radiation exposure, 522
  oxygen transport in, 434–436
  radionuclide effects, 522–524
  see also Respiratory tract
Lung cancer:
  radiation-induced, 523, 549
  unique detection method, 524
Lymph nodes, radionuclide effects, 527
Lymphocytes:
  in radiation bioassays, 545
  microwave effects on, 629

Machine-paced work, ergonomics of, 415–416
Machines, *vs.* humans, 421–422
Macrophages, particle ingestion by, 152
Malathion, 250
  detoxication, 161
Male workers, 110–111
  chemical toxicity and, 93
Mammals, late effects of radiation, 515
Mandelic acid, effect of conjugate hydrolysis on, 105
Mantoux test, for biological agents, 455
Manual material handling, ergonomic factors, 394–398
Maple bark disease, 467–468
Mastoiditis, and middle ear disease, 566
Maximum breathing capacity (MBC), test for byssinosis, 460–461
Maximum permissible limits, for lifting work, 396–397
Meal timing, and chronobiologic adjustment, 284
Meat packing, brucellosis in, 473
Medical practice:
  examinations for biological agents, 455
  history of past exposures, 362–363
  implications of chronobiology, 225–263
  programs for biological agents, 455
  recordkeeping in toxicologic studies, 13
  work exposure evaluation, 355–356
Medications:
  age and metabolism, 91–92
  biological rhythms and, 235–245
  blood binding of, 87–88
  in breast milk, 134
  for chronobiologic adjustments, 283–284
  concurrent use of, 92–93
  enzyme-stimulating, 87
  urinary kinetics of, 237
  see also Drugs

Melatonin, circadian rhythms of, 210
Membranes:
  absorption-distribution defenses, 145–148
  conceptual depiction, 145
  lipid layers, 145–146
  transfer factors, 145–146
Memorization, and recall limitations, 401
Memory:
  types of aids, 401–402
  typical span, 401
Meniere's disease, treatment of, 566
Menstrual cycle, and epileptic seizures, 234
Menstruation:
  and perspiration, 495
  vibration-induced disorders, 677
Mental disorders, in shiftworkers, 223
Mequitazine, chronokinetics of, 237, 238
Mercapturic acid in glutathione conjugation, 163
Mercuric chloride, circadian-susceptibility rhythms, 250
Mercury:
  in contaminated fish, 126
  exposure studies, 351, 360
  protein-binding detoxication, 155
  urine levels for, 115
Mesor, defined, 273
Metabolism:
  acetylation and, 85
  age and, 91–92
  biological monitoring, 81–94
  chemical exposure dose effect, 89–90
  in cold environments, 483
  concurrent use of medicinals, 92–93
  diet and dietary state, 86
  disease and, 92
  elimination half-time, 85
  enzymatic variations, 85–86, 90–91
  factors affecting, 84
  microwave effects on, 626
  obesity and fat/muscle ratio, 91
  pregnancy and, 94
  screening for abnormalities, 79
  sex differences, 93
  smoking and, 94
  of toxicants, 157
  unexpected products of, 85
  variations in, 86
Metal-binding proteins, 155
Metallo enzymes, in detoxication, 168–169
Metallothionein, 155
Metals:
  biological threshold limit values for, 115
  in breast milk, 134

INDEX

content of, in hair, 125
in human nails, 136
in saliva, 136
in scalp hair, 100
urine analysis for, 114–115
Meter readers, shiftwork performance, 288
Methanol, excreted *vs.* exposure levels, 94–95
Methylation reactions, in detoxication, 165–166
Micropolyspora faeni:
in air ducts, 459
in farmer's lung, 462
Microwave cataracts, 630
Microwave ovens, emission standards, 638–639
Microwave radiation, 619–640
animal studies, 640
biological effects, 620–621
cellular and genetic effects, 622–624
characteristics of, 620–621
critical organs, 621
dosimetry, 624–626
eastern European reports, 633
epidemiologic studies, 626–629
exposure standards, 634–638
growth and development effects, 631–632
human exposures, 626–629
industrial frequencies, 620
nervous system and cardiovascular effects, 631
ocular effects of, 629–631
problems and recommendations, 639–640
protection guides and standards, 634–640
"scaling" studies, 640
Middle ear:
anatomy and physiology, 559
diseases and treatment, 565–566
Milk analysis, of industrial exposure, 133–135
Military personnel:
chronobiologic adjustments in, 286
infrared radiation among, 602
Mining:
chemical carcinogens in, 550–551
immunologic studies, 550
Minute ventilation:
reduction in, 153, 154
time-response curves, 154
Mist, inhalation of, 353–354
Mixed-function monooxygenases, in detoxication, 158–159
Modified work schedules, popularity of, 187–188
Monday morning fever, 469
Monge's disease, 442

Morphine, enzyme stimulation of, 87
Motion effects, 709–715
human response to, 711–713
and postural function, 712
preventive measures, 713–715
and task disruption, 712–713
see also Vibration
Motion sickness, 414, 660, 711
acute *vs.* chronic, 442–443
Mucociliary clearance, as defense mechanism, 150–151
Mucus, as protective barrier, 152–153
Multiplace hyperbaric chambers, 446–448
Mumps, and hearing loss, 563–565
Muscular insufficiency, in working dimensions, 380
Musculoskeletal injuries, from vibration syndrome, 699–700
Mushrooms, biological agents in, 468–469
Mutagenesis, threshold concept, 43
Mutagenic chemicals, identification of, 358
Mycobacterial infections:
from swimming pools, 475
water-based, 453
Myocardial infarction, temporal distribution, 228

Nail analysis, of industrial exposure, 135–136
Nasal cavity, receptors in, 153
National Academy of Sciences, 368
on ultraviolet radiation, 587
National Cancer Institute, 65
National Center for Devices and Radiological Health, 580
National Center for Health Statistics, 210
National Council on Radiation Protection (NCRP), 530, 531
radiation standards and guidelines, 530, 533
National Institute of Occupational Safety and Health (NIOSH):
on manual lifting, 396–397
on organic mercury, 360
sampling strategy, 336
Navy, radar exposure studies, 628
Nervous system:
membrane barriers in, 148
microwave effects, 633
microwave-radiofrequency effects, 623–631
vibration effects on, 668
Neurological effects, of vibration syndrome, 699
Neurological tests, for vibration syndrome, 704–705

Neuromuscular effects, of whole-body vibration, 666–668
Neutron damage, in external radiation, 503
Neutron dosimetry, 507–508
Neutron radiation, 505
  biological effects of, 511–518
  detection and measurement, 505
  dosimetry, 507–508
Neutrons, thermal *vs.* fast, 505
New York University Medical Center, 512
Nicotine, circadian-susceptibility rhythms, 248
Night shift:
  accident proneness during, 289
  biological rhythm adjustments, 276–279
  categories of, 182
  natural aversion to, 178
  physiological aspects of, 364
  time-qualified reference values for, 292
  worker lifestyles, 191
  see also Shiftwork
p-Nitrophenol, glucuronic acid conjugation, 162
Nocturnal asthma, occupationally induced, 233
Noise:
  accident studies, 411
  defined, 558
  ergonomic factors, 407–414
  health and safety issues, 411
  and hearing loss, 408, 575
  levels, 570
  and productivity, 409–411
  sleep problems, 196
  and speech interference, 408–409
  and vibration, 677–678
Noise control:
  engineering measures, 412–413
  and hearing conservation, 411–413
  self-protective measures, 413
Noise exposure, 557–578
  auditory effects of, 574–576
  biological effects, 574–577
  character of, 570–571
  classification methods, 568
  duration of, 571–572
  early studies, 557–558
  ergonomic factors, 407–414
  hair cell damage, 575–576
  "index," 568
  job classification index, 572
  long-term studies, 575
  measurement of, 568–574
  nonauditory health effects, 576–577

occupational standards, 408
  permanent effects, 574
  severity of, 407
  temporary effects, 571–572
  types of, 567–570
Nonfiber molds, in warm air, 459
Nonionizing electromagnetic energies, 579–640
  biophysics of, 580–583
  dose-effect relationship, 584
  exposure limits, 583–584
  protection guides and standards, 583–585
  types of, 579
Nonliving materials:
  biological agents in, 458–470
  exposure modes, 458–459
  pharmacologic effects, 458
Nonmembrane defenses, to absorption and distribution, 148–155
Nonmetals, hair content, 126
Nonshivering thermogenesis, 483
Nonworkplace progenitors, of index chemicals, 100–104
Norepinephrine, and cold acclimatization, 488
Normal hearing:
  defined, 560–561
  frequencies, 561
Noxious stimulation, 609
Nuclear facilities, control-display problems at, 390
Nuclear reactors, late effects of radiation, 514–515

Obesity, and metabolism, 91
Occupational heath, 175–300
  biological rhythms and, 175–300
  circadian chronotoxicity and, 255–258
  early regulations, 14
  environment and, 2–4
  hazards, 18–19, 21–22
  noise and, 411
  nursing, 15
  organizations recommending exposure limits, 369
  position descriptions, 337–338
  shiftwork and, 175–300
  stresses, 8–9
  toxicologic studies, 12–13
  vibration-related disorders, 673–674
  whole-body vibration, 673–677
  work schedule and, 364
Occupational hygienists, Soviet, 663–664
Occupational physicians, 294

Occupational Safety and Health Act (1970), 14, 579, 584
Occupational Safety and Health Administration (OSHA):
  exposure limits, 368
  on noise levels, 570
  on unusual workshifts, 190
Octave band analyzer, in noise exposure measurement, 573
Ocular damage, infrared-induced, 611
Ocular effects, of microwave-radiofrequency radiation, 624, 629–631
Odorous substances, employee complaints of, 415
Office work, health hazards of, 420
Off-the-job stresses, 6–7
  industrial hygiene programs, 25
Oil refinery, shiftwork tolerance study, 279
Oil shale mining, chemical carcinogens in, 550–551
Older workers, metabolism of, 91–92
Ophthalmologic tests, for vibration syndrome, 704
Oral temperature, and circadian rhythms, 211–212, 270–280
Organic acid, ionization of, 146
Organ retention:
  localized, 510–511
  and particulate radiation, 509–510
Ornithosis, 453–474
Osteogenic cells, radiation dosimetry, 511
Otitis, chronic external, 565
Otosclerosis:
  hereditary, 563
  surgical treatment, 566
Ovaries, radiation sensitivity of, 512
Oxidant defense mechanisms, to toxicant exposure, 166–170
Oxygen:
  toxicity seizures, 249
  transport in lung, 434–436
Oxygen uptake:
  in cold environments, 483–486
  in fit and unfit individuals, 317
  indirect measurement of, 316–317
  test procedures, 318–324
  treadmill tests, 318–321
  for various work levels, 329
Ozone, 585
  toxic injury to lung, 152

Pain:
  drug effects, 610
  infrared-induced, 608–610
  sensation threshold, 625
Paired displays, design factors, 388
Pallesthesiometry, and vibration syndrome, 704–705
Paraquat, circadian-susceptibility rhythms, 250
Parasitic diseases, off-the-job, 364
Particulate radiation:
  accidental exposure, 506
  anatomical defenses, 149–150
  dosimetry, 508–511
  "hot spot" problem, 510–511
  inhalation studies, 353
  localized organ retention, 510–511
  Lockport exposures, 506
  measurement of, 506
  uniform organ retention, 509–510
Penicillin, toxicity in animals, 39
Pennsylvania, University of, 190
Pennsylvania Department of Health, 369
Peptic ulcers:
  in night workers, 364
  vibration-induced, 673
Perceptual demands:
  ergonomic factors, 398–401
  sensory overloading, 399–400
  time-sharing, 400–401
  vigilance studies, 398
Performance:
  levels of evaluation, 313
  secondary factors influencing, 329–330
  vibration effects on, 669–672
Periodogram, in chronobiological research, 272
Peripheral nerve function tests, for vibration white finger, 704
Peripheral vascular function tests, for vibration syndrome, 704
Permanent Commission and International Association on Occupational Health, 177
Permanent threshold shift, noise-induced, 572
Permissible exposure limits (PEL), and rapid rotation shift schedules, 294
Peroxidase, as defense mechanism, 169–170
Perspiration analysis, of industrial exposure, 135–136
Peru, altitude acclimatization in, 440
Pesticides:
  in breast milk, 134
  enzyme-stimulating, 87
Petrochemical industry, 12-hour workshift, 188

Pharmaceutical industry, 12-hour workshift, 187–188
Pharmacogenetics, 85
Pharmacokinetics:
    biological rhythms and, 225–226
    and cataractogenic activity, 47–48
    dose-interspecies response, 45–47
    linear and nonlinear, 48–57
    in toxicity data extrapolations, 45–57
Phase shifting:
    of biological rhythms, 265–269
    human models, 267–269
    rodent models, 266–267
Phenobarbital:
    circadian-sensitivity rhythms, 248
    enzyme stimulation of, 87
Phenols:
    normal urinary levels, 102–104
    sulfate conjugation of, 162
    urinary levels, 107–108
Phenylbutazone, and chemical binding, 88
Phosphorus exposure, recovery time from, 351
Photic seizures, 597
Photoluminescence, 595
Photons, defined, 580
Photophobia, 601
Physical fitness:
    and aerobic capacity, 318
    as antivibration factor, 708
    in cold environments, 485
    tests, 314–330
    training programs, 323
    of various populations, 313–314
Physical work:
    excessive, 392–393
    tolerance time for, 393
Physician, guidance in risk therapy, 79
Physiology, of prolonged work, 327–329
Pilot error, analysis of, 389
Pilots, vibration effects on, 669, 674
Placental barrier, to toxicant exposure, 148
Plastics operations, field study, 423
Plutonium exposure, chelation therapy, 540–541
Plutonium workers, radiation effects, 519
Pneumatic tools, harmful effects of, 684
Pneumoconioses, radiologic classification, 363
Poland, microwave exposure study, 628–636
Polarity:
    in circadian system, 267
    endogenous, 268

Position descriptions, in surveillance data acquisition, 337–338
Postlunch dip, in human performance, 288
Postnatal growth, microwave effects on, 622
Postural sway, vibration-induced, 667
Posture:
    motion effects, 712
    and vibration-related disorders, 686
    workspace design factors, 381–382
Power tools:
    harmful effects of, 683
    musculoskeletal injuries from, 699–700
    vibration reduction in, 706
    vibration white finger from, 693
Preemployment monitoring:
    for metabolic abnormalities, 79
    urine analysis in, 107
Pregnancy, and metabolism, 94
Presbycusis, 566–567
Production lines:
    ergonomic considerations, 423
    worker-robot interactions, 419
Productivity, noise effects on, 409–411
Proficiency analytical testing (PAT), of chemical agents, 336
Prolactin, circadian rhythms of, 210
Prolonged work, physiology of, 327–329
Propranolol, chronokinetics of, 236, 237
Protein binding, as barrier to toxicant distribution, 154–155
Proteins, and chronobiologic adjustment, 284
Psychologist, in industrial hygiene, 15
Psychosocial problems, of shiftworkers, 191–193, 416–418
Pulling, 394
    safe limits for, 398
Pulmonary capillary, schematic model, 434
Pulmonary disorders, autorhythmometry and, 294
Pulmonary edema, high-altitude, 442
Pulmonary lavage, in radiation therapy, 541–543
Pulmonary tests, for biological agents, 455
Puncture wounds, radiation bioassays, 547–548
Pushing, 394
    safe limits for, 398

Q fever, 453
    in packing house employees, 474–475
Q-switched laser, radiation hazards of, 613, 615

INDEX

Quality control, government regulations, 336
Quanta, defined, 580

Radar:
　exposure effects, 628
　frequency bands of, 619
Radiation:
　animal experiments, 515
　cataracts, 605, 606
　detectors, 504
　dosage range, 512
　exposure, sources of, 501
　exposure types, 502–503
　by high local electrical fields, 595
　humans, effects on, 513
　internal sources, 501
　"leakage," 504
　measurement, 504–507
　mixed potential hazards, 549
　public opinion, 548
　sources, 548–551
　standards and guidelines, 530–536
　stimulating effects of, 515
　as toxic agent, 501
　see also Ionizing radiation
Radiation Control for Health and Safety Act (1968), 579, 584
Radiation syndrome, acute, 512–513
Radiobiology:
　controversial issues in, 516
　electromagnetic radiation exposure, 504
　external radiation effects, 502
Radiofrequency radiation, 619–640
　biological effects, 620–621
　cellular and genetic effects, 622–624
　characteristics of, 620–621
　chromosome effects, 622–624
　defined, 619
　dosimetry, 624–626
　epidemiologic studies, 626–629, 633–634
　exposure standards, 634–638
　human exposures, 626–629, 631–632
　industrial frequencies, 620
　in medical applications, 632
　nervous system and cardiovascular effects, 631
　ocular effects of, 629–631
　problems and recommendations, 639–640
　protection guides and standards, 634–640
Radioimmune assays, in hair analysis, 126
Radiological examination, for vibration syndrome, 705

Radionuclide radiation:
　animal studies, 522–523
　biological effects, 519–530
　biological modeling, 520–522
　vs. electromagnetic radiation, 519–520
　human studies, 523
　industrial importance of, 519
　inhalation studies, 522–523
　kinetic model, 520
　soft tissue effects, 525–530
Radiowave sickness, 631
Radium workers:
　bone sarcomas in, 528
　epidemiologic studies, 519
Rapid rotation:
　and circadian rhythms, 189
　employee effects, 185–187
　schedules, 189
　of workshifts, 185–187
Raynaud's phenomenon, 685
　disorders arising from, 696–697
　etiology of, 700–702
　medical treatment for, 708–709
　of occupational origin, 689–698
　severe cases, 691–692
　symptomatology and progression, 692–698
　symptoms and differential diagnosis, 689–692
　vasomotor symptoms other than, 698
　see also Vibration syndrome
Reaction time, studies of, 401
Recordkeeping, government regulations, 336–337
Recovery time, in industrial exposure, 351
Recreation, 6
　unusual work schedules and, 179
Reflectance:
　defined, 402
　recommended ratios, 405
Relative work loads, 314–329
Renal excretion:
　of toxicants, 156
　transport mechanisms, 156
Repetitive manipulations, and system-equipment design, 391–392
Replication errors, and cytotoxic carcinogenesis, 59–60
Respiration, reflex inhibition of, 153
Respirators, program requirements, 340
Respiratory quotient, 433
Respiratory tract:
　alveolar macrophages, 152

Respiratory tract (*Continued*)
  anatomical defenses to particle deposition, 149–150
  animal studies, 153–154
  biological agents, 452
  as defense mechanism, 148–154
  index chemical concentration and, 119
  inhaled toxicant defenses of, 149
  mucus as protective barrier, 152–153
  nerve receptors in, 153
  particle deposition and retention, 509
  radiation studies, 522
  reflex responses to inhaled toxicants, 153–154
Respirometer, in physical energy measurement, 325
Rest allowance, calculations, 394
Retention factor, in toxicology studies, 38
Reticuloendothelial system, radionuclide effects, 526–528
Retinal disability, infrared-induced, 607
Retired workers, circadian rhythms of, 283
Rhythmic contractions, in workspace design, 382
Risk assessment, carcinogenic mechanisms in, 57
Robotics, ergonomic considerations, 419
Robots, 418
Rochester, University of, 545
Rock-drilling, harmful effects of, 682–684
Rodents:
  biological rhythms of, 215
  chronotoxicology evaluations, 259–261
  susceptibility-resistance rhythms in, 246–251
  toxicology evaluation of, 246–251
Room sampling, 349
Rubella, and hearing loss, 563–565

Safety, 27–70
  chemical evaluation, 27–70
  dose response, 28–32
  exposure route, 32–34
  noise and, 411
  toxicologic data, 27–70
Salicylic acid, and chemical binding, 89
Saliva analysis, of industrial exposure, 135–136
Salt depletion, in hot environments, 492
Sampling:
  "ceiling value," 338
  data acquisition, strategy in, 334–335
  equipment calibration, 335
  government regulations, 336
Sarcomas, circadian-dependent, 247
Schizophrenia, chemical monitoring problems, 95
Scotomatic glare, 599
Sensory effects, of whole-body vibration, 666–668
Sensory overloading, and perceptual demands, 399–400
Sequoiosis, 467
Sex:
  index chemical concentration and, 119
  and metabolism, 93
Shampoos, phenol content of, 104
Shiftwork, 181–182
  adverse effects of, 179
  age factor in, 274
  categories of systems, 181
  and chronobiology, 177, 204–225, 283–286
  circadian rhythms and, 257, 264, 286–289
  cosinor summary of tolerance, 279
  defined, 181
  detection and description of biological rhythms, 271–273
  different schedules, 193–194
  "dropouts," 198–199, 202–204
  early studies, 176, 178
  employee dissatisfaction, 191–204
  endogenous *vs.* exogenous intolerance factors, 274–276
  epidemiologic investigations, 197–199
  family problems, 192–193
  and fatigue, 193–197
  forward *vs.* backward rotation, 268
  free-running rhythms in, 223
  health hazards of, 176
  improving conditions of, 299
  industry studies, 277
  instrumentation for research, 270–271
  international conferences, 177
  intolerance of, 175–176, 274–276, 280–283
  long-range predictions, 273
  management view of research, 270
  methodological difficulties, 270–271
  morbidity and mortality, 176
  and occupational health, 175–300
  percentage of workforce, 182
  performance studies, 286–288
  permanent systems, 181–182
  psychosocial problems, 176–177
  rotating systems, 182, 294
  simulated conditions, 269
  sleep disruption, 193–197, 264–265

unusual schedules, 295–296
Shiftworkers:
    chronobiological field studies, 270–273
    vs. dayworkers, 192
    morbidity of, 197–198
    psychosocial problems of, 191–193
    retired, 283
    selection criteria, 273–289
    social activities of, 192
    steel mill study, 191–192
    tolerant vs. nontolerant, 277, 280
Ship vibration, human effects of, 670
Shivering, 483
Silica:
    administrative controls, 351
    exposure, 341, 344
    lung toxicity from, 152, 353
Sisal, biological agents in, 461–462
Sitting grasp reach envelope, in workspace design, 380, 382
Skin:
    absorption of toxic substances, 34, 452
    as defense mechanism, 146–147
    hydration, 147
    infrared burns, 603
    laser damage to, 615
    microwave perception, 624
    tests for biological agents, 455
    toxicant passage through, 147
    ultraviolet radiation effects, 589–590
    visible light radiation, 596–597
Skin absorption, 34–35
    of chemical substances, 77, 120, 355
    individual variations in, 95
Skin cancer:
    DNA repair mechanisms, 171
    radiation-induced, 515
    ultraviolet-induced, 589
Sleep, 7
    effect on health, 199
    by industry and shift, 198
    rhythms, 223
    schedules, 264–265
    shiftwork and, 193–197
Sliding caliper, 379
Smoking:
    biological agents and, 454
    environmental exposure to, 367
    and lung cancer, 523–524
    and metabolism, 94
    and radon tumorigenesis, 550
    and xenobiotic metabolism, 94
Sneezing, 153

Snowbird Actinide Workshop, 519
Snow blindness, 591
Social Security Act (1935), 14
Sodium orthophenyl phenol (SOPP), 62–65
Soft tissues, radionuclide effects, 525–530
Soil, biological agents in, 452–453, 471
Solvents:
    in breast milk, 134
    defined, 75
    exhaled air analysis, 121–123
Somatic mutation theory, of chemical carcinogenesis, 57–58
Sound level A (dBA), and hearing conservation, 568
Sound level meters, in noise exposure measurement, 573
Sound pressure level, in noise exposure measurement, 573
Soviet Union:
    microwave exposure studies, 631
    microwave-radiofrequency exposure standards, 634–635, 637
    whole-body vibration studies, 663–664
Space vehicles, vibration in, 677–678
Specific absorption rate, of microwave-radiofrequency radiation, 624–625
Specific tissue localization, of radionuclides, 529–530
Spectral analyses, in chronobiological research, 272
Spectrophotometry, visible vs. ultraviolet, 106
Speech:
    frequencies and hearing loss, 562
    noise interference, 408–409
Spinal injury, vibration-induced, 672
Spleen, radionuclide effects, 527
Split-shift schedules, in merchant navy, 182
Spontaneous mutations, in cytotoxic carcinogenesis, 59–60
Spot sampling, of environmental exposure, 21
Spray painting, field study, 423
Spreading caliper, 379
Squat lifting, 395
Staggered work hours, 183
Stagnant hypoxia, 437
Standard workweek, 180–181
Stanford Medical Center, 367
Stapedectomy, 566
Static exercise, relative work loads, 315
Steady noise, 567, 569
Steel tape, 379
Sterility, from chemical agents, 358
Sterilization, of biological agents, 456

Stress:
　environmental, 4–9
　of machine-paced work, 416
　noise-induced, 411
　psychosocial, 416, 417–418
　of vibration and noise exposure, 677
Stress agents:
　data evaluation, 21–22
　direct measurement, 18–19
　worker exposure to, 18
Stretch reflex, vibration-induced, 667
Strychnine, circadian-susceptibility rhythms, 248
Student t-test, for biological rhythms, 272
Stuffy office syndrome, 420
Styrene exposure, 87, 100
　field study, 425
Subchronic toxicity, 37–39
Subtilin, in washing powders, 469
Sugar cane, biological agents in, 463
Sulfa drugs, toxicity differences, 85, 155
Sulfate conjugation, of toxicants, 162
Sulfur compounds, in radiation therapy, 536
Sunburn, 587
Sunlamps, radiation hazards of, 590
Superoxide dismutase, as toxicant defense, 168–169
Surface seekers, 511, 524
Surveillance data acquisition:
　chronological log, 338
　engineering controls, 339–340
　of exposure levels, 337–340
　position descriptions, 337–338
　sampling and analytical procedures, 337
　weather conditions, 338–339
　worker training and attitude, 340
Susceptibility-resistance rhythms:
　evaluation of, 259–261
　in humans, 261–263
　by occupation, 263
　in rodents, 246–252
　significance of, 255–258
Sweating, 491, 494
Sweden, microwave-radiofrequency exposure standards, 637–638
Swine, brucellosis from, 473
Sympathicoadrenal hormones, and cold exposure, 488
Synchronization, of biological rhythms, 215–221
Synchronizers, 215
　advancing vs. delaying schedules, 267
Syncope, in cold exposure, 489

System-equipment design, in ergonomics, 383–392
Systemic poisons, work exposure evaluation, 358
System responses, to control devices, 387, 389
Systems approaches:
　ergonomic guidelines, 421–422
　to work design, 421–422

Tall buildings, cantilever oscillations of, 710
Taylor's grading system, for vibration white finger, 698
Tea, and chronobiologic adjustment, 284
Tears analysis, of industrial exposure, 135–136
Technology, 9–10
　health hazards in, 420
Telephone operators, shiftwork performance of, 286–287
Temperature:
　cycles and free-running rhythms, 223
　ergonomic factors, 407
　regulation and physiology, 491
Tendon jerks, vibration-induced, 667
Tenosynovitis, prevention of, 392
Teratogenesis, threshold concept, 43
Testes, radiation sensitivity of, 515, 622
Tetanus infection, 455
　soil sources, 452
Textile fibers, biological agents in, 459–462
Thermal environments, 491–497. *See also* Cold environments; Hot environments
Thermal radiation, 595
　pain threshold of, 609
Thermoluminescent detectors, in neutron measurement, 505
Thesaurosis, 459
Thiol compounds, glucuronic acid detoxication, 161
Three Mile Island, 549
　control-display deficiency at, 389–390
Threshold, 42
　in dose-response curves, 42–44
Threshold limit value (TLV):
　and administrative controls, 351
　of chemical substances, 77
　chronotoxicity and, 294
　defined, 370
　handbook, 80
　organizations recommending, 368–370
　and skin absorption, 77
　of target organs, 80
　in testing procedures, 349

INDEX

for ultraviolet radiation, 593–594
and unusual workshifts, 190
in urine analysis, 108–109
validating for concurrent exposure, 80
Threshold shift, temporary, 572, 574
Throat lozenges, urinary phenol from, 109
Thyroid, radionuclide localization, 529–530
Time-motion studies, and energy cost measurement, 317
Time-qualified reference values:
  in biological monitoring, 290–292
  day vs. night workers, 291–292
Time-sharing:
  guidelines for, 400
  perceptual demands of, 400–401
Time-weighted average, in noise exposure measurement, 573
Tissue:
  index chemical solubility in, 118
  oxygen transport from lung, 434–436
Tobacco dust, biological agents in, 468
Tolvene, hippuric acid following exposure, 97, 102
Tool handles, system-equipment design, 390–391
Toxicant exposure:
  biochemical defenses, 156–171
  body defense mechanisms, 143–172
  DNA repair and, 170–171
  excretory defense mechanisms, 155–156
  immunological defense mechanisms, 171
  membrane defenses, 145–148
  nonmembrane defenses, 148–155
  oxidant defense mechanisms, 166–170
Toxicants:
  antidotal therapy, 155
  biotransformation of, 157
  body defense mechanisms, 143–172
  chronesthesy of, 241
  chronobiological studies, 245–246, 252–261
  covalent binding of, 158
  cutaneous absorption of, 34
  distribution of, 154–155
  exposure route, 32–34
  ingestion of, 34
  inhalation of, 32–33
  interaction with biological systems, 144
  lipophilicity and absorption, 156
  liver (biliary) excretion of, 156
  membrane defenses, 145–148
  metabolism of, 157
  mucociliary clearance of, 150–151
  passive diffusion of, 146

renal excretion of, 156
Toxic hearing loss, 563–567
Toxicity, acute, 34–36
Toxicologist, 298
Toxicology, 27–70
  animal studies, 39–42
  background, 1–2
  and biological rhythms, 247
  chemical safety data, 27–70
  chronic studies, 37
  dose-interspecies response data, 45–47
  dynamic factors, 41
  early studies, 12
  evaluation in rodents, 246–251
  human, 41
  subchronic studies, 38
  tests used in, 246–247
Toxic Substances Control Act (1976), 345
Tractor drivers, vibration-induced ailments, 673
Traumatic vasospastic disease, 695, 700. See also Raynaud's phenomenon; Vibration syndrome
Treadmills, 318–321
Tremor, vibration-induced, 667
Trichloroacetic acid (TCA), dose effect on metabolism, 89
Trichloroethane, toxicity, 31
1,1,2-Trichloroethylene (TRI):
  animal studies, 65–66
  carcinogenic studies, 65–69
  and DNA alkylation, 66–69
  hepatotoxicity levels, 68
2,4,5-Trichlorophenoxyacetic acid (2,4,5-T):
  animal studies, 49–55
  human studies, 55–57
  linear and nonlinear pharmacokinetics, 48–57
  species differences, 48–57
Tritium exposure, bioassay techniques, 546
Truck drivers:
  sleep problems of, 288
  vibration-related disorders, 674
Tuberculosis immunizations, 455
Tumorigenesis:
  experimental approaches, 61–69
  genetic vs. epigenetic mechanisms, 61
  mutational origin, 58
Tumors:
  chemical-induced, 58
  circadian variation in, 251
Turkeys, bird fancier's disease from, 466
Twins, circadian rhythms in, 224

Typical workday, work output during, 327–329

Ulcers, among shiftworkers, 202, 203
Ultradian rhythms, 207
Ultraviolet radiation, 585–594
  absorption effects, 587
  critical organs, 589
  dosimetry, 586, 591
  effects of, 586–587
  in medical practice, 586
  nonionizing portions, 585
  occupational exposure guides, 589, 594
  pathophysiology, 587–588
  protection guides, 592–594
  sources of, 585–586
Uniform organ retention, of particulate radiation, 509–510
Unions, and shiftwork research, 270
United States:
  compressed workweek in, 185
  exposure limits in, 368, 634
  shiftwork in, 175, 182
  unusual workweeks in, 184
  work force trends, 418
Unusual workshifts, 179–180, 183–187
  advantages and disadvantages, 189–190
  chronotoxicity and, 295–296
  defined, 183–184
  evaluation of, 190–191
  examples of, 184
  federal legislation and, 189–190
  history of, 187
  industry requirements, 184
  main factors affecting usage, 190
  special problems of, 295
Uranium exposure, off-the-job, 363
Uranium mining:
  and lung cancer, 523
  particle detectors in, 507
  radiation hazards, 549–551
Urine:
  index chemical concentrations in, 107–108
  osmolality of, 96
  sample dilution of factors, 96–98
  specific gravity of, 96
  spot vs. 24-hr samples, 112–113
Urine analysis, 76, 107–116
  biological threshold limit values, 108–109
  colorimetric method, 113–114
  of conjugated and unconjugated index chemicals, 112
  factors in methodology, 109–114
  for inorganic ions, 114–115
  nonlinear response, 112
  optimum conditions for use of, 115–116
  slope of dose-response curve, 109–112
  variations in, 99
  of various chemicals, 110–111
Urticaria, in cold exposure, 489
Utah, University of, 524

Vacationers, chronobiological adjustments of, 284
Vapors:
  animal studies, 36
  sampling of, 354
Vasoconstriction, from altitude exposure, 436–437
Vasomotor symptoms, of vibration syndrome, 698
Vector-borne diseases, 476
Vegetative organisms, diseases from, 471–472
Veiling glare, 598
Venom, *see* Envenomization
Venous lymphocytes, circadian rhythms in, 210, 212
Ventilation:
  controls in spray painting, 423
  local exhaust, 350–351
  systems evaluation, 339–340
  tests for biological agents, 470
Very low-frequency vibration, 709–715
  defined, 710
  engineering and biomedical approaches to, 713–715
  examples of, 710
  exposure standards, 715
  human response to, 711–713
  preventive measures, 713–715
Vibrating tools:
  harmful effects of, 683–684
  skill factor, 708
  and white finger disorder, 414
  see also Power tools
Vibration, 653–715
  American national standard, 660
  complexity of, 655–666
  defined, 653
  description and measurement, 654–658
  direction of, 658
  duration and time course, 658
  engineering, 662, 678, 680
  ergonomic factors, 413–414
  exposure modes, 653–654
  exposure standards, 679–680

INDEX 751

frequency of, 655
intensity of, 656–658
minimization of adverse effects, 681
morbidity studies, 414
noise and, 413–414, 677–678
performance effects, 669–672
receptor organs, 666
sensory and neuromuscular effects, 666–668
sinusoidal, 656
soporific effects, 668
source/individual controls, 680
stress, 677–678
tolerance studies, 658
trauma, 666
in vehicles and machines, 656
see also Whole-body vibration
Vibration syndrome, 687–709
biodynamic aspects of, 702–703
climatic factors, 701
defined, 682–683
diagnostic and screening tests, 703–705
engineering solutions, 706
etiology of, 700–702
exposure standards, 705–706
human factors engineering and operational solutions, 706–708
incidence by occupation, 703
medical treatment for, 708–709
musculoskeletal injuries, 699–700
neuromuscular reflex mechanisms, 701–702
occupational vs. personal factors, 690
physical measurements and electromyography, 705
prevalence of, 702
prevention and treatment, 705–709
radiological examination of bones and joints, 705
recent data, 709
risk factors, 690
screening tests, 694
varieties of, 688–689, 698–700
vasomotor symptoms, 698
Vibration white finger:
climatic factors, 695
etiology of, 701
latent period of exposure, 693
medical treatment for, 708
musculoskeletal injuries, 699–700
neurological effects, 699
nonworking effects of, 695–696
screening tests, 694

symptomatology and progression of, 692–698
Taylor's grading system for, 698
vasomotor symptoms, 698
see also Raynaud's phenomenon
Video display terminals (VDT):
control-display problems with, 390
health hazards of, 420
Vigilance studies, main factors, 398–399
Vinyl chloride monomer (VCM), dose effect on metabolism, 89–90
Virus diseases, and hearing loss, 563–565
Visible light radiation, 594–601
animal experiments, 597
critical organs, 596–598
defined, 594–595
design and layout, 385
nomenclature, 596
occupational exposures, 596
pathophysiology, 596–599
protection guides, 600–601
recommended indicators, 384
sources of, 595–596
system-equipment design, 383–384
threshold dosimetry, 600
warning or emergency, 384
Vitamin E, as oxidant defense mechanism, 170
Volume seekers, 510, 524

Walking test, of children and aged, 321
Walsh-Healy Act (1936), 14
Warning signals, shiftworkers and, 288
Watchkeeping, perceptual demands of, 398
Water:
biological agents in, 453
contaminated, 453
depletion in hot environments, 492
Weather conditions, in surveillance data acquisition, 338–339
Welding:
carcinogenic effect, 588
radiation hazards of, 590
Wharton Business School, 190
White-collar workers:
ergonomic considerations, 420
increase in, 418
White finger, see Vibration white finger
Whole-body vibration, 659–681
absorption and transmission factors, 662
biodynamics of, 661–663
chronic occupational exposure, 673–677
direction of, 658

Whole-body vibration (*Continued*)
  effects on humans, 661–672
  electroencephalograms, 668
  engineering controls, 680
  exposure guidelines, 663, 679
  frequency ranges of, 660–661, 663
  industrial concern, 659–660
  minimization of adverse effects, 681
  morbidity studies, 414
  noise and, 677–678
  pathology of, 672–678
  performance effects, 669–672
  physiology of, 663–666
  protection from effects of, 678–681
  resonant modes, 662–663
  safe levels, 414
  sensation threshold, 667
  sensory and neuromuscular effects, 666–668
  severe acute exposure, 672
  sex differences, 676
  source/individual controls, 680
  stressors associated with, 677–678
  types of, 659
Wilson's disease, chemical monitoring problems, 95
Wives, epidemiologic studies, 361
Women workers:
  ergonomic problems, 420–421
  hair lead levels of, 128, 132
  in nontraditional jobs, 420–421
  physiological responses to cold, 485
  vibration susceptibility of, 676
Wood dust disease, 467–468
Wool sorter's disease, 472
Work:
  behavioral procedures, 425
  capacity studies, 325–326
  costs, 313–331
  energy expenditure by task, 328
  ergonomic factors, 392–401
  exposure to carcinogens, 357–358
  functions, 416
  measurements, 313–331
  organization, 415–416
  output during typical workday, 327–329
  performance, 401
  physiological costs, 315
  physiology of, 326
  practices, 423–424
  psychosocial stresses, 416–418
  stress tests, 704
  unusual physical conditions, 420–421

Workbench, optimal height of, 381
Work design, 378–382
  behavioral approaches, 423–425
  body dimensions in, 381
  distances in, 380–381
  heights in, 380–381
  systems approaches, 421–422
  see also Ergonomics
Workers:
  age and metabolism, 91–92
  anthropometric data on, 378–379
  capacity studies, 325–326
  changes in, 420–421
  chemical carcinogenicity in, 40–41
  circadian rhythms and shiftwork performance, 286–289
  educational involvement of, 24
  energy consumption of, 393
  ergonomic considerations, 420–421
  vs. machines, 421–422
  male *vs.* female, 93
  microwave-radiofrequency exposure, 626–629
  physical energy measurements, 324–327
  physical performance capacity, 393
  predicting toxicity in, 39–42
  psychosocial factors and stress-strain outcomes, 417–418
  susceptibility-resistance rhythms, 261–263
  training of, 340
  vibration effects, 655–656, 661–672
  women, 420–421
Work Practices Guide for Manual Lifting, 396
Work schedules:
  biological rhythms and, 178–180
  diurnal *vs.* nocturnal, 178
  flexible or staggered, 183
  and occupational health, 364
  popular industry, 180–183
  sociological perspective, 179
  standard, 180–181
  traditional, 179
  unusual, 179–180, 183–187
Workspace, 378–382
  hazards, 420
  limits for exposure, 368–372
  psychosocial factors and stress–strain outcomes, 417–418
  stresses, 7
World Health Organization, 496, 638
World War II, impact on worker exposure, 20–21

Xenobiotics:
  biological monitoring of, 78
  blood concentrations of, 88
  body mechanisms for handling, 81–82
  in breast milk, 134–135
  defined, 76
  metabolism and, 81–83
  see also Chemical agents; Toxicants
X-rays:
  for biological agents, 470
  circadian-susceptibility rhythms, 249

Yawning, in cold environments, 483

Zeitgeber, see Synchronizers
Zinc:
  in chelation therapy, 540
  content of, in hair, 132
Zoonoses, 473–475